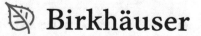 Birkhäuser

Xiaochun Liu • Bert-Wolfgang Schulze

Boundary Value Problems with Global Projection Conditions

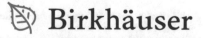 Birkhäuser

Xiaochun Liu
School of Mathematics and Statistics
Wuhan University
Wuhan, China

Bert-Wolfgang Schulze
Institut für Mathematik
Universität Potsdam
Potsdam, Germany

ISSN 0255-0156   ISSN 2296-4878   (electronic)
Operator Theory: Advances and Applications
ISBN 978-3-030-09933-6   ISBN 978-3-319-70114-1   (eBook)
https://doi.org/10.1007/978-3-319-70114-1

Mathematics Subject Classification (2010): 35S15, 58J20

This book is published under the imprint Birkhäuser, www.birkhauser-science.com by the registered company Springer Nature Switzerland AG.
The registered company address is: Gewerbestrasse 11, 6330 Cham, Switzerland

as to contain all "standard" elliptic boundary value problems for differential operators and to be closed under the construction of parametrices of elliptic elements. While the space of BVPs for differential operators with Shapiro–Lopatinskii (SL)-elliptic conditions just generates the above-mentioned algebra of BVPs with the transmission property at the boundary, cf. [10] or [34], [19], the Toeplitz analogue is designed to include the parametrices of Dirac operators with global projection conditions (especially APS-conditions in the sense of Atiyah, Patodi, and Singer [4, 5, 6]) as well as elliptic BVPs for geometric and other elliptic differential operators, with conditions of SL-elliptic or global projection (GP)-elliptic type. More precisely, every elliptic differential (and then also pseudo-differential operator with the transmission property at the boundary) on a smooth compact manifold with boundary belongs to the algebra, and, as we shall see, every such operator admits elliptic boundary conditions of that kind.

In that sense the Toeplitz algebra of BVPs unifies the concept of elliptic conditions of SL- and GP-elliptic type. Ellipticity in this context is equivalent with the Fredholm properties in the respective scales of spaces (standard Sobolev spaces in the SL case, spaces of Hardy type in the GP case). The 2×2 block matrices contain the above-mentioned algebra of Toeplitz operators on the boundary as a subalgebra, and hence also the Fredholm property of elliptic operators between Hardy-type Sobolev spaces is equivalent with the respective GP-ellipticity. In order to make the machinery transparent, we provide a concise introduction to the Boutet de Monvel algebra of BVPs with the transmission property at the boundary. In this framework, we consider cutting and pasting of elliptic BVPs and analogues of index formulas of Agranovich and Dynin, and we analyse the pseudo-differential nature of projections of Calderón–Seeley type.

Ellipticity on manifolds with boundary and approaches to treating solvability near the boundary constitute a prominent field of PDEs, geometric analysis, and index theory. Numerous papers and monographs are devoted to special cases and explicit computations, see, for instance, Booss-Bavnbek and Wojciechowski [8], Grubb and Seeley [20], Savin and Sternin [37], and the work of many other authors, in particular, joint work [53] with Seiler on elliptic complexes of BVPs in GP-framework, and references therein. Additional references and results can be found in the monograph [29] of Nazaikinskij et al.

Another part of this text studies a Toeplitz analogue of the edge algebra, see, in particular, the articles of Schulze and Seiler [52, 54]. The original edge pseudo-differential algebra was introduced in [43] as a calculus that contains all edge-degenerate differential operators on a manifold with edge, together with the parametrices of elliptic elements. The ellipticity first refers to an analogue of Shapiro–Lopatinskii conditions, i.e., a bijectivity condition for an operator-valued principal symbol structure which contains also trace and potential operators with respect to the edge, a substitute of the former boundary. In order to keep the material self-contained, we briefly outline the basic parts of the edge calculus. Again there is a topological obstruction to the existence of such edge conditions, and we complete the algebra by edge conditions of global projection type to an algebra referred

Preface

Boundary value problems (BVPs) for elliptic and other types of partial differential equations belong to the classical areas of mathematical analysis. Prototypes are the Dirichlet or the Neumann problem for the Laplace operator, and practically every textbook on PDEs treats problems of that type. However, it is by no means obvious whether (or that) the operators which are involved in the solvability process (such as Green's function, or potential operators) have a pseudo-differential structure, or how many other elliptic boundary value problems can be posed for the Laplacian. Another question is whether an elliptic differential operator admits (Shapiro–Lopatinskii)-elliptic boundary conditions at all, for instance, the Cauchy–Riemann operator in a smooth domain in the complex plane, or Dirac operators on a manifold with smooth boundary. The more we look in this direction, the more obscure the notion of ellipticity of a BVP becomes. And if there are also singularities on the boundary (even of a moderate complexity, such as conical points or edges), substantial difficulties arise in the search for a natural approach to reflect basic solvability properties.

The present text is devoted to developing general concepts of ellipticity of BVPs. First we introduce necessary tools on pseudo-differential operators. We define ellipticity on an open C^∞ manifold and construct parametrices within the algebra of standard pseudo-differential operators. Then we pass to Toeplitz operators on a closed compact C^∞ manifold based on pseudo-differential operators and pseudo-differential projections. Other essential topics are operators with operator-valued symbols with twisted symbol estimates, where we establish basic results. We also present some material on pseudo-differential operators on manifolds with conical exit to infinity, especially manifolds modelled on infinite cylinders (or half-cylinders). Then we study elliptic BVPs of Shapiro–Lopatinskii type in the framework of Boutet de Monvel's calculus [10]. In particular, we give a K-theoretic explanation of the existence of such conditions, cf. Atiyah and Bott [3], Harutyunyan and Schulze [23, Subsection 3.3.4], and elucidate the pseudo-differential structure of parametrices. After that we formulate a Toeplitz analogue of the algebras of pseudo-differential operators with the transmission property at the boundary. Such an operator algebra, first introduced in [47] (see also a more detailed version in [48]), is built for similar reasons as other pseudo-differential algebras, namely, so to as the Toeplitz analogue of the edge algebra. It contains Shapiro–Lopatinskii elliptic edge conditions as a special case. Similarly as in the calculus of BVPs, we obtain the Fredholm property in spaces on the boundary that are derived from standard Sobolev spaces and a subsequent pseudo-differential projection. We show that elliptic conditions of global projection type exist for arbitrary edge-degenerate elliptic operators in the top left corner.

We then pass to a special case of the edge calculus, namely, BVPs on a manifold with smooth boundary, cf. the article of Schulze and Seiler [51]. Clearly, all results of the general edge calculus developed before remain true. However, regarding a manifold with boundary as a special manifold with edge allows us to single out a specific subclass of pseudo-differential operators which are more in the focus of boundary value problems, namely, operators with standard symbols rather than edge-degenerate ones obtained by restriction of pseudo-differential operators of an ambient open manifold containing the considered embedded manifold with smooth boundary. It is by no means evident that these operators generate a subcalculus of the general edge algebra. However, we show that this is indeed the case, and we obtain an approach of BVPs without (or with) the transmission property at the boundary which is much more general than the calculus of [10]. Note in this connection that the theory of Vishik and Eskin also treats BVPs without the transmission property, cf. [62], [63] and Eskin's book [14], but the edge calculus approach is rather different, and it produces an algebra. Nevertheless, [14] contains an algebra of pseudo-differential operators on the half-line which became later on an important ingredient of the edge symbolic calulus. This part of the development is also commented in detail in the monograph [45], see also Rempel and Schulze [35]. BVPs without the transmission property at the boundary are of interest also in connection with mixed and transmission problems, cf. Harutyunyan and Schulze [23], Wong [56], Chang et al. [11]. Here we mainly focus on the feature that elliptic boundary problems with global projection conditions can be studied for similar reasons as in the corresponding Toeplitz variant of the general edge calculus. But it remains a remarkable effect that there are relevant subalgebras which are closed under the construction of parametrices.

Contents

Introduction

Boundary value problems are important topics in the analysis of partial differential equations. The motivation comes from physics and wide areas of the applied sciences, but also from index theory, geometry, and other fields of pure mathematics. Similarly to the case of open manifolds, where ellipticity plays a crucial role in understanding solvability properties for many types of equations, e.g., also parabolic ones, ellipticity on a manifold with boundary is not only interesting on its own right, but also for large classes of more general problems, where the control of phenomena up to the boundary is a specific aspect. Moreover, a (say, smooth) boundary can be interpreted as a special geometric singularity, namely, as an edge with the inner normal as the model cone of a corresponding wedge, here a collar neighbourhood of the boundary. It turns out that the analysis of elliptic BVPs yields methods and insight on ellipticity on manifolds with singularities, and many ideas in this field can be read off from the case of boundary value problems. Conversely, it turns out that the ideas for analysing equations on manifolds with singularities, especially for cones and wedges, shed a new light on the structure of solvability of BVPs, e.g., mixed problems, for instance, the Zaremba problem for the Laplacian, with mixed conditions of Dirichlet and Neumann type which have a jump on an interface of the boundary, cf. [23], or [11].

A smooth manifold X with boundary ∂X can be interpreted as a stratified space, i.e., a disjoint union of smooth open manifolds

$$X = s_0(X) \cup s_1(X),$$

with $s_0(X) := \operatorname{int} X = X \setminus \partial X$ and $s_1(X) := \partial X$. Parallel to the stratification of X, symbolised by the sequence of strata

$$s(X) = \big(s_0(X), s_1(X)\big),$$

the operators A under consideration have a principal symbol hierarchy

$$\sigma(A) = \big(\sigma_0(A), \sigma_1(A)\big) \tag{0.1}$$

with $\sigma_i(A)$ being associated with $s_i(X)$, $i = 0, 1$, where $\sigma_0(A)$ is also called the principal interior symbol and $\sigma_1(A)$ the principal boundary symbol of A. Later on, in Section 2.3, we prefer to write $(\sigma_\psi, \sigma_\partial)$ rather than (σ_0, σ_1).

If A is a differential operator of order $\mu \in \mathbb{N}$ ($= \{0, 1, 2, \dots\}$) written locally in coordinates $x \in \mathbb{R}^n$ in the form

$$A = \sum_{|\alpha| \leq \mu} a_\alpha(x) D_x^\alpha$$

with smooth coefficients, where X close to the boundary is identified with the half-space $\overline{\mathbb{R}}_+^n = \{x \in \mathbb{R}^n : x_n \geq 0\}$, we have

$$\sigma_0(A)(x, \xi) = \sum_{|\alpha| = \mu} a_\alpha(x) \xi^\alpha \tag{0.2}$$

and

$$\sigma_1(A)(x', \xi') = \sum_{|\alpha| = \mu} a_\alpha(x', 0) (\xi', D_{x_n})^\alpha, \tag{0.3}$$

where $x = (x', x_n)$, $\xi = (\xi', \xi_n)$. While (0.2) is interpreted as a smooth scalar function on $T^*X \setminus 0$, the cotangent bundle of X with the zero section removed, (0.3) is a smooth operator-valued function on $T^*(\partial X) \setminus 0$. The action of the symbol (0.3) can be regarded in standard Sobolev spaces on $\mathbb{R}_+$, namely,

$$\sigma_1(A)(x', \xi') : H^s(\mathbb{R}_+) \to H^{s-\mu}(\mathbb{R}_+), \tag{0.4}$$

for any $s \in \mathbb{R}$. The meaning of (0.1) in the pseudo-differential set-up will be explained below. If the operator A is elliptic with respect to σ_0, i.e., the symbol (0.2) does not vanish for $\xi \neq 0$, the operators (0.4) form a family of Fredholm operators for $s - \mu > -1/2$. Ellipticity of A with respect to (0.1) should include an invertibility condition on (0.4) for $\xi' \neq 0$. However, the operator family (0.4) is bijective only in exceptional cases (in the pseudo-differential set-up). For a differential operator A the boundary symbol $\sigma_1(A)$ is surjective but not injective. In order to have invertibility we should complete the Fredholm family (0.4) to a family of isomorphisms

$$\sigma_1(\mathcal{A})(x', \xi') := \begin{pmatrix} \sigma_1(A) \\ \sigma_1(T) \end{pmatrix} (x', \xi') : H^s(\mathbb{R}_+) \to \begin{matrix} H^{s-\mu}(\mathbb{R}_+) \\ \oplus \\ \mathbb{C}^{N_2} \end{matrix} \tag{0.5}$$

by an extra operator family $\sigma_1(T)(x', \xi') : H^s(\mathbb{R}_+) \to \mathbb{C}^{N_2}$ that maps the kernel of $\sigma_1(A)(x', \xi')$ isomorphically to $\mathbb{C}^{N_2}$. Globally $\mathbb{C}^{N_2}$ is interpreted as the fibre of a vector bundle G over $T^*(\partial X) \setminus 0$; in general, this bundle is not trivial. In addition, to interpret $\sigma_1(T)$ as the boundary symbol of an operator

$$T : H^s(X) \to H^{s-\mu}(\partial X, J_2) \tag{0.6}$$

for some smooth complex vector bundle J_2 over ∂X, we have to require that

$$G = \pi_{\partial X}^* J_2, \tag{0.7}$$

i.e., the above-mentioned G has to be the pull-back of such a J_2 under the canonical projection $\pi_{\partial X} : T^*(\partial X) \setminus 0 \to \partial X$. As for the shift μ of smoothness in (0.6), we are free to impose any real number; for convenience we took $\mu = \operatorname{ord} A$, the order of A. The property (0.7) is a topological condition on the behavior of $\sigma_0(A)$ close to the boundary, see [3] and Section 3.2 below, necessary and sufficient for the existence of a Fredholm operator of the form

$$\mathcal{A} = \begin{pmatrix} A & K \\ T & Q \end{pmatrix} : \begin{matrix} H^s(X) \\ \oplus \\ H^s(\partial X, J_1) \end{matrix} \to \begin{matrix} H^{s-\mu}(X) \\ \oplus \\ H^{s-\mu}(\partial X, J_2) \end{matrix} \tag{0.8}$$

belonging to the Boutet de Monvel's calculus, in the general pseudo-differential case also containing non-trivial entries K, Q and another smooth complex vector bundle J_1. The Fredholm property of (0.8) is equivalent to the σ_0-ellipticity of A together with the bijectivity of the boundary symbol

$$\sigma_1(\mathcal{A})(x', \xi') = \begin{pmatrix} \sigma_1(A) & \sigma_1(K) \\ \sigma_1(T) & \sigma_1(Q) \end{pmatrix} (x', \xi') : \begin{matrix} H^s(\mathbb{R}_+) \\ \oplus \\ (\pi_{\partial X}^* J_1)_{x', \xi'} \end{matrix} \to \begin{matrix} H^{s-\mu}(\mathbb{R}_+) \\ \oplus \\ (\pi_{\partial X}^* J_2)_{x', \xi'} \end{matrix} \tag{0.9}$$

for every $(x', \xi') \in T^*(\partial X) \setminus 0$ and sufficiently large s. The relation (0.9) is a block matrix generalisation of (0.5), meaningful in the pseudo-differential case, and formulated in global form. The bundles J_1, J_2 over ∂X appear at the same time as soon as both $\ker \sigma_1(A)(x', \xi')$ and $\operatorname{coker} \sigma_1(A)(x', \xi')$ are non-trivial.

The operators K, T, Q are said to satisfy the (pseudo-differential analogue of the) Shapiro–Lopatinskii condition with respect to A if (0.9) is a family of bijections. In Chapter 3 we will provide more details, in particular, on boundary conditions when the above-mentioned topological obstruction does not vanish. A general answer was first given in [47]. The corresponding extension of Boutet de Monvel's calculus to a Toeplitz calculus is presented here in Chapter 4. Special geometric differential operators have been studied before, including extra global boundary conditions that guarantee a finite Fredholm index, see the work of Atiyah, Patodi, and Singer [4, 5, 6] and of many other authors.

In Chapter 5 we study in detail the case of BVPs for differential operators. First, in Section 5.1, we establish general cutting and pasting constructions and compare the indices of elliptic operators on a closed manifold M with the indices of elliptic BVPs on submanifolds M_+, M_- with common smooth boundary Y that subdivide M as $M_+ \cup M_-$ with $M_+ \cap M_- = Y$. In Section 5.2 we study an extension of the concept of spectral BVPs to arbitrary elliptic differential operators of any order, following joint work [30], [28] with Nazaikinskij, Savin, Sternin and Shatalov. Section 5.3 is devoted to further observations on Calderón–Seeley projections.

In Part II we consider elliptic operators on a manifold with edge, here with edge conditions rather than boundary conditions, again under the unifying goal of understanding Shapiro–Lopatinskii and global projection edge conditions within a Toeplitz analogue of the edge calculus. This material refers to a paper of Schulze and Seiler [52], and again we focus on the operator algebra aspect and the construction of parametrices of elliptic elements within this calculus.

Part III is devoted to the case of Toeplitz calculus where the underlying space is a smooth manifold with boundary. This corresponds to the case where the model cone transverse to the edge, the boundary, is equal to $\overline{\mathbb{R}}_+$, the inner normal. We refer to the fact that the calculus contains all interior classical symbols, not necessarily with the transmission property, which are smooth up to the boundary. Such a subcalculus is much more general than that of Part I with the transmission property, though here we realize operators in weighted edge spaces. In Section 3.2 we briefly outline other approaches to symbols without the transmission property and make interesting observations, suggested by several applications. Moreover, we give a number of modifications and extensions of the calculus, such as to the truncation quantization, already occurring both in Vishik and Eskin's work as well in the Boutet de Monvel calculus, see also the joint paper [50, 51] with Seiler.

It is interesting to study operators on more general singular spaces with boundary, e.g., when the boundary itself has conical singularities, edges or higher corners. However, this problem seems to be open, at least as far as the operator algebra approach with global projection conditions is concerned. A "final" calculus answer should be fitted in the general strategy of establishing operator algebras with symbol hierarchies on stratified spaces with boundary, analogously to the program outlined in [49].

Acknowledgement: The first author appreciates her former PhD supervisor, Professor Chen Hua from Wuhan University for his steady encouragement and generous support and help. She would like to acknowledge the grant (No. 11371282, 11571259, 11771342) from the NNSF of China for support. Deepest gratitude goes to her family for permanent support for her scientific endeavours. The second author wants to thank Professor Chen Hua and many other colleagues from the Wuhan University for the excellent working conditions, hospitality and care during many years of cooperation.

Part I

Boundary Value Problems with Global Projection Conditions

Chapter 1

Pseudo-differential operators

1.1 Basics of the pseudo-differential calculus

We first outline some notation and well-known material on standard pseudo-differential operators. Proofs, as far as they are skipped here, can be found in textbooks on the pseudo-differential calculus. Pseudo-differential operators in their "most classical form" in an open set $\Omega \subseteq \mathbb{R}^n$ are defined in terms of oscillatory integrals, based on the Fourier transform, namely,

$$Au(x) := \mathrm{Op}_x(a)u(x) := \iint e^{i(x-x')\xi} a(x, x', \xi) u(x') dx' \,\bar{d}\xi, \tag{1.1}$$

$\bar{d}\xi = (2\pi)^{-n} d\xi$. Here we first assume $u \in C_0^\infty(\Omega)$; later on we extend the operators to more general function and distribution spaces. In order to specify the involved amplitude functions $a(x, x', \xi)$ we employ the abbreviation

$$\langle \xi \rangle := \left(1 + |\xi|^2\right)^{1/2}. \tag{1.2}$$

The function (1.2) belongs to $C^\infty(\mathbb{R}^n)$ and satisfies the estimates

$$c_1 |\xi| \le \langle \xi \rangle \le c_2 |\xi| \tag{1.3}$$

for all $\xi \in \mathbb{R}^n$, $|\xi| \ge 1$, and some constants $c_1, c_2 > 0$. We assume $a(x, x', \xi) \in S_{(\mathrm{cl})}^\mu(\Omega \times \Omega \times \mathbb{R}^n)$, $\mu \in \mathbb{R}$, according to the following definition.

Definition 1.1.1. (i) Let $U \subseteq \mathbb{R}^d$ be an open set; then $S^\mu(U \times \mathbb{R}^n)$, $\mu \in \mathbb{R}$, is defined to be the set of all $a(x, \xi) \in C^\infty(U \times \mathbb{R}^n)$ satisfying the symbol estimates

$$\left| D_x^\alpha D_\xi^\beta a(x, \xi) \right| \le c \, \langle \xi \rangle^{\mu - |\beta|}$$

for all $(x, \xi) \in K \times \mathbb{R}^n$, $K \Subset U$, $\alpha \in \mathbb{N}^d$, $\beta \in \mathbb{N}^n$, for constants $c = c(\alpha, \beta, K) > 0$; here $K \Subset U$ means that K is a compact subset of

© Springer Nature Switzerland AG 2018
X. Liu, B.-W. Schulze, *Boundary Value Problems with Global Projection Conditions*,
Operator Theory: Advances and Applications 265, https://doi.org/10.1007/978-3-319-70114-1_1

U, and $D_x^\alpha = \left(i^{-1}\partial_{x_1}\right)^{\alpha_1}\cdots\left(i^{-1}\partial_{x_d}\right)^{\alpha_d}$ for $\alpha = (\alpha_1,\ldots,\alpha_d)$. The elements of $S^\mu(U \times \mathbb{R}^n)$ are also called symbols of order μ.

(ii) Let $S^{(\nu)}(U \times (\mathbb{R}^n \setminus \{0\}))$, $\nu \in \mathbb{R}$, be the space of all $f_{(\nu)}(x,\xi) \in C^\infty(U \times (\mathbb{R}^n \setminus \{0\}))$ such that $f_{(\nu)}(x,\lambda\xi) = \lambda^\nu f_{(\nu)}(x,\xi)$ for all $\lambda \in \mathbb{R}_+$. A symbol $a(x,\xi) \in S^\mu(U \times \mathbb{R}^n)$ is called classical, if there are homogeneous components $a_{(\mu-j)}(x,\xi) \in S^{(\mu-j)}(U \times (\mathbb{R}^n\setminus\{0\}))$, $j \in \mathbb{N}$, such that

$$a(x,\xi) - \sum_{j=0}^{N} \chi(\xi)a_{(\mu-j)}(x,\xi) \in S^{\mu-(N+1)}(U \times \mathbb{R}^n) \qquad (1.4)$$

for every $N \in \mathbb{N}$; here χ is a so-called excision function (i.e., $\chi \in C^\infty(\mathbb{R}^n)$, $\chi(\xi) = 0$ for $|\xi| \le c_0$, $\chi(\xi) = 1$ for $|\xi| \ge c_1$, for some $0 \le c_0 \le c_1$). By $S_{\mathrm{cl}}^\mu(U \times \mathbb{R}^n)$ we denote the space of all classical symbols of order μ.

Note that $S^\mu(U \times \mathbb{R}^n)$ is a Fréchet space with a semi-norm system

$$\sup_{(x,\xi)\in K\times\mathbb{R}^n} \langle\xi\rangle^{-\mu+|\beta|}\left|D_x^\alpha D_\xi^\beta a(x,\xi)\right|, \qquad (1.5)$$

$K \Subset U$, $\alpha \in \mathbb{N}^d$, $\beta \in \mathbb{N}^n$. Also, $S_{\mathrm{cl}}^\mu(U \times \mathbb{R}^n)$ is Fréchet in a stronger topology than that induced from $S^\mu(U \times \mathbb{R}^n)$, namely, with the semi-norms (1.5) together with those from (1.4) and from the homogeneous components $a_{(\mu-j)}(x,\xi) \in C^\infty(U \times (\mathbb{R}^n\setminus\{0\}))$ which are uniquely determined by a. First we can recover $a_{(\mu)}(x,\xi)$ by

$$a_{(\mu)}(x,\xi) = \lim_{\lambda\to\infty} \lambda^{-\mu}a(x,\lambda\xi). \qquad (1.6)$$

Then, because $a(x,\xi) - \chi(\xi)a_{(\mu)}(x,\xi) \in S_{\mathrm{cl}}^{\mu-1}(U \times \mathbb{R}^n)$ for any excision function χ, we can apply the same procedure again which yields $a_{(\mu-1)}(x,\xi)$, and then, successively, $a_{(\mu-j)}(x,\xi)$ for every $j \in \mathbb{N}$.

If a consideration refers both to classical and general symbols (later on also in the operator-valued setup) we write subscript "(cl)". Let $S_{(\mathrm{cl})}^\mu(\mathbb{R}^n)$ denote the subspaces of x-independent symbols ("with constant coefficients"). The spaces $S_{(\mathrm{cl})}^\mu(\mathbb{R}^n)$ are closed in $S_{(\mathrm{cl})}^\mu(U \times \mathbb{R}^n)$, and we have

$$S_{(\mathrm{cl})}^\mu(U \times \mathbb{R}^n) = C^\infty\left(U, S_{(\mathrm{cl})}^\mu(\mathbb{R}^n)\right).$$

Set $S^{-\infty}(U \times \mathbb{R}^n) = \bigcap_{\mu\in\mathbb{R}} S^\mu(U \times \mathbb{R}^n)$; this space is isomorphic to $C^\infty(U, \mathcal{S}(\mathbb{R}^n))$.

In the future we will employ numerous variants of such symbol spaces; they will be defined when they first appear. Let us now observe a specific property of symbols, first with constant coefficients. We have

$$S^\mu(\mathbb{R}_\xi^n) \subset \mathcal{S}'(\mathbb{R}_\xi^n),$$

hence $F_{\xi\to\theta}^{-1}S^\mu(\mathbb{R}^n) \subset \mathcal{S}'(\mathbb{R}_\theta^n)$, where F^{-1} is the inverse Fourier transform in $\mathbb{R}^n$.

Proposition 1.1.2. *For every* $a(\xi) \in S^{\mu}_{(\text{cl})}(\mathbb{R}^n)$ *the distribution*

$$k_F(a)(\theta) := \int e^{i\theta\xi} a(\xi)\mathchar'26\mkern-12mu d\xi,$$

$k_F(a) \in \mathcal{S}'(\mathbb{R}^n)$, *has the property* $(1 - \psi(\theta))k_F(a)(\theta) \in \mathcal{S}(\mathbb{R}^n_\theta)$ *for every* $\psi \in C^{\infty}_0(\mathbb{R}^n)$, $\psi \equiv 1$ *close to 0, and hence*

$$a_0(\xi) := \int e^{-i\theta\xi}\psi(\theta)k_F(a)(\theta)d\theta \in S^{\mu}_{(\text{cl})}(\mathbb{R}^n).$$

More generally, for every $a(\xi, \eta) \in S^{\mu}_{(\text{cl})}(\mathbb{R}^{n+q})$ *the distribution*

$$k_F(a)(\theta, \eta) := \int e^{i\theta\xi} a(\xi, \eta)\mathchar'26\mkern-12mu d\xi \in \mathcal{S}'(\mathbb{R}^{n+q}_{\theta,\eta})$$

has the property $(1 - \psi(\theta))k_F(a)(\theta, \eta) \in \mathcal{S}(\mathbb{R}^{n+q}_{\theta,\eta})$. *This implies*

$$a_0(\theta, \eta) = \int e^{-i\theta\xi}\psi(\theta)k_F(a)(\theta, \eta)d\theta \in S^{\mu}_{(\text{cl})}(\mathbb{R}^{n+q}_{\xi,\eta}).$$

Later on the map

$$S^{\mu}_{(\text{cl})}(\mathbb{R}^n) \to S^{\mu}_{(\text{cl})}(\mathbb{R}^n), \quad a(\xi) \mapsto a_0(\xi),$$

is referred to as a kernel cut-off operator.

Henceforth, if φ, φ' are functions we write $\varphi \prec \varphi'$ if $\varphi' \equiv 1$ on supp φ.

Corollary 1.1.3. *Let* $\varphi \prec \varphi'$ *be arbitrary functions in* $C^{\infty}_0(\mathbb{R})$. *Then for every* $a(\xi) \in S^{\mu}(\mathbb{R})$ *we have*

$$\varphi(x)k_F(a)(x - x')(1 - \varphi'(x')) \in \mathcal{S}(\mathbb{R}_x \times \mathbb{R}_{x'}).$$

An analogous property holds for symbols $a(x, \xi) \in C^{\infty}_b(\mathbb{R}_{x'}, S^{\mu}(\mathbb{R}_\xi))$.

Example 1.1.4. (i) *Polynomials* $a(x, \xi) = \sum_{|\alpha| \le \mu} c_\alpha(x)\xi^\alpha$ *with coefficients* $c_\alpha \in C^{\infty}(\Omega)$ *belong to* $S^{\mu}_{\text{cl}}(\Omega \times \mathbb{R}^n)$.

(ii) $\langle\xi\rangle^\mu \in S^{\mu}_{\text{cl}}(\mathbb{R}^n)$ *for every* $\mu \in \mathbb{R}$; *however,* $\langle\xi\rangle^\mu + \langle\xi\rangle^{\mu-1/2} \in S^{\mu}(\mathbb{R}^n)\backslash S^{\mu}_{\text{cl}}(\mathbb{R}^n)$.

Theorem 1.1.5. *Let* $a_k(x, \xi) \in S^{\mu_k}_{(\text{cl})}(U \times \mathbb{R}^n)$, $k \in \mathbb{N}$, *be an arbitrary sequence, where* $\mu_k \to -\infty$ *as* $k \to \infty$ *(with* $\mu_k := \mu - k$ *in the classical case). Then there is an* $a(x, \xi) \in S^{\mu}_{\text{cl}}(U \times \mathbb{R}^n)$, $\mu = \max_{k\in\mathbb{N}}\{\mu_k\}$, *such that*

$$a(x, \xi) - \sum_{k=0}^{N} a_k(x, \xi) \in S^{\mu_{(N)}}(U \times \mathbb{R}^n),$$

where $\mu_{(N)} \to -\infty$ *as* $N \to \infty$. *The symbol* $a(x, \xi)$ *is unique mod* $S^{-\infty}(U \times \mathbb{R}^n)$.

We write $a \sim \sum_{k=0}^{N} a_k$ and call a an asymptotic sum of the a_k. An asymptotic sum $a \sim \sum_{j=0}^{\infty} a_j$ may be constructed as a convergent sum

$$a(x,\xi) = \sum_{j=0}^{\infty} \chi(\xi/c_j) a_j(x,\xi) \tag{1.7}$$

for an excision function χ and constants $c_j > 0$ tending to ∞ sufficiently fast as $j \to \infty$, where $\sum_{j=J}^{\infty} \chi(\xi/c_j) a_j(x,\xi)$ converges in $S^{m_J}(U \times \mathbb{R}^n)$ for $m_J = \max\{\mu_j : j \geq J\}$ for every $J \in \mathbb{N}$.

Remark 1.1.6.   (i) Let $f(\zeta,\eta) \in S^{(\mu)}(\mathbb{R}_{\zeta,\eta}^{d+q} \setminus \{0\})$; then for every fixed $\eta^1 \neq 0$ we have $f(\zeta,\eta^1) \in S_{\mathrm{cl}}^{\mu}(\mathbb{R}_\zeta^d)$, and the homogeneous component $f_{(\mu-j)}(\zeta,\eta^1) \in S^{(\mu-j)}(\mathbb{R}_\zeta^d \setminus \{0\})$ of order $\mu - j$ is a polynomial in $\eta^1 \in \mathbb{R}^q$ of degree j for every $j \in \mathbb{N}$. Clearly, we have similar relations for f smoothly depending on extra variables z, y.

(ii) Let $f(z,y,\zeta,\eta) \in S_{(\mathrm{cl})}^{\mu}(V_z \times U_y \times \mathbb{R}_{\zeta,\eta}^{d+q})$, for open $V \subseteq \mathbb{R}^e$, $U \subseteq \mathbb{R}^p$. Then for every fixed $y^1 \in \mathbb{R}^p$, $\eta^1 \in \mathbb{R}^q$ we have $f(z,y^1,\zeta,\eta^1) \in S_{(\mathrm{cl})}^{\mu}(V_z \times \mathbb{R}_\zeta^d)$. In the classical case the homogeneous principal symbol of $f(z,y^1,\zeta,\eta^1)$ in (z,ζ) is independent of η^1.

In fact, for (i) we write $f(\zeta,\eta^1) = |\zeta|^\mu f(\zeta/|\zeta|, \eta^1/|\zeta|)$ and $\sigma := \zeta/|\zeta|$, $\varepsilon := |\zeta|^{-1}$. Then Taylor expansion of $f(\sigma, \varepsilon\eta^1)$ in ε at $\varepsilon = 0$ gives

$$f(\sigma, \varepsilon\eta^1) \sim \sum_{j=0}^{\infty} \sum_{|\alpha|=j} b_{\alpha,j}(\sigma)(\eta^1)^\alpha \varepsilon^j$$

for coefficients $b_{\alpha,j}(\sigma) \in C^\infty(S^{d-1})$, which yields

$$f(\zeta,\eta^1) \sim \sum_{j=0}^{\infty} \sum_{|\alpha|=j} b_{\alpha,j}(\zeta/|\zeta|)(\eta^1)^\alpha |\zeta|^{\mu-j}.$$

The explanation for (ii) is simple as well and left to the reader. The latter expression can also be interpreted as an asymptotic expansion in the sense of Theorem 1.1.5 with respect to the covariables ζ.

Definition 1.1.7. Let $\Omega \subseteq \mathbb{R}^n$ be open, $\mu \in \mathbb{R}$; then

$$L_{(\mathrm{cl})}^{\mu}(\Omega) := \{\mathrm{Op}(a) : a(x,x',\xi) \in S_{(\mathrm{cl})}^{\mu}(\Omega \times \Omega \times \mathbb{R}^n)\}$$

is called the space of pseudo-differential operators on Ω of order μ (classical or general, according to a). Moreover, we set $L^{-\infty}(\Omega) = \bigcap_{\mu \in \mathbb{R}} L^\mu(\Omega)$.

An operator $A \in L^\mu(\Omega)$ induces a continuous operator

$$A : C_0^\infty(\Omega) \to C^\infty(\Omega).$$

As such it has a distributional kernel $K_A \in \mathcal{D}'(\Omega \times \Omega)$ such that

$$\langle Au, v \rangle = \langle K_A, u \otimes v \rangle$$

for all $u, v \in C_0^\infty(\Omega)$ (with $\langle \cdot, \cdot \rangle$ denoting the bilinear pairing $\int_\Omega u(x)v(x)dx$ or the application of a distribution to a test function). We have

$$\text{sing supp } K_A \subseteq \text{diag}(\Omega \times \Omega) \tag{1.8}$$

where $\text{diag}(\Omega \times \Omega) := \{(x,y) \in \Omega \times \Omega : x = y\}$. The property (1.8) is called the pseudo-locality of pseudo-differential operators.

By virtue of (1.8) we can write every $A \in L^\mu(\Omega)$ in the form

$$A = A_0 + C$$

for an $A_0 \in L^\mu(\Omega)$ which is properly supported, i.e., K_{A_0} has a proper support in the sense that $K_{A_0} \cap (M \times \Omega)$ and $K_{A_0} \cap (\Omega \times M)$ are compact for every $M \Subset \Omega$, and C is an operator with kernel $c(x, x') \in C^\infty(\Omega \times \Omega)$, i.e.,

$$Cu(x) = \int c(x, x')u(x')dx'. \tag{1.9}$$

Observe that for every $c(x, x') \in C^\infty(\Omega \times \Omega)$ we find an $a(x, x', \xi) \in S^{-\infty}(\Omega \times \Omega \times \mathbb{R}^n)$ such that

$$\int c(x, x')u(x')dx' = \text{Op}(a)u(x)$$

for all $u \in C_0^\infty(\Omega)$. In fact, for every $\psi(\xi) \in \mathcal{S}(\mathbb{R}^n)$ such that $\int \psi(\xi)d\xi = 1$ we can write

$$\int c(x, x')u(x')dx' = \int e^{i(x-x')\xi} \big(c(x, x')e^{-i(x-x')\xi}\psi(\xi) \big) u(x')dx'd\xi,$$

where $a(x, x', \xi) = c(x, x')e^{-i(x-x')\xi}\psi(\xi)$ is as asserted.

Remark 1.1.8. The space $L^{-\infty}(\Omega)$ coincides with the space of operators (1.9) with $c(x, x') \in C^\infty(\Omega \times \Omega)$.

A properly supported operator $A \in L^\mu(\Omega)$ induces continuous operators

$$A : C_0^\infty(\Omega) \to C_0^\infty(\Omega), \quad C^\infty(\Omega) \to C^\infty(\Omega) \tag{1.10}$$

and extends to

$$A : \mathcal{E}'(\Omega) \to \mathcal{E}'(\Omega), \quad \mathcal{D}'(\Omega) \to \mathcal{D}'(\Omega).$$

Relation (1.10) allows us to form

$$a(x, \xi) = e_{-\xi} A e_\xi \tag{1.11}$$

for $e_\xi(x) := e^{ix\xi}$ for a properly supported A.

Let $\boldsymbol{K} \Subset \Omega$ and set for the moment

$$L_{\mathrm{cl}}^\mu(\Omega)_{\boldsymbol{K}} := \big\{ A \in L_{\mathrm{cl}}^\mu(\Omega) : \mathrm{supp}\, K_A \subseteq \boldsymbol{K} \big\}.$$

Then using the Fourier inversion formula $u(x) = \int e^{ix\xi} \widehat{u}(\xi) d\xi$ it follows that

$$Au(x) = \int \big(Ae^{ix\xi}\big) \widehat{u}(\xi) d\xi = \int e^{ix\xi} a(x,\xi) \widehat{u}(\xi) d\xi = \mathrm{Op}(a)u(x).$$

Together with the fact that (1.11) belongs to $S_{\mathrm{cl}}^\mu(\Omega \times \mathbb{R}^n)$ for $A \in L_{\mathrm{cl}}^\mu(\Omega)_{\boldsymbol{K}}$ we obtain a space

$$S_{\mathrm{cl}}^\mu(\Omega \times \mathbb{R}^n)_{\boldsymbol{K}} := \big\{ e_{-\xi} A e_\xi : A \in L_{\mathrm{cl}}^\mu(\Omega)_{\boldsymbol{K}} \big\}$$

which is closed in $S_{\mathrm{cl}}^\mu(\Omega \times \mathbb{R}^n)$. Via the bijection

$$\mathrm{Op}(\cdot) : S_{\mathrm{cl}}^\mu(\Omega \times \mathbb{R}^n)_{\boldsymbol{K}} \to L_{\mathrm{cl}}^\mu(\Omega)_{\boldsymbol{K}}$$

the space $L_{\mathrm{cl}}^\mu(\Omega)_{\boldsymbol{K}}$ becomes a Fréchet space. Then, also

$$L_{\mathrm{cl}}^\mu(\Omega) = L_{\mathrm{cl}}^\mu(\Omega)_{\boldsymbol{K}} + L^{-\infty}(\Omega)$$

is Fréchet in the topology of the non-direct sum (which is independent of the choice of $\boldsymbol{K}$).

If an operator $A \in L^\mu(\Omega)$ is written in the form $A = \mathrm{Op}(a) \bmod L^{-\infty}(\Omega)$ for an $a(x, x', \xi) \in S^\mu(\Omega \times \Omega \times \mathbb{R}^n)$, we also call $a(x, x', \xi)$ a double symbol of A (which is not uniquely determined by the operator). If a is independent of x' (x) we call it a left (right) symbol, sometimes written $a_{\mathrm{L}}(x, \xi)$ ($a_{\mathrm{R}}(x', \xi)$). Incidentally, for double symbols we also write $a_{\mathrm{D}}(x, x', \xi)$ rather than $a(x, x', \xi)$.

Theorem 1.1.9. *Every* $A \in L^\mu(\Omega)$ *admits representations*

$$A = \mathrm{Op}(a_{\mathrm{L}}) \bmod L^{-\infty}(\Omega) \quad and \quad A = \mathrm{Op}(a_{\mathrm{R}}) \bmod L^{-\infty}(\Omega)$$

with left and right symbols $a_{\mathrm{L}}(x, \xi)$ *and* $a_{\mathrm{R}}(x', \xi)$, *respectively. If* $a_{\mathrm{D}}(x, x', \xi)$ *is any double symbol of* A, *then we have asymptotic expansions*

$$a_{\mathrm{L}}(x, \xi) \sim \sum_{\alpha \in \mathbb{N}^n} \frac{1}{\alpha!} \big(\partial_\xi^\alpha D_{x'}^\alpha a_{\mathrm{D}}\big)(x, x', \xi)\big|_{x'=x}$$

and

$$a_{\mathrm{R}}(x', \xi) \sim \sum_{\alpha \in \mathbb{N}^n} \frac{1}{\alpha!} (-1)^{|\alpha|} \big(\partial_\xi^\alpha D_x^\alpha a_{\mathrm{D}}\big)(x, x', \xi)\big|_{x=x'},$$

respectively. Here $\partial_\xi^\alpha := \partial_{\xi_1}^{\alpha_1} \cdots \partial_{\xi_n}^{\alpha_n}$ *for* $\alpha = (\alpha_1, \ldots, \alpha_n)$, *and* $\alpha! := \alpha_1! \cdots \alpha_n!$.

Theorem 1.1.10. *Let $\chi : \Omega \to \widetilde{\Omega}$ be a diffeomorphism between open sets $\Omega \subseteq \mathbb{R}_x^n$, $\widetilde{\Omega} \subseteq \mathbb{R}_{\widetilde{x}}^n$, and let*

$$\widetilde{A} := \chi_* A := \left(\chi^*\right)^{-1} \circ A \circ \chi^* \tag{1.12}$$

be the operator push forward in the sense of operators $A : C_0^\infty(\Omega) \to C^\infty(\Omega)$ and $\widetilde{A} : C_0^\infty(\widetilde{\Omega}) \to C^\infty(\widetilde{\Omega})$, respectively, with the function pull-back χ^*. Then χ_* induces an isomorphism

$$\chi_* : L_{(\mathrm{cl})}^\mu(\Omega) \to L_{(\mathrm{cl})}^\mu(\widetilde{\Omega})$$

for every $\mu \in \mathbb{R}$. In particular, $A = \mathrm{Op}_x(a)$ mod $L^{-\infty}(\Omega)$ for an $a(x, \xi) \in S_{(\mathrm{cl})}^\mu(\Omega \times \mathbb{R}^n)$ implies $\widetilde{A} = \mathrm{Op}_{\widetilde{x}}(\widetilde{a})$ mod $L^{-\infty}(\widetilde{\Omega})$ for an $\widetilde{a}(\widetilde{x}, \widetilde{\xi}) \in S_{(\mathrm{cl})}^\mu(\widetilde{\Omega} \times \mathbb{R}^n)$, and we have

$$\widetilde{a}(\widetilde{x}, \widetilde{\xi})|_{\widetilde{x}=\chi(x)} \sim \sum_{\alpha \in \mathbb{N}^n} \frac{1}{\alpha!} \left(\partial_\xi^\alpha a\right)\left(x, {}^t d\chi(x)\widetilde{\xi}\right) \Pi_\alpha\left(x, \widetilde{\xi}\right)$$

for

$$\Pi_\alpha\left(x, \widetilde{\xi}\right) := D_z^\alpha e^{i\delta(x,z)\widetilde{\xi}}\Big|_{z=x}, \quad \delta(x, z) := \chi(z) - \chi(x) - d\chi(x)(z - x).$$

Here $d\chi(x)$ denotes the Jacobi matrix of χ at x.

It can be easily checked that $\Pi_\alpha(x, \widetilde{\xi})$ is a polynomial in $\widetilde{\xi}$ of degree $\leq |\alpha|/2$. In particular, we see that

$$\widetilde{a}\left(\chi(x), \left({}^t d\chi(x)\right)^{-1}\xi\right) = a(x, \xi) \mod S^{\mu-1}(\Omega \times \mathbb{R}^n).$$

In the case of classical operators A we have the homogeneous principal symbol

$$\sigma_\psi(A)(x, \xi) := a_{(\mu)}(x, \xi)$$

with $a_{(\mu)}(x, \xi)$ being the homogeneous principal component of $a(x, \xi) \in S_{\mathrm{cl}}^\mu(\Omega \times \mathbb{R}^n)$ of order μ. Then $\sigma_\psi(\widetilde{A})(\widetilde{x}, \widetilde{\xi}) = \sigma(A)(x, \xi)$ for $(\widetilde{x}, \widetilde{\xi}) = (\chi(x), ({}^t d\chi(x))^{-1}\xi)$ (in the notation of Theorem 1.1.10).

Let us now turn to pseudo-differential operators on a C^∞ manifold M. Throughout this consideration we assume that M is a Riemannian manifold, $n = \dim M$. The fixed Riemannian metric gives rise to a measure dx on M and an identification between $C^\infty(M \times M)$ and the space $L^{-\infty}(M)$ of smoothing operators

$$Cu(x) = \int_M c(x, x')u(x')dx',$$

with $c(x, x') \in C^\infty(M \times M)$.

If

$$A : C_0^\infty(M) \to C^\infty(M) \tag{1.13}$$

is a continuous operator, then for every open $U \subseteq M$ we can consider its restriction

$$A_U : C_0^\infty(U) \to C^\infty(U) \tag{1.14}$$

defined by $(Au)\big|_U$ for $u \in C_0^\infty(U)$. For a coordinate neighbourhood U on M and a chart $\chi : U \to \Omega$, $\Omega \subseteq \mathbb{R}^n$ open, we have the operator push-forward

$$\chi_* A_U : C_0^\infty(\Omega) \to C^\infty(\Omega) \tag{1.15}$$

defined analogously to (1.12). The space of pseudo-differential operators $L_{(\mathrm{cl})}^\mu(M)$ is defined to be the set of all operators (1.13) that are mod $L^{-\infty}(M)$ determined by the restrictions (1.14) with $\chi_* A_U \in L_{(\mathrm{cl})}^\mu(\Omega)$ for all U belonging to an atlas of charts $\chi : U \to \Omega$ on M.

The space $L_{(\mathrm{cl})}^\mu(M)$ admits a straightforward extension to $(m \times k)$-systems $L_{(\mathrm{cl})}^\mu(M; \mathbb{C}^k, \mathbb{C}^m)$ of pseudo-differential operators. Let us now generalise pseudo-differential operators to the case of spaces of distributional sections of (smooth complex) vector bundles over the manifold M. Let $\mathrm{Vect}(M)$ denote the set of such vector bundles. By definition every $E \in \mathrm{Vect}(M)$ has a system of trivialisations

$$\tau_U : E\big|_U \to U \times \mathbb{C}^k,$$

where k is the fibre dimension of E, and U runs over an open covering of M (say, by contractible coordinate neighbourhoods). Let $F \in \mathrm{Vect}(M)$ be another vector bundle of fibre dimension l, with the system of trivialisations

$$\eta_U : F\big|_U \to U \times \mathbb{C}^m.$$

Consider a continuous operator

$$A : C_0^\infty(M, E) \to C^\infty(M, F). \tag{1.16}$$

Then similarly as before, we have restrictions

$$A_U : C_0^\infty\big(U, E\big|_U\big) \to C^\infty(U, F|_U). \tag{1.17}$$

Using the isomorphisms

$$\tau_U^* : C_0^\infty(U, \mathbb{C}^k) \to C_0^\infty\big(U, E\big|_U\big), \quad \eta_U^* : C^\infty(U, \mathbb{C}^m) \to C^\infty\big(U, F\big|_U\big),$$

induced by the trivialisations, we pass to the $(m \times k)$-systems of operators

$$\left(\eta_U^*\right)^{-1} \circ A_U \circ \tau_U^* : C_0^\infty(U, \mathbb{C}^k) \to C^\infty(U, \mathbb{C}^m).$$

The space $L^{-\infty}(M; E, F)$ of smoothing operators $C : C_0^\infty(M, E) \to C^\infty(M, F)$ is defined as the set of all integral operators

$$Cu(x) = \int_M \langle c(x, x'), u(x') \rangle_{E_{x'}} \, dx'$$

with $c(x, x') \in C^\infty(M \times M, F \boxtimes E')$. Here E' is the dual bundle, with $\langle \cdot, \cdot \rangle_E$ being the fibre-wise defined bilinear pairing between E and E'. Moreover, $F \boxtimes E' := \pi_1^* F \otimes \pi_2^* E'$ for the projections $\pi_i : M \times M \to M$, $\pi_1(x, x') := x$, $\pi_2(x, x') := x'$, and the respective bundle pull-backs π_i^*, and $\otimes$ the tensor product between bundles.

Let $L_{(\mathrm{cl})}^\mu(M; E, F)$ be the set of operators (1.16) that are mod $L^{-\infty}(M; E, F)$ determined by their restrictions (1.17) such that

$$\chi_*\left(\left(\eta_U^*\right)^{-1} \circ A_U \circ \tau_U^*\right) : C_0^\infty(\Omega, \mathbb{C}^k) \to C^\infty(\Omega, \mathbb{C}^m), \tag{1.18}$$

belong to $L_{(\mathrm{cl})}^\mu(\Omega; \mathbb{C}^k, \mathbb{C}^m)$, for all coordinate neighbourhoods U from charts $\chi : U \to \Omega$ on M.

Remark 1.1.11. The space $L_{(\mathrm{cl})}^\mu(M; E, F)$ can be equivalently defined as follows. Let $(U_j)_{j \in \mathbb{N}}$ be a locally finite open covering of M by coordinate neighbourhoods, $\chi_j : U_j \to \mathbb{R}^n$ charts, $(\varphi_j)_{j \in \mathbb{N}}$ a subordinate partition of unity, and $(\psi_j)_{j \in \mathbb{N}}$ any system of functions $\psi_j \in C_0^\infty(U_j)$, $\varphi_j \prec \psi_j$ for all j (here $f \prec g$ or $g \succ f$ means that the function g is equal to 1 on supp f). Then we have

$$L_{(\mathrm{cl})}^\mu(M; E, F) = \Big\{ A_0 + C : A_0 = \sum_{j \in \mathbb{N}} \varphi_j \{(\chi_j^{-1})_* \mathrm{Op}(a_j)\} \psi_j,$$

$$a_j(x, \xi) \in S_{(\mathrm{cl})}^\mu(\mathbb{R}^n \times \mathbb{R}^n; \mathbb{C}^k, \mathbb{C}^m), \ C \in L^{-\infty}(M; E, F) \Big\}. \tag{1.19}$$

In this notation $(\chi_j^{-1})_* \mathrm{Op}(a_j)$ is an abbreviation of $\eta_{U_j}^* (\chi_j^{-1})_* \mathrm{Op}(a_j) \tau_{U_j}^{*-1}$, cf. (1.18). Observe that A_0 is properly supported.

Remark 1.1.12. The spaces $L_{(\mathrm{cl})}^\mu(M; E, F)$ are Fréchet in a natural way.

Let us make a choice of an adequate countable semi-norm system. For $A \in L_{(\mathrm{cl})}^\mu(M; E, F)$ and every $j \in \mathbb{N}$ the operator push-forward

$$(\chi_j)_* (\varphi_j A \psi_j) \in L_{(\mathrm{cl})}^\mu(\mathbb{R}^n; \mathbb{C}^k, \mathbb{C}^m)$$

is properly supported. Using an analogue of the above-mentioned bijection

$$\mathrm{symb} := \mathrm{Op}(\cdot)^{-1} : L_{(\mathrm{cl})}^\mu(\mathbb{R}^n; \mathbb{C}^k, \mathbb{C}^m)_K \to S_{(\mathrm{cl})}^\mu(\mathbb{R}^n; \mathbb{C}^k, \mathbb{C}^m)_K,$$

for every semi-norm ρ from the Fréchet topology of $S_{(\mathrm{cl})}^\mu(\mathbb{R}^n; \mathbb{C}^k, \mathbb{C}^m)_K$ we obtain a semi-norm

$$\rho\left(\mathrm{symb}\left((\chi_j)_* (\varphi_j A \psi_j)\right)\right)$$

on $L_{(\mathrm{cl})}^\mu(M; E, F)$. Moreover, we have

$$R := A - \sum_{j \in \mathbb{N}} \varphi_j A \psi_j \in L^{-\infty}(M; E, F).$$

This gives a well-defined map

$$\mathrm{rem}: L_{(\mathrm{cl})}^{\mu}(M; E, F) \to L^{-\infty}(M; E, F), \quad A \mapsto R,$$

and every semi-norm λ from the Fréchet topology of $L^{-\infty}(M; E, F)$ gives rise to a semi-norm $\lambda(\mathrm{rem}(A))$ on $L_{(\mathrm{cl})}^{\mu}(M; E, F)$. Now ρ, j and λ run over countable sets, i.e., we defined a countable system of semi-norms on the space $L_{(\mathrm{cl})}^{\mu}(M; E, F)$. We leave it as an exercise for the reader to verify that the space $L_{(\mathrm{cl})}^{\mu}(M; E, F)$ is Fréchet and that another choice of the involved data leads to an equivalent semi-norm system.

For $L_1, L_2 \in \mathrm{Vect}(T^*M \setminus 0)$ by

$$S^{(\mu)}(T^*M \setminus 0; L_1, L_2), \quad \mu \in \mathbb{R}, \tag{1.20}$$

we denote the set of bundle morphisms $\sigma: L_1 \to L_2$ such that

$$\sigma(x, \lambda\xi) = \lambda^{\mu}\sigma(x, \xi) \text{ for all } \lambda \in \mathbb{R}_+. \tag{1.21}$$

Here $(x, \lambda\xi)$, $\lambda \in \mathbb{R}_+$, means the invariantly defined $\mathbb{R}_+$-action on the fibres of the cotangent bundle, while λ^{μ} on the right of (1.21) means the corresponding action in the fibres of L_2 over (x, ξ).

In the special case $L_1 = \pi_M^* E$, $L_2 = \pi_M^* F$ for $E, F \in \mathrm{Vect}(M)$ we simply write

$$S^{(\mu)}(T^*M \setminus 0; E, F) \tag{1.22}$$

rather than (1.20).

The analogue of the above-mentioned homogeneous principal symbol $\sigma_\psi(A)$ in the case $A \in L_{\mathrm{cl}}^{\mu}(M; E, F)$ is now a (uniquely determined) bundle morphism

$$\sigma_\psi(A) : \pi_M^* E \to \pi_M^* F, \tag{1.23}$$

$\pi_M : T^*M \setminus 0 \to M$, belonging to $S^{(\mu)}(T^*M \setminus 0; E, F)$.

Proposition 1.1.13. *The principal symbol map*

$$\sigma_\psi : L_{\mathrm{cl}}^{\mu}(M; E, F) \to S^{(\mu)}(T^*M \setminus 0; E, F)$$

is surjective with $\ker \sigma_\psi = L_{\mathrm{cl}}^{\mu-1}(M; E, F)$, *and there is a linear map*

$$\mathrm{op} : S^{(\mu)}(T^*M \setminus 0; E, F) \to L_{\mathrm{cl}}^{\mu}(M; E, F)$$

such that $\sigma_\psi \circ \mathrm{op} = \mathrm{id}$.

Next we fix some notation on Sobolev spaces. First, for $s \in \mathbb{R}$ we set

$$H^s(\mathbb{R}^n) := \left\{ u \in \mathcal{S}'(\mathbb{R}^n) : \widehat{u} \in L_{\mathrm{loc}}^1(\mathbb{R}^n), \langle\xi\rangle^s \widehat{u}(\xi) \in L^2(\mathbb{R}^n) \right\}$$

which is the same as the completion of $C_0^\infty(\mathbb{R}^n)$ (or $\mathcal{S}(\mathbb{R}^n)$) with respect to the norm

$$\|u\|_{H^s(\mathbb{R}^n)} = \left\{ \int \langle \xi \rangle^{2s} |\hat{u}(\xi)|^2 d\xi \right\}^{1/2}.$$

Similarly, we have the space $H^s(\mathbb{R}^n, \mathbb{C}^k) = H^s(\mathbb{R}^n) \otimes \mathbb{C}^k$ for any $k \in \mathbb{N}$. Now if M is a C^∞ manifold and $E \in \mathrm{Vect}(M)$ with fibre dimension k, we let $H^s_{\mathrm{loc}}(M, E)$ denote the completion of $C^\infty(M, E)$ with respect to the system of semi-norms

$$\left\| (\chi^{-1})^* \left((\tau_U^*)^{-1}(\varphi u) \right) \right\|_{H^s(\mathbb{R}^n, \mathbb{C}^k)}$$

for an atlas $\chi : U \to \mathbb{R}^n$ on M, arbitrary $\varphi \in C_0^\infty(U)$, and the above-mentioned isomorphism $\tau_U^* : C_0^\infty(U, \mathbb{C}^k) \to C_0^\infty(U, E|_U)$, here combined with

$$\left(\chi^{-1} \right)^* : C_0^\infty(U, \mathbb{C}^k) \to C_0^\infty(\mathbb{R}^n, \mathbb{C}^k).$$

By $H^s_{\mathrm{comp}}(M, E)$ we denote the subspace of all $u \in H^s_{\mathrm{loc}}(M, E)$ with compact support. If M is compact the spaces $H^s_{\mathrm{comp}}(M, E)$ and $H^s_{\mathrm{loc}}(M, E)$ coincide, and we write $H^s(M, E)$, $s \in \mathbb{R}$. In particular, we fix an identification $H^0(M, E) = L^2(M, E)$ with a corresponding scalar product in the L^2-space, based on the Riemannian metric on M and a Hermitean metric in the bundle M. If M is not necessarily compact, we can define the space $L^2_{\mathrm{loc}}(M, E)$ endowed with a corresponding local scalar product.

Theorem 1.1.14. *An $A \in L^\mu(M; E, F)$ induces continuous operators*

$$A : H^s_{\mathrm{comp}}(M, E) \to H^{s-\mu}_{\mathrm{loc}}(M, F) \tag{1.24}$$

for all $s \in \mathbb{R}$. For compact M we have

$$A : H^s(M, E) \to H^{s-\mu}(M, F). \tag{1.25}$$

Remark 1.1.15. Every $A \in L^\mu(M; E, F)$ can be written in the form $A = A_0 + C$ for a properly supported $A_0 \in L^\mu(M; E, F)$ and a $C \in L^{-\infty}(M; E, F)$. If an operator A is properly supported, it induces continuous operators

$$A : H^s_{\mathrm{comp}}(M, E) \to H^{s-\mu}_{\mathrm{comp}}(M, F), \quad H^s_{\mathrm{loc}}(M, E) \to H^{s-\mu}_{\mathrm{loc}}(M, F)$$

for all $s \in \mathbb{R}$.

Theorem 1.1.16. *Let $A \in L^\mu_{(\mathrm{cl})}(M; E, F)$; then for the formal adjoint defined by*

$$(u, A^* v)_{L^2_{\mathrm{loc}}(M, E)} = (Au, v)_{L^2_{\mathrm{loc}}(M, F)}$$

for all $u \in C_0^\infty(M, E)$ and $v \in C_0^\infty(M, F)$, we have $A^* \in L^\mu_{(\mathrm{cl})}(M; F, E)$, and in the classical case

$$\sigma_\psi(A^*) = \sigma_\psi(A)^*$$

with "$*$" on the right-hand side referring to the Hermitean structures in the involved bundles.

Theorem 1.1.17. *Let* $A \in L^{\mu}_{(\mathrm{cl})}(M; E_0, F)$, $B \in L^{\nu}_{(\mathrm{cl})}(M; E, E_0)$, *and let* A *be properly supported. Then* $AB \in L^{\mu+\nu}_{(\mathrm{cl})}(M; E, F)$, *and in the classical case*

$$\sigma_\psi(AB) = \sigma_\psi(A)\sigma_\psi(B).$$

Definition 1.1.18. *An operator* $A \in L^{\mu}_{\mathrm{cl}}(M; E, F)$ *is called elliptic (of order* μ) *if its homogeneous principal symbol*

$$\sigma_\psi(A) : \pi_M^* E \to \pi_M^* F, \tag{1.26}$$

where $\pi_M : T^*M \setminus 0 \to M$, *is an isomorphism.*

Clearly, there is also an adequate notion of ellipticity in the non-classical case, and most of the general assertions below hold in analogous form also for non-classical operators.

Theorem 1.1.19. *An elliptic operator* $A \in L^{\mu}_{\mathrm{cl}}(M; E, F)$ *has a properly supported parametrix* $B \in L^{-\mu}_{\mathrm{cl}}(M; F, E)$, *i.e.,*

$$1 - BA \in L^{-\infty}(M; E, E), \quad 1 - AB \in L^{-\infty}(M; F, F).$$

Theorem 1.1.20. *Let* M *be compact and* $A \in L^{\mu}_{\mathrm{cl}}(M; E, F)$; *then the following conditions are equivalent:*

(i) *A is elliptic.*

(ii) *The operator*

$$A : H^s(M, E) \to H^{s-\mu}(M, F) \tag{1.27}$$

is Fredholm for some $s = s_0 \in \mathbb{R}$.

Remark 1.1.21. Let M be compact and $A \in L^{\mu}_{\mathrm{cl}}(M; E, F)$ elliptic.

(i) The operator (1.27) is Fredholm for all $s \in \mathbb{R}$.

(ii) $\ker_s A = \{u \in H^s(M, E) : Au = 0\}$ is a finite-dimensional subspace $V \subset C^\infty(M, E)$ independent of s, and there is a finite-dimensional subspace $W \subset C^\infty(M, F)$ independent of s such that $\mathrm{im}_s A + W = H^{s-\mu}(M, F)$ for every $s \in \mathbb{R}$; here $\mathrm{im}_s A = \{Au : u \in H^s(M, E)\}$.

(iii) There is a parametrix $B \in L^{-\mu}_{\mathrm{cl}}(M; F, E)$ such that

$$1 - BA : L^2(M, E) \to V \text{ and } 1 - AB : L^2(M, F) \to W$$

are projections.

For future references we consider a generalisation of Definition 1.1.18.

Definition 1.1.22. *An operator* $A \in L^{\mu}_{\mathrm{cl}}(M; E, F)$ *is called overdetermined (resp. underdetermined) elliptic if* (1.26) *is surjective (resp. injective).*

In the following remark we assume for simplicity that M is compact.

Remark 1.1.23. (i) If $A \in L_{\mathrm{cl}}^{\mu}(M; E, F)$ is overdetermined (underdetermined) elliptic, then the formal adjoint A^* is underdetermined (overdetermined) elliptic.

(ii) If $A \in L_{\mathrm{cl}}^{\mu}(M; E, F)$ is overdetermined elliptic, then $AA^* \in L_{\mathrm{cl}}^{2\mu}(M; F, F)$ is elliptic; moreover, there is a right parametrix $B_{\mathrm{R}} \in L_{\mathrm{cl}}^{-\mu}(M; F, E)$, i.e., we have $1 - AB_{\mathrm{R}} \in L^{-\infty}(M; F, F)$. If $A \in L_{\mathrm{cl}}^{\mu}(M; E, F)$ is underdetermined elliptic then

$$A^*A \in L_{\mathrm{cl}}^{2\mu}(M; E, E)$$

is elliptic; moreover, there is a left parametrix $B_{\mathrm{L}} \in L_{\mathrm{cl}}^{-\mu}(M; F, E)$, i.e., we have $1 - B_{\mathrm{L}}A \in L^{-\infty}(M; E, E)$.

In fact, for Remark 1.1.23 (ii) in the overdetermined case we take a parametrix $P \in L_{\mathrm{cl}}^{-2\mu}(M; F, F)$ of AA^*, using Theorem 1.1.19, and we can set $B_{\mathrm{R}} := A^*P$. In the underdetermined case we choose a parametrix $Q \in L_{\mathrm{cl}}^{-2\mu}(M; E, E)$ of A^*A, and we can set $B_{\mathrm{L}} := QA^*$.

Let us now introduce pseudo-differential operators with parameters $\lambda \in \mathbb{R}^l$. First, the space $L^{-\infty}(M; E, F)$ is Fréchet, and we set

$$L^{-\infty}(M; E, F; \mathbb{R}^l) := \mathcal{S}\big(\mathbb{R}^l, L^{-\infty}(M; E, F)\big). \qquad (1.28)$$

Moreover, we define the space $L_{(\mathrm{cl})}^{\mu}(M, E, F, \mathbb{R}^l)$ in a similar manner as (1.19) for $C \in L^{-\infty}(M; E, F; \mathbb{R}^l)$ and $a_j(x, \xi, \lambda) \in S_{(\mathrm{cl})}^{\mu}\big(\mathbb{R}^m \times \mathbb{R}_{\xi,\lambda}^{n+l}; \mathbb{C}^k, \mathbb{C}^m\big)$ for all j.

Remark 1.1.24. There is an analogue of Remark 1.1.12 in the parameter-dependent case, i.e., $L_{(\mathrm{cl})}^{\mu}(M; E, F; \mathbb{R}^l)$ is Fréchet in a natural way.

Every $A \in L_{(\mathrm{cl})}^{\mu}(M; E, F; \mathbb{R}^l)$ has a parameter-dependent homogeneous principal symbol

$$\sigma_{\psi}(A) \in S^{(\mu)}\big((T^*M \times \mathbb{R}^l) \setminus 0; E, F\big),$$

where 0 stands for $(\xi, \lambda) = 0$. In the parameter-dependent case we have analogues of Proposition 1.1.13 as well as of Example 1.11 and Theorem 1.1.5.

Theorem 1.1.25. *Let M be a smooth closed manifold. For every operator $A \in L^{\mu}(M; E, F; \mathbb{R}^l)$ and $\nu \geq \mu$ we have*

$$\|A(\lambda)\|_{\mathcal{L}(H^s(M,E), H^{s-\nu}(M,F))} \leq c \, \langle \lambda \rangle^{\max\{\mu, \mu-\nu\}}$$

for a constant $c = c(s, \nu) > 0$.

Corollary 1.1.26. *For every $A \in L^{-1}(M; E, F; \mathbb{R}^l)$ we have*

$$\|A(\lambda)\|_{\mathcal{L}(L^2(M,E), L^2(M,F))} \leq c \, \langle \lambda \rangle^{-1}$$

for a constant $c \geq 0$.

Definition 1.1.27. An operator $A \in L^{\mu}_{\mathrm{cl}}(M; E, F; \mathbb{R}^l)$ is called parameter-dependent elliptic of order μ if $\sigma_{\psi}(A)(x, \xi, \lambda) : E_x \to F_x$ is a family of isomorphisms for all $(x, \xi, \lambda) \in T^*X \times \mathbb{R}^l \setminus 0$.

Theorem 1.1.28. *A parameter-dependent elliptic* $A \in L^{\mu}_{\mathrm{cl}}(M; E, F; \mathbb{R}^l)$ *has a properly supported parameter-dependent parametrix* $B \in L^{-\mu}_{\mathrm{cl}}(M; F, E; \mathbb{R}^l)$, *i.e.,*

$$1 - BA \in L^{-\infty}(M; E, E; \mathbb{R}^l), \quad 1 - AB \in L^{-\infty}(M; F, F; \mathbb{R}^l).$$

Theorem 1.1.29. *Let M be a smooth closed manifold, and let $A \in L^{\mu}_{\mathrm{cl}}(M; E, F; \mathbb{R}^l)$ be parameter-dependent elliptic. Then*

$$A(\lambda) : H^s(M, E) \to H^{s-\mu}(M, F) \tag{1.29}$$

is a family of Fredholm operators of index zero, and there is a $C > 0$ such that the operators (1.29) *are isomorphisms for all $|\lambda| \geq C$ and $s \in \mathbb{R}$.*

1.2 Projections and Toeplitz operators

We now generalise the concept of pseudo-differential operators and study the so-called Toeplitz operators. Those are defined as pseudo-differential operators composed with projections on both sides. In our considerations the Hilbert spaces H are assumed to be separable and complex. Let $\mathcal{L}(H)$ denote the space of linear continuous operators $H \to H$. An operator $P \in \mathcal{L}(H)$ is called a projection if $P^2 = P$.

Remark 1.2.1. A projection P in a Hilbert space H is orthogonal if and only if $P^* = P$.

In fact, orthogonality of P means $((1 - P)u, Pv) = 0$ for all $u, v \in H$. This is the case when $P^* = P$. Conversely, $((1 - P)u, Pv) = 0$ for all $u, v \in H$ yields $0 = (u, Pv) - (Pu, Pv) = (P^*u, v) - (Pu, Pv)$, which entails $P^* = P^*P$. By forming adjoints it follows that $P = P^*P$, i.e., $P = P^*$.

Let M be an oriented closed compact C^{∞} manifold.

Theorem 1.2.2. *Let $J \in \mathrm{Vect}(M)$, and let $p : \pi_M^* J \to \pi_M^* J$ be a smooth bundle morphism of homogeneity 0 in the covariable $\xi \neq 0$, i.e., $p(x, \lambda\xi) = p(x, \xi)$ for all $(x, \xi) \in T^*M \setminus 0$, $\lambda \in \mathbb{R}_+$, and $p^2 = p$. Then there exists a $P \in L^0_{\mathrm{cl}}(M; J, J)$ such that $\sigma_{\psi}(P) = p$ and $P^2 = P$. Moreover, if p satisfies the condition $p^* = p$, the operator P can be chosen in such a way that $P^* = P$.*

For the proof we prepare the following general result. Let H be a Hilbert space and $\mathcal{K}(H) \subset \mathcal{L}(H)$ the set of compact operators in H. Recall that the quotient space $\mathcal{L}(H)/\mathcal{K}(H)$ is called the Calkin algebra. Let $\pi : \mathcal{L}(H) \to \mathcal{L}(H)/\mathcal{K}(H)$ be the corresponding canonical map. One of the basic properties of the Calkin algebra is that an operator $A \in \mathcal{L}(H)$ is Fredholm if and only if πA is invertible in $\mathcal{L}(H)/\mathcal{K}(H)$.

A set D in a topological space will be called discrete if its intersection with every compact set is finite.

Lemma 1.2.3. *Let $p \in \mathcal{L}(H)/\mathcal{K}(H)$ with $p^2 = p$, and let $Q \in \mathcal{L}(H)$ be an element such that $\pi Q = p$. Then the spectrum $\mathrm{spec}_{\mathcal{L}(H)}(Q)$ of Q has the property that*

$$\mathrm{spec}_{\mathcal{L}(H)}(Q) \cap \big(\mathbb{C}\backslash(\{0\} \cup \{1\})\big)$$

is a discrete set.

Proof. Let us first verify that $p^2 = p$ implies $\mathrm{spec}_{\mathcal{L}(H)/\mathcal{K}(H)}(p) \subseteq \{0\}\cup\{1\}$. In fact, denoting by $e \in \mathcal{L}(H)/\mathcal{K}(H)$ the identity in the Calkin algebra (which is equal to πI for the identity operator I in $\mathcal{L}(H)$), for every $\lambda \in \mathbb{C}\backslash(\{0\} \cup \{1\})$ there exists the inverse $(\lambda e - p)^{-1} = p(\lambda - 1)^{-1} + (e - p)\lambda^{-1}$. Now, setting $U := \mathbb{C}\backslash(\{0\}\cup\{1\})$, the operators $\lambda I - Q \in \mathcal{L}(H)$ form a holomorphic family of Fredholm operators in the complex variable $\lambda \in U$. Moreover, the operator $\lambda I - Q$ is invertible for $|\lambda| > \|Q\|_{\mathcal{L}(H)}$, as we see from

$$\left\|\left(I - \frac{Q}{\lambda}\right)^{-1}\right\|_{\mathcal{L}(H)} = \sum_{j=0}^{\infty} \left\|\left(\frac{Q}{\lambda}\right)^j\right\|_{\mathcal{L}(H)} \leq \sum_{j=0}^{\infty} \left(\frac{\|Q\|}{|\lambda|}\right)^j < \infty.$$

To complete the proof it suffices to employ the fact that $\lambda I - Q$ is invertible for all $\lambda \in U\backslash D$, where $D \subset U$ is some discrete subset (for a proof see [44, Subsection 2.2.3]). □

Proof of Theorem 1.2.2. By Lemma 1.2.3, the spectrum $\mathrm{spec}_{\mathcal{L}(H)}(Q)$ intersects the set $\mathbb{C}\backslash(\{0\} \cup \{1\})$ in a discrete set. Therefore, there is a $0 < \delta < 1$ such that the circle $C_\delta := \{\lambda \in \mathbb{C} : |\lambda - 1| = \delta\}$ does not intersect $\mathrm{spec}_{\mathcal{L}(H)}(Q)$. Let us form

$$P := \frac{1}{2\pi i} \int_{C_\delta} (\lambda I - Q)^{-1} d\lambda. \tag{1.30}$$

Then we have $P \in L^0_{\mathrm{cl}}(M; J, J)$ and $P^2 = P$ as a consequence of the holomorphic functional calculus for $L^0_{\mathrm{cl}}(M; J, J)$. The relation $\sigma_\psi(P) = p$ is obtained from

$$\sigma_\psi(P) = \frac{1}{2\pi i} \int_{C_\delta} (\lambda e - p)^{-1} d\lambda$$

$$= \left\{\frac{1}{2\pi i} \int_{C_\delta} \frac{1}{\lambda - 1} d\lambda\right\} p + \left\{\frac{1}{2\pi i} \int_{C_\delta} \frac{1}{\lambda} d\lambda\right\} (e - p) = p.$$

To prove the first assertion it remains to apply the Residue Theorem and Cauchy's Theorem, which show that the first summand on the right is equal to p and the second one vanishes.

For the second part of Theorem 1.2.2 we assume $p^* = p$. Applying the first part of the proof, we find a $P_1 \in L^0_{\mathrm{cl}}(M; J, J)$ such that $P_1^2 = P_1$ and $\sigma_\psi(P_1) = p$. Then $Q := P_1^* P_1 \in L^0_{\mathrm{cl}}(M; J, J)$ satisfies the relation $\sigma_\psi(Q) = p^* p = p^2 = p$. The operator Q has the property $Q = Q^* \geq 0$. If η denotes the spectral measure of Q, the projection $P \in L^0_{cl}(M; J, J)$ defined by the formula (1.30) is equal to the spectral projection $\eta(B_\delta(1) \cap \mathrm{spec}_{\mathcal{L}(L^2(M,J))}(Q))$ for $B_\delta(1) = \{\lambda \in \mathbb{C} : |\lambda - 1| < \delta\}$. In particular, we have $P = P^* = P^2$, and $\sigma_\psi(P) = p$ as before. $\qquad\square$

Remark 1.2.4. Observe that for a given $p(x, \xi)$ as in Theorem 1.2.2 there are many different pseudo-differential projections $P \in L^0_{cl}(M; J, J)$ with p as the homogeneous principal symbol. In fact, if $G \in L^0_{cl}(M; J, J)$ is an arbitrary elliptic pseudo-differential operator that induces an isomorphism $G : L^2(M, J) \to L^2(M, J)$, it is known that $G^{-1} \in L^0_{cl}(M; J, J)$, and $\tilde{P} := GPG^{-1} \in L^0_{cl}(M; J, J)$ is again a projection having p as its homogeneous principal symbol.

The construction of Theorem 1.2.2 of pseudo-differential projections for a given principal symbol which is a projection has the following general background. If Ψ is a Fréchet operator algebra with a given ideal $\mathcal{I}$, it is known that there is a lifting of idempotent elements of $\Psi/\mathcal{I}$ to idempotent elements in Ψ, provided that some natural assumptions on the operator algebra are satisfied, cf., Gramsch [17]. Such a situation is $\Psi := L^0_{cl}(M; J, J)$, $\mathcal{I} := L^{-1}_{cl}(M; J, J)$, and the quotient space $\Psi/\mathcal{I}$ is isomorphic to the space of homogeneous principal symbols of order 0. The general theory gives us a characterisation of all idempotent elements $P \in L^0_{cl}(M; J, J)$ belonging to the connected component of a given idempotent element $P_1 \in L^0_{cl}(M; J, J)$ and having the same homogeneous principal symbol as P_1. The result says that those P have the form GPG^{-1}, where G varies over the connected components of the identity in the group $\{I + K : I + K \text{ invertible}, K \in L^{-1}_{cl}(M; J, J)\}$.

Let H be a Hilbert space; then an element $P \in \mathcal{L}(H)$ is called a projection if it satisfies the relation $P^2 = P$. Observe that also $1 - P$ is a projection, the so-called complementary projection.

Proposition 1.2.5. *Let P, Q be projections in H such that $P - Q$ is a compact operator. Then the restrictions of P to $\operatorname{im} Q$ and of Q to $\operatorname{im} P$ are Fredholm operators*

$$P_Q : \operatorname{im} Q \to \operatorname{im} P, \quad Q_P : \operatorname{im} P \to \operatorname{im} Q$$

between the respective closed subspaces of H, and Q_P is a parametrix of P_Q, and vice versa.

Proof. Since the operator Q acts as the identity on $\operatorname{im} Q$, we have

$$Q_P P_Q - 1_{\operatorname{im} Q} = Q_P P_Q - Q^2 = Q_P(P_Q - Q_P) : \operatorname{im} Q \to \operatorname{im} Q,$$

i.e., $Q_P P_Q - 1_{\operatorname{im} Q}$ is a compact operator on $\operatorname{im} Q$. It follows that Q_P is a left parametrix of P_Q. In an analogous manner we see that $P_Q Q_P - 1_{\operatorname{im} P} = P_Q(Q_P - P_Q) : \operatorname{im} P \to \operatorname{im} P$ is compact, i.e., Q_P is also a right parametrix of P_Q, which means that P_Q is a Fredholm operator. At the same time we see that Q_P is also Fredholm. $\qquad\square$

Remark 1.2.6. Let ind (P, Q) denote the index of $P_Q : \operatorname{im} Q \to \operatorname{im} P$. Then we have

$$\operatorname{ind}(P, Q) = -\operatorname{ind}(Q, P).$$

Concerning more material and other observations on pseudo-differential projections, see also Birman and Solomyak [7] or Solomyak [61].

Let $J \in \operatorname{Vect}(M)$ and

$$p : \pi_M^* J \to \pi_M^* J \tag{1.31}$$

be a projection as in Theorem 1.2.2. Then

$$L := \operatorname{im} p \in \operatorname{Vect}(T^* M \backslash 0), \tag{1.32}$$

which is a subbundle of $\pi_M^* J$. Conversely, for every $L \in \operatorname{Vect}(T^* M \backslash 0)$ there exists a $J \in \operatorname{Vect}(M)$ such that L is a subbundle of $\pi_M^* J$. In fact, there exist an N and an $L^\perp \in \operatorname{Vect}(T^* M \backslash 0)$ such that $L \oplus L^\perp = (T^* M \backslash 0) \times \mathbb{C}^N$.

Definition 1.2.7. A triple $\mathbb{L} := (P, J, L)$ will be called projection data on M when $P \in L_{\mathrm{cl}}^0(M; J, J)$ is a projection as in Theorem 1.2.2, and L defined by (1.32). Let $\mathbb{P}(M)$ denote the set of all such projection data.

Proposition 1.2.8. (i) *For every* $J \in \operatorname{Vect}(M)$ *we have* $(\operatorname{id}, J, \pi_M^* J) \in \mathbb{P}(M)$.

(ii) *For every* $\mathbb{L} = (P, J, L)$, $\widetilde{\mathbb{L}} = (\widetilde{P}, \widetilde{J}, \widetilde{L}) \in \mathbb{P}(M)$ *we have*

$$\mathbb{L} \cup \widetilde{\mathbb{L}} := (P \cup \widetilde{P}, J \oplus \widetilde{J}, L \oplus \widetilde{L}) \in \mathbb{P}(M).$$

(iii) *Every* $\mathbb{L} = (P, J, L) \in \mathbb{P}(M)$ *admits complementary projection data* $\mathbb{L}^\perp \in \mathbb{P}(M)$ *in the sense that* $\mathbb{L} \oplus \mathbb{L}^\perp = (\operatorname{id}, F, \pi_M^* F)$ *for some* $F \in \operatorname{Vect}(M)$.

(iv) *Every* $\mathbb{L} = (P, J, L) \in \mathbb{P}(M)$ *has an adjoint* $\mathbb{L}^* = (P^*, J, L^*) \in \mathbb{P}(M)$, *where* P^* *is the adjoint of* P *in* $L^2(M, J)$ *given by* $\operatorname{im} p^*$ *for* $p^* = \sigma_\psi(P^*)$ *and* $L^* \in \operatorname{Vect}(T^* M \backslash 0)$. *Note that* $(\mathbb{L}^*)^* = \mathbb{L}$.

(v) *For every subbundle* L *of* $\pi_M^* J$, $J \in \operatorname{Vect}(M)$, *there exist projection data* $\mathbb{L} = (P, J, L) \in \mathbb{P}(M)$.

The proof is straightforward.

Remark 1.2.9. For every $\mathbb{L} = (P, J, L) \in \mathbb{P}(M)$ we find complementary projection data $\mathbb{L}^\perp = (1 - P, J, L^\perp)$ by setting $L^\perp = \operatorname{im} \sigma_\psi(1 - P)$, where we have $L \oplus L^\perp = \pi_M^* J$.

Remark 1.2.10. For every $\mathbb{L} = (P, J, L)$ we can form continuous projections also in Sobolev spaces $H^s(M, J)$ of distributional sections in the bundle J,

$$P : H^s(M, J) \to H^s(M, J), \quad s \in \mathbb{R}.$$

Let us set

$$H^s(M, \mathbb{L}) := PH^s(M, J). \tag{1.33}$$

This is a closed subspace of $H^s(M,J)$, in fact, a Hilbert space with the scalar product induced by $H^s(M,J)$. Occasionally if P is regarded as an operator on $H^s(M,J)$ we also write P^s.

Remark 1.2.11. Let $\mathbb{L} = (P,J,L)$ and $\widetilde{\mathbb{L}} = (\widetilde{P},\widetilde{J},L) \in \mathbb{P}(M)$ where J is a subbundle of $\widetilde{J}$ (such that L is a subbundle also of $\pi_M^* \widetilde{J}$) and $P : H^s(M,J) \to H^s(M,\mathbb{L})$ the restriction of $\widetilde{P} : H^s(M,\widetilde{J}) \to H^s(M,\widetilde{\mathbb{L}})$ to $H^s(M,J)$. Then $H^s(M,\mathbb{L}) = H^s(M,\widetilde{\mathbb{L}})$.

Let us now formulate other properties of the spaces $H^s(M,\mathbb{L})$.

Proposition 1.2.12. *We have continuous embeddings*

$$H^{s'}(M,\mathbb{L}) \hookrightarrow H^s(M,\mathbb{L}) \tag{1.34}$$

for every $s' \geq s$, which are compact for $s' > s$.

Proof. First recall that the embeddings $H^{s'}(M,J) \hookrightarrow H^s(M,J)$ are continuous for $s' \geq s$ and compact for $s' > s$. The closed subspace $H^{s'}(M,\mathbb{L})$ of $H^{s'}(M,J)$ is continuously embedded in $H^s(M,J)$, namely, as the space $\{P^{s'} u' : u' \in H^{s'}(M,J)\} \subseteq \{P^s u : u \in H^s(M,J)\}$, since $P^s u' = P^{s'} u'$ for $u' \in H^{s'}(M,J)$; in other words, we have an inclusion $H^{s'}(M,\mathbb{L}) \subset H^s(M,\mathbb{L})$. Let $\{u_k'\}_{k \in \mathbb{N}}$ be a sequence that converges to a limit u in $H^{s'}(M,\mathbb{L})$. We have $u_k' = P^{s'} u_k' = P^s u_k'$ since $P^{s'} = P^s|_{H^{s'}(M,J)}$. The continuous embedding $H^{s'}(M,\mathbb{L}) \hookrightarrow H^s(M,J)$ entails the convergence of $\{u_k'\}_{k \in \mathbb{N}}$ in $H^s(M,J)$. At the same time, $u_k' \in H^s(M,\mathbb{L})$; since the latter space is closed in $H^s(M,J)$, we obtain $u \in H^s(M,\mathbb{L})$.

Now let $s' > s$, and let $B \subset H^{s'}(M,\mathbb{L})$ be a bounded set. Then every sequence $\{u_k'\}_{k \in \mathbb{N}} \subset B$ contains a subsequence $\{u_{k_j}'\}_{j \in \mathbb{N}}$ convergent in $H^s(M,J)$. As above we have $u_{k_j}' = P^{s'} u_{k_j}' = P^s u_{k_j}'$, and we obtain convergence of this sequence in $H^s(M,J)$. Since $P^s u_{k_j}' \in H^s(M,\mathbb{L})$, one obtains the convergence in this closed subspace, which shows the compactness of (1.34) for $s' > s$. $\qquad\square$

Proposition 1.2.13. (i) *The space $H^\infty(M,\mathbb{L}) = \bigcap_{s \in \mathbb{R}} H^s(M,\mathbb{L})$ is dense in the space $H^s(M,\mathbb{L})$ for every $s \in \mathbb{R}$.*

(ii) *Let $H^0(M,\mathbb{L})$ be endowed with the scalar product from $H^0(M,J)$, and let $V \subset H^\infty(M,\mathbb{L})$ be a subspace of finite dimension. Then the orthogonal projection $C_V : H^0(M,\mathbb{L}) \to V$ induces continuous operators $C_V : H^s(M,\mathbb{L}) \to V$ for all $s \in \mathbb{R}$, and C_V is a compact operator $H^s(M,\mathbb{L}) \to H^s(M,\mathbb{L})$ for every $s \in \mathbb{R}$.*

Proof. (i) First recall that $C^\infty(M,J)$ is dense in $H^s(M,J)$. Thus, writing $u \in PH^s(M,J)$ in the form $u = P^s u$ and choosing a sequence $\{\varphi_k\}_{k \in \mathbb{N}} \subset C^\infty(M,J)$ with $\varphi_k \to u$ in $H^s(M,J)$ as $k \to \infty$, it follows that $P^s \varphi_k \to P^s u = u$ in $H^s(M,J)$ as $k \to \infty$. It remains to show that $P^s C^\infty(M,J) \subset H^\infty(M,\mathbb{L})$ for every $s \in \mathbb{R}$. However, since $C^\infty(M,J) = \bigcap_{s \in \mathbb{R}} H^{s'}(M,J)$, and $P^s|_{H^{s'}(M,J)} = P^{s'}$, for

$u' \in H^{s'}(M, J)$ we obtain $P^s u' \in H^{s'}(M, J)$ for every $s' \geq s$, i.e., $P^s C^\infty(M, J) \subset$ $P^{s'} H^{s'}(M, J)$ for every $s' \geq s$. This yields $P^s C^\infty(M, J) \subset \bigcap_{s' \in \mathbb{R}} H^{s'}(M, \mathbb{L}) = H^\infty(M, \mathbb{L})$.

(ii) Let $N = \dim V$ and choose an orthogonal basis $\{v_j\}_{j=1,\dots,N}$ in V. Then the orthogonal projection $C_V : H^0(M, \mathbb{L}) \to V$ can be written in the form

$$C_V u = \sum_{j=1}^{N} (u, v_k) v_k, \tag{1.35}$$

where $(\cdot, \cdot)$ denotes the $H^0(M, J)$-scalar product. Now using the fact that $V \subset H^\infty(M, J)$ and the non-degenerate sesquilinear pairing

$$H^s(M, J) \times H^{-s}(M, J) \to \mathbb{C} \tag{1.36}$$

via $(\cdot, \cdot)$, we deduce that (u, v_k) exists for every $s \in \mathbb{R}$. Thus (1.35) defines a continuous operator $H^s(M, J) \to V$ for every $s \in \mathbb{R}$. Because of the inclusion $V \subset H^\infty(M, \mathbb{L})$ and Proposition 1.2.12, we have a continuous embedding $V \hookrightarrow H^{s'}(M, \mathbb{L})$ for every $s' \in \mathbb{R}$. This establishes the continuity of $C_V : H^s(M, \mathbb{L}) \to H^{s'}(M, \mathbb{L})$ for every $s' \in \mathbb{R}$. For $s' \geq s$ we also have a compact embedding $H^{s'}(M, \mathbb{L}) \to H^s(M, \mathbb{L})$, cf. Proposition 1.2.12. This gives us altogether the compactness of $C_V : H^s(M, \mathbb{L}) \to H^s(M, \mathbb{L})$. $\qquad\square$

Remark 1.2.14. Let $\mathbb{L} = (P, J, L) \subset \mathbb{P}(M)$ and let $\mathbb{L}^* = (\varGamma^*, J, L^*)$ be its adjoint in the sense of Proposition 1.2.8 (iv). Then we have the subspaces

$$H^s(M, \mathbb{L}) \subseteq H^s(M, J), \quad H^{-s}(M, \mathbb{L}^*) \subseteq H^{-s}(M, J)$$

for every $s \in \mathbb{R}$. Then the sesquilinear pairing $(\cdot, \cdot) : H^\infty(M, \mathbb{L}) \times H^\infty(M, \mathbb{L}^*) \to \mathbb{C}$ induced by (1.36) extends to a non-degenerate sesquilinear pairing

$$(\cdot, \cdot) : H^s(M, \mathbb{L}) \times H^{-s}(M, \mathbb{L}^*) \to \mathbb{C} \tag{1.37}$$

for every $s \in \mathbb{R}$ (which is nothing else than the restriction of (1.36) to the respective subspaces). Whenever necessary the pairing will also be denoted by

$$(\cdot, \cdot)_{H^s(M, \mathbb{L}) \times H^{-s}(M, \mathbb{L}^*)}.$$

It follows that for every $s \in \mathbb{R}$ we have an equivalence of norms

$$\|u\|_{H^s(M, \mathbb{L})} \sim \sup_{\substack{f \in H^{-s}(M, \mathbb{L}^*) \\ f \neq 0}} \frac{|(u, f)|}{\|f\|_{H^{-s}(M, \mathbb{L}^*)}} = \sup_{\substack{f \in H^\infty(M, \mathbb{L}^*) \\ f \neq 0}} \frac{|(u, f)|}{\|f\|_{H^{-s}(M, \mathbb{L}^*)}}$$

and analogously for $\|f\|_{H^{-s}(M, \mathbb{L}^*)}$.

The only point to show is that (1.37) is non-degenerate. So assume that for an element $v \in H^{-s}(M, \mathbb{L}^*)$ we have $(u, v) = 0$ for all $u \in H^s(M, \mathbb{L})$; as always,

$(\cdot, \cdot)$ means the scalar product in $H^0(M, J)$. Then since $H^s(M, \mathbb{L}) = PH^s(M, J)$, it follows that $(Pf, v) = 0$ for all $f \in H^s(M, J)$. This gives $(f, P^*v) = 0$ for all those f. Since $v \in H^{-s}(M, \mathbb{L}^*)$ means $P^*v = v$, we obtain $(f, v) = 0$ for all f, which entails $v = 0$ because of the non-degeneracy of (1.36).

Observe that for every $s \in \mathbb{R}$ we have

$$|(u, v)| \le \|u\|_{H^s(M,\mathbb{L})} \|v\|_{H^{-s}(M,\mathbb{L}^*)}, \qquad (1.38)$$

first for $u \in H^\infty(M, \mathbb{L})$, $v \in H^\infty(M, \mathbb{L}^*)$, and then for all $u \in H^s(M, \mathbb{L})$, $v \in H^{-s}(M, \mathbb{L}^*)$. In fact, from the corresponding property of the $H^0(M, J)$ scalar product we know that

$$|(f, g)| \le \|f\|_{H^s(M,J)} \|g\|_{H^{-s}(M,J)} \qquad (1.39)$$

for all $f \in H^s(M, J)$, $g \in H^{-s}(M, J)$. Then, since $H^s(M, \mathbb{L})$ and $H^{-s}(M, \mathbb{L}^*)$ are closed subspaces of $H^s(M, J)$ we may insert in (1.39) the corresponding elements in the subspaces, which gives us the estimate (1.38).

Let $\mathbb{L}_i = (P_i, J_i, L_i) \in \mathbb{P}(M)$, $i = 1, 2$, and let $A : H^\infty(M, \mathbb{L}_1) \to H^\infty(M, \mathbb{L}_2)$ be an operator that extends to a continuous operator

$$A : H^s(M, \mathbb{L}_1) \to H^{s-\mu}(M, \mathbb{L}_2)$$

for all $s \in \mathbb{R}$ and some $\mu \in \mathbb{R}$. Then there is a (unique) $A^* : H^\infty(M, \mathbb{L}_2^*) \to H^\infty(M, \mathbb{L}_1^*)$, defined by

$$(Au, v)_{H^0(M,\mathbb{L}_2) \times H^0(M,\mathbb{L}_2^*)} = (u, A^*v)_{H^0(M,\mathbb{L}_1) \times H^0(M,\mathbb{L}_1^*)} \qquad (1.40)$$

for all $u \in H^\infty(M, \mathbb{L}_1)$, $v \in H^\infty(M, \mathbb{L}_2^*)$, that extends to a continuous operator

$$A^* : H^s(M, \mathbb{L}_2^*) \to H^{s-\mu}(M, \mathbb{L}_1^*) \qquad (1.41)$$

for every $s \in \mathbb{R}$. This operator A^* is called the formal adjoint of the operator A.

Let us briefly give the arguments for the latter assertion. First, if A_1^* and A_2^* satisfy the relation (1.40) for the same A and all u, v as indicated, then

$$(u, (A_1^* - A_2^*)v)_{H^0(M,\mathbb{L}_1) \times H^0(M,\mathbb{L}_1^*)} = 0.$$

This means, since $f := (A_1^* - A_2^*)v$ belongs to $H^\infty(M, \mathbb{L}_1^*)$, that

$$(u, f)_{H^0(M,\mathbb{L}_1) \times H^0(M,\mathbb{L}_1^*)} = 0$$

for all $u \in H^\infty(M, \mathbb{L}_1)$. As noted before in the context of non-degeneracy of the pairing and since $H^\infty(M, \mathbb{L}_1)$ is dense in $H^s(M, \mathbb{L}_1)$ for every $s \in \mathbb{R}$, it follows that $f = 0$. Let us now verify that A^* induces a continuous operator (1.41) for every $s \in \mathbb{R}$.

By virtue of the above-mentioned equivalence of norms, we have

$$\|A^*v\|_{H^{-s}(M,\mathbb{L}_1^*)} \sim \sup_{\substack{u\in H^{\infty}(M,\mathbb{L}_1) \\ u\neq 0}} \frac{|(u, A^*v)|}{\|u\|_{H^s(M,\mathbb{L}_1)}}$$

$$= \sup \frac{|(Au, v)|}{\|u\|_{H^s(M,\mathbb{L}_1)}}$$

$$\leq \sup \frac{\|Au\|_{H^{s-\mu}(M,\mathbb{L}_2)}}{\|u\|_{H^s(M,\mathbb{L}_1)}} \|v\|_{H^{-s+\mu}(M,\mathbb{L}_2^*)}$$

$$= \|A\|_{\mathcal{L}(H^s(M,\mathbb{L}_1), H^{s-\mu}(M,\mathbb{L}_2))} \|v\|_{H^{-s+\mu}(M,\mathbb{L}_2^*)}.$$

This shows the continuity of $A^* : H^{-s+\mu}(M,\mathbb{L}_2^*) \to H^{-s}(M,\mathbb{L}_1^*)$. Since this holds for every s, we obtain (1.41).

Given data $\mathbb{L} = (P, J, L) \in \mathbb{P}(M)$ with the subspaces

$$H^s(M, \mathbb{L}) \hookrightarrow H^s(M, J), \quad s \in \mathbb{R}, \tag{1.42}$$

we consider the embedding operator E given by (1.42). Analogously we observe the embedding

$$e : L \to \pi_M^* J \tag{1.43}$$

as a subbundle, where (1.43) is assumed to be homogeneous of order 0 in ξ. More precisely, if S^*M denotes the unit cosphere bundle induced by $T^*M\backslash 0$ (with respect to a fixed Riemannian metric on M) and if $\pi_1 : T^*M\backslash 0 \to S^*M$ denotes the canonical projection, defined by $(x, \xi) \to (x, \xi/|\xi|)$, then we have $L = \pi_1^* L_1$ for $L_1 := L|_{S^*M} \in \mathrm{Vect}(S^*M)$. Similarly we have $\pi_M^* J = \pi_1^*((\pi_M^* J)_1)$; then we obtain an embedding

$$e_1(x, \xi) : (L_1)_{(x,\xi)} \to ((\pi_M^* J)_1)_{(x,\xi)}, \quad (x, \xi) \in S^*M, \tag{1.44}$$

which induces embeddings

$$e(x, \xi) : L_{(x,\xi)} \to (\pi_M^* J)_{(x,\xi)}, \quad (x, \xi) \in T^*M\backslash 0, \tag{1.45}$$

defined by the composition of linear mappings

$$e(x, \xi) : L_{(x,\xi)} \to L_{(x,\xi/|\xi|)} \to ((\pi_M^* J)_1)_{(x,\xi/|\xi|)} \to (\pi_M^* J)_{(x,\xi)},$$

where the first mapping is defined by the bundle pull back under the embedding $S^*M \hookrightarrow T^*M\backslash 0$, the second one by (1.44), and the third one by the identification $(\pi_M^* J)_{(x,\xi)} = J_x$, $x \in M$. Then we have $e(x, \lambda\xi) = e(x, \xi)$ for all $\lambda \in \mathbb{R}$, $(x, \xi) \in T^*M\backslash 0$.

Definition 1.2.15. Let $\mathbb{L}_i := (P_i, J_i, L_i)$, $i = 1, 2$, with the corresponding operators $E_1 : H^{\infty}(M, \mathbb{L}_1) \to H^{\infty}(M, J_1)$ and $P_2 : H^{\infty}(M, J_2) \to H^{\infty}(M, \mathbb{L}_2)$, respectively. An operator of the form

$$A = P_2 \widetilde{A} E_1$$

for some $\widetilde{A} \in L^\mu_{cl}(M; J_1, J_2)$, $\mu \in \mathbb{R}$, is called a Toeplitz operator of order $\mu \in \mathbb{R}$ associated with the projection data $\mathbb{L}_1$, $\mathbb{L}_2$. We denote by $T^\mu(M; \mathbb{L}_1, \mathbb{L}_2)$ the set of all Toeplitz operators on M of order μ. Moreover, we set

$$T^{-\infty}(M; \mathbb{L}_1, \mathbb{L}_2) := \{P_2\widetilde{C}E_1 : \widetilde{C} \in L^{-\infty}(M; J_1, J_2)\}. \tag{1.46}$$

Observe that $T^{\mu-j}(M; \mathbb{L}_1, \mathbb{L}_2) \subseteq T^\mu(M; \mathbb{L}_1, \mathbb{L}_2)$ for every $j \in \mathbb{N}$, and

$$T^{-\infty}(M; \mathbb{L}_1, \mathbb{L}_2) \subseteq T^\mu(M; \mathbb{L}_1, \mathbb{L}_2) \quad \text{for every } \mu.$$

Thus

$$T^{-\infty}(M; \mathbb{L}_1, \mathbb{L}_2) \subseteq \bigcap_{\mu\in\mathbb{R}} T^\mu(M; \mathbb{L}_1, \mathbb{L}_2).$$

Note that the space $T^{-\infty}(M; \mathbb{L}_1, \mathbb{L}_2)$ can be equivalently defined as the set of all $A \in T^\infty(M; \mathbb{L}_1, \mathbb{L}_2) := \bigcup_{\mu\in\mathbb{R}} T^\mu(M; \mathbb{L}_1, \mathbb{L}_2)$ such that there is an operator $\widetilde{A} \in L^\mu_{cl}(M; J_1, J_2)$ for some $\mu \in \mathbb{R}$ with $A = P_2\widetilde{A}E_1$, where

$$\widetilde{C} := P_2\widetilde{A}P_1 \in L^{-\infty}(M; J_1, J_2);$$

then $A = P_2\widetilde{C}E_1$. Moreover,

$$P_2\widetilde{C}E_1 \in T^{-\infty}(M; \mathbb{L}_1, \mathbb{L}_2) \iff P_2\widetilde{C}P_1 \in L^{-\infty}(M; J_1, J_2). \tag{1.47}$$

Proposition 1.2.16. *Given $\mathbb{L}_i \in \mathbb{P}(M)$, $i = 1, 2$, we have a canonical isomorphism*

$$T^\mu(M; \mathbb{L}_1, \mathbb{L}_2) \to \{P_2\widetilde{A}P_1 : \widetilde{A} \in L^\mu_{cl}(M; J_1, J_2)\}. \tag{1.48}$$

Proof. In $L^\mu_{cl}(M; J_1, J_2) \ni \widetilde{A}, \widetilde{B}$ we define an equivalence relation by

$$\widetilde{A} \sim \widetilde{B} \iff P_2\widetilde{A}P_1 = P_2\widetilde{B}P_1. \tag{1.49}$$

Both sides represent the same element in $T^\mu(M; \mathbb{L}_1, \mathbb{L}_2)$. In fact, since the operators vanish on $\mathrm{im}\,(1 - P_1)$, they coincide on $\mathrm{im}\,P_1$, i.e., $P_2\widetilde{A}E_1 = P_2\widetilde{B}E_1$. Conversely, we can identify $P_2\widetilde{A}E_1$ with $P_2\widetilde{A}P_1|_{\mathrm{im}\,P_1} : \mathrm{im}\,P_1 \to \mathrm{im}\,P_2$; and $P_2\widetilde{B}E_1$ with $P_2\widetilde{B}P_1|_{\mathrm{im}\,P_1} : \mathrm{im}\,P_1 \to \mathrm{im}\,P_2$. Now $P_2\widetilde{A}P_1|_{\mathrm{im}\,P_1}$ can be identified with $P_2\widetilde{A}P_1$, since this operator vanishes on $\mathrm{im}\,(1 - P_1)$. The same conclusion for $\widetilde{B}$ shows the relation (1.49). $\qquad\square$

Remark 1.2.17. The space $T^\mu(M; \mathbb{L}_1, \mathbb{L}_2)$ can be identified with the corresponding quotient space $L^\mu_{cl}(M; J_1, J_2)/\sim$.

Observe that for $\mathbb{L}_i \in \mathbb{P}(M)$, $i = 1, 2$, and $\widetilde{A} \in L^\mu_{cl}(M; J_1, J_2)$ we have $\widetilde{A} \sim P_2\widetilde{A}P_1$.

Theorem 1.2.18. *An operator $A \in \mathcal{T}^\mu(M; \mathbb{L}_1, \mathbb{L}_2)$ extends to a continuous operator*

$$A : H^s(M, \mathbb{L}_1) \to H^{s-\mu}(M, \mathbb{L}_2) \tag{1.50}$$

for every $s \in \mathbb{R}$.

Proof. The proof is an immediate consequence of the definition of a Toeplitz operator as $A = P_2 \widetilde{A} E_1$. Both the embedding $E_1 : H^s(M, \mathbb{L}_1) \to H^s(M, J_1)$ and the projection $P_2 : H^{s-\mu}(M, J_2) \to H^{s-\mu}(M, \mathbb{L}_2)$ are continuous, and also $\widetilde{A} : H^s(M, J_1) \to H^{s-\mu}(M, J_2)$ is a continuous operator. $\qquad \square$

Remark 1.2.19. For every pair of projection data $\mathbb{L}_i = (P_i, J_i, L_i) \in \mathbb{P}(M), i = 1, 2,$ there exist $\mathbb{M}_i = (Q_i, \mathbb{C}^m, L_i) \in \mathbb{P}(M)$, such that

$$\mathcal{T}^\mu(M; \mathbb{L}_1, \mathbb{L}_2) = \mathcal{T}^\mu(M; \mathbb{M}_1, \mathbb{M}_2). \tag{1.51}$$

In fact, every two bundles J_1, J_2 over M can be regarded as subbundles of a trivial bundle $\mathbb{C}^m$; it suffices to use the fact that $J_1 \oplus J_2$ has a complementary bundle $(J_1 \oplus J_2)^\perp$, where $J_1 \oplus J_2 \oplus (J_1 \oplus J_2)^\perp = \mathbb{C}^m$ for a resulting m. Let $J_i^\perp$ be the complementary bundle of J_i in $\mathbb{C}^m$, $i = 1, 2$. According to Theorem 1.2.2, with the projection $\pi_M^* \mathbb{C}^m \to \pi_M^* J_i$ along $\pi_M^* J_i^\perp$ we can associate pseudo-differential projections $\widetilde{P}_i \in L^0_{\mathrm{cl}}(M; \mathbb{C}^m, \mathbb{C}^m)$, $\widetilde{P}_i : H^s(M, \mathbb{C}^m) \to H^s(M, \mathbb{C}^m)$, $i = 1, 2$. Moreover, we have our original projections $P_i : \pi_M^* J_i \to L_i$, which gives us projections $\pi_M^* \mathbb{C}^m \to L_i$ and associated pseudo-differential projections $Q_i \in L^0_{\mathrm{cl}}(M; \mathbb{C}^m, \mathbb{C}^m)$, where $Q_i = P_i \widetilde{P}_i$, and $Q_i : H^s(M, \mathbb{C}^m) \to H^s(M, L_i)$, $i = 1, 2$. The relation (1.51) then follows from the fact that every $\widetilde{A} \in L^\mu_{\mathrm{cl}}(M; J_1, J_2)$ can be identified with some $\widetilde{\widetilde{A}} \in L^\mu_{\mathrm{cl}}(M; \mathbb{C}^m, \mathbb{C}^m)$ by setting $\widetilde{A} = \widetilde{\widetilde{A}} \widetilde{P}_1$.

Recall that whenever $A \in L^\mu_{\mathrm{cl}}(M; J_1, J_2)$, $A' \in L^\mu_{\mathrm{cl}}(M; J_1', J_2')$ are pseudo-differential operators, $J_i, J_i' \in \mathrm{Vect}(M)$, $i = 1, 2$, we have the direct sum

$$A \oplus A' := \mathrm{diag}(A, A') \in L^\mu_{\mathrm{cl}}\big(M; J_1 \oplus J_1', J_2 \oplus J_2'\big).$$

A similar operation is possible on the level of Toeplitz operators. In fact, let $\mathbb{L}_i = (P_i, J_i, L_i)$, $\mathbb{L}_i' = (P_i', J_i', L_i')$, $i = 1, 2$, be projection data on M; then for $A := P_2 \widetilde{A} E_1 \in \mathcal{T}^\mu(M; \mathbb{L}_1, \mathbb{L}_2)$ and $A' := P_2' \widetilde{A}' R_1' \in \mathcal{T}^\mu(M; \mathbb{L}_1', \mathbb{L}_2')$ we have the direct sum

$$A \oplus A' := \mathrm{diag}(A, A') \in \mathcal{T}^\mu\big(M; \mathbb{L}_1 \oplus \mathbb{L}_1', \mathbb{L}_2 \oplus \mathbb{L}_2'\big).$$

Proposition 1.2.20. *Let $A_j \in \mathcal{T}^{\mu-j}(M; \mathbb{L}_1, \mathbb{L}_2)$, $j \in \mathbb{N}$, be an arbitrary sequence. Then there exists an operator $A \in \mathcal{T}^\mu(M; \mathbb{L}_1, \mathbb{L}_2)$ called an asymptotic sum of the A_j, such that*

$$A - \sum_{j=1}^{N} A_j \in \mathcal{T}^{\mu-(N+1)}(M; \mathbb{L}_1, \mathbb{L}_2) \tag{1.52}$$

for every $N \in \mathbb{N}$, and A is unique mod $\mathcal{T}^{-\infty}(M; \mathbb{L}_1, \mathbb{L}_2)$.

Proof. By definition, we have representations $A_j = P_2 \widetilde{A}_j E_1$ for suitable $\widetilde{A}_j \in L_{\mathrm{cl}}^{\mu-j}(M; J_1, J_2)$. We apply the well-known property of standard pseudo-differential operators that there is an $\widetilde{A} \in L_{\mathrm{cl}}^{\mu}(M; J_1, J_2)$ such that

$$\widetilde{A} - \sum_{j=1}^{N} \widetilde{A}_j \in L_{\mathrm{cl}}^{\mu-(N+1)}(M; J_1, J_2)$$

for every $N \in \mathbb{N}$, where $\widetilde{A}$ is unique mod $L^{-\infty}(M; J_1, J_2)$. Setting $A := P_2 \widetilde{A} E_1$, we obviously obtain (1.52). Moreover, if we have an $A' \in \mathcal{T}^{\mu}(M; \mathbb{L}_1, \mathbb{L}_2)$ (which is always of the form $A' := P_2 \widetilde{A}' E_1$ for some $\widetilde{A}' \in L_{\mathrm{cl}}^{\mu}(M; J_1, J_2)$) satisfying an analogue of relation (1.52), then we have $A - A' = P_2(\widetilde{A} - \widetilde{A}')E_1$ and

$$P_2(\widetilde{A} - \widetilde{A}')P_1 \in \cap_{j \in \mathbb{N}} L_{\mathrm{cl}}^{\mu-j}(M; J_1, J_2) = L^{-\infty}(M; J_1, J_2).$$

This entails

$$A - A' = P_2(\widetilde{A} - \widetilde{A}')P_1 E_1 \in \mathcal{T}^{-\infty}(M; \mathbb{L}_1, \mathbb{L}_2),$$

cf. also the relation (1.47). □

Definition 1.2.21. Let $A \in \mathcal{T}^{\mu}(M; \mathbb{L}_1, \mathbb{L}_2)$ satisfy $A = P_2 \widetilde{A} E_1$ for some $\widetilde{A} \in L_{\mathrm{cl}}^{\mu}(M; J_1, J_2)$. We define the homogeneous principal symbol of A as the bundle morphism

$$\sigma_\psi(A) : L_1 \to L_2,$$

given fibrewise over $(x, \xi) \in T^*M \setminus 0$ as the composition

$$\sigma_\psi(A)(x, \xi) = \sigma_\psi(P_2)(x, \xi)\sigma_\psi(\widetilde{A})(x, \xi)\sigma_\psi(E_1)(x, \xi). \tag{1.53}$$

Here $\sigma_\psi(E_1)(x, \xi)$ is interpreted as (1.45) for L_1 and J_1 instead of L and J, respectively, while $\sigma_\psi(\widetilde{A})(x, \xi)$ and $\sigma_\psi(P_2)(x, \xi)$ are the standard homogeneous principal symbols of the corresponding classical pseudo-differential operators.

To simplify considerations, we occasionally identify $\sigma_\psi(A)$ with

$$\sigma_\psi(P_2)\sigma_\psi(\widetilde{A})\sigma_\psi(P_1),$$

where $\sigma_\psi(P_1)|_{L_1} : L_1 \to \pi_M^* J_1$ is identified with the embedding $\sigma_\psi(E_1)$.

Let $S^{(\mu)}(T^*M \setminus 0; L_1, L_2)$ for $L_1, L_2 \in \mathrm{Vect}(T^*M \setminus 0)$ denote the space of all bundle morphisms

$$\sigma : L_1 \to L_2$$

such that $\sigma(x, \lambda\xi) = \lambda^\mu \sigma(x, \xi)$, $\lambda \in \mathbb{R}_+$, as a linear mapping $L_{1,(x,\xi)} \to L_{2,(x,\xi)}$ for every $(x, \xi) \in T^*M \setminus 0$.

Then σ_ψ yields a linear map

$$\sigma_\psi : \mathcal{T}^{\mu}(M; \mathbb{L}_1, \mathbb{L}_2) \to S^{(\mu)}(T^*M \setminus 0; L_1, L_2). \tag{1.54}$$

Theorem 1.2.22. (i) *The principal symbol map* (1.54) *is surjective, and there is a right inverse, also called an operator convention,*

$$\operatorname{op}: S^{(\mu)}(T^*M \setminus 0; L_1, L_2) \to \mathcal{T}^\mu(M; \mathbb{L}_1, \mathbb{L}_2).$$

(ii) *The kernel of* (1.54) *coincides with* $\mathcal{T}^{\mu-1}(M; \mathbb{L}_1, \mathbb{L}_2)$.

Proof. (i) To show the surjectivity of (1.54) we first observe that for every

$$t_{(\mu)} \in S^{(\mu)}(T^*M \setminus 0; L_1, L_2)$$

there exists an $\tilde{a}_{(\mu)} \in S^{(\mu)}(T^*M \setminus 0; \pi_M^* J_1, \pi_M^* J_2)$ such that $\tilde{a}_{(\mu)}|_{L_1} = t_{(\mu)}$. In fact, there are complementary bundles $L_i^\perp$ such that $L_i \oplus L_i^\perp = \pi_M^* J_i$, $i = 1, 2$, and it suffices to set $\tilde{a}_{(\mu)}|_{L_1^\perp} = 0$. The standard pseudo-differential calculus provides an $\tilde{A} \in L_{\mathrm{cl}}^\mu(M; J_1, J_2)$ such that $\sigma_\psi(\tilde{A}) = \tilde{a}_{(\mu)}$. Then, setting $A = P_2 \tilde{A} E_1$ it follows that $\sigma_\psi(A) = \tilde{t}_{(\mu)}$.

(ii) Let $A \in \mathcal{T}^\mu(M; \mathbb{L}_1, \mathbb{L}_2)$ and $\sigma_\psi(A) = 0$. Write $A = P_2 \tilde{A} E_1$ for some $\tilde{A} \in L_{\mathrm{cl}}^\mu(M; J_1, J_2)$ which allows us to identify $\sigma_\psi(A)$ with $\sigma_\psi(P_2 \tilde{A} P_1)|_{L_1}$ which vanishes. Since

$$A = P_2(P_2 \tilde{A} P_1) E_1 \tag{1.55}$$

and since $\ker \sigma_\psi(1 - P_1) =: L_1$ is a complementary bundle to $\ker \sigma_\psi(P_1) =: L_1^\perp$, i.e., $L_1 \oplus L_1^\perp = T^*M \setminus 0$, it follows that $\sigma_\psi(P_2 \tilde{A} P_1)$ vanishes on $T^*M \setminus 0$. This yields $P_2 A P_1 \in L_{\mathrm{cl}}^{\mu-1}(M; J_1, J_2)$. Then equality (1.55) shows that $A \in \mathcal{T}^{\mu-1}(M; \mathbb{L}_1, \mathbb{L}_2)$. □

Remark 1.2.23. Let $A \in \mathcal{T}^\mu(M; \mathbb{L}_1, \mathbb{L}_2)$ be an operator such that $\sigma_\psi(A) = 0$. Then (1.50) is a compact operator for every $s \in \mathbb{R}$.

In fact, Theorem 1.2.22 and Theorem 1.2.18 show that $A \in \mathcal{T}^{\mu-1}(M; \mathbb{L}_1, \mathbb{L}_2)$ and $A : H^s(M, \mathbb{L}_1) \to H^{s-\mu+1}(M, \mathbb{L}_2)$ is continuous; then the compactness of (1.50) is a consequence of Proposition 1.2.12.

Theorem 1.2.24. *Let* $A \in \mathcal{T}^\mu(M; \mathbb{L}_0, \mathbb{L}_2)$ *and* $B \in \mathcal{T}^\nu(M; \mathbb{L}_1, \mathbb{L}_0)$ *for* $\mu, \nu \in \mathbb{R}$, *and* $\mathbb{L}_1, \mathbb{L}_0, \mathbb{L}_2 \in \mathbb{P}(M)$. *Then* $AB \in \mathcal{T}^{\mu+\nu}(M; \mathbb{L}_1, \mathbb{L}_2)$ *and*

$$\sigma_\psi(AB) = \sigma_\psi(A)\sigma_\psi(B). \tag{1.56}$$

Proof. Writing $\mathbb{L}_i = (P_i, J_i, L_i)$, $i = 0, 1, 2$, and denoting the respective embedding operators by E_i, we have elements $\tilde{A} \in L_{\mathrm{cl}}^\mu(M; J_0, J_2)$, $\tilde{B} \in L_{\mathrm{cl}}^\nu(M; J_1, J_0)$ such that $A = P_2 \tilde{A} E_0$, $B = P_0 \tilde{B} E_1$. Then

$$AB = P_2 \tilde{A} E_0 P_0 \tilde{B} E_1 = P_2 \tilde{A} P_0 \tilde{B} E_1. \tag{1.57}$$

Since $P_0 \in L_{\mathrm{cl}}^0(M; J_0, J_0)$, the composition rule of standard pseudo-differential operators yields that $\tilde{A} P_0 \tilde{B} \in L_{\mathrm{cl}}^{\mu+\nu}(M; J_1, J_2)$. Now (1.57) yields

$$AB \in \mathcal{T}^{\mu+\nu}(M; \mathbb{L}_1, \mathbb{L}_2).$$

The symbol rule (1.56) is a consequence of the relation

$$\sigma_\psi(\widetilde{A}\,P_0\widetilde{B}) = \sigma_\psi(\widetilde{A})\sigma_\psi(P_0)\sigma_\psi(\widetilde{B}),$$

which gives

$$\sigma_\psi(AB) = \sigma_\psi(P_2)\sigma_\psi(\widetilde{A})\sigma_\psi(P_0)\sigma_\psi(\widetilde{B})\sigma_\psi(E_1);$$

but in the middle one is allowed to replace $\sigma_\psi(P_0)$ by $\sigma_\psi(E_0)\sigma_\psi(P_0)$. $\square$

Theorem 1.2.25. *Given* $A \in T^\mu(M; \mathbb{L}_1, \mathbb{L}_2)$, $\mathbb{L}_i = (P_i, J_i, L_i) \in \mathbb{P}(M)$, $i = 1, 2$, *for the formal adjoint in the sense of* Remark 1.2.14 *we have* $A^* \in T^\mu(M; \mathbb{L}_2^*, \mathbb{L}_1^*)$, *where* $\mathbb{L}_i^* \in \mathbb{P}(M)$ (*see* Proposition 1.2.8 (iv)) *for* $i = 1, 2$, *and*

$$\sigma_\psi(A^*) = \sigma_\psi(A)^*.$$

Proof. By (1.40), $(Au, v) = (u, A^*v)$ for a uniquely defined operator

$$A^* : H^s(M, \mathbb{L}_2^*) \to H^{s-\mu}(M, \mathbb{L}_1^*).$$

Let us write $A = P_2 \widetilde{A}\, E_1$ for a certain $\widetilde{A} \in L_{cl}^\mu(M; J_1, J_2)$. The pairing $(\cdot, \cdot)$ is coming from the $H^0(M, J)$-scalar product and we have $(P_2 \widetilde{A} P_1 u, v) = (u, P_1^* \widetilde{A}^* P_2^* v)$ for the adjoint $\widetilde{A}^* \in L_{cl}^\mu(M; J_2, J_1)$. Restricting this relation to $u \in H^\infty(M, \mathbb{L}_1)$, $v \in H^\infty(M, \mathbb{L}_2^*)$, we have $(P_2 \widetilde{A}\, E_1 u, v) = (u, P_1^* \widetilde{A}^* E_2^* v)$. This shows that $A^* = P_1^* \widetilde{A}^* E_2^* \in T^\mu(M; \mathbb{L}_2^*, \mathbb{L}_1^*)$. To see the symbol rule for adjoints, we first identify $\sigma_\psi(A)$ with $\sigma_\psi(P_2 \widetilde{A} P_1)$ (see the explanation in connection with (1.53)). From the standard pseudo-differential calculus we have $\sigma_\psi((P_2 \widetilde{A} P_1)^*) = \sigma_\psi(P_1^* \widetilde{A}^* P_2^*) = \sigma_\psi(P_1^*)\sigma_\psi(\widetilde{A}^*)\sigma_\psi(P_2^*)$, and here we can again identify $\sigma_\psi(P_2^*)$ with $\sigma_\psi(E_2^*)$. $\square$

Definition 1.2.26. An operator $A \in T^\mu(M; \mathbb{L}_1, \mathbb{L}_2)$, $\mu \in \mathbb{R}$, for $\mathbb{L}_i \in \mathbb{P}(M)$, $i = 1, 2$, is called elliptic (of order μ) if $\sigma_\psi(A) : L_1 \to L_2$ is an isomorphism.

Example 1.2.27. Let $\mathbb{L} = (P, J, L) \in \mathbb{P}(M)$, and let

$$a_{(\mu)} : \pi_M^* J \to \pi_M^* J$$

for any fixed $\mu \in \mathbb{R}$ denote the unique smooth bundle morphism such that $a_{(\mu)} : \pi_1^* J \to \pi_1^* J$ for $\pi_1 : S^* M \to M$ is the identity map and $a_{(\mu)}(x, \lambda\xi) = \lambda^\mu a_{(\mu)}(x, \xi)$ for all $(x, \xi) \in T^* M \setminus 0$, $\lambda \in \mathbb{R}_+$. Let $\widetilde{A} \in L_{cl}^\mu(M; J, J)$ be any element with $\sigma_\psi(\widetilde{A}) = a_{(\mu)}$, and consider the composition $P\widetilde{A} P \in L_{cl}^\mu(M; J, J)$. Then $P\widetilde{A} P$ identified with $P\widetilde{A} P|_{H^\infty(M, \mathbb{L})}$ represents an elliptic operator in $T^\mu(M; \mathbb{L}, \mathbb{L})$.

Definition 1.2.28. Let $A \in T^\mu(M; \mathbb{L}_1, \mathbb{L}_2)$, $\mu \in \mathbb{R}$, $\mathbb{L}_i \in \mathbb{P}(M)$, $i = 1, 2$. Then an operator $B \in T^{-\mu}(M; \mathbb{L}_1, \mathbb{L}_2)$ is called a parametrix of A, if B satisfies the relations

$$C_{\mathrm{L}} := I - BA \in T^{-\infty}(M; \mathbb{L}_1, \mathbb{L}_1), \quad C_{\mathrm{R}} := I - AB \in T^{-\infty}(M; \mathbb{L}_2, \mathbb{L}_2); \quad (1.58)$$

here I denotes corresponding identity operators.

Remark 1.2.29. For every $\mathbb{L} := (P, J, L) \in \mathbb{P}(M)$ and every $\mu \in \mathbb{R}$ there exists an elliptic operator $R_\mathbb{L}^\mu \in \mathcal{T}^\mu(M; \mathbb{L}, \mathbb{L})$.

In fact, let $a_{(\mu)} \in S_{(\mathrm{cl})}^\mu(T^*M \setminus 0; J, J)$ be the unique element that restricts to the identity map on $\pi_1^* J$, where $\pi_1 : S^*M \to M$ is the canonical projection of the unit cosphere bundle S^*M induced by T^*M to M. Set $\widetilde{A} := \mathrm{op}(a_{(\mu)})$, cf. Proposition 1.1.13. Then $P\widetilde{A} E$ for the embedding $E : H^s(M, \mathbb{L}) \to H^s(M, J)$ is elliptic because $\sigma_\psi(P\widetilde{A} E) : L \to L$ is an isomorphism.

Proposition 1.2.30. *Let $A \in \mathcal{T}^\mu(M; \mathbb{L}_1, \mathbb{L}_2)$ be an elliptic operator, and represent A as an element $A \in \mathcal{T}^\mu(M; \mathbb{M}_1, \mathbb{M}_2)$ for $\mathbb{M}_i := (Q_i, \mathbb{C}^m, L_i)$, $i = 1, 2$, for a sufficiently large m (cf. Remark 1.2.19). Then there exists an elliptic operator $A^\perp \in \mathcal{T}^\mu(M; \mathbb{M}_1^\perp, \mathbb{M}_2^\perp)$ for suitable $\mathbb{M}_i^\perp \in \mathbb{P}(M)$ such that $A \oplus A^\perp \in L_{\mathrm{cl}}^\mu(M; \mathbb{C}^m, \mathbb{C}^m)$ is elliptic in the sense of* Definition 1.1.18.

Proof. Let $\mathbb{L}_i = (P_i, J_i, L_i)$, and realise L_i as subbundles of the trivial bundle $\mathbb{C}^m$ on $T^*M \setminus 0$ in such a way that $L_{1,(x,\xi)} \cap L_{2,(x,\xi)} = \{0\}$ for every $(x, \xi) \in T^*M \setminus 0$. To see that this is possible, it suffices to choose m so large that $\mathbb{C}^m = L_1 \oplus L_2 \oplus G$ for a complementary bundle $G = (L_1 \oplus L_2)^\perp$ of the direct sum. Using the isomorphisms

$$\sigma_\psi(A)(x, \xi) : L_{1,(x,\xi)} \to L_{2,(x,\xi)}$$

and

$$\sigma_\psi(A)^{-1}(x, \xi) : L_{2,(x,\xi)} \to L_{1,(x,\xi)}$$

for $(x, \xi) \in S^*M$ we can form an isomorphism

$$\mathrm{diag}\left(\sigma_\psi(A), \sigma_\psi(A)^{-1}, \mathrm{id}\right) : L_1 \oplus L_2 \oplus G\big|_{S^*M} \to L_2 \oplus L_1 \oplus G\big|_{S^*M}. \tag{1.59}$$

Set $L_1^\perp := L_2 \oplus G$, $L_2^\perp := L_1 \oplus G$, and define $\sigma_\psi(A^\perp) \in S^{(\mu)}(T^*M \setminus 0; L_1^\perp, L_2^\perp)$ as the unique element such that

$$\sigma_\psi(A^\perp)\big|_{S^*M} = \mathrm{diag}\left(\sigma_\psi(A)^{-1}, \mathrm{id}_G\right)\big|_{S^*M}.$$

Moreover, set

$$\mathbb{M}_i^\perp := (Q_i^\perp, \mathbb{C}^m, L_i^\perp), \quad i = 1, 2.$$

Our operator A is of the form $A = Q_2 \widetilde{A}_1 E_1$ for some $\widetilde{A}_1 \in L_{\mathrm{cl}}^\mu(M; \mathbb{C}^m, \mathbb{C}^m)$, with the embedding $E_1 : H^s(M, \mathbb{M}_1) \to H^s(M, \mathbb{C}^m)$. Moreover, let $\sigma_\psi(\widetilde{A}_2) \in S^{(\mu)}(T^*M \setminus 0; \mathbb{C}^m, \mathbb{C}^m)$ denote the unique element such that $\sigma_\psi(\widetilde{A}_2)\big|_{S^*M}$ coincides with (1.59). Then applying (1.22) to $\sigma_\psi(\widetilde{A}_2)$ we obtain an associated elliptic operator $\widetilde{A}_2 \in L_{\mathrm{cl}}^\mu(M; \mathbb{C}^m, \mathbb{C}^m)$. Then $Q_2^\perp \widetilde{A}_2 P_1^\perp \in L_{\mathrm{cl}}^\mu(M; \mathbb{C}^m, \mathbb{C}^m)$ has a symbol which restricts to $\sigma_\psi(\widetilde{A}_2)\big|_{L_1 \oplus G}$ and vanishes on L_2. We therefore set $A^\perp := Q_2^\perp \widetilde{A}_2 P_1^\perp$ and thus obtain

$$A \oplus A^\perp = \mathrm{diag}\left(Q_2 \widetilde{A}_1 E_1, Q_2^\perp \widetilde{A}_2 E_1^\perp\right).$$

The latter direct sum represents an element in $L_{\mathrm{cl}}^\mu(M; \mathbb{C}^m, \mathbb{C}^m)$, since it can be identified with a sum $Q_2 \widetilde{A}_1 P_1 + Q_2^\perp \widetilde{A}_2 P_1^\perp$ of standard pseudo-differential operators. The ellipticity both of $A^\perp$ and A is evident by construction. $\qquad\square$

Theorem 1.2.31. *Let* $A \in \mathcal{T}^\mu(M; \mathbb{L}_1, \mathbb{L}_2)$, $\mu \in \mathbb{R}$, $\mathbb{L}_i = (P_i, J_i, L_i) \in \mathbb{P}(M)$.

(i) *The operator* A *is elliptic* (*of order* μ) *if and only if*

$$A : H^s(M, \mathbb{L}_1) \to H^{s-\mu}(M, \mathbb{L}_2) \tag{1.60}$$

is a Fredholm operator for some $s = s_0 \in \mathbb{R}$.

(ii) *If* A *is elliptic, then* (1.60) *is Fredholm for all* $s \in \mathbb{R}$, *and* $\dim \ker A$ (*as well as the kernel itself*) *and* $\dim \operatorname{coker} A$ *are independent of* s.

(iii) *An elliptic operator* $A \in \mathcal{T}^\mu(M; \mathbb{L}_1, \mathbb{L}_2)$ *has a parametrix*

$$B \in \mathcal{T}^{-\mu}(M; \mathbb{L}_2, \mathbb{L}_1),$$

and B *can be chosen in such a way that the remainders in relation* (1.58) *are projections*

$$C_{\mathrm{L}} : H^s(M, \mathbb{L}_1) \to V, \quad C_{\mathrm{R}} : H^{s-\mu}(M, \mathbb{L}_2) \to W$$

for all $s \in \mathbb{R}$, *for* $V := \ker A \subset H^\infty(M, \mathbb{L}_1)$, *and a finite-dimensional subspace* $W \subset H^\infty(M, \mathbb{L}_2)$ *such that* $W + \operatorname{im} A = H^{s-\mu}(M, \mathbb{L}_2)$ *and* $W \cap \operatorname{im} A = \{0\}$ *for all* $s \in \mathbb{R}$.

Proof. Let $A \in \mathcal{T}^\mu(M; \mathbb{L}_1, \mathbb{L}_2)$ be elliptic. We first construct a parametrix as claimed in assertion (iii). Without loss of generality we assume $\mathbb{L}_i = (P_i, \mathbb{C}^m, L_i)$, $i = 1, 2$, for some sufficiently large m, cf. the relation (1.51). Proposition 1.2.30 yields an elliptic operator $A^\perp \in \mathcal{T}^\mu(M; \mathbb{L}_1^\perp, \mathbb{L}_2^\perp)$ such that $\widetilde{A} = A \oplus A^\perp$ is elliptic in $L_{\mathrm{cl}}^\mu(M; \mathbb{C}^m, \mathbb{C}^m)$. The standard pseudo-differential calculus then gives a parametrix $\widetilde{P} \in L_{\mathrm{cl}}{}^\mu(M; \mathbb{C}^m, \mathbb{C}^m)$, and we set $B_0 := P_1 \widetilde{P} E_2$, which belongs to $\mathcal{T}^{-\mu}(M; \mathbb{L}_2, \mathbb{L}_1)$. Writing $A = P_2 \widetilde{A} E_1$, it follows that

$$C_{\mathrm{L}}^0 := I - B_0 A = I - P_1 \widetilde{P} E_2 P_2 \widetilde{A} E_1 = I - P_1 \widetilde{P} P_2 \widetilde{A} E_1.$$

This yields $\sigma_\psi(C_{\mathrm{L}}^0) = 0$, i.e., $C_{\mathrm{L}}^0 \in \mathcal{T}^{-1}(M; \mathbb{L}_1, \mathbb{L}_1)$, cf. Theorem 1.2.22 (ii). In a similar manner it follows that $C_{\mathrm{R}}^0 := I - A B_0 \in \mathcal{T}^{-1}(M; \mathbb{L}_2, \mathbb{L}_2)$. From Remark 1.2.23 we know that the operators

$$C_{\mathrm{L}}^0 \in \mathcal{L}\big(H^s(M, \mathbb{L}_1)\big) \text{ and } C_{\mathrm{R}}^0 \in \mathcal{L}\big(H^{s-\mu}(M, \mathbb{L}_2)\big)$$

are compact. Thus (1.60) is a Fredholm operator for every $s \in \mathbb{R}$. We now improve C_{L}^0 and C_{R}^0 by a formal Neumann series argument. By Theorem 1.2.24, $(C_{\mathrm{L}}^0)^j \in \mathcal{T}^{-j}(M; \mathbb{L}_1, \mathbb{L}_1)$ for every j, and Proposition 1.2.20 allows us to form an asymptotic sum $K \sim -\sum_{j=1}^\infty (C_{\mathrm{L}}^0)^j$, $K \in \mathcal{T}^{-1}(M; \mathbb{L}_1, \mathbb{L}_1)$.

For $B_{\mathrm{L}} := (I - K) B_0 \in \mathcal{T}^{-\mu}(M; \mathbb{L}_2, \mathbb{L}_1)$ we then obtain $C_{\mathrm{L}} := I - B_{\mathrm{L}} A \in \mathcal{T}^{-\infty}(M; \mathbb{L}_1, \mathbb{L}_1)$. In an analogous manner we find a $B_{\mathrm{R}} \in \mathcal{T}^{-\mu}(M; \mathbb{L}_2, \mathbb{L}_1)$ such that $C_{\mathrm{R}} := I - A B_{\mathrm{R}} \in \mathcal{T}^{-\infty}(M; \mathbb{L}_2, \mathbb{L}_2)$. A standard algebraic argument shows that B_L is a two-sided parametrix of A. In fact, $B_L A = I - C_L$ and $AB_R = I - C_R$ imply that $B_L A B_R = B_L(I - C_R) = (I - C_L)B_R$ and $B_L - B_R = B_L C_R - C_L B_B =:$ $C' \in \mathcal{T}^{-\infty}(M; \mathbb{L}_2, \mathbb{L}_1)$. Then $AB_L = A(B_L + C') = I$ mod $\mathcal{T}^{-\infty}(M; \mathbb{L}_2, \mathbb{L}_2)$. Theorems 1.1.19 and 1.1.20, together with the first part of proof, allow us to apply Remark 1.1.21 to the present situation. This gives us the assertion of Theorem 1.2.31 (ii), (iii).

It remains to show the second part of Theorem 1.2.31, namely, that the Fredholm property of (1.60) for an $s = s_0 \in \mathbb{R}$ implies the ellipticity of A.

Without loss of generality, we may assume $s_0 = \mu = 0$. Remark 1.2.29 gives us elliptic operators

$$R_{\mathbb{L}_1}^{-s_0} : H^0(M, \mathbb{L}_1) \to H^{s_0}(M, \mathbb{L}_1), \quad R_{\mathbb{L}_2}^{s_0-\mu} : H^{s_0-\mu}(M, \mathbb{L}_2) \to H^0(M, \mathbb{L}_2).$$

These are Fredholm operators by the first part of the proof of Theorem 1.2.31. Thus, also $A_0 := R_{\mathbb{L}_2}^{s_0-\mu} A R_{\mathbb{L}_1}^{-s_0} : H^0(M, \mathbb{L}_1) \to H^0(M, \mathbb{L}_2)$ is Fredholm.

If we show the ellipticity of A_0, we obtain at once the ellipticity of A since ellipticity is preserved under compositions. In other words, it suffices to consider the case $A := A_0$ and

$$A : H^0(M, \mathbb{L}_1) \to H^0(M, \mathbb{L}_2). \tag{1.61}$$

Further, let $E : H^0(M, \mathbb{L}_1^\perp) \to H^0(M, \mathbb{L}_1^\perp)$ be the identity operator. Then

$$L^2(M, J_1) = H^0(M, \mathbb{L}_1) \oplus H^0(M, \mathbb{L}_1^\perp)$$

and there are continuous embeddings

$$E_2 : H^0(M, \mathbb{L}_2) \to L^2(M, J_2), \quad E_1^\perp : H^0(M, \mathbb{L}_1^\perp) \to L^2(M, J_1).$$

Then we can pass to the operator

$$B = \begin{pmatrix} E_2 & 0 \\ 0 & E_1^\perp \end{pmatrix} \begin{pmatrix} A & 0 \\ 0 & E \end{pmatrix} : L^2(M, J_1) \to L^2(M, J_2 \oplus J_1),$$

which is an element of $L_{\mathrm{cl}}^0(M; J_1, J_2 \oplus J_1)$. By assumption, the operator (1.61) is Fredholm. In particular, there is an operator $Q : H^0(M, \mathbb{L}_2) \to H^0(M, \mathbb{L}_1)$ such that $I - QA : H^0(M, \mathbb{L}_1) \to H^0(M, \mathbb{L}_1)$ is compact.

Let $S : L^2(M, J_2 \oplus J_1) \to H^0(M, \mathbb{L}_2) \oplus H^0(M, \mathbb{L}_1^\perp)$ be a projection. Then $T := \mathrm{diag}(Q, E) \circ S$ has the property that $I - TB =: K$ is compact in $L^2(M, J_1)$. Since $I - K$ is a Fredholm operator, $\dim \ker(I - K) < \infty$, and then $\dim \ker B < \infty$, since $Bu = 0$ implies $TBu = 0$, which yields $(I - K)u = 0$, i.e., $\ker B \subseteq \ker(I - K)$.

The operator $\widetilde{B} := B^* B : L^2(M, J_1) \to L^2(M, J_1)$ belongs to $L_{\mathrm{cl}}^0(M; J_1, J_1)$ and is self-adjoint and Fredholm.

From Remark 1.1.23 we know that $\widetilde{B}$ is elliptic. It follows that $\sigma_\psi(B)$ is injective and also $\sigma_\psi(A)$ is injective. In an analogous manner we can show that $\sigma_\psi(A)$ is also surjective.

This completes the proof of Theorem 1.2.31. $\qquad\square$

Remark 1.2.32. The ellipticity of an operator $A \in T^\mu(M; \mathbb{L}_1, \mathbb{L}_2)$, with $\mathbb{L}_i = (P_i, J_i, L_i)$, $i = 1, 2$, only depends on the bundles L_1, L_2, and not on the projections P_1, P_2 or the chosen bundles J_1, J_2 over M. Of course, the spaces $H^s(M, \mathbb{L}_1)$ and $H^{s-\mu}(M, \mathbb{L}_2)$ depend on the choice of the projections P_1, P_2. So the Fredholm index of (1.60) may change when projections are varied.

The corresponding effect can be illustrated in functional analytic terms.

Theorem 1.2.33. *Let H_i, $i = 1, 2$, be Hilbert spaces, and let $P_i, Q_i \in \mathcal{L}(H_i)$ be continuous projections such that $P_i - Q_i \in \mathcal{K}(H_i)$, $i = 1, 2$. Moreover, let $A \in \mathcal{L}(H_1, H_2)$ be an operator such that*

$$A := P_2 \widetilde{A} : \operatorname{im} P_1 \to \operatorname{im} P_2$$

is a Fredholm operator. Then also

$$B := Q_2 \widetilde{A} : \operatorname{im} Q_1 \to \operatorname{im} Q_2$$

is a Fredholm operator, and we have

$$\operatorname{ind} A - \operatorname{ind} B = \operatorname{ind}(P_1, Q_1) - \operatorname{ind}(P_2, Q_2), \tag{1.62}$$

cf. Remark 1.2.6.

Proof. In this proof we set $P_{jj} := P_j|_{\operatorname{im} Q_j}$, $Q_{jj} := Q_j|_{\operatorname{im} P_j}$ for $j = 1, 2$. From Proposition 1.2.5 we know that

$$P_{11} : \operatorname{im} Q_1 \to \operatorname{im} P_1, \quad Q_{22} : \operatorname{im} P_2 \to \operatorname{im} Q_2$$

are Fredholm operators, and Q_{11} is a parametrix of P_{11}. This shows that $Q_{11} \circ P_{11} : \operatorname{im} Q_1 \to \operatorname{im} Q_1$ is a Fredholm operator of index 0. An analogous relation holds for the operators P_{22} and Q_{22}, respectively. Thus, the composition

$$D : \operatorname{im} Q_1 \xrightarrow{P_{11}} \operatorname{im} P_1 \xrightarrow{A} \operatorname{im} P_2 \xrightarrow{Q_{22}} \operatorname{im} Q_2$$

is again a Fredholm operator, of index

$$\operatorname{ind} D = \operatorname{ind} A + \operatorname{ind} P_{11} + \operatorname{ind} Q_{22}.$$

In the notation of Remark 1.2.6, it follows that

$$\operatorname{ind} D = \operatorname{ind} A + \operatorname{ind}(P_1, Q_1) - \operatorname{ind}(P_2, Q_2). \tag{1.63}$$

We have the identity

$$
\begin{aligned}
D &= Q_{22} P_2 \widetilde{A} (Q_{11} + I_{\operatorname{im} P_1} - Q_{11}) P_{11} \\
&= Q_{22} P_2 \widetilde{A} Q_{11} P_{11} + Q_{22} P_2 \widetilde{A} (I_{\operatorname{im} P_1} - Q_{11}) P_{11} \\
&= Q_{22} Q_{22} P_2 \widetilde{A} Q_{11} P_{11} + Q_{22} P_2 \widetilde{A} (I_{\operatorname{im} P_1} - Q_{11}) P_{11} \\
&= Q_{22} P_{22} B Q_{11} P_{11} - Q_{22} [P_{22}, Q_{22}] \widetilde{A} Q_{11} P_{11} \\
&\quad + Q_{22} P_2 \widetilde{A} (I_{\operatorname{im} P_1} - Q_{11}) P_{11},
\end{aligned}
$$

where $[P_{22}, Q_{22}]$ is the commutator in H_2, which is compact, since

$$[P_{22}, Q_{22}] = P_{22}Q_{22} - Q_{22}P_{22} = (P_{22} - Q_{22})(I_{\operatorname{im} P_2} - Q_{22} - P_{22}).$$

Moreover, $(I_{\operatorname{im} P_1} - Q_{11})P_{11} = (P_{11} - Q_{11})P_{11} : H_1 \to H_1$ is compact. Hence, the operator $Q_{22}P_{22}BQ_{11}P_{11} - D$ is compact, i.e., $Q_{22}P_{22}BQ_{11}P_{11}$ is Fredholm, and we have

$$\operatorname{ind} D = \operatorname{ind}(Q_{22}P_{22}BQ_{11}P_{11}). \tag{1.64}$$

By Proposition 1.2.5, the operators $Q_{11}P_{11} : \operatorname{im} Q_1 \to \operatorname{im} Q_1$ and $Q_{22}P_{22} : \operatorname{im} Q_2 \to \operatorname{im} Q_2$ are Fredholm and of index 0. Therefore, we have

$$\operatorname{ind} B = \operatorname{ind}(Q_{22}P_{22}BQ_{11}P_{11}),$$

and the assertion is a consequence of relations (1.63) and (1.64). $\qquad\square$

Corollary 1.2.34. *Let* $A \in T^\mu(M; \mathbb{L}_1, \mathbb{L}_2)$, *with* $\mathbb{L}_j := (P_j, J_j, L_j)$, *and* $B \in T^\mu(M; \mathbb{M}_1, \mathbb{M}_2)$, *with* $\mathbb{M}_j := (Q_j, J_j, L_j)$, *and assume that* $\sigma_\psi(A) = \sigma_\psi(B)$. *Then the Fredholm indices of* A *and* B *as operators*

$$A : H^s(M, \mathbb{L}_1) \to H^{s-\mu}(M, \mathbb{L}_2), \quad B : H^s(M, \mathbb{M}_1) \to H^{s-\mu}(M, \mathbb{M}_2)$$

are related by the formula (1.62), *which is independent of* s.

In fact, the Fredholm indices of A and B are independent of s, cf. Theorem 1.2.31, and hence, we may apply (1.62) for any fixed s.

Remark 1.2.35. Let $\mathbb{L} := (P, J, L)$ and $\mathbb{M} := (Q, J, L)$ be projection data; we interpret the operators

$$P : H^s(M, \mathbb{M}) \to H^s(M, \mathbb{L}) \text{ and } Q : H^s(M, \mathbb{L}) \to H^s(M, \mathbb{M})$$

as elements of $T^0(M; \mathbb{M}, \mathbb{L})$ and $T^0(M; \mathbb{L}, \mathbb{M})$, respectively. Recall that, for instance, we have an identification

$$T^0(M; \mathbb{M}, \mathbb{L}) \cong \{P\widetilde{A}Q : \widetilde{A} \in L^\mu_{\mathrm{cl}}(M; J, J)\},$$

cf. Proposition 1.2.16. Inserting $\widetilde{A} = P$ we obtain $P\widetilde{A}Q = P : \operatorname{im} Q \to \operatorname{im} P$. For a similar reason we interpret Q as a Toeplitz operator $Q\widetilde{A}P = Q : \operatorname{im} P \to \operatorname{im} Q$ for $\widetilde{A} = Q$. Since $\sigma_\psi(P) = \sigma_\psi(Q) = \operatorname{id}_L$, the operators P and Q are elliptic in the respective classes. Moreover, we have

$$\operatorname{ind} P = \operatorname{ind}(P, Q), \quad \operatorname{ind} Q = \operatorname{ind}(Q, P),$$

cf. also Proposition 1.2.5.

Remark 1.2.36. Let $A, B \in T^\mu(M; \mathbb{L}_1, \mathbb{L}_2)$ be elliptic, and assume that the principal symbols

$$\sigma_\psi(A), \sigma_\psi(B) : L_1 \to L_2$$

coincide. Then we have

$$\operatorname{ind} A = \operatorname{ind} B.$$

In fact, Proposition 1.2.23 gives $\sigma_\psi(A - B) = 0$, i.e.,

$$A - B \in \mathcal{T}^{\mu-1}(M; \mathbb{L}_1, \mathbb{L}_2),$$

and hence A is equal to B modulo a compact operator.

Let us assume now that $L_i(t) := (P_i(t), J_i, L_i(t))$, $0 \leq t \leq 1$, $i = 1, 2$, is a family of elements in $\mathbb{P}(M)$, where $P_i(t) \in C([0,1], L^0_{\mathrm{cl}}(M; J_1, J_2))$ are families of projections, such that $L_i(t) = \sigma_\psi(P_i(t))\pi^*_M J_i$ are families of subbundles in $\pi^*_M J_i$. Suppose that

$$a_{(\mu)}(t) : L_1(t) \to L_2(t)$$

is a continuous family of isomorphisms, smooth in $(x, \xi) \in T^*M \setminus 0$ and homogeneous of degree of μ. We can complete $a_{(\mu)}(t)$ to a continuous family of morphisms

$$\tilde{a}_{(\mu)}(t) : \pi^*_M J_1 \to \pi^*_M J_2$$

such that $a_{(\mu)}(t) = \sigma_\psi(P_2(t))\tilde{a}_{(\mu)}(t)\sigma_\psi(P_1(t))$ for all t. Let us set $\tilde{A}_{(\mu)}(t) := \mathrm{op}(\tilde{a}_{(\mu)}(t))$, cf. Proposition 1.1.13, giving us an element of $C([0,1], L^\mu_{\mathrm{cl}}(M; J_1, J_2))$.

We then obtain a family

$$A_t := P_2(t)\tilde{A}(t)E_1(t) \in \mathcal{T}^\mu(M; \mathbb{L}_1(t), \mathbb{L}_2(t)),$$

where $E_1(t) : P_1(t)H^s(M, J_1) \to H^s(M, J_1)$ are the canonical embeddings. The operators A_t are elliptic for all $t \in [0, 1]$.

Theorem 1.2.37. *Under the condition on $A_t \in \mathcal{T}^\mu(M; \mathbb{L}_1(t), \mathbb{L}_2(t))$, $0 \leq t \leq 1$, mentioned above we have*

$$\mathrm{ind}\, A_0 = \mathrm{ind}\, A_1,$$

where the index of A_t refers to the Fredholm operator

$$A_t : P_1(t)H^s(M, J_1) \to P_2(t)H^{s-\mu}(M, J_2).$$

Theorem 1.2.37 has a more general functional analytic background. The following considerations up to the end of this section have been contributed by Thomas Krainer.

Theorem 1.2.38. *Let H_1 and H_2 be Hilbert spaces, and consider families of operators*

$$(A_t)_{0 \leq t \leq 1} \subset C([0,1], \mathcal{L}(H_1, H_2)),$$

$(P_t)_{0 \leq t \leq 1} \subset C([0,1], \mathcal{L}(H_2))$, and $(Q_t)_{0 \leq t \leq 1} \subset C([0,1], \mathcal{L}(H_1))$. Assume $P^2_t = P_t$, $Q^2_t = Q_t$ for all $t \in [0,1]$. Moreover, let

$$P_t A_t Q_t : \mathrm{im}\, Q_t \to \mathrm{im}\, P_t$$

be a Fredholm operator for every $t \in [0,1]$. Then we have

$$\mathrm{ind}\,(P_0 A_0 Q_0 : \mathrm{im}\, Q_0 \to \mathrm{im}\, P_0) = \mathrm{ind}\,(P_1 A_1 Q_1 : \mathrm{im}\, Q_1 \to \mathrm{im}\, P_1).$$

To prove this theorem we first show another result. Consider the set

$$\Pi(H_2) \times \mathcal{L}(H_1, H_2) \times \Pi(H_1), \tag{1.65}$$

where $\Pi(H)$ for a Hilbert space H denotes the set of all $P \in \mathcal{L}(H)$ such that $P^2 = P$. Let $\Phi_k(H_1, H_2)$ be the set of all triples (P, A, Q) in (1.65) such that

$$PAQ : \operatorname{im} Q \to \operatorname{im} P$$

is a Fredholm operator of index k.

Proposition 1.2.39. *For every $k \in \mathbb{Z}$ the set $\Phi_k(H_1, H_2)$ is open in (1.65).*

Proof. As is well known, the set of Fredholm operators of index k between Hilbert spaces H and $\widetilde{H}$ is open in $\mathcal{L}(H, \widetilde{H})$. Applying this to $H := \operatorname{im} Q$ and $\widetilde{H} := \operatorname{im} P$, it follows that for any triple $(P, A, Q) \in \Phi_k(H_1, H_2)$ there exists an $\varepsilon_0 > 0$ such that $(P, A + K, Q) \in \Phi_k(H_1, H_2)$ for every $K \in \mathcal{L}(H_1, H_2)$, $\|K\| < \varepsilon_0$.

Now we prove that for every $(P, A, Q) \in \Phi_k(H_1, H_2)$, there exist constants $\alpha > 0$, $\varepsilon > 0$, $\beta > 0$ such that

$$(P', B, Q') \in \Pi(H_2) \times \mathcal{L}(H_1, H_2) \times \Pi(H_1)$$

and

$$\|P' - P\| < \alpha, \quad \|B - A\| < \varepsilon, \quad \|Q' - Q\| < \beta$$

imply

$$(P', B, Q') \in \Phi_k(H_1, H_2). \tag{1.66}$$

Let $G_1 \in \mathcal{L}(H_1)$, $G_2 \in \mathcal{L}(H_2)$ be invertible elements such that

$$\|G_1 - I\| < \delta_1, \quad \|G_2 - I\| < \delta_2 \tag{1.67}$$

for sufficiently small $\delta_1, \delta_2 < 1$. Set $P' := G_2 P G_2^{-1}$ and $Q' = G_1 Q G_1^{-1}$. We can prove that relation (1.67) holds for ε and δ_1, δ_2 so small that

$$\frac{1 + \delta_1}{1 - \delta_2} \varepsilon + \frac{\delta_1 + \delta_2}{1 - \delta_2} \|A\| < \varepsilon_0 \tag{1.68}$$

holds. In fact, Neumann series arguments show that

$$\|G_2^{-1}\| < \frac{1}{1 - \delta_2}, \quad \|G_2^{-1} - I\| < \frac{\delta_2}{1 - \delta_2}, \quad \text{and} \quad \|G_1\| < 1 + \delta_1. \tag{1.69}$$

We now rewrite the operator $P'BQ'$ as follows:

$$P'BQ' = G_2 P G_2^{-1} B G_1 Q G_1^{-1} = G_2 P(A + K) Q G_1^{-1} = G_2 (PKQ + PAQ) G_1^{-1}, \tag{1.70}$$

with

$$K := G_2^{-1} B G_1 - A = G_2^{-1}(B - A)G_1 + (G_2^{-1} - I)A + A(G_1 - I) + (G_2^{-1} - I)A(G_1 - I).$$

Using (1.67), (1.68), and (1.69), we obtain

$$\|K\| \leq \|G_2^{-1}\|\|B - A\|\|G_1\| + \|G_2^{-1} - I\|\|A\|$$
$$+ \|A\|\|G_1 - I\| + \|G_2^{-1} - I\|\|A\|\|G_1 - I\|$$
$$\leq \frac{1 + \delta_1}{1 - \delta_2}\varepsilon + \left[\frac{\delta_2}{1 - \delta_2} + \delta_1 + \frac{\delta_1\delta_2}{1 - \delta_2}\right]\|A\| < \varepsilon_0.$$

Thus, from the first part of the proof it follows that $(P, A + K, Q) \in \Phi_k(H_1, H_2)$. Moreover, (1.70) together with the isomorphisms

$$G_1 : \operatorname{im} Q' \to \operatorname{im} Q, \quad G_2 : \operatorname{im} P \to \operatorname{im} P'$$

gives the relation $P'BQ' : \operatorname{im} Q' \to \operatorname{im} P'$.

Let us finally prove that P' and Q' are well-defined. For this we consider the map

$$s : \Pi(H_2) \to \mathcal{L}(H_2), \quad P' \mapsto P'P + (I - P')(I - P).$$

The map s is continuous, and we have $s(P) = I$. Let us choose $\delta_2' > 0$ such that $\|s(P') - s(P)\| < \delta_2$ when $\|P' - P\| < \delta_2'$. Since δ_2 is very small, we have the inverse $s(P')^{-1} \in \mathcal{L}(H_2)$ for $\|P' - P\| < \delta_2'$. In a similar manner it follows that $\|Q' - Q\| < \delta_1'$ for a suitable small $\delta_1' > 0$ implies $Q' = G_1QG_1^{-1}$ for an invertible $G_1 \in \mathcal{L}(H_2)$, $\|G_1 - I\| < \delta_1$. This completes the proof of Proposition 1.2.39. □

Proof of Theorem 1.2.38. By Proposition 1.2.39, the map $[0, 1] \to \mathbb{Z}$ defined by $t \mapsto \operatorname{ind} P_t A_t Q_t$ is continuous, and hence constant. □

1.3   Operator-valued symbols and abstract edge spaces

It will be necessary for several reasons to extend the concept of pseudo-differential operators to the set-up with vector- or operator-valued symbols. There are, in fact, many variants.

Definition 1.3.1. (i) If E is a Fréchet space with the semi-norm system $(\pi_j)_{j\in\mathbb{N}}$, then

$$S^\mu(U \times \mathbb{R}^q, E)$$

for an open set $U \subseteq \mathbb{R}^m$ and $\mu \in \mathbb{R}$ is defined to be the set of all $a(y, \eta) \in C^\infty(U \times \mathbb{R}^q, E)$ satisfying the symbol estimates

$$\pi_j\big(D_y^\alpha D_\eta^\beta a(y, \eta)\big) \leq c\langle\eta\rangle^{\mu - |\beta|}$$

for all $(y, \eta) \in K \times \mathbb{R}^q$, $K \Subset U$, and for all $\alpha \in \mathbb{N}^p$, $\beta \in \mathbb{N}^q$, $j \in \mathbb{N}$, with constants $c = c(\alpha, \beta, j, K) > 0$.

(ii) A symbol $a(y,\eta) \in S^{\mu}(U \times \mathbb{R}^{q}, E)$ is called classical, if there are homogeneous components $a_{(\mu-k)}(y,\eta) \in C^{\infty}(U \times (\mathbb{R}^{q} \setminus \{0\}), E)$, $k \in \mathbb{N}$, $a_{(\mu-k)}(y, \lambda\eta) = \lambda^{\mu-k}a_{(\mu-k)}(y,\eta)$ for all $\lambda \in \mathbb{R}_{+}$, $(y,\eta) \in U \times (\mathbb{R}^{q} \setminus \{0\}))$, such that

$$a(y,\eta) - \sum_{k=0}^{N} \chi(\eta)a_{(\mu-k)}(y,\eta) \in S^{\mu-(N+1)}(U \times \mathbb{R}^{q}, E)$$

for every $N \in \mathbb{N}$ and some excision function χ. By $S^{\mu}_{\mathrm{cl}}(U \times \mathbb{R}^{q}, E)$ we denote the space of all classical symbols of order μ.

Analogously to the scalar case, cf. Section 1.1, the space $S^{\mu}(U \times \mathbb{R}^{q}, E)$ is Fréchet with a semi-norm system

$$\sup_{(y,\eta)\in K \times \mathbb{R}^{q}} \langle\eta\rangle^{-\mu+|\beta|}\pi_{j}\big(D_{y}^{\alpha}D_{\eta}^{\beta}a(y,\eta)\big), \tag{1.71}$$

$K \Subset U$, $\alpha \in \mathbb{N}^{p}$, $\beta \in \mathbb{N}^{q}$, $j \in \mathbb{N}$. Also $S^{\mu}_{\mathrm{cl}}(U \times \mathbb{R}^{q}, E)$ is Fréchet in a stronger topology than that induced from $S^{\mu}(U \times \mathbb{R}^{q}, E)$, namely, with the semi-norms (1.71) together with those from (1.4) and from the homogeneous components $a_{(\mu-j)}(y,\eta) \in C^{\infty}(U \times (\mathbb{R}^{q} \setminus \{0\}), E)$ which are uniquely determined by a. First we can recover $a_{(\mu)}(x,\xi)$ by

$$a_{(\mu)}(y,\eta) = \lim_{\lambda\to\infty} \lambda^{-\mu}a(y, \lambda\eta). \tag{1.72}$$

Then, since $a(x,\xi) - \chi(\eta)a_{(\mu)}(y,\eta) \in S^{\mu-1}_{\mathrm{cl}}(U \times \mathbb{R}^{q}, E)$ for any excision function χ, we can apply the same procedure again, which yields $a_{(\mu-1)}(y,\eta)$, and then, successively, $a_{(\mu-j)}(y,\eta)$ for every $j \in \mathbb{N}$.

We have

$$S^{\mu}_{(\mathrm{cl})}(U \times \mathbb{R}^{q}, E) = C^{\infty}\big(U, S^{\mu}_{(\mathrm{cl})}(\mathbb{R}^{q}, E)\big)$$

with $S^{\mu}_{(\mathrm{cl})}(\mathbb{R}^{q}, E)$ being the (closed) subspace of $S^{\mu}_{(\mathrm{cl})}(U \times \mathbb{R}^{q}, E)$ of x-independent elements, and

$$S^{-\infty}(U \times \mathbb{R}^{q}, E) := \bigcap_{\mu\in\mathbb{R}} S^{\mu}(\mathbb{R}^{q}, E) = C^{\infty}(U, \mathcal{S}(\mathbb{R}^{q}, E)).$$

The generalisation of results on scalar symbols to the vector-valued case is straightforward. We will tacitly employ many results for scalar symbols that are true in analogous form in the vector-valued case as well. This concerns, in particular, asymptotic summations.

Example 1.3.2. Let H, $\widetilde{H}$ be Hilbert spaces and $E := \mathcal{L}(H, \widetilde{H})$, equipped with the operator-norm topology. Then $S^{\mu}_{(\mathrm{cl})}(U \times \mathbb{R}^{q}, \mathcal{L}(H, \widetilde{H}))$ consists of operator-valued symbols, i.e., operator functions $U \times \mathbb{R}^{q} \to \mathcal{L}(H, \widetilde{H})$, for which the absolute value in the symbol estimates of Definition 1.1.1 (i) is replaced by the operator norm.

Proposition 1.3.3. *Let E and F be Fréchet spaces with the semi-norm systems $(\pi_j)_{j\in\mathbb{N}}$ and $(\varphi_k)_{k\in\mathbb{N}}$, respectively. Let $T : E \to F$ be a continuous operator, i.e., for every $k \in \mathbb{N}$ there exists a $j \in \mathbb{N}$, such that*

$$\varphi_k(Te) \leq c\,\pi_{j(k)}(e)$$

for all $e \in E$, with some $c = c(k) > 0$. Moreover, let $a(y,\eta) \in S^\mu(U \times \mathbb{R}^q; E)$. Then the (y,η)-wise application of T on the values of $a(y,\eta)$ induces a continuous operator

$$T : S^\mu(U \times \mathbb{R}^q; E) \to S^\mu(U \times \mathbb{R}^q; F). \tag{1.73}$$

Proof. The application of T on the values of a yields

$$\varphi_k\big(TD_y^\alpha D_\eta^\beta a(y,\eta)\big) \leq c\,\pi_{j(k)}\big(D_y^\alpha D_\eta^\beta a(y,\eta)\big) \leq C\,\langle\eta\rangle^{\mu-|\beta|}$$

for all $(y,\eta) \in K \times \mathbb{R}^q$, $K \Subset U$, with constants $C = C(\alpha,\beta,k,K) > 0$. This shows the continuity of (1.73). $\qquad\square$

Corollary 1.3.4. *Let $E := S^\nu(\mathbb{R}_\tau^d)$, $\nu \in \mathbb{R}$, and*

$$T := \mathrm{Op}_t : S^\nu(\mathbb{R}_\tau^d) \to \mathcal{L}\big(H^s(\mathbb{R}^d), H^{s-\nu}(\mathbb{R}^d)\big)$$

be the continuous operator defined by the rule $p(\tau) \mapsto \mathrm{Op}_t(p)$, $s \in \mathbb{R}$. Then Op_t induces a continuous operator

$$\mathrm{Op}_t : S^\mu\big(U \times \mathbb{R}_\eta^q, S^\nu(\mathbb{R}_\tau^d)\big) \to S^\mu\big(U \times \mathbb{R}_\eta^q, \mathcal{L}\big(H^s(\mathbb{R}^d), H^{s-\nu}(\mathbb{R}^d)\big)\big)$$

for every $\mu, s \in \mathbb{R}$. In other words, for every $\mathfrak{p}(y,\eta,\tau) \in S^\mu\big(U \times \mathbb{R}_\eta^q, S^\nu(\mathbb{R}_\tau^d)\big)$ we have

$$\big\|D_y^\alpha D_\eta^\beta \mathrm{Op}_t(\mathfrak{p})(y,\eta)\big\|_{\mathcal{L}(H^s(\mathbb{R}^d), H^{s-\nu}(\mathbb{R}^d))} \leq c\,\langle\eta\rangle^{\mu-|\beta|}$$

for all $(y,\eta) \in K \times \mathbb{R}^q$, $K \Subset U$, $\alpha \in \mathbb{N}^p$, $\beta \in \mathbb{N}^q$, with constants $c = c(a,\beta,K) > 0$.

Let us turn to another generalisation of scalar symbols, referring to a Hilbert space with group action. The following definition has been introduced in [47] as a tool for formulating edge pseudo-differential operators.

Definition 1.3.5. A separable Hilbert space H is said to be endowed with a group action $\kappa = \{\kappa_\lambda\}_{\lambda\in\mathbb{R}_+}$ if κ is a group of isomorphisms

$$\kappa_\lambda : H \to H, \quad \kappa_\lambda \kappa_\nu = \kappa_{\lambda\nu} \ \text{ for all } \lambda,\nu \in \mathbb{R}_+, \ \kappa_1 = \mathrm{id},$$

such that $\kappa_\lambda h \in C\big(\mathbb{R}_{+,\lambda}, H\big)$ for every $h \in H$ (i.e., κ is strongly continuous).

More generally, a Fréchet space E which is written as a projective limit of Hilbert spaces E^j, $j \in \mathbb{N}$, with continuous embeddings $E^j \hookrightarrow E^0$ for all j, is said to be endowed with a group action $\kappa = \{\kappa_\lambda\}_{\lambda\in\mathbb{R}_+}$ if E^0 is endowed with κ in the above-mentioned sense and if $\kappa|_{E_j} := \{\kappa_\lambda|_{E^j}\}_{\lambda\in\mathbb{R}_+}$ is a group action on E^j for every j.

Remark 1.3.6. The set of adjoint operators $\kappa^* := \{\kappa_\lambda^*\}_{\lambda \in \mathbb{R}_+}$ is again a group action on H, cf. [32, Corollary 1. 10. 6].

Proposition 1.3.7. *Let H be a Hilbert space with group action $\kappa = \{\kappa_\lambda\}_{\lambda \in \mathbb{R}_+}$. Then there are constants $C, M > 0$ such that*

$$\|\kappa_\lambda\|_{\mathcal{L}(H)} \leq C\left(\max\left\{\lambda, \lambda^{-1}\right\}\right)^M. \tag{1.74}$$

Proof. By assumption, for every $h \in H$ the function $\lambda \mapsto \kappa_\lambda h$ belongs to $C(\mathbb{R}_+, H)$. Thus the set $\{\kappa_\lambda h : \lambda \in [\alpha, \beta]\}$ is bounded in H for every compact interval $[\alpha, \beta] \subset \mathbb{R}_+$. By the Banach–Steinhaus Theorem, $\sup_{\lambda \in [\alpha, \beta]} \|\kappa_\lambda\|_{\mathcal{L}(H)} \leq C$ for some $C = C(\alpha, \beta) > 0$. Thus, in particular, $\sup_{\lambda \in [e^{-1}, e]} \|\kappa_\lambda\|_{\mathcal{L}(H)} \leq C$, where we assume $C \geq 1$. It follows that $\|\kappa_{\lambda^n}\|_{\mathcal{L}(H)} \leq \|\kappa_\lambda\|_{\mathcal{L}(H)}^n \leq C^n$, i.e.,

$$\|\kappa_\lambda\|_{\mathcal{L}(H)} \leq C^n \quad \text{for every} \quad \lambda \in [e^{-n}, e^n], \, n \in \mathbb{N}.$$

Thus for $\lambda_n := e^n$, $n \in \mathbb{N}$, it follows that

$$\|\kappa_{\lambda_n}\|_{\mathcal{L}(H)} \leq C^n = e^{n \log C} = \lambda_n^{\log C}.$$

For $n \geq 1$ and $\lambda \in [e^{n-1}, e^n]$ there is a $\delta \in [1, e]$ such that $\lambda = \delta e^{n-1} = \delta \lambda_{n-1}$. For $\lambda \in [e^{n-1}, e^n]$ we obtain

$$\begin{aligned}
\|\kappa_\lambda\|_{\mathcal{L}(H)} = \|\kappa_{\delta\lambda_{n-1}}\|_{\mathcal{L}(H)} &\leq \|\kappa_\delta\|_{\mathcal{L}(H)} \|\kappa_{\lambda_{n-1}}\|_{\mathcal{L}(H)} \leq CC^{n-1} \\
&= C(\lambda_{n-1})^{\log C} \leq C(\delta\lambda_{n-1})^{\log C} = C\lambda^{\log C}.
\end{aligned} \tag{1.75}$$

Since the right-hand side of the last estimate is independent of n, we proved the asserted estimate for $M = \log C$, for all $\lambda \geq 1$. In a similar manner we can argue for $0 < \lambda \leq 1$. In fact, for $\lambda_{-n} := e^{-n}$, $n \geq 1$, we have

$$\|\kappa_{\lambda_{-n}}\|_{\mathcal{L}(H)} \leq C^n = e^{n \log C} = (\lambda_{-n}^{-1})^{\log C}.$$

Moreover, for $\lambda \in [e^{-n}, e^{-(n-1)}]$ there is a $\delta \in [e^{-1}, 1]$ such that $\lambda = \delta\lambda_{-(n-1)}$. This gives us

$$\begin{aligned}
\|\kappa_\lambda\|_{\mathcal{L}(H)} = \|\kappa_{\delta\lambda_{-(n-1)}}\|_{\mathcal{L}(H)} &\leq C(\lambda_{-(n-1)}^{-1})^{\log C} \\
&\leq C((\delta\lambda_{-(n-1)})^{-1})^{\log C} = C(\lambda^{-1})^{\log C},
\end{aligned}$$

which corresponds to the assertion for $0 < \lambda \leq 1$. $\qquad\square$

Clearly, in the Fréchet case the constants C and M in the estimates (1.74) for $\kappa_\lambda|_{E^j}$ may depend on j.

For a Hilbert space H it follows that

$$\left\|\kappa_{\langle\xi\rangle}^{-1}\kappa_{\langle\eta\rangle}\right\|_{\mathcal{L}(H)} \leq c\,\langle\xi - \eta\rangle^M \quad \text{for all } \xi, \eta \in \mathbb{R}^q$$

for some $c > 0$, by using Peetre's inequality.

Definition 1.3.8. (i) Let H and $\widetilde{H}$ be Hilbert spaces with group actions κ and $\widetilde{\kappa}$, respectively. Then $S^\mu(U \times \mathbb{R}^q; H, \widetilde{H})$ for $U \subseteq \mathbb{R}^p$ open, $\mu \in \mathbb{R}$, is defined to be the set of all $a(y, \eta) \in C^\infty(U \times \mathbb{R}^n, \mathcal{L}(H, \widetilde{H}))$ such that

$$\left\| \widetilde{\kappa}_{\langle \eta \rangle}^{-1} \{ D_y^\alpha D_\eta^\beta a(y, \eta) \} \kappa_{\langle \eta \rangle} \right\|_{\mathcal{L}(H, \widetilde{H})} \leq c \, \langle \eta \rangle^{\mu - |\beta|} \tag{1.76}$$

for all $(y, \eta) \in K \times \mathbb{R}^q$, $K \Subset U$, for all $\alpha \in \mathbb{N}^p$, $\beta \in \mathbb{N}^q$, with constants $c = c(\alpha, \beta, K) > 0$. The estimates (1.76) will be referred to as twisted symbol estimates.

(ii) Let $S^{(\nu)}(U \times (\mathbb{R}^q \setminus \{0\}); H, \widetilde{H})$, $\nu \in \mathbb{R}$, be the space of all

$$f_{(\nu)}(y, \eta) \in C^\infty(U \times (\mathbb{R}^q \setminus \{0\}), \mathcal{L}(H, \widetilde{H}))$$

such that

$$f_{(\nu)}(y, \lambda \eta) = \lambda^\nu \kappa_\lambda f_{(\nu)}(y, \eta) \kappa_\lambda^{-1} \tag{1.77}$$

for all $\lambda \in \mathbb{R}_+$. A symbol $a(y, \eta) \in S^\mu(U \times \mathbb{R}^q; H, \widetilde{H})$ is called classical, if there are homogeneous components

$$a_{(\mu-k)}(y, \eta) \in S^{(\mu-k)}(U \times (\mathbb{R}^q \setminus \{0\}); H, \widetilde{H}), \quad k \in \mathbb{N},$$

such that

$$a(y, \eta) - \sum_{k=0}^N \chi(\eta) a_{(\mu-k)}(y, \eta) \in S^{\mu-(N+1)}(U \times \mathbb{R}^q; H, \widetilde{H}) \tag{1.78}$$

for every $N \in \mathbb{N}$ and some excision function χ. By $S_{\mathrm{cl}}^\mu(U \times \mathbb{R}^q; H, \widetilde{H})$ we denote the space of all such classical symbols of order μ. As in Section 1.1, we write subscript "(cl)" when a consideration is valid both in the classical and the general case.

(iii) Let E and $\widetilde{E}$ be Fréchet spaces with group actions κ and $\widetilde{\kappa}$, respectively. Let $r : \mathbb{N} \to \mathbb{N}$ be a fixed mapping. Then

$$S_{(\mathrm{cl})}^\mu(U \times \mathbb{R}^q; E, \widetilde{E})_r$$

for $U \subseteq \mathbb{R}^q$ open, $\mu \in \mathbb{R}$, is defined to be the set of all

$$a(x, \xi) \in \bigcap_{j \in \mathbb{N}} C^\infty(U \times \mathbb{R}^q, \mathcal{L}(E^{r(j)}, \widetilde{E}^j))$$

such that

$$a(y, \eta) \in \bigcap_{j \in \mathbb{N}} S_{(\mathrm{cl})}^\mu(U \times \mathbb{R}^q; E^{r(j)}, \widetilde{E}^j).$$

We set $S_{(\mathrm{cl})}^\mu(U \times \mathbb{R}^q; E, \widetilde{E}) = \bigcup_r S_{(\mathrm{cl})}^\mu(U \times \mathbb{R}^q; E, \widetilde{E})_r$.

Analogously to the scalar case, cf. Section 1.1, $S^\mu(U \times \mathbb{R}^q; H, \widetilde{H})$ is a Fréchet space with a semi-norm system

$$\sup_{(y,\eta)\in K \times \mathbb{R}^q} \langle\eta\rangle^{-\mu+|\beta|} \big\| \widetilde{\kappa}_{\langle\eta\rangle}^{-1} \{ D_y^\alpha D_\eta^\beta a(y,\eta) \} \kappa_{\langle\eta\rangle} \big\|_{\mathcal{L}(H,\widetilde{H})}, \tag{1.79}$$

$K \Subset U$, $\alpha \in \mathbb{N}^p$, $\beta \in \mathbb{N}^q$. Also $S_{\mathrm{cl}}^\mu(U \times \mathbb{R}^q; H, \widetilde{H})$ is Fréchet in a stronger topology than that induced from $S^\mu(U \times \mathbb{R}^q; H, \widetilde{H})$, namely, with the semi-norms (1.79) together with those from (1.78) and from the homogeneous components $a_{(\mu-j)}(y,\eta) \in C^\infty(U \times (\mathbb{R}^q\backslash\{0\}), \mathcal{L}(H, \widetilde{H}))$ which are uniquely determined by a. First we can recover $a_{(\mu)}(y,\eta)$ by

$$a_{(\mu)}(y,\eta) = \lim_{\lambda\to\infty} \lambda^{-\mu} \widetilde{\kappa}_\lambda^{-1} a(y,\lambda\eta)\kappa_\lambda. \tag{1.80}$$

Then, since $a(y,\eta) - \chi(\eta)a_{(\mu)}(y,\eta) \in S_{\mathrm{cl}}^{\mu-1}(U \times \mathbb{R}^q; H, \widetilde{H})$ for any excision function χ, we can apply again the same procedure, which yields $a_{(\mu-1)}(y,\eta)$, and then, successively, $a_{(\mu-j)}(y,\eta)$ for every $j \in \mathbb{N}$.

If necessary, in order to indicate the dependence of symbol spaces on the group actions $\kappa, \widetilde{\kappa}$ in the involved spaces, we write

$$S_{\mathrm{cl}}^\mu(U \times \mathbb{R}^q; \cdot, \cdot)_{\kappa, \widetilde{\kappa}} \tag{1.81}$$

instead of $S_{\mathrm{cl}}^\mu(U \times \mathbb{R}^q; \cdot, \cdot)$. If H is a Hilbert space with two group actions $\kappa = \{\kappa_\lambda\}_{\lambda\in\mathbb{R}_+}$ and $\vartheta = \{\vartheta_\lambda\}_{\lambda\in\mathbb{R}_+}$, then κ and ϑ are said to be equivalent on H if

$$\sup_{\lambda\in\mathbb{R}_+} \|\kappa_\lambda \vartheta_\lambda^{-1}\|_{\mathcal{L}(H)} < \infty.$$

In the case of a Fréchet space $E = \mathrm{proj}\lim_{j\in\mathbb{N}} E^j$ with group actions κ and ϑ, equivalence on E means equivalence of $\kappa|_{E^j}$ and $\vartheta|_{E^j}$ on E^j for every j. By $\kappa \sim \vartheta$ we indicate equivalence.

We write

$$\kappa = \mathrm{id} \text{ if } \kappa_\lambda = \mathrm{id}_H \text{ for all } \lambda \in \mathbb{R}_+. \tag{1.82}$$

For $H = \widetilde{H} = \mathbb{C}$ and $\kappa = \widetilde{\kappa} = \mathrm{id}$, we have

$$S_{(\mathrm{cl})}^\mu(U \times \mathbb{R}^q; \mathbb{C}, \mathbb{C})_{\mathrm{id},\mathrm{id}} = S_{(\mathrm{cl})}^\mu(U \times \mathbb{R}^q) \tag{1.83}$$

for the scalar symbol spaces of Section 1.1.

Remark 1.3.9. (i) $S^\mu(U \times \mathbb{R}^q; \cdot, \cdot)_{\kappa, \widetilde{\kappa}} = S^\mu(U \times \mathbb{R}^q; \cdot, \cdot)_{\vartheta, \widetilde{\vartheta}}$ if $\kappa \sim \vartheta, \widetilde{\kappa} \sim \widetilde{\vartheta}$.

(ii) The spaces $S_{\mathrm{cl}}^\mu(U \times \mathbb{R}^q; \cdot, \cdot)_{\kappa, \widetilde{\kappa}}$ and $S_{\mathrm{cl}}^\mu(U \times \mathbb{R}^q; \cdot, \cdot)_{\vartheta, \widetilde{\vartheta}}$ can be different for $\kappa \sim \vartheta, \widetilde{\kappa} \sim \widetilde{\vartheta}$.

Theorem 1.3.10. *For every sequence of elements* $a_j \in S^{\mu_j}_{(\mathrm{cl})}(\Omega \times \mathbb{R}^q; H, \widetilde{H})$, $j \in \mathbb{N}$, $\mu_j \to -\infty$ *as* $j \to \infty$, $\mu_j = \mu - j$ *in the classical case, there is an asymptotic sum* $a \sim \sum_{j=0}^{\infty} a_j$, *i.e., an* $a \in S^{\mu}_{(\mathrm{cl})}(\Omega \times \mathbb{R}^q; H, \widetilde{H})$ *for* $\mu = \max\{\mu_j\}$, *such that for every* $N \in \mathbb{N}$ *there is a* $\nu_N \in \mathbb{R}$, $\nu_N \to -\infty$ *as* $N \to \infty$, *such that* $a - \sum_{j=0}^{N} a_j \in S^{\nu_N}(\Omega \times \mathbb{R}^q; H, \widetilde{H})$. *If* $b \in S^{\mu}_{(\mathrm{cl})}(\Omega \times \mathbb{R}^q; H, \widetilde{H})$ *is another such symbol, then* $a = b \bmod S^{-\infty}(\Omega \times \mathbb{R}^q; H, \widetilde{H})$.

A similar result holds for $a_j \in S^{\mu_j}_{(\mathrm{cl})}(\Omega \times \mathbb{R}^q; E, \widetilde{E})_f$ for any fixed $f : \mathbb{N} \to \mathbb{N}$; in this case we obtain $a \sim \sum_{j=0}^{\infty} a_j \in S^{\mu}_{(\mathrm{cl})}(\Omega \times \mathbb{R}^q; E, \widetilde{E})_f$. Note that the asymptotic sum in the operator-valued set-up may be obtained by a convergent sum similar to (1.7). However, later on in our main applications the symbol spaces will be more subtle, for instance, in the edge calculus, and then we will employ other arguments where the remainders of order $-\infty$ are particularly "precise", i.e., Green symbols of order $-\infty$ in the notation of Section 7.2.

Let us now add some general functional analytic considerations.

Definition 1.3.11. We call $\{H, H_0, H'\}$ a Hilbert space triple if H, H_0, H' are Hilbert spaces continuously embedded in a Hausdorff topological vector space V, such that $H \cap H_0 \cap H'$ is dense in H, H_0 and H', and if the H_0-scalar product $(\cdot, \cdot)$ extends from $(H \cap H_0) \times (H_0 \cap H') \to \mathbb{C}$ to a continuous non-degenerate sesquilinear form $H \times H' \to \mathbb{C}$ such that

$$\|h\| = \sup\left\{ \frac{|(h, h')|}{\|h'\|_{H'}} : h' \in H' \setminus \{0\} \right\}$$

and

$$\|h'\| = \sup\left\{ \frac{|(h, h')|}{\|h\|_H} : h \in H \setminus \{0\} \right\}$$

are equivalent norms in H and H', respectively, so that $(\cdot, \cdot)$ gives rise to antilinear isomorphisms from the dual of H to H' and vice versa.

Definition 1.3.12. By a Hilbert space triple $\{H, H_0, H'; \kappa\}$ with group action κ we understand a triple $\{H, H_0, H'\}$ in the sense of Definition 1.3.11 such that

(i) κ is a representation of the multiplicative group $\mathbb{R}_+$ in V which restricts to group actions in H, H_0 and H';

(ii) κ acts on H_0 as a unitary group.

Remark 1.3.13. Let $\{H, H_0, H'; \kappa\}$, $\{\widetilde{H}, \widetilde{H}_0, \widetilde{H}'; \widetilde{\kappa}\}$ be Hilbert triples with group action. Then:

(i) for every $a \in \mathcal{L}(H, \widetilde{H})$ there exists a unique $a^{(*)} \in \mathcal{L}(\widetilde{H}', H')$ such that

$$(au, \widetilde{u}')_{\widetilde{H}_0} = (u, a^{(*)}\widetilde{u}')_{H_0} \quad \text{for all} \quad u \in H, \widetilde{u}' \in \widetilde{H}'.$$

The mapping $a \mapsto a^{(*)}$ establishes an antilinear isomorphism $\mathcal{L}(H, \widetilde{H}) \to \mathcal{L}(\widetilde{H}', H')$;

(ii) the scalar product on $L^2(\mathbb{R}^q, H_0)$ induces a non-degenerate sesquilinear pairing

$$(\cdot, \cdot)_{L^2(\mathbb{R}^q, H_0)} : \mathcal{S}(\mathbb{R}^q, H) \times \mathcal{S}(\mathbb{R}^q, H') \to \mathbb{C}.$$

Theorem 1.3.14. *Let E, F be Fréchet spaces and $G = E \widehat{\otimes}_\pi F$ their projective tensor product. Then every $g \in G$ can be written as a convergent sum*

$$g = \sum_{j=0}^{\infty} \lambda_j e_j \otimes f_j$$

with $\lambda_j \in \mathbb{C}$, $\sum_{j=0}^{\infty} |\lambda_j| < \infty$ and sequences $e_j \in E$, $f_j \in F$, tending to zero in the respective spaces as $j \to \infty$.

Given Fréchet spaces E_0, E_1 continuously embedded in a Hausdorff topological vector space, we define the non-direct sum

$$E_0 + E_1 := \{e_0 + e_1 : e_0 \in E_0, e_1 \in E_1\}$$

endowed with the Fréchet topology of the identification

$$E_0 + E_1 \cong (E_0 \oplus E_1)/\Delta, \quad \text{where } \Delta = \{(e, -e) : e \in E_0 \cap E_1\}.$$

The semi-norm system of $E_0 + E_1$ may be generated as

$$r(e) := \inf_{e=e_0+e_1} \{p(e_0) + q(e_1) : e_0 \in E_0, e_1 \in E_1\}$$

where p and q run over the semi-norms in E_0 and E_1, respectively. Moreover, given a Fréchet space E which is a left module over an algebra A, we write for $a \in A$

$$[a]E := \text{completion of } \{ae : e \in E\} \text{ in } E. \tag{1.84}$$

In an analogous manner we define $E[a]$ when A acts from the right on E.

Remark 1.3.15. Note that there are canonical continuous embeddings

$$E_i \hookrightarrow E_0 + E_1 \quad \text{for } i = 0, 1. \tag{1.85}$$

More precisely, E_i may be interpreted as a subspace of $E_0 + E_1$ with a stronger topology than (or identical to) that induced by $E_0 + E_1$.

In fact, we may realise (1.85), say, for $i = 0$ as a composition of the continuous operators $E_0 \to E_0 \oplus E_1$, $e_0 \mapsto e_0 \oplus 0$ and $E_0 \to E_0 \oplus E_1 \to E_0 \to (E_0 \oplus E_1)/\Delta$. The composition is continuous and injective, since $e_0 \oplus 0$ cannot be mapped to 0 unless $e_0 = 0$.

Lemma 1.3.16. *Let $E = E_0 + E_1$ be a non-direct sum of Fréchet spaces and $K : E \to F$ an isomorphism from E to a Fréchet space F, then we have a non-direct sum $F = F_0 + F_1$ for the Fréchet spaces $F_i := KE_i$ with the Fréchet topology from the bijection $E_i \to F_i$.*

Proof. By virtue of Remark 1.3.15, the space E_i is a linear subspace of E. Applying K we obtain a linear subspace KE_i of F that we endow with the Fréchet topology induced by the bijection $K : E_i \to F_i$. It is clear that $F = F_0 + F_1$ as an algebraic non-direct sum. Moreover, K induces a bijection $K : E_0 \cap E_1 \to F_0 \cap F_1$, and hence $\Delta_E = \{(e, -e) : e \in E_0 \cap E_1\}$ correspondence to $\Delta_F = \{(f, -f) : f \in F_0 \cap F_1\}$, and it follows that $F = (F_0 \oplus F_1)/\Delta_F$. $\qquad\square$

Proposition 1.3.17. *Given Fréchet spaces E,F,G we have*

$$(E + F) \,\widehat{\otimes}_\pi\, G = E \,\widehat{\otimes}_\pi\, G + F \,\widehat{\otimes}_\pi\, G.$$

The following proposition is a consequence of Theorem 1.3.14.

Proposition 1.3.18. *We have*

$$S^\mu_{(\mathrm{cl})}\big(U \times \mathbb{R}^q; E, \widetilde{E}\big)_r = C^\infty\big(U, S^\mu_{(\mathrm{cl})}\big(\mathbb{R}^q; E, \widetilde{E}\big)_r\big) = C^\infty(U) \,\widehat{\otimes}_\pi\, S^\mu_{(\mathrm{cl})}\big(\mathbb{R}^q; E, \widetilde{E}\big)_r$$

for the projective tensor product $\widehat{\otimes}_\pi$ between the respective Fréchet spaces. In particular, every element $a(y, \eta) \in S^\mu_{(\mathrm{cl})}(\mathbb{R}^q; E, \widetilde{E})_r$ can be written as a convergent sum

$$a(y, \eta) = \sum_{j=0}^\infty \delta_j \alpha_j(y) a_j(\eta)$$

with null-sequences α_j and a_j in $C^\infty(U)$ and $S^\mu_{(\mathrm{cl})}(\mathbb{R}^q; E, \widetilde{E})_r$, respectively, and constants $\delta_j \in \mathbb{C}$ such that $\sum_{j=0}^\infty |\delta_j| < \infty$.

Remark 1.3.19. (i) Let $f(\zeta, \eta) \in S^{(\mu)}\big(\mathbb{R}^{d+q}_{\zeta,\eta} \setminus \{0\}; H, \widetilde{H}\big)$. Then for every fixed $\eta^1 \neq 0$ we have $f(\zeta, \eta^1) \in S^\mu_{\mathrm{cl}}(\mathbb{R}^d_\zeta; H, \widetilde{H})$, and the homogeneous component

$$f_{(\mu-j)}(\zeta, \eta^1) \in S^{(\mu-j)}\big(\mathbb{R}^d_\zeta \setminus \{0\}; H, \widetilde{H}\big)$$

of order $\mu - j$ is a polynomial in $\eta^1 \in \mathbb{R}^q$ of degree j for every $j \in \mathbb{N}$. Similar relations hold for f smoothly depending on extra variables z, y.

(ii) Let $f(z, y, \zeta, \eta) \in S^\mu_{(\mathrm{cl})}\big(V_z \times U_y \times \mathbb{R}^{d+q}_{\zeta,\eta}; H, \widetilde{H}\big)$, for open $V \subseteq \mathbb{R}^e$, $U \subseteq \mathbb{R}^p$. Then for every fixed $y^1 \in \mathbb{R}^p$, $\eta^1 \in \mathbb{R}^q$ we have

$$f(z, y^1, \zeta, \eta^1) \in S^\mu_{(\mathrm{cl})}\big(V_z \times \mathbb{R}^d_\zeta; H, \widetilde{H}\big).$$

In the classical case the homogeneous principal symbol of $f(z, y^1, \zeta, \eta^1)$ in (z, ζ) is independent of η^1.

The arguments are similar to those in the scalar case, cf. Remark 1.1.6.

Remark 1.3.20. Let $f(y, \eta) \in C^\infty\big(U \times \mathbb{R}^q_\eta, \mathcal{L}(H, \widetilde{H})\big)$ be a function such that

$$f(y, \lambda\eta) = \lambda^\mu f(y, \eta)$$

for all $y \in U$, $\lambda \geq 1$, $|\eta| \geq c$ for some $c > 0$. Then $f(y, \eta) \in S_{\mathrm{cl}}^{\mu}(U \times \mathbb{R}^q; H, \widetilde{H})$. In particular, for every $f_{(\mu)}(y, \eta) \in S^{(\mu)}(U \times (\mathbb{R}_{\eta}^q \setminus \{0\}); H, \widetilde{H})$ and any excision function $\chi(\eta)$ we have

$$\chi(\eta) f_{(\mu)}(y, \eta) \in S_{\mathrm{cl}}^{\mu}(U \times \mathbb{R}^q; H, \widetilde{H}).$$

Similar observations hold for Fréchet spaces.

Homogeneous operator functions occur in many applications. Consider the function $\lambda \mapsto b_0(\lambda)$ defined by

$$b_0(\lambda) := \widetilde{\kappa}_{\lambda} b \kappa_{\lambda}^{-1}, \quad \lambda \in \mathbb{R}_+, \tag{1.86}$$

for any fixed $b \in \mathcal{L}(H, \widetilde{H})$. Then we have

$$b_0(\delta\lambda) := \widetilde{\kappa}_{\delta} b_0(\lambda) \kappa_{\delta}^{-1} \tag{1.87}$$

for $\delta \in \mathbb{R}_+$; however, $b_0(\lambda)$ is not necessarily smooth in λ. An example is $H = \widetilde{H} = L^2(\mathbb{R}^n)$, $\kappa = \widetilde{\kappa}$, and b_0 the operator of multiplication by the characteristic function χ of the unit ball in $\mathbb{R}^n$. Then

$$b_0(\lambda) u(x) = \kappa_{\lambda} \chi(x) \kappa_{\lambda}^{-1} u(x) = \chi(\lambda x) u(x),$$

but the operator function $b_0(\lambda)$ is not smooth in λ. However,

$$y(\mu) \cdot \int_0^{\infty} \mu^{-1} \psi\left(\frac{\rho - \delta}{\rho}\right) b_0(\delta) d\delta$$

belongs to $C^{\infty}(\mathbb{R}_+, \mathcal{L}(H, \widetilde{H}))$ for any $\varphi \in C_0^{\infty}(\mathbb{R})$.

Remark 1.3.21. There is a $\varphi \in C_0^{\infty}(\mathbb{R})$, $\varphi \geq 0$, $\varphi \neq 0$, such that

$$c(\rho) := \int_0^{\infty} \rho^{-1} \varphi\left(\frac{\rho - \delta}{\rho}\right) d\delta > 0$$

for every $\rho \in \mathbb{R}_+$. In fact, it suffices to assume that $\varphi \in C_0^{\infty}(\mathbb{R}_-)$, $\varphi \geq 0$, and $\varphi(\beta) > 0$ for some $\beta \in \mathbb{R}_-$.

Let us now verify that $b(\rho)$ is twisted homogenous of order zero, i.e., $g(\lambda\rho) = \widetilde{\kappa}_{\lambda} g(\rho) \kappa_{\lambda}^{-1}$, $\lambda \in \mathbb{R}_+$. In fact, from (1.87) we obtain

$$\widetilde{\kappa}_{\lambda} g(\rho) \kappa_{\lambda}^{-1} = \int_0^{\infty} \rho^{-1} \varphi\left(\frac{\rho - \delta}{\rho}\right) \widetilde{\kappa}_{\lambda} b_0(\delta) \kappa_{\lambda}^{-1} d\delta$$

$$= \int_0^{\infty} \rho^{-1} \varphi\left(\frac{\rho - \delta}{\rho}\right) b_0(\delta\lambda) d\delta$$

$$= \int_0^{\infty} \rho^{-1} \varphi\left(\frac{\rho - \widetilde{\delta}/\lambda}{\rho}\right) b_0(\widetilde{\delta}) \lambda^{-1} d\widetilde{\delta}$$

$$= \int_0^{\infty} \rho^{-1} \lambda^{-1} \varphi\left(\frac{\lambda\rho - \widetilde{\delta}}{\lambda\rho}\right) b_0(\widetilde{\delta}) d\widetilde{\delta} = g(\lambda\rho).$$

Setting $a(\eta) := [\eta]^\mu g([\eta])$, $\mu \in \mathbb{R}$, Remark 1.3.20 shows that $a(\eta) \in S_{\mathrm{cl}}^\mu(\mathbb{R}^q; H, \widetilde{H})$, and

$$a_{(\mu)}(\eta) := |\eta|^\mu g(|\eta|) \in S^{(\mu)}(\mathbb{R}^q \setminus \{0\}; \mathcal{L}(H, \widetilde{H})).$$

Thus $a(\eta)$ is an example of a classical symbol of order μ. Summing up, we obtained the following result.

Proposition 1.3.22. *For any pair of Hilbert spaces H and $\widetilde{H}$ with group actions κ and $\widetilde{\kappa}$, respectively, and given $\mu \in \mathbb{R}$ there exists an $a(\eta) \in S_{\mathrm{cl}}^\mu(\mathbb{R}^q; H, \widetilde{H})$, $a(\eta) \neq 0$.*

Let us now establish an approximation property. Choose an ε-dependent family of functions $\varphi_\varepsilon \in C_0^\infty(\mathbb{R})$, $0 < \varepsilon < 1/2$, such that $\varphi_\varepsilon \geq 0$, $\int_{-\infty}^\infty \varphi_\varepsilon(t)dt = 1$, and $\operatorname{supp} \varphi_\varepsilon \subset [-\varepsilon, \varepsilon]$ for every ε.

Proposition 1.3.23. *Let $H, \widetilde{H}$ be as in Proposition 1.3.22, let $b \in \mathcal{L}(H, \widetilde{H})$ and $b_0(\lambda)$ defined by (1.86), and set*

$$g_\varepsilon(\lambda) := \int_0^\infty \lambda^{-1} \varphi_\varepsilon\left(\frac{\lambda - \delta}{\lambda}\right) b_0(\delta) d\delta.$$

Then $\|g_\varepsilon(\lambda)u - b_0(\lambda)u\|_{\widetilde{H}} \to 0$ as $\varepsilon \to 0$ for every $\lambda \in \mathbb{R}_+$, $u \in H$.

Proof. By construction, we have

$$g_\varepsilon(\lambda)u - b_0(\lambda)u = \int_0^\infty \lambda^{-1} \varphi_\varepsilon\left(\frac{\lambda - \delta}{\lambda}\right) b_0(\delta) u d\delta - b_0(\lambda)u$$

$$= \int_0^\infty \lambda^{-1} \varphi_\varepsilon\left(\frac{\lambda - \delta}{\lambda}\right) (b_0(\delta) - b_0(\lambda)) u d\delta.$$

Thus

$$\|g_\varepsilon(\lambda)u - b_0(\lambda)u\|_{\widetilde{H}} \leq \int_0^\infty \lambda^{-1} \varphi_\varepsilon\left(\frac{\lambda - \delta}{\lambda}\right) \|b_0(\delta)u - b_0(\lambda)u\|_{\widetilde{H}} d\delta$$

$$\leq \sup_{\delta \in I_\varepsilon} \|b_0(\delta)u - b_0(\lambda)u\|_{\widetilde{H}},$$

where I_ε is the support of $\varphi_\varepsilon\left(\frac{\lambda - \delta}{\lambda}\right)$ in δ for fixed λ. We use the fact that

$$\int_0^\infty \lambda^{-1} \varphi_\varepsilon\left(\frac{\lambda - \delta}{\lambda}\right) d\delta = 1.$$

Since $b_0(\lambda)u$ is continuous in λ with values in $\widetilde{H}$ and I_ε shrinks to λ as $\varepsilon \to 0$, we get our assertion. $\square$

The following properties of symbols, including Remark 1.3.29 below, may also be found in the article [50] jointly with J. Seiler in the context of applications. For

$$a(y, t, \eta, \tau) \in S^\mu(U \times \mathbb{R}^d \times \mathbb{R}_{\eta, \tau}^{q+d}) \tag{1.88}$$

we set

$$\mathfrak{a}(y,t,\eta,\tau) := a\big(y,\langle\eta\rangle^{-1}t,\eta,\langle\eta\rangle\tau\big), \tag{1.89}$$

referred to as the decoupled symbol associated with (1.88). We will show that for

$$a(y,t,\eta,\tau) \in S^{\mu}_{(\mathrm{cl})}\big(U_y \times \mathbb{R}^d_t \times \mathbb{R}^{q+d}_{\eta,\tau}\big) \tag{1.90}$$

the decoupled symbol belongs to the space $S^{\mu}_{(\mathrm{cl})}\big(U \times \mathbb{R}^q, S^{\mu}_{(\mathrm{cl})}(\mathbb{R}^d \times \mathbb{R}^d)\big)$.
In addition, we often form operator families

$$\mathrm{Op}_t(\mathfrak{a})(y,\eta)v(t) = \iint e^{i(t-t')\tau}\mathfrak{a}(y,\eta,t,\tau)v(t')dt'\,\bar{d}\tau$$

for (1.89), and then

$$\mathrm{Op}_t(\mathfrak{a})(y,\eta) = \kappa^{-1}_{\langle\eta\rangle}\mathrm{Op}_t(a)(y,\eta)\kappa_{\langle\eta\rangle}. \tag{1.91}$$

Lemma 1.3.24. *For $a(y,\eta,\tau) \in S^{\mu}(U_y \times \mathbb{R}^{q+d}_{\eta,\tau})$ we have*

$$\mathrm{Op}_t(a)(y,\eta) \in S^{\mu}\big(U \times \mathbb{R}^q; H^s(\mathbb{R}^d), H^{s-\mu}(\mathbb{R}^d)\big) \tag{1.92}$$

for every $s \in \mathbb{R}$. Here the group action on $H^s(\mathbb{R}^d)$ is as in Theorem 1.3.33 (ii). Moreover, $a(y,\eta,\tau) \in S^{\mu}_{\mathrm{cl}}(U_y \times \mathbb{R}^{q+d}_{\eta,\tau})$ yields

$$\mathrm{Op}_t(a)(y,\eta) \in S^{\mu}_{\mathrm{cl}}\big(U \times \mathbb{R}^q; H^s(\mathbb{R}^d), H^{s-\mu}(\mathbb{R}^d)\big).$$

Proof. Using

$$\langle\eta,\langle\eta\rangle\tau\rangle = \langle\eta\rangle\langle\tau\rangle \tag{1.93}$$

and the estimate $|\mathfrak{a}(y,\eta,\tau)| \le c\,\langle\eta,\langle\eta\rangle\tau\rangle^{\mu} = c\,\langle\eta\rangle^{\mu}\langle\tau\rangle^{\mu}$ uniformly in $y \in K$, $K \Subset U$, it follows that

$$\begin{aligned}
\big\|\kappa^{-1}_{\langle\eta\rangle}\mathrm{Op}_t(a)(y,\eta)\kappa_{\langle\eta\rangle}v\big\|^2_{H^{s-\mu}(\mathbb{R}^d)} &= \big\|\mathrm{Op}_t(\mathfrak{a})(y,\eta)v\big\|^2_{H^{s-\mu}(\mathbb{R}^d)} \\
&= \int \langle\tau\rangle^{2(s-\mu)}\big|\mathfrak{a}(y,\eta,\tau)\widehat{v}(\tau)\big|^2\bar{d}\tau \\
&\le c\int \langle\tau\rangle^{2(s-\mu)}\langle\tau\rangle^{2\mu}\langle\eta\rangle^{2\mu}\big|\widehat{v}(\tau)\big|^2\bar{d}\tau \\
&\le c\,\langle\eta\rangle^{2\mu}\int \langle\tau\rangle^{2s}\big|\widehat{v}(\tau)\big|^2\bar{d}\tau \\
&= c\,\langle\eta\rangle^{2\mu}\|v\|^2_{H^s(\mathbb{R}^d)},
\end{aligned}$$

i.e.,

$$\big\|\kappa^{-1}_{\langle\eta\rangle}\mathrm{Op}_t(a)(y,\eta)\kappa_{\langle\eta\rangle}\big\|_{\mathcal{L}(H^s(\mathbb{R}^d),H^{s-\mu}(\mathbb{R}^d))} \le c\,\langle\eta\rangle^{\mu}$$

with some $c = c(K) > 0$. In a similar manner we obtain the symbol estimates for the (y,η)-derivatives of $\mathrm{Op}_t(a)$.

Now let a be classical. Then for every $N \in \mathbb{N}$ we can write

$$a(y, \eta, \tau) = \sum_{j=0}^{N} \chi(\eta, \tau) a_{(\mu-j)}(y, \eta, \tau) + r_{N+1}(y, \eta, \tau)$$

for an excision function $\chi(\eta, \tau)$, homogeneous components $a_{(\mu-j)}(y, \eta, \tau)$ of a, i.e.,

$$a_{(\mu-j)}(y, \delta\eta, \delta\tau) = \delta^{\mu-j} a_{(\mu-j)}(y, \eta, \tau), \quad \delta \in \mathbb{R}_+,$$

and a remainder $r_{N+1}(y, \eta, \tau) \in S^{\mu-(N+1)}(U_y \times \mathbb{R}_{\eta,\tau}^{q+d})$. Let

$$a_{\mu-j}(y, \eta, \tau) := \chi(\eta, \tau) a_{(\mu-j)}(y, \eta, \tau).$$

Then we have $a_{\mu-j}(y, \delta\eta, \delta\tau) = \delta^{\mu-j} a_{\mu-j}(y, \eta, \tau)$ for all $\delta \geq 1$ and $|\eta, \tau| \geq$ const for a constant > 0. Let us verify that

$$\mathrm{Op}_t(a_{\mu-j})(y, \delta\eta) = \delta^{\mu-j} \kappa_\delta \, \mathrm{Op}_t(a_{\mu-j})(y, \eta) \kappa_\delta^{-1} \tag{1.94}$$

for all $\delta \geq 1$ and $|\eta| \geq$ const. This is a consequence of the relations

$$\kappa_\delta^{-1} \mathrm{Op}_t(a_{\mu-j})(y, \delta\eta) \kappa_\delta v(t) = \iota_\delta^{-1} \iint e^{i(t-t')\tau} a_{\mu-j}(y, \delta\eta, \tau) v(\delta t') dt' d\!\!\!{}^-\tau$$

$$= \iint e^{i(t-\tilde{t})\delta^{-1}\tau} a_{\mu-j}(y, \delta\eta, \tau) v(\tilde{t}) \delta^{-1} d\tilde{t} \, d\!\!\!{}^-\tau$$

$$= \iint e^{i(t-\tilde{t})\tilde{\tau}} a_{\mu-j}(y, \delta\eta, \delta\tilde{\tau}) v(\tilde{t}) d\tilde{t} \, d\!\!\!{}^-\tilde{\tau}$$

$$= \delta^{\mu-j} \mathrm{Op}_t(a_{\mu-j})(y, \eta).$$

Therefore, the operator functions $\mathrm{Op}_t(a_{\mu-j})(y, \eta)$ are classical symbols of order $\mu - j$, cf. Remark 1.3.20. The first part of the proof gives

$$\mathrm{Op}_t(a)(y, \eta) - \sum_{j=0}^{N} \mathrm{Op}_t(a_{\mu-j})(y, \eta)$$

$$= \mathrm{Op}_t(r_{N+1})(y, \eta) \in S^{\mu-(N+1)}\left(U \times \mathbb{R}^q; H^s(\mathbb{R}^d), H^{s-\mu}(\mathbb{R}^d)\right).$$

Thus, $\mathrm{Op}_t(a)(y, \eta)$ is a classical symbol.   □

Lemma 1.3.25. Let $a(t, \eta) \in S_{\mathrm{cl}}^{\mu}(\mathbb{R}_\eta^q, C^\infty(\mathbb{R}^d))$. Then for $\mathfrak{a}(t, \eta) := a([\eta]^{-1}t, \eta)$, we have

$$\mathfrak{a}(t, \eta) \in S_{\mathrm{cl}}^{\mu}(\mathbb{R}_\eta^q, C^\infty(\mathbb{R}^d)).$$

Proof. Taylor's formula for $\mathfrak{a}$ in t at 0 gives

$$\mathfrak{a}(t, \eta) = \sum_{|\alpha| \leq N} \frac{1}{\alpha!} (\partial_t^\alpha \mathfrak{a})(0, \eta) t^\alpha + r_{N+1}(t, \eta) \tag{1.95}$$

where $(\partial_t^\alpha \mathfrak{a})(0,\eta) = [\eta]^{-|\alpha|}(\partial_t^\alpha a)(0,\eta)$ and

$$r_{N+1}(t,\eta) = (N+1) \sum_{|\sigma|=N+1} \frac{t^\sigma}{\sigma!} \int_0^1 (1-\theta)^N (\partial_t^\sigma \mathfrak{a})(\theta t,\eta)d\theta.$$

The summands in (1.95) belong to $S_{cl}^{\mu-|\alpha|}(\mathbb{R}_\eta^q, C^\infty(\mathbb{R}^d))$, while the remainder is an element of $S^{\mu-(N+1)}(\mathbb{R}_\eta^q, C^\infty(\mathbb{R}^d))$. Those properties follow by checking the symbol estimates

$$\sup_{t\in K} \left| \partial_t^j (D_\eta^\alpha r_{N+1}(t,\eta)) \right| \le c \langle\eta\rangle^{\mu-(N+1)-|\alpha|}$$

for every j, with constants $c = c(j,K,\alpha) > 0$. The latter computations are straightforward. $\qquad\square$

Let us set $E := S^\mu(\mathbb{R}^d \times \mathbb{R}^d)$ with the above-mentioned Fréchet topology, with the semi-norms

$$\pi_{\gamma\delta,K}(e) = \sup_{t\in K, \tau\in\mathbb{R}^d} \langle\eta\rangle^{-\mu+|\delta|} \left| D_t^\gamma D_\tau^\delta e(t,\tau) \right|,$$

where $\gamma, \delta \in \mathbb{N}^d$, $K \Subset \mathbb{R}^d$.

Lemma 1.3.26. *For $a(y,t,\eta,\tau) \in S^\mu(U \times \mathbb{R}^d \times \mathbb{R}_{\eta,\tau}^{q+d})$ and*

$$\mathfrak{a}(y,t,\eta,\tau) = a(y, \langle\eta\rangle^{-1} t, \eta, \langle\eta\rangle \tau)$$

the map $a \mapsto \mathfrak{a}$ defines a continuous operator

$$S^\mu(U \times \mathbb{R}^d \times \mathbb{R}_{\eta,\tau}^{q+d}) \to S^\mu(U \times \mathbb{R}_\eta^q, S^\mu(\mathbb{R}^d \times \mathbb{R}_\tau^d))$$

for every $\mu \in \mathbb{R}$.

Proof. Let $a(y,t,\eta,\tau) \in S^\mu(U \times \mathbb{R}^d \times \mathbb{R}_{\eta,\tau}^{q+d})$, and set

$$p(y,\eta) := a(y, \langle\eta\rangle^{-1} t, \eta, \langle\eta\rangle \tau). \qquad (1.96)$$

The semi-norms in the space $E := S^\mu(\mathbb{R}^d \times \mathbb{R}^d)$ have the form

$$\pi_{\gamma\delta,K}(e) = \sup_{t\in K, \tau\in\mathbb{R}^d} \langle\tau\rangle^{-\mu+|\delta|} \left| D_t^\gamma D_\tau^\delta e(t,\tau) \right|, \qquad (1.97)$$

$\gamma, \delta \in \mathbb{N}^d$, $K \Subset \mathbb{R}^d$. We have to show that

$$\pi_{\gamma\delta,K}(D_y^\alpha D_\eta^\beta p(y,\eta)) \le c \langle\eta\rangle^{\mu-|\beta|} \qquad (1.98)$$

for all $\alpha \in \mathbb{N}^p$, $\beta \in \mathbb{N}^q$, $(y,\eta) \in A \times \mathbb{R}^q$, $A \Subset U$, with some constants $c = c(\alpha,\beta,A) > 0$. The symbol estimates for a itself read

$$\left| D_y^\alpha D_t^\gamma D_\eta^\beta D_\tau^\delta a(y,t,\eta,\tau) \right| \le c \langle\eta,\tau\rangle^{\mu-(|\beta|+|\delta|)}, \qquad (1.99)$$

with constants $c = c(\alpha, \gamma, \beta, \delta, K, A) > 0$, $K \Subset \mathbb{R}^d$, $A \Subset U$.

Let us confine ourselves to the case of y-independent symbols; the general case is completely analogous. We apply (1.97) to

$$e(t, \tau) := D_\eta^\beta p(\eta),$$

with p defined by (1.96). In this case $D_\eta^\beta p(\eta)$ is a linear combination of expressions

$$\langle \eta \rangle^{-|\iota|} \left(D_t^\iota D_\eta^\vartheta D_\tau^\zeta a \right) \left(\langle \eta \rangle^{-1} t, \eta, \langle \eta \rangle \tau \right) \tag{1.100}$$

for $|\iota| + |\vartheta| + |\zeta| = |\beta|$. In order to estimate (1.97) we apply $D_t^\gamma D_\tau^\delta$ to (1.100), which yields expressions like

$$\langle \eta \rangle^{-(|\iota| + |\gamma|)} \langle \eta \rangle^{|\delta|} \left(D_t^{\iota + \gamma} D_\eta^\vartheta D_\tau^{\zeta + \delta} a \right) \left(\langle \eta \rangle^{-1} t, \eta, \langle \eta \rangle \tau \right). \tag{1.101}$$

By (1.99), the absolute value of (1.101) can be estimated as

$$c \langle \eta \rangle^{-(|\iota| + |\gamma|)} \langle \eta \rangle^{|\delta|} \langle \eta, \langle \eta \rangle \tau \rangle^{\mu - (|\vartheta| + |\zeta| + |\delta|)}$$

uniformly in $t \in K$. Thus (1.93) yields the upper bound

$$c \langle \eta \rangle^{-(|\iota| + |\gamma|)} \langle \eta \rangle^{\mu - |\vartheta| - |\zeta|} \langle \tau \rangle^{\mu - (|\vartheta| + |\zeta| + |\delta|)} \leq c \langle \eta \rangle^{\mu - |\beta|} \langle \tau \rangle^{\mu - |\delta|},$$

which entails the estimate (1.98).   □

Lemma 1.3.27. *For $a(y, t, \eta, \tau) \in S_{cl}^\mu(U \times \mathbb{R}^d \times \mathbb{R}_{\eta, \tau}^{q+d})$ and any excision function $\chi(\tau)$, we have*

$$(1 - \chi(\tau)) a(y, t, \eta, \tau) \in S_{cl}^\mu \left(U \times \mathbb{R}^q, S^{-\infty} \left(\mathbb{R}^d \times \mathbb{R}^d \right) \right).$$

Proof. For simplicity we consider again the case of y-independent symbols. Set

$$p(t, \eta, \tau) := (1 - \chi(\tau)) a \left([\eta]^{-1} t, \eta, [\eta] \tau \right).$$

We have to show that for every $\beta \in \mathbb{N}^q$ and every semi-norm

$$\pi_{\gamma\delta, K}^M(e) = \sup_{t \in K, \tau \in \mathbb{R}^d} \langle \tau \rangle^M \left| D_t^\gamma D_\tau^\delta e(t, \tau) \right|,$$

where $\gamma, \delta \in \mathbb{N}^d$, $K \Subset \mathbb{R}^d$, $M \in \mathbb{N}$, on the space $S^{-\infty}(\mathbb{R}^d \times \mathbb{R}^d) \ni e(t, \tau)$ we have the estimates

$$\pi_{\gamma\delta, K} \left(D_\eta^\beta p(t, \eta, \tau) \right) \leq c \langle \eta \rangle^{\mu - |\beta|}$$

for all $\beta \in \mathbb{N}^q$, $c = c(\beta) > 0$. Let us first consider the case where a is independent of t. Then, we replace $\pi_{\gamma\delta, K}^M$ by $\pi_{\gamma\delta}^M$ and we show the symbol estimates

$$\pi_{\gamma\delta}^M \left(D_\eta^\beta p(\eta, \tau) \right) \leq c \langle \eta \rangle^{\mu - |\beta|}.$$

Let us write

$$a(\eta, \tau) = \widetilde{\chi}(\eta, \tau) \sum_{j=0}^{N} a_{(\mu-j)}(\eta, \tau) + r_{N+1}(\eta, \tau) \tag{1.102}$$

for some excision function $\widetilde{\chi}(\eta, \tau)$, where $r_{N+1}(\eta, \tau) \in S_{\mathrm{cl}}^{\mu-(N+1)}(\mathbb{R}^{q+d})$, $N \in \mathbb{N}$. The summands in the right-hand side of (1.102) give rise to functions

$$p_j(\eta, \tau) := (1 - \chi(\tau))\widetilde{\chi}(\eta, [\eta]\tau) \sum_{j=0}^{N} a_{(\mu-j)}(\eta, [\eta]\tau).$$

Because of the factor $(1 - \chi(\tau))$, the function $p_j(\eta, \tau)$ takes values in $S^{-\infty}(\mathbb{R}_\tau^d)$. Moreover, we have

$$p_j(\lambda\eta, \tau) = \lambda^{\mu-j} p_j(\eta, \tau)$$

for $\lambda \geq 1$, $|\eta| \geq$ const. In fact, for $|\eta| \geq$ const we have $\widetilde{\chi}(\eta, [\eta]\tau) = 1$ and $a_{(\mu-j)}(\lambda\eta, [\lambda\eta]\tau) = \lambda^{\mu-j} a_{(\mu-j)}(\eta, [\eta]\tau)$. Thus $p_j(\eta, \tau) \in C^\infty(\mathbb{R}_\eta^q, S^{-\infty}(\mathbb{R}_\tau^d))$, and the indicated homogeneity in η yields

$$p_j(\eta, \tau) \in S_{\mathrm{cl}}^{\mu-j}(\mathbb{R}^q, S^{-\infty}(\mathbb{R}^d)), \quad j = 1, \ldots, N.$$

Now for fixed $\beta \in \mathbb{N}^q$ let us show that

$$\pi_{\gamma\delta}^M((1 - \chi(\tau))D_\eta^\beta r_{N+1}(\eta, [\eta]\tau)) < c \langle\eta\rangle^{\mu-|\beta|}.$$

To this end consider the semi-norm

$$(1 - \chi(\tau))r_{N+1}(\eta, [\eta]\tau) \longrightarrow \sup_{\tau \in \mathbb{R}^d} \langle\tau\rangle^M \left| D_\tau^\delta(1 - \chi(\tau))r_{N+1}(\eta, [\eta]\tau)\right|.$$

Since $D_\eta^\beta r_{N+1}(\eta, \tau) \in S^{\mu-(N+1)-|\beta|}(\mathbb{R}^{q+d})$ we have the symbol estimates

$$\left|D_\eta^\beta D_\tau^\delta r_{N+1}(\eta, \tau)\right| \leq c \langle\eta, \tau\rangle^{\mu-(N+1)-|\delta|-|\beta|},$$

whence

$$\left| \langle\tau\rangle^M D_\tau^\delta(1 - \chi(\tau))D_\eta^\beta r_{N+1}(\eta, [\eta]\tau)\right|$$

$$\leq c \sup_{|\tau| \leq C} [\eta]^\delta \langle\eta, [\eta]\tau\rangle^{\mu-(N+1)-|\delta|-|\beta|}$$

$$\leq c \sup_{|\tau| \leq C} [\eta]^\delta \langle\eta\rangle^{\mu-(N+1)-|\delta|-|\beta|} \langle\tau\rangle^{\mu-(N+1)-|\delta|-|\beta|}$$

$$\leq c \langle\eta\rangle^{\mu-(N+1)-|\beta|}$$

where C is determined by $(1 - \chi(\tau)) = 0$ for $|\tau| \geq C$. This shows altogether that our symbol belongs to $S_{\mathrm{cl}}^\mu(U \times \mathbb{R}^q, S^{-\infty}(\mathbb{R}^d \times \mathbb{R}^d))$, and the proof for t-independent symbols a is complete. The general case is dealt with by combining the arguments from the first part of the proof with Lemma 1.3.25. $\qquad\square$

Proposition 1.3.28. *The map* $a \mapsto \mathfrak{a}$ *for* (1.90) *and* (1.89) *defines a continuous operator*

$$S^\mu_{\mathrm{cl}}\big(U \times \mathbb{R}^d \times \mathbb{R}^{q+d}_{\eta,\tau}\big) \to S^\mu_{\mathrm{cl}}\big(U \times \mathbb{R}^q, S^\mu_{\mathrm{cl}}(\mathbb{R}^d \times \mathbb{R}^d)\big) \qquad (1.103)$$

for every $\mu \in \mathbb{R}$.

Proof. For convenience, in this proof we replace $\langle \eta \rangle$ by $[\eta]$; the result itself is not affected by this modification. We first choose an excision function $\chi(\tau)$ such that for a $c > 0$ we have $\chi(\tau) = 0$ for $|\tau| \le c$. Then $\chi([\eta]^{-1}\tau) = 0$ for $|\tau| \le c[\eta]$. Thus there is an excision function $\widetilde{\chi}(\eta, \tau)$ such that $\widetilde{\chi}(\eta, \tau)\chi([\eta]^{-1}\tau) = \chi([\eta]^{-1}\tau)$ for all η, τ, or, equivalently,

$$\widetilde{\chi}(\eta, [\eta]\tau)\chi(\tau) = \chi(\tau) \qquad (1.104)$$

for all η, τ. Throughout the ensuing argument we drop y, and then we have

$$\mathfrak{a}(t, \eta, \tau) = a\big([\eta]^{-1}t, \eta, [\eta]\tau\big).$$

Let us write

$$a(t, \eta, \tau) = \widetilde{\chi}(\eta, \tau)\sum_{l=0}^{N} a_{(\mu-l)}(t, \eta, \tau) + c_N(t, \eta, \tau) \qquad (1.105)$$

for a remainder $c_N(t, \eta, \tau) \in S^{\mu-(N+1)}_{\mathrm{cl}}(\mathbb{R}^d \times \mathbb{R}^q \times \mathbb{R}^d)$. In order to show that

$$\mathfrak{a}(t, \eta, \tau) \in S^\mu_{\mathrm{cl}}\big(\mathbb{R}^q, S^\mu_{\mathrm{cl}}(\mathbb{R}^d \times \mathbb{R}^d)\big),$$

we verify that

$$\widetilde{\chi}(\eta, [\eta]\tau)\mathfrak{a}_l(t, \eta, \tau) \in S^{\mu-l}_{\mathrm{cl}}\big(\mathbb{R}^q, S^{\mu-l}_{\mathrm{cl}}(\mathbb{R}^d \times \mathbb{R}^d)\big)$$

for

$$\mathfrak{a}_l(t, \eta, \tau) := a_{(\mu-l)}\big([\eta]^{-1}t, \eta, [\eta]\tau\big),$$

and that

$$\mathfrak{c}_N(t, \eta, \tau) \in S^{\mu-(N+1)}\big(\mathbb{R}^q, S^{\mu-(N+1)}_{\mathrm{cl}}(\mathbb{R}^d \times \mathbb{R}^d)\big),$$

where $\mathfrak{c}_N(t, \eta, \tau) := c_N([\eta]^{-1}t, \eta, [\eta]\tau)$ is the decoupled version of $c_N(t, \eta, \tau)$. In the first step of the proof we show that

$$\mathfrak{a}(t, \eta, \tau) \in S^\mu\big(\mathbb{R}^q, S^\mu_{\mathrm{cl}}(\mathbb{R}^d \times \mathbb{R}^d)\big). \qquad (1.106)$$

Let $(\pi_\iota)_{\iota \in \mathbb{N}}$ denote a system of semi-norms for the Fréchet topology of $S^\mu_{\mathrm{cl}}(\mathbb{R}^d \times \mathbb{R}^d)$, cf. the discussion after Definition 1.1.1. Then (1.106) means that for every ι

$$\pi_\iota(D^\beta_\eta \mathfrak{a}(t, \eta, \tau)) \le c \langle \eta \rangle^{\mu-|\beta|} \qquad (1.107)$$

for all $\beta \in \mathbb{N}^q$, $\eta \in \mathbb{R}^q$, with constants $c_\iota(\beta) > 0$. Consider the decomposition

$$\mathfrak{a}(t, \eta, \tau) = \chi(\tau)\mathfrak{a}(t, \eta, \tau) + (1 - \chi(\tau))\mathfrak{a}(t, \eta, \tau).$$

We obtain

$$
\mathfrak{a}(t,\eta,\tau) - \chi(\tau)\sum_{l=0}^{N}\mathfrak{a}_l(t,\eta,\tau)
$$

$$
= \chi(\tau)\left\{\mathfrak{a}(t,\eta,\tau) - \sum_{l=0}^{N}\mathfrak{a}_l(t,\eta,\tau)\right\} + (1-\chi(\tau))\mathfrak{a}(t,\eta,\tau).
$$

Lemma 1.3.27 gives

$$
(1-\chi(\tau))\mathfrak{a}(t,\eta,\tau) \in S_{\mathrm{cl}}^{\mu}\big(\mathbb{R}^q, S^{-\infty}\big(\mathbb{R}^d \times \mathbb{R}^d\big)\big).
$$

Applying Lemma 1.3.26 to (1.105) one obtains the relation

$$
\mathfrak{a}(t,\eta,\tau) - \widetilde{\chi}(\eta,[\eta]\tau)\sum_{l=0}^{N}\mathfrak{a}_l(t,\eta,\tau) \in S^{\mu-(N+1)}\big(\mathbb{R}^q, S^{\mu-(N+1)}\big(\mathbb{R}^d \times \mathbb{R}^d\big)\big). \quad (1.108)
$$

Now multiplying (1.108) by $\chi(\tau)$ and using (1.104) we see that

$$
\chi(\tau)\{\mathfrak{a}(t,\eta,\tau) - \sum_{l=0}^{N}\mathfrak{a}_l(t,\eta,\tau)\} \in S^{\mu-(N+1)}\big(\mathbb{R}^q, S^{\mu-(N+1)}\big(\mathbb{R}^d \times \mathbb{R}^d\big)\big).
$$

We obtain

$$
\mathfrak{a}(t,\eta,\tau) - \chi(\tau)\sum_{l=0}^{N}a_{(\mu-l)}([\eta]^{-1}t,\eta,[\eta]\tau) = \mathfrak{r}_N(t,\eta,\tau) + (1-\chi(\tau))\mathfrak{a}(t,\eta,\tau), \quad (1.109)
$$

where

$$
\mathfrak{r}_N(t,\eta,\tau) = \chi(\tau)\left(\mathfrak{a}(t,\eta,\tau) - \widetilde{\chi}(\eta,\tau[\eta])\sum_{l=0}^{N}a_{(\mu-l)}([\eta]^{-1}t,\eta,[\eta]\tau)\right).
$$

Thanks to Lemma 1.3.27, the second right-hand term in (1.109) belongs to the space $S_{\mathrm{cl}}^{\mu}\big(\mathbb{R}^q, S^{-\infty}\big(\mathbb{R}^d \times \mathbb{R}^d\big)\big)$. Relation (1.106) will be established once we show that

$$
\mathfrak{r}_N(t,\eta,\tau) \in S^{\mu-(N+1)}\big(\mathbb{R}^q, S^{\mu-(N+1)}\big(\mathbb{R}^d \times \mathbb{R}^d\big)\big), \quad (1.110)
$$

$$
\mathfrak{b}(t,\eta,\tau) := \mathfrak{a}(t,\eta,\tau) - \chi(\tau)\sum_{l=0}^{N}a_{(\mu-l)}([\eta]^{-1}t,\eta,[\eta]\tau)
$$
$$
\in S^{\mu}\big(\mathbb{R}^q, S^{\mu-(N+1)}\big(\mathbb{R}^d \times \mathbb{R}^d\big)\big), \quad (1.111)
$$

and

$$
\chi(\tau)\sum_{l=0}^{N}a_{(\mu-l)}([\eta]^{-1}t,\eta,[\eta]\tau) \in S^{\mu}\big(\mathbb{R}^q, S_{\mathrm{cl}}^{\mu}\big(\mathbb{R}^d \times \mathbb{R}^d\big)\big). \quad (1.112)
$$

By Lemma 1.3.26, if we denote

$$\mathfrak{f}(t,\eta,\tau) := \mathfrak{a}(t,\eta,\tau) - \tilde{\chi}(\eta,[\eta]\tau) \sum_{l=0}^{N} a_{(\mu-l)}\big([\eta]^{-1}t,\eta,[\eta]\tau\big),$$

then $\mathfrak{f}(t,\eta,\tau) \in S^{\mu-(N+1)}\big(\mathbb{R}^q, S^{\mu-(N+1)}\big(\mathbb{R}^d \times \mathbb{R}^d\big)\big)$. Let us show that

$$\mathfrak{r}_N(t,\eta,\tau) = \chi(\tau)\mathfrak{f}(t,\eta,\tau)$$

belongs to $S^{\mu-(N+1)}\big(\mathbb{R}^q, S^{\mu-(N+1)}\big(\mathbb{R}^d \times \mathbb{R}^d\big)\big)$. To this end we insert $D_\eta^\beta \mathfrak{r}_N$ into the semi-norm $\pi_{\gamma\delta,K}(\cdot)$ in the space $S^{\mu-(N+1)}(\mathbb{R}^d \times \mathbb{R}^d)$, related to the order $\mu - (N+1) =: \nu$, cf. the expression (1.5). Then similarly to (1.98), we have to verify that

$$\pi_{\gamma\delta,K}\big(D_\eta^\beta \mathfrak{r}_N(t,\eta,\tau)\big) \le c \langle\eta\rangle^{\nu-|\beta|} \tag{1.113}$$

for every $\beta \in \mathbb{N}^q$, $K \Subset \mathbb{R}^d$. We have

$$\pi_{\gamma\delta,K}\big(\chi(\tau)D_\eta^\beta \mathfrak{f}(t,\eta,\tau)\big) = \sup_{t\in K, \tau\in\mathbb{R}^d} \langle\tau\rangle^{-\nu+|\delta|} \big|D_\tau^\delta\big(\chi(\tau)D_t^\gamma D_\eta^\beta \mathfrak{f}(t,\eta,\tau)\big)\big|$$

$$\le c \sup_{t\in K, \tau\in\mathbb{R}^d} \langle\tau\rangle^{-\nu+|\delta|} \big|D_\tau^\delta D_t^\gamma D_\eta^\beta \mathfrak{f}(t,\eta,\tau)\big|$$

$$+ c\sum_\zeta \sup_{t\in K, |\tau|\le C} \langle\tau\rangle^{-\nu+|\delta|} \big|D_\tau^\zeta D_t^\gamma D_\eta^\beta \mathfrak{f}(t,\eta,\tau)\big|,$$

where $\sum_\zeta$ is taken over the multi-indices $|\zeta| < |\delta|$. We obtain (1.113), similarly as in the proof of Lemma 1.3.26. Thus, we proved the relation (1.110). Relation (1.111) immediately follows from (1.109), i.e., from Lemma 1.3.27 and (1.110). In order to prove (1.112), we consider one of the summands, namely,

$$\chi(\tau)a_{(\mu-l)}([\eta]^{-1}t,\eta,[\eta]\tau), \quad l = 0,\ldots,N.$$

Using Taylor's formula for $a_{(\mu-l)}$ in the variable η for $\tau \neq 0$, we obtain

$$a_{(\mu-l)}(t,\eta,\tau) = \sum_{|\alpha|\le N} \frac{1}{\alpha!} a_{l\alpha}(t,\tau)\eta^\alpha$$

$$+ (N+1) \sum_{|\sigma|=N+1} \frac{\eta^\sigma}{\sigma!} \int_0^1 (1-\theta)^N \big(\partial_\eta^\sigma a_{(\mu-l)}\big)(t,\theta\eta,\tau)d\theta,$$

where

$$a_{l\alpha}(t,\tau) = \partial_\eta^\alpha a_{(\mu-l)}(t,0,\tau)$$

is (positively) homogeneous in τ of order $\mu - l - |\alpha|$. This gives us

$$a_{(\mu-k)}([\eta]^{-1}t,\eta,[\eta]\tau) = \sum_{|\alpha|\le N} \frac{1}{\alpha!} a_{l\alpha}([\eta]^{-1}t,[\eta]\tau)\eta^\alpha$$

$$+ (N+1) \sum_{|\sigma|=N+1} \frac{\eta^\sigma}{\sigma!} \int_0^1 (1-\theta)^N \big(\partial_\eta^\sigma a_{(\mu-l)}\big)([\eta]^{-1}t,\theta\eta,[\eta]\tau)d\theta.$$

Since
$$a_{l\alpha}([\eta]^{-1}t, [\eta]\tau) = a_{l\alpha}([\eta]^{-1}t, \tau)[\eta]^{\mu-l-|\alpha|},$$

it follows that

$$a_{(\mu-l)}([\eta]^{-1}t, \eta, [\eta]\tau) = \sum_{|\alpha| \leq N} \frac{1}{\alpha!} a_{l\alpha}([\eta]^{-1}t, [\eta]\tau)[\eta]^{\mu-l-|\alpha|}\eta^\alpha + R_{Nl}(t, \eta, \tau),$$

with

$$R_{Nl}(t, \eta, \tau) = (N+1) \sum_{|\sigma|=N+1} \frac{\eta^\sigma}{\sigma!} \int_0^1 (1-\theta)^N \left(\partial_\eta^\sigma a_{(\mu-l)}\right)([\eta]^{-1}t, \theta\eta, [\eta]\tau)d\theta.$$

For

$$b_{l\alpha}(t, \eta, \tau) := \frac{1}{\alpha!} a_{l\alpha}([\eta]^{-1}t, \tau)[\eta]^{\mu-l-|\alpha|}\eta^\alpha$$

we have

$$\chi(\tau)b_{l\alpha}(t, \eta, \tau) \in S^{\mu-(l+|\alpha|)}\left(\mathbb{R}^q, S_{cl}^\mu(\mathbb{R}^d \times \mathbb{R}^d)\right),$$

as implied by Lemma 1.3.24 and the homogeneity of $b_{l\alpha}(t, \eta, \tau)$ in τ of order $\mu - (l + |\alpha|)$. In order to complete the proof of (1.112) it remains to show that

$$\chi(\tau)R_{Nl}(t, \eta, \tau) \in S^\mu\left(\mathbb{R}^q, S^{\mu-(N+1)}(\mathbb{R}^d \times \mathbb{R}^d)\right).$$

This will be a consequence of the fact that

$$\mathfrak{g}_l(t, \eta, \tau) := \chi(\tau)\eta^\sigma \int_0^1 (1-\theta)^N \left(\partial_\eta^\sigma a_{(\mu-l)}\right)([\eta]^{-1}t, \theta\eta, [\eta]\tau)d\theta$$
$$\in S^\mu(\mathbb{R}_\eta^q, S^{\mu-(N+1)}(\mathbb{R}^d \times \mathbb{R}^d))$$

for every σ, $|\sigma| = N + 1$. We have

$$\chi(\rho)\left|\left(\partial_\eta^\sigma a_{(\mu-l)}\right)([\tau, \eta]^{-1}t, \theta\eta, [\eta]\tau)\right| \leq c\chi(\tau)\langle\theta\eta, [\eta]\tau\rangle^{\mu-l-|\sigma|}$$
$$= c\chi(\tau)[\eta]^{\mu-l-|\sigma|}(\tau^2 + \langle\theta\eta\rangle^2 [\eta]^{-2})^{(\mu-l-|\sigma|)}.$$
$$\tag{1.114}$$

For $\mu - l - |\sigma| \geq 0$ the right-hand side of (1.114) can be estimated by

$$c[\eta]^{\mu-l-|\sigma|} \langle\tau\rangle^{\mu-l-|\sigma|},$$

since $\langle\theta\eta\rangle^2 [\eta]^{-2}$ is uniformly bounded in $\eta \in \mathbb{R}^q$ for $0 \leq \theta \leq 1$. For $\mu - l - |\sigma| < 0$ the left-hand side of (1.114) can be estimated by

$$c\chi(\tau) \langle\theta\eta, [\eta]\tau\rangle^{\mu-l-|\sigma|} \leq c\chi(\tau) \langle[\eta]\tau\rangle^{\mu-l-|\sigma|}$$
$$\leq c\chi(\tau)[\eta]^{\mu-l-|\sigma|}|\tau|^{\mu-l-|\sigma|}$$
$$= c[\eta]^{\mu-l-|\sigma|} \langle\tau\rangle^{\mu-l-|\sigma|} \chi(\rho)\frac{|\tau|^{\mu-l-|\sigma|}}{\langle\tau\rangle^{\mu-l-|\sigma|}}$$
$$\leq c[\eta]^{\mu-l-|\sigma|} \langle\tau\rangle^{\mu-l-|\sigma|}.$$

In the following we assume, without loss of generality, that $\mu - l - |\sigma| < 0$. In much the same way as above we can prove

$$\pi_{\gamma\delta,K}\left(D_\eta^\beta \mathfrak{g}_l\right) \le c \langle\eta\rangle^{\mu-|\beta|}$$

for all $\beta \in \mathbb{N}^q$, $c = c(\beta) > 0$, where $\pi_{\gamma\delta,K}$ again refers to $S^{\mu-(N+1)}(\mathbb{R}^d \times \mathbb{R}^d)$. Summing up, we have proved the relation (1.106). In the final step we show

$$\mathfrak{a}(t,\eta,\tau) \in S_{\mathrm{cl}}^\mu\big(\mathbb{R}^q, S_{\mathrm{cl}}^\mu(\mathbb{R}^d \times \mathbb{R}^d)\big). \tag{1.115}$$

To this end we employ (1.105). In decoupled form, we have

$$\mathfrak{a}(t,\eta,\tau) = \sum_{l=0}^N \widetilde{\chi}(\eta,[\eta]\tau)a_{(\mu-l)}\big([\eta]^{-1}t,\eta,[\eta]\tau\big) + \mathfrak{r}_N(t,\eta,\tau).$$

From the first part of the proof we know that

$$\mathfrak{r}_N(t,\eta,\tau) \in S^{\mu-(N+1)}\big(\mathbb{R}^q, S_{\mathrm{cl}}^{\mu-(N+1)}(\mathbb{R}^d \times \mathbb{R}^d)\big).$$

Thus it remains to verify that

$$\widetilde{\chi}(\eta,[\eta]\tau)a_{(\mu-l)}\big([\eta]^{-1}t,\eta,[\eta]\tau\big) \in S_{\mathrm{cl}}^{\mu-l}(\mathbb{R}^q, S_{\mathrm{cl}}^{\mu-l}(\mathbb{R}^d \times \mathbb{R}^d))$$

for every $l = 0,\ldots,N$. From Taylor's formula in t at $t = 0$ we obtain

$$\widetilde{\chi}(\eta,[\eta]\tau)a_{(\mu-l)}\big([\eta]^{-1}t,\eta,[\eta]\tau\big) = \sum_{k=0}^N \frac{1}{k!}[\eta]^{-k}n_{lk}(t,\eta,\tau) + [\eta]^{-(N+1)}r_{N+1}(t,\eta,\tau), \tag{1.116}$$

with

$$n_{lk}(t,\eta,\tau) = t^k\widetilde{\chi}(\eta,[\eta]\tau)\partial_t^k a_{(\mu-l)}(0,\eta,[\eta]\tau) \in S^{\mu-l}\big(\mathbb{R}^q, S_{\mathrm{cl}}^{\mu-l}(\mathbb{R}^d \times \mathbb{R}^d)\big) \tag{1.117}$$

and

$$r_{N+1}(t,\eta,\tau) = \frac{t^{N+1}}{N!}\widetilde{\chi}(\eta,[\eta]\tau)\int_0^1 (1-\theta)^N \big(\partial_t^{N+1} a_{(\mu-l)}\big)\big(\theta[\eta]^{-1}t,\eta,[\eta]\tau\big)d\theta$$
$$\in S^{\mu-l}\big(\mathbb{R}^q, S_{\mathrm{cl}}^{\mu-l}(\mathbb{R}^d \times \mathbb{R}^d)\big). \tag{1.118}$$

The relations (1.117) and (1.118) are obtained in much the same way as in the first part of the proof. Now for $|\eta| \ge C$ for a sufficiently large $C > 0$ we have $\widetilde{\chi}(\eta,[\eta]\tau) = 1$ and $[\eta] = |\eta|$. Thus

$$[\eta]^{-k}n_{lk}(t,\eta,\tau) = |\eta|^{\mu-l-k}t^k\big(\partial_t^k a_{(\mu-l)}\big)(0,\eta/|\eta|,\tau) \quad \text{for } |\eta| \ge C,$$

cf. formula (1.116). It follows that

$$[\lambda\eta]^{-k} n_{lk}(t, \lambda\eta, \tau) = \lambda^{\mu-l-k} [\eta]^{-k} n_{lk}(t, \eta, \tau)$$

for $|\eta| \geq C$, $\lambda \geq 1$. Thus $[\eta]^{-k} n_{lk}(t, \eta, \tau) \in S^{\mu-l-k}(\mathbb{R}^q, S_{\mathrm{cl}}^{\mu-l}(\mathbb{R}^d \times \mathbb{R}^d))$ is homogeneous in η of order $\mu - l - k$ for $|\eta| \geq C$, i.e., it belongs to the space $S_{\mathrm{cl}}^{\mu-l-k}(\mathbb{R}^q, S_{\mathrm{cl}}^{\mu-l}(\mathbb{R}^d \times \mathbb{R}^d))$. Moreover, we have

$$[\eta]^{-(N+1)} r_{N+1}(t, \eta, \tau) \in S^{\mu-l-(N+1)}(\mathbb{R}^q, S_{\mathrm{cl}}^{\mu-l}(\mathbb{R}^d \times \mathbb{R}^d))$$

as a consequence of (1.118), cf. also formula (1.116). This yields the desired relation (1.115). As for the continuity of (1.103), it suffices to apply Lemma 1.3.26 together with the closed graph theorem. □

Remark 1.3.29. Lemma 1.3.26 and Proposition 1.3.28 can be easily extended to symbols $a(y, t, \eta, \tau) \in S_{(\mathrm{cl})}^{\mu}(U \times \overline{\mathbb{R}}_+ \times \mathbb{R}^{d-1} \times \mathbb{R}_{\eta,\tau}^{q+d})$ for $d \geq 1$, i.e., symbols that are smooth in $t \in \overline{\mathbb{R}}_+ \times \mathbb{R}^{d-1}$ up to 0 in the first component, i.e., the mapping $a \mapsto \mathfrak{a}$ defines a continuous operator

$$S_{(\mathrm{cl})}^{\mu}(U \times \overline{\mathbb{R}}_+ \times \mathbb{R}^{d-1} \times \mathbb{R}_{\eta,\tau}^{q+d}) \to S_{(\mathrm{cl})}^{\mu}(U \times \mathbb{R}^q, S_{(\mathrm{cl})}^{\mu}(\overline{\mathbb{R}}_+ \times \mathbb{R}^{d-1} \times \mathbb{R}_\tau^d)).$$

Let us now turn to pseudo-differential operators associated with symbols $a(y, y', \eta) \in S_{(\mathrm{cl})}^{\mu}(\Omega \times \Omega \times \mathbb{R}^q; H, \widetilde{H})$, $\Omega \subseteq \mathbb{R}^q$ open. For simplicity the generalities will be formulated for the case of Hilbert spaces H and $\widetilde{H}$ with group actions κ and $\widetilde{\kappa}$, respectively. The case of Fréchet spaces is analogous and will tacitly be used below.

Definition 1.3.30. Given Hilbert spaces H and $\widetilde{H}$ with group actions κ and $\widetilde{\kappa}$, respectively, for any open set $\Omega \subseteq \mathbb{R}^q$ we set

$$L_{\mathrm{cl}}^{\mu}(\Omega; H, \widetilde{H}) := \{\mathrm{Op}_y(a) : a(y, y', \eta) \in S_{\mathrm{cl}}^{\mu}(\Omega \times \Omega \times \mathbb{R}^q)\}.$$

If necessary, we write $L_{\mathrm{cl}}^{\mu}(\Omega; H, \widetilde{H})_{\kappa,\widetilde{\kappa}}$ to indicate the chosen group actions.

The space $L^{-\infty}(\Omega; H, \widetilde{H}) := \bigcap_{\mu \in \mathbb{R}} L^{\mu}(\Omega; H, \widetilde{H})_{\kappa,\widetilde{\kappa}}$ can be characterised as the set of integral operators with kernel $c(x, x') \in C^{\infty}(\Omega \times \Omega, \mathcal{L}(H, \widetilde{H}))$; it is independent of the specific choice of $\kappa, \widetilde{\kappa}$. The isomorphism $L^{-\infty}(\Omega; H, \widetilde{H}) \cong C^{\infty}(\Omega \times \Omega, \mathcal{L}(H, \widetilde{H}))$ yields a Fréchet topology in the space $L^{-\infty}(\Omega; H, \widetilde{H})$.

An element $A \in L^{\mu}(\Omega; H, \widetilde{H})$ induces a continuous operator

$$A : C_0^{\infty}(\Omega, H) \to C^{\infty}(\Omega, \widetilde{H}).$$

There are analogues of Sobolev spaces and corresponding continuity properties of pseudo-differential operators.

Definition 1.3.31. (i) Let H be a Hilbert space with group action $\kappa = \{\kappa_\lambda\}_{\lambda \in \mathbb{R}_+}$. Then $\mathcal{W}^s(\mathbb{R}^q, H)$ for $s \in \mathbb{R}$ is defined to be the completion of $\mathcal{S}(\mathbb{R}^q, H)$ (or, equivalently, of $C_0^\infty(\mathbb{R}^q, H)$) with respect to the norm

$$\|u\|_{\mathcal{W}^s(\mathbb{R}^q, H)} = \left\{ \int \langle \eta \rangle^{2s} \left\| \kappa_{\langle \eta \rangle}^{-1} \widehat{u}(\eta) \right\|_H^2 \, d\eta \right\}^{1/2}, \tag{1.119}$$

where $\widehat{u}(\eta) = Fu(\eta) = \int e^{-iy\eta} u(y) dy$ is the Fourier transform in $\mathbb{R}^q$ and the action of $\kappa_{\langle \eta \rangle}^{-1}$ concerns the values of $\widehat{u}$ in the space H.

(ii) Let $E = \varprojlim_{j \in \mathbb{N}} E^j$ be a Fréchet space with group action $\kappa = \{\kappa_\lambda\}_{\lambda \in \mathbb{R}_+}$. We set

$$\mathcal{W}^s(\mathbb{R}^q, E) := \varprojlim_{j \in \mathbb{N}} \mathcal{W}^s(\mathbb{R}^q, E^j).$$

If necessary, we also write $\mathcal{W}^s(\mathbb{R}^q, H)_\kappa$ rather than $\mathcal{W}^s(\mathbb{R}^q, H)$.

The spaces $\mathcal{W}^s(\mathbb{R}^q, \cdot)$ have been introduced in [43] in connection with operators on manifolds with edges, see also [36], and their properties are studied in [44, 46, 24, 59]. Incidentally, for convenience, we employ equivalent norms in $\mathcal{W}^s(\mathbb{R}^q, H)$ rather than (1.119), e.g., defined with $\eta \to [\eta]$ instead of $\langle \eta \rangle$, where $[\eta]$ is any strictly positive smooth function in $\mathbb{R}^q$ such that $[\eta] = |\eta|$ for $|\eta| \geq C$ for some constant $C > 0$.

Remark 1.3.32. The spaces $\mathcal{W}^s(\mathbb{R}^q, H)$ depend on the choice of κ; if necessary, we write

$$\mathcal{W}^s(\mathbb{R}^q, H)_\kappa = \mathcal{W}^s(\mathbb{R}^q, H). \tag{1.120}$$

Note that $\mathcal{W}^\infty(\mathbb{R}^q, H) := \bigcap_{s \in \mathbb{R}} \mathcal{W}^s(\mathbb{R}^q, H)_\kappa$ is independent of κ.

The following theorems summarise other essential properties of edge spaces.

Theorem 1.3.33. *Let H be a Hilbert space with group action κ, and let $s \in \mathbb{R}$.*

(i) *The operator $F^{-1} \kappa_{\langle \eta \rangle}^{-1} F : \mathcal{S}(\mathbb{R}^q, H) \to \mathcal{S}(\mathbb{R}^q, H)$ extends to an isomorphism*

$$F^{-1} \kappa_{\langle \eta \rangle}^{-1} F : \mathcal{W}^s(\mathbb{R}^q, H)_\kappa \to \mathcal{W}^s(\mathbb{R}^q, H)_{\mathrm{id}}.$$

Moreover, if κ, ϑ are group actions on H, then the relation $\kappa \sim \vartheta$ gives rise to an isomorphism

$$F^{-1} \vartheta_{\langle \eta \rangle} \kappa_{\langle \eta \rangle}^{-1} F : \mathcal{W}^s(\mathbb{R}^q, H)_\kappa \to \mathcal{W}^s(\mathbb{R}^q, H)_\vartheta$$

for every s.

(ii) *([46, Example 1.3.23]) For $H = H^s(\mathbb{R}^q)$ and $(\kappa_\lambda u)(y) = \lambda^{q/2} u(\lambda y)$ we have*

$$\mathcal{W}^s(\mathbb{R}^p, H^s(\mathbb{R}^q)) = H^s(\mathbb{R}^{p+q}).$$

(iii) ([24]) *We have*

$$\mathcal{W}^s(\mathbb{R}^q, H) = \left\{ u \in \mathcal{S}'(\mathbb{R}^q, H) : \langle \eta \rangle^s \kappa_{\langle \eta \rangle}^{-1} \widehat{u}(\eta) \in L^2(\mathbb{R}^q_\eta, H) \right\}$$

where $\mathcal{S}'(\mathbb{R}^q, H) = \mathcal{L}(\mathcal{S}(\mathbb{R}^q), H)$. *The space* $\mathcal{W}^s(\mathbb{R}^q, H)$ *is a Hilbert space with the scalar product*

$$(u, v)_{\mathcal{W}^s(\mathbb{R}^q, H)} = \left(\langle \eta \rangle^s \kappa_{\langle \eta \rangle}^{-1} \widehat{u}(\eta), \langle \eta \rangle^s \kappa_{\langle \eta \rangle}^{-1} \widehat{v}(\eta) \right)_{L^2(\mathbb{R}^q_\eta, H)}.$$

(iv) ([46, Theorem 1.3.34]) *The operator of multiplication* M_φ *by a function* $\varphi \in \mathcal{S}(\mathbb{R}^q)$ *induces a continuous operator*

$$M_\varphi : \mathcal{W}^s(\mathbb{R}^q, H) \to \mathcal{W}^s(\mathbb{R}^q, H),$$

and the mapping $\varphi \mapsto M_\varphi$ *represents a continuous operator*

$$\mathcal{S}(\mathbb{R}^q) \to S^0(\mathbb{R}^q; H, H)).$$

(v) ([44, Subsection 3.1.2, relations (24), (25)], [46, Proposition 1.3.44]) *The transformation*

$$(\chi_\lambda u)(y) := \lambda^{q/2} \kappa_\lambda u(\lambda y),$$

where $\lambda \in \mathbb{R}_+$, $u \in \mathcal{S}(\mathbb{R}^q, H)$, *extends to a group action on* $\mathcal{W}^s(\mathbb{R}^q, H)$, *and we have*

$$\mathcal{W}^s(\mathbb{R}^p, \mathcal{W}^s(\mathbb{R}^q, H)_\kappa)_\chi = \mathcal{W}^s(\mathbb{R}^{p+q}, H)_\kappa.$$

For an open $\Omega \subseteq \mathbb{R}^q$ and $s \in \mathbb{R}$ we set

$$\mathcal{W}^s_{\mathrm{loc}}(\Omega, H) := \left\{ u \in \mathcal{D}'(\Omega, H) : \varphi u \in \mathcal{W}^s(\mathbb{R}^q, H) \text{ for every } \varphi \in C_0^\infty(\Omega) \right\} \quad (1.121)$$

and

$$\mathcal{W}^s_{\mathrm{comp}}(\Omega, H) := \left\{ u \in \mathcal{W}^s_{\mathrm{loc}}(\Omega, H) : \mathrm{supp}\, u \text{ compact} \right\}. \qquad (1.122)$$

The space (1.121) is Fréchet, while (1.122) is an inductive limit of Fréchet spaces. Let us set, in particular,

$$H^s_{\mathrm{loc}(y)}(\Omega \times \mathbb{R}) := \mathcal{W}^s_{\mathrm{loc}}(\Omega, H^s(\mathbb{R})), \quad H^s_{\mathrm{comp}(y)}(\Omega \times \mathbb{R}) := \mathcal{W}^s_{\mathrm{comp}}(\Omega, H^s(\mathbb{R})),$$
$$(1.123)$$

and

$$H^s_{\mathrm{loc}(y)}(\Omega \times \mathbb{R}_\pm) := \mathcal{W}^s_{\mathrm{loc}}(\Omega, H^s(\mathbb{R}_\pm)), \quad H^s_{\mathrm{comp}(y)}(\Omega \times \mathbb{R}_\pm) := \mathcal{W}^s_{\mathrm{comp}}(\Omega, H^s(\mathbb{R}_\pm)).$$
$$(1.124)$$

The group actions in $H^s(\mathbb{R})$ and $H^s(\mathbb{R}_\pm)$ are defined by $u(t) \mapsto \lambda^{1/2} u(\lambda t)$, $\lambda \in \mathbb{R}_+$. The notation (1.124) is motivated by the fact that similarly to Theorem 1.3.33 (ii), we have

$$\mathcal{W}^s(\mathbb{R}^p, H^s(\mathbb{R}_\pm)) = H^s(\mathbb{R}^{p+1}_\pm).$$

Theorem 1.3.34. *Let* $a(y, y', \eta) \in S^\mu(\Omega \times \Omega \times \mathbb{R}^q; H, \widetilde{H})$, $\Omega \subseteq \mathbb{R}^q$ *open,* $\mu \in \mathbb{R}$, *where* H *and* $\widetilde{H}$ *are Hilbert spaces with group action* κ *and* $\widetilde{\kappa}$, *respectively. Then* $\mathrm{Op}(a) : C_0^\infty(\Omega, H) \to C^\infty(\Omega, \widetilde{H})$ *extends to a continuous operator*

$$\mathrm{Op}(a) : \mathcal{W}_{\mathrm{comp}}^s(\Omega, H) \to \mathcal{W}_{\mathrm{loc}}^{s-\mu}(\Omega, \widetilde{H}), \tag{1.125}$$

$s \in \mathbb{R}$. *For* $a(\eta) \in S^\mu(\mathbb{R}^q; H, \widetilde{H})$ *we obtain a continuous operator*

$$\mathrm{Op}(a) : \mathcal{W}^s(\mathbb{R}^q, H) \to \mathcal{W}^{s-\mu}(\mathbb{R}^q, \widetilde{H}), \tag{1.126}$$

and the mapping $a \mapsto \mathrm{Op}(a)$ *induces a continuous operator*

$$S^\mu(\mathbb{R}^q; H, \widetilde{H}) \to \mathcal{L}(\mathcal{W}^s(\mathbb{R}^q, H), \mathcal{W}^{s-\mu}(\mathbb{R}^q, \widetilde{H})) \tag{1.127}$$

for every $s \in \mathbb{R}$.

Proof. Let us first consider $a(\eta) \in S^\mu(\mathbb{R}^q; H, \widetilde{H})$. In this case we have

$$\|\mathrm{Op}(a)u\|_{\mathcal{W}^{s-\mu}(\mathbb{R}^q, \widetilde{H})}^2 = \int \langle \eta \rangle^{2(s-\mu)} \|\widetilde{\kappa}_{\langle \eta \rangle}^{-1} a(\eta) \widehat{u}(\eta)\|_{\widetilde{H}}^2 d\eta$$

$$\leq \sup_{\eta \in \mathbb{R}^q} \langle \eta \rangle^{-2\mu} \|\widetilde{\kappa}_{\langle \eta \rangle}^{-1} a(\eta) \kappa_{\langle \eta \rangle}\|_{\mathcal{L}(H, \widetilde{H})}^2 \int \langle \eta \rangle^{2s} \|\kappa_{\langle \eta \rangle}^{-1} \widehat{u}\|_H^2 d\eta. \tag{1.128}$$

This shows that (1.126) is continuous and

$$\|\mathrm{Op}(a)\|_{\mathcal{L}(\mathcal{W}^s(\mathbb{R}^q, H), \mathcal{W}^{s-\mu}(\mathbb{R}^q, \widetilde{H}))} \leq \sup_{\eta \in \mathbb{R}^q} \langle \eta \rangle^{-\mu} \|\widetilde{\kappa}_{\langle \eta \rangle}^{-1} a(\eta) \kappa_{\langle \eta \rangle}\|_{\mathcal{L}(H, \widetilde{H})}.$$

The last relation establishes the continuity of (1.127). In order to show the continuity of (1.125), we first pass from $a(y, y', \eta)$ to a left symbol $a_{\mathrm{L}}(y, \eta) \in S^\mu(\Omega \times \mathbb{R}^q; H, \widetilde{H})$,

$$a_{\mathrm{L}}(y, \eta) \sim \sum_{\alpha \in \mathbb{N}^q} \frac{1}{\alpha!} \partial_\eta^\alpha D_{y'}^\alpha a(y, y', \eta)|_{y'=y}.$$

Then

$$\mathrm{Op}(a) = \mathrm{Op}(a_{\mathrm{L}}) + C,$$

with an operator C with kernel $c(y, y') \in C^\infty(\Omega \times \Omega, \mathcal{L}(H, \widetilde{H}))$. The continuity of $C : \mathcal{W}_{\mathrm{comp}}^s(\Omega, H) \to \mathcal{W}_{\mathrm{loc}}^\infty(\Omega, \widetilde{H})$ is a simple exercise, left to the reader. Then it remains to show the continuity (1.125) for a_{L} rather than a. Since

$$S^\mu(\Omega \times \mathbb{R}^q; H, \widetilde{H}) = C^\infty(\Omega, S^\mu(\mathbb{R}^q; H, \widetilde{H})) = C^\infty(\Omega) \widehat{\otimes}_\pi S^\mu(\mathbb{R}^q; H, \widetilde{H}),$$

it suffices to assume $a_{\mathrm{L}}(y, \eta) \in C_0^\infty(K) \widehat{\otimes}_\pi S^\mu(\mathbb{R}^q; H, \widetilde{H}))$ for any $K \Subset \Omega$. We can write a_{L} as a convergent sum

$$a_{\mathrm{L}}(y, \eta) = \sum_{j=0}^\infty \lambda_j \varphi_j(y) a_j(\eta)$$

for $\lambda_j \in \mathbb{C}$, $\sum_{j=0}^\infty |\lambda_j| < \infty$ and $\varphi_j(y) \in C_0^\infty(K)$, $a_j(\eta) \in S^\mu(\mathbb{R}^q; H, \widetilde{H}))$, tending to zero in the respective spaces as $j \to \infty$, cf. also Proposition 1.3.14. This allows us to write

$$\mathrm{Op}(a_{\mathrm{L}}) = \sum_{j=0}^\infty \lambda_j \mathcal{M}_{\varphi_j} \mathrm{Op}(a_j).$$

Since $\mathcal{M}_{\varphi_j} \to 0$ in $S^0(\mathbb{R}^q; \widetilde{H}, \widetilde{H})$ and the mapping in (1.127) is continuous, it follows that $\mathrm{Op}(\mathcal{M}_{\varphi_j}) \to 0$ in $\mathcal{L}(\mathcal{W}^s(\mathbb{R}^q, \widetilde{H}), \mathcal{W}^s(\mathbb{R}^q, \widetilde{H}))$ as $j \to \infty$, for every $s \in \mathbb{R}$. Thus

$$\|\mathrm{Op}(a_{\mathrm{L}})\|_{\mathcal{L}(\mathcal{W}^s(\mathbb{R}^q, H), \mathcal{W}^{s-\mu}(\mathbb{R}^q, \widetilde{H}))}$$

$$\leq \sum_{j=0}^\infty |\lambda_j| \|\mathrm{Op}(\mathcal{M}_{\varphi_j})\|_{\mathcal{L}(\mathcal{W}^{s-\mu}(\mathbb{R}^q, \widetilde{H}), \mathcal{W}^{s-\mu}(\mathbb{R}^q, \widetilde{H}))} \|\mathrm{Op}(a_j)\|_{\mathcal{L}(\mathcal{W}^s(\mathbb{R}^q, H), \mathcal{W}^{s-\mu}(\mathbb{R}^q, \widetilde{H})}$$

$$< \omega.$$

$$(1.129)$$

$\square$

Theorem 1.3.35 ([59], [23, Theorem 2.2.20]). *Let H and $\widetilde{H}$ be Hilbert spaces with group action κ and $\widetilde{\kappa}$, respectively. Let $a(y, \eta) \in C^\infty(\mathbb{R}^{2q}, \mathcal{L}(H, \widetilde{H}))$ satisfy the estimates*

$$\pi(a) := \sup_{\substack{(y, \eta) \in \mathbb{R}^{2q} \\ \alpha \leq \boldsymbol{\alpha}, \beta \leq \boldsymbol{\beta}}} \|\widetilde{\kappa}_{\langle \eta \rangle}^{-1} \{D_y^\alpha D_\eta^\beta a(y, \eta)\} \kappa_{\langle \eta \rangle}\|_{\mathcal{L}(H, \widetilde{H})} < \infty$$

for $\boldsymbol{\alpha} := (M+1, \ldots, M+1)$, $\boldsymbol{\beta} := (1, \ldots, 1)$, with $M \in \mathbb{N}$ being a constant belonging to $\widetilde{\kappa}$ in the sense of Proposition 1.3.7. Then $\mathrm{Op}(a)$ induces a continuous operator

$$\mathrm{Op}(a) : \mathcal{W}^0(\mathbb{R}^q, H) \to \mathcal{W}^0(\mathbb{R}^q, \widetilde{H})$$

and we have

$$\|\mathrm{Op}(a)\|_{\mathcal{L}(\mathcal{W}^0(\mathbb{R}^q, H), \mathcal{W}^0(\mathbb{R}^q, \widetilde{H}))} \leq c\, \pi(a)$$

for some $c > 0$ independent of a.

1.4   Oscillatory integrals based on the Fourier transform

The starting point of this section consists of ideas of Kumano-go [26] on oscillatory integrals and pseudo-differential operators globally in $\mathbb{R}^q$. We employ elements of this approach here in various generalised settings, e.g., for operator-valued amplitude functions with twisted symbol estimates, parameter-dependent variants, or symbols taking values in operator algebras with symbol structures, and symbols with complex covariables and holomorphic dependence. We outline here the necessary results, not always with proofs when they are accessible elsewhere, up to simple modifications. In some cases we also sketch the proofs.

First we consider oscillatory integrals for symbols taking values in Fréchet spaces.

Definition 1.4.1. Let V be a Fréchet space with the system of semi-norms $(\pi_\iota)_{\iota \in \mathbb{N}}$. Then the space

$$S^{\boldsymbol{\mu};\boldsymbol{\nu}}(\mathbb{R}^{2q}, V) \tag{1.130}$$

for sequences of reals $\boldsymbol{\mu} := (\mu_\iota)_{\iota \in \mathbb{N}}$, $\boldsymbol{\nu} := (\nu_\iota)_{\iota \in \mathbb{N}}$ is defined to be the set of all $a(y, \eta) \in C^\infty(\mathbb{R}^{2q}, V)$ such that

$$\sup_{(y,\eta) \in \mathbb{R}^{2q}} \langle \eta \rangle^{-\mu_\iota} \langle y \rangle^{-\nu_\iota} \pi_\iota \big(D_y^\alpha D_\eta^\beta a(y, \eta) \big) < \infty \tag{1.131}$$

for all $\alpha, \beta \in \mathbb{N}^q$, $\iota \in \mathbb{N}$. Moreover, we set

$$S^{\infty;\infty}(\mathbb{R}^{2q}, V) := \bigcup_{\boldsymbol{\mu},\boldsymbol{\nu}} S^{\boldsymbol{\mu};\boldsymbol{\nu}}(\mathbb{R}^{2q}, V).$$

Remark 1.4.2. For every $\boldsymbol{\mu}, \boldsymbol{\nu}$ the space (1.130) is Fréchet with the system of semi-norms (1.131).

Remark 1.4.3. Note that $a(y, \eta) \in S^{\infty;\infty}(\mathbb{R}^{2q}, V)$ entails $\langle y \rangle^N \langle \eta \rangle^M a(y, \eta) \in S^{\infty;\infty}(\mathbb{R}^{2q}, V)$ for every $N, M \in \mathbb{R}$.

The spaces (1.130) have a number of natural properties that are summarised as follows.

Proposition 1.4.4.   (i) $a \in S^{\boldsymbol{\mu};\boldsymbol{\nu}}(\mathbb{R}^{2q}, V)$ *implies* $D_y^\alpha D_\eta^\beta a \in S^{\boldsymbol{\mu};\boldsymbol{\nu}}(\mathbb{R}^{2q}, V)$ *for every* $\alpha, \beta \in \mathbb{N}^q$.

(ii) *If* $T : V \to \widetilde{V}$ *is a continuous operator between Fréchet spaces* $V, \widetilde{V}$, *then we have*

$$a \in S^{\infty;\infty}(\mathbb{R}^{2q}, V) \implies Ta := ((y, \eta) \mapsto T(a(y, \eta))) \in S^{\infty;\infty}(\mathbb{R}^{2q}, \widetilde{V}).$$

More precisely, the mapping $a \mapsto Ta$ *defines a continuous operator*

$$S^{\boldsymbol{\mu};\boldsymbol{\nu}}(\mathbb{R}^{2q}, V) \to S^{\widetilde{\boldsymbol{\mu}};\widetilde{\boldsymbol{\nu}}}(\mathbb{R}^{2q}, \widetilde{V})$$

for every $\boldsymbol{\mu}, \boldsymbol{\nu}$ *and a resulting pair* $\widetilde{\boldsymbol{\mu}}, \widetilde{\boldsymbol{\nu}}$ *of orders.*

(iii) *Let V be the projective limit of Fréchet spaces V_j with respect to linear maps $T_j : V \to V_j$, $j \in I$. Then $a \in S^{\infty;\infty}(\mathbb{R}^{2q}, V)$ is equivalent to $T_j a \in S^{\infty;\infty}(\mathbb{R}^{2q}, V_j)$ for every $j \in I$.*

(iv) *If V_0, V_1, V are Fréchet spaces and $\langle \cdot, \cdot \rangle : V_0 \times V_1 \to V$ a continuous bilinear map, then $a_k \in S^{\infty;\infty}(\mathbb{R}^{2q}, V_k)$, $k = 0, 1$, implies $\langle a_0, a_1 \rangle \in S^{\infty;\infty}(\mathbb{R}^{2q}, V)$. More precisely, the mapping $(a_0, a_1) \mapsto \langle a_0, a_1 \rangle$ induces a continuous map*

$$S^{\mu_0;\nu_0}(\mathbb{R}^{2q}, V_0) \times S^{\mu_1;\nu_1}(\mathbb{R}^{2q}, V_1) \to S^{\mu;\nu}(\mathbb{R}^{2q}, V)$$

for any pairs of orders μ_0, ν_0 and μ_1, ν_1, for some resulting μ, ν.

(v) *Let $W \subseteq V$ be a closed subspace of V and let $[\cdot] : V \to V/W$ denote the quotient map. Then $a \in S^{\infty;\infty}(\mathbb{R}^{2q}, V)$ implies $[a] \in S^{\infty;\infty}(\mathbb{R}^{2q}, V/W)$.*

The proof immediately follows from Definition 1.4.1, see also [26].
Oscillatory integrals with amplitude functions $a \in S^{\mu;\nu}(\mathbb{R}^{2q}, V)$ are expressions of the form

$$\mathrm{Os}[a] := \iint e^{-iy\eta} a(y, \eta) dy đ\eta, \tag{1.132}$$

which are not always convergent in the standard sense. The notion of oscillatory integral indicates a way to give (1.132) a meaning. There are two equivalent methods of regularising the integrals (1.132).
One is based on a limit, using any $\chi(y, \eta) \in \mathcal{S}(\mathbb{R}^{2q})$ such that $\chi(0, 0) = 1$. It turns out that

$$\mathrm{Os}[a] = \lim_{\varepsilon \to 0} \iint e^{-iy\eta} \chi(\varepsilon y, \varepsilon \eta) a(y, \eta) dy đ\eta \tag{1.133}$$

is finite and the limit is independent of χ. The argument is as follows.
Assume first that $a(y, \eta) \in \mathcal{S}(\mathbb{R}^{2q}, V)$ and write

$$e^{-iy\eta} = \langle \eta \rangle^{-2M}(1 - \Delta_y)^M e^{-iy\eta}, \quad e^{-iy\eta} = \langle y \rangle^{-2N}(1 - \Delta_\eta)^N e^{-iy\eta}.$$

Then integration by parts gives us

$$
\begin{aligned}
\iint & e^{-iy\eta} a(y, \eta) dy đ\eta \\
&= \iint e^{-iy\eta} \langle y \rangle^{-2N}(1 - \Delta_\eta)^N \langle \eta \rangle^{-2M}(1 - \Delta_y)^M a(y, \eta) dy đ\eta
\end{aligned}
\tag{1.134}
$$

for every $M, N \in \mathbb{N}$. Since the expressions (1.131) are finite, by choosing $N = N_\iota$, $M = M_\iota$ sufficiently large for any fixed $\iota \in \mathbb{N}$ the right-hand side of (1.134) converges with respect to the semi-norm π_ι also in $S^{\mu;\nu}(\mathbb{R}^{2q}, V)$. We obtain

$$
\begin{aligned}
\lim_{\varepsilon \to 0} & \iint e^{-iy\eta} \chi(\varepsilon y, \varepsilon \eta) a(y, \eta) dy đ\eta \\
&= \lim_{\varepsilon \to 0} \iint e^{-iy\eta} \langle y \rangle^{-2N}(1 - \Delta_\eta)^N \langle \eta \rangle^{-2M}(1 - \Delta_y)^M \chi(\varepsilon y, \varepsilon \eta) a(y, \eta) dy đ\eta
\end{aligned}
$$

with convergence with respect to π_ι. By virtue of Lebesgue's dominated convergence theorem, the right-hand side converges to an element of V for arbitrary $a(y, \eta) \in S^{\infty;\infty}(\mathbb{R}^{2q}, V)$. Thus the left-hand side also exists for any such $a(y, \eta)$. At the same time, we see for any π from the system of semi-norms for V that

$$\pi(\mathrm{Os}[a]) \leq \iint \pi(\langle y \rangle^{-2N} (1 - \Delta_\eta)^N \langle \eta \rangle^{-2M} (1 - \Delta_y)^M a(y, \eta)) dy d\eta.$$

This yields the following result.

Theorem 1.4.5. *For $a(y, \eta) \in S^{\infty;\infty}(\mathbb{R}^{2q}, V)$ the oscillatory integral (1.132) defined as the limit (1.133) defines an element of V which is independent of χ. Moreover, the association $a \mapsto \mathrm{Os}[a]$ induces a continuous map*

$$\mathrm{Os}[\,\cdot\,] : S^{\mu;\nu}(\mathbb{R}^{2q}, V) \to V \quad \text{for every } \mu, \nu.$$

Proposition 1.4.6. *For $a(x, y, \xi, \eta) \in S^{\mu;\nu}(\mathbb{R}^{2q}_{x,y} \times \mathbb{R}^{2q}_{\xi,\eta}, V)$, we have*

$$b(x, \xi) := \big((y, \eta) \mapsto a(x, y, \xi, \eta) \big) \in S^{\infty;\infty}\big(\mathbb{R}^{2q}_{x,\xi}, S^{\mu;\nu}(\mathbb{R}^{2q}_{y,\eta}, V) \big).$$

In particular, Theorem 1.4.5 *shows that $\mathrm{Os}[b] \in S^{\infty;\infty}(\mathbb{R}^{2q}_{y,\eta}, V)$. Moreover, we have*

$$\mathrm{Os}[a] = \mathrm{Os}[\mathrm{Os}[b]]. \tag{1.135}$$

The interpretation of the left-hand side of (1.135) is analogous to (1.133) for a regularising function $\chi_\varepsilon(x, y, \xi, \eta) : (0, 1] \times \mathbb{R}^{2q} \times \mathbb{R}^{2q} \to \mathbb{C}$, while on the right-hand side of (1.135) the oscillatory integral $c(y, \eta) = \mathrm{Os}[b]$ is defined with some $\chi_\varepsilon(x, \xi) : (0, 1] \times \mathbb{R}^{2q} \to \mathbb{C}$ and $\mathrm{Os}[c]$ with some $\chi_\varepsilon(y, \eta) : (0, 1] \times \mathbb{R}^{2q} \to \mathbb{C}$.

Definition 1.4.7. Fix some $m \in \mathbb{N}$, $m \geq 1$. A function $\chi_\varepsilon : (0, 1] \times \mathbb{R}^n \to \mathbb{C}$ is called regularising, if

(i) $\chi_\varepsilon(y) \in \mathcal{S}(\mathbb{R}^m)$ for every $0 < \varepsilon \leq 1$;

(ii) $\sup_{(\varepsilon, y) \in (0,1] \times \mathbb{R}^m} |D_y^\alpha \chi_\varepsilon(y)| < \infty$ for every $\alpha \in \mathbb{N}^m$;

(iii) $D_y^\alpha \chi_\varepsilon(y) \to \begin{cases} 1 & \text{for } \alpha = 0, \\ 0 & \text{for } \alpha \neq 0, \end{cases}$ with pointwise convergence in $\mathbb{R}^m$ for $\varepsilon \to 0$.

In particular, $\chi(\varepsilon y)$ for any $\chi \in \mathcal{S}(\mathbb{R}^m)$ with $\chi(0) = 1$ is regularising in the sense of Definition 1.4.7.

Remark 1.4.8. If $\chi(\varepsilon y, \varepsilon \eta)$ is any regularising function $(0, 1] \times \mathbb{R}^{2q} \to \mathbb{C}$ and $a(y, \eta) \in S^{\infty;\infty}(\mathbb{R}^{2q}, V)$, then $\mathrm{Os}[a]$ coincides with (1.133).

Proposition 1.4.9. *For every $a(y, \eta) \in S^{\infty;\infty}(\mathbb{R}^{2q}, V)$ we have*

(i) $\mathrm{Os}[y^\alpha a] = \mathrm{Os}[D_\eta^\alpha a]$, $\mathrm{Os}[\eta^\beta a] = \mathrm{Os}[D_y^\beta a]$ *for every $\alpha, \beta \in \mathbb{N}^q$;*

(ii) $\mathrm{Os}[a] = \mathrm{Os}\big[e^{-i(y\eta_0 + y_0\eta + y_0\eta_0)} a(y + y_0, \eta + \eta_0) \big]$ *for every $(y_0, \eta_0) \in \mathbb{R}^{2q}$;*

(iii) *if $T : V \to \widetilde{V}$ is a linear continuous operator, then $T(\mathrm{Os}[a]) = \mathrm{Os}[Ta]$.*

Let us now present crucial elements of the pseudo-differential calculus glob-ally in $\mathbb{R}^n$ for operator-valued symbols satisfying twisted symbol estimates.

For any Fréchet space E with the semi-norm system $(e_j)_{j \in \mathbb{N}}$ we define

$$C_{\mathrm{b}}^\infty(\mathbb{R}^q, E) := \left\{ u \in C^\infty(\mathbb{R}^q, E) : \sup_{y \in \mathbb{R}^q} e_j(D_y^\alpha u) < \infty \text{ for all } \alpha \in \mathbb{N}^q, j \in \mathbb{N} \right\}.$$

Definition 1.4.10. Let H and $\widetilde{H}$ be Hilbert spaces with group actions κ and $\widetilde{\kappa}$, respectively. We define

$$S_{(\mathrm{cl})}^\mu\left(\mathbb{R}^p \times \mathbb{R}^q; H, \widetilde{H}\right)_{\mathrm{b}} := C_{\mathrm{b}}^\infty\left(\mathbb{R}^p, S_{(\mathrm{cl})}^\mu\left(\mathbb{R}^q; H, \widetilde{H}\right)\right).$$

If necessary, we write $S_{(\mathrm{cl})}^\mu\left(\mathbb{R}^p \times \mathbb{R}^q; H, \widetilde{H}\right)_{\kappa, \widetilde{\kappa}, \mathrm{b}}$.

For any $a(y, y', \eta) \in S_{(\mathrm{cl})}^\mu\left(\mathbb{R}_{y,y'}^{2q} \times \mathbb{R}_\eta^q; H, \widetilde{H}\right)_{\mathrm{b}}$ and $u \in C_{\mathrm{b}}^\infty(\mathbb{R}^q, H)$, we define

$$\begin{aligned}
\mathrm{Op}(a)u(y) &:= \iint e^{i(y-y')\eta} a(y, y', \eta) u(y') dy' d\eta \\
&= \iint e^{-iy'\eta} a(y, y' + y, \eta) u(y' + y) dy' d\eta.
\end{aligned} \tag{1.136}$$

This corresponds to the standard operator convention in connection with the Fourier transform. The right-hand side of (1.136) is interpreted as an oscillatory integral $\mathrm{Os}[\cdot]$ with the amplitude function

$$f_y(y', \eta) := a(y, y' + y, \eta) u(y' + y) \in S^{\infty;\infty}\left(\mathbb{R}_{y',\eta}^{2q}, \widetilde{H}\right)$$

for every fixed $y \in \mathbb{R}$. Clearly, $\mathrm{Op}(a) \in L_{(\mathrm{cl})}^\mu\left(\mathbb{R}^q; H, \widetilde{H}\right)$ in the sense of Definition 1.3.30. Note that (the scalar) Kumano-go calculus also admits double symbols in the sense that there are not only (y, y')-variables, but also covariables (η, η'). The definition cannot so easily generalised to the operator-valued case. However, the only occasions where we need such symbols are considerations like

$$p(y, y'\eta, \eta') = a(y, \eta) b(y', \eta')$$

for

$$a(y, \eta) \in S_{(\mathrm{cl})}^\mu\left(\mathbb{R}^q \times \mathbb{R}_\eta^q; H_0, \widetilde{H}\right)_{\mathrm{b}}, \quad b(y', \eta') \in S_{(\mathrm{cl})}^\mu\left(\mathbb{R}^q \times \mathbb{R}_\eta^q; H, H_0\right)_{\mathrm{b}}.$$

In that case instead of (1.136) we form the oscillatory integral

$$\mathrm{Op}(p)u(y) = \iint e^{-i(x\eta - x'\eta')} p(y, y + x, \eta, \eta') u(y + x + x') dx dx' d\eta d\eta'$$

for $u \in C_{\mathrm{b}}^\infty(\mathbb{R}^q, H)$. The amplitude function for the $\mathrm{Os}[\cdot]$-expression is in this case

$$f_y(x, x', \eta, \eta') \in S^{\infty;\infty}\left(\mathbb{R}^{2q} \times \mathbb{R}^{2q}, \widetilde{H}\right),$$

for every fixed y.

Theorem 1.4.11. *Let* $a(y, y', \eta) \in S^\mu (\mathbb{R}^{2q} \times \mathbb{R}^{2q}; H, \widetilde{H})_b$. *Then* $A = \mathrm{Op}(a)$ *defines a continuous operator*

$$A : \mathcal{S}(\mathbb{R}^q, H) \to \mathcal{S}(\mathbb{R}^q, \widetilde{H}).$$

Proof. Let us first assume that a is independent of y'. Moreover, since the order μ is arbitrary, we may assume that H and $\widetilde{H}$ are endowed with the trivial group actions, since

$$S^\mu (\mathbb{R}^q \times \mathbb{R}^q; H, \widetilde{H})_{\kappa, \widetilde{\kappa}, b} \subseteq S^{\mu + M + \widetilde{M}} (\mathbb{R}^q \times \mathbb{R}^q; H, \widetilde{H})_{\mathrm{id}, \mathrm{id}, b}$$

with M and $\widetilde{M}$ being the constants belonging to κ and $\widetilde{\kappa}$, respectively, occurring in Proposition 1.3.7. We consider $\mathrm{Op}(a)$ in the form

$$\mathrm{Op}(a)u(y) = \int e^{iy\eta} a(y, \eta) \widehat{u}(\eta) \, d\eta.$$

For the Schwartz space $\mathcal{S}(\mathbb{R}^q, H)$ we consider the system of semi-norms

$$\pi_m(u) := \sup_{y \in \mathbb{R}^q} \max_{|\alpha| + |\beta| \le m} \left\| y^\alpha D_y^\beta u(y) \right\|_H, \quad m \in \mathbb{N}.$$

If necessary, we also write $\pi_{m,y}(u)$. If the semi-norm concerns $\widetilde{H}$ we write $\widetilde{\pi}_m$. We have to show that for every $\widetilde{m} \in \mathbb{N}$ there is an $m \in \mathbb{N}$ such that

$$\widetilde{\pi}_{\widetilde{m}}(Au) \le c\, \pi_m(u) \tag{1.137}$$

for all $u \in \mathcal{S}(\mathbb{R}^q, H)$ and some constant $c > 0$. The Fourier transform $F_{y \to \eta}$ induces an isomorphism $F : \mathcal{S}(\mathbb{R}^q_y, H) \to \mathcal{S}(\mathbb{R}^q_\eta, H)$, and for every $m \in \mathbb{N}$ there exists a $C > 0$ such that

$$\pi_{m,y}(Fu) \le C \pi_{m+q+1,y}(u) \tag{1.138}$$

for all $u \in \mathcal{S}(\mathbb{R}^q_y, H)$. For $h = \mathbb{C}$ this may be found, e.g., in [26, Chapter 1]. The simple generalisation to the H-valued case is left to the reader. We have for any $y \in \mathbb{R}^q$,

$$\|Au(y)\|_{\widetilde{H}} = \left\| \int e^{iy\eta} a(y, \eta) \widehat{u}(\eta) \, d\eta \right\|_{\widetilde{H}} \le \int \left\| \langle \eta \rangle^{-N} a(y, \eta) \langle \eta \rangle^N \widehat{u}(\eta) \right\|_{\widetilde{H}} d\eta$$

$$\le c \int \left\| \langle \eta \rangle^N \widehat{u}(\eta) \right\|_{\widetilde{H}} d\eta$$

for $c := \sup_{(y,\eta) \in \mathbb{R}^{2q}} \left(\langle \eta \rangle^{-N} \|a(y, \eta)\|_{\mathcal{L}(H, \widetilde{H})} \right)$. By virtue of the symbol estimates for a, we may choose N so large that $c < \infty$. Moreover,

$$\int \left\| \langle \eta \rangle^N \widehat{u}(\eta) \right\|_{\widetilde{H}} d\eta \le \sup_{\eta \in \mathbb{R}^q} \left(\langle \eta \rangle^{N+q+1} \|\widehat{u}(\eta)\|_{\widetilde{H}} \right) \int \langle \eta \rangle^{-(q+1)} d\eta$$

$$\le c\, \pi_{N+q+1, \eta}(\widehat{u})$$

$$\le c\, \pi_{N+2(q+1), y}(\widehat{u}),$$

cf. (1.138). It follows that

$$\widetilde{\pi}_0(Au) = \sup_{y \in \mathbb{R}^q} \|Au(y)\|_{\widetilde{H}} \leq \pi_{N+2(q+1)}(u)$$

for all $u \in \mathcal{S}(\mathbb{R}^q, H)$. Moreover, we have

$$\partial_{y_j} Au(y) = \int e^{iy\eta} \{i\eta_j + \partial_{y_j}\} a(y,\eta)\widehat{u}(\eta)\,d\eta$$

$$= \int e^{iy\eta} a(y,\eta) \widehat{(\partial_{y_j} u)}(\eta)\,d\eta + \int e^{iy\eta} \partial_{y_j} a(y,\eta)\widehat{u}(\eta)\,d\eta,$$

$$y_k Au(y) = -\int e^{iy\eta} (i\partial_{\eta_k} a(y,\eta))\widehat{u}\,d\eta - \int e^{iy\eta} ia(y,\eta)\partial_{\eta_k} a(y,\eta)\widehat{u}\,d\eta.$$

By virtue of $\partial_{y_j} a \in S^\mu$, $\partial_{\eta_k} a \in S^{\mu-1}$ it follows an estimate of the desired type for $\widetilde{\pi}_1(Au)$, and by iterating the procedure we can estimate $\widetilde{\pi}_{\widetilde{m}}(Au)$ for any $\widetilde{m}$. This gives us the estimates (1.137) in the y'-independent case.

Finally, assume $a = a(y, y', \eta)$. In this case

$$\mathrm{Op}(a)u(y) = F^{-1}_{\eta \to y}\left\{F_{y' \to \eta} a(y, y', \eta) u(y')\right\}$$

is a composition of three continuous operators:

$$B : \mathcal{S}\left(\mathbb{R}^q_{y'}, H\right) \to \mathcal{S}(\mathbb{R}^q_{y'}, S^\mu(\mathbb{R}^q_y \times \mathbb{R}^q_\eta, \widetilde{H})_{\mathrm{b}}),$$

defined by $Bu := a(y, y', \eta)u(y')$ for

$$S^\mu(\mathbb{R}^q_y \times \mathbb{R}^q_\eta, \widetilde{H})_{\mathrm{b}} := \Big\{ f(y,\eta) \in C^\infty(\mathbb{R}^q_y \times \mathbb{R}^q_\eta, \widetilde{H}) :$$

$$\sup_{(y,\eta) \in \mathbb{R}^q \times \mathbb{R}^q} \langle\eta\rangle^{\mu - |\beta|} \left\| D^\alpha_y D^\beta_\eta f(y,\eta) \right\|_{\widetilde{H}} < \infty \Big\},$$

cf. Definition 1.3.1 (ii), for all $\alpha, \beta \in \mathbb{N}^q$,

$$F_{y' \to \eta} : \mathcal{S}(\mathbb{R}^q_{y'}, S^\mu(\mathbb{R}^q_y \times \mathbb{R}^q_\eta, \widetilde{H})_{\mathrm{b}}) \to \mathcal{S}(\mathbb{R}^q_\eta, S^\mu(\mathbb{R}^q_y \times \mathbb{R}^q_\eta, \widetilde{H})_{\mathrm{b}})$$

$$= \mathcal{S}(\mathbb{R}^q_\eta, C^\infty_{\mathrm{b}}(\mathbb{R}^q_y, \widetilde{H})),$$

and

$$F^{-1}_{\eta \to y} : \mathcal{S}(\mathbb{R}^q_\eta, C^\infty_{\mathrm{b}}(\mathbb{R}^q_y, \widetilde{H})) \to \mathcal{S}(\mathbb{R}^q_\eta, C^\infty_{\mathrm{b}}(\mathbb{R}^q_y, \widetilde{H})) = \mathcal{S}(\mathbb{R}^q, \widetilde{H}). \qquad \square$$

Observe that the proof in the case of symbols $a(y, y', \eta)$ is independent of the first part of the proof and is also valid for symbols $a(y, \eta)$. Let us set

$$L^\mu_{(\mathrm{cl})}(\mathbb{R}^q; H, \widetilde{H})_{\mathrm{b}} := \left\{ \mathrm{Op}(a) : a(y, y', \eta) \in S^\mu_{(\mathrm{cl})}(\mathbb{R}^{2q} \times \mathbb{R}^{2q}; H, \widetilde{H})_{\mathrm{b}} \right\}.$$

Proposition 1.4.12. *The mapping*

$$\mathrm{Op} : S^{\mu}_{(\mathrm{cl})}\big(\mathbb{R}^q \times \mathbb{R}^q; H, \widetilde{H}\,\big)_{\mathrm{b}} \to L^{\mu}_{(\mathrm{cl})}\big(\mathbb{R}^q; H, \widetilde{H}\,\big)_{\mathrm{b}} \tag{1.139}$$

defined by (1.136) gives rise to an isomorphism (1.139), and $a(y, \eta)$ can be recovered from $A = \mathrm{Op}(a)$ by the formula $a(y, \eta)h = e^{-iy\eta} A e^{iy'\eta}$, $h \in H$.

Proof. We can write

$$\mathrm{Op}(a)u(y) = \iint e^{i(y-y')\eta} a(y, \eta)u(y')dy'đ\eta = \iint e^{-iy'\eta} a(y, \eta)u(y'+y)dy'đ\eta.$$

The right-hand side can be interpreted as an oscillatory integral with $a(y, \eta)u(y' + y)$ as the amplitude function in (y', η) for every fixed y. This allows us to assume $u(y') \in C^{\infty}_{\mathrm{b}}(\mathbb{R}^q_{y'}, H)$, in particular $u(y') = e^{iy'\xi}h$ for any $h \in H$, with a parameter $\xi \in \mathbb{R}^q$. This gives rise to

$$e^{-iy\xi} A e^{iy'\xi} h = \iint e^{iy'(\eta-\xi)} a(y, \eta)h\, dy'đ\eta$$

$$= \lim_{\varepsilon \to 0} \int \left\{ \int e^{-iy'\eta'} \chi(\varepsilon y')dy' \right\} a(y, \xi + \eta')h\, đ\eta'$$

$$= \lim_{\varepsilon \to 0} \int \varepsilon^{-q} \widehat{\chi}(\varepsilon^{-1}\eta')a(y, \xi + \eta')h\, đ\eta' = a(y, \xi)h$$

for any $\chi \in C^{\infty}_0(\mathbb{R}^q)$, $\chi(0) = 1$. Here we employed the fact that for any $\varphi(\eta) \in C^{\infty}_0(\mathbb{R}^q)$ with $\int \varphi(\eta)d\eta = 1$ and $\varphi_{\varepsilon}(\eta) := \varepsilon^{-q}\varphi(\varepsilon^{-1}\eta)$ we have $\lim_{\varepsilon \to 0} \varphi_{\varepsilon} * f = f$ for any $f \in \mathcal{S}'(\mathbb{R}^q)$. □

Theorem 1.4.13. *Let H and $\widetilde{H}$ be Hilbert spaces with group action κ and $\widetilde{\kappa}$, respectively, and let $A = \mathrm{Op}(a) \in L^{\mu}_{(\mathrm{cl})}\big(\mathbb{R}^q; H, \widetilde{H}\,\big)_{\mathrm{b}}$, $a(y, y', \eta) \in S^{\mu}_{(\mathrm{cl})}\big(\mathbb{R}^{2q} \times \mathbb{R}^q; H, \widetilde{H}\,\big)_{\mathrm{b}}$. Then*

(i) *there exist unique (so-called) left or right symbols*

$$a_{\mathrm{L}}(y, \eta), a_{\mathrm{R}}(y', \eta) \in S^{\mu}_{(\mathrm{cl})}\big(\mathbb{R}^q \times \mathbb{R}^q; H, \widetilde{H}\,\big)_{\mathrm{b}} \tag{1.140}$$

   such that

$$\mathrm{Op}(a) = \mathrm{Op}(a_{\mathrm{L}}) = \mathrm{Op}(a_{\mathrm{R}});$$

(ii) *the symbols (1.140) can be expressed by oscillatory integrals*

$$a_{\mathrm{L}}(y, \eta) = \iint e^{-ix\xi} a(y, y+x, \eta+\xi)dx đ\xi,$$

$$a_{\mathrm{R}}(y', \eta) = \iint e^{-ix\xi} a(y'+x, y', \eta-\xi)dx đ\xi$$

which have asymptotic expansions

$$a_{\mathrm{L}}(y,\eta) \sim \sum_{\alpha \in \mathbb{N}^q} \frac{1}{\alpha!} (D_{y'}^\alpha \partial_\eta^\alpha a)(y,y',\eta)\big|_{y'=y},$$

$$a_{\mathrm{R}}(y',\eta) \sim \sum_{\alpha \in \mathbb{N}^q} \frac{1}{\alpha!} (-1)^{|\alpha|} (D_y^\alpha \partial_\eta^\alpha a)(y,y',\eta)\big|_{y=y'};$$

(iii) *the mappings $a \mapsto a_{\mathrm{L}}$ and $a \mapsto a_{\mathrm{R}}$ are continuous;*

(iv) *writing*

$$a_{\mathrm{L}}(y,\eta) = \sum_{|\alpha| \le N} \frac{1}{\alpha!} (D_{y'}^\alpha \partial_\eta^\alpha a)(y,y',\eta)\big|_{y'=y} + r_{\mathrm{L},N+1}(y,\eta),$$

$$a_{\mathrm{R}}(y',\eta) = \sum_{|\alpha| \le N} \frac{1}{\alpha!} (-1)^{|\alpha|} D_y^\alpha \partial_\eta^\alpha a(y,y',\eta)\big|_{y=y'} + r_{\mathrm{R},N+1}(y,\eta),$$

for any $N \in \mathbb{N}$, we have $r_{\mathrm{L},N+1}, r_{\mathrm{R},N+1} \in S^{\mu-(N+1)}\big(\mathbb{R}^q \times \mathbb{R}^q; H, \widetilde{H}\big)_{\mathrm{b}}$ and

$$r_{\mathrm{L},N+1}(y,\eta) = (N+1) \sum_{|\alpha|=N+1} \int_0^1 \frac{(1-t)^N}{\alpha!} \tag{1.141}$$
$$\times \iint e^{-ix\xi} (D_{y'}^\alpha \partial_\eta^\alpha a)(y, y+x, \eta+t\xi) dx \,\text{đ}\xi dt,$$

$$r_{\mathrm{R},N+1}(y',\eta) = (N+1) \sum_{|\alpha|=N+1} (-1)^{N+1} \int_0^1 \frac{(1-t)^N}{\alpha!} \tag{1.142}$$
$$\times \iint e^{-ix\xi} (D_y^\alpha \partial_\eta^\alpha a)(y'+x, y', \eta-t\xi) dx \,\text{đ}\xi dt;$$

(v) *the mappings $a \mapsto r_{\mathrm{L},N+1}$ and $a \mapsto r_{\mathrm{R},N+1}$ are continuous.*

The proof can be obtained by methods analogous to those in Kumano-go's monograph [26, Chapter 2].

Theorem 1.4.14. *Let $H, \widetilde{H}$ and H_0 be Hilbert spaces with group action $\kappa, \widetilde{\kappa}$ and κ_0, respectively.*

(i) *$A = \mathrm{Op}(a) \in L_{(\mathrm{cl})}^\mu \big(\mathbb{R}^q; H_0, \widetilde{H}\big)_{\mathrm{b}}$, $B = \mathrm{Op}(b) \in L_{(\mathrm{cl})}^\nu \big(\mathbb{R}^q; H, H_0\big)_{\mathrm{b}}$ for $a(y,\eta) \in S_{(\mathrm{cl})}^\mu \big(\mathbb{R}^q \times \mathbb{R}^q; H_0, \widetilde{H}\big)_{\mathrm{b}}$, $b(y,\eta) \in S_{(\mathrm{cl})}^\nu \big(\mathbb{R}^q \times \mathbb{R}^q; H, H_0\big)_{\mathrm{b}}$, entails*

$$AB \in L_{(\mathrm{cl})}^{\mu+\nu} \big(\mathbb{R}^q; H, \widetilde{H}\big)_{\mathrm{b}},$$

and we have

$$AB = \mathrm{Op}(a\#b)$$

for a unique (so-called Leibniz product)

$$(a\#b)(y,\eta) \in S_{(\mathrm{cl})}^{\mu+\nu}\big(\mathbb{R}^q \times \mathbb{R}^q; H, \widetilde{H}\,\big)_{\mathrm{b}},$$

$$(a\#b)(y,\eta) = \iint e^{-ix\xi} a(y,\eta+\xi) b(y+x,\eta) dx d\xi.$$

There is an asymptotic expansion

$$(a\#b)(y,\eta) \sim \sum_{\alpha \in \mathbb{N}^q} \frac{1}{\alpha!} \big(\partial_\eta^\alpha a(y,\eta)\big) D_y^\alpha b(y,\eta).$$

(ii) *The mapping $(a,b) \mapsto a\#b$ is continuous bilinear between the respective spaces of symbols.*

(iii) *Writing*

$$(a\#b)(y,\eta) = \sum_{|\alpha| \le N} \frac{1}{\alpha!} \big(\partial_\eta^\alpha a(y,\eta)\big) D_y^\alpha b(y,\eta) + r_{N+1}(y,\eta), \quad N \in \mathbb{N},$$

we have $r_{N+1} \in S_{(\mathrm{cl})}^{\mu+\nu-(N+1)}\big(\mathbb{R}^q \times \mathbb{R}^q; H, \widetilde{H}\,\big)_{\mathrm{b}}$ and

$$r_{N+1}(y,\eta)$$
$$= (N+1) \sum_{|\alpha|=N+1} \int_0^1 \frac{(1-t)^N}{\alpha!} \iint e^{-ix\xi} \big(\partial_\eta^\alpha a\big)(y,\eta+t\xi)\big(D_y^\alpha b\big)(y+x,\eta) dx d\xi dt.$$

(iv) *The mapping $(a,b) \mapsto r_{N+1}$ is continuous bilinear between the respective spaces of symbols.*

Theorem 1.4.15. *Let $H, \widetilde{H}$ be Hilbert spaces with group action. Then $\mathrm{Op}(a)$ for $a(y,\eta) \in S_{(\mathrm{cl})}^{\mu}\big(\mathbb{R}^q \times \mathbb{R}^q; H, \widetilde{H}\,\big)_{\mathrm{b}}$ extends the mapping $\mathcal{S}(\mathbb{R}^q, H) \to \mathcal{S}(\mathbb{R}^q, \widetilde{H})$, cf. Theorem 1.4.11, to a continuous operator*

$$\mathrm{Op}(a): \mathcal{W}^s(\mathbb{R}^q, H) \to \mathcal{W}^{s-\mu}\big(\mathbb{R}^q, \widetilde{H}\,\big)$$

for every $s \in \mathbb{R}$. The mapping $a \mapsto \mathrm{Op}(a)$ induces a continuous operator

$$S_{(\mathrm{cl})}^{\mu}\big(\mathbb{R}^q \times \mathbb{R}^q; H, \widetilde{H}\,\big)_{\mathrm{b}} \to \mathcal{L}\big(\mathcal{W}^s(\mathbb{R}^q, H), \mathcal{W}^{s-\mu}(\mathbb{R}^q, \widetilde{H}\,)\big) \qquad (1.143)$$

for every $s \in \mathbb{R}$.

Proof. We fix s and reduce orders to zero by composing with

$$P^{\mu-s} := \mathrm{Op}(\langle\eta\rangle^{\mu-s}\mathrm{id}_{\widetilde{H}}), \quad Q^{-s} := \mathrm{Op}(\langle\eta\rangle^{-s}\mathrm{id}_H).$$

These mappings give rise to isomorphisms

$$Q^{-s}: \mathcal{W}^0(\mathbb{R}^q, H) \to \mathcal{W}^s(\mathbb{R}^q, H), \quad P^{\mu-s}: \mathcal{W}^{s-\mu}(\mathbb{R}^q, H) \to \mathcal{W}^0(\mathbb{R}^q, H).$$

By Theorem 1.4.11, the operator

$$P^{\mu-s}\mathrm{Op}(a)Q^{-s} : \mathcal{S}(\mathbb{R}^q, H) \to \mathcal{S}(\mathbb{R}^q, \widetilde{H})$$

is continuous. From Theorem 1.4.14 we obtain

$$P^{\mu-s}\mathrm{Op}(a)Q^{-s} = \mathrm{Op}(a_0)$$

for an $a_0(y,\eta) \in S^0\big(\mathbb{R}^q \times \mathbb{R}^q; H, \widetilde{H}\big)_{\mathrm{b}}$. Thus, from Theorem 1.3.35 it follows that $\mathrm{Op}(a_0)$ extends to a continuous operator

$$\mathrm{Op}(a_0) : \mathcal{W}^0(\mathbb{R}^q, H) \to \mathcal{W}^0(\mathbb{R}^q, \widetilde{H}),$$

with the estimate

$$\|\mathrm{Op}(a_0)\|_{\mathcal{L}(\mathcal{W}^0(\mathbb{R}^q,H),\mathcal{W}^0(\mathbb{R}^q,\widetilde{H}))} \le c\,\pi(a_0).$$

This immediately establishes the continuity of the operator

$$\mathrm{Op}(a) = P^{-s+\mu}\mathrm{Op}(a_0)Q^s : \mathcal{W}^s(\mathbb{R}^q, H) \to \mathcal{W}^{s-\mu}(\mathbb{R}^q, \widetilde{H}),$$

with the estimate

$$\|\mathrm{Op}(a_0)\|_{\mathcal{L}(\mathcal{W}^s(\mathbb{R}^q,H),\mathcal{W}^{s-\mu}(\mathbb{R}^q,\widetilde{H}))} \le \widetilde{c}\,\pi(a_0),$$

for another $\widetilde{c}$. This yields (1.143). $\qquad\square$

Let $\{H, H_0, H'; \kappa\}$ and $\{\widetilde{H}, \widetilde{H}_0, \widetilde{H}'; \widetilde{\kappa}\}$ be Hilbert space triples with group action, cf. Definition 1.3.12. Recall that the pointwise application of adjoints $\mathcal{L}(H, \widetilde{H}) \to \mathcal{L}(\widetilde{H}', H')$ gives rise to an antilinear isomorphism

$$(*) : S^\mu_{(\mathrm{cl})}\big(\mathbb{R}^q \times \mathbb{R}^q; H, \widetilde{H}\big)_{\mathrm{b}} \to S^\mu_{(\mathrm{cl})}\big(\mathbb{R}^q \times \mathbb{R}^q; \widetilde{H}', H'\big)_{\mathrm{b}}, \quad a(y,\eta) \mapsto a^{(*)}(y,\eta). \tag{1.144}$$

Theorem 1.4.16. *Let* $A \in L^\mu_{\mathrm{cl}}\big(\mathbb{R}^q; H, \widetilde{H}\big)_{\mathrm{b}}$, *and consider the formal adjoint* A^*, *defined by*

$$(Au, v)_{L^2(\mathbb{R}^q, \widetilde{H}_0)} = (u, A^*v)_{L^2(\mathbb{R}^q, H_0)}$$

for all $u \in \mathcal{S}(\mathbb{R}^q, H)$, $v \in \mathcal{S}(\mathbb{R}^q, \widetilde{H}')$.

(i) *We have* $A^* \in L^\mu_{(\mathrm{cl})}\big(\mathbb{R}^q; \widetilde{H}', H'\big)_{\mathrm{b}}$. *Moreover, for* $A = \mathrm{Op}(a)$, *with* $a(y,\eta) \in S^\mu_{(\mathrm{cl})}\big(\mathbb{R}^q \times \mathbb{R}^q; H, \widetilde{H}\big)_{\mathrm{b}}$ *we have*

$$A^* = \mathrm{Op}(a^*)$$

for $a^*(y,\eta) \in S^\mu_{(\mathrm{cl})}\big(\mathbb{R}^q \times \mathbb{R}^q; \widetilde{H}', H'\big)_{\mathrm{b}}$,

$$a^*(y,\eta) = \iint e^{-ix\xi} a^{(*)}(y+x, \eta+\xi)\,dx\,d\!\!\!/\xi.$$

There is an asymptotic expansion

$$a^*(y,\eta) \sim \sum_{\alpha \in \mathbb{N}^q} \frac{1}{\alpha!} D_y^\alpha \partial_\eta^\alpha a^{(*)}(y,\eta).$$

(ii) *The mapping $a \mapsto a^*$ is antilinear and continuous between the respective spaces of symbols.*

(iii) *Writing*

$$a^*(y,\eta) = \sum_{|\alpha| \leq N} \frac{1}{\alpha!} \left(\partial_\eta^\alpha D_y^\alpha a^{(*)}(y,\eta)\right) + r_{N+1}^*(y,\eta), \quad N \in \mathbb{N},$$

we have $r_{N+1}^* \in S_{(\mathrm{cl})}^{\mu-(N+1)}\left(\mathbb{R}^q \times \mathbb{R}^q; \widetilde{H}', \widetilde{H}\right)_\mathrm{b}$ and

$$r_{N+1}^*(y,\eta)$$
$$= (N+1) \sum_{|\alpha|=N+1} \int_0^1 \frac{(1-t)^N}{\alpha!} \iint e^{-ix\xi} \left(\partial_\eta^\alpha D_y^\alpha a^{(*)}\right)(y+x, \eta+t\xi) dx\, d\!\!\!/\xi\, dt.$$

(iv) *The mapping $a \mapsto r_{N+1}^*$ is antilinear continuous between the respective spaces of symbols.*

We now return to the kernel cut-off construction already mentioned after Proposition 1.1.2 in a simple case.
We mainly apply the kernel cut-off with respect to a one-dimensional covariable $\tau \in \mathbb{R}$ within $\xi = (\tau, \eta) \in \mathbb{R}^{1+q}$. For $a(\tau, \eta) \in S_{(\mathrm{cl})}^\mu(\mathbb{R}^{1+q})$ we first form

$$k_F(a)(\theta, \eta) := \int e^{i\theta\tau} a(\tau, \eta) d\!\!\!/\tau, \tag{1.145}$$

which is an element of $\mathcal{S}'(\mathbb{R}^{1+q})$. Let us fix an arbitrary cut-off function $\psi(\theta) \in C_0^\infty(\mathbb{R})$ (i.e., $\psi(\theta) = 1$ close to $\theta = 0$) and set $\chi(\theta) := 1 - \psi(\theta)$, which is an excision function. Then

$$a(\tau, \eta) = F_{\theta \to \tau}(k_F(a)(\theta, \eta)) = a_0(\tau, \eta) + c(\tau, \eta),$$

where $c(\tau, \eta) := F_{\theta \to \tau}(\chi(\theta) k_F(a)(\theta, \eta)) \in \mathcal{S}(\mathbb{R}^{1+q})$, and thus

$$a_0(\tau, \eta) := F_{\theta \to \tau}(\psi(\theta) k_F(a)(\theta, \eta)) \in S_{(\mathrm{cl})}^\mu(\mathbb{R}^{1+q}),$$

cf. Proposition 1.1.2. The resulting operator

$$S_{(\mathrm{cl})}^\mu(\mathbb{R}^{1+q}) \to S_{(\mathrm{cl})}^\mu(\mathbb{R}^{1+q}), \quad a(\tau, \eta) \mapsto a_0(\tau, \eta)$$

will be called a kernel cut-off operator. For the ensuing discussion it is essential to observe that $a_0(\tau, \eta)$ extends to an entire function in $v = \tau + i\sigma \in \mathbb{C}$ with values in $S^{\mu}_{(\text{cl})}(\mathbb{R}^q_\eta)$. It will be necessary to consider the kernel cut-off for any $\varphi \in C^\infty_0(\mathbb{R})$ rather than a cut-off function and also to admit functions in $C^\infty_{\text{b}}(\mathbb{R})$. A simple computation, first for $\varphi \in C^\infty_0(\mathbb{R})$, gives

$$\mathcal{V}(\varphi)a(\tau, \eta) := F_{\theta \to \tau}(k_F(a)(\theta, \eta)) = \iint e^{-i\theta\tilde{\tau}} \varphi(\theta) a(\tau - \tilde{\tau}, \eta) d\theta d\tilde{\tau}, \quad (1.146)$$

which is an oscillatory integral in the sense of (1.132) with respect to $(\theta, \tilde{\tau})$ for every fixed η. The above results show that $\mathcal{V}(\varphi)$ extends to $C^\infty_{\text{b}}(\mathbb{R})$. The following theorem generalizes the construction to operator-valued symbols with twisted symbol estimates, and it also states a corresponding continuity in φ, as well as in a.

Theorem 1.4.17. *The kernel cut-off operator* $(\varphi, a) \mapsto \mathcal{V}(\varphi)a$ *defines a bilinear continuous mapping*

$$\mathcal{V} : C^\infty_{\text{b}}(\mathbb{R}) \times S^{\mu}_{(\text{cl})}(\mathbb{R}^{1+q}_{\tau,\eta}; H, \tilde{H}) \to S^{\mu}_{(\text{cl})}(\mathbb{R}^{1+q}_{\tau,\eta}; H, \tilde{H}) \quad (1.147)$$

and $\mathcal{V}(\varphi)a(\tau, \eta)$ *admits an asymptotic expansion*

$$\mathcal{V}(\varphi)a(\tau, \eta) \sim \sum_{k=0}^{\infty} \frac{(-1)^k}{k!} (D^k_\theta \varphi)(0) \partial^k_\tau a(\tau, \eta). \quad (1.148)$$

In particular, if $\varphi = \psi$ *is a cut-off function, then*

$$\mathcal{V}(\psi)a(\tau, \eta) = a(\tau, \eta) \mod S^{-\infty}(\mathbb{R}^{1+q}; H, \tilde{H}).$$

Proof. The correspondence $(\varphi, a) \mapsto \varphi(\theta)a(\tau - \tilde{\tau}, \eta)$ defines a bilinear continuous operator

$$C^\infty_{\text{b}}(\mathbb{R}_\theta) \times S^{\mu}_{(\text{cl})}(\mathbb{R}^{1+q}_{\tau,\eta}; H, \tilde{H}) \to C^\infty(\mathbb{R}^{1+q}_{\tau,\eta}, S^{\mu}_{(\text{cl})}(\mathbb{R}_\theta \times \mathbb{R}_{\tilde{\tau}}; H, \tilde{H})_{\text{b}});$$

recall that $S^{\mu}_{(\text{cl})}(\mathbb{R}_\theta \times \mathbb{R}_{\tilde{\tau}}; H, \tilde{H})_{\text{b}} = C^\infty(\mathbb{R}_\theta, S^{\mu}_{(\text{cl})}(\mathbb{R}_{\tilde{\tau}}; H, \tilde{H})_{\text{b}})$. In order to show the continuity of (1.147) it suffices to verify that $\mathcal{V}(\varphi)a \in S^{\mu}_{(\text{cl})}(\mathbb{R}^{1+q}_{\tau,\eta}; H, \tilde{H})$ and apply the closed graph theorem. Since

$$D^\beta_{\tau,\eta}\mathcal{V}(\varphi)a(\tau, \eta) = \mathcal{V}(\varphi)\big(D^\beta_{\tau,\eta}a(\tau, \eta)\big),$$

when $\beta \in \mathbb{N}^{1+q}$, it suffices to show that

$$\big\|\tilde{\kappa}^{-1}_{\langle\tau,\eta\rangle}\mathcal{V}(\varphi)a(\tau, \eta)\tilde{\kappa}_{\langle\tau,\eta\rangle}\big\|_{\mathcal{L}(H,\tilde{H})} \le c\langle\tau, \eta\rangle^\mu \quad (1.149)$$

for all $(\tau, \eta) \in \mathbb{R}^{1+q}$. According to (1.134), we regularize the oscillatory integral (1.146) as

$$\mathcal{V}(\varphi)a(\tau, \eta) = \iint e^{-i\theta\tilde{\tau}} \langle\theta\rangle^{-2} \left\{(1 - \partial^2_\theta)^N \varphi(\theta)\right\} a_N(\tau, \tilde{\tau}, \eta) d\theta d\tilde{\tau}$$

for

$$a_N(\tau, \widetilde{\tau}, \eta) := \left(1 - \partial_{\widetilde{\tau}}^2\right)\left\{\langle\widetilde{\tau}\rangle^{-2N} a(\tau - \widetilde{\tau}, \eta)\right\}, \tag{1.150}$$

N sufficiently large. The function (1.150) is a linear combination of the terms

$$\left(\partial_{\widetilde{\tau}}^j \langle\widetilde{\tau}\rangle^{-2N}\right)\left(\partial_\tau^k a\right)(\tau - \widetilde{\tau}, \eta)$$

for $0 \leq j, k \leq 2$. From Peetre's inequality $\langle\xi' + \xi''\rangle^s \leq c^{|s|}\langle\xi'\rangle^{|s|}\langle\xi''\rangle^s$ for all $\xi', \xi'' \in \mathbb{R}$, $s \in \mathbb{R}$, for some $c > 0$, we obtain

$$\left\|\widetilde{\kappa}_{\langle\tau - \widetilde{\tau}, \eta\rangle\langle\tau, \eta\rangle^{-1}}\right\|_{\mathcal{L}(\widetilde{H})} \leq C\langle\widetilde{\tau}\rangle^{\widetilde{M}}, \quad \left\|\kappa_{\langle\tau - \widetilde{\tau}, \eta\rangle^{-1}\langle\tau, \eta\rangle}\right\|_{\mathcal{L}(H)} \leq C\langle\widetilde{\tau}\rangle^M$$

for a suitable $C > 0$ and constants $M, \widetilde{M}$ belonging to $\kappa, \widetilde{\kappa}$ in the sense of (1.74). Moreover, writing $(\tau - \widetilde{\tau}, \eta) = (\tau, \eta) - (\widetilde{\tau}, 0)$, again by Peetre's inequality we have $\langle\tau - \widetilde{\tau}, \eta\rangle^\mu \leq c^{|s|}\langle\widetilde{\tau}\rangle^{|\mu|}\langle\tau, \eta\rangle^\mu$. It follows that

$$\left\|\widetilde{\kappa}_{\langle\tau, \eta\rangle}^{-1}\left\{\partial_{\widetilde{\tau}}^j\langle\widetilde{\tau}\rangle^{-2N}\left(\partial_\tau^k a\right)(\tau - \widetilde{\tau}, \eta)\right\}\kappa_{\langle\tau, \eta\rangle}\right\|_{\mathcal{L}(H, \widetilde{H})}$$

$$\leq \left|\partial_{\widetilde{\tau}}^j\langle\widetilde{\tau}\rangle^{-2N}\right|\left\|\widetilde{\kappa}_{\langle\tau - \widetilde{\tau}, \eta\rangle\langle\tau, \eta\rangle^{-1}}\right\|_{\mathcal{L}(\widetilde{H})}$$

$$\cdot \left\|\widetilde{\kappa}_{\langle\tau - \widetilde{\tau}, \eta\rangle}^{-1}\left(\partial_\tau^k a\right)(\tau - \widetilde{\tau}, \eta)\kappa_{\langle\tau - \widetilde{\tau}, \eta\rangle}\right\|_{\mathcal{L}(H, \widetilde{H})}\left\|\kappa_{\langle\tau - \widetilde{\tau}, \eta\rangle^{-1}\langle\tau, \eta\rangle}\right\|_{\mathcal{L}(H)}$$

$$\leq c\langle\widetilde{\tau}\rangle^{M + \widetilde{M} - 2N}\langle\tau - \widetilde{\tau}, \eta\rangle^\mu \leq c\langle\widetilde{\tau}\rangle^{M + \widetilde{M} + \mu - 2N}\langle\tau, \eta\rangle^\mu$$

for some $c > 0$. This implies analogous estimates for the function (1.150). We then obtain the estimate (1.149) by taking N so large that $M + \widetilde{M} + \mu - 2N \leq 0$. For (1.148) we employ the Taylor expansion

$$\varphi(\theta) = \sum_{k=0}^N \frac{1}{k!}\left(\partial_\theta^k\varphi\right)(0)\theta^k + \theta^{N+1}\varphi_{N+1}(\theta),$$

with $\varphi_{N+1}(\theta) = \frac{1}{N!}\int_0^1 (1 - t)^N\left(\partial_\theta^{N+1}\varphi\right)(t\theta)dt$. The function $\varphi_{N+1}(\theta)$ belongs to $C_{\mathrm{b}}^\infty(\mathbb{R})$. Integration by parts in (1.146) yields

$$\mathcal{V}(\varphi)a(\tau, \eta) = \sum_{k=0}^N \frac{1}{k!}\left(\partial_\theta^k\varphi\right)(0)\iint e^{-i\theta\widetilde{\tau}}\theta^k a(\tau - \widetilde{\tau}, \eta)d\theta\đ\widetilde{\tau}$$

$$+ \iint e^{-i\theta\widetilde{\tau}}\theta^{N+1}\varphi_{N+1}(\theta)a(\tau - \widetilde{\tau}, \eta)d\theta\đ\widetilde{\tau}$$

$$= \sum_{k=0}^N \frac{1}{k!}\left(D_\theta^k\varphi\right)(0)\left(\partial_\tau^k a\right)(\tau, \eta) + i^{N+1}\mathcal{V}(\varphi_{N+1})\partial_\tau^{N+1}a(\tau, \eta).$$

Here we employed the identity

$$\left(\partial_\tau^k a\right)(\tau, \eta) = \iint e^{-i\theta\widetilde{\tau}}\left(\partial_\tau^k a\right)(\tau - \widetilde{\tau}, \eta)d\theta\đ\widetilde{\tau}$$

and the expression (1.146), applied to φ_{N+1}. In the first part of the proof we saw that $\mathcal{V}(\varphi_{N+1})(\partial_\tau^{N+1} a)(\tau, \eta)$ belongs to $S^{\mu-(N+1)}(\mathbb{R}^{1+q}; H, \widetilde{H})$. This completes the proof of (1.148). □

Theorem 1.4.18. *Let* $\psi(\theta) \in C_0^\infty(\mathbb{R})$ *be a cut-off function, and set* $\psi_\varepsilon(\theta) := \psi(\varepsilon\theta)$ *for* $0 < \varepsilon \leq 1$. *Then for every* $a(\tau, \eta) \in S_{(\mathrm{cl})}^\mu(\mathbb{R}^{1+q}; H, \widetilde{H})$ *we have*

$$\lim_{\varepsilon \to 0} \mathcal{V}(\psi_\varepsilon) a(\tau, \eta) = a(\tau, \eta), \tag{1.151}$$

with convergence in $S^\mu(\mathbb{R}^{1+q}; H, \widetilde{H})$.

Proof. We have $\lim_{\varepsilon \to 0} \psi_\varepsilon = 1$ in the topology of $C_\mathrm{b}^\infty(\mathbb{R})$. Then (1.151) is a consequence of the continuity of (1.147) and of $\mathcal{V}(1)a = a$. □

Remark 1.4.19. *Let* $\varphi \in C_\mathrm{b}^\infty(\mathbb{R})$ *and* $\partial_\theta^k \varphi(0) = 0$ *for all* $0 \leq k \leq N$; *then* $\mathcal{V}(\varphi)$ *defines a continuous operator*

$$\mathcal{V}(\varphi) : S_{(\mathrm{cl})}^\mu(\mathbb{R}^{1+q}; H, \widetilde{H}) \to S_{(\mathrm{cl})}^{\mu-(N+1)}(\mathbb{R}^{1+q}; H, \widetilde{H}). \tag{1.152}$$

Moreover, if $\chi(\theta)$ *is an excision function, then* $\mathcal{V}(\chi)$ *defines a continuous operator*

$$\mathcal{V}(\chi) : S^\mu(\mathbb{R}^{1+q}; H, \widetilde{H}) \to S^{-\infty}(\mathbb{R}^{1+q}; H, \widetilde{H}). \tag{1.153}$$

In fact, we have the expansion (1.148), and the first N summands vanish. Thus, applying the closed graph theorem establishes the continuity of (1.152). Moreover, the continuity of (1.152) is a consequence of the continuity of (1.152) for every N. Then (1.153) follows from $\chi(\theta) k_F(a)(\theta, \eta) = F_{\theta \to \tau}^{-1}(\mathcal{V}(\chi)a(\tau, \eta))$ and $S^{-\infty}(\mathbb{R}^{1+q}; H, \widetilde{H}) = \mathcal{S}(\mathbb{R}^{1+q}, \mathcal{L}(H, \widetilde{H}))$.

Definition 1.4.20. Set

$$\mathbf{\Gamma}_\delta := \{\zeta \in \mathbb{C} : \mathrm{Im}\, \zeta = \delta\}$$

and define

$$S_{(\mathrm{cl}),\mathcal{O}}^\mu(\mathbb{R}^q; H, \widetilde{H})$$

to be the set of all $h(\zeta, \eta) \in \mathcal{A}(\mathbb{C}, S_{(\mathrm{cl})}^\mu(\mathbb{R}_\eta^q; H, \widetilde{H}))$ such that

$$h|_{\mathbf{\Gamma}_\delta} \in S_{(\mathrm{cl})}^\mu(\mathbf{\Gamma}_\delta \times \mathbb{R}_\eta^q; H, \widetilde{H})$$

for every $\delta \in \mathbb{R}$, uniformly on compact δ-intervals.

Definition 1.4.20 induces a natural Fréchet topology in $S_{(\mathrm{cl}),\mathcal{O}}^\mu(\mathbb{R}^q; H, \widetilde{H})$.

Theorem 1.4.21. *The kernel cut-off operator* $\mathcal{V} : (\varphi, a) \mapsto \mathcal{V}(\varphi)a$ *induces a separately continuous mapping*

$$\mathcal{V} : C_0^\infty(\mathbb{R}) \times S_{(\mathrm{cl})}^\mu(\mathbb{R}_{\tau,\eta}^{1+q}; H, \widetilde{H}) \to S_{(\mathrm{cl}),\mathcal{O}}^\mu(\mathbb{R}^q; H, \widetilde{H}). \tag{1.154}$$

Proof. Let us write $\mathcal{V}(\varphi)a$ in the form

$$\mathcal{V}(\varphi)a(\tau,\eta) = \int e^{-i\theta\tau}\varphi(\theta)\left\{\int e^{i\theta\tau'}a(\tau',\eta)d\!\!\!/\tau'\right\}d\theta.$$

In this case we have

$$k_F(a)(\theta,\eta) = \int e^{i\theta\tau'}a(\tau',\eta)d\!\!\!/\tau' \in \mathcal{S}'\big(\mathbb{R}^{1+q}_{\theta,\eta},\mathcal{L}(H,\widetilde{H})\big),$$

and $\varphi(\theta)k_F(a)(\theta,\eta)$ has compact support with respect to θ. Thus

$$\int e^{-i\theta\tau}\varphi(\theta)k_F(a)(\theta,\eta)d\theta$$

extends to a function

$$(\mathcal{V}(\varphi)a)(\tau+i\sigma,\eta) = \int e^{-i\theta\tau}e^{\theta\sigma}\varphi(\theta)k_F(a)(\theta,\eta)d\theta$$

which is holomorphic in $\zeta = \tau + i\sigma \in \mathbb{C}$. We have

$$(\mathcal{V}(\varphi)a)(\tau+i\sigma,\eta) = \mathcal{V}(\varphi_\sigma)a(\tau,\eta) \tag{1.155}$$

for $\varphi_\sigma(\theta) := e^{\theta\sigma}\varphi(\theta) \in C_0^\infty(\mathbb{R})$. By Theorem 1.4.17,

$$\mathcal{V}(\varphi)a(\tau+i\sigma,\eta) \in S^\mu_{(\mathrm{cl})}\big(\mathbb{R}^{1+q}_{\tau,\eta}; H,\widetilde{H}\big)$$

for every $\sigma \in \mathbb{R}$. The map $\mathbb{R} \to C_0^\infty(\mathbb{R})$ defined by $\sigma \mapsto \varphi_\sigma$ is continuous. Thus the mapping $a \mapsto \mathcal{V}(\varphi)a$, $a \in S^\mu_{(\mathrm{cl})}\big(\mathbb{R}^{1+q}_{\tau,\eta}; H,\widetilde{H}\big)$, is uniformly continuous on compact σ-intervals. The closed graph theorem gives us also the continuity of (1.154) with respect to the Fréchet topology of $S^\mu_{(\mathrm{cl}),\mathcal{O}}\big(\mathbb{R}^q; H,\widetilde{H}\big)$. □

Proposition 1.4.22. *Let* $a(\tau,\eta) \in S^\mu_{(\mathrm{cl})}\big(\mathbb{R}^{1+q}_{\tau,\eta}; H,\widetilde{H}\big)$. *There exists an*

$$h(\zeta,\eta) \in S^\mu_{(\mathrm{cl}),\mathcal{O}}\big(\mathbb{R}^q; H,\widetilde{H}\big)$$

such that

$$a(\tau,\eta) = h(\zeta,\eta)|_{\Gamma_0} \bmod S^{-\infty}\big(\mathbb{R}^{1+q}_{\tau,\eta}; H,\widetilde{H}\big). \tag{1.156}$$

Proof. We choose an arbitrary cut-off function ψ and define $h := \mathcal{V}(\psi)a$. Then, by Theorem 1.4.21, $h(\zeta,\eta) \in S^\mu_{(\mathrm{cl}),\mathcal{O}}\big(\mathbb{R}^q; H,\widetilde{H}\big)$. The relation (1.156) is a consequence of the asymptotic expansion (1.148). □

Remark 1.4.23. The kernel cut-off operator may be introduced also with respect to any other weight line Γ_β rather than Γ_0. In other words, for every $a(\tau+i\beta,\eta) \in S^\mu_{(\mathrm{cl})}\big(\Gamma_\beta \times \mathbb{R}^q; H,\widetilde{H}\big)$ there exists an $h(\zeta,\eta) \in S^\mu_{(\mathrm{cl}),\mathcal{O}}\big(\mathbb{R}^q; H,\widetilde{H}\big)$ such that

$$a(\tau+i\beta,\eta) = h(\zeta,\eta)|_{\Gamma_\beta} \bmod S^{-\infty}\big(\Gamma_\beta \times \mathbb{R}^q; H,\widetilde{H}\big).$$

In fact, it suffices to apply Proposition 1.4.22 to $a_\beta(\tau, \eta) := a(\tau + i\beta, \eta)$ as an element of $S^\mu_{(cl)}\big(\mathbb{R}^{1+q}_{\tau,\eta}; H, \widetilde{H}\big)$ which yields a corresponding $h_0(\zeta, \eta)$, and then we simply set $h(\zeta, \eta) := h_0(\zeta + i\beta, \eta)$, which belongs $S^\mu_{(cl),\mathcal{O}}\big(\mathbb{R}^q_{\tau,\eta}; H, \widetilde{H}\big)$.

Thus in connection with kernel cut-off constructions we often focus on the weight line Γ_0.

Proposition 1.4.24. (i) *Let* $a(\tau, \eta) \in S^\mu_{(cl)}\big(\mathbb{R}^{1+q}; H, \widetilde{H}\big)$ *and* $h = \mathcal{V}(\varphi)$ *for a* $\varphi \in C_0^\infty(\mathbb{R})$. *Then for any* σ *we have an asymptotic expansion*

$$h(\tau + i\sigma, \eta) \sim \sum_{k=0}^\infty \frac{(-1)^k}{k!} (D_\theta^k \varphi_\sigma)(0) \partial_\tau^k a(\tau, \eta), \qquad (1.157)$$

for $\varphi_\sigma(\theta) = e^{\theta\sigma}\varphi(\theta)$.

(ii) $a(\tau, \eta) \in S^{-\infty}\big(\mathbb{R}^{1+q}; H, \widetilde{H}\big)$ *entails* $\mathcal{V}(\varphi)a \in S^{-\infty}_{\mathcal{O}}\big(\mathbb{R}^q; H, \widetilde{H}\big)$ *for any* $\varphi \in C_0^\infty(\mathbb{R})$.

Proof. (i) Applying the relation (1.155) to φ the expansion (1.157) is an obvious consequence of (1.148).
(ii) immediately follows from (1.157). □

Proposition 1.4.25. *Let* $h(\zeta, \eta) \in S^\mu_{(cl),\mathcal{O}}\big(\mathbb{R}^q; H, \widetilde{H}\big)$; *then we have*

(i)

$$h(\tau + i\sigma, \eta) \sim \sum_{k=0}^\infty \frac{(-1)^k}{k!} (D_\theta^k \psi_\sigma)(0) \partial_\tau^k h(\tau, \eta) \qquad (1.158)$$

for every fixed $\sigma \in \mathbb{R}$.

(ii) $h(\tau + i\sigma_0, \eta) \in S^{\mu-\varepsilon}_{(cl)}\big(\mathbb{R}^{1+q}; H, \widetilde{H}\big)$ *for some fixed* σ_0 *and* $0 < \varepsilon \le 1$ ($\varepsilon = 1$ *in the classical case*) *implies* $h(\zeta, \eta) \in S^{\mu-\varepsilon}_{(cl),\mathcal{O}}\big(\mathbb{R}^q; H, \widetilde{H}\big)$.

Proof. (i) We apply Proposition 1.4.22 to $a(\tau, \eta) := h(\tau, \eta)$. Then we obtain an $\widetilde{h}(\zeta, \eta) \in S^\mu_{(cl),\mathcal{O}}\big(\mathbb{R}^q; H, \widetilde{H}\big)$, and we have

$$\widetilde{h}(\tau, \eta) = h(\tau, \eta) \mod S^{-\infty}_{\mathcal{O}}\big(\mathbb{R}^q_\eta; H, \widetilde{H}\big).$$

Now, from Proposition 1.4.24 (ii) it follows that

$$\mathcal{V}(\psi)\big(\widetilde{h} - h\big)(\tau, \eta) \in S^{-\infty}_{\mathcal{O}}\big(\mathbb{R}^q_\eta; H, \widetilde{H}\big).$$

In other words, in

$$\widetilde{h}(\tau + i\sigma, \eta) \sim \sum_{k=0}^\infty \frac{(-1)^k}{k!} (D_\theta^k \psi_\sigma)(0) \partial_\tau^k h(\tau, \eta)$$

on the left we may replace $\widetilde{h}$ by h, i.e., we proved (i).

Claim (ii) is an immediate consequence of (i), applied to the translated function $h'(\zeta, \eta) := h(\zeta + i\sigma_0, \eta)$ which again belongs to $S^\mu_{(\mathrm{cl}),\mathcal{O}}(\mathbb{R}^q; H, \widetilde{H})$. In fact, the asymptotic expansion (1.158) contains an element in $S^{\mu-\varepsilon}_{(\mathrm{cl})}$ on the original weight line and then the values on other weight lines belong to $S^{\mu-\varepsilon}_{(\mathrm{cl})}$. $\qquad\square$

Corollary 1.4.26. *$a, b \in S^\mu_{(\mathrm{cl}),\mathcal{O}}(\mathbb{R}^q; H, \widetilde{H})$ and $(a-b)|_{\mathbf{\Gamma}_\beta} \in S^{-\infty}(\mathbf{\Gamma}_\beta \times \mathbb{R}^q; H, \widetilde{H})$ for some fixed $\beta \in \mathbb{R}$ implies $a = b \bmod S^{-\infty}_{\mathcal{O}}(\mathbb{R}^q; H, \widetilde{H})$.*

In fact, Proposition 1.4.25 (ii) shows $a = b \bmod S^{\mu-1}_{(\mathrm{cl}),\mathcal{O}}(\mathbb{R}^q; H, \widetilde{H})$, and then successively, $a = b \bmod S^{\mu-N}_{(\mathrm{cl}),\mathcal{O}}(\mathbb{R}^q; H, \widetilde{H})$ for every $N \in \mathbb{N}$.

Theorem 1.4.27. *For every sequence $a_j \in S^{\mu}_{(\mathrm{cl}),\mathcal{O}}(\mathbb{R}^q; H, \widetilde{H})$, $\mu_j \to -\infty$ as $j \to \infty$ ($\mu_j = \mu - j$, in the classical case) there is an asymptotic sum $a \in S^\mu_{(\mathrm{cl}),\mathcal{O}}(\mathbb{R}^q; H, \widetilde{H})$ for $\mu := \max_{j \in \mathbb{N}}\{\mu_j\}$, i.e., for every $N \in \mathbb{N}$ there is a $\nu_N \in \mathbb{R}$, $\nu_N \to -\infty$ as $N \to \infty$, such that $a - \sum_{j=0}^N a_j \in S^{\nu_N}_{\mathcal{O}}(\mathbb{R}^q; H, \widetilde{H})$. If $b \in S^\mu_{(\mathrm{cl}),\mathcal{O}}(\mathbb{R}^q; H, \widetilde{H})$ is another such symbol, then $a = b \bmod S^{-\infty}_{\mathcal{O}}(\mathbb{R}^q; H, \widetilde{H})$.*

Proof. By definition, we have $a_j|_{\mathbf{\Gamma}_0} \in S^{\mu_j}_{(\mathrm{cl})}(\mathbf{\Gamma}_0 \times \mathbb{R}^q; H, \widetilde{H})$, and Theorem 1.3.10 allows us to form an asymptotic sum $f \sim \sum_{j=0}^\infty a_j|_{\mathbf{\Gamma}_0} \in S^\mu_{(\mathrm{cl})}(\mathbf{\Gamma}_0 \times \mathbb{R}^q; H, \widetilde{H})$. Then for any cut-off function ψ we have $a := \mathcal{V}(\psi)f \in S^\mu_{(\mathrm{cl}),\mathcal{O}}(\mathbb{R}^q; H, \widetilde{H})$. From (1.156) it follows that $f(\tau, \eta) = a(\zeta, \eta)|_{\mathbf{\Gamma}_0} \in S^{-\infty}(\mathbf{\Gamma}_0 \times \mathbb{R}^q; H, \widetilde{H})$. The relation $(a - \sum_{j=0}^N a_j)|_{\mathbf{\Gamma}_0} \in S^{\nu_N}(\mathbf{\Gamma}_0 \times \mathbb{R}^q; H, \widetilde{H})$ together with Proposition 1.4.25 give $a - \sum_{j=0}^N a_j \in S^{\nu_N}_{\mathcal{O}}(\mathbb{R}^q; H, \widetilde{H})$, which is the first part of our assertion. If $b \in S^\mu_{(\mathrm{cl}),\mathcal{O}}(\mathbb{R}^q; H, \widetilde{H})$ is another asymptotic sum, we first have

$$(a - b)_{\mathbf{\Gamma}_0} \in S^{-\infty}(\mathbf{\Gamma}_0 \times \mathbb{R}^q; H, \widetilde{H}).$$

From Corollary 1.4.26 it follows that $a - b \in S^{-\infty}_{\mathcal{O}}(\mathbb{R}^q; H, \widetilde{H})$. $\qquad\square$

We now discuss the kernel cut-off on formal adjoints in the following sense. We have the antilinear operation $(*)$ between symbols, cf. (1.144). Then from (1.145) we can pass to

$$k_F(a)^{(*)}(\theta, \eta) = \int e^{-i\theta\tau'} a^{(*)}(\tau', \eta)\,d̄\tau'$$

and

$$(\mathcal{V}(\varphi)a)^{(*)}(\tau, \eta) = \int e^{i\theta\tau} \bar{\varphi}(\theta) k_F(a)^{(*)}(\theta, \eta)\,d\theta.$$

Analogously to (1.146), we interpret the latter expression as an oscillatory integral

$$(\mathcal{V}(\varphi)a)^{(*)} = \iint e^{-i\theta\widetilde{\tau}} \bar{\varphi}(\theta) a^{(*)}(\tau + \widetilde{\tau})\,d\theta\,d̄\widetilde{\tau}.$$

Remark 1.4.28. (i) There is a modification of Theorem 1.4.17 which yields an antibilinear continuous mapping

$$C_{\mathrm{b}}^{\infty}(\mathbb{R}) \times S_{(\mathrm{cl})}^{\mu}\left(\mathbb{R}^{1+q}; H, \widetilde{H}\right) \to S_{(\mathrm{cl})}^{\mu}\left(\mathbb{R}^{1+q}; \widetilde{H}', H'\right).$$

(ii) There is a modification of Theorem 1.4.21 which gives a separately continuous antibilinear mapping

$$C_0^{\infty}(\mathbb{R}) \times S_{(\mathrm{cl})}^{\mu}\left(\mathbb{R}^{1+q}; H, \widetilde{H}\right) \to S_{(\mathrm{cl})}^{\mu}\left(\mathbb{R}^{1+q}; \widetilde{H}', H'\right).$$

Remark 1.4.29. So far, kernel cut-off has been formulated for symbols with constant coefficients. The procedure is valid in analogous form for symbols with variable coefficients, e.g., in

$$S_{(\mathrm{cl})}^{\mu}\left(U \times \Omega \times \mathbb{R}^{1+q}; H, \widetilde{H}\right) = C^{\infty}\left(U \times \Omega \times \mathbb{R}^{1+q}; S_{(\mathrm{cl})}^{\mu}\left(\mathbb{R}^{1+q}; H, \widetilde{H}\right)\right)$$

for some open $U \subseteq \mathbb{R}$, $\Omega \subseteq \mathbb{R}^q$. In other words, the kernel cut-off operator works on covariables and is carried over for symbols that also depend on variables.

Remark 1.4.30. In kernel cut-off constructions we may have also symbols $a(\tau, \eta, \lambda)$ with an additional parameter $\lambda \in \mathbb{R}^d$, involved as an extra covariable. Then the corresponding kernel cut-off operator with respect to λ gives rise to holomorphic symbols in $\mathbb{C}^d$ with similar properties as before.

1.5 Operators on manifolds with conical exit to infinity

We now present some material on pseudo-differential operators on manifolds with conical exit to infinity. It is not our intention to develop the corresponding pseudo-differential calculus in maximal generality. The material here has auxiliary character; more details can be found in [46]. The simplest case of a manifold with conical exit is the cylinder

$$X^{\asymp} := \mathbb{R} \times X \ni (r, x),$$

where X is a smooth closed manifold, $n = \dim X$, and the superscript "$\asymp$" indicates a specific choice of charts for $r \to \pm\infty$. In the case $X = S^n$, the unit sphere in $\mathbb{R}^{n+1}$, such charts may have the form $\chi_+ : (1, +\infty) \times X \to \{\widetilde{x} \in \mathbb{R}^{n+1} : |\widetilde{x}| > 1\}$, where $\chi_+(r, x) := rx$, $x \in S^n$, and analogously

$$\chi_- : (-\infty, -1) \times X \to \{\widetilde{x} \in \mathbb{R}^{n+1} : |\widetilde{x}| > 1\},$$

where $\chi_-(r, x) := -rx$. In this way $X^{\asymp}$ is equipped with the metric of a straight cone for $r \to \pm\infty$. In the following we consider the case $r \to +\infty$. The analogue for $r \to -\infty$ can be studied in an analogous manner. If X is arbitrary we choose charts of the form

$$\chi_+ : (1, +\infty) \times U \to \left\{\widetilde{x} \in \mathbb{R}^{n+1} : |\widetilde{x}| > 1, \frac{\widetilde{x}}{|\widetilde{x}|} \in V_1\right\},$$

where U is a coordinate neighbourhood on X and $\chi_+(r,x) := r\chi_1(x)$ for a diffeomorphism $\chi_1 : U \to V_1$ onto an open subset $V_1 \subset S^n$. Analogously we have

$$\chi_- : (-\infty, -1) \times U \to \left\{ \widetilde{x} \in \mathbb{R}^{n+1} : |\widetilde{x}| > 1, \frac{\widetilde{x}}{|\widetilde{x}|} \in V_1 \right\}.$$

Let us now define

$$H^{s;0}(X^{\asymp}) := \left\{ u \in H^s_{\mathrm{loc}}(X^{\asymp}) : \left((1 - \omega_\pm)\varphi u \right) \circ \chi_\pm^{-1} \in H^s\left(\mathbb{R}^{n+1} \right) \right\},$$

for any $\varphi \in C_0^\infty(U)$ and cut-off functions $\omega_\pm$ on $r \gtrless 0$ such that $1 - \omega_\pm$ is supported in $r > 1$ and $r < -1$, respectively. Moreover, let

$$H^{s;g}(X^{\asymp}) := \langle r \rangle^{-g} H^{s;0}(X^{\asymp}), \quad s, g \in \mathbb{R}.$$

Observe that an equivalent definition is based on charts

$$\chi_+ : (1, +\infty) \times U \to \left\{ \widetilde{x} \in \mathbb{R}^{n+1} : \widetilde{x} = r\chi_0(x), r > 1, x \in U \right\}$$

for a diffeomorphism $\chi_0 : U \to V_0$ onto an open subset $V_0 \subset \mathbb{R}^n_{\widetilde{x}'}$. Here $\mathbb{R}^{n+1} = \mathbb{R}_r \times \mathbb{R}^n_{\widetilde{x}'}$, i.e., $\widetilde{x} = (r, \widetilde{x}')$, and analogously for χ.

Finally, a manifold M with conical exit to infinity is a C^∞ manifold such that there exists a smooth closed manifold X and a decomposition

$$M = M_0 \cup M_\infty \tag{1.159}$$

where M_0 is a smooth compact manifold with boundary $\partial M_0 \cong X$ and $M_\infty = (1, +\infty) \times X$ (the latter equality means the identification via a fixed diffeomorphism), where $(1, +\infty) \times X$ has a conical exit for $r \to +\infty$ as described before. Clearly, X may have different connected components. For instance, for $M := X^{\asymp}$ we can set $M_0 = [-1, +1] \times X$, and $M_\infty = (1, +\infty) \times \left(X \bigcup_{\mathrm{disj}} X \right)$, the disjoint union of two copies of X.

The space $H^{s;g}(M)$ for M in general is defined as the set of all $u \in H^s_{\mathrm{loc}}(M)$ such that

$$u\big|_{(1,\infty) \times X} \in H^{s;g}(X^{\asymp})\big|_{(1,\infty) \times X}.$$

Pseudo-differential operators on M will belong to $L^\mu_{\mathrm{cl}}(M)$ with a specific behaviour for $r \to \infty$. This can be described in local terms in $\mathbb{R}^{n+1}$. Since the features are invariant under the transition maps for different charts χ_+, the corresponding notions make sense globally on M.

We now consider the case $M = \mathbb{R}^m \ni x$, interpreted as a manifold with conical exit to infinity $|x| \to \infty$. According to (1.159) in this case we can set

$$M_0 := \{|x| \le 1\}, \quad M_\infty := \{|x| \ge 1\}.$$

Definition 1.5.1. (i) The space $S^{\mu;\nu}(\mathbb{R}^m \times \mathbb{R}^m)$ for $\mu, \nu \in \mathbb{R}$ is defined to be the set of all $a(x, \xi) \in C^\infty(\mathbb{R}^m_x \times \mathbb{R}^m_\xi)$ such that

$$\sup_{x, \xi \in \mathbb{R}^m} \langle x \rangle^{-\nu + |\alpha|} \langle \xi \rangle^{-\mu + |\beta|} |D^\alpha_x D^\beta_\xi a(x, \xi)| < \infty \qquad (1.160)$$

for all $\alpha, \beta \in \mathbb{N}^m$.

(ii) The space $S^{\mu;\nu,\nu'}(\mathbb{R}^m \times \mathbb{R}^m \times \mathbb{R}^m)$ for $\mu, \nu, \nu' \in \mathbb{R}$ is defined to be the set of all $a(x, x', \xi) \in C^\infty(\mathbb{R}^m_x \times \mathbb{R}^m_{x'} \times \mathbb{R}^m_\xi)$ such that

$$\sup_{x, x', \xi \in \mathbb{R}^m} \langle x \rangle^{-\nu + |\alpha|} \langle x' \rangle^{-\nu' + |\alpha'|} \langle \xi \rangle^{-\mu + |\beta|} |D^\alpha_x D^{\alpha'}_{x'} D^\beta_\xi a(x, x', \xi)| < \infty \quad (1.161)$$

for all $\alpha, \alpha', \beta \in \mathbb{N}^m$.

Remark 1.5.2. The expressions (1.160) for $\alpha, \beta \in \mathbb{N}^m$ form a countable system of semi-norms that turns $S^{\mu;\nu}(\mathbb{R}^m \times \mathbb{R}^m)$ into a Fréchet space. Analogously $S^{\mu;\nu,\nu'}(\mathbb{R}^m \times \mathbb{R}^m \times \mathbb{R}^m)$ is a Fréchet space with the system of semi-norms (1.161).

Remark 1.5.3. We have $a^{t,h}(x, \xi) := \langle x \rangle^h \langle \xi \rangle^t \in S^{t;h}(\mathbb{R}^m \times \mathbb{R}^m)$, $t, h \in \mathbb{R}$. The operator of multiplication by $a^{t,h}$ induces an isomorphism

$$a^{t,h} : S^{\mu;\nu}(\mathbb{R}^m \times \mathbb{R}^m) \to S^{\mu+t;\nu+h}(\mathbb{R}^m \times \mathbb{R}^m)$$

for all $\mu, \nu \in \mathbb{R}$.

Example 1.5.4. We have $S^\mu_{(cl)}(\mathbb{R}^m_\xi) \subset S^{\mu;0}(\mathbb{R}^m \times \mathbb{R}^m)$, $S^\nu_{(cl)}(\mathbb{R}^m_x) \subset S^{0;\nu}(\mathbb{R}^m \times \mathbb{R}^m)$. The classical symbol spaces are nuclear. The space

$$S^\mu_{cl}(\mathbb{R}^m_\xi) \widehat{\otimes}_\pi S^\nu_{cl}(\mathbb{R}^m_x) =: S^{\mu;\nu}_{cl_{\xi;x}}(\mathbb{R}^m \times \mathbb{R}^m)$$

consists of symbols in $S^{\mu;\nu}(\mathbb{R}^m \times \mathbb{R}^m)$ that are classical both in ξ and x (of order μ in ξ and ν in x, respectively). For example, if $\widetilde{p}(\xi) \in S^\mu_{cl}(\mathbb{R}^m)$ is arbitrary, then we have

$$p(x, \xi) := \chi(x)\widetilde{p}(|x|\xi) \in S^{\mu;\mu}_{cl_{\xi;x}}(\mathbb{R}^m \times \mathbb{R}^m)$$

for any excision function χ in $x \in \mathbb{R}^m$.

Proposition 1.5.5. *For every sequence $a_j(x, \xi) \in S^{\mu_j;\nu_j}(\mathbb{R}^m \times \mathbb{R}^m)$, $j \in \mathbb{N}$, $\mu_j \to -\infty$, $\nu_j \to -\infty$ as $j \to \infty$, there exists an $a(x, \xi) \in S^{\mu;\nu}(\mathbb{R}^m \times \mathbb{R}^m)$ for $\mu = \max\{\mu_j\}$, $\nu = \max\{\nu_j\}$, written $a \sim \sum_{j=0}^\infty a_j$, such that for every $k \in \mathbb{N}$ there exists an $N = N(k) \in \mathbb{N}$ with $a(x, \xi) - \sum_{j=0}^N a_j(x, \xi) \in S^{\mu-k;\nu-k}(\mathbb{R}^m \times \mathbb{R}^m)$, and every such $a(x, \xi)$ is unique mod $S^{-\infty;-\infty}(\mathbb{R}^m \times \mathbb{R}^m)$. Similarly, for $a_j(x, \xi) \in S^{\mu_j;\nu}(\mathbb{R}^m \times \mathbb{R}^m)$, $j \in \mathbb{N}$, $\mu_j \to -\infty$ as $j \to \infty$, $\nu \in \mathbb{R}$ fixed, there exists an $a(x, \xi) \in S^{\mu;\nu}(\mathbb{R}^m \times \mathbb{R}^m)$ for $\mu = \max\{\mu_j\}$, such that for every $k \in \mathbb{N}$ there is an $N = N(k) \in \mathbb{R}$ with $a(x, \xi) - \sum_{j=0}^N a_j(x, \xi) \in S^{\mu-k;\nu}(\mathbb{R}^m \times \mathbb{R}^m)$, and every such $a(x, \xi)$ is unique mod $S^{-\infty;\nu}(\mathbb{R}^m \times \mathbb{R}^m)$. By interchanging the roles of x and ξ we obtain an analogous result when only the second order tends to $-\infty$.*

For more material on this topic, cf. [46, Subsection 1.4.1].

Lemma 1.5.6. *There is a function* $\omega(x, x') \in S^{0;0,0}(\mathbb{R}^m \times \mathbb{R}^m \times \mathbb{R}^m)$ *(independent of the covariables ξ) such that, for a given $\rho > 0$,*

$$\omega(x, x') = 1 \text{ for } |x - x'| \leq \frac{\rho}{2}\langle x \rangle, \quad \omega(x, x') = 0 \text{ for } |x - x'| > \rho\langle x \rangle.$$

It suffices to set $\omega(x, x') := \psi(|x - x'|)/\rho\langle x \rangle)$ *for any* $\psi \in C^\infty(\mathbb{R})$ *such that* $\psi(t) = 1$ *for* $|t| \leq 1/2$ *and* $\psi(t) = 0$ *for* $|t| > 1$.

Remark 1.5.7. $a(x, \xi) \in S^{\mu;\nu}(\mathbb{R}^m \times \mathbb{R}^m)$ *implies* $\omega(x, x')a(x, \xi) \in S^{\mu;\nu,0}(\mathbb{R}^m \times \mathbb{R}^m \times \mathbb{R}^m)$ *for any* ω *as in Lemma 1.5.6.*

Let

$$\begin{aligned} \mathrm{Op}(a)u(x) &:= \iint e^{i(x-x')\xi}a(x, x', \xi)u(x')dx'd\xi \\ &= \iint e^{-ix'\xi}a(x, x' + x, \xi)u(x' + x)dx'd\xi \end{aligned} \tag{1.162}$$

for $a(x, x', \xi) \in S^{\mu;\nu,\nu'}(\mathbb{R}^m \times \mathbb{R}^m \times \mathbb{R}^m)$ and $u \in \mathcal{S}(\mathbb{R}^m)$. Then, using standard techniques of pseudo-differential operators (especially, oscillatory integral arguments), we obtain a continuous operator

$$\mathrm{Op}(a) : \mathcal{S}(\mathbb{R}^m) \to \mathcal{S}(\mathbb{R}^m). \tag{1.163}$$

Set

$$L^{\mu;\nu}(\mathbb{R}^m) := \{\mathrm{Op}(a) : a(x, \xi) \in S^{\mu;\nu}(\mathbb{R}^m \times \mathbb{R}^m)\}, \tag{1.164}$$

and

$$L^{-\infty;-\infty}(\mathbb{R}^m) := \bigcap_{\mu,\nu \in \mathbb{R}} L^{\mu;\nu}(\mathbb{R}^m). \tag{1.165}$$

Proposition 1.5.8. *The space* $L^{-\infty;-\infty}(\mathbb{R}^m)$ *coincides with the set of all integral operators of the form* $Cu(x) = \int c(x, x')u(x')dx'$ *with* $c(x, x') \in \mathcal{S}(\mathbb{R}^m \times \mathbb{R}^m)$.

Proof. Let $C \in L^{-\infty;-\infty}(\mathbb{R}^m)$; then for every $\mu, \nu \in \mathbb{R}$ we have a representation $C = \mathrm{Op}(a^{\mu;\nu})$ with some symbol $a^{\mu;\nu}(x, \xi) \in S^{\mu;\nu}(\mathbb{R}^m \times \mathbb{R}^m)$. The (uniquely determined) kernel $c(x, x')$ of C has the form

$$c(x, x') = \int e^{i(x-x')\xi}a^{\mu;\nu}(x, \xi)d\xi \tag{1.166}$$

for every μ, ν. Because of (1.165) it suffices to take $\mu, \nu \leq -N$ for some $N \in \mathbb{N}$. To verify that $c(x, x') \in \mathcal{S}(\mathbb{R}^m \times \mathbb{R}^m)$, we have to show that for every fixed semi-norm on the space $\mathcal{S}(\mathbb{R}^m \times \mathbb{R}^m) \ni \varphi(x, x')$,

$$\pi_k(\varphi) := \sup_{\substack{x,x' \in \mathbb{R}^m \\ |\alpha|+|\alpha'|+|\beta|+|\beta'| \leq k}} |x^\alpha x'^{\alpha'} D_x^\beta D_{x'}^{\beta'} \varphi(x, x')|,$$

there exists a pair of orders $\mu(k), \nu(k) \in \mathbb{R}$ such that $\pi_k(c) < \infty$ when $c(x, x')$ is represented in the form (1.166) with $\mu = \mu(k)$, $\nu = \nu(k)$. However, this is quite elementary; so we see that $c(x, x') \in \mathcal{S}(\mathbb{R}^m \times \mathbb{R}^m)$. Conversely, given $c(x, x') \in \mathcal{S}(\mathbb{R}^m \times \mathbb{R}^m)$ we set $a(x, \xi) := e^{-ix\xi} \int c(x, x') e^{ix'\xi} dx'$. Since $c(x, x')$ is a Schwartz function with respect to x', the function $d(x, \xi) = \int c(x, x') e^{ix'\xi} dx'$ belongs to $\mathcal{S}(\mathbb{R}^m \times \mathbb{R}^m)$, and then also $e^{-ix\xi} d(x, \xi) \in \mathcal{S}(\mathbb{R}^m \times \mathbb{R}^m)$. We finally have $u(x') = \int e^{ix'\xi} \hat{u}(\xi) d\xi$, and hence

$$Cu(x) = \int C(e^{ix'\xi}) \hat{u}(\xi) d\xi = \int e^{ix\xi} e^{-ix\xi} C(e^{ix'\xi}) \hat{u}(\xi) d\xi = \int e^{ix\xi} a(x, \xi) \hat{u}(\xi) d\xi,$$

i.e., $a(x, \xi)$ is a symbol of C, and it follows that $C = \mathrm{Op}(a)$. $\qquad\square$

Theorem 1.5.9. (i) *For every* $a(x, x', \xi) \in S^{\mu;\nu,\nu'}(\mathbb{R}^{2m} \times \mathbb{R}^m; H, \widetilde{H})$ *we have*

$$\mathrm{Op}(a) \in L^{\mu;\nu+\nu'}(\mathbb{R}^m).$$

(ii) *For every* $a(x, x', \xi) \in S^{\mu;\nu,\nu'}(\mathbb{R}^{2m} \times \mathbb{R}^m)$ *there are unique left and right symbols*

$$a_{\mathrm{L}}(x, \xi) \in S^{\mu;\nu+\nu'}(\mathbb{R}^m \times \mathbb{R}^m), \quad a_{\mathrm{R}}(x', \xi) \in S^{\mu;\nu+\nu'}(\mathbb{R}^m \times \mathbb{R}^m),$$

such that $\mathrm{Op}(a) = \mathrm{Op}(a_{\mathrm{L}}) = \mathrm{Op}(a_{\mathrm{R}})$, *where*

$$a_{\mathrm{L}}(x, \xi) = \iint e^{-ix'\eta} a(x, x' + x, \xi + \eta) dx' d\eta \qquad (1.167)$$

and

$$a_{\mathrm{R}}(x', \xi) = \iint e^{-ix\eta} a(x' + x, x', \xi - \eta) dx\, d\eta. \qquad (1.168)$$

The mappings $a \mapsto a_{\mathrm{L}}, a \mapsto a_{\mathrm{R}}$ *are continuous acting as*

$$S^{\mu;\nu,\nu'}(\mathbb{R}^{2m} \times \mathbb{R}^m) \to S^{\mu;\nu+\nu'}(\mathbb{R}^m \times \mathbb{R}^m).$$

Moreover, we have asymptotic expansions

$$a_{\mathrm{L}}(x, \xi) \sim \sum_{\alpha \in \mathbb{N}^m} \frac{1}{\alpha!} D_{x'}^\alpha \partial_\xi^\alpha a(x, x', \xi)\big|_{x'=x},$$

$$a_{\mathrm{R}}(x', \xi) \sim \sum_{\alpha \in \mathbb{N}^m} \frac{1}{\alpha!} (-1)^{|\alpha|} D_x^\alpha \partial_\xi^\alpha a(x, x', \xi)\big|_{x=x'}.$$

(iii) *We have* $L^{\mu;\nu}(\mathbb{R}^m) = \{\mathrm{Op}(a) : a(x, x', \xi) \in S^{\mu;\nu,0}(\mathbb{R}^m \times \mathbb{R}^m \times \mathbb{R}^m)\}$. *The map*

$$\mathrm{Op} : S^{\mu;\nu}(\mathbb{R}^m \times \mathbb{R}^m) \to L^{\mu;\nu}(\mathbb{R}^m) \qquad (1.169)$$

is an isomorphism.

Remark 1.5.10. Using Remark 1.5.2 and Theorem 1.5.9 (iii) we obtain a Fréchet space structure on $L^{\mu;\nu}(\mathbb{R}^m)$.

According to the notation introduced above, we have

$$H^{s;g}(\mathbb{R}^m) = \langle x \rangle^{-g} H^s(\mathbb{R}^m), \quad s, g \in \mathbb{R}, \tag{1.170}$$

with $H^s(\mathbb{R}^m)$ being the standard Sobolev space in $\mathbb{R}^m$ of smoothness $s \in \mathbb{R}$.

Theorem 1.5.11. *An $A \in L^{\mu;\nu}(\mathbb{R}^m)$ induces continuous operators*

$$A : H^{s;g}(\mathbb{R}^m) \to H^{s-\mu;g-\nu}(\mathbb{R}^m)$$

for all $s, g \in \mathbb{R}$.

Theorem 1.5.12. *If $A \in L^{\mu;\nu}(\mathbb{R}^m)$ and $B \in L^{\rho;\sigma}(\mathbb{R}^m)$, then $AB \in L^{\mu+\rho;\nu+\sigma}(\mathbb{R}^m)$. If A and B are given in the form*

$$A = \mathrm{Op}(a), \quad B = \mathrm{Op}(b) \tag{1.171}$$

for symbols $a(x,\xi) \in S^{\mu;\nu}(\mathbb{R}^m)$, $b(x,\xi) \in S^{\rho;\sigma}(\mathbb{R}^m)$, then there is a unique (so-called Leibniz product) $(a\#b)(x,\xi) \in S^{\mu+\rho;\nu+\sigma}(\mathbb{R}^m \times \mathbb{R}^m)$ such that $AB = \mathrm{Op}(a\#b)$, where

$$(a\#b)(x,\xi) = \iint e^{-iy\eta} a(x,\xi+\eta) b(x+y,\xi) dy \mathchar'26\mkern-12mu d\eta, \tag{1.172}$$

interpreted as an oscillatory integral. Moreover,

$$(a\#b)(x,\xi) \sim \sum_{\alpha \in \mathbb{N}^m} \frac{1}{\alpha!} \left(\partial_\xi^\alpha a(x,\xi) \right) D_x^\alpha b(x,\xi).$$

Theorem 1.5.13. *Let $A \in L^{\mu;\nu}(\mathbb{R}^m)$, and define the formal adjoint A^* by*

$$(Au, v)_{L^2(\mathbb{R}^m)} = (u, A^* v)_{L^2(\mathbb{R}^m)}$$

for $u, v \in \mathcal{S}(\mathbb{R}^m)$. Then $A^* \in L^{\mu;\nu}(\mathbb{R}^m)$. If A is given in the form $A = \mathrm{Op}(a)$ for a symbol $a(x,\xi) \in S^{\mu;\nu}(\mathbb{R}^m)$, then there is a unique $a^*(x,\xi) \in S^{\mu;\nu}(\mathbb{R}^m)$ such that $A^* = \mathrm{Op}(a^*)$, where

$$a^*(x,\xi) = \iint e^{-iy\eta} \overline{a}(x+y,\xi+\eta) dy \mathchar'26\mkern-12mu d\eta, \tag{1.173}$$

interpreted as an oscillatory integral. Moreover,

$$a^*(x,\xi) \sim \sum_{\alpha \in \mathbb{N}^m} \frac{1}{\alpha!} \partial_\xi^\alpha D_x^\alpha \overline{a}(x,\xi).$$

Remark 1.5.14. The operator $A^{t,h} := \mathrm{Op}(a^{t,h})$ with $a^{t,h}(x,\xi) = \langle x \rangle^h \langle \xi \rangle^t \in S^{t;h}(\mathbb{R}^m \times \mathbb{R}^m)$, $t, h \in \mathbb{R}$, induces isomorphisms $A^{t,h} : H^{s;g}(\mathbb{R}^m) \to H^{s-t;g-h}(\mathbb{R}^m)$ for all $s, g \in \mathbb{R}$, and we have $(A^{t,h})^{-1} = \mathrm{Op}(p^{-t,-h})$ for some $p^{-t,-h}(x,\xi) \in S^{-t;-h}(\mathbb{R}^m \times \mathbb{R}^m)$.

In fact, the operator $\mathrm{Op}(a^{t,h}) = \mathcal{M}_{\langle x \rangle^h}\mathrm{Op}(\langle \xi \rangle^t)$, first acting on $\mathcal{S}(\mathbb{R}^m)$, obviously induces an isomorphism $A := \mathrm{Op}(a^{t,h}) : \mathcal{S}(\mathbb{R}^m) \to \mathcal{S}(\mathbb{R}^m)$, with the inverse $\left(\mathrm{Op}(\langle \xi \rangle^t)\right)^{-1}(\mathcal{M}_{\langle x \rangle^h})^{-1} = \mathrm{Op}(\langle \xi \rangle^{-t})\mathcal{M}_{\langle x \rangle^{-h}} = \mathrm{Op}(\langle \xi \rangle^{-t}\#\langle x \rangle^{-h})$ for a unique symbol $p^{-t;-h}(x,\xi) := \langle \xi \rangle^{-t}\#\langle x \rangle^{-h}$ in $S^{-t;-h}(\mathbb{R}^m \times \mathbb{R}^m)$, see Theorem 1.5.12. In other words, $\mathrm{Op}(a^{t,h})^{-1} = \mathrm{Op}(p^{-t,-h}) =: P$ on $\mathcal{S}(\mathbb{R}^m)$. The operators

$$A : H^{s,-g}(\mathbb{R}^m) \to H^{s-t;g-h}(\mathbb{R}^m), \quad P : H^{s-t;g-h}(\mathbb{R}^m) \to H^{s;g}(\mathbb{R}^m) \qquad (1.174)$$

are continuous for all $s, g \in \mathbb{R}$, see Theorem 1.5.11. Then, since

$$PA = \mathrm{Op}(p^{-t,-h}\#a^{t,h}) = 1,$$

the operators PA and AP extend from $\mathcal{S}(\mathbb{R}^m)$ to the identity maps on the spaces $H^{s;g}(\mathbb{R}^m)$ and $H^{s-t;g-h}(\mathbb{R}^m)$, respectively; thus, (1.174) are isomorphisms.

Theorem 1.5.15. *We have a canonical continuous embedding*

$$\iota : H^{s';g'}(\mathbb{R}^m) \hookrightarrow H^{s;g}(\mathbb{R}^m) \qquad (1.175)$$

for any $s' \geq s$, $g' \geq g$, and ι is compact when $s' > s$, $g' > g$.

Let us now return to classical symbols in x, ξ. First the space $S^\mu(\mathbb{R}^m_\xi)$ can be identified with the set of all elements of $S^{\mu;0}(\mathbb{R}^m \times \mathbb{R}^m)$ that are independent of x. Recall that an $a(\xi) \in S^\mu(\mathbb{R}^m)$ is said to be classical if there is a sequence of functions $a_{(\mu-j)}(\xi) \in C^\infty(\mathbb{R}^m \setminus \{0\})$, $j \in \mathbb{N}$, with the property $a_{(\mu-j)}(\lambda\xi) = \lambda^{\mu-j}a(\xi)$ for all $\lambda \in \mathbb{R}_+$, $\xi \in \mathbb{R}^m \setminus \{0\}$, such that

$$a(\xi) - \sum_{j=0}^{N} \chi(\xi)a_{(\mu-j)}(\xi) \in S^{\mu-(N+1)}(\mathbb{R}^m) \qquad (1.176)$$

for every $N \in \mathbb{N}$ and any excision function χ. As before, $S^\mu_{\mathrm{cl}}(\mathbb{R}^m)$ denotes the space of classical symbols in that sense. The homogeneous components $a_{(\mu-j)}$, $j \in \mathbb{N}$, are uniquely determined by a. This gives us a sequence of linear operators

$$S^\mu_{\mathrm{cl}}(\mathbb{R}^m) \to C^\infty(S^{m-1}), \quad a \mapsto a_{(\mu-j)}|_{S^{m-1}}, \; j \in \mathbb{N}. \qquad (1.177)$$

Moreover, (1.176) defines a sequence of linear operators

$$S^\mu_{\mathrm{cl}}(\mathbb{R}^m) \to S^{\mu-(N+1)}(\mathbb{R}^m), \quad N \in \mathbb{N}. \qquad (1.178)$$

We endow $S^\mu_{\mathrm{cl}}(\mathbb{R}^m)$ with the Fréchet topology of the projective limit with respect to the mappings (1.177), (1.178). This turns $S^\mu_{\mathrm{cl}}(\mathbb{R}^m)$ into a nuclear Fréchet space.

Let $S^{(\mu)}(\mathbb{R}^m \setminus \{0\})$ denote the space of all $a_{(\mu)} \in C^\infty(\mathbb{R}^m \setminus \{0\})$ satisfying the homogeneity condition $a_{(\mu)}(\lambda\xi) = \lambda^\mu a_{(\mu)}(\xi)$ for all $\lambda \in \mathbb{R}_+$, $\xi \in \mathbb{R}^m \setminus \{0\}$. We now consider the spaces $S^\mu_{\mathrm{cl}}(\mathbb{R}^m_\xi)$ and $S^\nu_{\mathrm{cl}}(\mathbb{R}^m_x)$ in the variables ξ and x, respectively, and

$$S^{\mu;\nu}_{\mathrm{cl}_{(\xi;x)}}(\mathbb{R}^m \times \mathbb{R}^m) = S^\mu_{\mathrm{cl}}(\mathbb{R}^m_\xi) \,\widehat{\otimes}_\pi\, S^\nu_{\mathrm{cl}}(\mathbb{R}^m_x),$$

see Example 1.5.4. We then set

$$L^{\mu;\nu}_{\mathrm{cl}}(\mathbb{R}^m) := \left\{ \mathrm{Op}(a) : a(x,\xi) \in S^{\mu;\nu}_{\mathrm{cl}_{(\xi;x)}}(\mathbb{R}^m \times \mathbb{R}^m) \right\}. \tag{1.179}$$

If a consideration is valid both for classical and general symbols or operators, we write subscripts "$(\mathrm{cl}_{(\xi;x)})$" and "(cl)", respectively.

Recall that $E \,\widehat{\otimes}_\pi\, F$ denotes the projective tensor product of the Fréchet spaces E and F. Then, if G is another Fréchet space and $\sigma : E \to G$ a linear continuous operator, we also obtain a continuous operator

$$\sigma \otimes \mathrm{id}_F : E \,\widehat{\otimes}_\pi\, F \to G \,\widehat{\otimes}_\pi\, F.$$

A similar relation holds with respect to the second factor. In particular, let

$$\sigma_\psi : S^\mu_{\mathrm{cl}}(\mathbb{R}^m_\xi) \to S^{(\mu)}(\mathbb{R}^m_\xi \setminus \{0\}) \;\text{ and }\; \sigma_{\mathrm{e}} : S^\nu_{\mathrm{cl}}(\mathbb{R}^m_x) \to S^{(\nu)}(\mathbb{R}^m_x \setminus \{0\})$$

denote the operators that map a symbol to its homogeneous principal component of order μ and ν in the corresponding variables ξ and x, respectively. Then we obtain operators

$$\sigma_\psi : S^{\mu;\nu}_{\mathrm{cl}_{(\xi;x)}}(\mathbb{R}^m \times \mathbb{R}^m) \to S^{(\mu)}(\mathbb{R}^m \setminus \{0\}) \,\widehat{\otimes}_\pi\, S^\nu_{\mathrm{cl}}(\mathbb{R}^m), \tag{1.180}$$

$$\sigma_{\mathrm{e}} : S^{\mu;\nu}_{\mathrm{cl}_{(\xi;x)}}(\mathbb{R}^m \times \mathbb{R}^m) \to S^\mu_{\mathrm{cl}}(\mathbb{R}^m) \,\widehat{\otimes}_\pi\, S^{(\nu)}(\mathbb{R}^m \setminus \{0\}). \tag{1.181}$$

For simplicity, we omitted corresponding identity maps for the other factors. Now we can apply σ_{e} in (1.181) with respect to x and σ_ψ in (1.181) with respect to ξ. It is well known that the resulting maps coincide and define a map

$$\sigma_{\psi,\mathrm{e}} := \sigma_\psi \otimes \sigma_{\mathrm{e}} : S^{\mu;\nu}_{\mathrm{cl}_{(\xi;x)}}(\mathbb{R}^m \times \mathbb{R}^m) \to S^{(\mu)}(\mathbb{R}^m \setminus \{0\}) \,\widehat{\otimes}_\pi\, S^{(\nu)}(\mathbb{R}^m \setminus \{0\}).$$

We call $\sigma(a) := (\sigma_\psi(a),\, \sigma_{\mathrm{e}}(a),\, \sigma_{\psi,\mathrm{e}}(a))$ the principal symbol of the classical symbol $a(x,\xi) \in S^{\mu;\nu}_{\mathrm{cl}_{(\xi;x)}}(\mathbb{R}^m \times \mathbb{R}^m)$, and we set

$$\sigma_{\mathrm{E}}(a) := (\sigma_{\mathrm{e}}(a), \sigma_{\psi,\mathrm{e}}(a)).$$

Observe that $a \in S^{\mu;\nu}_{\mathrm{cl}_{(\xi;x)}}(\mathbb{R}^m \times \mathbb{R}^m))$, $b \in S^{\rho;\sigma}_{\mathrm{cl}_{(\xi;x)}}(\mathbb{R}^m \times \mathbb{R}^m)$ implies $ab \in S^{\mu+\rho;\nu+\sigma}_{\mathrm{cl}_{(\xi;x)}}(\mathbb{R}^m \times \mathbb{R}^m)$ and

$$\sigma_\psi(ab) = \sigma_\psi(a)\sigma_\psi(b), \quad \sigma_{\mathrm{e}}(ab) = \sigma_{\mathrm{e}}(a)\sigma_{\mathrm{e}}(b), \quad \sigma_{\psi,\mathrm{e}}(ab) = \sigma_{\psi,\mathrm{e}}(a)\sigma_{\psi,\mathrm{e}}(b). \tag{1.182}$$

For $A = \mathrm{Op}(a)$ we also write $\sigma(A) := \sigma(a)$, $\sigma_\psi(A) := \sigma_\psi(a)$, $\sigma_{\mathrm{e}}(A) := \sigma_{\mathrm{e}}(a)$, $\sigma_{\psi,\mathrm{e}}(A) := \sigma_{\psi,\mathrm{e}}(a)$, and

$$\sigma_{\mathrm{E}}(A) := (\sigma_{\mathrm{e}}(A), \sigma_{\psi,\mathrm{e}}(A)). \tag{1.183}$$

Remark 1.5.16. $a(x,\xi) \in S^{\mu;\nu}_{\mathrm{cl}(\xi;x)}(\mathbb{R}^m \times \mathbb{R}^m)$ and $\sigma_\psi(a) = 0$, $\sigma_{\mathrm{e}}(a) = 0$, $\sigma_{\psi,\mathrm{e}}(a) = 0$ imply $a(x,\xi) \in S^{\mu-1;\nu-1}_{\mathrm{cl}(\xi,x)}(\mathbb{R}^m \times \mathbb{R}^m)$. Concerning a proof, see [44, Proposition 1.4.27].

Remark 1.5.17.   (i) Theorem 1.5.9 applied to classical symbols

$$a(x,x',\xi) \in S^{\mu;\nu,\nu'}_{\mathrm{cl}(\xi;x,x')}(\mathbb{R}^m \times \mathbb{R}^m \times \mathbb{R}^m)$$

gives rise to $a_{\mathrm{L}}(x,\xi) \in S^{\mu;\nu+\nu'}_{\mathrm{cl}(\xi,x)}(\mathbb{R}^m \times \mathbb{R}^m)$ and $a_{\mathrm{R}}(x',\xi) \in S^{\mu}_{\mathrm{cl}(\xi,x')}(\mathbb{R}^m \times \mathbb{R}^m)$, respectively.

(ii) Theorem 1.5.12 holds in analogous form for $A \in L^{\mu;\nu}_{\mathrm{cl}}(\mathbb{R}^m)$, $B \in L^{\rho;\sigma}_{\mathrm{cl}}(\mathbb{R}^m)$ and yields $AB \in L^{\mu+\rho;\nu+\sigma}_{\mathrm{cl}}(\mathbb{R}^m)$. In the representation of A and B of the form (1.171) with classical symbols a and b, respectively, $a\#b$ is also classical, and we have

$$\sigma_\psi(AB) = \sigma_\psi(A)\sigma_\psi(B), \quad \sigma_{\mathrm{e}}(AB) = \sigma_{\mathrm{e}}(A)\sigma_{\mathrm{e}}(B),$$
$$\sigma_{\psi,\mathrm{e}}(AB) = \sigma_{\psi,\mathrm{e}}(A)\sigma_{\psi,\mathrm{e}}(B).$$

(iii) Theorem 1.5.13 holds in analogous form for $A \in L^{\mu;\nu}_{\mathrm{cl}}(\mathbb{R}^m)$; we then have $A^* \in L^{\mu;\nu}_{\mathrm{cl}}(\mathbb{R}^m)$, and

$$\sigma_\psi(A^*)(x,\xi) = \overline{\sigma_\psi(A)(x,\xi)}, \quad \sigma_{\mathrm{e}}(A^*)(x,\xi) = \overline{\sigma_{\mathrm{e}}(A)(x,\xi)},$$
$$\sigma_{\psi,\mathrm{e}}(A^*)(x,\xi) = \overline{\sigma_{\psi,\mathrm{e}}(A)(x,\xi)}.$$

Definition 1.5.18. An operator $A \in L^{\mu;\nu}(\mathbb{R}^m)$ is called elliptic, if for the symbol $a(x,\xi) \in S^{\mu;\nu}(\mathbb{R}^m \times \mathbb{R}^m)$ in the representation $A = \mathrm{Op}(a)$ (cf. Theorem 1.5.8 (iii)) there is a $p(x,\xi) \in S^{-\mu;-\nu}(\mathbb{R}^m \times \mathbb{R}^m)$ such that

$$1 - p(x,\xi)a(x,\xi) \in S^{-1;-1}(\mathbb{R}^m \times \mathbb{R}^m). \tag{1.184}$$

Observe that $a(x,\xi) \in S^{\mu;\nu}(\mathbb{R}^m \times \mathbb{R}^m)$ is elliptic if and only if there are constants $R > 0$, $c > 0$ such that $a(x,\xi) \neq 0$ for all $(x,\xi) \in \mathbb{R}^m \times \mathbb{R}^m$, $|x| + |\xi| \geq R$, and $|a^{-1}(x,\xi)| \leq c\langle\xi\rangle^{-\mu}\langle x\rangle^{-\nu}$ for all $(x,\xi) \in \mathbb{R}^m \times \mathbb{R}^m$, $|x| + |\xi| \geq R$.

Remark 1.5.19. An operator $A \in L^{\mu;\nu}_{\mathrm{cl}}(\mathbb{R}^m)$ is elliptic if and only if

$$\sigma_\psi(A)(x,\xi) \neq 0 \text{ for all } (x,\xi) \in \mathbb{R}^m \times (\mathbb{R}^m \setminus \{0\}),$$
$$\sigma_{\mathrm{e}}(A)(x,\xi) \neq 0 \text{ for all } (x,\xi) \in (\mathbb{R}^m \setminus \{0\}) \times \mathbb{R}^m,$$
$$\sigma_{\psi,\mathrm{e}}(A)(x,\xi) \neq 0 \text{ for all } (x,\xi) \in (\mathbb{R}^m \setminus \{0\}) \times (\mathbb{R}^m \setminus \{0\}).$$

A proof is given in [46, Proposition 1.4.37]. These conditions are independent of each other. For instance, $A = \mathrm{Op}(a)$ for $a(x,\xi) = \langle\xi\rangle^\mu + \langle x\rangle^\nu$ satisfies the first two conditions, but the third one is violated.

Theorem 1.5.20. *For an operator $A \in L^{\mu;\nu}(\mathbb{R}^m)$ the following conditions are equivalent:*

(i) *the operator*

$$A : H^{s;g}(\mathbb{R}^m) \to H^{s-\mu;g-\nu}(\mathbb{R}^m) \tag{1.185}$$

is Fredholm for some $s = s_0,\ g = g_0 \in \mathbb{R}$;

(ii) *the operator A is elliptic.*

The Fredholm property (1.185) of $A \in L^{\mu;\nu}(\mathbb{R}^m)$ for $s_0, g_0 \in \mathbb{R}$ implies the same for all $s, g \in \mathbb{R}$.

Theorem 1.5.21.   (i) *An elliptic operator $A \in L^{\mu;\nu}_{(\mathrm{cl})}(\mathbb{R}^m)$ has a parametrix $P \in L^{-\mu;-\nu}_{(\mathrm{cl})}(\mathbb{R}^m)$, i.e., the remainders*

$$C := I - PA, \quad \widetilde{C} := I - AP \tag{1.186}$$

belong to $L^{-\infty;-\infty}(\mathbb{R}^m)$; in the classical case we have

$$\sigma_\psi(P) = \sigma_\psi(A)^{-1}, \quad \sigma_{\mathrm{e}}(P) = \sigma_{\mathrm{e}}(A)^{-1}, \quad \sigma_{\psi,\mathrm{e}}(P) = \sigma_{\psi,\mathrm{e}}(A)^{-1}.$$

The parametrix P can be chosen in such a way that $C, \widetilde{C}$ are projections to finite-dimensional subspaces $V, \widetilde{V}$ of $\mathcal{S}(\mathbb{R}^m)$, and, if A is realised as (1.185), we have

$$V = \ker A, \quad \widetilde{V} + \operatorname{im} A = H^{s-\mu;g-\nu}(\mathbb{R}^m) \tag{1.187}$$

for all $s, g \in \mathbb{R}$;

(ii) *Let $A \in L^{\mu;\nu}(\mathbb{R}^m)$ be elliptic; then*

$$Au = f \in H^{s-\mu;g-\nu}(\mathbb{R}^m), \quad u \in \mathcal{H}^{-\infty;-\infty}(\mathbb{R}^m) \tag{1.188}$$

implies $u \in H^{s;g}(\mathbb{R}^m)$ for every $s, g \in \mathbb{R}$.

Theorem 1.5.22. *Assume that $A \in L^{\mu;\nu}_{(\mathrm{cl})}(\mathbb{R}^m)$ induces an isomorphism*

$$A : H^{s;g}(\mathbb{R}^m) \to H^{s-\mu;g-\nu}(\mathbb{R}^m) \tag{1.189}$$

for some $s = s_0,\ g = g_0 \in \mathbb{R}$. Then A is elliptic and induces isomorphisms (1.189) for all $s, g \in \mathbb{R}$, and we have $A^{-1} \in L^{-\mu;-\nu}_{(\mathrm{cl})}(\mathbb{R}^m)$.

Let M be a C^∞ manifold, $m = \dim M$, equipped with a free $\mathbb{R}_+$-action $m \to \delta_\lambda m$, $\lambda \in \mathbb{R}_+$, $m \in M$. Then the orbit space X is a C^∞ manifold, and we have a diffeomorphism

$$e : M \to \mathbb{R}_+ \times X, \quad e(m) =: (r, x), \tag{1.190}$$

such that $e(\delta_\lambda m) = (\lambda r, x)$ for all $\lambda \in \mathbb{R}_+$, $m \in M$. We will say that M is endowed with the structure of an infinite cone, if there is fixed an open covering $\mathfrak{V}$ of M

by neighbourhoods of the form $V = e^{-1}(\mathbb{R}_+ \times U)$, where U runs over an open covering $\mathfrak{U}$ of X by coordinate neighbourhoods, together with a system of charts

$$\chi : V \to \Gamma \qquad (1.191)$$

such that $\chi(m) = (r, r\chi_1(x))$, $m \in V$, for charts $\chi_1 : U \to B := \{y \in \mathbb{R}^n : |y| < 1\}$, where $\Gamma = \{(r, \widetilde{y}) \in \mathbb{R}^{1+n} : r > 0, \widetilde{y}/r \in B\}$. The manifold X will also be called the base of the cone.

In our notation we mainly focus on what happens at infinity, i.e., over $e^{-1}((R, \infty) \times X)$ for any $R > 0$. We assume in this context that X is compact.

Definition 1.5.23. A C^∞ manifold M is said to be a manifold with conical exit (to infinity), if M contains a submanifold M_∞ endowed with the structure of an infinite cone such that when $e : M_\infty \to \mathbb{R}_+ \times X$ is a map as in (1.190), the set

$$M_0 := M \setminus e^{-1}((R, \infty) \times X) \qquad (1.192)$$

is a C^∞ manifold with boundary $\partial M_0 \cong X$, for some $R > 0$.

On $M_{[R,\infty)} := M_\infty \setminus e^{-1}((0, R) \times X)$ for some $R \geq 1$ we define dilations $\delta_\lambda : M_{[R,\infty)} \to M_{[R,\infty)}$ for $\lambda \geq 1$ by $\delta_\lambda m := e^{-1}(\lambda r, x)$ for $e(m) = (r, x)$. By definition, a manifold M with conical exit can be written as a union

$$M = M_0 \cup M_\infty.$$

Let us fix a partition of unity

$$\{\varphi_0, \varphi_\infty\}$$

subordinate to $\{\text{int } M_0, M_\infty\}$, $\varphi_0 \in C_0^\infty(\text{int } M_0)$, $\varphi_\infty \in C^\infty(M_\infty)$.

Remark 1.5.24. Manifolds with conical exit form a category in a natural way. A morphism $\chi : M \to \widetilde{M}$ in that category is a differentiable map such that there is an $R \geq 1$, for which $\chi(\delta_\lambda m) = \delta_\lambda \chi(m)$ for all $m \in M_\infty \setminus e^{-1}((0, R) \times X)$ and $\lambda \geq 1$.

In the following notation we assume, for simplicity, that the base X of the cone is a compact closed C^∞ manifold.

Let $H^{s;g}(M)$ for $s, g \in \mathbb{R}$ denote the space of all $u \in H^s_{\text{loc}}(M)$ such that

$$((\varphi_\infty u) \circ e^{-1})(r, x) \in H^{s;g}_{\text{cone}}(X^\wedge)$$

for $H^{s;g}_{\text{cone}}(X^\wedge) := \langle r \rangle^{-g} H^s_{\text{cone}}(X^\wedge)$, cf. the formula (1.170). We can set

$$\|u\|_{H^{s;g}(M)} = \left\{ \|\varphi_0 u\|^2_{H^s(\text{int } M_0)} + \|(\varphi_\infty u) \circ e^{-1}\|^2_{H^{s;g}_{\text{cone}}(X^\wedge)} \right\}^{\frac{1}{2}}, \qquad (1.193)$$

where $H^s(\text{int } \cdot)$ means the standard Sobolev space of smoothness s for a compact C^∞ manifold with boundary (indicated by $\cdot$). The space $H^{s;g}(M)$ can also be equipped with a Hilbert space scalar product such that the associated norm is equivalent to (1.193).

Setting $\mathcal{S}(M) := \varprojlim_{k \in \mathbb{N}} H^{k;k}(M)$, we obtain an analogue of the Schwartz space on a manifold M with conical exit.

This space is a nuclear Fréchet space. Let us set

$$\mathcal{S}(M \times M) := \mathcal{S}(M) \,\widehat{\otimes}_\pi\, \mathcal{S}(M).$$

On M we fix a Riemannian metric dm which on $e^{-1}((R, \infty) \times X)$ is the pull-back of the cone metric $dr^2 + r^2 g_X$ under the mapping e for some Riemannian metric g_X on X. Observe that in $M := \mathbb{R}^m$ we can take the standard Euclidean metric $d\widetilde{x}$ in $\mathbb{R}^m$. This allows us to endow the space $H^{0;0}(M)$ with the scalar product

$$(u, v)_M := \int_M u(m)\overline{v(m)}dm.$$

Moreover, with a kernel $c(m, m') \in \mathcal{S}(M \times M)$ we associate an integral operator $Cu(m) := \int_M c(m, m')u(m')dm'$. Let

$$L^{-\infty;-\infty}(M) \tag{1.194}$$

denote the space of those operators.

For further use we fix a locally finite open covering by coordinate neighbourhoods

$$U_\iota, \ \iota \in I, \quad V_\kappa, \ \kappa \in K, \tag{1.195}$$

and we assume $M_0 \subset \bigcup_{\iota \in I} U_\iota$, $e^{-1}((R_1 \times \infty) \times X) \cap U_\iota = \emptyset$, $\iota \in I$, for some $R_1 > R_0$, and that $e(V_\kappa)$, $\kappa \in K$, is of the form $(R_0 \times \infty) \times V_\kappa'$ for a coordinate neighbourhood V_κ' on X and some $0 < R_0 < R$.

A function $\varphi \in C^\infty(M)$ is called homogeneous of order 0 far from M_0 if there exists an $R_1 \geq 1$ such that

$$(\varphi \circ e^{-1})(\lambda r, m) = (\varphi \circ e^{-1})(r, m)$$

for all $r \geq R_1$ and all $\lambda \geq 1$. In a similar vein we define homogeneity of order 0 of a function in $C^\infty(V_\kappa)$ far from M_0.

We fix a partition of unity

$$\Phi_{0,\iota}, \ \iota \in I, \quad \Phi_{\infty,\kappa}, \ \kappa \in K, \tag{1.196}$$

by functions $\Phi_{0,\iota} \in C_0^\infty(U_\iota)$, $\Phi_{\infty,\kappa} \in C^\infty(V_\kappa)$, such that $\Phi_{\infty,\kappa}$ is homogeneous of order zero far from M_0. In addition we choose functions

$$\Psi_{0,\iota}, \ \iota \in I, \quad \Psi_{\infty,\kappa}, \ \kappa \in K, \tag{1.197}$$

$\Psi_{0,\iota} \in C_0^\infty(U_\iota), \Psi_{\infty,\kappa} \in C^\infty(V_\kappa)$, such that also $\Psi_{\infty,\kappa}$ is homogeneous of order zero far from M_0, and $\Psi_{0,\iota} \equiv 1$ on supp $\Phi_{0,\iota}$, $\Psi_{\infty,\kappa} \equiv 1$ on supp $\Phi_{\infty,\kappa}$ for all ι, κ.

Choose a system of charts

$$\chi_{0,\iota} : U_\iota \to \mathbb{R}^n, \ \iota \in I, \quad \chi_{\infty,\kappa} : V_\kappa \to \Gamma, \ \kappa \in K, \qquad (1.198)$$

such that $\chi_{\infty,\kappa}(\delta_\lambda m) = \lambda r \chi_{\kappa,1}(x)$ for all $\lambda \in \mathbb{R}_+$ and $m \in V_\kappa$, cf. the formula (1.191) and the subsequent notation.

Definition 1.5.25. The space $L^{\mu;\nu}_{(\mathrm{cl})}(M)$ of (classical) pseudo-differential operators of order $\mu \in \mathbb{R}$ and exit order $\nu \in \mathbb{R}$ is defined to be the subspace of all $A = A_0 + C \in L^\mu_{(\mathrm{cl})}(M)$ for $C \in L^{-\infty;-\infty}(M)$ and

$$(\chi_{0,\iota})_* \Phi_{0,\iota} A_0 \Psi_{0,\iota} \in L^\mu_{(\mathrm{cl})}(\mathbb{R}^n), \ \iota \in I, \quad (\chi_{\infty,\kappa})_* \Phi_{\infty,\kappa} A_0 \Psi_{\infty,\kappa} \in L^{\mu;\nu}_{(\mathrm{cl})}(\mathbb{R}^n), \ \kappa \in K.$$

Next we give a list of generalisations of the above calculus of exit pseudo-differential operators in $\mathbb{R}^m$ to the case of manifolds M with conical exit to infinity. Recall that $X = \partial M_0$ is assumed to be compact.

Theorem 1.5.26. *An operator* $A \in L^{\mu;\nu}_{(\mathrm{cl})}(M)$ *induces continuous operators*

$$A : \mathcal{S}(M) \to \mathcal{S}(M)$$

and

$$A : H^{s;g}(M) \to H^{s-\mu;g-\nu}(M)$$

and for any $s, g \in \mathbb{R}$.

Theorem 1.5.27. *We have canonical continuous embeddings*

$$\iota : H^{s';g'}(M) \hookrightarrow H^{s;g}(M)$$

for any $s' \geq s$, $g' \geq g$; *moreover,* ι *is compact when* $s' > s$, $g' > g$.

Theorem 1.5.28. $A \in L^{\mu;\nu}_{(\mathrm{cl})}(M)$, $B \in L^{\rho;\sigma}_{(\mathrm{cl})}(M)$ *implies* $AB \in L^{\mu+\rho;\nu+\sigma}_{(\mathrm{cl})}(M)$.

Theorem 1.5.29. *Let* $A \in L^{\mu;\nu}_{(\mathrm{cl})}(M)$ *and define the formal adjoint* A^* *by*

$$(Au, v)_{H^{0,0}(M)} = (u, A^* v)_{H^{0,0}(M)} \quad \text{for all } u, v \in C_0^\infty(M).$$

Then $A^* \in L^{\mu;\nu}(M)$.

Definition 1.5.30. An operator $A \in L^{\mu;\nu}_{(\mathrm{cl})}(M)$ is called elliptic if it is elliptic in the standard sense as a pseudo-differential operator on the open manifold M and if, in addition, for any chart $\chi : V \to \Gamma$ like (1.191) for the push-forward $\chi_*(A|_V) \in L^\mu_{(\mathrm{cl})}(\Gamma)$ and every pair of functions $\varphi_\infty, \psi_\infty \in C^\infty(\Gamma)$, homogeneous of order zero far from the origin, and vanishing close to the origin and near $\partial\Gamma$, there is an elliptic operator $\widetilde{A} \in L^{\mu;\nu}_{(\mathrm{cl})}(\mathbb{R}^{n+1})$ in the sense of Definition 1.5.18, such that $\varphi_\infty \chi_*(A|_V) \psi_\infty = \varphi_\infty \widetilde{A} \psi_\infty$.

Theorem 1.5.31. *For an operator* $A \in L^{\mu;\nu}_{(cl)}(M)$ *the following conditions are equivalent:*

(i) *The operator*

$$A : H^{s;g}(M) \rightarrow H^{s-\mu;g-\nu}(M) \tag{1.199}$$

is Fredholm for some $s = s_0, g = g_0 \in \mathbb{R}$.

(ii) *The operator* A *is elliptic.*

The Fredholm property (1.199) *of* $A \in L^{\mu;\nu}(\mathbb{R}^m)$ *for* $s_0, g_0 \in \mathbb{R}$ *implies the same for all* $s, g \in \mathbb{R}$.

Theorem 1.5.32. (i) *An elliptic operator* $A \in L^{\mu;\nu}_{(cl)}(M)$ *has a parametrix* $P \in L^{-\mu;-\nu}_{(cl)}(M)$, *i.e.,* $C := I - PA, \widetilde{C} := I - AP \in L^{-\infty;-\infty}(M)$. *The parametrix can be chosen in such a way that* $C, \widetilde{C}$ *are projections to finite-dimensional subspaces* $V, \widetilde{V} \subset \mathcal{S}(M)$, *and* $V = \ker A, \widetilde{V} + \operatorname{im} A = H^{s-\mu;g-\nu}(M)$ *for every* $s, g \in \mathbb{R}$.

(ii) *Let* $A \in L^{\mu;\nu}(M)$ *be elliptic; then* $Au = f \in H^{s-\mu;g-\nu}(M)$, $u \in H^{-\infty;-\infty}(M)$ *implies* $u \in H^{s;g}(M)$ *for every* $s, g \in \mathbb{R}$.

Theorem 1.5.33. *Assume that* $A \in L^{\mu;\nu}_{(cl)}(M)$ *induces an isomorphism* (1.199) *for some* $s = s_0, g = g_0 \in \mathbb{R}$. *Then* A *induces isomorphisms for all* $s, g \in \mathbb{R}$, *and we have* $A^{-1} \in L^{-\mu;-\nu}_{(cl)}(M)$.

The theory of pseudo-differential operators on a manifold M with conical exit also makes sense for non-compact X. In that case we have continuity results in $H^{s;g}_{comp}(M)$ and $H^{s;g}_{loc}(M)$ spaces, for a straightforward generalisation of the $H^{s;g}$-spaces with respect to localisations "in X-direction", but with a behaviour as before when we approach the conical exit.

Let us finally recall that $X^{\asymp}$ is a manifold with conical exits to infinity. The spaces $L^{\mu;\nu}_{(cl)}(X^{\asymp})$ will play a crucial role in the cone pseudo-differential calculus over $X^{\wedge} \ni (r,x)$ for large r.

Chapter 2

BVPs with the transmission property

2.1 Symbols with the transmission property

Let us first give a motivation for the transmission property of a symbol at the boundary. Given a smooth manifold X of dimension n with boundary Y, we can form the double $2X$; which is an open manifold obtained by gluing together two copies $X_{\pm}$ of X along the common boundary (here we identify X with X_{+}, the plus side). Local considerations close to the boundary will be performed in $\Omega \times \overline{\mathbb{R}}_{+}$ in the splitting of variables $x = (y, t)$, and the double is equal to $\Omega \times \mathbb{R}$ (where $\Omega \times \overline{\mathbb{R}}_{-}$ has the meaning of the minus side, and $\Omega \subseteq \mathbb{R}^{n-1}$ open corresponds to a chart on Y). A differential operator $A = \sum_{|\alpha| \leq \mu} a_{\alpha}(x) D_x^{\alpha}$, say, with coefficients $a_{\alpha} \in C_0^{\infty}(\Omega \times \overline{\mathbb{R}}_{+})$, can always be obtained from an extension $\widetilde{A} = \sum_{|\alpha| \leq \mu} \widetilde{a}_{\alpha}(x) D_x^{\alpha}$ to the double $\Omega \times \mathbb{R}$, with coefficients $\widetilde{a}_{\alpha} \in C_0^{\infty}(\Omega \times \mathbb{R})$. Now if A is elliptic we obtain ellipticity of $\widetilde{A}$ in a neighbourhood of the boundary Ω. Thus, when we start the program of constructing parametrices of boundary value problems (BVPs) for elliptic operators A, the first step will be to analyse parametrices of $\widetilde{A}$ close to the boundary and then take the restriction to the plus side. What we have is a very specific behaviour of the resulting symbols; we just obtain symbols with the transmission property. In order to see some features, we first consider the (complete) symbol of A

$$a(x, \xi) = \sum_{|\alpha| \leq \mu} a_{\alpha}(x) \xi^{\alpha}.$$

It can be written as a sum of components $a_{(\mu-j)}(x, \xi)$ of homogeneity $\mu - j$, $j = 0, \ldots, \mu$, and we see that, apart from the standard homogeneity,

$$a_{(\mu-j)}(x, \lambda \xi) = \lambda^{\mu-j} a_{(\mu-j)}(x, \xi)$$

© Springer Nature Switzerland AG 2018
X. Liu, B.-W. Schulze, *Boundary Value Problems with Global Projection Conditions*,
Operator Theory: Advances and Applications 265, https://doi.org/10.1007/978-3-319-70114-1_2

for all $\lambda \in \mathbb{R}_+$, we have such a relation even for all $\lambda \in \mathbb{R}$. In particular, it follows that

$$a_{(\mu-j)}(x, \xi) = (-1)^{\mu-j} a_{(\mu-j)}(x, -\xi)$$

for all x, ξ. It can be easily verified that such a stronger homogeneity is preserved for the homogeneous components of the classical symbols of parametrices (when A is elliptic). Also other natural operations, for instance, compositions, preserve such a property. The symbols that we just described are a special case of Definition 2.1.1 below.

Another important motivation lies in the nature of the operator convention close to the boundary. When we have a differential operator $\widetilde{A}$ in $\Omega \times \mathbb{R}$ and define $A = \widetilde{A}|_{\Omega \times \mathbb{R}_+}$, which induces continuous operators between the comp/loc-variants of standard Sobolev spaces over $\Omega \times \mathbb{R}$, restricted to $\Omega \times \mathbb{R}_+$, then we can write

$$A = \mathrm{r}^+ \widetilde{A} \mathrm{e}^+,$$

where e^+ denotes the extension of the respective Sobolev distribution over $\Omega \times \mathbb{R}_+$ by zero for $t < 0$ (which gives a distribution over $\Omega \times \mathbb{R}$ when $s > -1/2$) and r^+ the restriction to $\Omega \times \mathbb{R}_+$. For instance, if $\widetilde{a}_\alpha \in C_0^\infty(\mathbb{R}^{n-1} \times \mathbb{R})$, we obtain a continuous operator

$$\mathrm{r}^+ \widetilde{A} \mathrm{e}^+ : H^s(\mathbb{R}_+^n) \to H^{s-\mu}(\mathbb{R}_+^n)$$

for $H^s(\mathbb{R}_+^n) := H^s(\mathbb{R}^n)|_{\mathbb{R}_+^n}$. Proceeding in the same way for an operator $\widetilde{A} \in L_{\mathrm{cl}}^\mu(\mathbb{R}^n)$ we will not always yield continuity in Sobolev spaces in such a sense, i.e., with control of Sobolev smoothness up to the boundary which is the case when the transmission property holds. A counter-example is any $\widetilde{A} \in L_{\mathrm{cl}}^1(\mathbb{R}^n)$ with $|\xi|$ as its homogeneous principal symbol.

Definition 2.1.1. A symbol $a(y, t, \eta, \tau) \in S_{\mathrm{cl}}^\mu(\Omega \times \mathbb{R} \times \mathbb{R}_{\eta,\tau}^n)$ for $\mu \in \mathbb{Z}$ is said to have the transmission property at $t = 0$ if it satisfies the conditions

$$\left(D_t^k D_\xi^\alpha a_{(\mu-j)}\right)(y, 0, 0, 1) = (-1)^{\mu-j-|\alpha|}\left(D_t^k D_\xi^\alpha a_{(\mu-j)}\right)(y, 0, 0, -1) \qquad (2.1)$$

for all $y \in \Omega$ and $k, j \in \mathbb{N}, \alpha \in \mathbb{N}^n$; here $x = (y, t), \xi = (\eta, \tau)$ (recall that $a_{(\mu-j)}$ is the homogeneous component of a of order $\mu - j$). Let

$$S_{\mathrm{tr}}^\mu(\Omega \times \mathbb{R} \times \mathbb{R}^n) \qquad (2.2)$$

denote the space of all symbols of this kind. Moreover, set

$$S_{\mathrm{tr}}^\mu(\Omega \times \overline{\mathbb{R}}_\pm \times \mathbb{R}^n) := \left\{a|_{\Omega \times \overline{\mathbb{R}}_\pm \times \mathbb{R}^n} : a \in S_{\mathrm{tr}}^\mu(\Omega \times \mathbb{R} \times \mathbb{R}^n)\right\}.$$

Remark 2.1.2. The spaces $S_{\mathrm{tr}}^\mu(\Omega \times \mathbb{R} \times \mathbb{R}^n)$ and $S_{\mathrm{tr}}^\mu(\Omega \times \overline{\mathbb{R}}_\pm \times \mathbb{R}^n)$ are closed in the Fréchet topologies induced by $S_{\mathrm{cl}}^\mu(\Omega \times \mathbb{R} \times \mathbb{R}^n)$ and $S_{\mathrm{cl}}^\mu(\Omega \times \overline{\mathbb{R}}_\pm \times \mathbb{R}^n)$, respectively. Also the subspaces $S_{\mathrm{tr}}^\mu(\mathbb{R}^n)$ of symbols depending only on $\xi = (\eta, \tau)$ are closed in the Fréchet topologies of $S_{\mathrm{cl}}^\mu(\mathbb{R}^n)$.

Remark 2.1.3. The transmission property is preserved under natural operations with symbols:

(i) $a \in S_{\mathrm{tr}}^{\mu}$, $b \in S_{\mathrm{tr}}^{\nu}$ $\implies$ $ab \in S_{\mathrm{tr}}^{\mu+\nu}$ (for the (x, ξ)-wise product).

(ii) $a \in S_{\mathrm{tr}}^{\mu}$, $b \in S_{\mathrm{tr}}^{\nu}$ $\implies$ $a \sharp b \in S_{\mathrm{tr}}^{\mu+\nu}$ (for the Leibniz product).

(iii) $a_j \in S_{\mathrm{tr}}^{\mu-j}$, $j \in \mathbb{N}$ $\implies$ $\sum_{j=0}^{\infty} a_j \in S_{\mathrm{tr}}^{\mu}$ (asymptotic summation).

(iv) $a \in S_{\mathrm{tr}}^{\mu}$ elliptic entails $a^{\sharp^{-1}} \in S_{\mathrm{tr}}^{-\mu}$ (for the inverse $\sharp^{-1}$ under Leibniz multiplication).

(v) If χ is a diffeomorphism, smooth up to the boundary, then $a \in S_{\mathrm{tr}}^{\mu}$ implies $\chi_* a \in S_{\mathrm{tr}}^{\mu}$ (for the symbol push-forward χ_* belonging to the push-forward of associated pseudo-differential operators).

The latter remark provides many examples of symbols with the transmission property, starting from specific ones, e.g., polynomials in the covariables.

Example 2.1.4. Let $\varphi(\tau) \in \mathcal{S}(\mathbb{R})$ be a function such that $\varphi(0) = 1$ and

$$\operatorname{supp}\left(F^{-1}\varphi\right) \subset \mathbb{R}_-,$$

where $F = F_{t \to \tau}$ is the one-dimensional Fourier transform (e.g.,

$$\psi(\imath) - \imath^{-1} \int_{-\infty}^{0} e^{-it\tau}\psi(\imath)d\imath$$

for any $\psi \in C_0^{\infty}(\mathbb{R}_-)$ such that $c = \int_{-\infty}^{0} \psi(t)dt \neq 0$). For any $\mu \in \mathbb{Z}$ we have

$$r_{\pm}^{\mu}(\eta, \tau) := \left(\varphi\left(\frac{\tau}{C\langle \eta \rangle}\right)\langle \eta \rangle \pm i\tau\right)^{\mu} \in S_{\mathrm{tr}}^{\mu}(\mathbb{R}^n) \tag{2.3}$$

for any constant $C > 0$.

Symbols of this structure have been considered from different viewpoints in [18] and [39] (for more see also [40]).

It is instructive to consider separately symbols with the transmission property in dimension one, i.e., $S_{\mathrm{tr}}^{\mu}(\mathbb{R} \times \mathbb{R})$, and then interpret symbols in arbitrary dimension as a kind of standard symbols tangential to the boundary with values in $S_{\mathrm{tr}}^{\mu}(\mathbb{R} \times \mathbb{R})$. Later on, by analysing the structure of symbols in one dimension, we will see the differences between symbols with and without the transmission property.

Proposition 2.1.5. *The space $S_{\mathrm{tr}}^{\mu}(\mathbb{R} \times \mathbb{R})$ can be characterised as the set of all $a(t, \tau) \in S_{\mathrm{cl}}^{\mu}(\mathbb{R} \times \mathbb{R})$ such that the coefficients $a_{j,\pm}(t) \in C^{\infty}(\mathbb{R})$ in the asymptotic expansion*

$$a(t, \tau) \sim \sum_{j=0}^{\infty} a_{j,\pm}(t)(i\tau)^{\mu-j} \quad \text{as } \tau \to \pm\infty \tag{2.4}$$

satisfy the conditions

$$(D_t^k a_{j,+})(0) = (D_t^k a_{j,-})(0) \quad \text{for all } j, k \in \mathbb{N}. \tag{2.5}$$

Proof. For simplicity we consider the case of symbols with constant coefficients (the general case is similar and left to the reader). For $a(\tau) \in S_{\text{cl}}^\mu(\mathbb{R})$ the property (2.5) means that

$$a_{j,+} = a_{j,-} \quad \text{for all } j \in \mathbb{N}. \tag{2.6}$$

The homogeneous components of $a(\tau)$ are the functions

$$a_{(\mu-j)}(\tau) = \{a_{j,+}\theta^+(\tau) + a_{j,-}\theta^-(\tau)\}(i\tau)^{\mu-j}, \tag{2.7}$$

where $\theta^\pm$ denotes the characteristic functions of the $\pm$-half-line in τ. The condition $a(\tau) \in S_{\text{tr}}^\mu(\mathbb{R})$ means that

$$a_{(\mu-j)}(\tau) = (-1)^{\mu-j} a_{(\mu-j)}(-\tau)$$

for all $\tau \in \mathbb{R} \setminus \{0\}$ and all $j \in \mathbb{N}$, cf. Definition 2.1.1. Because of (2.7) one thus has

$$(a_{j,+}\theta^+(\tau) + a_{j,-}\theta^-(\tau))(i\tau)^{\mu-j} = (-1)^{\mu-j}(a_{j,+}\theta^+(-\tau) + a_{j,-}\theta^-(-\tau))(-i\tau)^{\mu-j},$$

which is equivalent to

$$a_{j,+}\theta^+(\tau) + a_{j,-}\theta^-(\tau) = a_{j,+}\theta^+(-\tau) + a_{j,-}\theta^-(-\tau) \tag{2.8}$$

for all $\tau \neq 0$. For $\tau > 0$ we have $\theta^-(\tau) = \theta^+(-\tau) = 0$ and $\theta^+(\tau) = \theta^-(-\tau) = 1$ (similar relations hold for $\tau < 0$). In any case (2.8) is equivalent to (2.6). $\quad\square$

Proposition 2.1.6. *For every $a(y, t, \eta, \tau) \in S_{\text{tr}}^\mu(\Omega \times \mathbb{R} \times \mathbb{R}_{\eta,\tau}^n)$ and fixed $(y, \eta) \in \Omega \times \mathbb{R}^{n-1}$, we have*

$$a_{y,\eta}(t, \tau) := a(y, t, \eta, \tau) \in S_{\text{tr}}^\mu(\mathbb{R}_t \times \mathbb{R}_\tau).$$

Proof. The symbol $\chi(\eta, \tau)a_{\mu-j}(y, t, \eta, \tau)$ for an excision function χ and fixed j belongs to $S_{\text{tr}}^{\mu-j}(\Omega \times \mathbb{R} \times \mathbb{R}^n)$. This easily follows from Proposition 2.1.5. Now it suffices to note that for fixed $(y, \eta) \in \Omega \times \mathbb{R}^{n-1}$ we also have $a \sim \sum_{j=0}^\infty \chi a_{\mu-j}$ in the space $S_{\text{tr}}^\mu(\mathbb{R}_t \times \mathbb{R}_\tau)$. $\quad\square$

In order to analyse the mapping properties of pseudo-differential operators in local coordinates $x = (y, t) \in \mathbb{R}_+^n$, we first ignore the half-space and the transmission property and consider a symbol

$$p(y, t, \eta, \tau) \in S^\mu(\Omega \times \mathbb{R}_t \times \mathbb{R}_{\eta,\tau}^n),$$

$\Omega \subseteq \mathbb{R}^{n-1}$ open. The associated operator

$$\text{Op}_x(p) : C_0^\infty(\Omega \times \mathbb{R}) \to C^\infty(\Omega \times \mathbb{R})$$

can be written in iterated form, namely,

$$\mathrm{Op}_x(p) = \mathrm{Op}_y(\mathrm{Op}_t(p)),$$

where $\mathrm{Op}_t(p)(y,\eta)$ is regarded as an operator-valued symbol. In order to simplify the considerations we first assume p to be independent of t. In this case

$$\mathrm{Op}_t(p)(y,\eta) : H^s(\mathbb{R}) \to H^{s-\mu}(\mathbb{R})$$

is a family of continuous operators for every $s \in \mathbb{R}$, parametrised by $(y,\eta) \in \Omega \times \mathbb{R}^{n-1}$. With the family of isomorphisms $\kappa_\lambda : H^s(\mathbb{R}) \to H^s(\mathbb{R})$, $\lambda \in \mathbb{R}_+$, $(\kappa_\lambda u)(t) := \lambda^{1/2} u(\lambda t)$, we form

$$\kappa_{\langle\eta\rangle}^{-1} \mathrm{Op}_t(p)(y,\eta)\kappa_{\langle\eta\rangle} = \mathrm{Op}_t(\mathfrak{p})(y,\eta) \tag{2.9}$$

for

$$\mathfrak{p}(y,\eta,\tau) := p(y,\eta,\langle\eta\rangle\tau).$$

From the symbol estimates for p, in particular, $|p(y,\eta,\tau)| \le c\langle\eta,\tau\rangle^\mu$ for a constant $c > 0$ that is uniform on compact subsets of $\mathbb{R}^{n-1}$, we obtain

$$|\mathfrak{p}(y,\eta,\tau)| = |p(y,\eta,\langle\eta\rangle\tau)| \le c\langle\eta,\langle\eta\rangle\tau\rangle^\mu = c\langle\eta\rangle^\mu\langle\tau\rangle^\mu. \tag{2.10}$$

This corresponds to a special case for the symbol estimates in the operator-valued set up, cf. Definition 1.3.8.

Lemma 2.1.7. *For every* $p(y,\eta,\tau) \in S^\mu(\Omega \times \mathbb{R}^n_{\eta,\tau})$ *we have*

$$\left\| \kappa_{\langle\eta\rangle}^{-1} \{ D_y^\alpha D_\eta^\beta \mathrm{Op}_t(p)(y,\eta) \} \kappa_{\langle\eta\rangle} \right\|_{\mathcal{L}(H^s(\mathbb{R}), H^{s-\mu+|\beta|}(\mathbb{R}))} \le c\langle\eta\rangle^{\mu-|\beta|} \tag{2.11}$$

for all $(y,\eta) \in K \times \mathbb{R}^{n-1}$, $K \Subset \Omega$, *all* $\alpha, \beta \in \mathbb{N}^{n-1}$, *and every* $s \in \mathbb{R}$, *with constants* $c = c(\alpha,\beta,K,s) > 0$.

Proof. Let first $\alpha = \beta = 0$, and set $a(y,\eta) = \mathrm{Op}_t(p)(y,\eta)$. Then the relation (2.10) and the estimates (2.11) yield

$$\left\| \kappa_{\langle\eta\rangle}^{-1} a(y,\eta)\kappa_{\langle\eta\rangle} u \right\|_{H^{s-\mu}(\mathbb{R})}^2 = \int \langle\tau\rangle^{2(s-\mu)} |p(y,\eta,\langle\eta\rangle\tau)\widehat{u}(\tau)|^2 d\tau$$

$$\le \sup_{\tau \in \mathbb{R},\, y \in K} \langle\tau\rangle^{-2\mu} |p(y,\eta,\langle\eta\rangle\tau)|^2 \int \langle\tau\rangle^{2\mu} |\widehat{u}(\tau)|^2 d\tau$$

$$\le c\langle\eta\rangle^{2\mu} \|u\|_{H^s(\mathbb{R})}^2.$$

This implies (2.11) for $\alpha = \beta = 0$. The assertion for arbitrary α, β follows in an analogous manner from the relation

$$D_y^\alpha D_\eta^\beta \mathrm{Op}_t(p)(y,\eta) = \mathrm{Op}_t(D_y^\alpha D_\eta^\beta p)(y,\eta)$$

and $D_y^\alpha D_\eta^\beta p(y,\eta,\tau) \in S^{\mu-|\beta|}(\Omega \times \mathbb{R}^n_{\eta,\tau})$. $\square$

Corollary 2.1.8. *For any $p(y, \eta, \tau)$ as in* Lemma 2.1.7 *we have*

$$\mathrm{Op}_t(p)(y, \eta) \in S^\mu(\Omega \times \mathbb{R}^{n-1}; H^s(\mathbb{R}), H^{s-\mu}(\mathbb{R})) \tag{2.12}$$

for every $s \in \mathbb{R}$.

Observe that we even have

$$\mathrm{Op}_t(p)(y, \eta) \in S^\mu_{\mathrm{cl}}(\Omega \times \mathbb{R}^{n-1}; H^s(\mathbb{R}), H^{s-\mu}(\mathbb{R})), \tag{2.13}$$

$s \in \mathbb{R}$, when $p(y, \eta, \tau) \in S^\mu_{\mathrm{cl}}(\Omega \times \mathbb{R}^n_{\eta, \tau})$, because p is independent of t.

Remark 2.1.9. Let $S^\mu(\Omega \times \mathbb{R} \times \mathbb{R}^n_{\eta, \tau})_C$ for some $C > 0$ be the set of all $p(y, t, \eta, \tau) \in S^\mu(\Omega \times \mathbb{R} \times \mathbb{R}^n_{\eta, \tau})$ that are independent of t for $|t| \geq C$. Relation (2.12) also holds for symbols $p(y, t, \eta, \tau) \in S^\mu(\Omega \times \mathbb{R} \times \mathbb{R}^n_{\eta, \tau})_C$ for

$$\mathfrak{p}(y, t, \eta, \tau) := p\big(y, \langle\eta\rangle^{-1}t, \eta, \langle\eta\rangle\tau\big). \tag{2.14}$$

In fact, it suffices to apply Lemma 1.3.26.

2.2   Operators on the half-line

As noted in the preceding section a pseudo-differential operator in the half-space can be written as a composition of an operator normal to the boundary (i.e., on the half-line) and some object tangent to the boundary. The latter will be an operator with operator-valued symbol in the sense of Section 1.3. Here we study the action in normal direction which is specific for BVPs. For convenience we freeze (and then suppress) the variables (y, η) and consider symbols $a(t, \tau) \in S^\mu_{\mathrm{tr}}(\mathbb{R} \times \mathbb{R})$. Let us first assume $a(\tau) \in S^\mu_{\mathrm{tr}}(\mathbb{R})$, which will be a main feature for homogeneous boundary symbols.

Remark 2.2.1. Every $a(\tau) \in S^\mu_{\mathrm{tr}}(\mathbb{R})$ can be written in the form

$$a(\tau) = \sum_{j=0}^\mu a_j \tau^j + b(\tau) \tag{2.15}$$

with coefficients $a_j \in \mathbb{C}$, and a remainder $b(\tau) \in S^{-1}_{\mathrm{tr}}(\mathbb{R})$ (clearly, the sum in (2.15) is non-trivial only for $\mu \in \mathbb{N}$).

In fact, the relation (2.6) shows that from $a(\tau)$ we can split a polynomial in τ of order μ, and be left with a symbol of order -1.

The first summand on the right of (2.15) represents a differential operator, and the definition of its action on distributions on $\mathbb{R}_+$ is canonically defined. As we shall see below, cf. Proposition 2.2.16, the space $S^{-1}_{\mathrm{tr}}(\mathbb{R})$ can be characterised as

$$S^{-1}_{\mathrm{tr}}(\mathbb{R}) = F_{t \to \tau} \mathcal{S}(\overline{\mathbb{R}}_-) + F_{t \to \tau} \mathcal{S}(\overline{\mathbb{R}}_+)$$

for the one-dimensional Fourier transform $Fu(\tau) = \int e^{-it\tau} u(t) dt$ and $\mathcal{S}(\overline{\mathbb{R}}_\pm) = \mathcal{S}(\mathbb{R})|_{\overline{\mathbb{R}}_\pm}$.

Concerning $b(\tau)$, the choice of a natural operator convention on the half-line is much less evident. The operator $\mathrm{op}_t(b)$ on $\mathbb{R}$ is non-local, and we need a rule to truncate it to $\mathbb{R}_+$ in such a way that we do not lose the control on the Sobolev smoothness under the action close to 0. It turns out that in the case of symbols with the transmission property the operator convention

$$S_{\mathrm{tr}}^\mu(\mathbb{R}) \ni a(\tau) \to \mathrm{r}^+ \mathrm{op}(a) \mathrm{e}^+ =: \mathrm{op}^+(a) \qquad (2.16)$$

works, with $\mathrm{op}(a)u(t) = \iint e^{i(t-t')\tau} a(\tau) u(t') dt' d\tau$ being the standard pseudo-differential action on $\mathbb{R}$, based on the Fourier transform, and

$$\mathrm{r}^+ u := \begin{cases} u & \text{on } \mathbb{R}_+, \\ 0 & \text{on } \mathbb{R}_-, \end{cases} \qquad \text{for } u \in H^s(\mathbb{R}_+),\ s > -1/2,$$

$\mathrm{r}^+ f := f|_{\mathbb{R}_+}$ (the restriction of distributions on $\mathbb{R}$ to $\mathbb{R}_+$). For future use we also introduce the notation e^- (extension by 0 from $\mathbb{R}_-$ to $\mathbb{R}$) and r^- (restriction to $\mathbb{R}_-$). Since $\mathrm{e}^+ H^s(\mathbb{R}_+) \subset \mathcal{S}'(\mathbb{R})$ for $s > -1/2$ for every $a(\tau) \in S_{\mathrm{cl}}^\mu(\mathbb{R})$, we have

$$\mathrm{op}^+(a) : H^s(\mathbb{R}_+) \to \mathcal{S}'(\mathbb{R})|_{\mathbb{R}_+}$$

(more precisely, the image belongs to $H_{\mathrm{loc}}^{s-\mu}(\mathbb{R}_+)$). However, for $a(\tau) \in S_{\mathrm{tr}}^\mu(\mathbb{R})$, as we shall see below, we in fact have the continuity of

$$\mathrm{op}^+(a) : H^s(\mathbb{R}_+) \to H^{s-\mu}(\mathbb{R}_+)$$

for $s > -1/2$. For $a(\tau) \notin S_{\mathrm{tr}}^\mu(\mathbb{R})$ this is not the case, cf. Part III.

Apart from the transmission property it will be useful to have the notion of plus/minus-symbols.

By $\mathcal{A}(U)$ for open $U \subseteq \mathbb{C}$ we denote the space of all holomorphic functions in U. Denote $\mathbb{C}_\pm := \{\zeta \in \mathbb{C} : \mathrm{Im}\,\zeta \gtrless 0\}$.

Definition 2.2.2. Let $\zeta = \tau + i\vartheta$. A plus-symbol of order $\mu \in \mathbb{R}$ is a symbol $a(\tau) \in S_{\mathrm{cl}}^\mu(\mathbb{R})$ such that there is an $a_+(\zeta) \in C^\infty(\overline{\mathbb{C}}_-) \cap \mathcal{A}(\mathbb{C}_-)$ such that $a(\tau) = a_+(\tau)$ and

$$|a_+(\zeta)| \le c\,(1 + |\zeta|^2)^{\mu/2} \qquad (2.17)$$

for all $\zeta \in \overline{\mathbb{C}}_-$. A minus-symbol of order $\mu \in \mathbb{R}$ is a symbol $a(\tau) \in S_{\mathrm{cl}}^\mu(\mathbb{R})$ such that there is an $a_-(\zeta) \in C^\infty(\overline{\mathbb{C}}_+) \cap \mathcal{A}(\mathbb{C}_+)$ such that $a(\tau) = a_-(\tau)$ and (2.17) holds for all $\zeta \in \overline{\mathbb{C}}_+$.

Example 2.2.3. (i) A polynomial $a(\tau) = \sum_{j=0}^\mu a_j \tau^j$ is both a plus- and minus-symbol of order μ.

(ii) The function $(\delta \pm i\tau)^j$ for $\delta \in \mathbb{R}_+$, $j \in \mathbb{Z}$, are plus/minus-symbols (according to the sign at $i\tau$) of order j.

Clearly, the symbols in Example 2.2.3 also have the transmission property. As usual, for $s \in \mathbb{R}$ we set

$$H^s(\mathbb{R}_\pm) := \left\{ u|_{\mathbb{R}_\pm} : u \in H^s(\mathbb{R}) \right\}$$

and

$$H_0^s(\overline{\mathbb{R}}_\pm) := \left\{ u \in H^s(\mathbb{R}) : u|_{\mathbb{R}_\mp} = 0 \right\} = \left\{ u \in H^s(\mathbb{R}) : \operatorname{supp} u \subseteq \overline{\mathbb{R}}_\pm \right\}. \qquad (2.18)$$

Observe that we have canonical identifications

$$H^s(\mathbb{R}_+) = H^s(\mathbb{R})/H_0^s(\overline{\mathbb{R}}_-), \quad H^s(\mathbb{R}_-) = H^s(\mathbb{R})/H_0^s(\overline{\mathbb{R}}_+). \qquad (2.19)$$

The sets $H_0^s(\overline{\mathbb{R}}_\pm)$ are closed subspaces of $H^s(\mathbb{R})$. In fact, e.g., in the plus-case and for $s \geq 0$, the $H^s(\mathbb{R})$-convergence of a sequence $\{u_k\}_{k\in\mathbb{N}} \subset H_0^s(\overline{\mathbb{R}}_+)$ to a $u \in H^s(\mathbb{R})$ entails the convergence to u in $L^2(\mathbb{R})$, and it is clear that vanishing of $u_k|_{\mathbb{R}_-}$ for all $k \in \mathbb{N}$ entails $u|_{\mathbb{R}_-} = 0$ for the $L^2(\mathbb{R})$-limit u. For $s \leq 0$ we can argue that the $H^s(\mathbb{R})$-convergence of $\{u_k\}_{k\in\mathbb{N}} \subset H_0^s(\overline{\mathbb{R}}_+)$ to a $u \in H^s(\mathbb{R})$ implies the convergence $u_k \to u$ in $\mathcal{S}'(\mathbb{R})$, i.e., $\langle u_k, \varphi \rangle \to \langle u, \varphi \rangle$ for every $\varphi \in \mathcal{S}(\mathbb{R})$. This is true, in particular, for all φ with compact support in $\mathbb{R}_-$. Then $\langle u_k, \varphi \rangle = 0$ for all k implies $\langle u, \varphi \rangle = 0$ for every φ and it follows again that $u|_{\mathbb{R}_-} = 0$ (of course, the latter argument is valid for all $s \in \mathbb{R}$).

Since $H^s(\mathbb{R})$ is a Hilbert space, $H_0^s(\overline{\mathbb{R}}_\pm)$ are Hilbert subspaces, and the spaces (2.19) can be identified with the respective orthogonal complements in $H^s(\mathbb{R})$, i.e.,

$$H^s(\mathbb{R}_\pm) \cong H_0^s(\overline{\mathbb{R}}_\pm)^\perp.$$

In other words, $H^s(\mathbb{R}_\pm)$ are Hilbert spaces as well.

As a consequence of (2.19) we obtain that there are continuous (extension) operators

$$e_s^+ : H^s(\mathbb{R}_+) \to H^s(\mathbb{R}), \quad e_s^- : H^s(\mathbb{R}_-) \to H^s(\mathbb{R})$$

such that $e_s^+ u|_{\mathbb{R}_+} = u$ for all $u \in H^s(\mathbb{R}_+)$, $e_s^- v|_{\mathbb{R}_-} = v$ for all $v \in H^s(\mathbb{R}_-)$. In fact, setting for the moment $H^s(\mathbb{R})_1 := H_0^s(\overline{\mathbb{R}}_-)^\perp$, every $u \in H^s(\mathbb{R})$ has a unique decomposition $u = u_0 + u_1$ with $u_0 \in H_0^s(\overline{\mathbb{R}}_-)$ and $u_1 \in H^s(\mathbb{R})_1$. This shows that the restriction operator

$$r_s^+ : H^s(\mathbb{R})_1 \to H^s(\mathbb{R}_+) \qquad (2.20)$$

is surjective. It is also injective, since $\ker r_s^+ = H_0^s(\overline{\mathbb{R}}_-)$. Thus we can set $e_s^+ := (r_s^+)^{-1}$. The minus case is analogous.

Proposition 2.2.4. *For $|s| < 1/2$ we have*

$$H_0^s(\overline{\mathbb{R}}_\pm) = H^s(\mathbb{R}_\pm)$$

and the operators $e^\pm : H^s(\mathbb{R}_\pm) \to H^s(\mathbb{R})$ are continuous.

For a proof, see Eskin's book [14, Chapter 5].

Theorem 2.2.5. *Let $u_+(t) \in H_0^s(\overline{\mathbb{R}}_+), s \in \mathbb{R}$. Then the Fourier transform*

$$\widehat{u}_+(\tau) = \int e^{-it\tau} u_+(t) dt$$

extends to a function

$$h_+(\zeta) = \int e^{-it\tau}(e^{t\vartheta}u_+(t)) dt \in C(\operatorname{Im}\zeta \le 0) \cap \mathcal{A}(\operatorname{Im}\zeta < 0), \qquad (2.21)$$

$\zeta = \tau + i\vartheta$, *such that*

$$\int (1 + |\tau| + |\vartheta|)^{2s} |h_+(\tau + i\vartheta)|^2 d\tau \le C \qquad (2.22)$$

for all $\vartheta \le 0$, with a constant $C > 0$ independent of ϑ.

Conversely, let $h_+(\tau + i\vartheta)$ be a locally integrable function in τ for $-\infty < \vartheta < 0$ satisfying the estimate (2.22) for some $C > 0$ independent of ϑ and belonging to $\mathcal{A}(\operatorname{Im}\zeta < 0)$. Then there is an $u_+(t) \in H_0^s(\overline{\mathbb{R}}_+), s \in \mathbb{R}$, such that

$$h_+(\zeta) = \int e^{-it\tau}(e^{t\vartheta}u_+(t)) dt.$$

A proof of this version of the Paley–Wiener Theorem in higher dimensions can be found in Eskin's book [14], cf. also Section 6.12 below.

Remark 2.2.6. There is an analogue of Theorem 2.2.5 on the characterisation of the Fourier transform of $H_0^s(\overline{\mathbb{R}}_-)$. The modifications, including interchanging the sign of $\operatorname{Im}\zeta$, are straightforward, and in the following they will be tacitly used.

Let us briefly comment the case $s = 0$, where we employ the identifications

$$H^0(\mathbb{R}) = L^2(\mathbb{R}), \quad H_0^0(\overline{\mathbb{R}}_\pm) = L^2(\mathbb{R}_\pm).$$

The restriction operators

$$\mathrm{r}^\pm : L^2(\mathbb{R}) \to L^2(\mathbb{R}_\pm)$$

determine complementary projections. Those may be also identified with the operators of multiplication by $\theta^\pm(t)$, the characteristic functions of $\mathbb{R}_\pm$,

$$\theta^\pm : L^2(\mathbb{R}_t) \to L^2(\mathbb{R}_{\pm,t}). \qquad (2.23)$$

The operators

$$\Pi^\pm := F\theta^\pm F^{-1} : L^2(\mathbb{R}_\tau) \to V^\pm(\mathbb{R}_\tau). \qquad (2.24)$$

also represent complementary projections in the image under the Fourier transform $F = F_{t \to \tau}$, for certain closed subspaces $V^\pm(\mathbb{R}) \subset L^2(\mathbb{R})$,

$$L^2(\mathbb{R}) = V^-(\mathbb{R}) \oplus V^+(\mathbb{R}).$$

Remark 2.2.7. The operators (2.24) can be interpreted as pseudo-differential projections $\Pi^\pm \in L^0_{\mathrm{cl}}(\mathbb{R})$ on the τ-axis.

In fact, for instance, in the plus case, we can write $\Pi^+ u(\tau) = \mathrm{op}_\tau(p^+)u(\tau)$ for $p_+(t) := \theta^+(-t)$, with t being interpreted as a covariable. For an excision function $\chi(t)$ we have $\chi(t)p_+(t) \in S^0_{\mathrm{cl}}(\mathbb{R})$, $\mathrm{op}_\tau(p^+) = \mathrm{op}_\tau(\chi p^+) + \mathrm{op}_\tau((1-\chi)p^+)$, where $\mathrm{op}_\tau((1-\chi)p^+)$ is a smoothing operator. In the minus case we can argue in a similar manner.

Theorem 2.2.8. (i) *Let $a(\tau) \in S^\mu_{\mathrm{cl}}(\mathbb{R})$ be a plus-symbol. Then $\mathrm{op}(a)$ induces a continuous operator*

$$\mathrm{op}(a) : H^s_0\big(\overline{\mathbb{R}}_+\big) \to H^{s-\mu}_0\big(\overline{\mathbb{R}}_+\big) \tag{2.25}$$

for every $s \in \mathbb{R}$. Analogously, if $a(\tau) \in S^\mu_{\mathrm{cl}}(\mathbb{R})$ is a minus-symbol, then

$$\mathrm{op}(a) : H^s_0\big(\overline{\mathbb{R}}_-\big) \to H^{s-\mu}_0\big(\overline{\mathbb{R}}_-\big) \tag{2.26}$$

is continuous for every $s \in \mathbb{R}$.

(ii) *Let $a(\tau) \in S^\mu_{\mathrm{cl}}(\mathbb{R})$ be a minus-symbol and let $\mathrm{e}^+_s : H^s(\mathbb{R}_+) \to H^s(\mathbb{R})$ be any continuous extension operator for fixed $s \in \mathbb{R}$. Then*

$$\mathrm{r}^+\mathrm{op}(a)\,\mathrm{e}^+_s : H^s(\mathbb{R}_+) \to H^{s-\mu}(\mathbb{R}_+) \tag{2.27}$$

is continuous and this operator is independent of the choice of e^+_s. Moreover, we have $\mathrm{r}^+\mathrm{op}(a)\,\mathrm{e}^+_s = \mathrm{op}^+(a)$ for all $s \in \mathbb{R}, s > 1/2$.

Proof. (i) Since $\mathrm{op}(a)u = F^{-1}a(\tau)Fu$, the assertion is equivalent to the continuity of the operator of multiplication

$$a(\tau) : F\big(H^s_0(\mathbb{R}_+)\big) \to F\big(H^{s-\mu}_0(\mathbb{R}_+)\big). \tag{2.28}$$

Since $a(\tau)$ is a plus-symbol, cf. Definition 2.2.2, there is a symbol $a_+(\zeta)$ that extends $a(\tau)$ to $\mathrm{Im}\,\zeta < 0$ as a holomorphic function with the estimate (2.17). Moreover, by virtue of Theorem 2.2.5, any $\widehat{u}(\tau)$ for $u \in H^s_0(\overline{\mathbb{R}}_+)$ extends to an $h_+(\zeta)$ which is holomorphic in $\mathrm{Im}\,\zeta < 0$ with the estimate (2.22). Now

$$\sup_{\vartheta \le 0} \int (1 + |\tau| + |\vartheta|)^{2(s-\mu)} |a_+(\tau + i\vartheta)h_+(\tau + i\vartheta)|^2 d\tau$$

$$\le c_1^2 \sup_{\vartheta \le 0} \int (1 + |\tau| + |\vartheta|)^{2s} |h_+(\tau + i\vartheta)|^2 d\tau \le c,$$

where $c_1 = \sup_{\zeta \in \overline{\mathbb{C}}_-}(1 + |\tau| + |\vartheta|)^{-\mu}|a_+(\zeta)|$. Moreover,

$$a_+(\zeta)h_+(\zeta) \in C(\mathrm{Im}\,\zeta \le 0) \cap \mathcal{A}(\mathrm{Im}\,\zeta < 0).$$

Thus $a_+(\zeta)h_+(\zeta)$ belongs to $\widehat{H}^{s-\mu}_0(\overline{\mathbb{R}}_+)$, and hence (2.25) is continuous. The continuity of (2.26) can be proved in an analogous manner.

(ii) The operator (2.27) is obviously continuous for any $a \in S^\mu(\mathbb{R})$. However, if a is a minus-symbol, and if e_s^+, l_s^+ are two continuous extension operators $H^s(\mathbb{R}_+) \to H^s(\mathbb{R})$, then

$$r^+ \mathrm{op}(a)(e_s^+ - l_s^+)u = 0,$$

since $(e_s^+ - l_s^+)u \in H_0^s(\overline{\mathbb{R}}_-)$, $\mathrm{op}(a)(e_s^+ - l_s^+)u \in H_0^{s-\mu}(\overline{\mathbb{R}}_-)$, cf. (2.26), and

$$r^+ H_0^{s-\mu}(\overline{\mathbb{R}}_-) = \{0\} \tag{2.29}$$

for every $s \in \mathbb{R}$. Now if we replace e_s^+ by e^+ for $s > -1/2$, then

$$(e_s^+ - e^+)u \in H_0^{\tilde{s}}(\overline{\mathbb{R}}_+) \tag{2.30}$$

for some $\tilde{s} \in \mathbb{R}$. In fact, we have $e^+ u \in H^{s_1}(\mathbb{R})$ for some $s_1 \in \mathbb{R}$, moreover, $e_s^+ u \in H^s(\mathbb{R})$, and

$$e_s^+ u|_{\mathbb{R}_+} = e^+ u|_{\mathbb{R}_+}.$$

This implies (2.30) for $\tilde{s} = \min\{s, s_1\}$. It follows that

$$r^+ \mathrm{op}(a)e^+ u = r^+ \mathrm{op}(a)e_s^+ u,$$

since $r^+ \mathrm{op}(a)(e^+ - e_s^+)u$ vanishes because of (2.29), and hence the continuity of $r^+ \mathrm{op}(a)e^+ : H^s(\mathbb{R}_+) \to H^{s-\mu}(\mathbb{R}_+)$ is a consequence of (2.27). $\qquad\square$

Proposition 2.2.9. *For every $a(\tau) \in S_{\mathrm{tr}}^\mu(\mathbb{R})$ and every $N \in \mathbb{N}$ there exist a minus-symbol $m_N(\tau) \in S_{\mathrm{tr}}^\mu(\mathbb{R})$ and a plus-symbol $p_N(\tau) \in S_{\mathrm{tr}}^\mu(\mathbb{R})$ such that*

$$a(\tau) - m_N(\tau) \in S_{\mathrm{tr}}^{-(N+1)}(\mathbb{R}), \quad a(\tau) - p_N(\tau) \in S_{\mathrm{tr}}^{-(N+1)}(\mathbb{R}). \tag{2.31}$$

Proof. Because of Remark 2.2.1 and Example 2.2.3 (i), without loss of generality we may assume $\mu \leq -1$. By Proposition 2.1.5, there are constants $a_j \in \mathbb{C}$ such that

$$a(\tau) - \chi(\tau) \sum_{j=1}^N a_j (i\tau)^{-j} \in S_{\mathrm{cl}}^{-(N+1)}(\mathbb{R}).$$

Here χ is any fixed excision function. In order to show the existence of $m_N(\tau)$ it suffices to write

$$\chi(\tau)(i\tau)^{-j} = m_{N,j}(\tau) \bmod S_{\mathrm{cl}}^{-(N+1)}(\mathbb{R}) \tag{2.32}$$

for a minus-symbol $m_{N,j}(\tau) \in S_{\mathrm{tr}}^\mu(\mathbb{R})$, for every $1 \leq j \leq N$. We have

$$(i\tau)^{-1} = -(1-i\tau)^{-1} + (i\tau)^{-1}(1-i\tau)^{-1} = \cdots = -\sum_{k=1}^N (1-i\tau)^{-k} + (i\tau)^{-1}(1-i\tau)^{-N},$$

which yields

$$\chi(\tau)(i\tau)^{-1} = -\sum_{k=1}^N (1-i\tau)^{-k} + \chi(\tau)(i\tau)^{-1}(1-i\tau)^{-N} \bmod S_{\mathrm{cl}}^{-(N+1)}(\mathbb{R}),$$

and more generally,

$$\chi(\tau)(i\tau)^{-j} = \left(-\sum_{k=1}^{N}(1 - i\tau)^{-k}\right)^{-j} \mod S_{\text{cl}}^{-(N+1)}(\mathbb{R})$$

for every $1 \leq j \leq N$.

From Example 2.2.3 (ii) we thus obtain the decomposition (2.31). The remainder belongs to $S_{\text{tr}}^{-(N+1)}(\mathbb{R})$, since both a and m_N have the transmission property. A similar computation, using

$$(i\tau)^{-j} = (1 + i\tau)^{-1} + (i\tau)^{-1}(1 + i\tau)^{-1} = \cdots,$$

gives us a plus-symbol $p_N(\tau)$ with the desired property. $\square$

Remark 2.2.10. It is easily verified that the decompositions of Proposition 2.2.9, namely, $a(\tau) \mapsto a(\tau) - p_N(\tau)$, $a(\tau) \mapsto a(\tau) - m_N(\tau)$, etc., represent continuous operators

$$S_{\text{tr}}^{\mu}(\mathbb{R}) \to S_{\text{tr}}^{\mu-(N+1)}(\mathbb{R}), \quad S_{\text{tr}}^{\mu}(\mathbb{R}) \to S_{\text{tr}}^{-(N+1)}(\mathbb{R})$$

in the respective Fréchet topologies.

Theorem 2.2.11. *Let* $a(\tau) \in S_{\text{tr}}^{\mu}(\mathbb{R})$; *then* $\text{op}^+(a)$ *induces a continuous operator*

$$\text{op}^+(a) : H^s(\mathbb{R}_+) \to H^{s-\mu}(\mathbb{R}_+) \tag{2.33}$$

for every $s \in \mathbb{R}$, $s > -1/2$. *Moreover, the correspondence* $a \mapsto \text{op}^+(a)$ *yields a continuous operator*

$$S_{\text{tr}}^{\mu}(\mathbb{R}) \to \mathcal{L}\big(H^s(\mathbb{R}_+), H^{s-\mu}(\mathbb{R}_+)\big) \tag{2.34}$$

for all those s. *In addition,* $\text{op}^+(a)$ *induces a continuous operator*

$$\text{op}^+(a) : \mathcal{S}\big(\overline{\mathbb{R}}_+\big) \to \mathcal{S}\big(\overline{\mathbb{R}}_+\big), \tag{2.35}$$

and $a \mapsto \text{op}^+(a)$ *yields a continuous operator* $S_{\text{tr}}^{\mu}(\mathbb{R}) \to \mathcal{L}\big(\mathcal{S}\big(\overline{\mathbb{R}}_+\big)\big)$.

Proof. To show the continuity (2.33), we write $a(\tau) = m_N(\tau) + c_N(\tau)$ for a minus symbol $m_N(\tau)$ and a remainder $c_N(\tau) \in S_{\text{cl}}^{-(N+1)}(\mathbb{R})$. Then it suffices to show the continuity of $\text{op}^+(m_N)$ and $\text{op}^+(c_N)$ separately. For m_N we apply Theorem 2.2.8 (ii). As for c_N, we may choose N as large as we want. Since $e^+ H^s(\mathbb{R}_+) \subset \mathcal{S}'(\mathbb{R})$ for every $k \in \mathbb{N}$ we find an N such that $\text{op}(c_N)$ has an integral kernel in $C^k(\mathbb{R} \times \mathbb{R})$. For any $\psi, \tilde{\psi} \in C_0^{\infty}(\mathbb{R})$, $\psi = 1$ near 0, $\tilde{\psi} \equiv 1$ on $\text{supp}\,\psi$, we obtain $\tilde{\psi}\,\text{op}(c_N)e^+\psi : H^s(\mathbb{R}_+) \to C_0^k(\mathbb{R})$, $(1 - \tilde{\psi})\,\text{op}(c_N)e^+\psi : H^s(\mathbb{R}_+) \to \mathcal{S}(\mathbb{R})$, which gives us a continuous operator $\text{r}^+\text{op}(c_N)e^+\psi : H^s(\mathbb{R}_+) \to H^{s-\mu}(\mathbb{R}_+)$ when k is sufficiently large. Moreover, $\text{r}^+\text{op}(c_N)e^+(1 - \psi) : H^s(\mathbb{R}_+) \to H^{s-\mu}(\mathbb{R}_+)$ is continuous anyway since $(1 - \psi)H^s(\mathbb{R}_+) \subset H^s(\mathbb{R}_+)$. The simple proof of the continuity of the operation (2.34) is left to the reader. The property (2.35) follows from the characterization of $S^\mu_{\mathrm{tr}}(\mathbb{R})$ as

$$S^\mu_{\mathrm{tr}}(\mathbb{R}) = \{\text{polynomials in } \tau \text{ of order } \mu\} + F\mathcal{S}(\overline{\mathbb{R}}_-) + F\mathcal{S}(\overline{\mathbb{R}}_+), \tag{2.36}$$

see Proposition 2.2.16 below. Then, because of $\mathrm{op}^+(a) = \mathrm{r}^+ F^{-1}a(\tau)Fe^+$ and relation (2.36), we immediately obtain (2.35), since the multiplication operator between elements of the space on the right-hand side of (2.36) and subsequent restriction r^+ of distributions to $\mathbb{R}_+$ preserves $\mathcal{S}(\mathbb{R})$ and is continuous. $\qquad\square$

Remark 2.2.12. Let $S^\mu_{\mathrm{tr}}(\mathbb{R} \times \mathbb{R})_C$ be the space of all $a(t,\tau) \in S^\mu_{\mathrm{tr}}(\mathbb{R} \times \mathbb{R})$ that are independent of t for $|t| \geq C$. Then the operator (2.33) is continuous for every $a(t,\tau) \in S^\mu_{\mathrm{tr}}(\mathbb{R} \times \mathbb{R})_C$, $s > -1/2$, and the correspondence $a \mapsto \mathrm{op}^+(a)$ defines continuous operators

$$S^\mu_{\mathrm{tr}}(\mathbb{R} \times \mathbb{R})_C \to \mathcal{L}\big(H^s(\mathbb{R}_+), H^{s-\mu}(\mathbb{R}_+)\big) \tag{2.37}$$

for every $s \in \mathbb{R}$, $s > -1/2$, and $S^\mu_{\mathrm{tr}}(\mathbb{R} \times \mathbb{R})_C \to \mathcal{L}\big(\mathcal{S}(\overline{\mathbb{R}}_+), \mathcal{S}(\overline{\mathbb{R}}_+)\big)$.

Remark 2.2.12 is a simple generalisation of Theorem 2.2.11; details are left to the reader.

Proposition 2.2.13. *For* $p(y,t,\eta,\tau) \subset S^\mu_{\mathrm{tr}}(\Omega \times \mathbb{R} \times \mathbb{R}^n_{\eta,\tau})_C$ *we have*

$$\mathrm{r}^+\mathrm{Op}_t(p)(y,\eta)\,e^+ \in S^\mu\big(\Omega \times \mathbb{R}^q; H^s(\mathbb{R}_+), H^{s-\mu}(\mathbb{R}_+)\big) \tag{2.38}$$

for every real $s > -1/2$, *and*

$$\mathrm{r}^+\mathrm{Op}_t(p)(y,\eta)\,e^+ \in S^\mu\big(\Omega \times \mathbb{R}^q; \mathcal{S}(\overline{\mathbb{R}}_+), \mathcal{S}(\overline{\mathbb{R}}_+)\big). \tag{2.39}$$

Proof. Applying Remark 2.1.9 for (2.14) we have relation (2.9). Moreover, Proposition 1.3.28, applied to the subspace

$$S^\mu_{\mathrm{tr}}\big(\Omega \times \mathbb{R} \times \mathbb{R}^n_{\eta,\tau}\big)_C \subset S^\mu_{\mathrm{cl}}\big(\Omega \times \mathbb{R}^q, S^\mu_{\mathrm{cl}}(\mathbb{R} \times \mathbb{R})\big)$$

for $q = n - 1$ shows that

$$\mathfrak{p}(y,t,\eta,\tau) \in S^\mu_{\mathrm{cl}}\big(\Omega \times \mathbb{R}^q_\eta, S^\mu_{\mathrm{tr}}(\mathbb{R} \times \mathbb{R}_\tau)_C\big).$$

The continuity of (2.37) together with Corollary 1.3.4 applied to $\mathrm{r}^+\mathrm{Op}_t(\,\cdot\,)e^+$ rather than $\mathrm{Op}_t(\,\cdot\,)$ yield relation (2.38). The property (2.39) can be reduced to the case of symbols with constant coefficients combined with a tensor product argument. In fact, we can write

$$p(y,t,\eta,\tau) = \sum_{j=0}^{\infty} \lambda_j \varphi_j(y,t) p_j(\eta,\tau)$$

for a sequence $\lambda_j \in \mathbb{C}$, $\sum_{j=0}^{\infty} |\lambda_j| < \infty$, functions $\varphi_j(y,t) \in C^{\infty}(\Omega \times \mathbb{R})_C$ where C indicates vanishing of functions for $|t| \geq C$, and symbols $p_j(\eta, \tau) \in S^{\mu}_{tr}(\mathbb{R}^n_{\eta,\tau})$, $\varphi_j \to 0$ and $p_j \to 0$ as $j \to \infty$ in the respective spaces. Then, using Theorem 2.2.11 concerning mappings in $\mathcal{S}(\overline{\mathbb{R}}_+)$ we see that

$$\mathrm{r}^+ \mathrm{Op}_t(p_j)\mathrm{e}^+ \to 0 \ \text{as} \ j \to \infty.$$

Those are specific symbols in $S^{\mu}(\mathbb{R}^q; \mathcal{S}(\overline{\mathbb{R}}_+), \mathcal{S}(\overline{\mathbb{R}}_+))$ also tending to zero in the corresponding symbol space. Then we have altogether

$$\mathrm{r}^+ \mathrm{Op}_t(p)(y,\eta)\mathrm{e}^+ = \sum \lambda_j \varphi_j(y,t) \mathrm{r}^+ \mathrm{Op}_t(p_j)\mathrm{e}^+ \tag{2.40}$$

which belongs to $S^{\mu}(\Omega \times \mathbb{R}^q; \mathcal{S}(\overline{\mathbb{R}}_+), \mathcal{S}(\overline{\mathbb{R}}_+))$, since the multiplication by $\varphi_j(y,t)$ enjoys the desired properties, and tends to zero in the sense of symbols. Thus the sum on the right-hand side of (2.40) converges within the claimed symbol space. $\qquad \square$

Proposition 2.2.14. *For a function $a(\tau) \in L^2(\mathbb{R}_{\tau})$ the following conditions are equivalent:*

(i) *$a(\tau) \in F(\mathrm{e}^+ \mathcal{S}(\overline{\mathbb{R}}_+))$, where $F = F_{t \to \tau}$ is the Fourier transform on the real line.*

(ii) *$a(\tau) \in S^{-1}_{\mathrm{cl}}(\mathbb{R}_{\tau})$ is a plus-symbol in the sense of Definition 2.2.2, the extension to an element $a_+(\zeta) \in C^{\infty}(\mathrm{Im}\,\zeta \leq 0) \cap \mathcal{A}(\mathrm{Im}\,\zeta < 0)$ has an asymptotic expansion*

$$a_+(\zeta) \sim \sum_{k \leq 1} a_k \zeta^k \quad \text{for } |\zeta| \to \infty, \ \mathrm{Im}\,\zeta \leq 0, \tag{2.41}$$

and all derivatives $(\partial_{\tau}^l a)(\tau)$, $l \in \mathbb{N}$, extend to $(\partial_{\zeta}^l a_+)(\zeta)$ with analogous asymptotic expansions, obtained by formally differentiating (2.41) on both sides with respect to ζ.

An analogous characterisation holds for the space $F(\mathrm{e}^- \mathcal{S}(\overline{\mathbb{R}}_-))$.

Proof. (i) $\implies$ (ii) We have $\mathrm{e}^+ \mathcal{S}(\overline{\mathbb{R}}_+) \subset L^2(\mathbb{R}_+)$, and from Theorem 2.2.5 for $s = 0$ we already know that $a(\tau) = Fu(\tau)$ for $u(t) \in \mathrm{e}^+ \mathcal{S}(\overline{\mathbb{R}}_+)$ has an analytic extension $a_+(\zeta) \in \mathcal{A}(\mathrm{Im}\,\zeta < 0)$. In the present case it can be easily verified that also $a_+(\zeta) \in C^{\infty}(\mathrm{Im}\,\zeta \leq 0)$. Integration by parts in $a(\tau) = \int_0^{\infty} \mathrm{e}^{-it\tau} u(t)dt$ for $\tau \neq 0$ gives

$$a(\tau) = -\frac{1}{i\tau} \mathrm{e}^{-it\tau} u(t)\big|_0^{\infty} + \frac{1}{i\tau} \int_0^{\infty} \mathrm{e}^{-it\tau} \partial_t u(t)dt = \cdots =$$

$$= \frac{1}{i\tau} u(0) + \frac{1}{(i\tau)^2} \partial_t u(0) + \cdots + \frac{1}{(i\tau)^{k+1}} \partial_t^k u(0)$$

$$+ \frac{1}{(i\tau)^{k+1}} \int_0^{\infty} \mathrm{e}^{-it\tau} \partial_t^{k+1} u(t)dt.$$

Consequently,

$$a(\tau) \sim \sum_{j=0}^{\infty} \partial_t^j u(0)(i\tau)^{-(j+1)}.\tag{2.42}$$

In a similar manner we can show that

$$a'(\tau) = \int_0^{\infty} e^{-it\tau}(-it)u(t)dt \sim -i \sum_{k=0}^{\infty} \partial_t^k(tu)(0)(i\tau)^{-(k+1)}.$$

Using that $\partial_t^l(tu)(0) = l\partial_t^{l-1}(u)(0)$ we obtain

$$a'(\tau) \sim -i \sum_{k=1}^{\infty} k\partial_t^{k-1}(u)(0)(i\tau)^{-(k+1)}.$$

Thus the expansion for $a'(\tau)$ follows by formally differentiating (2.42). The higher derivatives can be treated in a similar way. It is evident that the asymptotic expansions also hold for $\operatorname{Im} \zeta < 0$.

(ii) $\Longrightarrow$ (i) Let $u(t) = (F^{-1}a)(t)$ for a function $a(\tau)$ with the properties (ii). From $a \in L^2(\mathbb{R})$ it follows that $u \in L^2(\mathbb{R})$. By Theorem 2.2.5, the function $u(t)$ vanishes for almost all $t < 0$. From the properties in (ii) it follows that the function $\tau^k D_\tau^l a(\tau) \in C^{\infty}(\mathbb{R})$ is the sum of a polynomial and a function h_{kl} satisfying the relations of (ii), for all $k, l \in \mathbb{N}$. Thus $(F^{-1}\tau^k D_\tau^l a)(t)$ is a sum of derivatives of the Dirac distribution at the origin and a function in $L^2(\mathbb{R}_+)$. It follow that $D_t^k t^l u(t)|_{t>0} \in L^2(\mathbb{R}_+)$ for every $k, l \in \mathbb{N}$, i.e., $u \in \mathcal{S}(\overline{\mathbb{R}}_+)$.

After the first part of the proof the characterisation of $F(e^{-}\mathcal{S}(\overline{\mathbb{R}}_-))$ is straightforward. Analogously as (2.42) for $a(\tau) = \int e^{-it\tau}u(t)dt$, $u(t) \in e^{-}\mathcal{S}(\overline{\mathbb{R}}_-)$, it follows that

$$a(\tau) \sim -\sum_{j=1}^{\infty} \partial_t^j u(0)(i\tau)^{-(j+1)}.\tag{2.43}$$

$\square$

Remark 2.2.15. We have $F(e^{\pm}\mathcal{S}(\overline{\mathbb{R}}_\pm)) \subseteq S_{\mathrm{tr}}^{-1}(\mathbb{R})$, and the Fourier transform induces continuous operators

$$F : e^{\pm}\mathcal{S}(\overline{\mathbb{R}}_\pm) \to S_{\mathrm{tr}}^{-1}(\mathbb{R}).\tag{2.44}$$

The first inclusion is an immediate consequence of the fact that the transmission property of $a(\tau) \in S_{\mathrm{tr}}^{-1}(\mathbb{R})$ is characterised by asymptotic expansions $a(\tau) \sim \sum_{j=0}^{\infty} a_j(i\tau)^{-(j+1)}$, $\tau \to \pm\infty$, with the same coefficients a_j on both sides.

The continuity of (2.44) follows from the continuity of $F : e^{\pm}\mathcal{S}(\overline{\mathbb{R}}_\pm) \to S_{\mathrm{cl}}^{-1}(\mathbb{R})$, a consequence of Proposition 2.2.14. At the same time we see that the coefficients a_j depend continuously on the argument function. This establishes the continuous dependence of the resulting homogeneous components $a_j(i\tau)^{-(j+1)}$ which also contribute to the Fréchet topology of $S_{\mathrm{tr}}^{-1}(\mathbb{R})$.

Proposition 2.2.16. *Every symbol $a(\tau) \in S_{\mathrm{tr}}^{-1}(\mathbb{R})$ has a representation*

$$a(\tau) = a_+(\tau) + a_-(\tau)$$

with plus/minus-symbols $a_\pm(\tau) \in S_{\mathrm{tr}}^{-1}(\mathbb{R})$, where

$$a_+(\tau) \in F_{t\to\tau}\big(\mathrm{e}^+ \mathcal{S}(\overline{\mathbb{R}}_+)\big), \quad a_-(\tau) \in F_{t\to\tau}\big(\mathrm{e}^- \mathcal{S}(\overline{\mathbb{R}}_-)\big),$$

$\mathcal{S}(\overline{\mathbb{R}}_\pm) := \mathcal{S}(\mathbb{R})|_{\overline{\mathbb{R}}_\pm}$.

Proof. Let $a(\tau) \in S_{\mathrm{tr}}^{-1}(\mathbb{R})$, and consider the expansion

$$a(\tau) \sim \sum_{j=0}^{\infty} a_j (i\tau)^{-(j+1)}, \quad \tau \to \pm\infty. \tag{2.45}$$

We use the fact that there is a $u \in \mathcal{S}(\overline{\mathbb{R}}_+)$ such that $a_j = \partial_t^j u(0)$ for all j; this is a consequence of Borel's theorem. By Proposition 2.2.14, the symbol $b(\tau) := (F_{t\to\tau} u)(\tau)$ is a plus-symbol, and we have

$$c(\tau) := a(\tau) - b(\tau) \in S^{-\infty}(\mathbb{R}_\tau).$$

Therefore, it suffices to decompose $c(\tau)$ into a plus- and a minus symbol. Here we can take

$$c_\pm(\tau) := \Pi^\pm c(\tau)$$

for the operators $\Pi^\pm$ in (2.24). In order to verify that, consider, for instance, the plus-case. Then $c_+(\tau) = \int_0^\infty e^{-it\tau} \widehat{c}(t)\, dt$. Again by Proposition 2.2.14 we see that c_+ is a plus-symbol. It follows altogether the desired decomposition, namely, $a = b + c_+ + c_-$, i.e.,

$$a_+ = b + c_+, \quad a_- = c_-. \qquad \square$$

A refinement of Proposition 2.2.9 is the following observation.

Remark 2.2.17. Every $a(\tau) \in S_{\mathrm{tr}}^{-1}(\mathbb{R})$ can be written in the forms

$$a(\tau) = b_+(\tau) + c(\tau)$$

and

$$a(\tau) = b_-(\tau) + d(\tau)$$

with plus/minus-symbols $b_\pm \in S_{\mathrm{tr}}^{-1}(\mathbb{R})$ and $c, d \in S^{-\infty}(\mathbb{R})$.

In fact, such a decomposition in the plus case was constructed in the proof of Proposition 2.2.16. The arguments for the minus case are analogous.

Theorem 2.2.18. *Let $a(t,\tau) \in S_{\mathrm{tr}}^\mu(\mathbb{R} \times \mathbb{R})_{\mathrm{C}}$. Then $\mathrm{op}^+(a)$ induces a continuous operator*

$$\mathrm{op}^+(a) : \mathcal{S}(\overline{\mathbb{R}}_+) \to \mathcal{S}(\overline{\mathbb{R}}_+), \tag{2.46}$$

and the correspondence $a \mapsto \mathrm{op}^+(a)$ defines a continuous operator

$$S_{\mathrm{tr}}^\mu(\mathbb{R} \times \mathbb{R})_{\mathrm{C}} \to \mathcal{L}\big(\mathcal{S}(\overline{\mathbb{R}}_+), \mathcal{S}(\overline{\mathbb{R}}_+)\big). \tag{2.47}$$

Proof. Let us consider the case of t-independent symbols; the general case is similar and left to the reader. The assertion is clear when $a(\tau)$ is a polynomial in τ. Therefore, by Remark 2.2.1, we may assume $a(\tau) \in S_{\mathrm{tr}}^{-1}(\mathbb{R})$. First we have continuous operators (2.44). Moreover, the multiplication of symbols in $S_{\mathrm{tr}}^{-1}(\mathbb{R})$ is bilinear and continuous. Thus, because of $\mathrm{op}^+(a) = \mathrm{r}^+ F^{-1} a(\tau) F \mathrm{e}^+$ our operator is a composition of continuous operators

$$\mathcal{S}(\overline{\mathbb{R}}_+) \to F\mathrm{e}^+ \mathcal{S}(\overline{\mathbb{R}}_+) \to \left(F(\mathrm{e}^+ \mathcal{S}(\overline{\mathbb{R}}_+) + \mathrm{e}^- \mathcal{S}(\overline{\mathbb{R}}_-))\right) \to \mathcal{S}(\overline{\mathbb{R}}_+).$$

The proof of the continuity of (2.47) is straightforward. □

Proposition 2.2.19. *The operator* $\mathrm{op}^+(l_-^\nu)$, $\nu \in \mathbb{R}$, *where* $l_-^\nu(\tau) := (\delta - i\tau)^\nu$, *induces a continuous operator*

$$\mathrm{op}^+(l_-^\nu) : \mathcal{S}(\overline{\mathbb{R}}_+) \to \mathcal{S}(\overline{\mathbb{R}}_+) \tag{2.48}$$

which is invertible, with $(\mathrm{op}^+(l_-^\nu))^{-1} = \mathrm{op}^+(l_-^{-\nu})$. *Moreover,*

$$\mathrm{r}^+ \mathrm{op}(l_-^\nu)\, \mathrm{e}_s^+ : H^s(\mathbb{R}_+) \to H^{s-\nu}(\mathbb{R}_+) \tag{2.49}$$

is an isomorphism for every $s \in \mathbb{R}$ *and* $(\mathrm{r}^+ \mathrm{op}(l_-^\nu)\, \mathrm{e}_s^+)^{-1} = \mathrm{r}^+ \mathrm{op}(l_-^{-\nu})\, \mathrm{e}_{s-\nu}^+$ *for arbitrary continuous extension operators* $\mathrm{e}_\sigma^+ : H^\sigma(\mathbb{R}_+) \to H^\sigma(\mathbb{R})$. *The involved operators are independent of the choice of* e_σ^+, *and the latter may be equivalently replaced by* e^+ *whenever* $\sigma > -1/2$.

Proof. By virtue of Seeley's extension theorem, cf. [57], there is a continuous extension operator $E : \mathcal{S}(\overline{\mathbb{R}}_+) \to \mathcal{S}(\mathbb{R})$. For $u \in \mathcal{S}(\overline{\mathbb{R}}_+)$ we have $\mathrm{op}^+(l_-^\nu)u = \mathrm{r}^+ \mathrm{op}(l_-^\nu)\mathrm{e}^+ u = \mathrm{r}^+ \mathrm{op}(l_-^\nu)Eu$ independently of the choice of the operator E. In fact, we have $\mathrm{e}^+ u = (\mathrm{e}^+ u - Eu) + Eu$. But $\mathrm{e}^+ u - Eu \in L^2(\mathbb{R}_-) = H^0(\overline{\mathbb{R}}_-)$ and the minus property of the symbol $l_-(\tau)$ allows us to apply Theorem 2.2.8 (ii). Thus $\mathrm{r}^+ \mathrm{op}(l_-^\nu)\mathrm{e}^+ u$ can be interpreted as $\mathrm{r}^+ \mathrm{op}(l_-^\nu)\,Eu$, which is a composition of continuous operators $E : \mathcal{S}(\overline{\mathbb{R}}_+) \to \mathcal{S}(\mathbb{R})$, $F : \mathcal{S}(\mathbb{R}) \to \mathcal{S}(\mathbb{R})$, $l_-^\nu : \mathcal{S}(\mathbb{R}) \to \mathcal{S}(\mathbb{R})$, $F^{-1} : \mathcal{S}(\mathbb{R}) \to \mathcal{S}(\mathbb{R})$, $\mathrm{r}^+ : \mathcal{S}(\mathbb{R}) \to \mathcal{S}(\overline{\mathbb{R}}_+)$. In order to check the form of the inverse we observe that

$$(\mathrm{r}^+ \mathrm{op}(l_-^{-\nu})\, \mathrm{e}^+)(\mathrm{r}^+ \mathrm{op}(l_-^\nu)\, \mathrm{e}^+) = \mathrm{r}^+ \mathrm{op}(l_-^{-\nu})\, \mathrm{op}(l_-^\nu)\, \mathrm{e}^+$$
$$- \mathrm{r}^+ \mathrm{op}(l_-^{-\nu}(1 - \mathrm{e}^+ \mathrm{r}^+)\, \mathrm{op}(l_-^\nu)\, \mathrm{e}^+.$$

The first summand on the right is the identity while the second one vanishes.

The assertion on (2.49) is also easy and is left to the reader. □

Proposition 2.2.20. *Let* $A = \mathrm{op}^+(a)$ *for* $a(\tau) \in S_{\mathrm{tr}}^0(\mathbb{R})$, *regarded as a continuous operator*

$$\mathrm{op}^+(a) : L^2(\mathbb{R}_+) \to L^2(\mathbb{R}_+)$$

cf. also Theorem 2.2.18 for $s = 0$. *Then for the* $L^2(\mathbb{R}_+)$-*adjoint we have*

$$\left(\mathrm{op}^+(a)\right)^* = \mathrm{op}^+(\overline{a})$$

for the complex conjugate $\overline{a}(\tau) \in S_{\mathrm{tr}}^0(\mathbb{R})$.

Proof. The proof is straightforward and left to the reader.   □

Remark 2.2.21. Clearly, $\mathrm{op}^+(a)^* = \mathrm{op}^+(\overline{a})$ also holds for arbitrary $a \in S^0(\mathbb{R})$. The adjoint refers to the sesquilinear scalar product $(u,v)_{L^2(\mathbb{R}_+)} = \int_0^\infty u(t)\overline{v}(t)dt$. It makes sense also to consider the bilinear pairing $\langle u,v \rangle = \int_0^\infty u(t)v(t)dt$. Then the transpose operator, defined by $\langle \mathrm{op}^+(a))u,v \rangle = \langle u,^t\mathrm{op}^+(a))v \rangle$, $u,v \in L^2(\mathbb{R}_+)$, $a \in S^0(\mathbb{R})$, has the form $^t\mathrm{op}^+(a)) = \mathrm{op}^+(a^\vee)$ for $(a^\vee)(\tau) := a(-\tau)$.

Theorem 2.2.22. *Let $\varepsilon : \mathbb{R}_\pm \to \mathbb{R}_\mp$ be defined by $\varepsilon(t) := -t$ and $\varepsilon^* : L^2(\mathbb{R}_\mp) \to L^2(\mathbb{R}_\pm)$ the corresponding function pull-back. Then*

$$\mathrm{r}^+\mathrm{op}(a)\,\mathrm{e}^-\varepsilon^*, \;\; \varepsilon^*\mathrm{r}^-\mathrm{op}(a)\mathrm{e}^+ : L^2(\mathbb{R}_+) \to L^2(\mathbb{R}_+) \tag{2.50}$$

for $a(\tau) \in S^0_{\mathrm{tr}}(\mathbb{R})$ induce continuous operators $L^2(\mathbb{R}_+) \to \mathcal{S}(\overline{\mathbb{R}}_+)$, and $a \mapsto \mathrm{r}^+\mathrm{op}(a)\,\mathrm{e}^-\varepsilon^*$, $a \mapsto \varepsilon^*\mathrm{r}^-\mathrm{op}(a)\,\mathrm{e}^+$ define continuous mappings

$$S^0_{\mathrm{tr}}(\mathbb{R}) \to \mathcal{L}(L^2(\mathbb{R}_+), \mathcal{S}(\overline{\mathbb{R}}_+)).$$

Proof. The assertion is true for symbols which are a constant (then both operators vanish). Thus by virtue of Remark 2.2.1 we may assume $a \in S^{-1}_{\mathrm{tr}}(\mathbb{R})$. Let us consider the operator $\mathrm{r}^+\mathrm{op}(a)\,\mathrm{e}^-\varepsilon^*$. If a is a minus-symbol we have $\mathrm{r}^+\mathrm{op}(a)\,\mathrm{e}^- = 0$. In fact, for $u \in L^2(\mathbb{R}_+)$ we have $\mathrm{e}^-\varepsilon^*u \in \mathrm{e}^-L^2(\mathbb{R}_-) = H^0_0(\overline{\mathbb{R}}_-)$. Using the decomposition

$$a(\tau) = m_N(\tau) + r_{N+1}(\tau)$$

from Proposition 2.2.9, we can write

$$\mathrm{r}^+\mathrm{op}(a)\,\mathrm{e}^- = \mathrm{r}^+\mathrm{op}(m_N)\,\mathrm{e}^- + \mathrm{r}^+\mathrm{op}(r_{N+1})\,\mathrm{e}^- = \mathrm{r}^+\mathrm{op}(r_{N+1})\,\mathrm{e}^-.$$

This allows us to assume that our symbol is of order ≤ -2. Now for $a(\tau) \in S^{-2}_{\mathrm{tr}}(\mathbb{R})$ we can write

$$\mathrm{r}^+\mathrm{op}(a)\mathrm{e}^-\varepsilon^*v(t) = \mathrm{r}^+\int_{\mathbb{R}}\int_0^\infty e^{i(t+t')\tau}a(\tau)v(-t')dt'đ\tau$$

$$= \mathrm{r}^+\int_0^\infty\left\{\int e^{i(t+t')\tau}a(\tau)đ\tau\right\}v(-t')dt'.$$

From Proposition 2.2.16 we have $\int e^{ir\tau}a(\tau)đ\tau \in \mathrm{e}^+\mathcal{S}(\overline{\mathbb{R}}_+) + \mathrm{e}^-\mathcal{S}(\overline{\mathbb{R}}_-)$. Since r has the meaning of $t + t'$ for $t > 0$, $t' > 0$, we obtain

$$\mathrm{r}^+\mathrm{op}(a)\mathrm{e}^-\varepsilon^*v(t) = \mathrm{r}^+\int f(t+t')v(-t')dt' \tag{2.51}$$

for some $f(r) \in \mathrm{e}^+\mathcal{S}(\overline{\mathbb{R}}_+)$. The right-hand side of (2.51) represents a continuous operator $L^2(\mathbb{R}_+) \to \mathcal{S}(\overline{\mathbb{R}}_+)$, and it is easy to verify that it depends continuously on a. As for the second operator in (2.50), we assume again $a \in S^{-1}_{\mathrm{tr}}(\mathbb{R})$ and observe that when a is a plus-symbol we have $\mathrm{r}^-\mathrm{op}(a)\mathrm{e}^+ = 0$. Thus, similarly as in the first part of the proof, we may assume $\mu \leq -2$. The rest of the proof is of analogous structure as before.   □

Corollary 2.2.23. *Let g denote one of the operators in (2.50), and let g^* be its $L^2(\mathbb{R}_+)$-adjoint. Then g and g^* induce continuous operators*

$$g, g^* : L^2(\mathbb{R}_+) \to S(\overline{\mathbb{R}}_+). \tag{2.52}$$

Proof. The assertion for $g = r^+\mathrm{op}(a)e^-\varepsilon^*$ is contained in Theorem 2.2.22. Moreover, since $g^* = \varepsilon^* r^-\mathrm{op}(\bar{a})e^+$ the result for g^* also follows from Theorem 2.2.22. □

Theorem 2.2.24. *For an operator $g \in \mathcal{L}(L^2(\mathbb{R}_+))$ the following properties are equivalent:*
(i) *g induces continuous operators (2.52).*
(ii) *There exists a $c(t,t') \in S(\overline{\mathbb{R}}_+ \times \overline{\mathbb{R}}_+)\big(= S(\mathbb{R} \times \mathbb{R})|_{\overline{\mathbb{R}}_+ \times \overline{\mathbb{R}}_+} = S(\overline{\mathbb{R}}_+) \hat{\otimes}_\pi S(\overline{\mathbb{R}}_+)\big)$ such that*

$$gu(t) = \int_0^\infty c(t,t')u(t')dt'.$$

Proof. The proof of (ii) $\Longrightarrow$ (i) is evident. For (i) $\Longrightarrow$ (ii) we argue as follows. By the continuity of (2.52), the kernel $c(t,t')$ belongs to

$$S(\overline{\mathbb{R}}_{+,t}) \hat{\otimes}_\pi L^2(\mathbb{R}_{+,t'}) \cap L^2(\mathbb{R}_{+,t}) \hat{\otimes}_\pi S(\overline{\mathbb{R}}_{+,t'}).$$

In particular, g is a Hilbert–Schmidt operator, $c(t,t') \in L^2(\mathbb{R}_+ \times \mathbb{R}_+)$. From (2.52) we also see that the operators

$$\langle t \rangle^l g : L^2(\mathbb{R}_+) \to S(\overline{\mathbb{R}}_{+,t}), \quad \langle t' \rangle^{l'} g^* : L^2(\mathbb{R}_+) \to S(\overline{\mathbb{R}}_{+,t'}), \quad l, l' \in \mathbb{N}, \tag{2.53}$$

are continuous. This implies $\langle t \rangle^l c(t,t'), \langle t' \rangle^{l'} c(t,t') \in L^2(\mathbb{R}_+ \times \mathbb{R}_+)$, $l, l' \in \mathbb{N}$. Since for every $N \in \mathbb{N}$ there are $l, l' \in \mathbb{N}$ such that

$$\langle t, t' \rangle^N \leq c\left(\langle t \rangle^l + \langle t' \rangle^{l'}\right)$$

for some $c > 0$, it follows that

$$\langle t, t' \rangle^N c(t,t') \in L^2(\mathbb{R}_+ \times \mathbb{R}_+), \quad N \in \mathbb{N}. \tag{2.54}$$

Another consequence of (2.52) is that the operators with kernels $\partial_t^k c(t,t')$ and $\partial_{t'}^{k'} c(t,t')$ are continuous as operators from $L^2(\mathbb{R}_+)$ to $S(\overline{\mathbb{R}}_+)$. After (2.54) we can say the same about the kernels $\partial_t^k \langle t,t' \rangle^N c(t,t')$ and $\partial_{t'}^{k'} \langle t,t' \rangle^N c(t,t')$ for all k, k', N. This implies

$$\partial_t^k \langle t,t' \rangle^N c(t,t'), \; \partial_{t'}^{k'} \langle t,t' \rangle^N c(t,t') \in L^2(\mathbb{R}_+ \times \mathbb{R}_+) \quad \text{for all } k, k', N \in \mathbb{N}. \tag{2.55}$$

It remains to conclude

$$\partial_t^m \partial_{t'}^{m'} \langle t,t' \rangle^N c(t,t') \in L^2(\mathbb{R}_+ \times \mathbb{R}_+) \quad \text{for all } m, m', N \in \mathbb{N}. \tag{2.56}$$

Let us show this first for $m = m' = 1$. Here we employ the fact that $(\partial_t + \partial_{t'})^2 = \partial_t^2 + \partial_{t'}^2 + 2\partial_t\partial_{t'}$. From (2.55) for $k = k' = 1$ we conclude that

$$c_1(t, t') := (\partial_t + \partial_{t'})\langle t, t'\rangle^N c(t, t') \in L^2(\mathbb{R}_+ \times \mathbb{R}_+).$$

Since the mapping properties for the operator with kernel $c_1(t, t')$ are again as in (2.52), we can apply again the operator $(\partial_t + \partial_{t'})$ to $c_1(t, t')$ and see that $c_2(t, t') := (\partial_t + \partial_{t'})c_1(t, t') = (\partial_t + \partial_{t'})^2\langle t, t'\rangle^N c(t, t') \in L^2(\mathbb{R}_+ \times \mathbb{R}_+)$. Because of (2.55) we then obtain (2.56) for $m = m' = 1$. By iterating the argument we finally obtain (2.56), which means that $c(t, t') \in \mathcal{S}(\overline{\mathbb{R}}_+ \times \overline{\mathbb{R}}_+)$.   □

Definition 2.2.25. (i) An operator $g \in \mathcal{L}(L^2(\mathbb{R}_+))$ as in Theorem 2.2.24 (i) or (ii) is called a Green operator (on the half-line) of type 0.

  (ii) An operator of the form $\sum_{j=0}^d g_j\partial_t^j$ with Green operators g_j of type 0 is called a Green operator (on the half-line) of type d.
  Let $\Gamma^d(\mathbb{R}_+)$ denote the space of all Green operators of type $d \in \mathbb{N}$.

  Later on, in the full calculus of 2×2 block matrices of operators on the half-line we use the notation $\mathcal{B}_G^d(\mathbb{R}_+)$.

Proposition 2.2.26. *Let $W \subset \mathcal{S}(\overline{\mathbb{R}}_+)$ be a finite-dimensional subspace. Then the orthogonal projection $g : L^2(\mathbb{R}_+) \to W$ belongs to $\Gamma^0(\mathbb{R}_+)$.*

Proof. The space W can be described as the linear span of elements $w_1, \ldots, w_m \in \mathcal{S}(\overline{\mathbb{R}}_+)$, $m = \dim W$, such that $\|w_j\|_{L^2(\mathbb{R}_+)} = 1$ and $(w_i, w_j)_{L^2(\mathbb{R}_+)} = \delta_{ij}$ for $j = 1, \ldots, m$. We can write

$$gu(t) = \sum_{j=1}^m (u, w_j)_{L^2(\mathbb{R}_+)} w_j = \int_0^\infty w_j(t)\overline{w}_j(t')u(t')dt'$$

for the kernel $c(t, t') = \sum_{j=1}^m w_j(t)\overline{w}_j(t') \in \mathcal{S}(\overline{\mathbb{R}}_+) \,\widehat{\otimes}_\pi\, \mathcal{S}(\overline{\mathbb{R}}_+)$.   □

Remark 2.2.27. For every $s > d - 1/2$, any $g \in \Gamma^d(\mathbb{R}_+)$ induces a compact operator

$$g : H^s(\mathbb{R}_+) \to H^s(\mathbb{R}_+) \tag{2.57}$$

and a continuous operator

$$g : H^s(\mathbb{R}_+) \to \mathcal{S}(\overline{\mathbb{R}}_+). \tag{2.58}$$

  In fact (2.57) is a consequence of (2.58), while (2.58) follows from the continuity of $\partial_t^j : H^s(\mathbb{R}_+) \to H^{s-j}(\mathbb{R}_+)$ and $g_0 : H^{s-j}(\mathbb{R}_+) \to \mathcal{S}(\overline{\mathbb{R}}_+)$ for every $g_0 \in \Gamma^0(\mathbb{R}_+)$, $s - j > -1/2$.

Proposition 2.2.28. *Every $g \in \Gamma^d(\mathbb{R}_+)$, $d \geq 1$, can be written in a unique way as*

$$g = g_0 + \sum_{j=0}^{d-1} l_j \gamma_j \tag{2.59}$$

with $g_0 \in \Gamma^0(\mathbb{R}_+)$, $l_j \in \mathcal{S}(\overline{\mathbb{R}}_+)$, $\gamma_j u := \partial_t^j u(0)$.

Proof. Let us first show the uniqueness of representation (2.59). If $g = h_0 + \sum_{j=0}^{d-1} m_j \gamma_j$ is another such representation, then we have

$$h_0 - g_0 = \sum_{j=0}^{d-1} (l_j - m_j) \gamma_j.$$

It follows that

$$\int_0^\infty (h_0 - g_0)(t, t') u(t') dt' = 0$$

for all $u \in C_0^\infty(\mathbb{R}_+)$, and hence $h_0 = g_0$. This yields $\sum_{j=0}^{d-1}(l_j - m_j)\gamma_j \omega(t) t^k = 0$ for a cut-off function ω and $0 \leq k \leq d-1$. It follows that $l_k = m_k$ for all k, and so the uniqueness is established.

Now an operator $gu(t) := \int_0^\infty c(t, t') \partial_{t'}^j u(t') dt'$ for $c(t, t') \in \mathcal{S}(\overline{\mathbb{R}}_+ \times \overline{\mathbb{R}}_+)$, $j \geq 1$, can be reformulated by integration by parts as

$$gu(t) = \int_0^\infty \left(- \partial_{t'} c(t, t')\right) \partial_{t'}^{j-1} u(t') dt' + c(t, t') \partial_{t'}^{j-1} u(t') \big|_0^\infty$$

$$= \int_0^\infty \left(- \partial_{t'} c(t, t')\right) \partial_{t'}^{j-1} u(t') dt' - c(t, 0) \gamma_{j-1} u.$$

Iterating this construction we obtain (2.59). □

Proposition 2.2.29. *Let $c(t, t') \in \mathcal{S}(\overline{\mathbb{R}}_+) \widehat{\otimes}_\pi \mathcal{S}(\overline{\mathbb{R}}_+)$ and $\mu \in \mathbb{R}, s \in \mathbb{N}, s > j - 1/2$, for some $j \in \mathbb{N}$. Then*

$$g : u \mapsto \operatorname{op}^+(l_-^{s-\mu}) \int_0^\infty c(t, t') \partial_{t'}^j \operatorname{op}^+(l_-^{-s}) u(t') dt'$$

defines an operator in $\Gamma^0(\mathbb{R}_+)$.

Proof. First note that the continuity (2.48) allows us to replace $c(t, t')$ by

$$c_1(t, t') := \operatorname{op}^+(l_-^{s-\mu}) c(t, t') \in \mathcal{S}(\overline{\mathbb{R}}_+) \widehat{\otimes}_\pi \mathcal{S}(\overline{\mathbb{R}}_+).$$

For the proof we check the mapping properties (2.52), cf. Theorem 2.2.24 and Definition 2.2.25. From Proposition 2.2.19 we have the continuity

$$\partial_{t'}^j \operatorname{op}^+(l_-^{-s}) : L^2(\mathbb{R}_+) \to H^{s-j}(\mathbb{R}_+).$$

This yields $\int_0^\infty c_1(t,t')\partial_{t'}^j \mathrm{op}^+(l_-^{-s})u(t')dt' \in \mathcal{S}(\overline{\mathbb{R}}_+)$ and also the continuity of $g : L^2(\mathbb{R}_+) \to \mathcal{S}(\overline{\mathbb{R}}_+)$. Moreover, we have $g^*v(t') = \mathrm{op}^+(\bar{b})\int_0^\infty \bar{c}_1(t,t')v(t)dt$ for $b(\tau) := l_-^{-s}(\tau)(i\tau)^j$. For the desired continuity of $g^* : L^2(\mathbb{R}_+) \to \mathcal{S}(\overline{\mathbb{R}}_+)$ it suffices now to observe that $\mathrm{op}^+(\bar{b}) : \mathcal{S}(\overline{\mathbb{R}}_+) \to \mathcal{S}(\overline{\mathbb{R}}_+)$ is continuous, which is a consequence of $\bar{b} \in S_{\mathrm{tr}}^{-s+j}(\mathbb{R})$, and Theorem 2.2.18. $\qquad\square$

Symbols with the transmission property also generate other operators which are typical in BVPs, for instance, potential operators.

Remark 2.2.30. Let δ_0 be the Dirac distribution at the origin, $a(\tau) \in S_{\mathrm{tr}}^{-1}(\mathbb{R})$, and apply $\mathrm{op}(a)$ to δ_0 in the distributional sense. Then

$$(\mathrm{op}(a)\delta_0)(t) \in \mathrm{e}^+\mathcal{S}(\overline{\mathbb{R}}_+) + \mathrm{e}^-\mathcal{S}(\overline{\mathbb{R}}_-). \tag{2.60}$$

Indeed, we have $(F_{t\to\tau}\delta_0)(\tau) = 1$, i.e.,

$$(\mathrm{op}(a)\delta_0)(t) = \int e^{it\tau}a(\tau)\hat{\delta}_0(\tau)d\!\!\!/\tau = \int e^{it\tau}a(\tau)d\!\!\!/\tau;$$

then (2.60) is a consequence of Proposition 2.2.16. This gives us a linear operator

$$k : \mathbb{C} \to \mathcal{S}(\overline{\mathbb{R}}_+), \quad c \mapsto \mathrm{r}^+\mathrm{op}(a)(c\delta_0), \tag{2.61}$$

which is an example of a potential operator in the calculus on the half-line, cf. Definition 2.2.32 below.

Theorem 2.2.31. (i) *Let $a(\tau) \in S_{\mathrm{tr}}^\mu(\mathbb{R})$, $b(\tau) \in S_{\mathrm{tr}}^\nu(\mathbb{R})$. Then we have*

$$\mathrm{op}^+(a)\mathrm{op}^+(b) = \mathrm{op}^+(ab) + g \tag{2.62}$$

for a Green operator $g \in \Gamma^{\max\{\nu,0\}}(\mathbb{R}_+)$.

(ii) *We have*

$$a(\tau) \in S_{\mathrm{tr}}^\mu(\mathbb{R}), g \in \Gamma^e(\mathbb{R}_+) \implies \mathrm{op}^+(a)\,g \in \Gamma^e(\mathbb{R}_+), \tag{2.63}$$

$$k \in \Gamma^d(\mathbb{R}_+), b(\tau) \in S_{\mathrm{tr}}^\nu(\mathbb{R}) \implies k\,\mathrm{op}^+(b) \in \Gamma^{\max\{\nu+d,0\}}(\mathbb{R}_+), \tag{2.64}$$

and

$$k \in \Gamma^d(\mathbb{R}_+), g \in \Gamma^e(\mathbb{R}_+) \implies kg \in \Gamma^e. \tag{2.65}$$

Proof. (i) We write

$$a(\tau) = a_0(\tau) + p(\tau), \quad b(\tau) = b_0(\tau) + q(\tau)$$

for $a_0, b_0 \in S_{\mathrm{tr}}^{-1}(\mathbb{R})$ and polynomials p and q of order μ and ν, respectively, cf. the formula (2.15). We have

$$\mathrm{op}^+(a)\,\mathrm{op}^+(b) = \mathrm{op}^+(a_0)\,\mathrm{op}^+(b_0) + \mathrm{op}^+(a_0)\,\mathrm{op}^+(q)$$
$$+ \mathrm{op}^+(p)\,\mathrm{op}^+(b_0) + \mathrm{op}^+(p)\,\mathrm{op}^+(q). \tag{2.66}$$

The first summand in (2.66) can be written as

$$\text{op}^+(a_0)\,\text{op}^+(b_0) = \text{op}^+(a_0 b_0) + \text{r}^+\text{op}(a_0)\,(1 - \text{e}^+\text{r}^+)\,\text{op}(b_0)\,\text{e}^+.$$

We show that $g_0 := \text{r}^+\text{op}(a_0)\,(1 - \text{e}^+\text{r}^+)\,\text{op}(b_0)\,\text{e}^+$ belongs to Γ^0, i.e., induces continuous operators

$$g_0,\, g_0^* : L^2(\mathbb{R}_+) \to \mathcal{S}(\overline{\mathbb{R}}_+).$$

Using the isomorphisms $\varepsilon^* : L^2(\mathbb{R}_\pm) \to L^2(\mathbb{R}_\mp)$ coming from $\varepsilon : \mathbb{R}_\mp \to \mathbb{R}_\pm$, $\varepsilon(t) = -t$, we obtain

$$g_0 = \text{r}^+\text{op}(a_0)\,\text{e}^-\varepsilon^*\varepsilon^*\text{r}^-\,\text{op}(b_0)\,\text{e}^+, \quad g_0^* = \text{r}^+\text{op}(\bar{b}_0)\,\text{e}^-\varepsilon^*\varepsilon^*\text{r}^-\,\text{op}(\bar{a}_0)\,\text{e}^+.$$

The desired mapping properties now follow from Theorem 2.2.22. Moreover, if $q(\tau)$ is a polynomial of order ν we have $\text{op}^+(q)\text{op}^+(a_0) = \text{op}^+(qa_0)$, and

$$\text{op}^+(a_0)\text{op}^+(q) = \text{op}^+(a_0 q) + h \quad \text{for some } h \in \Gamma^\nu(\mathbb{R}_+). \tag{2.67}$$

For the proof of (2.67) we assume $q(\tau) := \tau$. The general case easily follows by iterating the argument, while the case $\nu = 0$ is trivial, with $h = 0$. In the computation we may assume $u \in C_0^\infty(\overline{\mathbb{R}}_+)$; then

$$\text{op}^+(a_0)\text{op}^+(q)u(t) = \text{r}^+ \int e^{it\tau} a_0(\tau) \left\{ \int_0^\infty e^{-it'\tau} i^{-1} \partial_{t'} u(t')\,dt' \right\} d\!\!\!/\tau$$

$$= \text{r}^+ \int e^{it\tau} a_0(\tau) \left\{ e^{-it'\tau} u(t')\big|_0^\infty + \int_0^\infty e^{-it'\tau} \tau u(t')\,dt' \right\} d\!\!\!/\tau$$

$$= \text{op}^+(a_0\tau)u(t) + \text{r}^+ \int e^{it\tau} i^{-1} a_0(\tau)\,d\!\!\!/\tau\,\gamma_0 u.$$

From Proposition 2.2.28 and Remark 2.2.30 it follows that the second term on the right-hand side of the last relation represents an element of $\Gamma^1(\mathbb{R}_+)$. Finally, we obviously have $\text{op}^+(p)\text{op}^+(q) = \text{op}^+(pq)$.

(ii) Let a and g be as in (2.63). Then Theorem 2.2.24 and Definition 2.2.25 (ii) shows that

$$gu(t) = \sum_{j=0}^d \int_0^\infty c_j(t, t')\partial_{t'}^j u(t')\,dt'$$

for kernels $c_j(t, t') \in \mathcal{S}(\overline{\mathbb{R}}_+)\,\hat{\otimes}_\pi\,\mathcal{S}(\overline{\mathbb{R}}_+)$. Moreover, by Theorem 2.2.18, we have a continuous operator (2.46), and it follows that

$$\text{op}^+(a)\,c_j(t, t') \in \mathcal{S}(\overline{\mathbb{R}}_+)\,\hat{\otimes}_\pi\,\mathcal{S}(\overline{\mathbb{R}}_+).$$

This implies $\text{op}^+(a)\,g \in \Gamma^d(\mathbb{R}_+)$.

Next let k and b be as in (2.64), and assume for simplicity that

$$kv(t) = \int_0^\infty h(t, t')\partial_{t'}^d v(t')\,dt', \quad h(t, t') \in \mathcal{S}(\overline{\mathbb{R}}_+)\,\hat{\otimes}_\pi\,\mathcal{S}(\overline{\mathbb{R}}_+).$$

Then

$$k \operatorname{op}^+(b)v(t) = \int_0^\infty h(t,t')\operatorname{op}^+((i\tau)^d b)v(t')dt'.$$

If $d + \nu \le 0$, we have $\operatorname{op}^+((i\tau)^d b) : L^2(\mathbb{R}_+) \to L^2(\mathbb{R}_+)$, and it follows that $k \operatorname{op}^+(b) \in \Gamma^0(\mathbb{R}_+)$. In the case $d + \nu \ge 0$ we can write $(i\tau)^d b(\tau) = f_0(\tau) + r(\tau)$ for some $f_0(\tau) \in S_{\mathrm{tr}}^{-1}(\mathbb{R})$ and a polynomial $r(\tau)$ of order $d + \nu$. We then obtain

$$k \operatorname{op}^+(b)v(t) = \int_0^\infty h(t,t')\operatorname{op}^+(f_0)v(t')dt' + k \operatorname{op}^+(r)v.$$

The first summand defines an element of $\Gamma^0(\mathbb{R}_+)$ and the second one an element of $\Gamma^{d+\nu}(\mathbb{R}_+)$.

The relation (2.65) is a direct consequence of Theorem 2.2.24 and Definition 2.2.25 (ii). □

Definition 2.2.32. Let $\mathcal{B}^{\mu,d}(\overline{\mathbb{R}}_+; j_1, j_2)$ for $\mu \in \mathbb{Z}$, $d \in \mathbb{N}$, $j_1, j_2 \in \mathbb{N}$, denote the set of all block matrix operators of the form

$$\boldsymbol{a} := \begin{pmatrix} \operatorname{op}^+(a) + g_{11} & g_{12} \\ g_{21} & g_{22} \end{pmatrix} : \begin{matrix} \mathcal{S}(\overline{\mathbb{R}}_+) \\ \oplus \\ \mathbb{C}^{j_1} \end{matrix} \to \begin{matrix} \mathcal{S}(\overline{\mathbb{R}}_+) \\ \oplus \\ \mathbb{C}^{j_2} \end{matrix} \tag{2.68}$$

for any $a(\tau) \in S_{\mathrm{tr}}^\mu(\mathbb{R})$, $g_{11} \in \Gamma^d(\mathbb{R}_+)$, $g_{12} := (k_1, \ldots, k_{j_1})$, $g_{21} := {}^t(b_1, \ldots, b_{j_2})$ and $g_{22} \in \mathbb{C}^{j_2} \otimes \mathbb{C}^{j_1}$, for

$$k_n : \mathbb{C} \to \mathcal{S}(\overline{\mathbb{R}}_+), \quad n = 1, \ldots, j_1, \quad b_m : \mathcal{S}(\overline{\mathbb{R}}_+) \to \mathbb{C}, \quad m = 1, \ldots, j_2,$$

$b_m u := \sum_{l=0}^d \int_0^\infty c_{ml}(t)(\partial_t^l u)(t)dt$, for certain $c_{ml} \in \mathcal{S}(\overline{\mathbb{R}}_+)$, $l = 0, \ldots, d$, $m = 1, \ldots, j_2$.

The operator g_{21} is also called a trace (or boundary) operator of type d and g_{12} a potential operator (in the calculus of BVPs on the half-line). Moreover, let $\mathcal{B}_G^d(\overline{\mathbb{R}}_+; j_1, j_2)$ denote the set of all operators (2.68) of the form $\boldsymbol{g} = (g_{ij})_{i,j=1,2}$.

Remark 2.2.33. A trace operator g_{21} of type $d \ge 1$ (say, for $j_2 = 0$) has a unique representation

$$g_{21} = g_{0,21} + \sum_{j=0}^{d-1} m_j \gamma_j \tag{2.69}$$

with a trace operator $g_{0,21}$ of type 0, constants m_j, and $\gamma_j u = \partial_t^j u(0)$.

This can be verified in a similar manner as Proposition 2.2.28.

Remark 2.2.34. An operator $\boldsymbol{a} \in \mathcal{B}^{\mu,d}(\overline{\mathbb{R}}_+; j_0, j_2)$ extends to a continuous operator

$$\boldsymbol{a} : \begin{matrix} H^s(\mathbb{R}_+) \\ \oplus \\ \mathbb{C}^{j_1} \end{matrix} \to \begin{matrix} H^{s-\mu}(\mathbb{R}_+) \\ \oplus \\ \mathbb{C}^{j_2} \end{matrix} \tag{2.70}$$

for every $s > d - 1/2$.

Theorem 2.2.35. *If* $a \in \mathcal{B}^{\mu,d}\left(\overline{\mathbb{R}}_+; j_0, j_2\right)$ *and* $b \in \mathcal{B}^{\nu,e}\left(\overline{\mathbb{R}}_+; j_1, j_0\right)$ *then*

$$ab \in \mathcal{B}^{\mu+\nu,h}\left(\overline{\mathbb{R}}_+; j_1, j_2\right)$$

for $h = \max\{\nu + d, e\}$. *Moreover,* $a \in \mathcal{B}_G^{\mu,d}$ *or* $b \in \mathcal{B}_G^{\nu,e}$ *implies* $ab \in \mathcal{B}_G^{\mu+\nu,h}$.

Proof. Write

$$a := \begin{pmatrix} \operatorname{op}^+(a) + g_{11} & g_{12} \\ g_{21} & g_{22} \end{pmatrix}, \quad b := \begin{pmatrix} \operatorname{op}^+(b) + h_{11} & h_{12} \\ h_{21} & h_{22} \end{pmatrix}. \qquad (2.71)$$

The multiplication of the top left corners has been characterised in Theorem 2.2.31. For the other entries we assume for simplicity that $j_0 = j_1 = j_2 = 1$. It remains to consider

$$g_{12}h_{21}, \quad (\operatorname{op}^+(a) + g_{11})h_{12}, \quad g_{12}h_{22}, \quad g_{21}(\operatorname{op}^+(b) + h_{11}), \quad g_{22}h_{21}, \quad g_{22}h_{22}.$$

From Definition 2.2.32 we easily obtain that $g_{12}h_{21} \in \Gamma^e(\mathbb{R}_+)$, $g_{12}h_{22}$ is a potential operator, $g_{22}h_{21}$ is a trace operator of type e, and $g_{22}h_{22}$ is of the type of a bottom right corner. Moreover, $g_{11}h_{12}$ is a potential operator, and $g_{21}h_{11}$ is a trace operator of type e. Finally, $\operatorname{op}^+(a)h_{12}$ is a potential operator as a consequence of Theorem 2.2.18, and $g_{21}\operatorname{op}^+(b)$ is a trace operator of type $\max\{\nu+d, 0\}$ for similar reasons as (2.64). $\qquad \square$

Remark 2.2.36. The function $l_-^\mu(\tau) := (1 - i\tau)^\mu$ is a minus-symbol of order $\mu \in \mathbb{R}$, and

$$\operatorname{op}^+(l_-^\mu) : H^s(\mathbb{R}_+) \to H^{s-\mu}(\mathbb{R}_+)$$

is an isomorphism for every $s \in \mathbb{R}$, $s > \max\{-1/2, \mu-1/2\}$, where $\left(\operatorname{op}^+(l_-^\mu)\right)^{-1} = \operatorname{op}^+(l_-^{-\mu})$. Moreover, $l_+^\mu(\tau) := (1 + i\tau)^\mu$ is a plus-symbol of order $\mu \in \mathbb{R}$, and

$$\operatorname{op}(l_+^\mu) : H_0^s(\mathbb{R}_+) \to H_0^{s-\mu}(\mathbb{R}_+)$$

is an isomorphism for every $s \in \mathbb{R}$, where $\left(\operatorname{op}(l_+^\mu)\right)^{-1} = \operatorname{op}(l_+^{-\mu})$.

Proposition 2.2.37. *For every* $a(\tau) \in S^0(\mathbb{R})$ *and* $j \in \mathbb{N}$ *the following relations hold as operators* $L^2(\mathbb{R}_+) \to L^2(\mathbb{R}_+)$:

(i)

$$\operatorname{op}^+(l_-^{-j}a) = \operatorname{op}^+(l_-^{-j})\operatorname{op}^+(a),$$

(ii)

$$\operatorname{op}^+(al_+^{-j}) = \operatorname{op}^+(a)\operatorname{op}^+(l_+^{-j}).$$

Proof. Consider, for instance, (ii). In this case we employ the identity

$$
\begin{aligned}
\mathrm{op}^+(a l_+^{-j}) &= \mathrm{r}^+\mathrm{op}(a l_+^{-j})\,\mathrm{e}^+ \\
&= \mathrm{r}^+\mathrm{op}(a)\,\mathrm{op}(l_+^{-j})\,\mathrm{e}^+ \\
&= \mathrm{r}^+\mathrm{op}(a)\,\theta^+(t)\,\mathrm{op}(l_+^{-j})\,\mathrm{e}^+ + \mathrm{r}^+\mathrm{op}(a)\,\theta^-(t)\,\mathrm{op}(l_+^{-j})\,\mathrm{e}^+ \\
&= \mathrm{op}^+(a)\,\mathrm{op}^+(l_+^{-j}),
\end{aligned}
$$

where $\theta^\pm(t)$ are the characteristic functions of $\mathbb{R}_\pm$. The last equality on the right-hand side follows from the fact that $\mathrm{op}^+(l_+^{-j})$ acts from $L^2(\mathbb{R}_+)$ to $L^2(\mathbb{R}_+)$, which entails $\theta^-(t)\,\mathrm{op}(l_+^{-j})\,\mathrm{e}^+ = 0$. The case (i) can be treated in an analogous manner. □

Proposition 2.2.38. (i) *Let* $a(\tau) \in S_{\mathrm{tr}}^\mu(\mathbb{R})$, $s \in \mathbb{N}$, *and consider the operator* $\mathrm{op}^+(a) : H^s(\mathbb{R}_+) \to H^{s-\mu}(\mathbb{R}_+)$. *Then*

$$
\boldsymbol{a} := \mathrm{r}^+\mathrm{op}(l_-^{s-\mu})\,\mathrm{e}_{s-\mu}^+\mathrm{op}^+(a)\,\mathrm{op}^+(l_-^{-s}) = \mathrm{op}^+(l_-^{-\mu}a) + g_0 \tag{2.72}
$$

for a $g_0 \in \Gamma^0(\mathbb{R}_+)$. *For* $s - \mu \geq 0$ *the operator* $\boldsymbol{a}$ *coincides with*

$$
\mathrm{op}^+(l_-^{s-\mu})\,\mathrm{op}^+(a)\,\mathrm{op}^+(l_-^{-s}).
$$

(ii) *For* $g \in \Gamma^d(\mathbb{R}_+)$, $s \in \mathbb{N}$, $s > d - 1/2$, *we have*

$$
g_1 := \mathrm{op}^+(l_-^{s-\mu})\,g\,\mathrm{op}^+(l_-^{-s}) \in \Gamma^0(\mathbb{R}_+).
$$

Proof. (i) We have

$$
\mathrm{r}^+\mathrm{op}(l_-^{s-\mu})\,\mathrm{e}_{s-\mu}^+\mathrm{r}^+\mathrm{op}(a)\,\mathrm{e}^+ = \mathrm{r}^+\mathrm{op}(l_-^{s-\mu}a)\,\mathrm{e}^+ - \mathrm{r}^+\mathrm{op}(l_-^{s-\mu})\big(1-\mathrm{e}_{s-\mu}^+\mathrm{r}^+\big)\,\mathrm{op}(a)\,\mathrm{e}^+.
$$

Since $\big(1-\mathrm{e}_{s-\mu}^+\mathrm{r}^+\big)\mathrm{op}(a)\mathrm{e}^+ \in H_0^{s-\mu}(\overline{\mathbb{R}}_-)$ the right-hand side in the above equation is equal to $\mathrm{r}^+\mathrm{op}(l_-^{s-\mu}a)\mathrm{e}^+$, cf. Theorem 2.2.8. For $s - \mu \geq 0$ we may replace $\mathrm{e}_{s-\mu}^+$ by e^+ with the same result. Applying Theorem 2.2.35 it follows that

$$
\boldsymbol{a} = \mathrm{op}^+(l_-^{s-\mu}a)\,\mathrm{op}^+(l_-^{-s}) = \mathrm{op}^+(l_-^{-\mu}a) + g_0
$$

for a $g_0 \in \Gamma^0(\mathbb{R}_+)$, $h = \max\{\nu + d, e\}$.
   (ii) is a consequence of Proposition 2.2.29. □

Theorem 2.2.39. *Let* $\boldsymbol{a} \in \mathcal{B}^{\mu,d}(\overline{\mathbb{R}}_+)$ *and* $s \in \mathbb{N}$, $s \geq \max\{\mu, d\}$. *Then*

$$
R : \boldsymbol{a} \mapsto \mathrm{op}^+(l_-^{s-\mu})\,\boldsymbol{a}\,\mathrm{op}^+(l_-^{-s}) =: \boldsymbol{b} \tag{2.73}
$$

defines an isomorphism

$$
R : \mathcal{B}^{\mu,d}(\overline{\mathbb{R}}_+) \to \mathcal{B}^{0,0}(\overline{\mathbb{R}}_+) \tag{2.74}
$$

with the inverse

$$
R^{-1} : \boldsymbol{b} \mapsto \mathrm{op}^+(l_-^{-s+\mu})\,\boldsymbol{b}\,\mathrm{op}^+(l_-^{s}). \tag{2.75}
$$

Proof. That the map (2.74) is an isomorphism is a consequence of Proposition 2.2.38. To show that $RR^{-1} = \mathrm{id}$, we simply observe that

$$\mathrm{op}^+\left(l_-^{-s+\mu}\right)\left\{\mathrm{op}^+\left(l_-^{s-\mu}\right)\boldsymbol{a}\,\mathrm{op}^+\left(l_-^{-s}\right)\right\}\mathrm{op}^+\left(l_-^s\right) = \boldsymbol{a},$$

since $\mathrm{op}^+\left(l_-^{-s+\mu}\right)\mathrm{op}^+\left(l_-^{s-\mu}\right) = 1$ and $\mathrm{op}^+\left(l_-^{-s}\right)\mathrm{op}^+\left(l_-^s\right) = 1$, cf. Proposition 2.2.19. The relation $R^{-1}R = \mathrm{id}$ can be proved in a similar manner. $\qquad\square$

We now discuss invertibility of operators in the set $\mathcal{B}^{\mu,d}\left(\overline{\mathbb{R}}_+;j_1,j_2\right)$.

Proposition 2.2.40. *Let $g \in \Gamma^0(\mathbb{R}_+)$, and assume that the operator*

$$1 + g : L^2(\mathbb{R}_+) \to L^2(\mathbb{R}_+) \tag{2.76}$$

is an isomorphism. Then $(1+g)^{-1} = 1 + h$ for some $h \in \Gamma^0(\mathbb{R}_+)$.

Proof. First it is clear that $h := (1+g)^{-1} - 1$ and $1 + h$ belong to $\mathcal{L}\left(L^2(\mathbb{R}_+)\right)$. We employ the fact that $h \in \Gamma^0(\mathbb{R}_+)$ is equivalent to the continuity of h, h^* : $L^2(\mathbb{R}_+) \to \mathcal{S}\left(\overline{\mathbb{R}}_+\right)$, cf. formula (2.52) . The relation $(1+g)(1+h) = 1$ entails the identity $h = -g(1+h)$. Thus the continuity of $g : L^2(\mathbb{R}_+) \to \mathcal{S}\left(\overline{\mathbb{R}}_+\right)$ implies that of h. Moreover, $(1+g^*)(1+h^*) = 1$ yields $h^* = -g^*(1+h^*)$, and we also obtain the continuity of $h^* : L^2(\mathbb{R}_+) \to \mathcal{S}\left(\overline{\mathbb{R}}_+\right)$. $\qquad\square$

Remark 2.2.41. The operator (2.76) is Fredholm and of index zero, and there are finite-dimensional subspaces $V, W \subset \mathcal{S}\left(\overline{\mathbb{R}}_+\right)$ such that

$$V = \ker(1+g), \quad W \oplus \mathrm{im}\,(1+g) = L^2(\mathbb{R}_+). \tag{2.77}$$

In fact, the operator $g : L^2(\mathbb{R}_+) \to L^2(\mathbb{R}_+)$ is compact since it is Hilbert–Schmidt, and hence $1+g$ is of index zero. Moreover, $u \in \ker(1+g)$ entails $u = -gu$, and $g : L^2(\mathbb{R}_+) \to \mathcal{S}\left(\overline{\mathbb{R}}_+\right)$ shows $V \subset \mathcal{S}\left(\overline{\mathbb{R}}_+\right)$. Applying a similar conclusion to $\mathrm{coker}\,(1 + g) \cong \ker(1 + g^*) =: W$, we obtain the second relation of (2.77) and $W \subset \mathcal{S}\left(\overline{\mathbb{R}}_+\right)$.

Proposition 2.2.42. *Let $\boldsymbol{g} \in \mathcal{B}_G^d\left(\overline{\mathbb{R}}_+;j,j\right)$, and assume that the operator*

$$\begin{pmatrix} 1 & 0 \\ 0 & 0 \end{pmatrix} + \boldsymbol{g} : \begin{matrix} H^s(\mathbb{R}_+) \\ \oplus \\ \mathbb{C}^j \end{matrix} \to \begin{matrix} H^s(\mathbb{R}_+) \\ \oplus \\ \mathbb{C}^j \end{matrix} \tag{2.78}$$

is an isomorphism for some $s = s_0 > d - 1/2$. Then (2.78) is an isomorphism for every $s > d - 1/2$, and the inverse of (2.78) has the form

$$\begin{pmatrix} 1 & 0 \\ 0 & 0 \end{pmatrix} + \boldsymbol{k} \tag{2.79}$$

for some $\boldsymbol{k} \in \mathcal{B}_G^d\left(\overline{\mathbb{R}}_+;j,j\right)$.

Proof. Let us write

$$g =: \begin{pmatrix} f & k \\ t & q \end{pmatrix}$$

with obvious meaning of notation. Choose an invertible $(j \times j)$-matrix r such that the block matrix operator $\begin{pmatrix} 1+f & k \\ t & r \end{pmatrix}$ is invertible. Such a matrix r can be found by a small perturbation of q, using the fact that the isomorphisms in a Hilbert space form an open dense set. Then, once we have computed

$$\begin{pmatrix} 1+f & k \\ t & r \end{pmatrix}^{-1} := \begin{pmatrix} d_{11} & d_{12} \\ d_{21} & d_{22} \end{pmatrix}$$

we obtain

$$\begin{pmatrix} 1+f & k \\ t & q \end{pmatrix} \begin{pmatrix} d_{11} & d_{12} \\ d_{21} & d_{22} \end{pmatrix} = \begin{pmatrix} 1 & 0 \\ b & m \end{pmatrix},$$

where the right-hand side is invertible since both factors on the left-hand side are invertible. Since the right-hand side is a triangular matrix, the operator m is an invertible $(j \times j)$-matrix, and we have $b = td_{11} + qd_{21}$. Then, since

$$\begin{pmatrix} 1 & 0 \\ b & m \end{pmatrix} \begin{pmatrix} 1 & 0 \\ -m^{-1}b & m \end{pmatrix} = \begin{pmatrix} 1 & 0 \\ 0 & 1 \end{pmatrix}$$

we obtain

$$\begin{pmatrix} 1+f & k \\ t & q \end{pmatrix}^{-1} = \begin{pmatrix} d_{11} & d_{12} \\ d_{21} & d_{22} \end{pmatrix} \begin{pmatrix} 1 & 0 \\ -m^{-1}b & m \end{pmatrix}.$$

The invertibility of the matrix r allows us to form

$$\begin{pmatrix} 1 & -kr^{-1}t \\ 0 & 1 \end{pmatrix} \begin{pmatrix} 1+f & k \\ t & r \end{pmatrix} \begin{pmatrix} 1 & 0 \\ -r^{-1}t & r^{-1} \end{pmatrix} = \begin{pmatrix} 1+g & 0 \\ 0 & 1 \end{pmatrix} \tag{2.80}$$

for $g := f - kr^{-1}t \in \mathcal{B}_G^d(\overline{\mathbb{R}}_+)$, because $f \in \mathcal{B}_G^d(\overline{\mathbb{R}}_+)$ and $-kr^{-1}t \in \mathcal{B}_G^d(\overline{\mathbb{R}}_+)$. Then (2.80) gives

$$\begin{pmatrix} d_{11} & d_{12} \\ d_{21} & d_{22} \end{pmatrix} = \begin{pmatrix} 1 & 0 \\ -r^{-1}t & r^{-1} \end{pmatrix} \begin{pmatrix} (1+g)^{-1} & 0 \\ 0 & 1 \end{pmatrix} \begin{pmatrix} 1 & -kr^{-1} \\ 0 & 1 \end{pmatrix}, \tag{2.81}$$

and the next step is to compute $(1+g)^{-1}$. From (2.81) we see that

$$1 + g = 1 + f - kr^{-1}t : H^s(\mathbb{R}_+) \to H^s(\mathbb{R}_+) \tag{2.82}$$

is an isomorphism. We can write

$$1 + g = 1 + g_0 + \sum_{i=0}^{d-1} k_i \circ \gamma^i$$

for a unique $g_0 \in \mathcal{B}_{\mathrm{G}}^0(\overline{\mathbb{R}}_+)$, potential operators k_i, $i = 0, \ldots, d - 1$, and trace operators γ^i which are derivatives of order i composed with restrictions to the origin of the half-line. We now choose some $g_1 \in \mathcal{B}_{\mathrm{G}}^0(\overline{\mathbb{R}}_+)$ such that

$$1 + g_0 + g_1 : L^2(\mathbb{R}_+) \to L^2(\mathbb{R}_+) \tag{2.83}$$

is an isomorphism.

The construction of g_1 is as follows. We first observe that

$$1 + g_0 : L^2(\mathbb{R}_+) \to L^2(\mathbb{R}_+)$$

is Fredholm and of index 0, since g_0 is a compact operator in $L^2(\mathbb{R}_+)$. The spaces $V := \ker(1 + g_0)$, $W := \ker(1 + g_0^*)$ are of the same dimension, say, e, and are contained in $\mathcal{S}(\overline{\mathbb{R}}_+)$. This allows us to form an isomorphism

$$\begin{pmatrix} 1 + g_0 & w \\ v & 0 \end{pmatrix} : \begin{matrix} L^2(\mathbb{R}_+) \\ \oplus \\ \mathbb{C}^e \end{matrix} \to \begin{matrix} L^2(\mathbb{R}_+) \\ \oplus \\ \mathbb{C}^e \end{matrix}$$

for linear maps $v : V \to \mathbb{C}^e$, $w : \mathbb{C}^e \to W$. Thus for some sufficiently small $\varepsilon > 0$ the operator

$$\begin{pmatrix} 1 + g_0 & w \\ v & \varepsilon \cdot \mathrm{id}_{\mathbb{C}^e} \end{pmatrix} : \begin{matrix} L^2(\mathbb{R}_+) \\ \oplus \\ \mathbb{C}^e \end{matrix} \to \begin{matrix} L^2(\mathbb{R}_+) \\ \oplus \\ \mathbb{C}^e \end{matrix} \tag{2.84}$$

is also an isomorphism. In a similar manner as we saw that (2.82) is an isomorphism, we obtain an isomorphism (2.83) for $g_1 = -w\varepsilon^{-1}\mathrm{id}_{\mathbb{C}^e}v$. In addition, we have

$$(1 + g_0 + g_1)^{-1} = 1 + h$$

for some $\in \mathcal{B}_{\mathrm{G}}^0(\overline{\mathbb{R}}_+)$. Therefore, writing

$$1 + g = 1 + g_0 + g_1 - g_1 + \sum_{i=0}^{d-1} k_i \circ \gamma^i$$

it follows that

$$(1 + h)(1 + g) = 1 + (1 + h)\left(-g_1 + \sum_{i=0}^{d-1} k_i \circ \gamma^i\right) : H^s(\mathbb{R}_+) \to H^s(\mathbb{R}_+).$$

By construction, the operator $-g_1$ is of finite rank, namely,

$$-g_1 = \sum_{i=1}^{e} l_i m_i$$

for some potential operators $l_i : \mathbb{C} \to \mathcal{S}(\overline{\mathbb{R}}_+)$ and trace operators $m_i : L^2(\mathbb{R}_+) \to \mathbb{C}$ of type 0. Then

$$(1+h)\left(-g_1 + \sum_{i=0}^{d-1} k_i \circ \gamma^i\right) = \sum_{i=1}^{e} ((1+h)l_i m_i) + \sum_{i=0}^{d-1}(1+h)k_i \circ \gamma^i \qquad (2.85)$$

where $(1+h)l_i$ and $(1+h)k_i$ are potential operators. Thus (2.85) can be written as

$$(1+h)(1+g) = 1 + \sum_{j=1}^{e+d} p_j s_j = 1 + \mathcal{P}\mathcal{S} : H^s(\mathbb{R}_+) \to H^s(\mathbb{R}_+) \qquad (2.86)$$

for vectors of operators

$$\mathcal{P} := (p_1, \dots, p_{e+d}), \quad \mathcal{S} := (s_1, \dots, s_{e+d}),$$

and $1 + \mathcal{P}\mathcal{S}$ is invertible. Consider the operators

$$1 + \mathcal{P}\mathcal{S} : \mathbb{C}^{e+d} \to \mathbb{C}^{e+d}. \qquad (2.87)$$

Let us show that (2.87) is invertible if an only if $1 + \mathcal{S}\mathcal{P}$ is invertible. In fact, writing

$$\mathfrak{P} := \begin{pmatrix} 1 & \mathcal{P} \\ 0 & 1 \end{pmatrix}, \quad \mathfrak{S} := \begin{pmatrix} 1 & 0 \\ -\mathcal{S} & 1 \end{pmatrix}, \quad \mathfrak{F} := \begin{pmatrix} 1 & -\mathcal{P} \\ \mathcal{S} & 1 \end{pmatrix},$$

we have

$$\mathfrak{P}\mathfrak{F}\mathfrak{S} = \begin{pmatrix} 1+\mathcal{P}\mathcal{S} & 0 \\ 0 & 1 \end{pmatrix}, \quad \mathfrak{S}\mathfrak{F}\mathfrak{P} = \begin{pmatrix} 1 & 0 \\ 0 & 1+\mathcal{S}\mathcal{P} \end{pmatrix}.$$

We see that $1 + \mathcal{P}\mathcal{S}$ is invertible exactly when $1 + \mathcal{S}\mathcal{P}$ is invertible, and we have

$$(1+\mathcal{P}\mathcal{S})^{-1} = 1 - \mathcal{P}(1+\mathcal{S}\mathcal{P})^{-1}\mathcal{S}$$

where $g_3 := -\mathcal{P}(1+\mathcal{S}\mathcal{P})^{-1}\mathcal{S} \in \Gamma^d(\mathbb{R}_+)$. Thus the inverse of (2.86) has the form $1 + g_3 = ((1+h)(1+g))^{-1} = (1+g)^{-1}(1+h)^{-1}$. In other words, we proved that

$$(1+g)^{-1} = (1+g_3)(1+h),$$

i.e., $(1+g)^{-1} = 1 + k$ for $k = g_3 + h + g_3 h \in \Gamma^d(\mathbb{R}_+)$.

By Remark 2.2.27, the operator (2.57) is compact for any $s > d - 1/2$. Thus $\operatorname{ind}(1+g) = 0$ in $H^s(\mathbb{R}_+)$. However, $\ker(1+g) \subset \mathcal{S}(\overline{\mathbb{R}}_+)$ is independent of s, i.e., $\dim \ker(1+g) = 0$. This implies $\dim \operatorname{coker}(1+g) = 0$, and hence (2.78) is an isomorphism for all $s > d - 1/2$. □

Definition 2.2.43.   (i) A symbol $a(\tau) \in S^\mu_{\mathrm{tr}}(\mathbb{R})$ is called elliptic, if $a(\tau) \neq 0$ for all $\tau \in \mathbb{R}$ and $a_{0,+} = a_{0,-} \neq 0$, cf. the notation in (2.6).

(ii) An operator

$$a := \begin{pmatrix} \mathrm{op}^+(a) + g & k \\ b & q \end{pmatrix} \in \mathcal{B}^{\mu,d}(\overline{\mathbb{R}}_+; j_1, j_2) \tag{2.88}$$

is called elliptic, if $a(\tau) \in S_{\mathrm{tr}}^\mu(\mathbb{R})$ is elliptic in the sense of (i).

(iii) An operator

$$p := \begin{pmatrix} \mathrm{op}^+(p) + h & c \\ s & r \end{pmatrix} \in \mathcal{B}^{-\mu,e}(\overline{\mathbb{R}}_+; j_2, j_1) \tag{2.89}$$

is called a parametrix of $a \in \mathcal{B}^{\mu,d}(\overline{\mathbb{R}}_+; j_1, j_2)$, if $pa = \mathrm{diag}\,(1,0) + g_{\mathrm{L}}$, $ap = \mathrm{diag}\,(1,0) + g_{\mathrm{R}}$ with operators $g_{\mathrm{L}} \in \mathcal{B}_G^{d_{\mathrm{L}}}(\overline{\mathbb{R}}_+; j_1, j_1)$, $g_{\mathrm{R}} \in \mathcal{B}_G^{d_{\mathrm{R}}}(\overline{\mathbb{R}}_+; j_2, j_2)$ for some $d_{\mathrm{L}}, d_{\mathrm{R}} \in \mathbb{N}$.

Proposition 2.2.44. *Let* $\mathrm{op}^+(b) + k \in \mathcal{B}^{0,0}(\overline{\mathbb{R}}_+)$ *for* $b(\tau) \in S_{\mathrm{tr}}^0(\mathbb{R})$, $k \in \Gamma^0(\mathbb{R}_+)$ *be realised as a continuous operator*

$$b : L^2(\mathbb{R}_+) \to L^2(\mathbb{R}_+) \tag{2.90}$$

or

$$b_S : \mathcal{S}(\overline{\mathbb{R}}_+) \to \mathcal{S}(\overline{\mathbb{R}}_+). \tag{2.91}$$

Then, if b *is elliptic in the sense of* Definition 2.2.43, *the operators* (2.90) *and* (2.91) *are Fredholm, and there is a parametrix* $q := \mathrm{op}^+(b^{-1})$, *such that*

$$qb = 1 + k_{\mathrm{L}}, \quad bq = 1 + k_{\mathrm{R}} \tag{2.92}$$

with operators $k_{\mathrm{L}}, k_{\mathrm{R}} \in \Gamma^0(\mathbb{R}_+)$. *There are finite-dimensional subspaces* $K, L \subset \mathcal{S}(\overline{\mathbb{R}}_+)$ *such that*

$$K = \ker b, \quad L \cap \mathrm{im}\, b = \{0\}, \quad L + \mathrm{im}\, b = L^2(\mathbb{R}_+), \tag{2.93}$$
$$K = \ker b_S, \quad L \cap \mathrm{im}\, b_S = \{0\}, \quad L + \mathrm{im}\, b_S = \mathcal{S}(\overline{\mathbb{R}}_+), \tag{2.94}$$

and hence, $\mathrm{ind}\, b = \mathrm{ind}\, b_S$. *In particular,* (2.90) *is an isomorphism if and only if* (2.91) *is an isomorphism, and we have* $b^{-1} \in \mathcal{B}^{0,0}(\overline{\mathbb{R}}_+)$.

Proof. If $b(\tau)$ is elliptic, then the relations (2.92) hold for $q := \mathrm{op}^+(b^{-1})$ with Green operators $k_{\mathrm{L}}, k_{\mathrm{R}} \in \Gamma^0(\mathbb{R}_+)$. Since Green operators are compact in $L^2(\mathbb{R}_+)$, the operator (2.90) is Fredholm. We have $\ker b := K \subset \mathcal{S}(\overline{\mathbb{R}}_+)$, since $u \in L^2(\mathbb{R}_+)$, $u \in \ker b$ implies $\mathrm{op}^+(b)u = -ku \in \mathcal{S}(\overline{\mathbb{R}}_+)$ and $\mathrm{op}^+(b^{-1})\mathrm{op}^+(b)u = (1 + k_{\mathrm{L}})u$ yields $(1 + k_{\mathrm{L}})u = -\mathrm{op}^+(b^{-1})ku = -hu$ for some $h \in \Gamma^0(\mathbb{R}_+)$, which yields $u = -(k_{\mathrm{L}} + h)u \in \mathcal{S}(\overline{\mathbb{R}}_+)$. For the cokernel we can proceed in an analogous manner. It suffices to observe that $b^* = \mathrm{op}^+(\overline{b}) + k^* \in \mathcal{B}^{0,0}(\overline{\mathbb{R}}_+)$, and then $L := \ker b^* \subset \mathcal{S}(\overline{\mathbb{R}}_+)$ gives the relations (2.93) concerning L. Moreover, we also obtain (2.94), since the kernel of b in $L^2(\mathbb{R}_+)$ is already contained in $\mathcal{S}(\overline{\mathbb{R}}_+)$. For the cokernel we can argue in an analogous manner; this gives the second part of (2.94). Thus the bijectivity of (2.90) implies the one of (2.91) and vice versa.

Now assume that (2.90) is invertible. Then, since $k : L^2(\mathbb{R}_+) \to L^2(\mathbb{R}_+)$ is compact, we have $\operatorname{ind} \operatorname{op}^+(b) = 0$. Since $\operatorname{op}^+(b^{-1}) \operatorname{op}^+(b) u = (1 + k_{\mathrm{L}})$, it follows that also $\operatorname{ind} \operatorname{op}^+(b^{-1}) = 0$. Analogously to (2.93) we have

$$K_1 = \ker \operatorname{op}^+(b^{-1}), \quad L_1 \cap \operatorname{im} \operatorname{op}^+(b^{-1}) = \{0\}, \quad L_1 + \operatorname{im} \operatorname{op}^+(b^{-1}) = L^2(\mathbb{R}_+),$$
$$(2.95)$$

for certain finite-dimensional subspaces $K_1, L_1 \subset \mathcal{S}(\overline{\mathbb{R}}_+)$. Those allow us the construction of an $l \in \Gamma^0(\mathbb{R}_+)$ such that $\boldsymbol{q}_1 := \operatorname{op}^+(b^{-1}) + l : L^2(\mathbb{R}_+) \to L^2(\mathbb{R}_+)$ is invertible. In fact, choosing isomorphisms $k_1 : \mathbb{C}^m \to L_1$, $d_1 : K_1 \to \mathbb{C}^m$ for $m := \dim L_1 = \dim K_1$, we can pass to an isomorphism

$$\begin{pmatrix} \operatorname{op}^+(b^{-1}) & k_1 \\ d_1 & \varepsilon \end{pmatrix} : \begin{matrix} L^2(\mathbb{R}_+) & & L^2(\mathbb{R}_+) \\ \oplus & \to & \oplus \\ \mathbb{C}^m & & \mathbb{C}^m \end{matrix} \qquad (2.96)$$

first for $\varepsilon = 0$, and then for sufficiently small $\varepsilon > 0$, cf. analogously (2.84). Then it suffices to set $l := -k_1 \varepsilon^{-1} d_1$. It follows that $\boldsymbol{q}_1 \boldsymbol{b} = 1 + n : L^2(\mathbb{R}_+) \to L^2(\mathbb{R}_+)$ is invertible, where $n \in \Gamma^0(\mathbb{R}_+)$, and then Proposition 2.2.40 shows that there is an $h_1 \in \Gamma^0(\mathbb{R}_+)$ such that $(1 + n)^{-1} = 1 + h_1$. We finally obtain $(1 + h_1) \boldsymbol{q}_1 \boldsymbol{b} = 1$, i.e., $(1 + h_1) \boldsymbol{q}_1 = \boldsymbol{b}^{-1}$, and, according to Theorem 2.2.35, we have $\boldsymbol{b}^{-1} \in \mathcal{B}^{0,0}(\overline{\mathbb{R}}_+)$. □

Proposition 2.2.45. *Let* $\operatorname{op}^+(b) + k \in \mathcal{B}^{0,0}(\overline{\mathbb{R}}_+)$ *be as in* Proposition 2.2.44 *elliptic, now realised as a continuous operator*

$$\boldsymbol{b} : H^s(\mathbb{R}_+) \to H^s(\mathbb{R}_+) \qquad (2.97)$$

for $s \in \mathbb{R}$, $s > -1/2$. *The operator* (2.97) *is Fredholm, and* $\boldsymbol{q} = \operatorname{op}^+(b^{-1})$ *is a parametrix of* $\operatorname{op}^+(b) + k$ *also in the sense of* (2.97). *Moreover, the finite-dimensional subspaces* $K, L \subset \mathcal{S}(\overline{\mathbb{R}}_+)$ *of* Proposition 2.2.44 *have analogous properties with respect to* (2.97), *namely,*

$$K = \ker \boldsymbol{b}, \quad L \cap \operatorname{im} \boldsymbol{b} = \{0\}, \quad L + \operatorname{im} \boldsymbol{b} = H^s(\mathbb{R}_+), \qquad (2.98)$$

and the relation (2.94) *for the same* K, L. *Thus* $\operatorname{ind} \boldsymbol{b} = \operatorname{ind} \boldsymbol{b}_{\mathcal{S}}$. *Finally,* (2.97) *is an isomorphism if and only if* (2.91) *is an isomorphism.*

Proof. The operator $\operatorname{op}^+(b) + k$ induces a continuous operator (2.97) for any $s > -1/2$, cf. Theorem 2.2.11 and Remark 2.2.27. If $b(\tau)$ is elliptic, then the operator (2.97) is Fredholm. In fact, $\boldsymbol{q}$ is a parametrix and the remainders in the analogue of (2.94) are the same as in Proposition 2.2.44, namely k_{L} and k_{R}. Those are also compact in $H^s(\mathbb{R}_+)$ for $s > 1/2$, see Remark 2.2.27. Similarly as in the proof of Proposition 2.2.44, it follows that $K = \ker \boldsymbol{b} \subset \mathcal{S}(\overline{\mathbb{R}}_+)$, and we obtain the first relation in (2.98). The second one follows by using the shape of the formal adjoint, namely, $(\operatorname{op}^+(b) + k)^* = \operatorname{op}^+(\overline{b}) + k^*$. This shows that (2.97) is an isomorphism exactly when $\boldsymbol{b}_{\mathcal{S}} : \mathcal{S}(\overline{\mathbb{R}}_+) \to \mathcal{S}(\overline{\mathbb{R}}_+)$ is an isomorphism. Moreover, it follows that the inverse $(\mathrm{op}^+(b)+k)^{-1}$ computed in the proof of Proposition 2.2.44 also induces the inverse of (2.97). □

Proposition 2.2.46. *Let* $\mathrm{op}^+(a) + g \in \mathcal{B}^{\mu,d}(\overline{\mathbb{R}}_+)$, *for* $a(\tau) \in S_{\mathrm{tr}}^{\mu}(\mathbb{R})$, $g \in \Gamma^d(\mathbb{R}_+)$ *realised as a continuous operator*

$$\boldsymbol{a} : H^s(\mathbb{R}_+) \to H^{s-\mu}(\mathbb{R}_+) \tag{2.99}$$

for an $s > \max\{\mu, d\} - 1/2$, *or*

$$\boldsymbol{a}_{\mathcal{S}} : \mathcal{S}(\overline{\mathbb{R}}_+) \to \mathcal{S}(\overline{\mathbb{R}}_+). \tag{2.100}$$

Then, if a is elliptic in the sense of Definition 2.2.43, *the operators* (2.99) *and* (2.100) *are Fredholm, and there is a parametrix* $\boldsymbol{p} := \mathrm{op}^+(a^{-1})$, *such that*

$$\boldsymbol{p}\boldsymbol{a} = 1 + g_{\mathrm{L}}, \quad \boldsymbol{a}\boldsymbol{p} = 1 + g_{\mathrm{R}} \tag{2.101}$$

for operators $g_{\mathrm{L}} \in \Gamma^{\max\{\mu,d\}}(\mathbb{R}_+), g_{\mathrm{R}} \in \Gamma^{\max\{d-\mu,0\}}(\mathbb{R}_+)$. *There exist finite-dimensional subspaces* $V, W \subset \mathcal{S}(\overline{\mathbb{R}}_+)$ *such that*

$$V = \ker \boldsymbol{a}, \quad W \cap \mathrm{im}\,\boldsymbol{a} = \{0\}, \quad W + \mathrm{im}\,\boldsymbol{a} = H^{s-\mu}(\mathbb{R}_+), \tag{2.102}$$

$$V = \ker \boldsymbol{a}_{\mathcal{S}}, \quad W \cap \mathrm{im}\,\boldsymbol{a}_{\mathcal{S}} - \{0\}, \quad W + \mathrm{im}\,\boldsymbol{a}_{\mathcal{S}} = \mathcal{S}(\overline{\mathbb{R}}_+). \tag{2.103}$$

In particular, (2.99) *is an isomorphism if and only if* (2.100) *is an isomorphism, and we have* $\boldsymbol{a}^{-1} \in \mathcal{B}^{-\mu,\max\{d-\mu,0\}}(\overline{\mathbb{R}}_+)$.

Proof. Applying Theorem 2.2.39 we can pass to the operator

$$\boldsymbol{b}_{s_1} := \mathrm{op}^+\big(l_-^{s_1-\mu}\big)\boldsymbol{a}\,\mathrm{op}^+\big(l_-^{-s_1}\big) \in \mathcal{B}^{0,0}(\overline{\mathbb{R}}_+), \quad \boldsymbol{b}_{s_1} : H^{s_0}(\mathbb{R}_+) \to H^{s_0}(\mathbb{R}_+),$$

$s_0 > -1/2$, for $s_1 := \max\{\mu, d\}$, which satisfies the assumptions of Proposition 2.2.45. In the above notation we obtain $b(\tau) = l_-^{s_1-\mu}(\tau)a(\tau)l_-^{-s_1}(\tau)$, which is elliptic, and we have the relations

$$\boldsymbol{q}\boldsymbol{b}_{s_1} = 1 + k_{\mathrm{L}}, \quad \boldsymbol{b}_{s_1}\boldsymbol{q} = 1 + k_{\mathrm{R}} \tag{2.104}$$

for $\boldsymbol{q} := \mathrm{op}^+(b^{-1})$. In other words, it follows that

$$\boldsymbol{q}\,\mathrm{op}^+\big(l_-^{s_1-\mu}\big)\boldsymbol{a}\,\mathrm{op}^+\big(l_-^{-s_1}\big) = 1 + k_{\mathrm{L}}, \quad \mathrm{op}^+\big(l_-^{s_1-\mu}\big)\boldsymbol{a}\,\mathrm{op}^+\big(l_-^{-s_1}\big)\boldsymbol{q} = 1 + k_{\mathrm{R}}.$$

Thus for $\boldsymbol{p} := \mathrm{op}^+\big(l_-^{-s_1}\big)\boldsymbol{q}\,\mathrm{op}^+\big(l_-^{s_1-\mu}\big)$ we obtain

$$\boldsymbol{p}\boldsymbol{a} = \mathrm{op}^+\big(l_-^{-s_1}\big)(1 + k_{\mathrm{L}})\,\mathrm{op}^+\big(l_-^{s_1}\big) = 1 + g_{\mathrm{L}}$$

for $g_{\mathrm{L}} := \mathrm{op}^+\big(l_-^{-s_1}\big)k_{\mathrm{L}}\mathrm{op}^+\big(l_-^{s_1}\big) \in \Gamma^{\max\{\mu,d\}}(\mathbb{R}_+)$, cf. (2.64). In an analogous manner it follows that

$$\boldsymbol{a}\boldsymbol{p} = \mathrm{op}^+\big(l_-^{-s_1+\mu}\big)(1 + k_{\mathrm{R}})\,\mathrm{op}^+\big(l_-^{s_1-\mu}\big) = 1 + g_{\mathrm{R}}$$

for $g_{\mathrm{R}} := \mathrm{op}^+ \big(l_-^{-s+\mu} \big) k_{\mathrm{R}} \mathrm{op}^+ \big(l_-^{s-\mu} \big) \in \Gamma^{\max\{d-\mu,0\}}(\overline{\mathbb{R}}_+)$. Since

$$\mathrm{op}^+ \big(l_-^{-s_1} \big) : H^{s_0}(\mathbb{R}_+) \to H^s(\mathbb{R}_+), \quad \mathrm{op}^+ \big(l_-^{-s_1} \big) : H^{s_0}(\mathbb{R}_+) \to H^{s-\mu}(\mathbb{R}_+)$$

are isomorphisms for $s := s_0 + s_1$ it follows that $\boldsymbol{a} = \mathrm{op}^+ \big(l_-^{-s_1+\mu} \big) \boldsymbol{b}_{s_1} \mathrm{op}^+ \big(l_-^{s_1} \big)$ defines Fredholm operators (2.99) and (2.100). At the same time we obtain the relations (2.102) and (2.103) for the spaces $V := \mathrm{op}^+ (l_-^{-s_1}) K$ and $W := \mathrm{op}^+ \big(l_-^{-s_1} \big) L$. $\qquad\square$

Lemma 2.2.47 ([23, Lemma 3.3.23]). *Let $H, \widetilde{H}$ be Hilbert spaces.*

(i) *Let*

$$\begin{pmatrix} a & k \\ t & q \end{pmatrix} : \begin{matrix} H \\ \oplus \\ \mathbb{C}^{j_1} \end{matrix} \to \begin{matrix} \widetilde{H} \\ \oplus \\ \mathbb{C}^{j_2} \end{matrix} \tag{2.105}$$

be an isomorphism. Then $a : H \to \widetilde{H}$ is a Fredholm operator, and we have $\mathrm{ind}\, a = j_2 - j_1$.

(ii) *Let $a : H \to \widetilde{H}$ be a Fredholm operator. Then there are operators $k : \mathbb{C}^{j_1} \to \widetilde{H}$, $t : H \to \mathbb{C}^{j_2}$, $q : \mathbb{C}^{j_1} \to \mathbb{C}^{j_2}$ for certain j_1, j_2 such that (2.105) is an isomorphism.*

Theorem 2.2.48. *Let*

$$\begin{pmatrix} \mathrm{op}^+(a) + g_{11} & g_{12} \\ g_{21} & g_{22} \end{pmatrix} \in \mathcal{B}^{\mu,d} \big(\overline{\mathbb{R}}_+; j_1, j_2 \big), \tag{2.106}$$

realised as a continuous map

$$\boldsymbol{a} : \begin{matrix} H^s(\mathbb{R}_+) \\ \oplus \\ \mathbb{C}^{j_1} \end{matrix} \to \begin{matrix} H^{s-\mu}(\mathbb{R}_+) \\ \oplus \\ \mathbb{C}^{j_2} \end{matrix} \tag{2.107}$$

for $s > \max\{\mu, d\} - 1/2$, or

$$\boldsymbol{a}_{\mathcal{S}} : \begin{matrix} \mathcal{S}(\overline{\mathbb{R}}_+) \\ \oplus \\ \mathbb{C}^{j_1} \end{matrix} \to \begin{matrix} \mathcal{S}(\overline{\mathbb{R}}_+) \\ \oplus \\ \mathbb{C}^{j_2} \end{matrix}. \tag{2.108}$$

Let $a(\tau)$ be elliptic in the sense of Definition 2.2.43. *Then (2.107) is an isomorphism if and only if (2.108) is an isomorphism, and we have*

$$\begin{pmatrix} \mathrm{op}^+(a) + g_{11} & g_{12} \\ g_{21} & g_{22} \end{pmatrix}^{-1} \in \mathcal{B}^{-\mu,\max\{d-\mu,0\}} \big(\overline{\mathbb{R}}_+; j_2, j_1 \big). \tag{2.109}$$

In that case the operators

$$\mathrm{op}^+(a) + g_{11} : H^s(\mathbb{R}_+) \to H^{s-\mu}(\mathbb{R}_+), \ s > \max\{\mu, d\} - 1/2, \ \text{or} \ \mathcal{S}(\overline{\mathbb{R}}_+) \to \mathcal{S}(\overline{\mathbb{R}}_+) \tag{2.110}$$

are Fredholm, and we have

$$\text{ind}\left(\text{op}^+(a) + g_{11}\right) = j_2 - j_1. \tag{2.111}$$

Proof. From Lemma 2.2.47 and Proposition 2.2.46 it follows that the opera-tors (2.110) are Fredholm of index $j_2 - j_1$. Moreover, the operator $\text{op}^+\left(a^{-1}\right)$ is a parametrix of $\text{op}^+(a) + g_{11}$ and hence $\text{ind} \, \text{op}^+\left(a^{-1}\right) = j_1 - j_2$. Applying Lemma 2.2.47 to $\text{op}^+\left(a^{-1}\right)$ we find numbers l_1, l_2 such that

$$\boldsymbol{p} := \begin{pmatrix} \text{op}^+(a^{-1}) & c \\ b & r \end{pmatrix} : \begin{array}{c} H^{s-\mu}(\mathbb{R}_+) \\ \oplus \\ \mathbb{C}^{l_2} \end{array} \to \begin{array}{c} H^s(\mathbb{R}_+) \\ \oplus \\ \mathbb{C}^{l_1} \end{array} \tag{2.112}$$

is an isomorphism. From Proposition 2.2.46 we easily see that the entries c, b, r in (2.112) can be chosen in such a way that (2.112) will belong to $\mathcal{B}^{-\mu,0}\left(\overline{\mathbb{R}}_+; l_2, l_1\right)$. Lemma 2.2.47 gives us $\text{ind} \, \boldsymbol{p} = l_1 - l_2$ which is equal to $j_1 - j_2$. Therefore, there is an $n \in \mathbb{N}$ such that $l_i = j_i + n$ or $l_i + n = j_i$ for $i = 1, 2$. In the first case we show the invertibility of $\boldsymbol{a} \oplus \text{id}_{\mathbb{C}^n}$ and in the second case we replace $\boldsymbol{p}$ by $\boldsymbol{p} \oplus \text{id}_{\mathbb{C}^n}$. This allows us to assume $l_i = j_i$, $i = 1, 2$, without loss of generality. In this case we can form the composition $\boldsymbol{pa}$, which is of the form $\begin{pmatrix} 1 & 0 \\ 0 & 0 \end{pmatrix} + \boldsymbol{g}$ for some $\boldsymbol{g} \in \mathcal{B}_G^e\left(\overline{\mathbb{R}}_+; j_1, j_1\right)$ where 1 means $\text{id}_{H^s(\mathbb{R}_+)}$. Proposition 2.2.42 and Theorem 2.2.35 show that

$$\boldsymbol{a}^{-1} = \left(\begin{pmatrix} 1 & 0 \\ 0 & 0 \end{pmatrix} + \boldsymbol{g}\right)^{-1} \boldsymbol{p} \in \mathcal{B}^{-\mu, \max\{d-\mu, 0\}}\left(\overline{\mathbb{R}}_+; j_2, j_1\right). \qquad \square$$

2.3 A relationship between boundary symbols and Toeplitz operators

Let us return once again to the operators

$$\text{op}^\pm(a) = \text{r}^\pm \, \text{op}(a) \, \text{e}^\pm : L^2\left(\mathbb{R}_\pm\right) \to L^2\left(\mathbb{R}_\pm\right)$$

for a symbol $a(\tau) \in S_{\text{tr}}^0(\mathbb{R})$, cf. relation (2.16), where $\text{r}^\pm$ denotes the restriction to $\mathbb{R}_+$ and $\text{e}^\pm$ the operator of extension by zero from $\mathbb{R}_+$ to the opposite side. Recall that we have projections

$$\theta^\pm : L^2(\mathbb{R}_t) \to L^2(\mathbb{R}_{\pm,t}) \quad \text{and} \quad \Pi^\pm := F\theta^\pm F^{-1} : L^2(\mathbb{R}_\tau) \to V^\pm(\mathbb{R}_\tau), \tag{2.113}$$

cf. the formulas (2.23), (2.24), and corresponding direct decompositions

$$L^2(\mathbb{R}_t) = L^2(\mathbb{R}_{+,t}) \oplus L^2(\mathbb{R}_{-,t}) \quad \text{and} \quad L^2(\mathbb{R}_\tau) = V^+(\mathbb{R}_\tau) \oplus V^-(\mathbb{R}_\tau). \tag{2.114}$$

We often identify the spaces $\theta^\pm L^2(\mathbb{R}_t)$ with $L^2(\mathbb{R}_{\pm,t})$. In the following we mainly focus on the plus side; the minus case is analogous. Denoting the operator of multiplication by a function φ also by $\mathcal{M}_\varphi$ we write

$$\text{op}^+(a) = \mathcal{M}_{\theta^+} F^{-1} \mathcal{M}_a F \mathcal{M}_{\theta^+} \tag{2.115}$$

for the Fourier transform $F = F_{t \to \tau}$. From $\Pi^+|_{V^+(\mathbb{R})} = \mathrm{id}_{V^+(\mathbb{R})}$ together with (2.113) we obtain the operator

$$F \, \mathrm{op}^+(a) \, F^{-1} = \Pi^+ \mathcal{M}_a : V^+(\mathbb{R}) \to V^+(\mathbb{R}). \tag{2.116}$$

For future use we formulate a lemma on the Mellin transform of some special function, see also Eskin's book [14]. A basic tool on the half-line is the Mellin transform

$$Mu(z) := \int_0^\infty t^{z-1} u(t) dt. \tag{2.117}$$

Assuming $u(t) \in C_0^\infty(\mathbb{R}_+)$ we obtain an entire function in $z \in \mathbb{C}$. Writing $z = \tau + i\sigma$ the substitution $\boldsymbol{t} := -\log t$ gives a relationship between (2.117) and the Fourier transform

$$F_{\boldsymbol{t} \to \sigma}\big(e^{-\tau \boldsymbol{t}} v(\boldsymbol{t})\big)(\sigma) = \int_{-\infty}^\infty e^{-i\sigma \boldsymbol{t}} e^{-\tau \boldsymbol{t}} v(\boldsymbol{t}) d\boldsymbol{t}$$

for $v(\boldsymbol{t}) := u\big(e^{-\boldsymbol{t}}\big)$.

Lemma 2.3.1. *The Mellin transform of $u(t) := t^\delta/(t-a)$, $|\delta| < 1/2$, $a \in \mathbb{C} \setminus \overline{\mathbb{R}}_+$, has the form*

$$Mu(z) = 2\pi i \, \frac{e^{(\delta+z-1)\log a}}{1 - e^{2\pi i(\delta+z)}}. \tag{2.118}$$

Proof. The integral

$$Mu(z) = \int_0^\infty \frac{t^{\delta+z-1}}{t-a} dt$$

is absolutely convergent for all z in the strip $-\delta < \mathrm{Re}\, z < 1 - \delta$. In fact, the absolute convergence of the integral in the indicated strip is guaranteed by the relations

$$\int_0^1 t^\alpha dt < \infty \ \text{ for } \alpha > -1, \quad \int_1^\infty t^{\alpha-1} dt < \infty \ \text{ for } \alpha - 1 < -1$$

when we set $\alpha = \delta + \mathrm{Re}\, z - 1$.

Let $C_{N,\varepsilon} \subset \mathbb{C}_w$ denote the piecewise smooth curve $C_{N,\varepsilon} = S_\varepsilon \cup S_N \cup I_{\varepsilon N} \cup I_{N\varepsilon}$ for $0 < \varepsilon < N < \infty$, $S_\varepsilon := \{w : |w| = \varepsilon\} \setminus \{\varepsilon\}$ with clockwise orientation, $S_N := \{w : |w| = N\} \setminus \{N\}$ with counter-clockwise orientation,

$$I_{\varepsilon N} := \{w : \varepsilon \le \mathrm{Re}\, w \le N, \ \mathrm{Im}\, w = 0\},$$

$I_{\varepsilon N}$ with the orientation induced by $\mathbb{R}_+$, and

$$I_{N\varepsilon} := \{w : N \ge \mathrm{Re}\, w \ge \varepsilon, \ \mathrm{Im}\, w = 0\}$$

with the opposite orientation. We assume $\varepsilon > 0$ to be sufficiently small and $N > \varepsilon$ sufficiently large. Then the curve $C_{N,\varepsilon}$ counter-clockwise surrounds the point $a \in \mathbb{C} \setminus \overline{\mathbb{R}}_+$. In the following computation we assume $z \in \mathbb{C}$, $-\delta < \mathrm{Re}\, z < 1 - \delta$. Set

$$f(w) := \frac{w^{\delta+z-1}}{w-a} = \frac{e^{(\delta+z-1)\log w}}{w-a}$$

with the branch of the logarithm defined with argument 0 on $I_{\varepsilon N}$ and 2π on $I_{N\varepsilon}$. Thus the function $f(w)$ is equal to $e^{2\pi i(\delta+z-1)}t^{\delta+z-1}/(t-a)$ on $I_{N\varepsilon}$ and $t^{\delta+z-1}/(t-a)$ on $I_{\varepsilon N}$. By the residue theorem,

$$\int_{C_{N,\varepsilon}} \frac{w^{\delta+z-1}}{w-a}\,dw = \operatorname*{Res}_{w=a}\left(\frac{w^{\delta+z-1}}{w-a}\right) = 2\pi i\, e^{(\delta+z-1)\log a}.$$

Since

$$\lim_{\varepsilon\to 0}\int_{C_{N,\varepsilon}} \frac{w^{\delta+z-1}}{w-a}\,dw = 0 \quad\text{and}\quad \lim_{N\to\infty}\int_{C_{N,\varepsilon}} \frac{w^{\delta+z-1}}{w-a}\,dw = 0,$$

it follows that

$$\lim_{\substack{\varepsilon\to 0\\ N\to\infty}}\int_{C_{N,\varepsilon}} \frac{w^{\delta+z-1}}{w-a}\,dw = \lim_{\substack{\varepsilon\to 0\\ N\to\infty}}\left\{\int_{I_{\varepsilon N}} \frac{w^{\delta+z-1}}{w-a}\,dw + \int_{I_{N\varepsilon}} \frac{w^{\delta+z-1}}{w-a}\,dw\right\}$$

$$= \int_0^\infty \frac{t^{\delta+z-1}}{t-a}\,dt - \int_\infty^0 e^{2\pi i(\delta+z-1)}\frac{t^{\delta+z-1}}{t-a}\,dt$$

$$= \left(1+e^{2\pi i(\delta+z-1)}\right)\int_0^\infty \frac{t^{\delta+z-1}}{t-a}\,dt.$$

Thus, we proved formula (2.118). $\qquad\square$

Remark 2.3.2. The function (2.118) is holomorphic in the strip $-\delta < \operatorname{Re} z < 1-\delta$, since the numerator is an entire function, and $Mu(z)$ itself is meromorphic with simple poles at those z where $e^{2\pi i(\delta+z)} = 1$, i.e., where $2\pi i(\delta + \operatorname{Re} z) = 2\pi i k$ for some $k \in \mathbb{Z}$, i.e., $\delta + \operatorname{Re} z = k$. The denominator is holomorphic for $-\delta < \operatorname{Re} z < 1-\delta$.

Let us now observe that $\Pi^\pm$ are classical pseudo-differential operators of order zero on $\mathbb{R}_\tau$. For instance, in the plus-case the expression $\Pi^+ f = F_{t\to\tau}M_{\theta+}F^{-1}_{\tau'\to t}f$ can be written as

$$\Pi^+ f(\tau) = \iint e^{i(\tau'-\tau)t}\theta^+(t)f(\tau')\,d̄\tau'dt,$$

$f \in C_0^\infty(\mathbb{R}_\tau)$. Since $\theta^+(t)$ is discontinuous and not a symbol in t (treated here as a covariable), we write

$$\theta^+(t) = \chi(t)\theta^+(t) + (1-\chi(t))\theta^+(t)$$

for an excision function $\chi(t)$. Then we have $\chi(t)\theta^+(t) \in S_{\mathrm{cl}}^0(\mathbb{R})$, while $(1-\chi(t))\theta^+(t)$ only contributes a smoothing operator. The assertion for the minus-case follows from the relation $\Pi^- = 1-\Pi^+$. More generally, we have $\Pi^+ M_a \in L_{\mathrm{cl}}^0(\mathbb{R}_\tau)$,

$$\Pi^+ M_a f(\tau) = \iint e^{i(\tau'-\tau)t}\theta^+(t)a(\tau')f(\tau')\,d̄\tau'dt.$$

Proposition 2.3.3. *The operators* $\Pi^\pm = F M_{\theta^\pm} F^{-1}$ *can be written in the form*

$$\Pi^+ f(\tau) = \lim_{\varepsilon \to +0} \frac{1}{2\pi i} \int_{-\infty}^{\infty} \frac{f(\nu)}{\tau - i\varepsilon - \nu} d\nu = \frac{1}{2}\left(f(\tau) + \frac{1}{\pi i} \text{ p.v.} \int \frac{f(\nu) d\nu}{\tau - \nu} \right) \quad (2.119)$$

and

$$\Pi^- f(\tau) = \lim_{\varepsilon \to +0} \frac{-1}{2\pi i} \int_{-\infty}^{\infty} \frac{f(\nu)}{\tau + i\varepsilon - \nu} d\nu = \frac{1}{2}\left(f(\tau) - \frac{1}{\pi i} \text{ p.v.} \int \frac{f(\nu) d\nu}{\tau - \nu} \right) \quad (2.120)$$

for $f \in \mathcal{S}(\mathbb{R})$.

Proof. We consider (2.119); the arguments for the minus-case are analogous. The assertion will follow from the relation $(\Pi^+ F u)(\tau) = (F\theta^+ u)(\tau)$, first for $u \in \mathcal{S}(\mathbb{R})$, and then by continuous extension for any $u \in L^2(\mathbb{R})$. First we have to show that

$$F\big(\theta^+(t)u(t)\big)(\tau) = \lim_{\varepsilon \to +0} \frac{1}{2\pi i} \int \frac{f(\nu)}{\tau - i\varepsilon - \nu} d\nu \quad \text{for} \quad u \in \mathcal{S}(\mathbb{R}), \quad f(\tau) = (Fu)(\tau).$$

Consider the following absolutely convergent integral

$$\int_0^\infty e^{-it(\tau - i\varepsilon)} dt = \int_0^\infty e^{-t\varepsilon - it\tau} dt = F(\theta^+ e^{-t\varepsilon}),$$

where $\varepsilon > 0$. Computing this integral, we get

$$F(\theta^+ e^{-t\varepsilon}) = \frac{-i}{\tau - i\varepsilon}, \quad \varepsilon > 0.$$

Since $F(vu) = (2\pi)^{-1} \int (Fv)(\tau - \nu)(Fu)(\nu) d\nu$, $u, v \in L^1(\mathbb{R})$, it follows that

$$F(\theta^+ e^{-\varepsilon t} u(t))(\tau) = \frac{1}{2\pi i} \int \frac{f(\nu) d\nu}{\tau - i\varepsilon - \nu}. \quad (2.121)$$

Further, since $\theta^+(t) e^{-\varepsilon t}$ converges to $\theta^+(t)$ as $\varepsilon \to +0$, Lebesgue's theorem gives

$$F(\theta^+ u(t)) = \lim_{\varepsilon \to +0} F(\theta^+ e^{-\varepsilon t} u(t)) = \lim_{\varepsilon \to +0} \frac{1}{2\pi i} \int \frac{f(\nu) d\nu}{\tau - i\varepsilon - \nu},$$

which is the first equation of (2.119). In order to prove the second one, we set

$$g(\tau - i\varepsilon) := \int \frac{f(\nu) d\nu}{\tau - i\varepsilon - \nu}, \quad \varepsilon > 0,$$

and

$$g_1(\tau, \varepsilon) := \int_{|\tau - \varepsilon| \leq 1} \frac{f(\nu) - f(\tau)}{\tau - i\varepsilon - \nu} d\nu + \int_{|\tau - \varepsilon| > 1} \frac{f(\nu) d\nu}{\tau - i\varepsilon - \nu},$$

$$g_2(\tau, \varepsilon) := f(\tau) \int_{|\tau - \varepsilon| < 1} \frac{d\nu}{\tau - i\varepsilon - \nu};$$

then $g = g_1 + g_2$. We have

$$g_2(\tau, \varepsilon) = f(\tau) \int_{|\rho| < 1} \frac{d\rho}{\rho - \varepsilon} = f(\tau) \left\{ \int_{-1}^{1} \frac{\rho d\rho}{\rho^2 + \varepsilon^2} + i\varepsilon \int_{-1}^{1} \frac{d\rho}{\rho^2 + \varepsilon^2} \right\}$$

$$= \frac{2}{i} \arctan \frac{1}{-\varepsilon} \to i\pi f(\tau) \quad \text{for } \varepsilon \to +0. \tag{2.122}$$

In the expression of $g_1(\tau, \varepsilon)$ we may pass to the limit $\varepsilon \to +0$ under the integrals. Note that

$$\int_{\varepsilon < |\tau - \nu| < 1} \frac{f(\nu)}{\tau - \nu} d\nu = 0,$$

since the integrand is an odd function. Thus

$$g_1(\tau, 0) = \lim_{\varepsilon \to +0} \int_{|\tau - \nu| > \varepsilon} \frac{f(\nu)}{\tau - \nu} d\nu = \text{p.v.} \int \frac{f(\nu)}{\tau - \nu} d\nu. \qquad \Box$$

The operator

$$H : f(\tau) \to \frac{1}{i\pi} \text{ p.v.} \int \frac{f(\nu)}{\tau - \nu} d\nu$$

is just the Hilbert transform.

As an immediate consequence of (2.119)–(2.120) we have

$$\Pi^+ f + \Pi^- f = f \quad \text{for } f \in \mathcal{S}(\mathbb{R}). \tag{2.123}$$

Since θ^+ is the homogeneous principal symbol of $\Pi^+ \in L^0_{\text{cl}}(\mathbb{R})$, formula (2.119) shows that the Hilbert transform H is also an operator in $L^0_{\text{cl}}(\mathbb{R})$, and its homogeneous principal symbol is equal to $\theta^+ - \theta^-$. The operator H is even elliptic; however, it has the anti-transmission property, in the terminology we introduce in Section 10.3 below. We now transfer the consideration to subspaces of $L^2(S^1)$ on the unit circle $S^1 = \{|z| = 1\}$ in the complex z-plane. We consider $L^2(S^1)$ with the scalar product

$$(u, v)_{L^2(S^1)} = \frac{1}{2\pi} \int_{-\pi}^{\pi} u(e^{i\varphi}) \overline{v}(e^{i\varphi}) d\varphi, \quad z = e^{i\varphi}.$$

The functions $z^j, j \in \mathbb{Z}$, form an orthonormal basis of the space $L^2(S^1)$, and we have the orthogonal decomposition

$$L^2(S^1) = W^+(S^1) \oplus W^-(S^1)$$

for the subspaces

$$W^+(S^1) := [z^j : j \in \mathbb{N}], \quad W^-(S^1) := [z^{-j-1} : j \in \mathbb{N}]. \tag{2.124}$$

Here $[\,\cdot\,]$ denotes the closure of the linear span of the functions in $[\,\cdot\,]$. The space W^+ is called the Hardy space, often denoted by $H^2(S^1)$. Our notation is only motivated by the context of other $\pm$-spaces. Let

$$\pi^\pm : L^2(S^1) \to W^\pm(S^1)$$

denote the corresponding orthogonal projections.

Now let us consider the diffeomorphism

$$\kappa : \mathbb{R} \to S^1 \setminus \{1\}, \quad \tau \mapsto z,$$

given by

$$\kappa(\tau) := \frac{1 - i\tau}{1 + i\tau} = z, \quad \text{where} \quad \kappa^{-1}(z) = \frac{1}{i}\frac{1 - z}{1 + z} = \tau.$$

We have

$$\kappa(0) = 1, \quad \text{and} \quad \kappa(\tau) \to -1 \quad \text{as} \quad \tau \to \pm\infty.$$

Moreover, $z = (1 + \tau^2)\{(1 - \tau^2) - 2i\tau\}$ implies

$$\arg \kappa(\tau) \gtreqless 0 \quad \text{for} \quad \tau \lesseqgtr 0.$$

Proposition 2.3.4. *The operator*

$$B : g \mapsto \kappa^*\big((1 + z)g(z)\big) = \frac{2}{1 + i\tau}g(\kappa(\tau))$$

induces an isomorphism

$$B : L^2(S^1) \to L^2(\mathbb{R}), \tag{2.125}$$

where

$$B(z^j) = 2\frac{(1 - i\tau)^j}{(1 + i\tau)^{j+1}}, \quad B(z^{-j-1}) = 2\frac{(1 + i\tau)^j}{(1 - i\tau)^{j+1}}, \quad j \in \mathbb{N}, \tag{2.126}$$

and we have

$$\pi^{\pm} = B^{-1}\Pi^{\pm}B \in L^0_{\mathrm{cl}}(S^1). \tag{2.127}$$

Moreover, the pull-back κ^* induces an isomorphism

$$\kappa^* : C^{\infty}(S^1) \to S^0_{\mathrm{tr}}(\mathbb{R}), \tag{2.128}$$

and for $a := \kappa^*(b)$ we have

$$\Pi^+ M_a = B\pi^+ M_b B^{-1} : V^+(\mathbb{R}) \to V^+(\mathbb{R}) \tag{2.129}$$

and

$$\pi^+ M_b : W^+(S^1) \to W^+(S^1). \tag{2.130}$$

Proof. The relations (2.126) can be verified by a direct computation. Since

$$V^+(\mathbb{R}) := \left[2\frac{(1 - i\tau)^j}{(1 + i\tau)^{j+1}} : j \in \mathbb{N}\right] \quad \text{and} \quad V^-(\mathbb{R}) := \left[2\frac{(1 + i\tau)^j}{(1 - i\tau)^{j+1}} : j \in \mathbb{N}\right], \tag{2.131}$$

it follows that B induces isomorphisms

$$B : W^{\pm}(S^1) \to V^{\pm}(\mathbb{R}), \tag{2.132}$$

which yield the isomorphism (2.125). At the same time we obtain (2.127). Let us now show (2.129). From (2.127) we conclude that

$$\Pi^{\pm} = B\pi^{\pm}B^{-1}.$$

For the plus-case it follows that

$$\Pi^{+}\mathcal{M}_a = B\pi^{+}B^{-1}\mathcal{M}_a = B\pi^{+}\left(B^{-1}\mathcal{M}_aB\right)B^{-1} = B\pi^{+}\mathcal{M}_bB^{-1}$$

for $\mathcal{M}_b = B^{-1}\mathcal{M}_aB$ and

$$b(z) = \left((\kappa^*)^{-1}a\right)(z). \tag{2.133}$$

$\square$

Instead of (2.130) we also write

$$T_b : W^{+}(S^1) \to W^{+}(S^1). \tag{2.134}$$

For the continuity of (2.134) it suffices to assume $b \in C(S^1)$; T_b is called a Toeplitz operator.

Corollary 2.3.5. *For a symbol $a(\tau) \in S^0_{\mathrm{tr}}(\mathbb{R}_\tau)$ and b as in (2.133) we have*

$$\left(B^{-1}F\right)\mathrm{op}^{+}(a)\left(F^{-1}B\right) = T_b : W^{+}(S^+) \to W^{+}(S^+). \tag{2.135}$$

In fact, it suffices to combine the relation (2.116) with (2.129) and (2.130) for $T_b = \pi^{+}\mathcal{M}_b$.

Theorem 2.3.6. *Let $b \in C(S^1)$ and $b(z) \neq 0$ for every $z \in S^1$. Then there exist a unique $m \in \mathbb{Z}$ and a function $\psi \in C(S^1)$ such that*

$$b(z) = z^m e^{\psi(z)}. \tag{2.136}$$

A proof can be found in [38, Satz 4.4.8]. We set

$$m := \mathrm{wind}\,(b),$$

called the winding number of b.

Definition 2.3.7. The operator T_b, with $b \in C(S^1)$, is called elliptic if $b(z) \neq 0$ for all $z \in S^1$.

Theorem 2.3.8. *Let $b \in C(S^1)$. The following conditions are equivalent:*

(i) *The operator T_b is elliptic.*

(ii) *The operator $T_b : W^{+} \to W^{+}$ is Fredholm.*

The ellipticity of T_b ensures the existence of a parametrix, namely, $(T_b)^{(-1)} = T_{b^{-1}}$. Moreover, we have

$$\mathrm{ind}\,T_b = -\mathrm{wind}\,(b). \tag{2.137}$$

For a proof, see [38, Satz 4.4.7].

Corollary 2.3.9. *For the operator*

$$\mathrm{op}^+(a) : L^2(\mathbb{R}_+) \to L^2(\mathbb{R}_+) \tag{2.138}$$

with $a(\tau) \in S^0_{\mathrm{tr}}(\mathbb{R})$, the following conditions are equivalent:

(i) *a is elliptic, i.e., $a(\tau) \neq 0$ for all $\tau \in \mathbb{R}$, and $a_{0,+} = a_{0,+} \neq 0$, cf. notation in formula (2.4) and Definition 2.2.43;*

(ii) *the operator (2.138) is Fredholm.*

In fact, condition (i) is equivalent to $b(z) \neq 0$ for all $z \in S^1$. Moreover, by virtue of relation (2.135), using $L^2(\mathbb{R}_+) = F^{-1}BW^+(S^1)$ and

$$\mathrm{op}^+(a) : F^{-1}BW^+(S^1) \to F^{-1}BW^+(S^1),$$

we see that (i) and (ii) are equivalent.

2.4   The algebra of boundary value problems

Let X be a compact C^∞ manifold with boundary Y, and let $2X$ denote the double of X, obtained by gluing together two copies X_+ and X_- of X along their common boundary Y by the identity map. We then identify X with X_+. Let e^+ denote the operator of extension of functions on $\mathrm{int}\, X_+$ by zero to the opposite side X_-, and r^+ denote the operator of restriction of distributions on $2X$ to $\mathrm{int}\, X_+$. In an analogous manner we define the operators e^- and r^- with respect to the minus-side of $2X$. On $2X$ we choose a Riemannian metric that is equal to the product metric of $Y \times (-1, 1)$ in a neighbourhood of Y for some Riemannian metric on Y.

For a given $E \in \mathrm{Vect}(X)$ we fix any $\widetilde{E} \in \mathrm{Vect}(X)$ and $E = \widetilde{E}|_X$. Now let E and F in $\mathrm{Vect}(X)$ with fibre dimensions k and m, respectively. Consider the space $L^\mu_{\mathrm{cl}}(2X; \widetilde{E}, \widetilde{F})$, cf. the notation in Section 1.1. For every chart $\chi : V \to \Omega$ on $2X$, $\Omega \subset \mathbb{R}^n$ open, and trivialisations $\widetilde{E}|_V \cong \Omega \times \mathbb{C}^k$, $\widetilde{F}|_V \cong \Omega \times \mathbb{C}^m$, the push-forward $\chi_* A$ of an operator $A \in L^\mu_{\mathrm{cl}}(2X; \widetilde{E}, \widetilde{F})$ belongs to $L^\mu_{\mathrm{cl}}(\Omega; \mathbb{C}^k, \mathbb{C}^m)$. The notation for the push-forward χ_* shows also the chosen trivialisations of $\widetilde{E}|_V$ and $\widetilde{F}|_V$; for simplicity those are not explicitly indicated here, cf. also the notation in Remark 1.1.11. This should not cause any confusion.

Let $V \cap Y \neq \emptyset$, $V := V' \times (-1, 1)$, where V' is a coordinate neighbourhood on the boundary Y, and assume that χ restricts to charts $\chi_\pm : V_\pm \to \Omega \times \overline{\mathbb{R}}_+$ on $X_\pm$ for $V_\pm := X_\pm \cap V$, and to a chart $\chi' := \chi|_{V'} : V' \to \Omega$ on Y, $\Omega \subset \mathbb{R}^{n-1}$.

Definition 2.4.1. For $\mu \in \mathbb{Z}$ we define $L^\mu_{\mathrm{tr}}(2X; \widetilde{E}, \widetilde{F})$ to be the space of all elements $\widetilde{A} \in L^\mu_{\mathrm{cl}}(2X; \widetilde{E}, \widetilde{F})$ such that for every chart $\chi : V \to \Omega \times \mathbb{R}$ of the described kind and $\varphi, \psi \in C^\infty_0(\Omega \times \mathbb{R})$ the symbol

$$\widetilde{a}(x, \xi) := e_{-\xi}\{\varphi(\chi_*)\widetilde{A}\psi\}e_\xi \quad \text{for } e_\xi := e^{ix\xi}$$

is an $m \times k$ matrix of elements in $S_{\mathrm{tr}}^{\mu}(\Omega \times \mathbb{R} \times \mathbb{R}^n)$, cf. formula (1.11) and Definition 2.1.1. Moreover, we set

$$L_{\mathrm{tr}}^{\mu}(X; E, F) := \{\mathrm{r}^+ \widetilde{A} \mathrm{e}^+ : \widetilde{A} \in L_{\mathrm{tr}}^{\mu}(2X; \widetilde{E}, \widetilde{F})\}. \tag{2.139}$$

There is a natural Fréchet topology in the space (2.139). This property is left as an exercise for the reader.

Let us now establish the continuity of operators in $L_{\mathrm{tr}}^{\mu}(X; E, F)$ between Sobolev spaces.

Proposition 2.4.2. *Let* $\Omega \subset \mathbb{R}^q$ *be an open set, and assume that* $a(y, t, \eta, \tau) \in S_{\mathrm{tr}}^{\mu}(\Omega \times \overline{\mathbb{R}}_+ \times \mathbb{R}^n)$ *is independent of* t *for* $t > c$ *for some constant* $c > 0$. *Then* $\mathrm{Op}^+(a) := \mathrm{r}^+ \mathrm{Op}(\widetilde{a}) \mathrm{e}^+$ *(for any* $\widetilde{a}(y, t, \eta, \tau) \in S_{\mathrm{tr}}^{\mu}(\Omega \times \mathbb{R} \times \mathbb{R}^n)$ *such that* $a = \widetilde{a}|_{\Omega \times \overline{\mathbb{R}}_+ \times \mathbb{R}^n}$) *induces a continuous operator*

$$\mathrm{Op}^+(a) : H_{\mathrm{comp}(y)}^s(\Omega \times \mathbb{R}_+) \to H_{\mathrm{loc}(y)}^{s-\mu}(\Omega \times \mathbb{R}_+)$$

for every $s \in \mathbb{R}$, $s > -1/2$.

Proof. It suffices to apply Theorem 1.3.34 for $H = H^s(\mathbb{R}_+)$, $\widetilde{H} = H^{s-\mu}(\mathbb{R}_+)$, in combination with relation $\mathrm{op}^+(a)(y, \eta) \in S^{\mu}(\Omega \times \mathbb{R}^{n-1}; H^s(\mathbb{R}_+), H^{s-\mu}(\mathbb{R}_+))$ for every real $s > -1/2$, cf. Proposition 2.2.13. $\qquad \square$

Theorem 2.4.3. *An* $A \in L_{\mathrm{tr}}^{\mu}(X; E, F)$ *induces a continuous operator*

$$A : H^s(\mathrm{int}\, X, E) \to H^{s-\mu}(\mathrm{int}\, X, F)$$

for every $s \in \mathbb{R}$, $s > -1/2$.

Proof. Let $\omega'' \prec \omega \prec \omega'$ (cf. notation in Remark 1.1.11) be functions in $C_0^{\infty}(Y \times [0, 1))$ that are equal to 1 in a neighbourhood of Y. Write the operator A in the form $\omega A \omega' + (1 - \omega) A (1 - \omega'') + C$. Then $C \in L^{-\infty}(X; E, F)$, and both C and $(1 - \omega) A (1 - \omega'')$ are continuous as asserted, cf. also Theorem 1.1.14. The continuity of $\omega A \omega' : H^s(X, E) \to H^{s-\mu}(X, F)$ for $s > -1/2$ follows from Proposition 2.4.2 by a simple partition of unity argument. $\qquad \square$

Given an $A \in L_{\mathrm{tr}}^{\mu}(X; E, F)$, $A = \mathrm{r}^+ \widetilde{A} \mathrm{e}^+$ for an $\widetilde{A} \in L_{\mathrm{tr}}^{\mu}(2X; \widetilde{E}, \widetilde{F})$, cf. formula (2.139), we first have the homogeneous principal symbol $\sigma_{\psi}(\widetilde{A}) : \pi_{2X}^* \widetilde{E} \to \pi_{2X}^* \widetilde{F}$, cf. the notation (1.23), and we set $\sigma_{\psi}(A) := \sigma_{\psi}(\widetilde{A})|_{T^*X \setminus 0}$,

$$\sigma_{\psi}(A) : \pi_X^* E \to \pi_X^* F, \tag{2.140}$$

$\pi_X : T^*X \setminus 0 \to X$ $\left(T^*X = T^*(2X)|_X\right)$. With (2.140) we associate a family of operators

$$\sigma_{\partial}(A)(y, \eta) := \mathrm{r}^+ \sigma_{\psi}(A)(y, 0, \eta, D_t) \mathrm{e}^+ = \mathrm{r}^+ \mathrm{op}\left(\sigma_{\psi}(A)|_{t=0}\right)(y, \eta) \mathrm{e}^+ \tag{2.141}$$

for $(y, \eta) \in T^*Y \setminus 0$. This refers to the variables $(y, t) \in Y \times [0, 1)$ of a collar neighbourhood of the boundary. The family of operators (2.141) represents a bundle morphism

$$\sigma_\partial(A) : \pi_Y^* E' \otimes H^s(\mathbb{R}_+) \to \pi_Y^* F' \otimes H^{s-\mu}(\mathbb{R}_+)$$

for every fixed $s \in \mathbb{R}$, $s > -1/2$, $\pi_Y : T^*Y \setminus 0 \to Y$, $E' := E\big|_Y$, $F' := F\big|_Y$. Alternatively, we interpret $\sigma_\partial(A)$ as a morphism

$$\sigma_\partial(A) : \pi_Y^* E' \otimes \mathcal{S}(\overline{\mathbb{R}}_+) \to \pi_Y^* F' \otimes \mathcal{S}(\overline{\mathbb{R}}_+),$$

$\mathcal{S}(\overline{\mathbb{R}}_+) := \mathcal{S}(\mathbb{R})\big|_{\overline{\mathbb{R}}_+}$. We call $\sigma_\psi(A)$ the principal interior symbol and $\sigma_\partial(A)$ the principal boundary symbol of the operator A.

Example 2.4.4. Let $A = \sum_{|\alpha| \le \mu} a_\alpha(x) D_x^\alpha$ be a differential operator in

$$\mathbb{R}_+^n = \{x = (y, t) \in \mathbb{R}^n : t > 0\}, \quad a_\alpha \in C^\infty(\overline{\mathbb{R}}_+^n).$$

Then

$$\sigma_\psi(A)(x, \xi) = \sum_{|\alpha| = \mu} a_\alpha(x) \xi^\alpha,$$

$$\sigma_\partial(A)(y, \eta) = \sum_{|\alpha| = \mu} a_\alpha(y, 0)(\eta, D_t)^\alpha.$$

Remark 2.4.5. For $(\kappa_\lambda u)(t) := \lambda^{1/2} u(\lambda t)$, $\lambda \in \mathbb{R}_+$, we have

$$\sigma_\partial(A)(y, \lambda \eta) = \lambda^\mu \kappa_\lambda \sigma_\partial(A)(y, \eta) \kappa_\lambda^{-1}$$

for all $\lambda \in \mathbb{R}_+$.

We now define a 2×2 block matrix algebra of boundary value problems on X with trace and potential conditions. The motivation is similar to that for classical pseudo-differential operators on an open manifold, where we complete the algebra of differential operators to an algebra that contains the parametrices of elliptic elements. In the present case it is the set of differential boundary value problems with differential boundary conditions (up to an order reduction on the boundary) that we complete to an algebra of pseudo-differential boundary value problems that contains the parametrices of elliptic elements.

The spaces $L_{\mathrm{tr}}^\mu(X; E, F)$ belong to the top left corners. However, if we compose two elements of that kind there appear remainders terms, here called Green operators. In addition, boundary operators as in classical BVPs (like Dirichlet or Neumann for Laplacians) have to be designed. This is precisely the topic of the following discussion. Also the respective 2×2 block matrices will be called Green operators since the 12- and 21-entries bear some similarity with the above-mentioned Green operators in the top left corners.

First we need the smoothing Green operators. They will also have a so-called type $d \in \mathbb{N}$; and we begin with the case $d = 0$.

Let $\mathcal{B}^{-\infty,0}(X;\boldsymbol{v})$ for $\boldsymbol{v} := (E,F,J_1,J_2)$ for E, $F \in \mathrm{Vect}(X)$, J_1 $J_2 \in \mathrm{Vect}(Y)$, denote the space of all operators

$$
\mathcal{G} := \begin{pmatrix} G & K \\ T & Q \end{pmatrix} : \begin{array}{c} C^\infty(X,E) \\ \oplus \\ C^\infty(Y,J_1) \end{array} \longrightarrow \begin{array}{c} C^\infty(X,F) \\ \oplus \\ C^\infty(Y,J_2) \end{array}
$$

such that $\mathcal{G}$ and $\mathcal{G}^*$ extend to continuous operators

$$
\mathcal{G} : \begin{array}{c} H^s(\mathrm{int} X, E) \\ \oplus \\ H^s(Y,J_1) \end{array} \longrightarrow \begin{array}{c} C^\infty(X,F) \\ \oplus \\ C^\infty(Y,J_2) \end{array}, \quad \mathcal{G}^* : \begin{array}{c} H^s(\mathrm{int} X, F) \\ \oplus \\ H^s(Y,J_2) \end{array} \longrightarrow \begin{array}{c} C^\infty(X,E) \\ \oplus \\ C^\infty(Y,J_1) \end{array}
$$

for all $s \in \mathbb{R}$, $s > -1/2$. Here $\mathcal{G}^*$ is the formal adjoint of $\mathcal{G}$ in the sense that

$$
(u, \mathcal{G}^* v)_{L^2(X,E)\oplus L^2(Y,J_1)} = (\mathcal{G}u, v)_{L^2(X,F)\oplus L^2(Y,J_2)} \tag{2.142}
$$

for all $u \in C^\infty(X,E) \oplus C^\infty(Y,J_1)$, $v \in C^\infty(X,F) \oplus C^\infty(Y,J_2)$; the L^2-scalar products refers to the chosen Riemannian metrics on X and Y and to the Hermitean metrics in the respective vector bundles.

In order to pass to operators of type $d \in \mathbb{N} \setminus \{0\}$ for every $E \in \mathrm{Vect}(X)$ we fix an operator $T : C^\infty(X,E) \to C^\infty(X,E)$ that is equal to $\mathrm{id}_{E'} \otimes \partial_t$ in a collar neighbourhood of Y, in the splitting of variables $(y,t) \in Y \times [0,1)$.

The space $\mathcal{B}^{-\infty,d}(X;\boldsymbol{v})$ of smoothing "BVPs" of type $d \in \mathbb{N} \setminus \{0\}$ is defined to be the set of all operators

$$
\mathcal{G} = \mathcal{G}_0 + \sum_{j=1}^{d} \mathcal{G}_j \mathrm{diag}\,(T^j, 0) \tag{2.143}
$$

with arbitrary $\mathcal{G}_j \in \mathcal{B}^{-\infty,0}(X;\boldsymbol{v})$, $j = 0,\ldots,d$.

In order to formulate the case of order $\mu \in \mathbb{R}$, $d \in \mathbb{N}$, we first introduce corresponding operator-valued symbols. The notion refers to the spaces

$$
H := L^2(\mathbb{R}_+, \mathbb{C}^k) \oplus \mathbb{C}^{j_1}, \quad \widetilde{H} := L^2(\mathbb{R}_+, \mathbb{C}^m) \oplus \mathbb{C}^{j_2},
$$

or Fréchet subspaces

$$
\mathcal{S}(\overline{\mathbb{R}}_+, \mathbb{C}^k) \oplus \mathbb{C}^{j_1}, \quad \mathcal{S}(\overline{\mathbb{R}}_+, \mathbb{C}^m) \oplus \mathbb{C}^{j_2},
$$

where the group actions are defined by

$$
u(t) \oplus c \to \lambda^{1/2} u(\lambda t) \oplus c, \quad \lambda \in \mathbb{R}_+,
$$

cf. also Definition 1.3.8.

Definition 2.4.6. Let k, m, j_1, $j_2 \in \mathbb{N}$, $\mu \in \mathbb{R}$, $\Omega \subset \mathbb{R}^q$ open, $q = \dim Y$. The space $\mathcal{R}_G^{\mu,0}(\Omega \times \mathbb{R}^q; \boldsymbol{w})$, $\boldsymbol{w} = (k, m; j_1, j_2)$, of Green symbols of order μ and type 0 is defined to be the set of all

$$g(y, \eta) \in S_{\mathrm{cl}}^{\mu}\big(\Omega \times \mathbb{R}^q; L^2(\mathbb{R}_+, \mathbb{C}^k) \oplus \mathbb{C}^{j_1}, \mathcal{S}(\overline{\mathbb{R}}_+, \mathbb{C}^m) \oplus \mathbb{C}^{j_2}\big)$$

such that

$$g^*(y, \eta) \in S_{\mathrm{cl}}^{\mu}\big(\Omega \times \mathbb{R}^q; L^2(\mathbb{R}_+, \mathbb{C}^m) \oplus \mathbb{C}^{j_2}, L^2(\mathbb{R}_+, \mathbb{C}^k) \oplus \mathbb{C}^{j_1}\big).$$

Here $g^*(y, \eta)$ is the pointwise adjoint in the sense

$$\big(u, g^*(y, \eta)v\big)_{L^2(\mathbb{R}_+, \mathbb{C}^k) \oplus \mathbb{C}^{j_1}} = \big(g(y, \eta)u, v\big)_{L^2(\mathbb{R}_+, \mathbb{C}^m) \oplus \mathbb{C}^{j_2}} \qquad (2.144)$$

for all $u \in L^2(\mathbb{R}_+, \mathbb{C}^k) \oplus \mathbb{C}^{j_1}$, $v \in L^2(\mathbb{R}_+, \mathbb{C}^m) \oplus \mathbb{C}^{j_2}$.

Remark 2.4.7. It can be proved that every $g(y, \eta) \in \mathcal{R}_G^{\mu,0}(\Omega \times \mathbb{R}^q; \boldsymbol{w})$ induces elements

$$g(y, \eta) \in S_{\mathrm{cl}}^{\mu}\big(\Omega \times \mathbb{R}^q; H^s(\mathbb{R}_+, \mathbb{C}^k) \oplus \mathbb{C}^{j_1}, \mathcal{S}(\overline{\mathbb{R}}_+, \mathbb{C}^m) \oplus \mathbb{C}^{j_2}\big)$$

for all $s \in \mathbb{R}$, $s > -1/2$.

Remark 2.4.8. The operator $\mathrm{id}_{\mathbb{C}^k} \otimes \partial_t^j$ represents an operator-valued symbol

$$\mathrm{id}_{\mathbb{C}^k} \otimes \partial_t^j \in S_{\mathrm{cl}}^{j}\big(\Omega \times \mathbb{R}^q; H^s(\mathbb{R}_+, \mathbb{C}^k), H^{s-j}(\overline{\mathbb{R}}_+, \mathbb{C}^k)\big)$$

for every $s \in \mathbb{R}$ (in this case there is no dependence on $(y, \eta) \in \Omega \times \mathbb{R}^q$).

In fact, the operator $T^j := \partial_t^j : H^s(\mathbb{R}_+) \to H^{s-j}(\mathbb{R}_+)$ belongs to the space $C^{\infty}\big(\Omega \times \mathbb{R}^q, \mathcal{L}(H^s(\mathbb{R}_+), \mathcal{H}^{s-j}(\mathbb{R}_+))\big)$ and satisfies the relation

$$T^j = \lambda^j \kappa_\lambda T^j \kappa_\lambda^{-1} \text{ for all } \lambda \in \mathbb{R}_+,$$

cf. Remark 1.3.20.

Definition 2.4.9. (i) By $\mathcal{R}_G^{\mu,d}(\Omega \times \mathbb{R}^q; \boldsymbol{w})$ for $\mu \in \mathbb{R}$, $d \in \mathbb{N}$, we denote the space of all operator functions

$$g(y, \eta) := g_0(y, \eta) + \sum_{j=1}^{d} g_j(y, \eta) \, \mathrm{diag}\,(\partial_t^j, 0)$$

with arbitrary $g_j(y, \eta) \in \mathcal{R}^{\mu-j,0}(\Omega \times \mathbb{R}^q; \boldsymbol{w})$. The elements of $\mathcal{R}_G^{\mu,d}(\Omega \times \mathbb{R}^q; \boldsymbol{w})$ are called Green symbols of order μ and type d.

(ii) By $\mathcal{R}^{\mu,d}(\Omega \times \mathbb{R}^q; \boldsymbol{w})$ for $\mu \in \mathbb{Z}$, $d \in \mathbb{N}$, we denote the space of all operator functions

$$a(y, \eta) := \mathrm{op}^+(p)(y, \eta) + g(y, \eta)$$

with arbitrary $p(y, t, \eta, \tau) \in S_{\mathrm{tr}}^{\mu}(\Omega \times \mathbb{R} \times \mathbb{R}^n; \mathbb{C}^k, \mathbb{C}^m)$ and

$$g(y, \eta) \in \mathcal{R}_G^{\mu,d}(\Omega \times \mathbb{R}^q; \boldsymbol{w}).$$

Remark 2.4.10. (i) Observe that elements of $\mathcal{R}_G^{\mu,d}(\Omega \times \mathbb{R}^q; w)$ or $\mathcal{R}^{\mu,d}(\Omega \times \mathbb{R}^q; w)$ can be composed by functions in $C^\infty(\Omega)$.

(ii) For $a(y,\eta) \in \mathcal{R}^{\mu,d}(\Omega \times \mathbb{R}^q; w)$ we have $D_y^\alpha D_\eta^\beta a(y,\eta) \in \mathcal{R}^{\mu-|\beta|,d}(\Omega \times \mathbb{R}^q; w)$ for every $\alpha, \beta \in \mathbb{N}^q$. Combining this with Theorem 2.2.35 for

$$a(y,\eta) \in \mathcal{R}^{\mu,d}(\Omega \times \mathbb{R}^q; v_0), \quad b(y,\eta) \in \mathcal{R}^{\nu,e}(\Omega \times \mathbb{R}^q; w_0),$$

$$v_0 := (k_0, m; j_0, j_2), \quad w_0 := (k, k_0; j_1, j_0), \tag{2.145}$$

we have

$$D_\eta^\alpha a(y,\eta) D_y^\alpha b(y,\eta) \in \mathcal{R}^{\mu+\nu-|\alpha|,h}(\Omega \times \mathbb{R}^q; v_0 \circ w_0) \tag{2.146}$$

for $v_0 \circ w_0 = (k, m; j_1, j_2)$,

Observe that $g(y,\eta) \in \mathcal{R}_G^{\mu,d}(\Omega \times \mathbb{R}^q; w)$ implies

$$g(y,\eta) \in S_{\mathrm{cl}}^\mu \big(\Omega \times \mathbb{R}^q; H^s(\mathbb{R}_+, \mathbb{C}^k) \oplus \mathbb{C}^{j_1}, \mathcal{S}(\overline{\mathbb{R}}_+, \mathbb{C}^m) \oplus \mathbb{C}^{j_2}\big) \tag{2.147}$$

for every $s \in \mathbb{R}$, $s > d - 1/2$.

The following statement is an immediate consequence of the definition.

Proposition 2.4.11. (i) *Let* $g_l(y,\eta) \in \mathcal{R}_G^{\mu-l,d}(\Omega \times \mathbb{R}^q; w)$, $l \in \mathbb{N}$, *be an arbitrary sequence. Then there is a* $g(y,\eta) \in \mathcal{R}_G^{\mu,d}(\Omega \times \mathbb{R}^q; w)$ *such that*

$$g(y,\eta) - \sum_{l=0}^N g_l(y,\eta) \in \mathcal{R}_G^{\mu-(N+1),d}(\Omega \times \mathbb{R}^q; w)$$

for every $N \in \mathbb{N}$, *and* $g(y,\eta)$ *is unique modulo* $\mathcal{R}_G^{-\infty,d}(\Omega \times \mathbb{R}^q; w)$.

(ii) *For arbitrary* $a_l(y,\eta) \in \mathcal{R}^{\mu-l,d}(\Omega \times \mathbb{R}^q; w)$, $l \in \mathbb{N}$, *there exists an* $a(y,\eta) \in \mathcal{R}^{\mu,d}(\Omega \times \mathbb{R}^q; w)$ *such that*

$$a(y,\eta) - \sum_{l=0}^N a_l(y,\eta) \in \mathcal{R}^{\mu-(N+1),d}(\Omega \times \mathbb{R}^q; w)$$

for every $N \in \mathbb{N}$, *and* $a(y,\eta)$ *is unique modulo* $\mathcal{R}_G^{-\infty,d}(\Omega \times \mathbb{R}^q; w)$.

With Green symbols we now associate so-called Green operators of order μ and type d, namely,

$$\mathcal{G} = \mathrm{Op}(g), \quad \mathrm{Op}(g)u(y) = \iint e^{i(y-y')\eta} g(y,\eta) u(y') dy' d\eta,$$

first for $u \in C_0^\infty(\Omega, H^s(\mathbb{R}_+, \mathbb{C}^k)) \oplus C_0^\infty(\Omega, \mathbb{C}^{j_1})$; then

$$\mathrm{Op}(g)u \in C^\infty(\Omega, H^{s-\mu}(\mathbb{R}_+, \mathbb{C}^m)) \oplus C^\infty(\Omega, \mathbb{C}^{j_2}).$$

Theorem 2.4.12. *For every* $g(y, \eta) \in \mathcal{R}_G^{\mu,d}(\Omega \times \mathbb{R}^q; \boldsymbol{w})$, $\boldsymbol{w} = (k, m; j_1, j_2)$, *the operator* $\mathcal{G} = \mathrm{Op}(g)$ *extends to a continuous operator*

$$\mathcal{G} : \begin{array}{c} H^s_{\mathrm{comp}(y)}(\Omega \times \mathbb{R}_+, \mathbb{C}^k) \\ \oplus \\ H^s_{\mathrm{comp}}(\Omega, \mathbb{C}^{j_1}) \end{array} \longrightarrow \begin{array}{c} H^{s-\mu}_{\mathrm{loc}(y)}(\Omega \times \mathbb{R}_+, \mathbb{C}^m) \\ \oplus \\ H^{s-\mu}_{\mathrm{loc}}(\Omega, \mathbb{C}^{j_2}) \end{array}$$

for every $s \in \mathbb{R}$, $s > d - 1/2$.

Proof. We have

$$g(y, \eta) \in S_{\mathrm{cl}}^{\mu}(\Omega \times \mathbb{R}^q; H^s(\mathbb{R}_+, \mathbb{C}^k) \oplus \mathbb{C}^{j_1}, H^{s-\mu}(\mathbb{R}_+, \mathbb{C}^m) \oplus \mathbb{C}^{j_2})$$

for $s > d - 1/2$ which is a consequence of (2.147). Then

$$\mathrm{Op}(g) : \mathcal{W}^s_{\mathrm{comp}}(\Omega, H^s(\mathbb{R}_+, \mathbb{C}^k) \oplus \mathbb{C}^{j_1}) \to \mathcal{W}^{s-\mu}_{\mathrm{loc}}(\Omega, H^{s-\mu}(\mathbb{R}_+, \mathbb{C}^m) \oplus \mathbb{C}^{j_2})$$

is continuous, cf. Theorem 1.3.34. Hence the assertion follows from the identities (1.124). $\qquad\square$

The symbolic structure of a Green operator $\mathcal{G} = \mathrm{Op}(g)$ allows us to define its boundary symbol

$$\sigma_\partial(\mathcal{G})(y, \eta) : H^s(\mathbb{R}_+, \mathbb{C}^k) \oplus \mathbb{C}^{j_1} \to \mathcal{S}(\overline{\mathbb{R}}_+, \mathbb{C}^m) \oplus \mathbb{C}^{j_2} \qquad (2.148)$$

for $(y, \eta) \in T^*\Omega \setminus 0$, namely, as $\sigma_\partial(\mathcal{G})(y, \eta) = g_{(\mu)}(y, \eta)$, the homogeneous principal component of $g(y, \eta)$, cf. Definition 1.3.8 (ii). Alternatively, we also write

$$\sigma_\partial(\mathcal{G})(y, \eta) : H^s(\mathbb{R}_+, \mathbb{C}^k) \oplus \mathbb{C}^{j_1} \to H^{s-\mu}(\overline{\mathbb{R}}_+, \mathbb{C}^m) \oplus \mathbb{C}^{j_2},$$

or

$$\sigma_\partial(\mathcal{G})(y, \eta) : \mathcal{S}(\overline{\mathbb{R}}_+, \mathbb{C}^k) \oplus \mathbb{C}^{j_1} \to \mathcal{S}(\overline{\mathbb{R}}_+, \mathbb{C}^m) \oplus \mathbb{C}^{j_2}.$$

Now we pass to global Green operators on a smooth compact manifold X with boundary Y. Let $Y \times [0, 1)$ be a collar neighbourhood of Y and $V' \subset Y$ a coordinate neighbourhood, $V := V' \times [0, 1)$, and $\chi : V \to \Omega \times \overline{\mathbb{R}}_+$ a chart that restricts to a chart $\chi' : V \to \Omega$. For our vector bundles $E, F \in \mathrm{Vect}(X)$ and $J_{1,2} \in \mathrm{Vect}(Y)$ we have trivialisations

$$E|_V \cong \Omega \times \overline{\mathbb{R}}_+ \times \mathbb{C}^k, \quad F|_V \cong \Omega \times \overline{\mathbb{R}}_+ \times \mathbb{C}^m, \quad \text{and} \quad J_{1,2}|_{V'} \cong \Omega \times \mathbb{C}^{j_{1,2}}.$$

Green operators $\mathcal{G}$ can be interpreted as operators between sections in the corresponding bundles over V and V', respectively, namely,

$$\mathcal{G}_V : C_0^\infty(V, E|_V) \oplus C_0^\infty(V', J_1|_{V'}) \to C^\infty(V, F|_V) \oplus C^\infty(V', J_2|_{V'}).$$

Let us write $\mathcal{G}_V = (\chi^{-1})_* \mathrm{Op}(g)$, where the push-forward $(\chi^{-1})_*$ is an abbreviation of the notation $\mathrm{diag}((\chi^{-1})_*, ((\chi')^{-1})_*)$, which also includes the cocycles of transition maps of the involved bundles (cf. analogously, Remark 1.1.11).

Let us fix a finite system of coordinate neighbourhoods $\{V_j\}_{j=1,...,L}$, $V_j = V_j' \times [0,1)$, for an open cover $\{V_j'\}_{j=1,...,L}$ of Y by coordinate neighbourhoods. Moreover, choose functions $\varphi_j \prec \psi_j$ in $C_0^\infty(V_j)$ and set $\varphi_j' := \varphi_j|_{V_j'}$, $\psi_j' := \psi_j|_{V_j'}$, and assume that $\{\varphi_j'\}_{j=1,...,L}$ is a partition of unity subordinate to the cover $\{V_j'\}_{j=1,...,L}$.

Definition 2.4.13. The space of $\mathcal{B}_G^{\mu,d}(X; \boldsymbol{v})$, $\boldsymbol{v} = (E, F; J_1, J_2)$, of Green operators of order μ and type d is defined to be the set of all operators

$$\mathcal{G} := \sum_{j=0}^{L} \operatorname{diag}\left(\varphi_j, \varphi_j'\right)\left(\chi_j^{-1}\right)_* \operatorname{Op}(g_j) \operatorname{diag}\left(\psi_j, \psi_j'\right) + \mathcal{C}$$

with arbitrary $g_j(y,\eta) \in \mathcal{R}_G^{\mu,d}(\Omega \times \mathbb{R}_+; \boldsymbol{w})$, $\boldsymbol{w} = (k, m; j_1, j_2)$, $1 \leq j \leq L$, and $\mathcal{C} \in \mathcal{B}_G^{-\infty,d}(X; \boldsymbol{v})$. The space of top left corners of elements in $\mathcal{B}_G^{\mu,d}(X; \boldsymbol{v})$ will also be denoted by $\mathcal{B}_G^{\mu,d}(X; E, F)$.

The families of maps (2.148) have an invariant meaning as bundle morphisms

$$\sigma_\partial(\mathcal{G}) : \pi_Y^* \begin{pmatrix} E' \otimes H^s(\mathbb{R}_+) \\ \oplus \\ J_1 \end{pmatrix} \longrightarrow \pi_Y^* \begin{pmatrix} F' \otimes H^{s-\mu}(\mathbb{R}_+) \\ \oplus \\ J_2 \end{pmatrix}, \tag{2.149}$$

$\pi_Y : T^*Y \setminus 0 \to Y$, $s > d - 1/2$ (alternatively, we may write $\mathcal{S}(\overline{\mathbb{R}}_+)$ instead of $H^{s-\mu}(\mathbb{R}_+)$ on the right of (2.149), or $\mathcal{S}(\overline{\mathbb{R}}_+)$ on both sides).

Let us now define the spaces of pseudo-differential BVPs for operators with the transmission property at the boundary, also referred to as the Boutet de Monvel's calculus.

Definition 2.4.14. The space $\mathcal{B}^{\mu,d}(X; \boldsymbol{v})$, $\mu \in \mathbb{Z}$, $d \in \mathbb{N}$, $\boldsymbol{v} = (E, F; J_1, J_2)$, is defined to be the set of operators of the form

$$\mathcal{A} = \begin{pmatrix} A & 0 \\ 0 & 0 \end{pmatrix} + \mathcal{G} \tag{2.150}$$

with arbitrary $A \in L_{\mathrm{tr}}^\mu(X; E, F)$, cf. notation (2.139), and $\mathcal{G} \in \mathcal{B}_G^{\mu,d}(X; \boldsymbol{v})$, cf. Definition 2.4.13. The elements of $\mathcal{B}^{\mu,d}(X; \boldsymbol{v})$ are called pseudo-differential BVPs of order μ and type d. The space of top left corners of elements in $\mathcal{B}^{\mu,d}(X; \boldsymbol{v})$ will also be denoted by $\mathcal{B}^{\mu,d}(X; E, F)$.

For $\mathcal{A} \in \mathcal{B}^{\mu,d}(X; \boldsymbol{v})$, we set

$$\sigma(\mathcal{A}) := \left(\sigma_\psi(\mathcal{A}), \sigma_\partial(\mathcal{A})\right),$$

where $\sigma_\psi(\mathcal{A}) := \sigma_\psi(A)$, cf. formula (2.140), called the (homogeneous principal) interior symbol of $\mathcal{A}$ of order μ, and

$$\sigma_\partial(\mathcal{A}) := \begin{pmatrix} \sigma_\partial(A) & 0 \\ 0 & 0 \end{pmatrix} + \sigma_\partial(\mathcal{G}),$$

called the (homogeneous principal) boundary symbol of $\mathcal{A}$ of order μ.

The homogeneity of $\sigma_\psi(\mathcal{A})$ is as usual, i.e., $\sigma_\psi(\mathcal{A})(x, \lambda\xi) = \lambda^\mu \sigma_\psi(\mathcal{A})(x, \xi)$ for all $\lambda \in \mathbb{R}_+$, $(x, \xi) \in T^*X \setminus 0$. For $\sigma_\partial(\mathcal{A})$ we have

$$\sigma_\partial(\mathcal{A})(y, \lambda\eta) = \lambda^\mu \operatorname{diag}(\kappa_\lambda, \operatorname{id}) \sigma_\partial(\mathcal{A})(y, \eta) \operatorname{diag}(\kappa_\lambda^{-1}, \operatorname{id}) \qquad (2.151)$$

for all $\lambda \in \mathbb{R}_+$, $(y, \eta) \in T^*Y \setminus 0$, cf. also relation (1.77).

Remark 2.4.15.   (i) We have

$$\mathcal{B}^{\mu-1, d}(X; \boldsymbol{v}) = \{\mathcal{A} \in \mathcal{B}^{\mu, d}(X; \boldsymbol{v}) : \sigma(\mathcal{A}) = 0\}.$$

Setting $\sigma(\mathcal{B}^{\mu, d}(X; \boldsymbol{v})) := \{\sigma(\mathcal{A}) : \mathcal{A} \in \mathcal{B}^{\mu, d}(X; \boldsymbol{v})\}$ there is an operator correspondence

$$\operatorname{op} : \sigma(\mathcal{B}^{\mu, d}(X; \boldsymbol{v})) \to \mathcal{B}^{\mu, d}(X; \boldsymbol{v})$$

in the form of a linear operator (non-canonical) such that $\sigma \circ \operatorname{op} = \operatorname{id}_{\sigma(\mathcal{B}^{\mu, d}(X; \boldsymbol{v}))}$. The principal symbolic map

$$\sigma : \mathcal{B}^{\mu, d}(X; \boldsymbol{v}) \to \sigma(\mathcal{B}^{\mu, d}(X; \boldsymbol{v}))$$

gives rise to an exact sequence

$$0 \to \mathcal{B}^{\mu-1, d}(X; \boldsymbol{v}) \to \mathcal{B}^{\mu, d}(X; \boldsymbol{v}) \xrightarrow{\sigma} \sigma(\mathcal{B}^{\mu, d}(X; \boldsymbol{v})) \to 0.$$

(ii) If in (i) we replace $\mathcal{B}^{\mu, d}(X; \boldsymbol{v})$ by $\mathcal{B}_G^{\mu, d}(X; \boldsymbol{v})$ and σ by σ_∂, then we obtain an analogue of (i) for Green operators.

Let $\mathcal{B}^{\mu, d}(X; E, F)$ and $\mathcal{B}_G^{\mu, d}(X; E, F)$ denote the spaces of top left corners of $\mathcal{B}^{\mu, d}(X; \boldsymbol{v})$ and $\mathcal{B}_G^{\mu, d}(X; \boldsymbol{v})$, respectively. By definition, we have

$$\mathcal{B}^{\mu, d}(X; E, F) = L_{\operatorname{tr}}^\mu(X; E, F) + \mathcal{B}_G^{\mu, d}(X; E, F).$$

Observe that $\mathcal{B}_G^{\mu, d}(X; E, F) \subset L^{-\infty}(\operatorname{int} X; E, F)$ and hence $\mathcal{B}^{\mu, d}(X; E, F) \subset L_{\operatorname{cl}}^\mu(\operatorname{int} X; E, F)$.

Remark 2.4.16. The statement $A \in \mathcal{B}^{\mu, d}(X; E, F) \cap L^{-\infty}(\operatorname{int} X; E, F)$ is equivalent to $A \in \mathcal{B}_G^{\mu, d}(X; E, F)$.

Writing an $\mathcal{A} \in \mathcal{B}^{\mu, d}(X; \boldsymbol{v})$ in the form $\mathcal{A} = (\mathcal{A}_{ij})_{i, j=1, 2}$, we also call $\mathcal{A}_{21}$ a trace operator and $\mathcal{A}_{12}$ a potential operator. For the bottom right corner $\mathcal{A}_{22}$ we simply have $\mathcal{A}_{22} \in L_{\operatorname{cl}}^\mu(Y; J_1, J_2)$.

Proposition 2.4.17. *Every* $G \in \mathcal{B}_G^{\mu, d}(X; E, F)$ *has a unique representation*

$$G = G_0 + \sum_{j=0}^{d-1} K_j \circ T^j$$

with a $G_0 \in \mathcal{B}_G^{\mu, 0}(X; E, F)$, *potential operators* $K_j \in \mathcal{B}^{\mu-j-1/2, 0}(X; (0, F; E', 0))$, *and trace operators* T^j *of the same form as in* (2.143).

The proof of Proposition 2.4.17 is straightforward, using Proposition 2.2.28, see also [45].

Theorem 2.4.18. *An $\mathcal{A} \in \mathcal{B}^{\mu,d}(X; \boldsymbol{v})$ for $\boldsymbol{v} = (E, F; J_1, J_2)$ induces a continuous operator*

$$\mathcal{A}: \begin{matrix} H^s(\text{int } X, E) \\ \oplus \\ H^s(Y, J_1) \end{matrix} \longrightarrow \begin{matrix} H^{s-\mu}(\text{int } X, F) \\ \oplus \\ H^{s-\mu}(Y, J_2) \end{matrix} \tag{2.152}$$

for every $s > d - 1/2$.

Proof. The assertion is a consequence of Theorem 2.4.3 and Theorem 2.4.12. ☐

Remark 2.4.19. Under the conditions of Theorem 2.4.18, the operator (2.152) is compact when $\mathcal{A} \in \mathcal{B}^{\mu-1,d}(X; \boldsymbol{v})$.

In fact, the compactness is a consequence of Theorem 2.4.18, applied for $\mu-1$, and the compactness of the embeddings

$$H^{s-(\mu+1)}(\text{int } X, F) \oplus H^{s-(\mu+1)}(Y, J_2) \hookrightarrow H^{s-\mu}(\text{int } X, F) \oplus H^{s-\mu}(Y, J_2)$$

Theorem 2.4.20. *Let $\mathcal{A}_j \in \mathcal{B}^{\mu-j,d}(X; \boldsymbol{v})$, $j \in \mathbb{N}$, be an arbitrary sequence. Then there exists an asymptotic sum $\mathcal{A} \sim \sum_{j=0}^{\infty} \mathcal{A}_j$ in $\mathcal{B}^{\mu,d}(X; \boldsymbol{v})$, i.e., an $\mathcal{A} \in \mathcal{B}^{\mu,d}(X; \boldsymbol{v})$ such that*

$$\mathcal{A} - \sum_{j=0}^{\infty} \mathcal{A}_j \in \mathcal{B}^{\mu-(N+1),d}(X; \boldsymbol{v})$$

for every $N \in \mathbb{N}$; moreover, $\mathcal{A}$ is unique mod $\mathcal{B}^{-\infty,d}(X; \boldsymbol{v})$.

Proof. The result can be easily reduced to local asymptotic summations of interior symbols, cf. Theorem 1.1.5, Remark 2.1.3 (iii), and boundary amplitude functions, cf. Proposition 2.4.11. ☐

Theorem 2.4.21. (i) *Consider $\mathcal{A} \in \mathcal{B}^{\mu,d}(X; \boldsymbol{v})$ with $\boldsymbol{v} = (E_0, F; J_0, J_2)$ and $\mathcal{B} \in \mathcal{B}^{\nu,e}(X; \boldsymbol{w})$ with $\boldsymbol{w} = (E, E_0; J_1, J_0)$. Then $\mathcal{AB} \in \mathcal{B}^{\mu+\nu,h}(X; \boldsymbol{v} \circ \boldsymbol{w})$ with $\boldsymbol{v} \circ \boldsymbol{w} := (E, F; J_1, J_2)$ and $h := \max(\nu + d, e)$, and we have $\sigma(\mathcal{AB}) = \sigma(\mathcal{A})\sigma(\mathcal{B})$ (with componentwise multiplication). If $\mathcal{A}$ or $\mathcal{B}$ is Green, then so is $\mathcal{AB}$.*

(ii) *Let $\mathcal{A} \in \mathcal{B}^{0,0}(X; \boldsymbol{v})$ with $\boldsymbol{v} = (E, F; J_1, J_2)$. Then for the adjoint (analogously defined as (2.142)) we have $\mathcal{A}^* \in \mathcal{B}^{0,0}(X; \boldsymbol{v}^*)$ with $\boldsymbol{v}^* = (F, E; J_2, J_1)$, and $\sigma(\mathcal{A}^*) = \sigma(\mathcal{A})^*$ (with componentwise adjoint, cf. also Theorem 1.1.16 and (2.144)).*

Proof. (i) The non-trivial part of the proof concerns the behaviour of local compositions

$$\text{Op}_y(a) \, \varphi \, \text{Op}_y(b) = \text{Op}_y(a \#_y \varphi b)$$

for any $\varphi \in C_0^\infty(\mathbb{R}^q)$ modulo smoothing operators of type h. In local compositions it is convenient to assume that

$$a(y, \eta) \in \mathcal{R}^{\mu,d}(\mathbb{R}^q \times \mathbb{R}^q; \boldsymbol{v}_0), \quad b(y, \eta) \in \mathcal{R}^{\nu,e}(\mathbb{R}^q \times \mathbb{R}^q; \boldsymbol{w}_0),$$

cf. (2.145), with k_0 and j_0 being the fibre dimensions of E_0 and J_0, respectively. The Leibniz product

$$a(y, \eta) \#_y \varphi b(y, \eta) \sim \sum_{\alpha \in \mathbb{N}^q} \frac{1}{\alpha!} \big(\partial_\eta^\alpha a(y, \eta) \big) D_y^\alpha \varphi(y) b(y, \eta)$$

can be carried out in $\mathcal{R}^{\mu+\nu,h}(\mathbb{R}^q \times \mathbb{R}^q; k, m; j_1, j_2)$, because of Theorem 2.2.35, Remark 2.4.10, in particular, relation (2.146), and Proposition 2.4.11. The remaining elements of the proof are straightforward.

(ii) The essential part of the proof consists of computing local formal adjoints $\mathrm{Op}_y(a)^*$ for $a(y, \eta) \in \mathcal{R}^{0,0}(\mathbb{R}^q \times \mathbb{R}^q; \boldsymbol{v})$, $\boldsymbol{v} = (k, m; j_1, j_2)$. The (y, η)-wise formal adjoint $a^{[*]}(y, \eta)$ of $a(y, \eta)$ belongs to

$$\mathcal{R}^{0,0}(\mathbb{R}^q \times \mathbb{R}^q; \boldsymbol{v}^*) \tag{2.153}$$

for $\boldsymbol{v}^* = (m, k; j_2, j_1)$. Analogously as in the standard pseudo-differential calculus we have $\mathrm{Op}_y(a)^* = \mathrm{Op}_y(a^{(*)})$ modulo smoothing operators of type 0, where

$$a^{(*)}(y, \eta) \sim \sum_{\alpha \in \mathbb{N}^q} \frac{1}{\alpha!} \partial_\eta^\alpha D_y^\alpha a^{[*]}(y, \eta) \tag{2.154}$$

can be carried out in (2.153), cf. Proposition 2.2.20 and its simple generalisation to the block matrix case. In (2.154) we also used Remark 2.4.10 (ii). $\square$

Chapter 3

Shapiro–Lopatinskii ellipticity

3.1 SL-ellipticity, parametrices, and the Fredholm property

We now define the notion of Shapiro–Lopatinskii ellipticity (also known as SL-ellipticity) of boundary conditions for an operator in Boutet de Monvel's calculus on a smooth manifold X with boundary Y. The results can be found, for instance, in the monograph [34] of Rempel and Schulze, and of course, also in the work [9] of Boutet de Monvel; see also the monograph of Grubb [19]. Therefore, here we only sketch the proofs.

Definition 3.1.1. Let $\mathcal{A} \in \mathcal{B}^{\mu,d}(X; \boldsymbol{v})$, $\mu \in \mathbb{Z}$, $d \in \mathbb{N}$, $\boldsymbol{v} = (E, F; J_1, J_2)$ for E, $F \in \mathrm{Vect}(X)$, $J_1, J_2 \in \mathrm{Vect}(Y)$. The operator $\mathcal{A}$ is called elliptic if

(i) $\mathcal{A}$ is σ_ψ-elliptic, i.e.,

$$\sigma_\psi(\mathcal{A}) : \pi_X^* E \to \pi_X^* F \tag{3.1}$$

for $\pi_X : T^*X \setminus 0 \to X$ is an isomorphism.

(ii) $\mathcal{A}$ is σ_∂-elliptic, i.e.,

$$\sigma_\partial(\mathcal{A}) : \pi_Y^* \begin{pmatrix} E' \otimes H^s(\mathbb{R}_+) \\ \oplus \\ J_1 \end{pmatrix} \longrightarrow \pi_Y^* \begin{pmatrix} F' \otimes H^{s-\mu}(\mathbb{R}_+) \\ \oplus \\ J_2 \end{pmatrix} \tag{3.2}$$

for $\pi_Y : T^*Y \setminus 0 \to Y$ is an isomorphism for some $s > \max\{\mu, d\} - 1/2$.

Condition (ii) for some $s = s_0 > \max\{\mu, d\} - 1/2$ is equivalent to this property for all $s > \max\{\mu, d\} - 1/2$. This in turn is equivalent to the bijectivity of

$$\sigma_\partial(\mathcal{A}) : \pi_Y^* \begin{pmatrix} E' \otimes \mathcal{S}(\overline{\mathbb{R}}_+) \\ \oplus \\ J_1 \end{pmatrix} \longrightarrow \pi_Y^* \begin{pmatrix} F' \otimes \mathcal{S}(\overline{\mathbb{R}}_+) \\ \oplus \\ J_2 \end{pmatrix}. \tag{3.3}$$

© Springer Nature Switzerland AG 2018
X. Liu, B.-W. Schulze, *Boundary Value Problems with Global Projection Conditions*,
Operator Theory: Advances and Applications 265, https://doi.org/10.1007/978-3-319-70114-1_3

This is a consequence of the considerations on operators on the half-line in Section 2.2.

Definition 3.1.2. Let $\mathcal{A} \in \mathcal{B}^{\mu,d}(X; \boldsymbol{v})$, $\boldsymbol{v} = (E, F; J_1, J_2)$; then a $\mathcal{P} \in \mathcal{B}^{-\mu,e}(X; \boldsymbol{v}^{-1})$ with $\boldsymbol{v}^{-1} = (F, E; J_2, J_1)$ and some $e \in \mathbb{N}$ is called a parametrix of $\mathcal{A}$, if

$$\mathcal{C}_{\mathrm{L}} := \mathcal{I} - \mathcal{P}\mathcal{A} \in \mathcal{B}^{-\infty,d_{\mathrm{L}}}(X; \boldsymbol{v}_{\mathrm{L}}), \quad \mathcal{C}_{\mathrm{R}} := \mathcal{I} - \mathcal{A}\mathcal{P} \in \mathcal{B}^{-\infty,d_{\mathrm{R}}}(X; \boldsymbol{v}_{\mathrm{R}})$$

for certain $d_{\mathrm{L}}, d_{\mathrm{R}} \in \mathbb{N}$ and $\boldsymbol{v}_{\mathrm{L}} := (E, E; J_1, J_1)$, $\boldsymbol{v}_{\mathrm{R}} := (F, F; J_2, J_2)$, where $\mathcal{I}$ denotes the respective identity operators.

Theorem 3.1.3. *Let $\mathcal{A} \in \mathcal{B}^{\mu,d}(X; \boldsymbol{v})$, $\boldsymbol{v} = (E, F; J_1, J_2)$, be elliptic. Then there is a parametrix $\mathcal{B} \in \mathcal{B}^{-\mu,(d-\mu)^+}(X; \boldsymbol{v}^{-1})$ with $\boldsymbol{v}^{-1} := (F, E; J_2, J_1)$, where*

$$\sigma(\mathcal{B}) = \sigma(\mathcal{A})^{-1}$$

with componentwise inverses, and

$$\mathcal{C}_{\mathrm{L}} := 1 - \mathcal{B}\mathcal{A} \in \mathcal{B}^{-\infty,d_{\mathrm{L}}}(X, \boldsymbol{v}_{\mathrm{L}}), \quad \mathcal{C}_{\mathrm{R}} := 1 - \mathcal{A}\mathcal{B} \in \mathcal{B}^{-\infty,d_{\mathrm{R}}}(X, \boldsymbol{v}_{\mathrm{R}})$$

for $\boldsymbol{v}_{\mathrm{L}} := (E, E; J_1, J_1)$, $\boldsymbol{v}_{\mathrm{R}} := (F, F; J_2, J_2)$, $d_{\mathrm{L}} = \max\{\mu, d\}$, $d_{\mathrm{R}} = (d - \mu)^+$ (where $\nu^+ := \max\{\nu, 0\}$). More precisely, $\mathcal{B}$ may be found in such a way that

$$\mathcal{C}_{\mathrm{L}} : H^s(\mathrm{int}\, X, E) \oplus H^s(Y, J_1) \to \mathcal{V}, \quad \mathcal{C}_{\mathrm{R}} : H^{s-\mu}(\mathrm{int}\, X, F) \oplus H^{s-\mu}(Y, J_2) \to \mathcal{W}$$

are projections for $s > \max\{\mu, d\} - 1/2$.

The proof follows by applying an operator correspondence to $\sigma(\mathcal{A})^{-1}$, namely, we form $\mathcal{B}_0 = \mathrm{op}(\sigma(\mathcal{A})^{-1}) \in \mathcal{B}^{-\mu,e}(X; \boldsymbol{v}^{-1})$ with $\boldsymbol{v}^{-1} = (F, E; J_2, J_1)$. Then, by Theorem 2.4.21, it follows that $\sigma(\mathcal{B}_0\mathcal{A}) = \mathrm{id}$, and hence $\mathcal{B}_0\mathcal{A} = \mathcal{I} + \mathcal{C}_0$ for a $\mathcal{C}_0 \in \mathcal{B}^{-1,h_0}(X; \boldsymbol{v}_{\mathrm{L}})$. Then a simple formal Neumann series argument allows us to form a $\mathcal{D}_0 \in \mathcal{B}^{-1,h_1}(X; \boldsymbol{v}_{\mathrm{L}})$ such that $(\mathcal{I} + \mathcal{D}_0)(\mathcal{I} + \mathcal{C}_0) = \mathcal{I} + \mathcal{C}_{\mathrm{L}}$ for a smoothing remainder $\mathcal{C}_{\mathrm{L}}$. Thus $\mathcal{B} = (\mathcal{I} + \mathcal{D}_0)\mathcal{B}_0$ is a left parametrix of $\mathcal{A}$. In a similar manner we obtain a right parametrix. The involved types can be computed via the rules of Theorem 2.4.21. The way of obtaining projections $\mathcal{C}_{\mathrm{L}}$ and $\mathcal{C}_{\mathrm{R}}$ is explained in Kapanadze and Schulze [25].

Theorem 3.1.4. *Let X be compact. For an operator $\mathcal{A} \in \mathcal{B}^{\mu,d}(X; \boldsymbol{v})$, with $\boldsymbol{v} = (E, F; J_1, J_2)$, the following conditions are equivalent:*

(i) *$\mathcal{A}$ is elliptic.*

(ii) *The operator*

$$\mathcal{A} : H^s(\mathrm{int}\, X, E) \oplus H^s(Y, J_1) \to H^{s-\mu}(\mathrm{int}\, X, F) \oplus H^{s-\mu}(Y, J_2) \qquad (3.4)$$

is Fredholm for some $s = s_0 > \max\{\mu, d\} - 1/2$.

The implication (i) $\Longrightarrow$ (ii) of Theorem 3.1.4 is a corollary of Theorem 3.1.3, since for compact X the remainders $\mathcal{C}_L$ and $\mathcal{C}_R$ are compact operators in the respective Sobolev spaces, and hence $\mathcal{A}$ is Fredholm. Concerning (ii) $\Longrightarrow$ (i), cf. Theorem 4.2.8 below.

Remark 3.1.5. Let X be compact and $\mathcal{A} \in \mathcal{B}^{\mu,d}(X; \boldsymbol{v})$ elliptic.

(i) The operator $\mathcal{A}$ is Fredholm for all $s > \max\{\mu, d\} - 1/2$.

(ii) $\mathcal{V} := \ker_s \mathcal{A} = \{u \in H^s(\operatorname{int} X, E) \oplus H^s(Y, J_1) : \mathcal{A}u = 0\}$ is a finite-dimensional subspace $H^s(\operatorname{int} X, E) \oplus H^s(Y, J_1) \subset C^\infty(X, E) \oplus C^\infty(Y, J_1)$ independent of s, and there is a finite-dimensional $\mathcal{W} \subset C^\infty(X, F) \oplus C^\infty(Y, J_2)$ independent of s such that $\operatorname{im}_s \mathcal{A} + \mathcal{W} = H^{s-\mu}(\operatorname{int} X, F) \oplus H^{s-\mu}(Y, J_2)$ for every s; here $\operatorname{im}_s \mathcal{A} = \{\mathcal{A}u : u \in H^s(\operatorname{int} X, E) \oplus H^s(Y, J_1)\}$.

Theorem 3.1.6. *Let X be compact, and assume that $\mathcal{A} \in \mathcal{B}^{\mu,d}(X; \boldsymbol{v})$ induces an isomorphism (3.4) for some $s = s_0 > \max\{\mu, d\} - 1/2$. Then (3.4) is an isomorphism for all $s > \max\{\mu, d\} - 1/2$, and we have $\mathcal{A}^{-1} \in \mathcal{B}^{-\mu,(\mu-d)^+}(X; \boldsymbol{v}^{-1})$.*

In fact, assume first that $\mathcal{A}$ is elliptic. Then Remark 3.1.5 states elliptic regularity, i.e., kernels and cokernels of $\mathcal{A}$ are independent of s. The inverse $\mathcal{A}^{-1}$ can be obtained by first constructing a parametrix of index 0 and then $\mathcal{I} + \mathcal{C}_L$ is invertible as well, and its inverse $\mathcal{I} + \mathcal{D}$ can be constructed in an elementary way. This yields $\mathcal{A}^{-1} = (\mathcal{I} + \mathcal{D})\mathcal{B}$, which also belongs to Boutet de Monvel's calculus. The assertion in general is proved in [34]. The statement itself is also called spectral invariance.

3.2 Interior ellipticity and the Atiyah–Bott obstruction

Let us now discuss the question to what extent a σ_ψ-elliptic operator on a smooth (compact) manifold with boundary admits Shapiro–Lopatinskii elliptic boundary conditions.

Theorem 3.2.1. *Let $A \in \mathcal{B}^{\mu,d}(X; E, F)$ be σ_ψ-elliptic, cf. Definition 3.1.1. Then the boundary symbol*

$$\sigma_\partial(A) : \pi_Y^*(E' \otimes H^s(\mathbb{R}_+)) \to \pi_Y^*(F' \otimes H^{s-\mu}(\mathbb{R}_+)) \tag{3.5}$$

represents a family of Fredholm operators for every $s > \max\{\mu, d\} - 1/2$, which is parametrised by $(y, \eta) \in T^*Y \setminus 0$, and $\ker \sigma_\partial(A)(y, \eta)$, $\operatorname{coker} \sigma_\partial(A)(y, \eta)$ are independent of s.

Proof. The assertion means that

$$\sigma_\partial(A)(y, \eta) : E_y' \otimes H^s(\mathbb{R}_+) \to F_y' \otimes H^{s-\mu}(\mathbb{R}_+)$$

is Fredholm for every $(y, \eta) \in T^*Y \setminus 0$. However, this is a straightforward matrix-valued analogue of Proposition 2.2.46. $\qquad\square$

By virtue of (2.151), we have

$$\sigma_\partial(A)(y, \lambda\eta) = \lambda^\mu \kappa_\lambda \sigma_\partial(A)(y, \eta) \kappa_\lambda^{-1} \tag{3.6}$$

for all $\lambda \in \mathbb{R}_+$. It follows that

$$\sigma_\partial(A)(y, \eta/|\eta|) : E'_y \otimes H^s(\mathbb{R}_+) \to F'_y \otimes H^{s-\mu}(\mathbb{R}_+)$$

is a family of Fredholm operators parametrised by $(y, \eta) \in S^*Y$, the unit co-sphere bundle induced by $T^*Y \setminus 0$ (referring to the fixed Riemannian metric). This corresponds to a standard situation of K-theory, discussed in abstract terms in Harutyunyan and Schulze [23], see also Section 3.5 below. The space S^*Y is compact, and hence there is a K-theoretic index element

$$\mathrm{ind}_{S^*Y} \sigma_\partial(A) \in K(S^*Y). \tag{3.7}$$

Let

$$\pi : S^*Y \to Y \tag{3.8}$$

denote the canonical projection. Then the bundle pull-back induces a homomorphism

$$\pi^* : K(Y) \to K(S^*Y).$$

The following result was first established by Atiyah and Bott in [3] in the case of differential operators and later on formulated for pseudo-differential operators with the transmission property at the boundary by Boutet de Monvel [10].

Let X be a smooth manifold with compact boundary Y, and $E, F \in \mathrm{Vect}(Y)$.

Theorem 3.2.2. *A σ_ψ-elliptic operator $A \in \mathcal{B}^{\mu,d}(X; E, F)$ admits a Shapiro–Lopatinskii elliptic BVP $\mathcal{A} \in \mathcal{B}^{\mu,d}(X; \boldsymbol{v})$, $\boldsymbol{v} := (E, F; J_1, J_2)$, for certain $J_1, J_2 \in \mathrm{Vect}(Y)$ (i.e., the top left corner of $\mathcal{A}$ is of the form $A + G$, for some $G \in \mathcal{B}_G^{\mu,d}(X; E, F)$ and $\sigma_\partial(\mathcal{A})$ is an isomorphism) if and only if*

$$\mathrm{ind}_{S^*Y} \sigma_\partial(A) \in \pi^* K(Y), \tag{3.9}$$

where $\pi : S^*Y \to Y$ is the canonical projection.

The proof will employ the following auxiliary considerations.

Remark 3.2.3. Let $A \in \mathcal{B}^{\mu,d}(X; E, F)$ be σ_ψ-elliptic, and consider the family of Fredholm operators $\sigma_\partial(A)(y, \eta) : E'_y \otimes H^s(\mathbb{R}_+) \to F'_y \otimes H^{s-\mu}(\mathbb{R}_+)$, $s > \max\{\mu, d\} - 1/2$. Then there exists a subbundle $\widetilde{W} \subset \pi^*F' \otimes \mathcal{S}(\overline{\mathbb{R}}_+)$ of finite fibre dimension such that

$$\widetilde{W}_{y,\eta} + \mathrm{im}\, \sigma_\partial(A)(y, \eta) = F'_y \otimes H^{s-\mu}(\mathbb{R}_+) \quad \text{for all } (y, \eta) \in S^*Y.$$

In fact, the observation essentially corresponds to Proposition 3.5.9 below. The main difference here is that the subbundle $\widetilde{W}$ which is the image under some potential boundary symbol k is possible with fibres $F'_y \otimes \mathcal{S}(\overline{\mathbb{R}}_+)$. This follows from an analogue of relation (2.103).

Proposition 3.2.4. Let $A \in \mathcal{B}^{\mu,d}(X; E, F)$ be σ_ψ-elliptic and choose L_1, $L_2 \in$ Vect(S^*Y) such that

$$\text{ind}_{S^*Y} \sigma_\partial(A) = [L_2] - [L_1]. \tag{3.10}$$

Then there exists an element $G \in \mathcal{B}_G^{\mu,0}(X; E, F)$ such that

$$\ker_{S^*Y} \sigma_\partial(A + G) \cong L_2, \quad \text{coker}_{S^*Y}\sigma_\partial(A + G) \cong L_1. \tag{3.11}$$

Proof. Let us fix some real $s > \max\{\mu, d\} - 1/2$; the specific choice is inessential. Let

$$a : \pi^* E' \otimes H^s(\mathbb{R}_+) \to \pi^* F' \otimes H^{s-\mu}(\mathbb{R}_+),$$

$\pi : S^*Y \to Y$, denote the restriction of $\sigma_\partial(A)$ to $(y, \eta) \in S^*Y$. According to Remark 3.2.3, there is a surjective bundle morphism

$$\begin{pmatrix} a & k \end{pmatrix} : \begin{array}{c} \pi^* E' \otimes H^s(\mathbb{R}_+) \\ \oplus \\ W \end{array} \longrightarrow \pi^* F' \otimes H^{s-\mu}(\mathbb{R}_+)$$

for a $W \in \text{Vect}(S^*Y)$, where $k : W \to \widetilde{W}$ is an isomorphism to a subbundle $\widetilde{W}$ of $\pi^* F' \otimes \mathcal{S}(\mathbb{R}_+)$. Without loss of generality, we assume W to be trivial. Let $p : \pi^* F' \otimes H^{s-\mu}(\mathbb{R}_+) \to \widetilde{W}$ be a projection that is orthogonal with respect to the scalar products of $F'_y \otimes L^2(\mathbb{R}_+)$. By adding, if necessary, another finite-dimensional subbundle to W and denoting the new bundle again by W we obtain the following properties. There are subbundles $\widetilde{L}_1 \subset \widetilde{W}$ and

$$\widetilde{L}_2 \subset \widetilde{V} := \ker_{S^*Y}((1 - p)a) \subset \pi^* E' \otimes H^s(\mathbb{R}_+)$$

such that $\widetilde{L}_1 \cong L_1$, $\widetilde{L}_2 \cong L_2$. In addition, choosing complementary bundles $\widetilde{L}_1^\perp$ in $\widetilde{W}$, $\widetilde{L}_2^\perp$ in $\widetilde{V}$ we have $\widetilde{L}_1^\perp \cong \widetilde{L}_2^\perp$, provided that the fibre dimension of W is sufficiently large. If $\lambda : \widetilde{L}_2^\perp \to \widetilde{L}_1^\perp$ is an isomorphism and if

$$\iota^\perp : \widetilde{L}_1^\perp \to \pi^* F' \otimes H^{s-\mu}(\mathbb{R}_+)$$

is the canonical embedding, and

$$\pi^\perp : \pi^* E' \otimes H^s(\mathbb{R}_+) \to \widetilde{L}_2^\perp$$

(for $s = 0$) the orthogonal projection, the operator family $a_0 := (1 - p)a + q$ for $q := \iota^\perp \circ \lambda \circ \pi^\perp$ has the property

$$L_2 \cong \ker_{S^*Y} a_0, \quad L_1 \cong \text{coker}_{S^*Y} a_0. \tag{3.12}$$

The operator function $g := -pa + q : \pi^* E' \otimes H^s(\mathbb{R}_+) \to \pi^* F' \otimes H^{s-\mu}(\mathbb{R}_+)$ can be extended by (twisted) homogeneity μ to a morphism

$$g_{(\mu)} : \pi_Y^* E' \otimes H^s(\mathbb{R}_+) \to \pi_Y^* F' \otimes H^{s-\mu}(\mathbb{R}_+),$$

i.e., $g_{(\mu)}(y, \lambda\eta) = \lambda^\mu \kappa_\lambda g_{(\mu)}(y, \eta)\kappa_\lambda^{-1}$ for all $\lambda \in \mathbb{R}_+$, $(y, \eta) \in T^*Y \setminus 0$, where $g_{(\mu)}|_{S^*Y} = g$. Now we may set

$$G := \sum_{j=1}^{L} \varphi_j(\chi_j^{-1})_* \mathrm{Op}(g_j)\psi_j,$$

cf. Definition 2.4.13, where $g_j(y, \eta)$ are local Green symbols of order μ which have $g_{(\mu)}(y, \eta)$ as homogeneous principal components. Since $a_0 = \sigma_\partial(A + G)|_{S^*Y}$, the assertion follows from the relations (3.12). $\qquad\square$

Proposition 3.2.5. *Let $A \in \mathcal{B}^{\mu,d}(X; E, F)$ be a σ_ψ-elliptic operator. Then there exist vector bundles $J_1, J_2 \in \mathrm{Vect}(Y)$, $L_1, L_2 \in \mathrm{Vect}(T^*Y \setminus 0)$, such that $\pi_Y^* J_i$ is a subbundle of L_i, $i = 1, 2$, and an operator*

$$\mathcal{A} = \begin{pmatrix} A + G & K \\ T & 0 \end{pmatrix} \tag{3.13}$$

in $\mathcal{B}^{\mu,d}(X; \boldsymbol{v})$ for $\boldsymbol{v} := (E, F; J_1, J_2)$, such that $\sigma_\partial(\mathcal{A})$ restricts to an isomorphism

$$
\begin{array}{ccc}
\pi_Y^* E' \otimes H^s(\mathbb{R}_+) & & \pi_Y^* F' \otimes H^{s-\mu}(\mathbb{R}_+) \\
\oplus & \longrightarrow & \oplus \\
L_1 & & L_2
\end{array}
\tag{3.14}
$$

for every $s > \max\{\mu, d\} - 1/2$.

Proof. If $A \in \mathcal{B}^{\mu,d}(X; E, F)$ is σ_ψ-elliptic the boundary symbol $\sigma_\partial(A)$ induces a family of Fredholm operators $\sigma_\partial(A) : \pi^* E' \otimes H^s(\mathbb{R}_+) \to \pi^* F' \otimes H^{s-\mu}(\mathbb{R}_+)$, $\pi : S^*Y \to Y$, for any $s > \max\{\mu, d\} - 1/2$. Thus there is an index element $\mathrm{ind}_{S^*Y} \sigma_\partial(A) \in K(S^*Y)$. Choose any $L_1, L_2 \in \mathrm{Vect}(S^*Y)$ such that

$$\mathrm{ind}_{S^*Y} \sigma_\partial(A) = [L_2] - [L_1].$$

For brevity L_1, L_2 will also denote the pull backs of these bundles to $T^*Y \setminus 0$ under the canonical projection $T^*Y \setminus 0 \to S^*Y$. By virtue of Proposition 3.2.4 we find a Green operator $G \in \mathcal{B}_G^{\mu,d}(X; E, F)$ such that the relations (3.11) hold. Choose arbitrary bundle morphisms

$$\widetilde{k} : L_1 \to \pi^* F' \otimes \mathcal{S}(\overline{\mathbb{R}}_+), \quad \widetilde{t} : \pi^* E' \otimes \mathcal{S}(\overline{\mathbb{R}}_+) \to L_2,$$

such that $\widetilde{k}$ represents an isomorphism $L_1 \to \widetilde{L}_1$, and $\widetilde{t}$ restricts to an isomorphism $\widetilde{L}_2 \to L_2$, cf. the notation in the proof of Proposition 3.2.4. Then the block matrix

$$
\begin{pmatrix} \sigma_\partial(A + G) & \widetilde{k} \\ \widetilde{t} & 0 \end{pmatrix} :
\begin{array}{ccc}
\pi_Y^* E' \otimes H^s(\mathbb{R}_+) & & \pi_Y^* F' \otimes H^{s-\mu}(\mathbb{R}_+) \\
\oplus & \longrightarrow & \oplus \\
L_1 & & L_2
\end{array}
\tag{3.15}
$$

is an isomorphism for every $s > \max\{\mu, d\} - 1/2$. Now let $J_1, J_2 \in \mathrm{Vect}(Y)$ be arbitrary bundles such that L_i is a subbundle of $\pi^* J_i$, $i = 1, 2$ (it suffices to take $J_i = \mathbb{C}^{j_i}$ for any sufficiently large j_i). From (3.15) we pass to a bundle morphism

$$\begin{pmatrix} \sigma_\partial(A+G) & k \\ t & 0 \end{pmatrix} : \pi^* \begin{pmatrix} E' \otimes H^s(\mathbb{R}_+) \\ \oplus \\ J_1 \end{pmatrix} \longrightarrow \pi^* \begin{pmatrix} F' \otimes H^{s-\mu}(\mathbb{R}_+) \\ \oplus \\ J_2 \end{pmatrix}, \qquad (3.16)$$

where $k := \widetilde{k} \circ p_1$ with a bundle projection $p_1 : \pi^* J_1 \to L_1$ and $t := e_2 \circ \widetilde{t}$ for the canonical embedding $e_2 : L_2 \to \pi^* J_2$. Then (3.16) restricts to the isomorphism (3.15). Next we extend (3.16) by twisted homogeneity μ to a boundary symbol defined for all $(y, \eta) \in T^*Y \setminus 0$; it has the form

$$\begin{pmatrix} \sigma_\partial(A+G) & k_{(\mu)} \\ t_{(\mu)} & 0 \end{pmatrix}$$

for unique $t_{(\mu)}(y, \eta)$, $k_{(\mu)}(y, \eta)$ satisfying

$$t_{(\mu)}(y, \eta/|\eta|) = t(y, \eta/|\eta|), \quad k_{(\mu)}(y, \eta/|\eta|) = k(y, \eta/|\eta|).$$

Now we find trace and potential operators T and K, respectively, such that $\sigma_\partial(T) = t_{(\mu)}$, $\sigma_\partial(K) = k_{(\mu)}$. With these T and K the operator (3.13) is as desired. □

Proof of Theorem 3.2.2. Let $A + G \in \mathcal{B}^{\mu,d}(X; E, F)$ be σ_ψ-elliptic and let $\mathcal{A} \in \mathcal{B}^{\mu,d}(X; \boldsymbol{v})$ have $A + G$ as the top left corner, then the second condition of Definition 3.1.1 has the consequence that also

$$\sigma_\partial(\mathcal{A}) : \pi^* \begin{pmatrix} E' \otimes H^s(\mathbb{R}_+) \\ \oplus \\ J_1 \end{pmatrix} \longrightarrow \pi^* \begin{pmatrix} F' \otimes H^{s-\mu}(\mathbb{R}_+) \\ \oplus \\ J_2 \end{pmatrix}$$

is an isomorphism, which means that

$$\sigma_\partial(A) : \pi^* E' \otimes H^s(\mathbb{R}_+) \to \pi^* F' \otimes H^{s-\mu}(\mathbb{R}_+) \qquad (3.17)$$

is a family of Fredholm operators and $\mathrm{ind}_{S^*Y} \sigma_\partial(A) = [\pi^* J_2] - [\pi^* J_1] \in \pi^* K(Y)$. Thus it remains to show that when $A \in \mathcal{B}^{\mu,d}(X; E, F)$ is σ_ψ-elliptic and (3.9) is satisfied, then there exists an SL-elliptic $\mathcal{A} \in \mathcal{B}^{\mu,d}(X; \boldsymbol{v})$ containing $A + G$ in the top left corner for some $G \in \mathcal{B}_G^{\mu,d}(X; E, F)$. We first complete the Fredholm family (3.17) to a family of isomorphisms (3.2) that is given for $(y, \eta) \in S^*Y$. Setting $J_1 := \mathbb{C}^{j_1}$ for sufficiently large j_1, there is an injective bundle morphism

$$k : \pi^* J_1 \to \pi^* F' \otimes \mathcal{S}(\overline{\mathbb{R}}_+)$$

such that

$$\begin{pmatrix} \sigma_\partial(A) & k \end{pmatrix} : \begin{matrix} \pi^* E' \otimes H^s(\mathbb{R}_+) \\ \oplus \\ \pi^* J_1 \end{matrix} \longrightarrow \pi^* F' \otimes H^{s-\mu}(\mathbb{R}_+)$$

is surjective. Then $\ker_{S^*Y} \left(\sigma_\partial(A) \quad k \right)$ is a finite-dimensional subbundle of

$$
\begin{array}{c}
\pi^* E' \otimes H^s(\mathbb{R}_+) \\
\oplus \\
\pi^* J_1
\end{array}
\tag{3.18}
$$

In much the same way as in the proof of Proposition 3.2.5 we see that for sufficiently large j_1 the bundle $\ker_{S^*Y} \left(\sigma_\partial(A) \quad k \right)$ is isomorphic to $\pi^* J_2$ for some $J_2 \in \mathrm{Vect}(S^*Y)$. Let us choose an isomorphism

$$
l : \ker_{S^*Y} \left(\sigma_\partial(A) \quad k \right) \to \pi^* J_2
$$

and let

$$
p : \begin{array}{c}
\pi^* E' \otimes H^s(\mathbb{R}_+) \\
\oplus \\
\pi^* J_1
\end{array} \longrightarrow \ker_{S^*Y} \left(\sigma_\partial(A) \quad k \right)
$$

be the projection induced by the orthogonal projection with respect to the $(E'_y \otimes L^2(\mathbb{R}_+)) \oplus J_{1,y}$-scalar products in the fibres, first for $s \geq \max\{\mu, d\}$ and then extended by continuity to all $s > \max\{\mu, d\} - 1/2$. Setting $(t \quad q) := l \circ p$ we obtain an isomorphism

$$
\begin{pmatrix} \sigma_\partial(A) & k \\ t & q \end{pmatrix} : \pi^* \begin{pmatrix} E' \otimes H^s(\mathbb{R}_+) \\ \oplus \\ J_1 \end{pmatrix} \longrightarrow \pi^* \begin{pmatrix} F' \otimes H^{s-\mu}(\mathbb{R}_+) \\ \oplus \\ J_2 \end{pmatrix}.
$$

There is now a unique extension to an isomorphism

$$
\sigma_\partial(\mathcal{A}) = \begin{pmatrix} \sigma_\partial(A) & k_{(\mu)} \\ t_{(\mu)} & q_{(\mu)} \end{pmatrix} : \pi^* \begin{pmatrix} E' \otimes H^s(\mathbb{R}_+) \\ \oplus \\ J_1 \end{pmatrix} \longrightarrow \pi^* \begin{pmatrix} F' \otimes H^{s-\mu}(\mathbb{R}_+) \\ \oplus \\ J_2 \end{pmatrix}
$$

homogeneous in the sense that

$$
\sigma_\partial(\mathcal{A})(y, \lambda\eta) = \lambda^\mu \, \mathrm{diag}\,(1, \kappa_\lambda) \, \sigma_\partial(\mathcal{A})(y, \eta) \, \mathrm{diag}\,(1, \kappa_\lambda^{-1})
$$

for all $(y, \eta) \in T^*Y \setminus 0$ and all $\lambda \in \mathbb{R}_+$. In the final step we pass to an element

$$
\mathcal{A} = \begin{pmatrix} A & K \\ T & Q \end{pmatrix} \in \mathcal{B}^{\mu,d}(X; \boldsymbol{v}), \quad \boldsymbol{v} = (E, F; J_1, J_2),
$$

which has $\sigma_\partial(\mathcal{A})$ as the homogeneous principal edge symbol. The operator A is given anyway. The construction of the entries T, K, Q in terms of $t_{(\mu)}$, $k_{(\mu)}$, $q_{(\mu)}$ is analogous to (3.13). $\qquad\square$

The K-theoretic index element (3.7) is directly deduced from the interior symbol

$$
\sigma_\psi(A)\big|_{T^*X|_Y \setminus 0},
$$

i.e., the restriction of $\sigma_\psi(A)$ to the boundary Y. More precisely, it is completely determined by $\sigma_\psi(A)|_{S^*X|_Y}$, which is a consequence of the twisted homogeneity (3.6). Let us assume $\mu = 0$, which we can ensure by multiplying the symbol by an order reducing factor of opposite order, see, formula (3.21) below. Then we have

$$\sigma_\psi(A)(y, 0, \lambda\eta, \lambda\tau) = \sigma_\psi(A)(y, 0, \eta, \tau)$$

for all $(\eta, \tau) \neq 0$ and $\lambda > 0$. However, thanks to the transmission property,

$$\sigma_\psi(A)(y, 0, 0, \tau) = \sigma_\psi(A)(y, 0, 0, -\tau) \tag{3.19}$$

for all $-1 \leq \tau \leq 1$. Thus σ_ψ induces an isomorphism

$$\sigma_\psi(A) : \pi_{\Xi}^* E' \to \pi_{\Xi}^* F', \tag{3.20}$$

where

$$\Xi := S^*X\big|_Y \cup N^*$$

and N^* is the trivial $[-1, 1]$-bundle normal to the boundary Y. A simple geometric consideration shows that (3.20) represents an element $d(\sigma_\psi(A)) \in K(\mathbb{R}^2 \times S^*Y)$, see also [34]. Then an observation of Boutet de Monvel in [10] is

Remark 3.2.6. If $\beta : K(S^*Y) \to K(\mathbb{R}^2 \times S^*Y)$ denotes the Bott isomorphism, then

$$\beta \operatorname{ind}_{S^*Y} \sigma_\partial(A) = d(\sigma_\psi(A)).$$

Note that such a relation extends to BVPs without transmission property, i.e., when the numbers $a_- := \sigma_\psi(A)(y, 0, 0, -1)$, $a_+ := \sigma_\psi(A)(y, 0, 0, 1)$ do not coincide, but are connected by the line segment in the complex plane plus the values of a smoothing Mellin symbol parametrized by N^* (vanishing at $\tau = \pm 1$) as the diffeomorphic image of a weight line in the Mellin plane of the complex covariable, see, in particular [35] and Parts II, III below.

3.3 Boundary value problems with parameters

Let us now consider a parameter-dependent analogue of $\mathcal{B}^{\mu,d}(X; \boldsymbol{v})$ in Definition 2.4.14, with the parameter space $\mathbb{R}^l \ni \lambda$.

First we give a straightforward extension of Definition 2.1.1 to the case of symbols with the transmission property with parameters. It suffices to replace the covariable $\eta \in \mathbb{R}^{n-1}$ by $(\eta, \lambda) \in \mathbb{R}^{n-1+l}$. The corresponding analogue of (2.2) will be denoted by $S_{\mathrm{tr}}^\mu(\Omega \times \mathbb{R} \times \mathbb{R}^{n+l})$, $\mu \in \mathbb{Z}$, and we set

$$S_{\mathrm{tr}}^\mu(\Omega \times \overline{\mathbb{R}}_\pm \times \mathbb{R}^{n+l}) := \left\{ a|_{\Omega \times \overline{\mathbb{R}}_\pm \times \mathbb{R}^{n+l}} : a \in S_{\mathrm{tr}}^\mu(\Omega \times \mathbb{R} \times \mathbb{R}^{n+l}) \right\}.$$

Later on we will employ the parameter-dependent analogue of Example 2.1.4, namely, the symbols

$$r_\pm^\mu(\eta, \tau) := \left(\varphi(\tau/C\langle \eta, \lambda \rangle) \langle \eta, \lambda \rangle \pm i\tau \right)^\mu \in S_{\mathrm{tr}}^\mu(\mathbb{R}^{n+l}). \tag{3.21}$$

Similarly to Definition 2.4.1, let $L_{\mathrm{tr}}^{\mu}\big(2X; \widetilde{E}, \widetilde{F}; \mathbb{R}^l\big)$ denote the space of parameter-dependent pseudo-differential operators on the double of a smooth compact manifold X with boundary, with matrices of local symbols in $S_{\mathrm{tr}}^{\mu}\big(\Omega \times \mathbb{R} \times \mathbb{R}^{n+l}\big)$, $\widetilde{E}, \widetilde{F} \in \mathrm{Vect}(2X)$. Set

$$L_{\mathrm{tr}}^{\mu}\big(X; E, F; \mathbb{R}^l\big) := \big\{ \mathrm{r}^+ \widetilde{A}\, \mathrm{e}^+ : \widetilde{A} \in L_{\mathrm{tr}}^{\mu}\big(2X; \widetilde{E}, \widetilde{F}; \mathbb{R}^l\big) \big\}.$$

Definition 2.4.9 has an obvious analogue in the parameter-dependent case where it suffices to replace $\eta \in \mathbb{R}^q$, $q = n - 1 = \dim Y$, by $(\eta, \lambda) \in \mathbb{R}^{q+l}$. This gives us the space $\mathcal{R}_{\mathrm{G}}^{\mu,d}\big(\Omega \times \mathbb{R}^q; \boldsymbol{w}; \mathbb{R}^l\big)$ of parameter-dependent Green symbols of order $\mu \in \mathbb{R}$ and type $d \in \mathbb{N}$. Thus, analogously as in Definition 2.4.13, we obtain the class $\mathcal{B}_{\mathrm{G}}^{\mu,d}\big(X; \boldsymbol{v}; \mathbb{R}^l\big)$, $\boldsymbol{v} = (E, F; J_1, J_2)$, of parameter-dependent Green operators of order $\mu \in \mathbb{R}$ and type $d \in \mathbb{N}$. The substitute of $\mathcal{B}_{\mathrm{G}}^{-\infty,d}(X; \boldsymbol{v})$ in this case is $\mathcal{S}\big(\mathbb{R}^l, \mathcal{B}_{\mathrm{G}}^{-\infty,d}(X; \boldsymbol{v})\big) =: \mathcal{B}_{\mathrm{G}}^{-\infty,d}\big(X; \boldsymbol{v}; \mathbb{R}^l\big)$, the Schwartz space over $\mathbb{R}^l$ with values in the Fréchet space $\mathcal{B}_{\mathrm{G}}^{-\infty,d}(X; \boldsymbol{v})$. The space $\mathcal{B}_{\mathrm{G}}^{-\infty,d}\big(X; \boldsymbol{v}; \mathbb{R}^l\big)$ will also play the role of smoothing elements in the parameter-dependent analogue of Boutet de Monvel's calculus, i.e., for similar reasons as

$$\mathcal{B}^{-\infty,d}(X; \boldsymbol{v}) = \mathcal{B}_{\mathrm{G}}^{-\infty,d}(X; \boldsymbol{v})$$

we set

$$\mathcal{B}^{-\infty,d}\big(X; \boldsymbol{v}; \mathbb{R}^l\big) := \mathcal{S}\big(\mathbb{R}^l, \mathcal{B}^{-\infty,d}(X; \boldsymbol{v})\big) \qquad (3.22)$$

Definition 3.3.1. The space $\mathcal{B}^{\mu,d}\big(X; \boldsymbol{v}; \mathbb{R}^l\big)$, $\mu \in \mathbb{Z}$, $d \in \mathbb{N}$, $\boldsymbol{v} = (E, F; J_1, J_2)$, is defined to be the set of all operator families of the form

$$\mathcal{A}(\lambda) = \begin{pmatrix} A(\lambda) & 0 \\ 0 & 0 \end{pmatrix} + \mathcal{G}(\lambda),$$

with arbitrary $A(\lambda) \in L_{\mathrm{tr}}^{\mu}\big(X; E, F; \mathbb{R}^l\big)$ and $\mathcal{G}(\lambda) \in \mathcal{B}^{-\infty,d}\big(X; \boldsymbol{v}; \mathbb{R}^l\big)$.

The rest of the material of Section 2.4 can be generalised to the parameter-dependent case. Moreover, there is also an extension of SL-ellipticity to parameter-dependent SL-ellipticity, cf. Section 3.1. Concerning the principal symbols

$$\sigma(\mathcal{A}(\lambda)) = \big(\sigma_\psi(\mathcal{A}(\lambda)), \sigma_\partial(\mathcal{A}(\lambda))\big)$$

we have in the first component dependence on $(\xi, \lambda) \neq 0$. More precisely, the parameter-dependent interior symbol is a morphism

$$\sigma_\psi(\mathcal{A}(\lambda)) : \pi_X^* E \to \pi_X^* F, \qquad (3.23)$$

$\pi_X : T^*X \times \mathbb{R}^l \backslash 0 \to X$ (with 0 indicating $(\xi, \lambda) = 0$) while the parameter-dependent boundary symbol is a morphism

$$\sigma_\partial(\mathcal{A}(\lambda)) : \pi_Y^* \begin{pmatrix} E' \otimes H^s(\mathbb{R}_+) \\ \oplus \\ J_1 \end{pmatrix} \longrightarrow \pi_Y^* \begin{pmatrix} F' \otimes H^{s-\mu}(\mathbb{R}_+) \\ \oplus \\ J_2 \end{pmatrix}, \qquad (3.24)$$

$\pi_Y : T^*Y \times \mathbb{R}^l \backslash 0 \to Y$ (with 0 indicating $(\eta, \lambda) = 0$).

Remark 3.3.2. Observe that for every $\lambda_0 \in \mathbb{R}^l$ we have

$$\mathcal{A}(\lambda) \in \mathcal{B}^{\mu,d}(X; \boldsymbol{v}; \mathbb{R}^l) \implies \mathcal{A}(\lambda_0) \in \mathcal{B}^{\mu,d}(X; \boldsymbol{v}).$$

Thus, according to Theorem 2.4.18 every $\mathcal{A}(\lambda) \in \mathcal{B}^{\mu,d}(X; \boldsymbol{v}; \mathbb{R}^l)$ defines a family of continuous operators (2.152).

We also have analogues of Theorems 2.4.20 and 2.4.21 in the parameter-dependent case.

Let us now turn to parameter-dependent ellipticity of parameter-dependent BVPs.

Definition 3.3.3. An $\mathcal{A} \in \mathcal{B}^{\mu,d}(X; \boldsymbol{v}; \mathbb{R}^l)$, $\mu \in \mathbb{Z}$, $d \in \mathbb{N}$, $\boldsymbol{v} = (E, F; J_1, J_2)$, is elliptic if

(i) (3.23) is an isomorphism,

(ii) (3.24) is an isomorphism for some $s > \max\{\mu, d\} - 1/2$.

Theorem 3.3.4. *Let X be compact and $\mathcal{A} \in \mathcal{B}^{\mu,d}(X; \boldsymbol{v}; \mathbb{R}^l)$, $\boldsymbol{v} = (E, F; J_1, J_2)$, parameter-dependent elliptic.*

(i) *$\mathcal{A}$ has a parameter-dependent elliptic parametrix $\mathcal{P} \in \mathcal{B}^{-\mu,(d-\mu)^+}(X; \boldsymbol{v}^{-1}; \mathbb{R}^l)$ for $\boldsymbol{v}^{-1} = (F, E; J_2, J_1)$, in the sense of*

$$\mathcal{C}_{\mathrm{L}} = \mathcal{I} - \mathcal{P}\mathcal{A} \in \mathcal{B}^{-\infty,d_{\mathrm{L}}}(X; \boldsymbol{v}_{\mathrm{L}}; \mathbb{R}^l), \quad \mathcal{C}_{\mathrm{R}} = \mathcal{I} - \mathcal{A}\mathcal{P} \in \mathcal{B}^{-\infty,d_{\mathrm{R}}}(X; \boldsymbol{v}_{\mathrm{R}}; \mathbb{R}^l)$$

for $d_{\mathrm{L}} = \max\{\mu, d\}, d_{\mathrm{R}} = (d - \mu)^+$, cf. also the notation in Theorem 3.1.3.

(ii) *For $l \geq 1$ the operators*

$$\mathcal{A}(\lambda) : H^s(\mathrm{int}\, X, E) \oplus H^s(Y, J_1) \to H^{s-\mu}(\mathrm{int}\, X, F) \oplus H^{s-\mu}(Y, J_2) \quad (3.25)$$

are Fredholm of index zero for every $s > \max\{\mu, d\} - 1/2$, and there is a $C > 0$ such that (3.25) are isomorphisms for all $|\lambda| > C$.

Proof. The arguments for (i) are similar to those for Theorem 3.1.3. (ii) The Fredholm property of (3.25) follows from Theorem 3.1.4, using the fact that $\mathcal{A}(\lambda_0)$ is SL-elliptic for every fixed λ_0. However, the existence of a parameter-dependent parametrix in the sense of (i) shows that, for instance, the operator norm of $\mathcal{C}_{\mathrm{L}}(\lambda)$ in $H^s(\mathrm{int}\, X, E) \oplus H^s(Y, J_1)$ tends to zero as $|\lambda| \to \infty$. Therefore, a Neumann series argument applied to $\mathcal{P}\mathcal{A}(\lambda) = \mathcal{I} - \mathcal{C}_{\mathrm{L}}(\lambda)$ shows that $\mathcal{P}\mathcal{A}(\lambda)$ is invertible for large $|\lambda|$, which implies that $\mathcal{A}(\lambda)$ is injective for large $|\lambda|$. A similar argument for $\mathcal{A}\mathcal{P}(\lambda) = \mathcal{I} - \mathcal{C}_{\mathrm{R}}(\lambda)$ shows that $\mathcal{A}(\lambda)$ is surjective for large $|\lambda|$. $\square$

3.4 Order-reducing operators on a manifold with boundary

Theorem 3.4.1. *For every $\mu \in \mathbb{Z}$, $V \in \mathrm{Vect}(X)$ there exists a parameter-dependent elliptic $R^\mu_V(\lambda) \in \mathcal{B}^{\mu,0}(X; V, V; \mathbb{R}^l)$ which induces isomorphisms*

$$R^\mu_V(\lambda) : H^s(X, V) \to H^{s-\mu}(X, V)$$

for all $\lambda \in \mathbb{R}^l$, $s \in \mathbb{R}$.

Proof. Let us sketch the idea; then the details are straightforward. For convenience, we assume that V is trivial and of fibre dimension 1. We set $\widetilde{\lambda} := (\lambda_0, \lambda)$, $\lambda_0 \in \mathbb{R}$, consider a parameter-dependent operator on X of the form $\mathrm{r}^+ \widetilde{R^\mu}(\widetilde{\lambda})\mathrm{e}^+$ for a parameter-dependent elliptic operator $\widetilde{R^\mu}(\widetilde{\lambda}) \in L^\mu_{\mathrm{cl}}\left(\widetilde{X}; \mathbb{R}^{1+l}_{\widetilde{\lambda}}\right)$ of order μ on the double $\widetilde{X} = 2X$, constructed in a similar manner as the operator A_0 in (1.19), where local interior symbols in a tubular neighbourhood of Y are of the form

$$r^\mu_-\left(\eta, \tau, \widetilde{\lambda}\right) := \left(\varphi\left(\frac{\tau}{C\langle\eta, \widetilde{\lambda}\rangle}\right)\langle\eta, \widetilde{\lambda}\rangle - i\tau\right)^\mu \in S^\mu_{\mathrm{tr}}\left(\mathbb{R}^{n+1+l}_{\xi, \widetilde{\lambda}}\right) \tag{3.26}$$

for a sufficiently large constant $C > 0$, cf. notation in (2.1.4), and off some larger tubular neighbourhood of the form $\left(c + |\xi|^2 + |\widetilde{\lambda}|^2\right)^{\mu/2}$ for some $c > 0$. This operator is parameter-dependent elliptic with parameter λ, both with respect to σ_ψ and to σ_∂ when C and c and $|\lambda_0|$ are sufficiently large, where ellipticity with respect to the boundary symbol holds without extra trace and potential entries. It suffices then to apply Theorem 3.3.4 for sufficiently large $|\widetilde{\lambda}|$ and to define $R^\mu(\lambda) := \mathrm{r}^+ \widetilde{R^\mu}(\lambda_0, \lambda)\mathrm{e}^+$ for any sufficiently large $|\lambda_0|$. □

Remark 3.4.2. Let us fix $\lambda_0 \in \mathbb{R}^l$, and set $R^\mu_V := R^\mu_V(\lambda_0)$, which belongs to $\mathcal{B}^{\mu,0}(X; V, V)$. Then $\sigma_\partial(R^\mu_V) : \pi_Y^* V' \oplus H^s(\mathbb{R}_+) \to \pi_Y^* V' \oplus H^{s-\mu}(\mathbb{R}_+)$ is a family of isomorphisms for $s > \mu - 1/2$, and also $\sigma_\partial(R^\mu_V) : \pi_Y^* V' \oplus \mathcal{S}(\overline{\mathbb{R}}_+) \to \pi_Y^* V' \oplus \mathcal{S}(\overline{\mathbb{R}}_+)$ is a family of isomorphisms.

3.5 Families of Fredholm operators

We consider here properties of block matrices of linear continuous operators between direct sums of Hilbert spaces. The assumptions are made for convenience. Some assertions are of a purely algebraic nature or valid in analogous form for more general linear operators and vector spaces. In applications, i.e., with Fréchet spaces, it will be easy to modify the arguments.

Lemma 3.5.1. *Let* $H, \widetilde{H}$ *and* M *be Hilbert spaces.*

(i) *A continuous linear operator*

$$\boldsymbol{a} := \begin{pmatrix} a \\ b \end{pmatrix} : H \to \begin{array}{c} \widetilde{H} \\ \oplus \\ M \end{array} \tag{3.27}$$

is an isomorphism if and only if $a : H \to \widetilde{H}$ *is surjective and* $b|_{\ker a} : \ker a \to M$ *is an isomorphism.*

(ii) *A continuous linear operator*

$$\boldsymbol{p} := (p \quad k) : \begin{matrix} \widetilde{H} \\ \oplus \\ M \end{matrix} \to H \qquad (3.28)$$

is an isomorphism if and only if $p : H \to \widetilde{H}$ is injective and $k : M \to L$ is an isomorphism to a subspace $L \subseteq H$ such that

$$\operatorname{im} p \cap L = \emptyset, \quad \operatorname{im} p \oplus L = H. \qquad (3.29)$$

Proposition 3.5.2. *Let $H, \widetilde{H}$ and M be Hilbert spaces and (3.27) an isomorphism. Moreover, let p be a right inverse of a, i.e., $ap = \operatorname{id}_{\widetilde{H}}$, and choose an isomorphism*

$$l : M \to \ker a.$$

Then for the isomorphism $q := bl : M \to M$ we have

$$\begin{pmatrix} a \\ b \end{pmatrix}^{-1} = (p + g \quad k), \qquad (3.30)$$

where

$$g = -lq^{-1}bp, \quad k = lq^{-1}.$$

Proof. First note that $L := \ker a$ satisfies the relations (3.29). From Lemma 3.5.1 (ii) we know that

$$(p \quad l) : \begin{matrix} \widetilde{H} \\ \oplus \\ M \end{matrix} \to H$$

is an isomorphism. Thus also

$$\begin{pmatrix} a \\ b \end{pmatrix} (p \quad l) = \begin{pmatrix} 1 & 0 \\ bp & q \end{pmatrix} : \begin{matrix} \widetilde{H} \\ \oplus \\ M \end{matrix} \to \begin{matrix} \widetilde{H} \\ \oplus \\ M \end{matrix}$$

is an isomorphism. In order to verify (3.30) it suffices to compose $(p \quad l)$ from the right with

$$\begin{pmatrix} 1 & 0 \\ bp & q \end{pmatrix}^{-1} = \begin{pmatrix} 1 & 0 \\ -q^{-1}bp & q^{-1} \end{pmatrix}. \qquad \square$$

Remark 3.5.3. Let $H, \widetilde{H}, M$ be Hilbert spaces, $a : H \to \widetilde{H}$ a surjective continuous linear map, and $v : H \to \ker a$ a continuous projection. Moreover, let $m : \ker a \to M$ be an isomorphism. Then the operator (3.27) is an isomorphism for $b := m \circ v$.

Proposition 3.5.4. *Let $H, \widetilde{H}$ be Hilbert spaces and $a : H \to \widetilde{H}$ a Fredholm operator. Then there are $d, e \in \mathbb{N}$ and a block matrix isomorphism*

$$
\boldsymbol{a} := \begin{pmatrix} a & k \\ t & q \end{pmatrix} : \begin{matrix} H \\ \oplus \\ \mathbb{C}^e \end{matrix} \to \begin{matrix} \widetilde{H} \\ \oplus \\ \mathbb{C}^d \end{matrix} \tag{3.31}
$$

for suitable finite rank operators

$$
k : \mathbb{C}^e \to \widetilde{H}, \quad t : H \to \mathbb{C}^d, \quad q : \mathbb{C}^e \to \mathbb{C}^d, \tag{3.32}
$$

and we have

$$
\operatorname{ind} a = d - e. \tag{3.33}
$$

Proof. First we choose any $e \in \mathbb{N}$ such that $\operatorname{codim} \operatorname{im} a \leq e$. Then we easily find a $k : \mathbb{C}^e \to \widetilde{H}$ such that

$$
\boldsymbol{c} := (a \quad k) : \begin{matrix} H \\ \oplus \\ \mathbb{C}^e \end{matrix} \to \widetilde{H} \tag{3.34}
$$

is surjective. In fact, we have a direct decomposition $\widetilde{H} = \operatorname{im} a \oplus W$ for a finite-dimensional subspace $W \subset \widetilde{H}$. Then for $e_1 := \dim W$ we choose an isomorphism $k_1 : \mathbb{C}^{e_1} \to W$, write $\mathbb{C}^e = \mathbb{C}^{e_0} \oplus \mathbb{C}^{e_1}$ for $e = e_0 + e_1$, and define $k := k_0 \oplus k_1 : \mathbb{C}^e \to \widetilde{H}$ by setting $k_0 := 0$. Since k is of finite rank, the operator (3.34) is Fredholm, and we set $d := \dim \ker \boldsymbol{c}$. Now for any isomorphism $\boldsymbol{b}_0 : \ker \boldsymbol{c} \to \mathbb{C}^d$ we form $\boldsymbol{b} := \boldsymbol{b}_0 \circ \boldsymbol{v}$ for the orthogonal projection

$$
\boldsymbol{v} : \begin{matrix} H \\ \oplus \\ \mathbb{C}^e \end{matrix} \to \ker \boldsymbol{c}.
$$

Then Lemma 3.5.1 gives us an isomorphism

$$
\begin{pmatrix} \boldsymbol{c} \\ \boldsymbol{b} \end{pmatrix} : \begin{matrix} H \\ \oplus \\ \mathbb{C}^e \end{matrix} \to \begin{matrix} \widetilde{H} \\ \oplus \\ \mathbb{C}^d \end{matrix}.
$$

We define $th := \boldsymbol{b}(h \oplus 0)$, $h \in H$, and $qf := \boldsymbol{b}(0 \oplus f)$, $f \in \mathbb{C}^d$, so that $\boldsymbol{b} = (t \quad q)$. Summing up, we constructed an isomorphism (3.31).

For the proof of (3.33) we first choose a parametrix $p \in \mathcal{L}(\widetilde{H}, H)$ of a, i.e., an operator such that $pa = \operatorname{id}_H - c_L$, $ap = \operatorname{id}_{\widetilde{H}} - c_R$ with compact remainders $c_L \in \mathcal{L}(H, H)$ and $c_R \in \mathcal{L}(\widetilde{H}, \widetilde{H})$, respectively. Then we have

$$
\boldsymbol{h} := \operatorname{diag}(p, \operatorname{id}_{\mathbb{C}^d}) \boldsymbol{a} = \begin{pmatrix} \operatorname{id}_H - c_L & pk \\ t & q \end{pmatrix} : \begin{matrix} H \\ \oplus \\ \mathbb{C}^e \end{matrix} \to \begin{matrix} H \\ \oplus \\ \mathbb{C}^d \end{matrix}. \tag{3.35}
$$

The family of operators

$$h_\tau := \begin{pmatrix} \mathrm{id}_H - \tau c_L & \tau pk \\ \tau t & \tau q \end{pmatrix}, \quad 0 \le \tau \le 1,$$

defines a homotopy through Fredholm operators between $h = h_1$ and

$$h_0 = \begin{pmatrix} \mathrm{id}_H & 0 \\ 0 & 0 \end{pmatrix},$$

with 0 indicating finite-dimensional operators that are 0 between the respective spaces. Clearly, we have $\mathrm{ind}\, h_0 = d - e$. Thanks to, the homotopy invariance of the index of Fredholm operators we finally obtain the relation (3.33). $\qquad\square$

Proposition 3.5.5. *Let $H, \widetilde{H}$ and $M, \widetilde{M}$ be Hilbert spaces and*

$$a := \begin{pmatrix} a & k \\ t & q \end{pmatrix} : \begin{matrix} H \\ \oplus \\ M \end{matrix} \to \begin{matrix} \widetilde{H} \\ \oplus \\ \widetilde{M} \end{matrix} \tag{3.36}$$

a linear operator.

(i) *Let*

$$a : H \to \widetilde{H} \tag{3.37}$$

be a Fredholm operator and $p : \widetilde{H} \to H$ a parametrix of a. Then (3.36) is Fredholm if and only if

$$b := q - tpk : M \to \widetilde{M} \tag{3.38}$$

is a Fredholm operator, and we have

$$\mathrm{ind}\, a = \mathrm{ind}\, a + \mathrm{ind}\, b. \tag{3.39}$$

Moreover, if $r : \widetilde{M} \to M$ is a parametrix of (3.38), then

$$p := \begin{pmatrix} p + pkrtp & -pkr \\ -rtp & r \end{pmatrix} : \begin{matrix} \widetilde{H} \\ \oplus \\ \widetilde{M} \end{matrix} \to \begin{matrix} H \\ \oplus \\ M \end{matrix} \tag{3.40}$$

is a parametrix of a.

(ii) *Let (3.37) be an isomorphism. Then (3.36) is an isomorphism if and only if (3.38) is an isomorphism, and we have $p = a^{-1}$ for $p := a^{-1}$ and $r := b^{-1}$ in formula (3.40).*

Proof. (i) We have

$$\begin{pmatrix} a & 0 \\ 0 & b \end{pmatrix} = \begin{pmatrix} 1 & 0 \\ -tp & 1 \end{pmatrix} \begin{pmatrix} a & k \\ t & q \end{pmatrix} \begin{pmatrix} 1 & -pk \\ 0 & 1 \end{pmatrix} \tag{3.41}$$

modulo compact operators. Here 1 denotes identity operators in different spaces. The first and the third factor in the right-hand side of (3.41) are isomorphisms. Therefore, a and diag (a, b) are simultaneously Fredholm or not. The latter is equivalent to the Fredholm property of b. The claimed index identity (3.39) is a consequence of (3.41). (ii) immediately follows from (3.41) under the assumption that a is an isomorphism. $\qquad\square$

We now pass to families of Fredholm operators, continuously parametrised by a compact topological space X. For notational convenience we assume that X is arcwise connected, i.e., for every two points $x_0, x_1 \in X$ there is a continuous map $\tau : I \to X$, $I := [0, 1]$, such that $\tau(i) = x_i$, $i = 0, 1$. (The modifications of our considerations for the general case are straightforward and left to the reader.) Kernels and cokernels depending on x form vector bundles over X when they are of constant dimension. Otherwise, although the index is always constant with respect to x, we may have variable dimensions of kernels and cokernels, and then the index can be described in terms of the K-functor. In the following we recall some material on the relationship between Fredholm operators and K-theory.

The set of all locally trivial continuous complex vector bundles over X will be denoted by Vect(X). The Cartesian product $X \times \mathbb{C}^N$ with the projection $\pi :$ $X \times \mathbb{C}^N \to X$, $(x, v) \mapsto x$, $x \in X$, is an example of an element of Vect(X), called trivial in this case. The fibre over $x \in X$ is defined as $\pi^{-1}(x) \cong \mathbb{C}^N$. Instead of $X \times \mathbb{C}^N$ we also write $\mathbb{C}^N$ if X is known from the context.

Elements of Vect(X) in general are determined by a total space E and a projection $\pi_E : E \to X$, where the fibres $E_x := \pi_E^{-1}(x)$, $x \in X$, are isomorphic to $\mathbb{C}^k$, for some $k \in \mathbb{N}$. Locally trivial means that every $x \in X$ has a neighbourhood U such that there is a bijective map $\eta_U : E|_U := \bigcup_{x \in U} E_x \to U \times \mathbb{C}^k$, a trivialisation of E over U, such that $\pi_E = p \circ \eta_U$ for $p : U \times \mathbb{C}^k \to U$, $(x, v) \mapsto x$, and η_U restricts to an isomorphism $E_x \to \mathbb{C}^k$ for every $x \in U$. Continuous means that the transition functions belonging to different trivialisations $\eta_U, \zeta_U : E|_U \to U \times \mathbb{C}^k$ over open sets U of some covering of X, namely, $\eta_U \circ \zeta_U^{-1} : U \times \mathbb{C}^k \to U \times \mathbb{C}^k$, are continuous. The fibre $\mathbb{C}^k$ occurring in the trivialisations will also be called the fibre of E. This should not cause confusions when we also speak about the fibres E_x over x. We do not give a self-contained account of the essential material on vector bundles, but fix some notation and recall structures that occur in our applications. For more details we refer to textbooks such as Atiyah [2]. In any case, if X is a smooth manifold, then the transition maps are required to be smooth, and in this case Vect(X) is the set of all locally trivial smooth complex vector bundles over X.

In an analogous manner we can define locally trivial continuous/smooth real vector bundles over X; in this case $\mathbb{C}^k$ is replaced by $\mathbb{R}^k$ for some k. For instance, tangent and cotangent bundles TM and T^*M over a smooth manifold M are real vector bundles of fibre dimension $\dim M$. For convenience, we confine ourselves to the complex case.

So far we tacitly identified a bundle E with its total space $\bigcup_{x \in X} E_x$. It is also common to identify a vector bundle with the triple (E, π_E, X). However, we keep writing E both for the bundle itself and its total space. The latter is a topological space in a natural way, with a basis of open sets coming from the system of trivialisations.

A (bundle) morphism $E \to F$ for $E, F \in \text{Vect}(X)$ is a continuous map $\eta : E \to F$ such that $\pi_E = \pi_F \circ \eta$ and $\eta\big|_{E_x} : E_x \to F_x$ is a linear map for every $x \in X$. If there is a morphism $\zeta : F \to E$ such that $\zeta \circ \eta = \text{id}_E$, $\eta \circ \zeta = \text{id}_F$, then η is called an isomorphism. We then write $E \cong F$.

For any continuous map $f : Y \to X$ and $E \in \text{Vect}(X)$, there is an element $F := f^*E \in \text{Vect}(Y)$, called the (bundle) pull-back of E, such that $f \circ \pi_F = \pi_E$. We set $F := \{(y, e) \in Y \times E : e \in E_{f(y)}\}$.

Given $E, F \in \text{Vect}(X)$ we have the direct sum and the tensor product

$$E \oplus F := \bigcup_{x \in X} E_x \oplus F_x \in \text{Vect}(X), \quad E \otimes F := \bigcup_{x \in X} E_x \otimes F_x \in \text{Vect}(X).$$

Then $E \oplus F \cong F \oplus E$, $E \otimes F \cong F \otimes E$, and $(E \oplus F) \otimes G = (E \otimes G) \oplus (F \otimes G)$ for any $E, F, G \in \text{Vect}(X)$.

Proposition 3.5.6. *For every $E \in \text{Vect}(X)$ there exists an $N \in \mathbb{N}$ and an element $F \in \text{Vect}(X)$ such that $E \oplus F = X \times \mathbb{C}^N$.*

This is an elementary well-known result on vector bundles over a compact topological space.

Definition 3.5.7. A bundle $F \in \text{Vect}(X)$ is called a subbundle of $E \in \text{Vect}(X)$ if there is a morphism $p : E \to F$ which restricts to projections $E_x \to F_x$ for all $x \in X$.

Remark 3.5.8. By Proposition 3.5.6, every $E \in \text{Vect}(X)$ can be represented as a subbundle of $X \times \mathbb{C}^N$ for a suitable $N \in \mathbb{N}$ depending on E. In other words, there is a continuous family of projections $p_x : \mathbb{C}^N \to \mathbb{C}^N$, $x \in X$, such that $p_x \mathbb{C}^N = E_x$ for all x. We also have the family of complementary projections $\text{id}_{\mathbb{C}^N} - p_x : \mathbb{C}^N \to \mathbb{C}^N$, $x \in X$, which defines another subbundle $E^\perp$ of $X \times \mathbb{C}^N$, called the complementary bundle, which satisfies the relation

$$E \oplus E^\perp = X \times \mathbb{C}^N. \tag{3.42}$$

As before, we assume that X is an arcwise connected topological space. Moreover, let H and $\widetilde{H}$ be infinite-dimensional, and let $\mathcal{F}(H, \widetilde{H})$ be the space of Fredholm operators $f : H \to \widetilde{H}$ equipped with the operator norm topology induced by $\mathcal{L}(H, \widetilde{H})$. Then we have $C(X, \mathcal{F}(H, \widetilde{H}))$, the space of continuous maps $a : X \to \mathcal{F}(H, \widetilde{H})$.

Proposition 3.5.9. *For every* $a \in C(X, \mathcal{F}(H, \widetilde{H}))$ *there exists an* $n \in \mathbb{N}$ *and a linear map* $k : \mathbb{C}^n \to \widetilde{H}$ *that*

$$
(a(x) \quad k) : \begin{matrix} H \\ \oplus \\ \mathbb{C}^n \end{matrix} \to \widetilde{H} \tag{3.43}
$$

is surjective for every $x \in X$.

Proof. For every $x_0 \in X$ there is a direct decomposition $\widetilde{H} = \operatorname{im} a(x_0) \oplus V_{x_0}$ for a finite-dimensional subspace $V_{x_0} \subset \widetilde{H}$. Let $n_{x_0} := \dim V_{x_0}$, and choose any isomorphism $k_{x_0} : \mathbb{C}^{n_{x_0}} \to V_{x_0}$. Using the embedding $V_{x_0} \hookrightarrow \widetilde{H}$, we interpret k_{x_0} also as an operator $k_{x_0} : \mathbb{C}^{n_{x_0}} \to \widetilde{H}$. It is then clear that

$$
(a(x_0) \quad k_{x_0}) : \begin{matrix} H \\ \oplus \\ \mathbb{C}^{n_{x_0}} \end{matrix} \to \widetilde{H}
$$

is surjective. Since the surjective operators form an open subset of $\mathcal{L}\left(\begin{matrix} H \\ \oplus \\ \mathbb{C}^{n_{x_0}} \end{matrix}, \widetilde{H} \right)$ there is an open neighbourhood U_{x_0} of x_0 in X such that $(a(x) \quad k_{x_0}) : \begin{matrix} H \\ \oplus \\ \mathbb{C}^{n_0} \end{matrix} \to \widetilde{H}$ is surjective for all $x \in U_{x_0}$. We obtain an open cover $\{U_{x_0}\}_{x_0 \in X}$ of X. Choose a finite subcover $\{U_{x_i}\}_{i=1,\ldots,N}$ using the compactness of X, and form the operator families $(a(x) \quad k_{x_i}) : \begin{matrix} H \\ \oplus \\ \mathbb{C}^{n_i} \end{matrix} \to \widetilde{H}$, which are surjective for all $x \in U_{x_i}$, for corresponding dimensions n_i, $i = 1, \ldots, N$. Now let us identify $\bigoplus_{i=1}^{N} \mathbb{C}^{n_i}$ with $\mathbb{C}^n$ with $n := \sum_{i=1}^{N} n_i$ and form $k := \operatorname{diag}(k_{x_1}, k_{x_2}, \ldots, k_{x_N}) : \mathbb{C}^n \to \widetilde{H}$. Then the operator (3.43) is surjective for all $x \in X$. $\qquad\square$

Remark 3.5.10. The surjective operator family (3.43) belongs to the space

$$
C\left(X, \mathcal{F}\left(\begin{matrix} H \\ \oplus \\ \mathbb{C}^n \end{matrix}, \widetilde{H} \right) \right).
$$

The notion of a locally trivial vector bundle also makes sense in the case of an infinite-dimensional fibre, e.g., a Hilbert space H. In particular, we can form the trivial bundle $X \times H$ over X. Similarly to the case of finite-dimensional fibres, we also write H rather than $X \times H$ when X is known from the context. Kuiper's theorem tells us that there are no other Hilbert bundles than the trivial ones, based on the fact that the group of unitary operators in an infinite-dimensional

Hilbert space is contractible. We do not discuss further this aspect here. Note that much like in $\mathrm{Vect}(X)$, we have also morphisms, isomorphisms, subbundles, the direct sum operation, etc. In particular, for any $E \in \mathrm{Vect}(X)$ the direct sum $(X \times H) \oplus E$ is again a Hilbert bundle over X with fibres $H \oplus E_x$. In general the transition maps between different trivialisations $U \times F \to U \times F$ of a Hilbert bundle over X with fibre F are represented by functions $g \in C(U, \mathrm{GL}(F))$ (or $g \in C^\infty(U, \mathrm{GL}(F))$ in the case of a C^∞ manifold X).

Remark 3.5.11. Let H be an infinite-dimensional Hilbert space. Every $E \in \mathrm{Vect}(X)$ is isomorphic to a subbundle of $X \times H$.

In fact, for any $N \in \mathbb{N}$ we can write $X \times H = (X \times H_0) \oplus (X \times H_1)$, where H_0 is an N-dimensional subspace of H and H_1 its orthogonal complement. Then, by virtue of (3.42), we have an isomorphism $(X \times H_0) \oplus (X \times H_1) \cong E \oplus E^\perp \oplus (X \times H_1)$, and the preimage of E is precisely the desired subbundle.

Proposition 3.5.12. *Let $a \in C\big(X, \mathcal{F}(H, \widetilde{H})\big)$, and assume that $a(x) : H \to \widetilde{H}$ is surjective for every $x \in X$. Then the family of kernels*

$$\ker_X a := \bigcup_{x \in X} \ker a(x) \subset X \times H \tag{3.44}$$

belongs to $\mathrm{Vect}(X)$.

Proof. Fix an arbitrary $x_0 \in X$, and let $p_{x_0} : H \to \ker a(x_0)$ be the orthogonal projection to $\ker a(x_0)$. Then

$$\boldsymbol{a}(x) := \begin{pmatrix} a(x) \\ p_{x_0} \end{pmatrix} : H \to \begin{matrix} \widetilde{H} \\ \oplus \\ \ker a(x_0) \end{matrix} \tag{3.45}$$

depends continuously on x. By Lemma 3.5.1, the operator $\boldsymbol{a}(x_0)$ is an isomorphism, because the first component of $\boldsymbol{a}(x_0)$ is surjective, and the second one restricts to the identity map on $\ker a(x_0)$. Isomorphisms form an open set in $\mathcal{L}\left(H, \begin{matrix} \widetilde{H} \\ \oplus \\ \ker a(x_0) \end{matrix} \right)$. Thus there is a neighbourhood U_{x_0} of x_0 such that (3.45) are isomorphisms for all $x \in U_{x_0}$. Applying Lemma 3.5.1 in the opposite direction we obtain that

$$p_{x_0}|_{\ker a(x)} : \ker a(x) \to \ker a(x_0) \tag{3.46}$$

is an isomorphism for all $x \in U_{x_0}$. Choosing any isomorphism $l : \ker a(x_0) \to \mathbb{C}^k$ we obtain a continuously parametrised family of isomorphisms

$$\eta_x := l \circ p_{x_0}|_{\ker a(x)} : \ker a(x) \to \mathbb{C}^k, \quad x \in U_{x_0}.$$

This gives us as a trivialisation

$$\eta_{U_{x_0}} : \big(\ker_X a\big)\big|_{U_{x_0}} \to U_{x_0} \times \mathbb{C}^k$$

over U_{x_0}. Since $x_0 \in X$ is arbitrary, it follows that $\ker_X a \in \mathrm{Vect}(X)$.   □

Denote by

$$p : H \to H \tag{3.47}$$

the family of orthogonal projections $\{p_x\}_{x \in X}$ from H to $\ker a(x)$, $x \in X$.

Corollary 3.5.13. *Suppose* $a \in C\big(X, \mathcal{F}(H, \widetilde{H})\big)$ *and* $a(x) : H \to \widetilde{H}$ *surjective for every* $x \in X$. *Then there exists a* $G \in \mathrm{Vect}(X)$ *and an isomorphism*

$$f : H \to \begin{matrix} \widetilde{H} \\ \oplus \\ G \end{matrix} \tag{3.48}$$

In fact, by virtue of Proposition 3.5.12, $\ker_X a \in \mathrm{Vect}(X)$, and from the proof we see that we can define

$$f := \begin{pmatrix} a \\ m \circ p \end{pmatrix} \tag{3.49}$$

for the family of projections (3.47) and an isomorphism $m : \ker_X a \to G$.

Theorem 3.5.14. *For every* $a \in C(X, \mathcal{F}(H, \widetilde{H}))$, *regarded as a (bundle) morphism* $a : H \to \widetilde{H}$, *there are elements* $J, G \in \mathrm{Vect}(X)$ *and morphisms*

$$k : J \to \widetilde{H}, \quad t : H \to G, \quad q : J \to G, \tag{3.50}$$

such that

$$\boldsymbol{a} := \begin{pmatrix} a & k \\ t & q \end{pmatrix} : \begin{matrix} H \\ \oplus \\ J \end{matrix} \to \begin{matrix} \widetilde{H} \\ \oplus \\ G \end{matrix} \tag{3.51}$$

is an isomorphism. Moreover, we have

$$\mathrm{ind}\, a(x) = g - j \quad \text{for every } x \in X, \tag{3.52}$$

where j *and* g *are the fibre dimensions of* J *and* G, *respectively.*

Proof. Remark 3.5.10 allows us to form the surjective Fredholm family

$$(a \quad k) : \begin{matrix} H \\ \oplus \\ \mathbb{C}^n \end{matrix} \to \widetilde{H}.$$

We now apply Corollary 3.5.13 to $(a \quad k)$ instead of a and $\begin{matrix} H \\ \oplus \\ \mathbb{C}^n \end{matrix}$ instead of H. Then the second component $m \circ p$ of (3.49) corresponds to the desired second row of (3.51), namely,

$$(t \quad q) : H \to \begin{matrix} \widetilde{H} \\ \oplus \\ G \end{matrix}.$$

Here J is the trivial bundle $\mathbb{C}^n$. Relation (3.52) corresponds to (3.33), applied to the isomorphism

$$\boldsymbol{a}(x) := \begin{pmatrix} a(x) & k(x) \\ t(x) & q(x) \end{pmatrix} : \begin{matrix} H \\ \oplus \\ J_x \end{matrix} \to \begin{matrix} \widetilde{H} \\ \oplus \\ G_x \end{matrix} \qquad (3.53)$$

for any fixed $x \in X$. $\qquad\qquad \square$

Remark 3.5.15. It is obvious that for the same a there are many choices of $J, G \in \mathrm{Vect}(X)$ and morphisms (3.50) such that (3.51) is an isomorphism.

In fact, for any $L \in \mathrm{Vect}(X)$ from (3.51) we can pass to another isomorphism

$$\begin{pmatrix} a & 0 \\ 0 & \mathrm{id}_L \end{pmatrix} : \begin{matrix} H \\ \oplus \\ \widetilde{J} \end{matrix} \to \begin{matrix} \widetilde{H} \\ \oplus \\ \widetilde{G} \end{matrix}, \quad \text{where} \quad \widetilde{J} := \begin{matrix} \mathbb{C}^n \\ \oplus \\ L \end{matrix} \quad \text{and} \quad \widetilde{G} := \begin{matrix} G \\ \oplus \\ L \end{matrix}.$$

Proposition 3.5.16. *For every $J, G \in \mathrm{Vect}(X)$ there exists an $a \in C\big(X, \mathcal{F}(H, \widetilde{H})\big)$ and an isomorphism (3.51) for a suitable choice of morphisms $k : J \to \widetilde{H}$, $t : H \to G$, $q : J \to G$.*

Proof. By assumption, the spaces H and $\widetilde{H}$ are of infinite dimension, and H and $\widetilde{H}$ are identified with $X \times H$ and $X \times \widetilde{H}$, respectively. There are subbundles $H_G \subset X \times H$, $\widetilde{H}_J \subset X \times \widetilde{H}$ and bundle homomorphisms $t : H \to G$ and $k : J \to \widetilde{H}$ that induce isomorphisms

$$t_G := t\big|_{H_G} : H_G \to G \quad \text{and} \quad k : J \to \widetilde{H}_J.$$

Let $P_G : H \to H_G$, $\widetilde{P}_J : \widetilde{H} \to \widetilde{H}_J$ be bundle projections, fibrewise defined as orthogonal projections. Moreover, let $a_1 : H \to \widetilde{H}$ be a continuous family of isomorphisms, say, independent of x. Then the assertion holds for $a := \widetilde{P}_J a_1 P_G$ and $q := 0$. $\qquad \square$

Observe that the right-hand side of (3.52) is a constant, though $\dim \ker a(x)$ and $\dim \mathrm{coker}\, a(x)$ may be variable under varying x. The idea of the K-theoretic index of the family $\boldsymbol{a}$ is now to give the difference of the bundles G and L a meaning.

Two pairs of bundles $(E, F), (\widetilde{E}, \widetilde{F}) \in \mathrm{Vect}(X) \times \mathrm{Vect}(X)$ are said to be equivalent, written

$$(E, F) \sim (\widetilde{E}, \widetilde{F}), \qquad (3.54)$$

if

$$E \oplus \widetilde{F} \oplus G \cong \widetilde{E} \oplus F \oplus G \quad \text{for some} \quad G \in \mathrm{Vect}(X). \qquad (3.55)$$

Note that the condition (3.55) is equivalent to

$$E \oplus L \cong \widetilde{E} \oplus M \quad \text{and} \quad F \oplus L \cong \widetilde{F} \oplus M \quad \text{for some} \quad L, M \in \mathrm{Vect}(X). \quad (3.56)$$

In fact, to show that $(3.55) \Longrightarrow (3.56)$ it suffices to set $L := \widetilde{F} \oplus G$, $M := F \oplus G$. Conversely, we obtain the implication $(3.56) \Longrightarrow (3.55)$ for $G := L \oplus M$. The equivalence class represented by $(E, F) \in \mathrm{Vect}(X) \times \mathrm{Vect}(X)$ will be denoted by

$$[E] - [F]. \tag{3.57}$$

In particular, if 0 means the bundle of fibre dimension 0, then we write $[E] := [E] - [0]$, $-[F] := [0] - [F]$.

Definition 3.5.17. For a compact topological space X we set

$$K(X) := \big\{ [E] - [F] : E, F \in \mathrm{Vect}(X) \big\}.$$

The correspondence $X \to K(X)$ is called the K-functor. We have commutative algebraic operations, namely, sum and product, induced by direct sums and tensor products of the involved vector bundles. More precisely,

$$([E] - [F]) + ([G] - [L]) = [E \oplus G] - [F \oplus L],$$

and

$$([E] - [F]) \cdot ([G] - [L]) = [(E \otimes G) \oplus (F \otimes L)] - [(E \otimes L) \oplus (F \otimes G)],$$

$E, F, G, L \in \mathrm{Vect}(X)$. The commutativity is a consequence of the isomorphisms

$$E \oplus F \cong F \oplus E, \quad E \otimes F \cong F \otimes E, \quad \text{for} \quad E, F \in \mathrm{Vect}(X).$$

Thus the K-functor takes values in commutative rings $K(X)$.

Proposition 3.5.18. *Let* $f : X \to Y$ *be a continuous map between compact spaces* X, Y. *Then the bundle pull-back* f^* *induces a ring homomorphism*

$$f^* : K(Y) \to K(X), \quad f^* : [E] - [F] \mapsto [f^*E] - [f^*F].$$

Proposition 3.5.18 follows from the simple fact that $(E, F) \sim \big(\widetilde{E}, \widetilde{F}\big)$ for pairs of bundles over Y entails $(f^*E, f^*F) \sim \big(f^*\widetilde{E}, f^*\widetilde{F}\big)$ for the corresponding pairs of pull-backs over X.

Theorem 3.5.19. *Let* $a \in C\big(X, \mathcal{F}(H, \widetilde{H})\big)$ *be as in* Theorem 3.5.14, *and define* $J, G \in \mathrm{Vect}(X)$ *by the isomorphism* (3.50). *Then*

$$\mathrm{ind}_X\, a := [G] - [J] \in K(X) \tag{3.58}$$

is independent of the choice of the morphisms k, t, q *in* (3.50). *The element* (3.58) *is called the* K-*theoretic index of the Fredholm family* a, *or simply the family index.*

A proof is given in [23, Subsection 3.3.4].

Chapter 4

Toeplitz boundary value problems

4.1 Operators with global projection conditions

In this section we extend the results of Section 1.2 to the case of a compact C^∞ manifold X with boundary Y.

Definition 4.1.1. Let $\mathbb{L}_i = (P_i, J_i, L_i) \in \mathbb{P}(Y)$ be projection data (cf. Definition 1.2.7), $V_i \in \mathrm{Vect}(X)$, $J_i \in \mathrm{Vect}(Y)$, $i = 1, 2$, and set

$$\boldsymbol{v} := (V_1, V_2; J_1, J_2), \quad \boldsymbol{l} := (V_1, V_2; \mathbb{L}_1, \mathbb{L}_2),$$

$$\mathcal{P}_2 := \mathrm{diag}\,(1, P_2), \quad \mathcal{E}_1 := \mathrm{diag}\,(1, E_1). \tag{4.1}$$

Then $\mathcal{T}^{\mu,d}(X; \boldsymbol{l})$ for $\mu \in \mathbb{Z}$, $d \in \mathbb{N}$, is defined to be the set of all operators

$$\mathcal{A} := \mathcal{P}_2 \tilde{\mathcal{A}} \mathcal{E}_1 \tag{4.2}$$

with arbitrary $\tilde{\mathcal{A}} \in \mathcal{B}^{\mu,d}(X; \boldsymbol{v})$. The elements of $\mathcal{T}^{\mu,d}(X; \boldsymbol{l})$ will be called boundary value problems of order μ and type d with global projection (boundary) conditions. Moreover, set

$$\mathcal{T}^{-\infty,d}(X; \boldsymbol{l}) := \{\mathcal{P}_2 \tilde{\mathcal{C}} \mathcal{E}_1 : \tilde{\mathcal{C}} \in \mathcal{B}^{-\infty,d}(X; \boldsymbol{v})\}, \tag{4.3}$$

and $\mathcal{T}^{\infty,d}(X; \boldsymbol{l}) = \bigcup_{\mu \in \mathbb{Z}} \mathcal{T}^{\mu,d}(X; \boldsymbol{l})$.

Remark 4.1.2. The system of spaces $\mathcal{T}^{\mu,d}(X; \boldsymbol{l})$ represents an extension of the spaces $\mathcal{B}^{\mu,d}(X; \boldsymbol{v})$.

In fact, the special case of operators in Boutet de Monvel's calculus is obtained if one takes $J_i = L_i$ and $P_i = \mathrm{id}$ the identity operators in $L^0_{\mathrm{cl}}(Y; J_i)$, $i = 1, 2$. The spaces $\mathcal{T}^{\mu,d}(X; \boldsymbol{l})$ may be regarded as Toeplitz analogue of Boutet de Monvel's

© Springer Nature Switzerland AG 2018

X. Liu, B.-W. Schulze, *Boundary Value Problems with Global Projection Conditions*,

Operator Theory: Advances and Applications 265, https://doi.org/10.1007/978-3-319-70114-1_4

calculus, but they are unifying the two concepts. This point of view has been first introduced in [51]. Similarly as in connection with Definition 1.2.15, an operator (4.2) first represents a continuous operator

$$
\mathcal{A}: \begin{array}{c} H^\infty(\operatorname{int} X, V_1) \\ \oplus \\ H^\infty(Y, \mathbb{L}_1) \end{array} \longrightarrow \begin{array}{c} H^\infty(\operatorname{int} X, V_2) \\ \oplus \\ H^\infty(Y, \mathbb{L}_2) \end{array} ,
$$

using the respective continuity of operators in $\mathcal{B}^{\mu,d}(X; \boldsymbol{v})$ and of the involved embedding and projection operators.

Observe that the space (4.3) can be equivalently characterised as the set of all $\mathcal{A} \in T^{\infty,d}(X; \boldsymbol{l})$, $\mathcal{A} = \mathcal{P}_2 \widetilde{\mathcal{A}} \mathcal{E}_1$ for some $\widetilde{\mathcal{A}} \in \mathcal{B}^{\mu,d}(X; \boldsymbol{v})$ such that $\mathcal{P}_2 \widetilde{\mathcal{A}} \mathcal{P}_1 \in \mathcal{B}^{-\infty,d}(X; \boldsymbol{v})$; then $\mathcal{A} = \mathcal{P}_2(\mathcal{P}_2 \widetilde{\mathcal{A}} \mathcal{P}_1) \mathcal{E}_1$. Moreover,

$$
\mathcal{P}_2(\mathcal{P}_2 \widetilde{\mathcal{A}} \mathcal{P}_1) \mathcal{E}_1 \in T^{-\infty,d}(X; \boldsymbol{l}) \implies \mathcal{P}_2 \widetilde{\mathcal{A}} \mathcal{P}_1 \in \mathcal{B}^{-\infty,d}(X; \boldsymbol{v}).
$$

Proposition 4.1.3. *Given $V_i \in \operatorname{Vect}(X)$, $\mathbb{L}_i \in \mathbb{P}(Y)$, $i = 1, 2$, we have a canonical isomorphism*

$$
T^{\mu,d}(X; \boldsymbol{l}) \to \{ \mathcal{P}_2 \widetilde{\mathcal{A}} \mathcal{P}_1 : \widetilde{\mathcal{A}} \in \mathcal{B}^{\mu,d}(X; \boldsymbol{v}) \}.
$$

Proof. The proof is analogous to that of Proposition 1.2.16; in the present case we employ the equivalence relation

$$
\widetilde{\mathcal{A}} \sim \widetilde{\mathcal{B}} \iff \mathcal{P}_2 \widetilde{\mathcal{A}} \mathcal{P}_1 = \mathcal{P}_2 \widetilde{\mathcal{B}} \mathcal{P}_1. \tag{4.4}
$$

$\square$

Remark 4.1.4. We have an identification

$$
T^{\mu,d}(X; \boldsymbol{l}) = \mathcal{B}^{\mu,d}(X; \boldsymbol{v})/\sim
$$

for the equivalence relation (4.4).

Proposition 4.1.5. *Every $\mathcal{A} \in T^{\mu,d}(X; \boldsymbol{l})$ induces continuous operators*

$$
\mathcal{A}: \begin{array}{c} H^s(\operatorname{int} X, V_1) \\ \oplus \\ H^s(Y, \mathbb{L}_1) \end{array} \longrightarrow \begin{array}{c} H^{s-\mu}(\operatorname{int} X, V_2) \\ \oplus \\ H^{s-\mu}(Y, \mathbb{L}_2) \end{array} \tag{4.5}
$$

for every $s \in \mathbb{R}$, $s > d - 1/2$.

Proof. The result immediately follows from the representation of $\mathcal{A}$ in the form $\mathcal{A} = \mathcal{P}_2 \widetilde{\mathcal{A}} \mathcal{E}_1$, the continuity of $\widetilde{\mathcal{A}} : H^s(\operatorname{int} X, V_1) \oplus H^s(Y, J_1) \to H^{s-\mu}(\operatorname{int} X, V_2) \oplus H^{s-\mu}(Y, J_2)$ for $s > d - 1/2$, together with the continuity of $E_1 : H^s(Y, \mathbb{L}_1) \to H^s(Y, J_1)$, $P_2 : H^{s-\mu}(Y, J_2) \to H^{s-\mu}(Y, \mathbb{L}_2)$. $\square$

Let us now introduce the principal symbol structure of $T^{\mu,d}(X; \boldsymbol{l})$. In the 2×2 block matrix structure $\mathcal{A} = (A_{ij})_{i,j=1,2}$ of our operators we have $A_{11} \in \mathcal{B}^{\mu,d}(X; V_1, V_2)$ and the (homogeneous principal) symbol

$$
\sigma_\psi(\mathcal{A}) := \sigma_\psi(A_{11}) : \pi_X^* V_1 \to \pi_X^* V_2
$$

as in (2.140). Occasionally we also call $\sigma_\psi(\mathcal{A})$ the interior symbol of $\mathcal{A}$. Moreover, the family of operators

$$\sigma_\partial(\mathcal{A}) := \begin{pmatrix} 1 & 0 \\ 0 & p_2 \end{pmatrix} \sigma_\partial(\widetilde{\mathcal{A}}) \begin{pmatrix} 1 & 0 \\ 0 & e_1 \end{pmatrix} : \begin{array}{c} \pi_Y^* V_1' \otimes H^s(\mathbb{R}_+) \\ \oplus \\ L_1 \end{array} \longrightarrow \begin{array}{c} \pi_Y^* V_2' \otimes H^{s-\mu}(\mathbb{R}_+) \\ \oplus \\ L_2 \end{array}$$

is called the (homogeneous principal) boundary symbol of $\mathcal{A}$ (recall that $p_2(y, \eta)$ is the homogeneous principal symbol of order zero of the projection $P_2 \in L^0_{cl}(Y; J_2, J_2)$ while $e_1 : L_1 \to \pi_Y^* J_1$ is the canonical embedding). Similarly as in $\mathcal{B}^{\mu,d}$, we set

$$\sigma(\mathcal{A}) := (\sigma_\psi(\mathcal{A}), \sigma_\partial(\mathcal{A})).$$

Remark 4.1.6. Identifying an operator $\mathcal{A} = \mathcal{P}_2 \widetilde{\mathcal{A}} \mathcal{E}_1 \in \mathcal{T}^{\mu,d}(X; l)$ with $\widetilde{\mathcal{A}} := \mathcal{P}_2 \widetilde{\mathcal{A}} \mathcal{P}_1 \in \mathcal{B}^{\mu,d}(X; v)$ (cf. Proposition 4.1.3), the relation $\sigma\left(\widetilde{\mathcal{A}}\right) = 0$ in the sense of $\mathcal{B}^{\mu,d}$ is equivalent to $\sigma(\mathcal{A}) = 0$ in the sense of $\mathcal{T}^{\mu,d}$.

Theorem 4.1.7. (i) $\mathcal{A} \in \mathcal{T}^{\mu,d}(X; l)$ and $\sigma(\mathcal{A}) = 0$ imply $\mathcal{A} \in \mathcal{T}^{\mu-1,d}(X; l)$ and the operator (4.5) is compact for every $s > d - 1/2$.

(ii) $\mathcal{A} \in \mathcal{T}^{\mu,d}(X; l_0)$, $\mathcal{B} \in \mathcal{T}^{\nu,e}(X; l_1)$ implies $\mathcal{AB} \in \mathcal{T}^{\mu+\nu,h}(X; l_0 \circ l_1)$ (when the bundle and projection data in the middle fit together so that $l_0 \circ l_1$ makes sense), $h = \max\{\nu + d, e\}$, and we have $\sigma(\mathcal{AB}) = \sigma(\mathcal{A})\sigma(\mathcal{B})$ with componentwise multiplication.

(iii) $\mathcal{A} \in \mathcal{T}^{0,0}(X; l)$ for $l = (V_1, V_2; \mathbb{L}_1, \mathbb{L}_2)$ implies $\mathcal{A}^* \in \mathcal{T}^{0,0}(X; l^*)$ for $l^* = (V_2, V_1; \mathbb{L}_2^*, \mathbb{L}_1^*)$, with $\mathcal{A}^*$ being defined by

$$(u, \mathcal{A}^* v)_{L^2(X, V_1) \oplus H^0(Y, \mathbb{L}_1)} = (\mathcal{A}u, v)_{L^2(X, V_2) \oplus H^0(Y, \mathbb{L}_2)}$$

for all $u \in L^2(X, V_1) \oplus H^0(Y, \mathbb{L}_1)$, $v \in L^2(X, V_2) \oplus H^0(Y, \mathbb{L}_2)$, and we have $\sigma(\mathcal{A}^*) = \sigma(\mathcal{A})^*$ with componentwise adjoint (cf. Theorems 2.4.21 and 1.2.25).

Proof. (i) We write $\mathcal{A}$ in the form (4.2) and form $\widetilde{\mathcal{A}} := \mathcal{P}_2 \widetilde{\mathcal{A}} \mathcal{P}_1 \in \mathcal{B}^{\mu,d}(X; v)$ for $v = (V_1, V_2; J_1, J_2)$. Then $\mathcal{A} = \mathcal{P}_2 \widetilde{\mathcal{A}} \mathcal{E}_1$, and $\sigma\left(\widetilde{\mathcal{A}}\right) = 0$ in the sense of $\mathcal{B}^{\mu,d}(X; v)$. It follows that $\widetilde{\mathcal{A}} \in \mathcal{B}^{\mu-1,d}(X; v)$, and hence $\mathcal{A} \in \mathcal{T}^{\mu-1,d}(X; l)$. The compactness of (4.5) is a consequence of the compactness of $\widetilde{\mathcal{A}}$ in usual Sobolev spaces.

(ii) We identify $\mathcal{A} \in \mathcal{T}^{\mu,d}(X; l_0)$, $l_0 = (V_0, V_2; \mathbb{L}_0, \mathbb{L}_2)$, $\mathcal{B} \in \mathcal{T}^{\nu,e}(X; l_1)$, $l_1 = (V_1, V_0; \mathbb{L}_1, \mathbb{L}_0)$, $\mathbb{L}_i = (P_i, J_i, L_i)$, with operators of the form

$$\mathcal{A} = \mathcal{P}_2 \widetilde{\mathcal{A}} \mathcal{P}_0, \quad \mathcal{B} = \mathcal{P}_0 \widetilde{\mathcal{B}} \mathcal{P}_1$$

for $\mathcal{P}_2 = \text{diag}(1, P_2)$, etc., and $\widetilde{\mathcal{A}} \in \mathcal{B}^{\mu,d}(X; v_0)$, $v_0 = (V_0, V_2; J_0, J_2)$, $\widetilde{\mathcal{B}} \in \mathcal{B}^{\nu,e}(X; v_1)$, $v_1 = (V_1, V_0; J_1, J_0)$. From $\mathcal{AB} = \mathcal{P}_2 \widetilde{\mathcal{A}} \mathcal{P}_0 \widetilde{\mathcal{B}} \mathcal{P}_1$ and

$\widetilde{\mathcal{C}} := \widetilde{\mathcal{A}} \mathcal{P}_0 \widetilde{\mathcal{B}} \in \mathcal{B}^{\mu+\nu,h}(X; \boldsymbol{v}_0 \circ \boldsymbol{v}_1)$, cf. Theorem 2.4.21 (i), it follows that $\mathcal{A}\mathcal{B} \in \mathcal{T}^{\mu+\nu,h}(X; \boldsymbol{l}_0 \circ \boldsymbol{l}_1)$. We have $\sigma(\mathcal{A}\mathcal{B}) = \sigma(\mathcal{A})\sigma(\mathcal{B})$, which is evident. Moreover,

$$\sigma_\partial(\widetilde{\mathcal{C}}) = \sigma_\partial(\widetilde{\mathcal{A}}) \operatorname{diag}(1, p_0) \, \sigma_\partial(\widetilde{\mathcal{B}}) = \sigma_\partial(\widetilde{\mathcal{A}}) \operatorname{diag}(1, p_0) \operatorname{diag}(1, p_0) \, \sigma_\partial(\widetilde{\mathcal{B}})$$

with $p_0 : \pi_Y^* J_0 \to L_0$ being the principal symbol of P_0. It follows that

$$\begin{aligned} \sigma_\partial(\mathcal{A}\mathcal{B}) &= \left\{ \operatorname{diag}(1, p_2) \, \sigma_\partial(\widetilde{\mathcal{A}}) \operatorname{diag}(1, p_0) \right\} \left\{ \operatorname{diag}(1, p_0) \, \sigma_\partial(\widetilde{\mathcal{B}}) \operatorname{diag}(1, p_1) \right\} \\ &= \sigma_\partial(\mathcal{A})\sigma_\partial(\mathcal{B}). \end{aligned}$$

(iii) We write $\mathcal{A}$ in the form $\mathcal{A} = \mathcal{P}_2 \widetilde{\mathcal{A}} \mathcal{P}_1$ for $\widetilde{\mathcal{A}} \in \mathcal{B}^{0,0}(X; \boldsymbol{v})$, $\boldsymbol{v} = (V_1, V_2; J_1, J_2)$. Then

$$\mathcal{A}^* = \operatorname{diag}(1, P_1^*) \, \widetilde{\mathcal{A}}^* \operatorname{diag}(1, P_2^*)$$

with $\widetilde{\mathcal{A}}^* \in \mathcal{B}^{0,0}(X; \boldsymbol{v}^*)$ as in Theorem 2.4.21 (ii) and P_1^*, P_2^* as in Theorem 1.2.25. This yields $\mathcal{A}^* \in \mathcal{T}^{0,0}(X; \boldsymbol{l}^*)$, cf. also the notation in Theorem 1.2.25. Moreover, we have

$$\sigma_\partial(\mathcal{A}^*) = \operatorname{diag}(1, p_1^*) \, \sigma_\partial(\widetilde{\mathcal{A}}^*) \operatorname{diag}(1, p_2^*)$$

where p_i^* is the homogeneous principal symbol of the projection P_i^*, $i = 1, 2$. From $\sigma_\partial(\mathcal{A}^*) = \sigma_\partial(\mathcal{A})^*$ we then obtain the assertion. $\square$

Theorem 4.1.8. *Let $\mathcal{A}_j \in \mathcal{T}^{\mu-j,d}(X; \boldsymbol{l})$, $j \in \mathbb{N}$, be an arbitrary sequence. Then there exists an $\mathcal{A} \in \mathcal{T}^{\mu,d}(X; \boldsymbol{l})$ such that*

$$\mathcal{A} - \sum_{j=0}^{N} \mathcal{A}_j \in \mathcal{T}^{\mu-(N+1),d}(X; \boldsymbol{l})$$

for every $N \in \mathbb{N}$, and $\mathcal{A}$ is unique mod $\mathcal{T}^{-\infty,d}(X; \boldsymbol{l})$.

Proof. The proof is evident after Theorem 2.4.20 and Definition 4.1.1. In fact, we have $\mathcal{A}_j = \mathcal{P}_2 \widetilde{\mathcal{A}}_j \mathcal{E}_1$ for a certain $\widetilde{\mathcal{A}} \in \mathcal{B}^{\mu-j,d}(X; \boldsymbol{v})$ and $\mathcal{P}_2, \mathcal{E}_1$ independent of j, and it suffices to set $\mathcal{A} = \mathcal{P}_1 \widetilde{\mathcal{A}} \mathcal{E}_1$ for $\widetilde{\mathcal{A}} \sim \sum_{j=0}^{\infty} \widetilde{\mathcal{A}}_j$ in the sense of Theorem 2.4.20. $\square$

Similarly as in the Toeplitz operator calculus on a closed manifold, it will be useful to form direct sums. Given $\mathcal{A} \in \mathcal{T}^{\mu,d}(X; \boldsymbol{l})$, $\mathcal{B} \in \mathcal{T}^{\mu,d}(X; \boldsymbol{m})$ for bundle and projection data

$$\begin{aligned} \boldsymbol{l} &:= (V_1, V_2; \mathbb{L}_1, \mathbb{L}_2), \qquad \mathbb{L}_i = (P_i, J_i, L_i) \in \mathbb{P}(Y), \\ \boldsymbol{m} &:= (W_1, W_2; \mathbb{M}_1, \mathbb{M}_2), \quad \mathbb{M}_i = (Q_i, G_i, M_i) \in \mathbb{P}(Y), \quad i = 1, 2, \end{aligned}$$

we set

$$\boldsymbol{l} \oplus \boldsymbol{m} := \left(V_1 \oplus W_1, V_2 \oplus W_2; \mathbb{L}_1 \oplus \mathbb{M}_1, \mathbb{L}_2 \oplus \mathbb{M}_2 \right).$$

Then for the direct sum of operators we obtain

$$\mathcal{A} \oplus \mathcal{B} \in \mathcal{T}^{\mu,d}(X; \boldsymbol{l} \oplus \boldsymbol{m})$$

and

$$\sigma_\psi(\mathcal{A} \oplus \mathcal{B}) = \sigma_\psi(\mathcal{A}) \oplus \sigma_\psi(\mathcal{B}), \quad \sigma_\partial(\mathcal{A} \oplus \mathcal{B}) = \sigma_\partial(\mathcal{A}) \oplus \sigma_\partial(\mathcal{B}),$$

with obvious meaning of notation.

4.2 Ellipticity, parametrices, and the Fredholm property

Let us study ellipticity in our Toeplitz calculus of boundary value problems.

Definition 4.2.1. An $\mathcal{A} \in \mathcal{T}^{\mu,d}(X; \boldsymbol{l})$ for $\boldsymbol{l} := (V_1, V_2; \mathbb{L}_1, \mathbb{L}_2)$, $\mathbb{L}_i = (P_i, J_i, L_i) \in \mathbb{P}(Y), i = 1, 2$, is said to be elliptic if

(i) the interior symbol

$$\sigma_\psi(\mathcal{A}) : \pi_X^* V_1 \to \pi_X^* V_2 \tag{4.6}$$

is an isomorphism;

(ii) the boundary symbol

$$\sigma_\partial(\mathcal{A}) : \begin{matrix} \pi_Y^* V_1' \otimes H^s(\mathbb{R}_+) & & \pi_Y^* V_2' \otimes H^{s-\mu}(\mathbb{R}_+) \\ \oplus & \longrightarrow & \oplus \\ L_1 & & L_2 \end{matrix} \tag{4.7}$$

is an isomorphism for every $s > \max\{\mu, d\} - 1/2$.

Remark 4.2.2. Similarly as in the $\mathcal{B}^{\mu,d}$-case, the condition (ii) in Definition 4.2.1 is equivalent to the bijectivity of

$$\sigma_\partial(\mathcal{A}) : \begin{matrix} \pi_Y^* V_1' \otimes \mathcal{S}(\overline{\mathbb{R}}_+) & & \pi_Y^* V_2' \otimes \mathcal{S}(\overline{\mathbb{R}}_+) \\ \oplus & \longrightarrow & \oplus \\ L_1 & & L_2 \end{matrix}. \tag{4.8}$$

Theorem 4.2.3. *For every $A \in L^\mu_{\mathrm{tr}}(X; V_1, V_2)$ elliptic with respect to σ_ψ (i.e., such that (4.6) is an isomorphism) there exist projection data $\mathbb{L}_1, \mathbb{L}_2 \in \mathbb{P}(Y)$ and an element $\mathcal{A} \in \mathcal{T}^{\mu,0}(X; \boldsymbol{l})$ with A as the top left corner, $\boldsymbol{l} := (V_1, V_2; \mathbb{L}_1, \mathbb{L}_2)$, which is elliptic in the sense of Definition 4.2.1.*

Proof. Let us choose elements $L_1, L_2 \in \mathrm{Vect}(S^*Y)$ such that

$$\mathrm{ind}_{S^*Y} \sigma_\partial(A) = [L_2] - [L_1]$$

holds in $K(S^*Y)$. Then using the operator $G \in \mathcal{B}_G^{\mu,d}(X; V_1, V_2)$ of Proposition 3.2.4 applied in the proof of Proposition 3.2.5 we obtain an operator (3.13) that we now denote by $\widetilde{\mathcal{A}}$, and it suffices to set $\mathcal{A} := \mathcal{P}_2 \widetilde{\mathcal{A}} \mathcal{E}_1$. $\qquad\square$

Proposition 4.2.4. *For every* $\mu \in \mathbb{Z}$, $V \in \mathrm{Vect}(X)$ *and* $\mathbb{L} \in \mathbb{P}(Y)$ *there exists an elliptic element* $\mathcal{R}^{\mu}_{V,\mathbb{L}} \in \mathcal{T}^{\mu,0}(X; l)$ *for* $l := (V, V; \mathbb{L}, \mathbb{L})$ *which induces a Fredholm operator*

$$\mathcal{R}^{\mu}_{V,\mathbb{L}} : \quad \begin{matrix} H^s(\mathrm{int}\, X, V) \\ \oplus \\ H^s(Y, \mathbb{L}) \end{matrix} \quad \longrightarrow \quad \begin{matrix} H^{s-\mu}(\mathrm{int}\, X, V) \\ \oplus \\ H^{s-\mu}(Y, \mathbb{L}) \end{matrix}$$

for every $s > \max\{\mu, 0\} - 1/2$.

Proof. It suffices to set

$$\mathcal{R}^{\mu}_{V,\mathbb{L}} := \mathrm{diag}\,(R^{\mu}_V, R^{\mu}_{\mathbb{L}})$$

with R^{μ}_V from Theorem 3.4.1 and $R^{\mu}_{\mathbb{L}}$ from Remark 1.2.29.   □

Remark 4.2.5. By virtue of the ellipticity of $R^{\mu}_V \in \mathcal{B}^{\mu,0}(X; V, V)$, the boundary symbol

$$\sigma_{\partial}(R^{\mu}_V) : \pi^*_Y V' \otimes H^s(\mathbb{R}_+) \to \pi^*_Y V' \otimes H^{s-\mu}(\mathbb{R}_+)$$

is an isomorphism for $s > \max\{\mu, 0\} - 1/2$. Thus $\mathrm{ind}_{S^*Y} \sigma_{\partial}(R^{\mu}_V) = 0$.

Similarly to R^{μ}_V, we can form an operator $S^{\mu}_V \in \mathcal{B}^{\mu,0}(X; V, V)$, the local symbol of which close to Y is equal to $\mathrm{r}^{\mu}_+(\eta, \tau) := \overline{\mathrm{r}^{\mu}_-(\eta, \tau)}$ (the complex conjugate). This operator can be chosen in such a way that

$$\sigma_{\partial}(S^{\mu}_V) : \pi^*_Y V' \otimes H^s(\mathbb{R}_+) \to \pi^*_Y V' \otimes H^{s-\mu}(\mathbb{R}_+)$$

is surjective and $\ker_{S^*Y} \sigma_{\partial}(S^{\mu}_V) = \mu[\pi^*_1 V']$; thus,

$$\mathrm{ind}_{S^*Y} \sigma_{\partial}(S^{\mu}_V) = \mu[\pi^* V'], \quad \pi : S^*Y \to Y. \tag{4.9}$$

Theorem 4.2.6. *For every elliptic operator* $\mathcal{A} \in \mathcal{T}^{\mu,d}(X; l)$, $l := (V_1, V_2; \mathbb{L}_1, \mathbb{L}_2)$, *there exists an elliptic operator* $\mathcal{B} \in \mathcal{T}^{\mu,d}(X; m)$, $m := (V_2, V_1; \mathbb{M}_1, \mathbb{M}_2)$, *for certain projection data* $\mathbb{M}_1, \mathbb{M}_2 \in \mathbb{P}(Y)$ *of the form* $\mathbb{M}_i := (Q_i, \mathbb{C}^N, M_i)$, $i = 1, 2$, *for some* $N \in \mathbb{N}$, *such that* $\mathcal{A} \oplus \mathcal{B} \in \mathcal{B}^{\mu,d}(X; v)$ *for* $v = (V_1 \oplus V_2, V_2 \oplus V_1; \mathbb{C}^N, \mathbb{C}^N)$ *is Shapiro–Lopatinskii elliptic.*

Proof. Let A denote the top left corner of $\mathcal{A}$, which belongs to $\mathcal{B}^{\mu,d}(X; V_1, V_2)$ and is σ_{ψ}-elliptic, cf. Definition 3.1.1 (i). By assumption, the boundary symbol

$$\sigma_{\partial}(A) : \pi^* V'_1 \otimes H^s(\mathbb{R}_+) \to \pi^* V'_2 \otimes H^{s-\mu}(\mathbb{R}_+), \quad \pi : S^*Y \to Y,$$

represents a family of Fredholm operators on S^*Y for every $s > \max\{\mu, d\} - 1/2$. The specific choice of s is not essential at this moment, but in connection with reductions of orders below we assume $s \in \mathbb{N}$ sufficiently large. Choose any $B \in \mathcal{B}^{\mu,d}(X; V_2, V_1)$ with the property that

$$\mathrm{ind}_{S^*Y} \sigma_{\partial}(B) = -\mathrm{ind}_{S^*Y} \sigma_{\partial}(A).$$

A way to find such a B is as follows. First consider the case $\mu = d = 0$. Then we can set $B := A^*$, cf. Theorem 2.4.21 (ii). In fact, we have $A^*A \in \mathcal{B}^{0,0}(X; V_1, V_1)$, and $\sigma_\partial(A^*A) = \sigma_\partial(A^*)\sigma_\partial(A)$, cf. Theorem 2.4.21 (i). From self-adjointness it follows that

$$\text{ind}_{S^*Y}\sigma_\partial(A^*A) = 0 = \text{ind}_{S^*Y}\sigma_\partial(A^*) + \text{ind}_{S^*Y}\sigma_\partial(A).$$

Now for arbitrary μ, d we write $A = A_\mu + G$ for $A_\mu \in \mathcal{B}^{\mu,0}(X; V_1, V_2)$, $G \in \mathcal{B}_G^{\mu,g}(X; V_1, V_2)$. We realise A_μ as a continuous operator $A_\mu : H^s(\text{int}\, X, V_1) \to H^{s-\mu}(\text{int}\, X, V_1)$ for some fixed sufficiently large $s \in \mathbb{N}$. Since $\sigma_\partial(G)$ is a family of compact operators, we have $\text{ind}_{S^*Y}\sigma_\partial(A) = \text{ind}_{S^*Y}\sigma_\partial(A_\mu)$. Thus we may ignore d, i.e., assume $d = 0$. Form the operator

$$A_0 := R_{V_2}^{s-\mu} A_\mu R_{V_1}^{-s} : L^2(X, V_1) \to L^2(X, V_2)$$

for order reducing operators $R_{V_1}^{-s}, R_{V_2}^{s-\mu}$, obtained from Theorem 3.4.1 for the corresponding choice of V and fixed $\lambda = \lambda_0$. We have $R_{V_1}^{-s} \in \mathcal{B}^{-s,0}(X; V_1, V_1)$, $R_{V_2}^{s-\mu} \in \mathcal{B}^{s-\mu,0}(X; V_2, V_2)$, and isomorphisms

$$R_{V_1}^{-s} : L^2(X, V_1) \to H^s(X, V_1), \quad R_{V_2}^{s-\mu} : H^{s-\mu}(\text{int}\, X, V_2) \to L^2(X, V_2).$$

Then $A_0 \in \mathcal{B}^{0,0}(X; V_1, V_2)$ and

$$\text{ind}_{S^*Y}\sigma_\partial(A_0) = \text{ind}_{S^*Y}\sigma_\partial(A) = [L_2] - [L_1].$$

For the L^2-adjoint $A_0^* \in \mathcal{B}^{0,0}(X; V_2, V_1)$ it follows that

$$\text{ind}_{S^*Y}\sigma_\partial(A_0^*) = [L_1] - [L_2].$$

and $\text{ind}_{S^*Y}\sigma_\partial(B_1) = [L_1] - [L_2]$ for $B_1 := R_{V_1}^{-s+\mu} A_0^* R_{V_2}^{s} \in \mathcal{B}^{\mu,s}(X; V_2, V_1)$, cf. Theorem 2.4.21 (i). The operator B_1 can be written as $B_1 = B + G$ for a $B \in \mathcal{B}^{\mu,0}(X; V_2, V_1)$ and a $G \in \mathcal{B}_G^{\mu,s}(X; V_2, V_1)$. Then

$$\text{ind}_{S^*Y}\sigma_\partial(B) = [L_1] - [L_2],$$

since $\sigma_\partial(G_1)$ takes values in compact operators. There are bundles $M_1, M_2 \in \text{Vect}(S^*Y)$ such that $M_1 \oplus L_1 \cong M_2 \oplus L_2 \cong \mathbb{C}^N$. Since $[M_1] + [L_1] = [\mathbb{C}^N]$ and $[M_2] + [L_2] = [\mathbb{C}^N]$, we obtain $[L_1] - [L_2] = ([\mathbb{C}^N] - [M_1]) - ([\mathbb{C}^N] - [M_2])$, i.e.,

$$\text{ind}_{S^*Y}\sigma_\partial(B) = [M_2] - [M_1].$$

Applying the constructions of Section 3.2 and Theorem 4.2.3, we find an elliptic operator $\mathcal{B} \in \mathcal{T}^{\mu,0}(X; \boldsymbol{m})$ for $\boldsymbol{m} := (V_2, V_1; \mathbb{M}_1, \mathbb{M}_2)$, $\mathbb{M}_i := (Q_i, \mathbb{C}^N, M_i)$, and $i = 1, 2$, such that $\ker_{S^*Y}\sigma_\partial(\mathcal{B}) \cong M_2$, $\text{coker}_{S^*Y}\sigma_\partial(\mathcal{B}) \cong M_1$. Taking for Q_1 (resp. Q_2) the complementary projection to P_2 (resp. P_1) it follows that $\mathcal{A} \oplus \mathcal{B}$ is elliptic in $\mathcal{B}^{\mu,d}(X; \boldsymbol{v})$ for $\boldsymbol{v} = (V_1 \oplus V_2, V_2 \oplus V_1; \mathbb{C}^N, \mathbb{C}^N)$. The operator $\mathcal{B}$ is then as asserted. $\square$

Note that Grubb and Seeley [20] used a similar idea to embed an elliptic boundary value problem with projection conditions into a standard one by using the adjoint operator and the complementary projection.

Definition 4.2.7. Let $\mathcal{A} \in \mathcal{T}^{\mu,d}(X; l)$, $l := (V_1, V_2; \mathbb{L}_1, \mathbb{L}_2)$, $\mathbb{L}_1, \mathbb{L}_2 \in \mathbb{P}(Y)$. An operator $\mathcal{P} \in \mathcal{T}^{-\mu,e}(X; l^{-1})$ for $l^{-1} := (V_2, V_1; \mathbb{L}_2, \mathbb{L}_1)$ and some $e \in \mathbb{N}$ is called a parametrix of $\mathcal{A}$, if the operators

$$\mathcal{C}_\mathrm{L} := \mathcal{I} - \mathcal{P}\mathcal{A} \text{ and } \mathcal{C}_\mathrm{R} := \mathcal{I} - \mathcal{A}\mathcal{P} \tag{4.10}$$

belong to $\mathcal{T}^{-\infty,d_\mathrm{L}}(X; m_\mathrm{L})$ and $\mathcal{T}^{-\infty,d_\mathrm{R}}(X; m_\mathrm{R})$, respectively, for

$$m_\mathrm{L} := (V_1, V_1; \mathbb{L}_1, \mathbb{L}_1), \quad m_\mathrm{R} := (V_2, V_2; \mathbb{L}_2, \mathbb{L}_2),$$

and certain $d_\mathrm{L}, d_\mathrm{R} \in \mathbb{N}$.

Theorem 4.2.8. Let $\mathcal{A} \in \mathcal{T}^{\mu,d}(X; l)$, $\mu \in \mathbb{Z}$, $d \in \mathbb{N}$, $l := (V_1, V_2; \mathbb{L}_1, \mathbb{L}_2)$ for V_1, $V_2 \in$ Vect(X), $\mathbb{L}_1, \mathbb{L}_2 \in \mathbb{P}(Y)$.

(i) *The operator $\mathcal{A}$ is elliptic if and only if*

$$
\mathcal{A}: \begin{array}{c} H^s(\operatorname{int} X, V_1) \\ \oplus \\ H^s(Y, \mathbb{L}_1) \end{array} \longrightarrow \begin{array}{c} H^{s-\mu}(\operatorname{int} X, V_2) \\ \oplus \\ H^{s-\mu}(Y, \mathbb{L}_2) \end{array} \tag{4.11}
$$

is a Fredholm operator for an $s = s_0$, $s_0 > \max\{\mu, d\} - 1/2$.

(ii) *If $\mathcal{A}$ is elliptic, (4.11) is Fredholm for all $s > \max\{\mu, d\} - 1/2$, and $\dim \ker \mathcal{A}$ and $\dim \operatorname{coker} \mathcal{A}$ are independent of s.*

(iii) *An elliptic operator $\mathcal{A}$ has a parametrix $\mathcal{P} \in \mathcal{T}^{-\mu,(d-\mu)^+}(X; l^{-1})$ in the sense of Definition 4.2.7 for $d_\mathrm{L} = \max\{\mu, d\}$, $d_\mathrm{R} = (d-\mu)^+$, and $\mathcal{P}$ can be chosen in such a way that the remainders in (4.10) are projections*

$$\mathcal{C}_\mathrm{L} : H^s(\operatorname{int} X, V_1) \oplus H^s(Y, \mathbb{L}_1) \to \mathcal{V}_1 \text{ and}$$
$$\mathcal{C}_\mathrm{R} : H^{s-\mu}(\operatorname{int} X, V_2) \oplus H^{s-\mu}(Y, \mathbb{L}_2) \to \mathcal{V}_2$$

for all $s > \max\{\mu, d\} - 1/2$, for $\mathcal{V}_1 = \ker \mathcal{A} \subset C^\infty(X, V_1) \oplus H^\infty(Y, \mathbb{L}_1)$ and a finite-dimensional subspace $\mathcal{V}_2 \subset C^\infty(X, V_2) \oplus H^\infty(Y, \mathbb{L}_2)$ with the property $\mathcal{V}_2 + \operatorname{im} \mathcal{A} = H^{s-\mu}(\operatorname{int} X, V_2) \oplus H^{s-\mu}(Y, \mathbb{L}_2)$, $\mathcal{V}_2 \cap \operatorname{im} \mathcal{A} = \{0\}$ for every $s > \max\{\mu, d\} - 1/2$.

Proof. We first show that an elliptic operator $\mathcal{A} \in \mathcal{T}^{\mu,d}(X; l)$ has a parametrix

$$\mathcal{P} \in \mathcal{T}^{-\mu,(d-\mu)^+}(X; l^{-1}).$$

We apply Theorem 4.2.6 to $\mathcal{A}$ and choose a complementary operator

$$\mathcal{B} \in \mathcal{T}^{\mu,d}(X; m), \quad m = (V_2, V_1; \mathbb{M}_1, \mathbb{M}_2),$$

such that $\widetilde{\mathcal{A}} := \mathcal{A} \oplus \mathcal{B} \in \mathcal{B}^{\mu,d}(X; \boldsymbol{v})$ for $\boldsymbol{v} = (V_1 \oplus V_2, V_2 \oplus V_1; \mathbb{C}^N, \mathbb{C}^N)$ is Shapiro–Lopatinskii-elliptic. Then

$$\mathcal{A} = \operatorname{diag}(1, P_2)\, \widetilde{\mathcal{A}} \,\operatorname{diag}(1, E_1). \tag{4.12}$$

Theorem 3.1.3 yields a parametrix $\widetilde{\mathcal{P}} \in \mathcal{B}^{-\mu,(d-\mu)^+}(X; \boldsymbol{v}^{-1})$ where $\boldsymbol{v}^{-1} := (V_2 \oplus V_1, V_1 \oplus V_2; \mathbb{C}^N, \mathbb{C}^N)$ and $\sigma(\widetilde{\mathcal{P}}) = \sigma(\widetilde{\mathcal{A}})^{-1}$. Let us set

$$\mathcal{P}_0 := \operatorname{diag}(1, P_1)\, \widetilde{\mathcal{P}} \,\operatorname{diag}(1, E_2) \in \mathcal{T}^{-\mu,(d-\mu)^+}(X; \boldsymbol{l}^{-1}),$$

where $E_2 : H^{s-\mu}(Y, \mathbb{L}_2) \to H^{s-\mu}(Y, J_2)$ is the canonical embedding and $P_1 : H^s(Y, J_1) \to H^s(Y, \mathbb{L}_1)$ the projection involved in $\mathbb{L}_1$, cf. notation in Definition 4.2.1. This yields

$$\mathcal{P}_0 \mathcal{A} = \operatorname{diag}(1, P_1)\, \widetilde{\mathcal{P}} \,\operatorname{diag}(1, P_2)\, \widetilde{\mathcal{A}} \,\operatorname{diag}(1, E_1).$$

Thus for $\mathcal{C}_{\mathrm{L}} := \mathcal{I} - \mathcal{P}_0 \mathcal{A} \in \mathcal{T}^{0,h}(X; \boldsymbol{v}_{\mathrm{L}})\, \operatorname{diag}(1, E_1)$ for $\boldsymbol{v}_{\mathrm{L}} = (V_1, V_1; \mathbb{L}_1, \mathbb{L}_1)$, $h = \max\{\mu, d\}$ we have $\sigma(\mathcal{C}_{\mathrm{L}}) = 0$, i.e., $\mathcal{C}_{\mathrm{L}} \in \mathcal{T}^{-1,h}(X; \boldsymbol{v}_{\mathrm{L}})$, cf. Theorem 4.1.7 (i). Applying Theorem 4.1.8 we find an operator $\mathcal{D}_{\mathrm{L}} \in \mathcal{T}^{-1,h}(X; \boldsymbol{v}_{\mathrm{L}})$ such that $(\mathcal{I} + \mathcal{D}_{\mathrm{L}})(\mathcal{I} - \mathcal{C}_{\mathrm{L}}) = \mathcal{I} \bmod \mathcal{T}^{-\infty,h}(X; \boldsymbol{v}_{\mathrm{L}})$. We can define $\mathcal{D}_{\mathrm{L}}$ as an asymptotic sum $\sum_{j=1}^{\infty} \mathcal{C}_{\mathrm{L}}^j$. Thus $(\mathcal{I} + \mathcal{D}_{\mathrm{L}})\mathcal{P}_0 \mathcal{A} = \mathcal{I} \bmod \mathcal{T}^{-\infty,h}(X; \boldsymbol{v}_{\mathrm{L}})$, and hence $\mathcal{P}_{\mathrm{L}} := \mathcal{I} + \mathcal{D}_{\mathrm{L}}\mathcal{P}_0 \in \mathcal{T}^{-\mu,(d-\mu)^+}(X; \boldsymbol{l}^{-1})$ is a left parametrix of $\mathcal{A}$. In a similar manner we find a right parametrix. Thus we may take $\mathcal{P} := \mathcal{P}_{\mathrm{L}}$.

The Fredholm property of (4.11) is a direct consequence of the compactness of the remainders $\mathcal{C}_{\mathrm{L}}, \mathcal{C}_{\mathrm{R}}$ in relation (4.10), cf. also Theorem 4.1.7. The second part of (iii) is a consequence of general facts on elliptic operators that are always true when elliptic regularity holds in the respective scales of spaces, see, for instance, [25, Subsection 1.2.7]. This confirms, in particular, assertion (ii).

It remains to show that the Fredholm property of (4.11) for $s = s_0$, $s_0 > \max\{\mu, d\} - 1/2$ entails ellipticity. We reduce order and type to 0 by means of elliptic operators from Proposition 4.2.4, namely,

$$\mathcal{R}_{V_1, \mathbb{L}_1}^{-s_0} : \begin{array}{c} L^2(X, V_1) \\ \oplus \\ H^0(Y, \mathbb{L}_1) \end{array} \longrightarrow \begin{array}{c} H^{s_0}(\operatorname{int} X, V_1) \\ \oplus \\ H^{s_0}(Y, \mathbb{L}_1) \end{array} , \quad \mathcal{R}_{V_2, \mathbb{L}_2}^{s_0 - \mu} : \begin{array}{c} H^{s_0 - \mu}(\operatorname{int} X, V_2) \\ \oplus \\ H^{s_0 - \mu}(Y, \mathbb{L}_2) \end{array} \longrightarrow \begin{array}{c} L^2(X, V_2) \\ \oplus \\ H^0(Y, \mathbb{L}_2) \end{array} \tag{4.13}$$

which are both Fredholm, according to the first part of the proof. The composition

$$\mathcal{A}_0 := \mathcal{R}_{V_2, \mathbb{L}_2}^{s_0 - \mu} \mathcal{A} \, \mathcal{R}_{V_1, \mathbb{L}_1}^{-s_0} : \begin{array}{c} L^2(X, V_1) \\ \oplus \\ H^0(Y, \mathbb{L}_1) \end{array} \longrightarrow \begin{array}{c} L^2(X, V_2) \\ \oplus \\ H^0(Y, \mathbb{L}_2) \end{array} \tag{4.14}$$

is again a Fredholm operator. In addition, it belongs to $\mathcal{T}^{0,0}(X; (V_1, V_2; \mathbb{L}_1, \mathbb{L}_2))$ (the type in the top left corner is necessarily 0 since it is acting in L^2). It suffices to show the ellipticity of $\mathcal{A}_0$. We now employ the fact that every $\mathbb{L} \in \mathbb{P}(Y)$ admits

complementary projection data $\mathbb{L}^{\perp} \in \mathbb{P}(Y)$, cf. Proposition 1.2.8 (iii). In particular, for $\mathbb{L}_1 = (P_1, J_1, \sigma_\psi(P_1)J_1)$ we form $\mathbb{L}_1^{\perp} = (1 - P_1, J_1, \sigma_\psi(1 - P_1)J_1)$. Then
$L^2(Y, J_1) = H^0(Y, \mathbb{L}_1) \oplus H^0(Y, \mathbb{L}_1^{\perp})$. We define an operator

$$\mathcal{B} := \mathcal{I}_2 \mathcal{E} \mathcal{C} \mathcal{I}_1 : \begin{array}{c} L^2(X, V_1) \\ \oplus \\ L^2(Y, J_1)) \end{array} \longrightarrow \begin{array}{c} L^2(X, V_2) \\ \oplus \\ L^2(Y, J_2 \oplus J_1) \end{array}$$

where

$$\mathcal{I}_1 : \begin{array}{c} L^2(X, V_1) \\ \oplus \\ L^2(Y, J_1) \end{array} \longrightarrow \begin{array}{c} L^2(X, V_1) \\ \oplus \\ H^0(Y, \mathbb{L}_1), \\ \oplus \\ H^0(Y, \mathbb{L}_1^{\perp}) \end{array} \qquad \mathcal{I}_2 : \begin{array}{c} L^2(X, V_2) \\ \oplus \\ L^2(Y, J_2) \\ \oplus \\ L^2(Y, J_1) \end{array} \longrightarrow \begin{array}{c} L^2(X, V_2) \\ \oplus \\ L^2(Y, J_2 \oplus J_1) \end{array}$$

are canonical identifications, and

$$\mathcal{C} : \begin{array}{c} L^2(X, V_1) \\ \oplus \\ H^0(Y, \mathbb{L}_1) \\ \oplus \\ H^0(Y, \mathbb{L}_1^{\perp}) \end{array} \longrightarrow \begin{array}{c} L^2(X, V_2) \\ \oplus \\ H^0(Y, \mathbb{L}_2), \\ \oplus \\ H^0(Y, \mathbb{L}_1^{\perp}) \end{array} \qquad \mathcal{E} : \begin{array}{c} L^2(X, V_2) \\ \oplus \\ H^0(Y, \mathbb{L}_2) \\ \oplus \\ H^0(Y, \mathbb{L}_1^{\perp}) \end{array} \hookrightarrow \begin{array}{c} L^2(X, V_2) \\ \oplus \\ L^2(Y, J_2) \\ \oplus \\ L^2(Y, J_1) \end{array}$$

with $\mathcal{E}$ being a canonical embedding, and $\mathcal{C} := \mathrm{diag}\left(\mathcal{A}_0, \mathrm{id}_{H^0(Y, \mathbb{L}_1^{\perp})}\right)$. We obviously
have $\dim \ker \mathcal{B} = \dim \ker \mathcal{A}_0 < \infty$. Moreover, $\ker \mathcal{B}^*\mathcal{B} = \ker \mathcal{B} = \mathrm{im}(\mathcal{B}^*\mathcal{B})^{\perp}$, and
$\mathcal{B}^*\mathcal{B}$ has closed range since $\mathcal{C}^*\mathcal{C}$ has. Therefore, $\mathcal{B}^*\mathcal{B} \in \mathcal{B}^{0,0}(X; (V_1, V_1; J_1, J_1))$ is a
Fredholm operator, and hence elliptic by Theorem 3.1.4. Therefore, both $\sigma_\psi(\mathcal{A}_0)$
and $\sigma_\partial(\mathcal{A}_0)$ are injective. Analogous arguments for adjoint operators show that
$\sigma_\psi(\mathcal{A}_0)$ and $\sigma_\partial(\mathcal{A}_0)$ are also surjective. □

4.3   Reduction to the boundary

The operator algebra furnished by the spaces $\mathcal{T}^{\mu,d}(X; l)$ for

$$l := (V_1, V_2; \mathbb{L}_1, \mathbb{L}_2), \quad \mathbb{L}_i = (P_i, J_i, L_i) \in \mathbb{P}(Y), \quad i = 1, 2,$$

contains the subalgebra of right bottom corners, consisting of the spaces

$$\mathcal{T}^\mu(Y; \mathbb{L}_1, \mathbb{L}_2),$$

studied in Section 1.2, cf. Definition 1.2.15. For the spaces $\mathcal{T}^{\mu,d}(X; l)$ it is not
essential that $\mu \in \mathbb{Z}$. The principle of reducing a boundary value problem (BVP)
to the boundary by means of another BVP has been well-known since a very long
time, see, for instance, the monograph [23] and the references therein. Let us illustrate this in the case of the Neumann problem on a smooth manifold X with

boundary Y using the Dirichlet problem. In this case the result is that the potential operator K_0 contained in the inverse ${}^t(\Delta \ T_0)^{-1} =: (P_0 \ K_0)$ of the Dirichlet problem (with T_0 being the restriction of a function to the boundary) is composed from the left with T_1 the boundary operator of the Neumann problem. One obtains $T_1 K_0$, which is a first-order classical elliptic pseudo-differential operator on the boundary. In other words, we are dealing with the composition

$$\begin{pmatrix} \Delta \\ T_1 \end{pmatrix} (P_0 \ K_0) = \begin{pmatrix} 1 & 0 \\ T_1 P_0 & T_1 K_0 \end{pmatrix}.$$

Clearly, if we replace Δ by another elliptic operator A with the transmission property at the boundary and if $\mathcal{A}_i := \begin{pmatrix} A \\ T_i \end{pmatrix}$, $i = 0, 1$, are SL-elliptic BVPs for A, then for a parametrix $\mathcal{P}_0 =: (P_0 \ K_0)$ of $\mathcal{A}_0$ we can form the composition

$$\mathcal{A}_1 \mathcal{P}_0 = \begin{pmatrix} A \\ T_1 \end{pmatrix} (P_0 \ K_0) = \begin{pmatrix} 1 & 0 \\ T_1 P_0 & T_1 K_0 \end{pmatrix} + \mathcal{C}_{\mathrm{R}}, \tag{4.15}$$

where $\mathcal{C}_{\mathrm{R}}$ is a compact remainder in Boutet de Monvel's calculus. The operator $R := T_1 K_0$ lives on the boundary Y and is elliptic. From (4.15) one obtains the Agranovich–Dynin formula for the Fredholm indices:

$$\operatorname{ind} \mathcal{A}_1 - \operatorname{ind} \mathcal{A}_0 = \operatorname{ind} R. \tag{4.16}$$

Note that from a parametrix $\mathcal{P}_0$ of $\mathcal{A}_0$ we get a parametrix $\mathcal{P}_1 =: (P_1 \ K_1)$ of $\mathcal{A}_0$ by a simple algebraic consideration, namely, writing

$$(P_1 \ K_1) = (P_0 - K_0 R^{(-1)} T_1 P_0 \ \ K_0 R^{(-1)}), \tag{4.17}$$

where $R^{(-1)}$ denotes a parametrix of R which is obtained in a more elementary way compared with the parametrix construction in Boutet de Monvel's calculus.

Let us now consider elliptic operators

$$\mathcal{A}_i = \begin{pmatrix} A \\ T_i \end{pmatrix} \in \mathcal{T}^{\mu, d_i}(X; \boldsymbol{l}_i), \ i = 0, 1, \quad \mathcal{A}_i : H^s(\operatorname{int} X, V_1) \longrightarrow \begin{matrix} H^{s-\mu}(\operatorname{int} X, V_2) \\ \oplus \\ H^{s-\mu}(Y, \mathbb{L}_i) \end{matrix},$$
$$\tag{4.18}$$

for $\mu > \max\{\mu, d\} - 1/2$, $\boldsymbol{l}_i = (V_1, V_2; \mathbb{O}, \mathbb{L}_i)$, $i = 0, 1$, $\mathbb{L}_i = (Q_i, J, L_i) \in \mathbb{P}(Y)$, where $\mathbb{O}$ indicates the case where the fibre dimension of the bundle in the middle is zero. For convenience, we assume the trace operators to be of the same orders as A. However, a simple reduction of order allows us to pass to arbitrary orders, cf. Remark 1.2.29. By Theorem 4.2.8 (iii), the operators $\mathcal{A}_i$ have parametrices $\mathcal{P}_i \in \mathcal{T}^{-\mu, (d_i - \mu)^+}(X; \boldsymbol{l}_i^{-1})$ for $\boldsymbol{l}_i^{-1} = (V_2, V_1; \mathbb{L}_i, \mathbb{O})$, $i = 0, 1$,

$$\mathcal{P}_i =: (P_i \ K_i), \quad i = 0, 1.$$

Since

$$\mathcal{A}_0 \mathcal{P}_0 = \operatorname{diag}\left(\operatorname{id}_{H^{s-\mu}(X,V_2)}, \operatorname{id}_{H^{s-\mu}(Y,\mathbb{L}_0)}\right) \bmod \mathcal{T}^{-\infty,(d-\mu)^+}(X; V_2, V_2; \mathbb{L}_0, \mathbb{L}_0),$$

it follows that

$$\mathcal{A}_1 \mathcal{P}_0 = \begin{pmatrix} \operatorname{id}_{H^{s-\mu}(X;V_2)} & 0 \\ T_1 P_0 & T_1 K_0 \end{pmatrix} \bmod \mathcal{T}^{-\infty,(d-\mu)^+}(X; V_2, V_2; \mathbb{L}_0, \mathbb{L}_1).$$

Since the latter operator is elliptic, so is $R := T_1 K_0 \in \mathcal{T}^0(Y; \mathbb{L}_0, \mathbb{L}_1)$, now in the Toeplitz calculus on the boundary, developed in Section 1.2. In particular,

$$R : H^{s-\mu}(Y, \mathbb{L}_0) \to H^{s-\mu}(Y, \mathbb{L}_1) \tag{4.19}$$

is a Fredholm operator, and we have an analogue of the Agranovich–Dynin formula (4.16). Moreover, knowing a parametrix $\mathcal{P}_0$ of $\mathcal{A}_0$ we can easily express a parametrix $\mathcal{P}_1$ of $\mathcal{A}_1$ by applying the corresponding analogue of relation (4.17), here using a parametrix $R^{(-1)} \in \mathcal{T}^0(Y; \mathbb{L}_1, \mathbb{L}_0)$ of the operator R.

Remark 4.3.1. Reductions of boundary conditions to the boundary in the Toeplitz analogue of Boutet de Monvel's calculus are possible also for 2×2 block matrix operators, containing trace and potential operators at the same time. The corresponding algebraic arguments are similar to those in [34, pages 252–254], see also the construction in Section 9.3 below, and there is then also an analogue of the Agranovich–Dynin formula.

Chapter 5

Cutting and pasting of elliptic operators, Cauchy data spaces

5.1 Cutting and pasting constructions

Let M be a smooth closed manifold which is decomposed as

$$M = X_- \cup X_+$$

where $X_\pm$ are smooth compact manifolds with common boundary $\partial X_- = \partial X_+ =: Y = X_- \cap X_+$. An example of this kind is $M = 2X$, the double of a smooth compact manifold X with boundary, with two copies of X glued together along $Y = \partial X$.

Consider an elliptic operator $A \in L^\mu_{\mathrm{cl}}(M; V, W)$ for $V, W \in \mathrm{Vect}(M)$, and let

$$A_\pm := A\big|_{\mathrm{int}\, X_\pm}.$$

Then $A_\pm \in L^\mu_{\mathrm{cl}}(X_\pm; V_\pm, W_\pm)$ for $V_\pm := V\big|_{\mathrm{int}\, X_\pm}$, $W_\pm := W\big|_{\mathrm{int}\, X_\pm}$. We study the question whether the Fredholm index of the operator

$$A : H^s(M, V) \to H^{s-\mu}(M, W)$$

can be compared with the Fredholm indices of elliptic boundary value problems for $A_\pm$ over $X_\pm$ referring to the respective boundary Y. This problem is quite natural when A is an elliptic differential operator. However, we have to be aware that the Atiyah–Bott obstruction for the existence of Shapiro–Lopatinskii conditions for $A_\pm$ might be non-vanishing and then BVPs would concern global projection conditions. Not less basic is another point, namely, that $A_\pm$ might not have the transmission property at Y. As we know, the transmission property "only" holds for a narrow set of pseudo-differential operators over M, cf. Chapter 2 and Part III below. Clearly, differential operators have the transmission property, and this is a

© Springer Nature Switzerland AG 2018
X. Liu, B.-W. Schulze, *Boundary Value Problems with Global Projection Conditions*,
Operator Theory: Advances and Applications 265, https://doi.org/10.1007/978-3-319-70114-1_5

case of independent interest. Therefore, in this section we assume that our operator $A \in L_{cl}^{\mu}(M; V, W)$ has the transmission property at Y. Moreover, for convenience we confine ourselves to the case $M = 2X$, with X identified with X_+. This has the advantage that we have a global reflection map

$$\varepsilon : M \to M$$

that maps a point $x_+ \in X_+$ to its counterpart $x_- \in X_-$ and vice versa, and Y remains fixed. Otherwise, such a map would exist for a tubular neighbourhood of Y in M, and our constructions could be modified for the general case. We also employ ε as diffeomorphisms

$$\varepsilon : X_- \to X_+,$$

between the respective manifolds with boundary. We have

$$r^+ A e^+ \in \mathcal{B}^{\mu,0}(X_+; V_+, W_+), \quad r^- A e^- \in \mathcal{B}^{\mu,0}(X_-; V_-, W_-),$$

and

$$r^+ A e^- \varepsilon^*, \ \varepsilon^* r^- A e^+ \in \mathcal{B}_G^{\mu,0}(X_+; V_+, W_+), \tag{5.1}$$

$$r^- A e^+ \varepsilon^*, \ \varepsilon^* r^+ A e^- \in \mathcal{B}_G^{\mu,0}(X_-; V_-, W_-). \tag{5.2}$$

Let us first consider the case of an elliptic operator $A_0 \in L_{cl}^0(M; V, W)$. The ellipticity of A_0 is equivalent to the Fredholm property of

$$A_0 : L^2(M, V) \to L^2(M, W),$$

cf. Theorem 1.1.20. This is equivalent to the Fredholm property of the block-matrix operator

$$\begin{pmatrix} r^+ A_0 e^+ & r^+ A_0 e^- \\ r^- A_0 e^+ & r^- A_0 e^- \end{pmatrix} : \begin{matrix} L^2(X_+, V_+) \\ \oplus \\ L^2(X_-, V_-) \end{matrix} \longrightarrow \begin{matrix} L^2(X_+, W_+) \\ \oplus \\ L^2(X_-, W_-) \end{matrix}$$

For $X = X_+$, $E_1 := V_+$, $E_2 := \varepsilon^* V_-$, $F_1 := W_+$, $F_2 := \varepsilon^* W_- \in \text{Vect}(X)$ we equivalently obtain the Fredholm property of

$$\mathcal{A} := \begin{pmatrix} r^+ A_0 e^+ & r^+ A_0 e^- \\ r^- A_0 e^+ & r^- A_0 e^- \end{pmatrix} : \begin{matrix} L^2(X, E_1) \\ \oplus \\ L^2(X, E_2) \end{matrix} \longrightarrow \begin{matrix} L^2(X, F_1) \\ \oplus \\ L^2(X, W, F_2) \end{matrix}$$

By assumption, the operator A_0 has the transmission property at Y. This condition is symmetric with respect to both sides $X_\pm$. Then $\mathcal{A} \in \mathcal{B}^{0,0}(X; E_1 \oplus E_2, F_1 \oplus F_2)$. By virtue of (5.1) and (5.2), the elements of the diagonal are Green operators of type 0. Now the Fredholm property of $\mathcal{A}$ is equivalent to the ellipticity in Boutet de Monvel's calculus, cf. Theorem 3.1.4. Thus the symbols

$$\sigma(\mathcal{A}) : \pi_X^*(E_1 \oplus E_2) \to \pi_X^*(F_1 \oplus F_2)$$

and

$$\sigma_\partial(\mathcal{A}) : \pi_Y^* \left(E_1' \oplus E_2' \right) \otimes L^2(\mathbb{R}_+) \to \pi_Y^* \left(F_1' \oplus F_2' \right) \otimes L^2(\mathbb{R}_+)$$

are isomorphisms. Since $E' := E_1' = E_2'$ and $F' := F_1' = F_2'$, we have

$$\sigma_\partial(\mathcal{A}) : \pi_Y^* \left(E' \oplus E' \right) \otimes L^2(\mathbb{R}_+) \to \pi_Y^* \left(F' \oplus F' \right) \otimes L^2(\mathbb{R}_+).$$

Recall that we even have $E' \cong F' \cong V' \cong W'$ for $V' = V|_Y$, $W' = W|_Y$ which is a consequence of the transmission property. Writing $\mathcal{A} = (\mathcal{A}_{ij})_{i,j=1,2}$ and taking into account that

$$\mathcal{A}_{12} \in \mathcal{B}_G^{0,0}(X; E_1, F_2), \quad \mathcal{A}_{21} \in \mathcal{B}_G^{0,0}(X; E_2, F_1),$$

which entails that both

$$\sigma_\partial(\mathcal{A}_{12}) : \pi_Y^*(E') \otimes L^2(\mathbb{R}_+) \to \pi_Y^*(F') \otimes L^2(\mathbb{R}_+)$$

and

$$\sigma_\partial(\mathcal{A}_{21}) : \pi_Y^*(E') \otimes L^2(\mathbb{R}_+) \to \pi_Y^*(F') \otimes L^2(\mathbb{R}_+)$$

take values in compact operators, it follows that

$$\mathrm{ind}_{S^*Y}\sigma_\partial(\mathcal{A}_{11}) + \mathrm{ind}_{S^*Y}\sigma_\partial(\mathcal{A}_{22}) = 0.$$

The notation $\sigma_\partial(\mathcal{A}_{11})$ is related to an operator over X_+. In order to distinguish this from the notation of boundary symbols over X_- we also write

$$\sigma_\partial(\mathcal{A}_{11}) := \sigma_{\partial,(+)}(\mathcal{A}_{11}).$$

Similarly, since $\mathcal{A}_{22} = \mathrm{r}^+(\varepsilon^* A_0)\mathrm{e}^+$ for the operator push-forward $\varepsilon_* A_0$ under $\varepsilon :$ $M \to M$, we set $\sigma_{\partial,(-)}(\mathrm{r}^- A_0 \mathrm{e}^-) = \sigma_{\partial,(+)}(\mathrm{r}^+(\varepsilon^* A_0)\mathrm{e}^+)$. Summing up we have the following result.

Proposition 5.1.1. *Let $A_0 \in L_{\mathrm{cl}}^0(M; V, W)$ be an elliptic operator with the transmission property at Y. Then*

$$\mathrm{ind}_{S^*Y}\sigma_{\partial,(+)}(\mathrm{r}^+ A_0 \mathrm{e}^+) + \mathrm{ind}_{S^*Y}\sigma_{\partial,(-)}(\mathrm{r}^- A_0 \mathrm{e}^-) = 0. \tag{5.3}$$

Relation (5.3) is interpreted as the addition in $K(S^*Y)$.

Let now $A \in L_{\mathrm{cl}}^\mu(M; V, W)$ be an arbitrary elliptic operator, $\mu \in \mathbb{Z}$. There exists an elliptic operator $\widetilde{R}_V^\mu \in L_{\mathrm{tr}}^\mu(M; V, V)$ which has the property that

$$\widetilde{R}_V^\mu : H^s(M, V) \to H^{s-\mu}(M, V) \tag{5.4}$$

is an isomorphism for every $s \in \mathbb{R}$ and also that

$$R_{V_+}^\mu := \mathrm{r}^+ \widetilde{R}_V^\mu \mathrm{e}^+ : H^s(\mathrm{int}\, X_+, V_+) \to L^2(X_+, V_+)$$

is an isomorphism with the inverse $r^+ \widetilde{R}_V^{-\mu} e^+$, where $\widetilde{R}_V^{-\mu}$ is the inverse of (5.4), realised for $s = \mu$. Let us set $A_0 := A \widetilde{R}_V^{-\mu}$. Then, according to the composition behaviour of truncated operators in Boutet de Monvel's calculus, we have

$$r^+ A_0 e^+ = (r^+ A e^+)(r^+ \widetilde{R}_V^{-\mu} e^+) + G$$

for some $G \in \mathcal{B}_G^{0,0}(X_+; V_+, W_+)$. Since $\mathrm{ind}_{S^* Y} \sigma_\partial (r^+ \widetilde{R}_V^{-\mu} e^+) = 0$ and $\sigma_\partial(G)(y, \eta)$ is compact for every $(y, \eta) \in S^* Y$, it follows that

$$\mathrm{ind}_{S^* Y} \sigma_{\partial,(+)} (r^+ A_0 e^+) = \mathrm{ind}_{S^* Y} \sigma_{\partial,(+)} (r^+ A e^+). \tag{5.5}$$

Moreover, we have

$$r^- A_0 e^- = (r^- A e^-)(r^- \widetilde{R}_V^{-\mu} e^-),$$

whence

$$
\begin{aligned}
\mathrm{ind}_{S^* Y} \sigma_{\partial,(-)} (r^- A_0 e^-) &= \mathrm{ind}_{S^* Y} \sigma_{\partial,(-)} (r^- A e^-) + \mathrm{ind}_{S^* Y} \sigma_{\partial,(-)} (r^- \widetilde{R}_V^{-\mu} e^-) \\
&= \mathrm{ind}_{S^* Y} \sigma_{\partial,(-)} (r^- A e^-) - \mu[\pi_1^* E'].
\end{aligned}
\tag{5.6}
$$

Here we used the property $\mathrm{ind}_{S^* Y} \sigma_{\partial,(-)} (r^- \widetilde{R}_V^{-\mu} e^-) = -\mu[\pi_1^* E']$ with $\pi : S^* Y \to Y$ being the canonical projection and $[\pi^* E']$ the element of $K(S^* Y)$ represented by $\pi^* E' \in \mathrm{Vect}(S^* Y)$, cf. also relation (4.9). From (5.5) and (5.6) it follows that

$$
\begin{aligned}
\mathrm{ind}_{S^* Y} \sigma_{\partial,(+)} (r^+ A_0 e^+) &+ \mathrm{ind}_{S^* Y} \sigma_{\partial,(-)} (r^- A_0 e^-) \\
&= \mathrm{ind}_{S^* Y} \sigma_{\partial,(+)} (r^+ A e^+) + \mathrm{ind}_{S^* Y} \sigma_{\partial,(-)} (r^- A e^-) - \mu[\pi_1^* E'].
\end{aligned}
$$

Taking into account Proposition 5.1.1, we obtain the following result.

Theorem 5.1.2. *For an elliptic operator $A \in L_{\mathrm{tr}}^\mu(M; V, W)$ we have*

$$\mathrm{ind}_{S^* Y} \sigma_{\partial,(+)} (r^+ A e^+) + \mathrm{ind}_{S^* Y} \sigma_{\partial,(-)} (r^- A e^-) = \mu[\pi_1^* E'].$$

Let us now specify this theorem for the case of an elliptic differential operator

$$D : H^s(M, V) \to H^{s-\mu}(M, W) \tag{5.7}$$

of order μ on the manifold M. In this case we know that both

$$\sigma_{\partial,(+)}(D) : \pi_Y^* E' \otimes H^s(\mathbb{R}_+) \to \pi_Y^* F' \otimes H^{s-\mu}(\mathbb{R}_+)$$

and

$$\sigma_{\partial,(-)}(D) : \pi_Y^* E' \otimes H^s(\mathbb{R}_-) \to \pi_Y^* F' \otimes H^{s-\mu}(\mathbb{R}_-)$$

are surjective for $s - \mu > -1/2$ (although $E' \cong F'$, we keep writing E' and F' in order to distinguish the directions of the maps). The kernels

$$\ker_{S^* Y} \sigma_{\partial,(+)}(D), \quad \ker_{S^* Y} \sigma_{\partial,(-)}(D)$$

are then subbundles of $\pi_1^* J$ for $J := E' \oplus \cdots \oplus E'$ (μ summands).

Theorem 5.1.3. *The Cauchy data spaces*

$$L_+(y,\eta) := \left\{ \left(D_t^j u \big|_{t=0} \right)_{j=0,\dots,\mu-1} : u \in E' \otimes \mathcal{S}(\overline{\mathbb{R}}_+),\ \sigma_{\partial,(+)}(D)(y,\eta)u = 0 \right\}$$

for $D_t := -i\partial_t$, *and*

$$L_-(y,\eta) := \left\{ \left(D_t^j u \big|_{t=0} \right)_{j=0,\dots,\mu-1} : u \in E' \otimes \mathcal{S}(\overline{\mathbb{R}}_-),\ \sigma_{\partial,(-)}(D)(y,\eta)u = 0 \right\},$$

$(y,\eta) \in S^*Y$, *form complementary subbundles of* $\bigoplus_{j=0}^{\mu-1} E'$.

The proof of Theorem 5.1.3 will be a consequence of Lemma 5.1.4 below. In a tubular neighbourhood $Y \times (-1,1) \ni (y,t)$ of Y the operator D can be written in the form

$$D = \sum_{j=0}^{\mu} a_j(t) D_t^j$$

for $D_t := -i\partial_t$, for coefficients $a_j \in C^\infty((-1,1), \mathrm{Diff}^{\mu-j}(Y; E', F'))$, $E' = V|_Y$, $F' = W|_Y$. We have

$$\sigma_{\partial,(\pm)}(D)(y,\eta) = \sum_{j=0}^{\mu} a_j(0)_{(\mu-j)}(y,\eta) D_t^j : H^s(\mathbb{R}_\pm, E') \to H^{s-\mu}(\mathbb{R}_\pm, F'),$$

where $a_j(0)_{(\mu-j)}(y,\eta)$ is the homogeneous principal symbol of

$$a_j(0) \in \mathrm{Diff}^{\mu-j}(Y; E', F')$$

of order $\mu - j$, $(y,\eta) \in T^*Y \setminus 0$, $j = 0,\dots,\mu$.

Lemma 5.1.4. *Let* $A := \sum_{k=0}^{\mu} b_k D_t^k$ *be an* $m \times m$-*system of differential operators over* $\mathbb{R}$ *with constant coefficients. Assume that* $b_\mu \in \mathrm{GL}(m,\mathbb{C})$, *and let* $\sum_{k=0}^{\mu} b_k \tau^k$ *be invertible for all* $\tau \in \mathbb{R}$. *Then for*

$$L_\pm := \left\{ \left(D_t^j u(0) \right)_{j=0,\dots,\mu-1} \in \mathbb{C}^{m\mu} : u \in \mathcal{S}(\overline{\mathbb{R}}_\pm, \mathbb{C}^m),\ Au = 0 \right\}$$

we have $L_+ \oplus L_- = \mathbb{C}^{m\mu}$.

Proof. Without loss of generality we may assume $b_\mu = \mathrm{id}_{\mathbb{C}^m}$; otherwise one can pass to a new system with coefficients $b_\mu^{-1} b_k$. First note that $Au = 0$ is equivalent to the system

$$D_t u_{\mu-1} + \sum_{k=0}^{\mu-1} b_k u_k = 0,\quad D_t u_j - u_{j+1} = 0 \quad \text{for } j = 0,\dots,\mu-2,$$

or

$$(D_t - A)U = 0 \text{ for } U = {}^{\mathrm{t}}(u_0,\dots,u_{\mu-1}),$$

with

$$A = \begin{pmatrix} 0 & 1 & 0 & \cdots & 0 \\ 0 & 0 & 1 & \cdots & 0 \\ \vdots & \vdots & \vdots & \ddots & \vdots \\ 0 & 0 & 0 & \cdots & 1 \\ -b_0 & -b_1 & -b_2 & \cdots & -b_{\mu-1} \end{pmatrix}.$$

We then have

$$L_{\pm} = \left\{ U\big|_{t=0} \in \mathbb{C}^{m\mu} : U \in \mathcal{S}\big(\overline{\mathbb{R}}_{\pm}, \mathbb{C}^{m\mu}\big), (D_t - A)U = 0 \right\}.$$

The assumed invertibility of $\sum_{k=0}^{\mu} b_k \tau_k$ for all $\tau \in \mathbb{R}$ is equivalent to the invertibility of the matrix $\tau - A$ for all $\tau \in \mathbb{R}$, i.e., the condition $\mathrm{spec}(A) \cap \mathbb{R} = \emptyset$. Let $\lambda \in \mathrm{spec}(A)$ and $S_\rho(\lambda) := \{ \zeta \in \mathbb{C} : |\zeta - \lambda| = \rho \}$. We have $(D_t - A)U = 0$ and $U|_{t=0} =: U_0$ if and only if

$$U(t) = e^{itA} U_0$$

and

$$e^{itA} U_0 = \sum_{\lambda \in \mathrm{spec}(A)} \frac{1}{2\pi i} \int_{S_\rho(\lambda)} e^{it\zeta} (\zeta - A)^{-1} d\zeta \, U_0$$

for every sufficiently small $0 < \rho < 1$. Setting

$$\mathrm{spec}_{\pm}(A) = \{ \lambda \in \mathrm{spec}(A) : \mathrm{Im}\,\lambda \gtrless 0 \},$$

we have $\mathrm{spec}(A) = \mathrm{spec}_+(A) \cup \mathrm{spec}_-(A)$, because by assumption $\mathrm{spec}(A) \cap \mathbb{R} = \emptyset$. It follows that

$$U(t) = \sum_{\lambda \in \mathrm{spec}_+(A)} \frac{1}{2\pi i} \int_{S_\rho(\lambda)} e^{it\zeta} (\zeta - A)^{-1} d\zeta \, U_0$$

$$+ \sum_{\lambda \in \mathrm{spec}_-(A)} \frac{1}{2\pi i} \int_{S_\rho(\lambda)} e^{it\zeta} (\zeta - A)^{-1} d\zeta \, U_0$$

and

$$U(0) = U_0 = \sum_{\lambda \in \mathrm{spec}_+(A)} \Pi_\lambda U_0 + \sum_{\lambda \in \mathrm{spec}_-(A)} \Pi_\lambda U_0,$$

where

$$\Pi_\lambda := \frac{1}{2\pi i} \int_{S_\rho(\lambda)} (\zeta - A)^{-1} d\zeta$$

is the projection to the eigenspace of A to the eigenvalue λ. We have

$$(D_t - A) \left\{ \frac{1}{2\pi i} \int_{S_\rho(\lambda)} e^{it\zeta} (\zeta - A)^{-1} d\zeta \right\} = 0$$

for every $\lambda \in \mathrm{spec}(\mathcal{A})$, and

$$\frac{1}{2\pi i} \int_{S_\rho(\lambda)} e^{it\zeta} (\zeta - \mathcal{A})^{-1} d\zeta = e^{i\lambda t} p(t)$$

for a suitable polynomial $p(t)$ with $m\mu \times m\mu$ matrix-valued coefficients, of order $m(\lambda) - 1$, where $m(\lambda)$ is the multiplicity of λ. This gives

$$L_+ = \mathcal{P}_+ \mathbb{C}^{m\mu}, \quad L_- = \mathcal{P}_- \mathbb{C}^{m\mu}$$

for the complementary projections

$$\mathcal{P}_+ := \sum_{\lambda \in \mathrm{spec}_+(\mathcal{A})} \Pi_\lambda \quad \text{and} \quad \mathcal{P}_- := \sum_{\lambda \in \mathrm{spec}_-(\mathcal{A})} \Pi_\lambda,$$

i.e., $L_+ \oplus L_- = \mathbb{C}^{m\mu}$. $\qquad\qquad\qquad\qquad\qquad\qquad\qquad\qquad\qquad\qquad\square$

Remark 5.1.5. Observe that we also have

$$(D_t - \mathcal{A}) \left\{ \frac{1}{2\pi i} \int_{S_\rho(\lambda)} e^{it\zeta} (\zeta - \mathcal{A})^{-1} \frac{d\zeta}{\zeta} \right\} = 0$$

for every $\lambda \in \mathrm{spec}(\mathcal{A})$ and $0 < \rho < \mathrm{dist}(\lambda, 0)$, i.e.,

$$\Pi_\lambda = \frac{1}{2\pi i} \int_{S_\rho(\lambda)} (\zeta - \mathcal{A})^{-1} d\zeta = \frac{1}{2\pi i} \int_{S_\rho(\lambda)} (\zeta - \mathcal{A})^{-1} \frac{d\zeta}{\zeta}.$$

Thus, setting $\Gamma_\pm := \{\zeta \in \mathbb{C} : \mathrm{Im}\, \zeta = \pm\delta\}$ for some sufficiently small $\delta > 0$, with Γ_+ (resp. Γ_-) oriented with increasing (resp. decreasing) $\mathrm{Re}\, \zeta$, it follows that

$$\mathcal{P}_\pm = \frac{1}{2\pi i} \mathcal{A} \int_{\Gamma_\pm} (\zeta - \mathcal{A})^{-1} \frac{d\zeta}{\zeta}.$$

Moreover, every solution of

$$(D_t - \mathcal{A})U = 0, \quad U\big|_{t=0} = U_0,$$

can be written in the form

$$U(t) = \frac{1}{2\pi i} \mathcal{A} \int_{\Gamma_+} e^{it\zeta} (\zeta - \mathcal{A})^{-1} \frac{d\zeta}{\zeta} U_0 + \frac{1}{2\pi i} \mathcal{A} \int_{\Gamma_-} e^{it\zeta} (\zeta - \mathcal{A})^{-1} \frac{d\zeta}{\zeta} U_0.$$

Let us now return to our elliptic differential operator (5.7), regarded as a map

$$D : H^\mu(M, V) \to L^2(M, W),\qquad\qquad\qquad\qquad\qquad (5.8)$$

and form the operators $A_\pm := \mathrm{r}^\pm D e^\pm$ over $\mathrm{int}\, X_\pm$.

Choose arbitrary elliptic elements

$$\mathcal{D}_\pm := \begin{pmatrix} A_\pm \\ T_\pm \end{pmatrix} \in T^{\mu,\mu}(X_\pm; l_\pm)$$

for $l_+ := (V_+, W_+; 0, \mathbb{L}_+)$, $l_- := (V_-, W_-; 0, \mathbb{L}_-)$ for suitable projection data $\mathbb{L}_\pm = (P_\pm, J, L_\pm)$.

According to Theorem 4.2.3, there exist such operators $T_\pm$. Recall that trace operators $T_\pm$ in the Toeplitz calculus have the form $T_\pm = P_\pm \widetilde{T}_\pm$ for suitable trace operators $\widetilde{T}_\pm \in \mathcal{B}^{\mu,\mu}(X_\pm; (V_\pm, W_\pm; 0, J))$. By virtue of the ellipticity, we have Fredholm operators (5.8) as well as

$$\mathcal{D}_\pm : H^\mu(X_\pm, V_\pm) \to \begin{matrix} L^2(X_\pm, W_\pm) \\ \oplus \\ H^0(Y, \mathbb{L}_\pm) \end{matrix}.$$

Let us now derive a relation between their indices. Consider the diagram

$$
\begin{array}{ccccccccc}
0 & \longrightarrow & H^\mu(M,V) & \xrightarrow{i} & \mathcal{M}_+ \oplus \mathcal{M}_- & \xrightarrow{j} & L^2(Y,J) & \longrightarrow & 0 \\
& & \downarrow D & & \mathcal{D}\downarrow\uparrow\mathcal{B} & & \uparrow\mathcal{R} & & \\
0 & \longleftarrow & L^2(M,V) & \xleftarrow{b} & \mathcal{N}_+ \oplus \mathcal{N}_- & \xleftarrow{a} & L^2(Y,J) & \longleftarrow & 0
\end{array}
\quad (5.9)
$$

where $\mathcal{M}_\pm := H^\mu(X_\pm, V_\pm)$, $\mathcal{N}_\pm := L^2(X_\pm, W_\pm) \oplus H^0(Y, \mathbb{L}_\pm)$. The maps i and j are defined as

$$i(u) := u\big|_{\text{int } X_1} \oplus u\big|_{\text{int } X}, \quad j(u_+ \oplus u_-) := \bigoplus_{k=0}^{\mu-1} \Delta_{E'}^{-\mu+k+1/2}(\mathrm{r}_+^k u_+ - \mathrm{r}_-^k u_-)$$

where $\mathrm{r}_\pm^k f := \partial_t^k f\big|_{Y_\pm}$, and ∂_t^k the corresponding derivative in the normal direction to Y and $\big|_{Y_\pm}$ indicating the restriction to Y from the $\pm$ side. The operator $\Delta_{E'}^\nu$ stands for any element in $L^\nu_{\text{cl}}(Y; E', E')$ that induces isomorphisms $H^s(Y, E') \to H^{s-\nu}(Y, E')$ for all $s \in \mathbb{R}$. The map a in (5.9) is the canonical embedding, where we use

$$H^s(Y,J) = H^s(Y, \mathbb{L}_+) \oplus H^s(Y, \mathbb{L}_-) \quad \text{for } s = 0.$$

The map b is the canonical projection, using

$$L^2(M,W) = L^2(X_+, W_+) \oplus L^2(X_-, W_-).$$

Moreover, we set

$$\mathcal{D} := \mathcal{D}_+ \oplus \mathcal{D}_-, \quad \mathcal{B} := \mathcal{B}_+ \oplus \mathcal{B}_-,$$

where $\mathcal{B}_\pm \in T^{-\mu,0}(X_\pm; l_\pm^{-1})$ are parametrices of $\mathcal{D}_\pm$. Finally, we set

$$R := j \circ \mathcal{B} \circ a : L^2(Y,J) \to L^2(Y,J)$$

which is an elliptic pseudo-differential operator on Y and, as such, Fredholm. The rows of the diagram (5.8) are exact, and we have

$$D = b \circ \mathcal{D} \circ i.$$

The assumptions of an abstract lemma in [34, Subsection 3.1.13] are satisfied, and thus we obtain the following index formula.

Theorem 5.1.6. *We have*

$$\operatorname{ind} D = \operatorname{ind} \mathcal{D}_+ + \operatorname{ind} \mathcal{D}_- + \operatorname{ind} R.$$

5.2 Spectral boundary value problems

In this section we study (non-homogeneous) boundary value problems for an elliptic differential operator

$$A : C^\infty(X, E) \to C^\infty(X, F) \tag{5.10}$$

on a smooth compact manifold X with boundary Y, $n = \dim X$, $E, F \in \operatorname{Vect}(X)$, as a natural generalisation of homogeneous problems studied in an L^2 set-up by Atiyah, Patodi, and Singer [4, 5, 6].

Concerning the material here, see also the joint paper with Nazaikinskij, Sternin, and Shatalov [30].

As before, in a collar neighbourhood $Y \times [0, 1)$ of Y in X we write $x = (y, t)$ and express our operator in the form

$$A = \sum_{j=0}^{\mu} a_j(t) D_t^j$$

for $D_t = -i\partial_t$, where $a_j \in C^\infty\big([0, 1), \operatorname{Diff}^{\mu-j}(Y; E', F')\big)$ for $E' = E\big|_Y$, $F' = F\big|_Y$. Recall that the ellipticity of A gives rise to an isomorphism

$$a_\mu(0) : E' \to F'.$$

We have

$$\sigma_\partial(A)(y, \eta) = \sum_{j=0}^{\mu} \sigma_\psi\big(a_j(0)\big)(y, \eta) D_t^j : H^s(\mathbb{R}_+, E') \to H^{s-\mu}(\mathbb{R}_+, F'), \tag{5.11}$$

with $\sigma_\psi(a_j(0))(y, \eta)$ being the homogeneous principal symbol of

$$a_j(0) \in \operatorname{Diff}^{\mu-j}(Y; E', F')$$

of order $\mu - j$. The boundary symbol (5.11) is a surjective family of Fredholm operators, parametrised by $(y, \eta) \in T^*Y \setminus 0$. Thus the family of kernels $\ker \sigma_\partial(A)(y, \eta)$ constitutes a bundle

$$L_+ := \ker \sigma_\partial(A) \in \mathrm{Vect}(T^*Y \setminus 0).$$

Consider the family of differential operators

$$\sigma_c(A)(w) := \sum_{j=0}^{\mu} a_j(0) w^j$$

parametrised by $w \in \mathbb{C}$.

Remark 5.2.1. We have

$$\sigma_c(A)(\beta + i\gamma) \in L_{\mathrm{cl}}^{\mu}(Y; E', F'; \mathbb{R}_\beta)$$

for every $\gamma \in \mathbb{R}$, and $\sigma_c(A)(\beta + i\gamma)$ is parameter-dependent elliptic with parameter $\beta \in \mathbb{R}$. Moreover, there exists a countable set $D \subset \mathbb{C}$ such that $D \cap \{w : c < \mathrm{Im}\, w < c'\}$ is finite for every $c \leq c'$, and

$$\sigma_c(A)(w) : H^s(Y, E') \to H^{s-\mu}(Y, F') \tag{5.12}$$

is an isomorphism for every $w \in \mathbb{C} \setminus D$ and $s \in \mathbb{R}$.

Such a situation is well-known in the context of operators on manifolds with conical singularities. In Chapter 6 below we return to considerations of that kind. Let

$$B := \sum_{j=0}^{\mu} a_j(0) D_t^j \tag{5.13}$$

be regarded as a differential operator on the infinite cylinder $Y \times \mathbb{R}$. As noted before the coefficient $a_\mu(0) : E' \to F'$ is an isomorphism. For convenience we set $J := E' = F'$ and assume $a_\mu(0) = \mathrm{id}_J$ (otherwise we pass to another operator by multiplying (5.13) from the left by $a_\mu^{-1}(0)$). Write B in the form

$$B = D_t^\mu + \sum_{j=0}^{\mu-1} b_j D_t^j, \tag{5.14}$$

where $b_j := a_j(0) \in \mathrm{Diff}^{\mu-j}(Y; J, J)$. The pull-back of the bundle J to $Y \times \mathbb{R}$ will be denoted again by J.

Let $H^s(Y \times \mathbb{R}, J)$ denote the cylindrical Sobolev space over $Y \times \mathbb{R}$ of smoothness $s \in \mathbb{R}$, defined as the completion of $C_0^\infty(Y \times \mathbb{R}, J)$ with respect to the norm

$$\left\{ \int \left\| R^s(\tau) \widehat{u}(\tau) \right\|_{L^2(Y,J)}^2 d\tau \right\}^{1/2},$$

where $R^s(\tau) \in L^\mu_{cl}(Y; J, J; \mathbb{R}_\tau)$ is any classical parameter-dependent elliptic pseudo-differential operator of order s with parameter τ which induces isomorphisms $R^s(\tau) : H^s(Y, J) \to L^2(Y, J)$ for every s. An equivalent definition for $s \in \mathbb{N}$ is

$$H^s(Y \times \mathbb{R}, J) = \{u(y, t) \in L^2(Y \times \mathbb{R}, J) : D_y^\alpha D_t^k u(y, t) \in L^2(Y \times \mathbb{R}, J)$$

$$\text{for all } |\alpha| + k \le s\},$$

where D_y^α runs over the set of all differential operators on Y of order $|\alpha|$, acting between sections of J. Moreover, set

$$H^{s,\gamma}(Y \times \mathbb{R}, J) := \{e^{\gamma t} u(y, t) : u \in H^s(Y \times \mathbb{R}, J)\}$$

for arbitrary $s, \gamma \in \mathbb{R}$. Let us reformulate the equation $Bu = f$ as

$$(D_t - \mathcal{A})U = F, \text{ for } U = {}^t(u_0, \dots, u_{\mu-1}) \tag{5.15}$$

and $u_j = D_t^j u$, $j = 0, \dots, \mu - 1$, $F := {}^t(0, \dots, 0, f)$,

$$\mathcal{A} := \begin{pmatrix} 0 & 1 & 0 & \cdots & 0 \\ 0 & 0 & 1 & \cdots & 0 \\ \vdots & \vdots & \vdots & \ddots & \vdots \\ 0 & 0 & 0 & \cdots & 1 \\ -b_0 & -b_1 & -b_2 & \cdots & -b_{\mu-1} \end{pmatrix}. \tag{5.16}$$

Observe that

$$\det(w - \mathcal{A}) = w^\mu + \sum_{j=0}^{\mu-1} b_j w^j =: \sigma_c(\mathcal{A})(w).$$

Lemma 5.2.2. *The operator*

$$w - \mathcal{A} : \bigoplus_{j=0}^{\mu-1} H^{s-j}(Y, J) \to \bigoplus_{k=1}^{\mu} H^{s-k}(Y, J) \tag{5.17}$$

is invertible for a $w \in \mathbb{C}$ if and only if

$$w^\mu + \sum_{j=0}^{\mu-1} b_j w^j : H^s(Y, J) \to H^{s-\mu}(Y, J) \tag{5.18}$$

is invertible for any $s \in \mathbb{R}$. In other words,

$$\text{spec}(\mathcal{A}) = \{w \in \mathbb{C} : \sigma_c(\mathcal{A})(w) \text{ is invertible}\}.$$

Moreover,

$$(w - \mathcal{A})^{-1} = \sigma_c(\mathcal{A})^{-1}(w) \mathcal{Q}(w) \tag{5.19}$$

for a matrix $\mathcal{Q}(w) = (Q_{jk}(w))_{1 \le k \le \mu, 0 \le j \le \mu-1}$ of differential operators, polynomially dependent on w, where $Q_{jk}(w)$ is parameter-dependent with parameter β (for $w = \beta + i\gamma$) of order $\mu - k + j$ for $k = 1, \dots, \mu$, $j = 0, \dots, \mu - 1$.

Proof. Formula (5.19) is a purely algebraic elementary fact, and the proof gives rise to the asserted characterisation of the entries of the matrix $\mathcal{Q}$. From (5.19) it follows that (5.17) is invertible if so is (5.18). Conversely, the invertibility of (5.17) entails that of (5.18) because $\sigma_c(A)^{-1}(w)f$ is equal to the first component of $(w - A)^{-1}({}^t(0, \ldots, 0, f))$.   □

Remark 5.2.3. The set spec $(\mathcal{A})$ is countable, and every strip

$$\{w \in \mathbb{C} : c < \operatorname{Im} w < c'\}$$

contains only finitely many elements of spec $(\mathcal{A})$ for arbitrary reals $c < c'$. The operator function $(w - \mathcal{A})^{-1}$ is meromorphic with poles of finite multiplicity at the points of spec$(\mathcal{A})$, and the Laurent coefficients of $(w - \mathcal{A})^{-1}$ at $(w - p)^{-(k+j)}$, $k \in \mathbb{N}$, are finite-rank operators in $L^{-\infty}(Y; J, J)$.

These observations go back to Agranovich and Vishik [1]. They are also systematically used in the calculus of operators on manifolds with conical singularities, see, for instance, [46, Subsection 1.2.4]. In particular, $\sigma_c(A)(w)$ is parameter-dependent elliptic with parameter $\operatorname{Re} w$ on every line $\operatorname{Im} w = \text{const}$ and invertible for large $|\operatorname{Re} w|$.

Now we interpret (5.14) as a continuous operator

$$B : H^s(Y \times \mathbb{R}_+, J) \to H^{s-\mu}(Y \times \mathbb{R}_+, J)$$

for $H^s(Y \times \mathbb{R}_+, J) := H^s(Y \times \mathbb{R}, J)|_{Y \times \mathbb{R}_+}$. For convenience, we assume that the set D of Remark 5.2.1 does not intersect the real line; otherwise we pass to a translated operator $(D_t - i\gamma)^\mu + \sum_{j=0}^{\mu-1} b_j(D_t - i\gamma)^j$ for a suitable real γ with a corresponding shifted set D_γ in the complex plane which does not intersect the real line.

For any integer $s \geq \mu$ we form the Cauchy data space

$$\mathcal{C}^{s,\mu}(Y, J) := \left\{ (D_t^k u(y, 0))_{k=0,\ldots,\mu-1} : u \in H^s(Y \times \mathbb{R}_+, J) \right\}.$$

Then

$$\mathcal{C}^{s,\mu}(Y, J) = \bigoplus_{k=0}^{\mu-1} H^{s-k-1/2}(Y, J),$$

and $\mathbb{T}^\mu := {}^t(r', r'D_t, \ldots, r'D_t^{\mu-1})$ for $r'u := u|_{t=0}$ defines a continuous operator

$$\mathbb{T}^\mu : H^s(Y \times \mathbb{R}_+) \to \mathcal{C}^{s,\mu}(Y, J).$$

We now investigate the solvability of the boundary value problem

$$Bu_+ = f_+ \in H^{s-\mu}(Y \times \mathbb{R}_+, J), \tag{5.20}$$

$$\mathbb{T}^\mu u_+ = g_+ \in \mathcal{C}^{s,\mu}(Y, J). \tag{5.21}$$

In general, solutions will not exist for all $g_+ \in C^{s,\mu}(Y, J)$, but only for g_+ in a subspace that is the image under a suitable pseudo-differential projection $\mathcal{P}_+$.

Analogously to (5.20)–(5.21), we can consider a boundary value problem on the negative half-cylinder

$$Bu_- = f_- \in H^{s-\mu}(Y \times \mathbb{R}_-, J), \tag{5.22}$$

$$\mathbb{T}^\mu u_- = g_- \in C^{s,\mu}(Y, J). \tag{5.23}$$

Then the admissible boundary data g_- will be determined by the complementary projection $\mathcal{P}_-$.

The projections $\mathcal{P}_\pm$ may be obtained as follows. We use that $\mathrm{spec}\,(\mathcal{A}) = D$, in the notation of Remark 5.2.1 and Lemma 5.2.2, does not intersect the real line $\mathrm{Im}\, w = 0$. Thus there is a $c > 0$ such that

$$\mathrm{spec}\,(\mathcal{A}) \cap \{w : -c \le \mathrm{Im}\, w \le c\} = \emptyset.$$

Let us fix some $0 < \varepsilon < c$ and set

$$\Gamma_\pm := \{w = \tau + i(c - \varepsilon) : \tau \in \mathbb{R}\},$$

oriented in direction of increasing τ on Γ_+ and decreasing τ on Γ_-. Then we form

$$\mathcal{P}_+ = \frac{1}{2\pi i} \mathcal{A} \int_{\Gamma_\pm} (w - \mathcal{A})^{-1} \frac{dw}{w}. \tag{5.24}$$

Remark 5.2.4. The operator $\mathcal{A}$ can be written in the form

$$\mathcal{A} = \mathcal{R} \mathcal{A}_1 \mathcal{R}^{-1} \tag{5.25}$$

with $\mathcal{R} := \mathrm{diag}(R_J^0, R_J^1, \ldots, R_J^{\mu-1})$, where $R_J^k : H^s(Y, J) \rightarrow H^{s-k}(Y, J)$ is an order-reducing isomorphism, $R_J^k \in L^k_{\mathrm{cl}}(Y; J, J)$, and $\mathcal{A}_1$ is a system of operators of order 1. Then $\mathcal{P}_\pm$ takes the form

$$\mathcal{P}_\pm = \mathcal{R} \left\{ \frac{1}{2\pi i} \mathcal{A}_1 \int_{\Gamma_\pm} (w - \mathcal{A}_1)^{-1} \frac{dw}{w} \right\} \mathcal{R}^{-1}. \tag{5.26}$$

Lemma 5.2.5. (i) *The integral* (5.24) *converges strongly in* $C^{s,\mu}(Y, J)$ *on the dense subset* $C^{s+1,\mu}(Y, J)$ *for every* $s \in \mathbb{R}$.

(ii) *The operators* $\mathcal{P}_\pm$ *form a matrix* $(\mathcal{P}_{\pm,jk})_{0 \le j \le \mu-1, 0 \le k \le \mu-1}$ *of elements in* $L^{j-k}_{\mathrm{cl}}(Y; J, J)$. *Thus* $\mathcal{P}_\pm$ *extend to continuous operators*

$$\mathcal{P}_\pm : C^{s,\mu}(Y, J) \rightarrow C^{s,\mu}(Y, J) \quad \text{for all } s \in \mathbb{R}.$$

Proof. (i) The entries of the matrix $(w - \mathcal{A})^{-1}$ belong to $L^{-1}_{\mathrm{cl}}(Y; J, J; \mathbb{R}_\tau)$, cf. Lemma 5.2.2, where the parameter $\tau \in \mathbb{R}$ runs over Γ_+ or Γ_- via $\tau \rightarrow w = \tau \pm i(c - \varepsilon)$. In fact, we have $\sigma_c(\mathcal{A})^{-1}(w) \in L^{-\mu}_{\mathrm{cl}}(Y; J, J; \mathbb{R})$ and the orders of the entries of $\mathcal{Q}(w)$ are $\leq \mu - 1$, cf. (5.19). By virtue of Theorem 1.1.25 from Section 1.1, the $\mathcal{L}(H^{s+1}(Y, J), H^{s+1}(Y, J))$-norm of every entry of $(w - \mathcal{A})^{-1}$ can be estimated by $c(1 + |w|)^{-1}$, $w \in \Gamma_\pm$, for a constant $c = c(s) > 0$. This gives us immediately the assertion (i).

(ii) Let us write $\mathcal{P}_\pm$ in the form (5.26). Then it suffices to observe that

$$\frac{1}{2\pi i} \int_{\Gamma_\pm} (w - \mathcal{A}_1)^{-1} \frac{dw}{w}$$

is a matrix of classical pseudo-differential operators of order -1. For the technique of the proof, see Seeley's paper [57]. $\square$

Let us write the operator (5.16) as $(\mathcal{A}_{kj})_{1 \leq k \leq \mu, 0 \leq j \leq \mu-1}$,

$$\mathcal{A} : \bigoplus_{j=0}^{\mu-1} H^{s-j}(Y, J) \to \bigoplus_{k=1}^{\mu} H^{s-k}(Y, J),$$

and interpret the orders $\operatorname{ord} \mathcal{A}_{kj} = k - j$ in the Douglis–Nirenberg sense, with homogeneous principal symbols $\sigma_\psi(\mathcal{A}_{kj})(y, \eta)$ of order $k - j$. For

$$\sigma_\psi(\mathcal{A})(y, \eta) := \big(\sigma_\psi(\mathcal{A}_{kj})(y, \eta)\big)_{1 \leq k \leq \mu, 0 \leq j \leq \mu-1}$$

it follows that

$$\sigma_\psi(\mathcal{A})(y, \eta) :=$$
$$\begin{pmatrix} 0 & 1 & 0 & \cdots & 0 \\ 0 & 0 & 1 & \cdots & 0 \\ \vdots & \vdots & \vdots & \ddots & \vdots \\ 0 & 0 & 0 & \cdots & 1 \\ -\sigma_\psi(b_0)(y, \eta) & -\sigma_\psi(b_1)(y, \eta) & -\sigma_\psi(b_2)(y, \eta) & \cdots & -\sigma_\psi(b_{\mu-1})(y, \eta) \end{pmatrix} \tag{5.27}$$

Moreover, let $N := \bigoplus_{k=0}^{\mu-1} J$ and

$$L_\pm(y, \eta) := \big\{ u(0) : u(t) \in \mathcal{S}\big(\overline{\mathbb{R}}_\pm, N_y\big), \big(D_t - \sigma_\psi(\mathcal{A})(y, \eta)\big)u = 0 \big\}.$$

Remark 5.2.6. We have a canonical one-to-one map

$$\ker \sigma_\partial(A)(y, \eta) \cong \ker \big(D_t - \sigma_\psi(\mathcal{A})(y, \eta)\big).$$

The right-hand side concerns $\mathcal{S}\big(\overline{\mathbb{R}}_+, N_y\big)$. A similar relation holds on the negative half-line.

Theorem 5.2.7. (i) *The operators $\mathcal{P}_\pm$ are complementary projections, i.e., $\mathcal{P}_\pm^2 = \mathcal{P}_\pm$, $\mathcal{P}_+ + \mathcal{P}_- = 1$, and we have*

$$\mathcal{P}_\pm \mathcal{A} = \mathcal{A} \mathcal{P}_\pm. \tag{5.28}$$

(ii) *The homogeneous principal symbols* $\sigma_\psi(\mathcal{P}_{\pm,jk})$ *of order* $j - k$ *of* $\mathcal{P}_{\pm,jk} \in$ $L_{\mathrm{cl}}^{j-k}(Y; J, J)$ (*cf.* Lemma 5.2.5 (ii)) *form projections*

$$(\sigma_\psi(\mathcal{P}_{\pm,jk})(y,\eta))_{0\leq j\leq\mu-1,0\leq k\leq\mu-1} =: \sigma_\psi(\mathcal{P}_\pm)(y,\eta) : J_y \to L_\pm(y,\eta)$$

along $L_\mp(y,\eta)$.

(iii) *The operator functions* $(w - \mathcal{A})^{-1}\mathcal{P}_\pm$ *are holomorphic in*

$$\mathrm{spec}\,(\mathcal{A}) \cap \{w : \mathrm{Im}\,w \gtrless 0\}.$$

Proof. The relation (5.28) is obvious. Moreover, $\mathcal{P}_+ + \mathcal{P}_- = 1$ follows from the formula

$$\mathcal{P}_+ + \mathcal{P}_- = \mathcal{A}\frac{1}{2\pi i}\int_{\Gamma_0}(w - \mathcal{A})^{-1}\frac{dw}{w},$$

for a small circle Γ_0 clockwise surrounding the origin. By Cauchy's residue theorem, the integral is equal to $2\pi i\mathcal{A}^{-1}$.

Next we calculate $\mathcal{P}_+^2$; the argument for $\mathcal{P}_-$ is analogous and left to the reader. Set $\Gamma_+' := \Gamma_+ + i\varepsilon'$ for some sufficiently small $\varepsilon' > 0$. Then (5.28) together with the resolvent identity $(w' - \mathcal{A})^{-1}(w - \mathcal{A})^{-1} = (w - w')^{-1}\{(w' - \mathcal{A})^{-1} - (w - \mathcal{A})^{-1}\}$ yield

$$\mathcal{P}_+^2 = \frac{1}{2\pi i}\mathcal{A}\int_{\Gamma_+}(w - \mathcal{A})^{-1}\frac{dw}{w}\frac{1}{2\pi i}\mathcal{A}\int_{\Gamma_+'}(w' - \mathcal{A})^{-1}\frac{dw'}{w'}$$

$$= -\left(\frac{\mathcal{A}}{2\pi}\right)^2\int_{\Gamma_+'}\left\{\int_{\Gamma_+}\frac{1}{w'w}(w' - \mathcal{A})^{-1}(w - \mathcal{A})^{-1}dw\right\}dw'$$

$$= -\left(\frac{\mathcal{A}}{2\pi}\right)^2\int_{\Gamma_+'}\left\{\int_{\Gamma_+}\frac{1}{w'w(w - w')}[(w' - \mathcal{A})^{-1} - (w - \mathcal{A})^{-1}]dw\right\}dw'$$

$$= I_0 + I_1,$$

where

$$I_0 = -\left(\frac{\mathcal{A}}{2\pi}\right)^2\int_{\Gamma_+'}\left\{\int_{\Gamma_+}(w - \mathcal{A})^{-1}\frac{dw}{w(w - w')}\right\}\frac{dw'}{w'}$$

$$= -\left(\frac{\mathcal{A}}{2\pi}\right)^2\int_{\Gamma_+}(w - \mathcal{A})^{-1}\left\{\int_{\Gamma_+'}\frac{dw'}{w'(w - w')}\right\}\frac{dw}{w} \qquad (5.29)$$

and

$$I_1 = -\left(\frac{\mathcal{A}}{2\pi}\right)^2\int_{\Gamma_+'}(w' - \mathcal{A})^{-1}\left\{\int_{\Gamma_+}\frac{dw}{w(w - w')}\right\}\frac{dw'}{w'}. \qquad (5.30)$$

We have $I_0 = 0$, since the inner integral on the right-hand side is zero. For I_1 we employ the relation

$$\int_{\Gamma_+}\frac{dw}{w(w - w')} = 2\pi i(w')^{-1}\int_{\Gamma_+}\frac{dw'}{(w')^2} = 0,$$

and obtain

$$I_1 = -\left(\frac{\mathcal{A}}{2\pi}\right)^2 2\pi i \int_{\Gamma'_+} (w' - \mathcal{A})^{-1} \frac{dw'}{(w')^2}$$

$$= \frac{\mathcal{A}}{2\pi i} \int_{\Gamma'_+} (\mathcal{A} - w' + w')(w' - \mathcal{A})^{-1} \frac{dw'}{(w')^2}$$

$$= \frac{\mathcal{A}}{2\pi i} \int_{\Gamma_+} (w' - \mathcal{A})^{-1} \frac{dw'}{w'} - \frac{\mathcal{A}}{2\pi i} \int_{\Gamma_+} \frac{dw'}{(w')^2} = \mathcal{P}_+,$$

cf. (5.24). Now (ii) follows from the considerations in the proof of Lemma 5.1.4. For proving (iii), let us consider the plus-case. Then

$$(w - \mathcal{A})^{-1} \mathcal{P}_+ = \frac{\mathcal{A}}{2\pi i} \int_{\Gamma_+} (w' - \mathcal{A})^{-1} (w - \mathcal{A})^{-1} \frac{dw'}{w'}$$

$$= \frac{\mathcal{A}}{2\pi i} \int_{\Gamma_+} (w' - \mathcal{A})^{-1} \frac{dw'}{w'(w - w')} - \frac{\mathcal{A}(w - \mathcal{A})^{-1}}{2\pi i} \int_{\Gamma_+} \frac{dw'}{w'(w - w')}.$$

The second integral on the right-hand side vanishes, and the first one is holomorphic in $\operatorname{spec}(\mathcal{A}) \cap \operatorname{Im} w > 0$. The case $\operatorname{spec}(\mathcal{A}) \cap \operatorname{Im} w < 0$ is analogous. □

For the next theorem we employ the Fourier–Laplace transform

$$(Fu)(w) = \widetilde{u}(w) = \int_0^\infty e^{-itw} u(t) dt$$

operating on functions over $\mathbb{R}_+$ and producing functions in $\operatorname{Im} w < 0$, with the inverse

$$(F^{-1}\widetilde{u})(t) = \frac{1}{2\pi} \int_{-\infty+i\delta}^{\infty+i\delta} e^{it\beta} \widetilde{u}(\beta) d\beta,$$

$\delta < 0$, $\beta = \operatorname{Re} w$. Let us define the maps

$$L : \mathbb{C} \to \mathbb{C}^\mu, \quad Lz := (0, \ldots, 0, z); \quad Q : \mathbb{C}^\mu \to \mathbb{C}, \quad Q(z_0, \ldots, z_{\mu-1}) := z_0.$$

Theorem 5.2.8. *The boundary value problem*

$$Bu = f \in H^{s-\mu}(Y \times \mathbb{R}_+, J),$$
$$\mathcal{P}_+ \mathbb{T}^\mu u = g \in \mathcal{P}_+ \mathcal{C}^{s,\mu}(Y, J)$$

has a unique solution $u \in H^s(Y \times \mathbb{R}_+, J)$, *which is of the form* $u = Sf + Kg$, *where*

$$Sf = QF^{-1}(w - \mathcal{A})^{-1} \left\{ L(\widetilde{f}(w)) - \frac{1}{2\pi} \int_{\Gamma_-} (z - \mathcal{A})^{-1} L(\widetilde{f}(w)) dz \right\},$$

$$Kg = -iQF^{-1}(w - \mathcal{A})^{-1} g.$$

The corresponding map $\mathcal{R} : (f, g) \mapsto u$ *is continuous as an operator*

$$\mathcal{R} : H^{s-\mu}(Y \times \mathbb{R}_+, J) \oplus \mathcal{P}_+ \mathcal{C}^{s,\mu}(Y, J) \to H^s(Y \times \mathbb{R}_+, J).$$

Proof. We show that $\mathcal{R}$ is a right inverse of the operator $^t(B, \mathcal{P}_+\mathbb{T}^\mu)$; the left inverse part is left to the reader. It will be convenient to pass from B to the operator $\mathcal{D} := D_t - \mathcal{A}$ for $\mathcal{A}$ as in (5.16) and the operator function U as in (5.15). Then we may omit the mappings Q and L which only single out the first component in U and replace $\mathbb{T}^\mu$ by r$'$, the restriction to $t = 0$. Then our operator corresponds to the column vector $^t(\mathcal{D}, \mathcal{P}_+r')$, and $\mathcal{R}$ is to be replaced by $\mathcal{R} = (\mathcal{S} \quad \mathcal{K})$ for

$$\mathcal{S}f := F^{-1}(w - \mathcal{A})^{-1}\left\{\widetilde{f}(w) - \frac{1}{2\pi}\int_{\Gamma_-}(z - \mathcal{A})^{-1}\widetilde{f}(z)dz\right\}, \qquad (5.31)$$

$$\mathcal{K}g := -iF^{-1}(w - \mathcal{A})^{-1}g. \qquad (5.32)$$

For the relation $\begin{pmatrix} \mathcal{D} \\ \mathcal{P}_+\text{r}' \end{pmatrix} (\mathcal{S} \quad \mathcal{K}) = \begin{pmatrix} 1 & 0 \\ 0 & 1 \end{pmatrix}$ we have to verify that

$$\mathcal{D}\mathcal{S}f = f, \qquad \mathcal{D}\mathcal{K}g = 0, \qquad (5.33)$$
$$\mathcal{P}_+\text{r}'\mathcal{S}f = 0, \qquad \mathcal{P}_+\text{r}'\mathcal{K}g = g \qquad (5.34)$$

for all f and g belonging to the respective spaces.

We have

$$\mathcal{D}\mathcal{S}f = \{F^{-1}(w - \mathcal{A})F\}$$
$$\cdot \left\{F^{-1}(w - \mathcal{A})^{-1}\widetilde{f}(w) - \frac{1}{2\pi}\mathcal{D}F^{-1}(w - \mathcal{A})^{-1}\int_{\Gamma_-}(z - \mathcal{A})^{-1}\widetilde{f}(z)dz\right\}.$$

The first summand on the right is equal to f. Moreover, we have

$$\mathcal{D}F^{-1}(w - \mathcal{A})^{-1}h = 0$$

for $h := \int_{\Gamma_-}(z - \mathcal{A})^{-1}\widetilde{f}(z)dz$ which is independent of w. For similar reasons we have $\mathcal{D}\mathcal{K}g = 0$. Thus we verified (5.33). For the second relation of (5.34) we write

$$\mathcal{P}_+\text{r}'\mathcal{K}g = -i\mathcal{P}_+ \lim_{t\to+0} F^{-1}(w - \mathcal{A})^{-1}g$$
$$= \mathcal{P}_+ \lim_{t\to+0} \frac{1}{2\pi i}\int_{\Gamma_-} e^{iwt}(w - \mathcal{A})^{-1}g\,dw. \qquad (5.35)$$

For $t > 0$ the exponent in e^{iwt} has a negative real part in the upper complex w-half-plane. Therefore, we may deform the contour of integration to a curve surrounding $\text{spec}_+(\mathcal{A}) = \{\lambda \in \text{spec}(\mathcal{A}) : \text{Im}\,\lambda > 0\}$ which, as noted in Remark 5.2.3, is a countable set which intersects every strip $\{w \in \mathbb{C} : c < \text{Im}\,w < c'\}$ in a finite set for arbitrary $c < c'$. It is now elementary to show that the limit on the right-hand side of (5.35) just equals $\mathcal{P}_+g = g$. By assumption we have $\mathcal{P}_+g = g$. Thus we verified the second relation of (5.34). Concerning the first relation of (5.34), we have

$$\mathcal{P}_+ \lim_{t \to +0} F^{-1}(w - \mathcal{A})^{-1} \left\{ \widetilde{f}(w) - \frac{1}{2\pi} \int_{\Gamma_-} (z - \mathcal{A})^{-1} \widetilde{f}(z) dz \right\}$$

$$= \left(F^{-1} \mathcal{P}_+ (w - \mathcal{A})^{-1} \widetilde{f}(w) \right) \big|_{t=0} - \frac{1}{2\pi} \int_{\Gamma_-} \mathcal{P}_+ (z - \mathcal{A})^{-1} \widetilde{f}(z) dz = 0. \quad \square$$

Remark 5.2.9. (i) If we first consider $\widetilde{\mathcal{S}} := F^{-1}(w - \mathcal{A})^{-1} \widetilde{f}(w)$ instead of the operator $\mathcal{S}$, then it suffices to verify the relations

$$\mathcal{D} \widetilde{\mathcal{S}} f = f, \quad \mathcal{D} \mathcal{K} g = 0, \quad \mathcal{P}_+ \mathrm{r}' \mathcal{K} g = g \tag{5.36}$$

for all f, g. Then

$$\begin{pmatrix} \mathcal{D} \\ \mathcal{P}_+ \mathrm{r}' \end{pmatrix} \begin{pmatrix} \widetilde{\mathcal{S}} & \mathcal{K} \end{pmatrix} = \begin{pmatrix} 1 & 0 \\ \mathcal{P}_+ \mathrm{r}' \widetilde{\mathcal{S}} & 1 \end{pmatrix}$$

yields $\mathcal{S} = \widetilde{\mathcal{S}} - \mathcal{K} \mathcal{P}_+ \mathrm{r}' \widetilde{\mathcal{S}}$. This provides an explanation of the second summand in (5.31).

(ii) The expression

$$\mathcal{S} \mathcal{D} f + \mathcal{K} \mathcal{P}_+ \mathrm{r}' f \tag{5.37}$$

runs through the image of all f under the Cauchy data map $\mathcal{P}_+ \mathrm{r}'$ for arbitrary f, and also the full space of images $\mathcal{D} f$ under vanishing Cauchy data, which is a consequence of (5.33), (5.34).

Now let us consider the elliptic operator (5.10) on a smooth compact manifold X with boundary Y. As before, we fix a collar neighbourhood $Y \times [0,1) \ni (y, t)$ of Y. Analogously to the cylindrical situation, we have the Cauchy data space $\mathcal{C}^{s,\mu}(Y, E')$ for any integer $s \geq \mu$, $E' = E|_Y$, a continuous operator

$$\mathbb{T}^\mu : H^s(\mathrm{int}\, X, E) \to \mathcal{C}^{s,\mu}(Y, E'), \tag{5.38}$$

and a pseudo-differential projection $\mathcal{P}_+ : \mathcal{C}^{s,\mu}(Y, E') \to \mathcal{C}^{s,\mu}(Y, E')$. Concerning the non-bijectivity points of the induced operator family (5.12), we again assume that there are no such points on $\mathrm{Re}\, w = 0$.

Theorem 5.2.10. *The spectral boundary value problem*

$$Au = f \in H^{s-\mu}(X, F), \tag{5.39}$$

$$\mathcal{P}_+ \mathbb{T}^\mu u = g \in \mathcal{P}_+ \mathcal{C}^{s,\mu}(Y, E') \tag{5.40}$$

defines a Fredholm operator

$$\begin{pmatrix} A \\ \mathcal{P}_+ \mathbb{T}^\mu \end{pmatrix} : H^s(X, E) \longrightarrow \begin{matrix} H^{s-\mu}(X, F) \\ \oplus \\ \mathcal{P}_+ \mathcal{C}^{s,\mu}(Y, E') \end{matrix} \tag{5.41}$$

for every $s \geq \mu$.

Proof. The operator (5.41) is of analogous structure as $\mathcal{A}$ in Theorem 4.2.8. The only modification here concerns the orders of the operators in the boundary conditions. In fact, we first form the operator $\mathcal{A}$, cf. (5.16), and set $\mathcal{A}_0 := \mathcal{R}_0^{-1}\mathcal{A}\mathcal{R}_0$ for a diagonal matrix $\mathcal{R}_0 = \mathrm{diag}\left(\mathcal{R}_{E'}^{-\mu+k+1/2}\right)_{k=0,\dots,\mu-1}$ of order-reducing operators

$$\mathcal{R}_{E'}^{-\mu+k+1/2} : H^{s-\mu}(Y,E') \to H^{s-(k+1/2)}(Y,E').$$

Since

$$C^{s,\mu}(Y,E') = \bigoplus_{k=0}^{\mu-1} H^{s-k-1/2}(Y,E'),$$

the operator $\mathcal{R}_0$ induces an isomorphism

$$\mathcal{R}_0 : \bigoplus_{k=0}^{\mu-1} H^{s-\mu}(Y,E') \to C^{s,\mu}(Y,E').$$

Similarly to (5.24), (5.25) and (5.26), we have

$$\mathcal{P}_\pm = \frac{1}{2\pi i}\mathcal{A}\int_{\Gamma_\pm}(w-\mathcal{A})^{-1}\frac{dw}{w} = \mathcal{R}_0\frac{1}{2\pi i}\int_{\Gamma_\pm}\mathcal{A}_0(w-\mathcal{A}_0)^{-1}\frac{dw}{w}\mathcal{R}_0^{-1}$$

and

$$P_\pm := \frac{1}{2\pi i}\int_{\Gamma_\pm}\mathcal{A}_0(w-\mathcal{A}_0)^{-1}\frac{dw}{w}$$

are pseudo-differential projections

$$P_\pm : \bigoplus_{k=0}^{\mu-1} H^{s-\mu}(Y,E') \to \bigoplus_{k=0}^{\mu-1} H^{s-\mu}(Y,E')$$

of the class $L^0_{\mathrm{cl}}\left(Y;\bigoplus_{0\le k\le\mu-1}E',\bigoplus_{0\le k\le\mu-1}E'\right)$. Now the boundary value problem (5.39), (5.40) representing the operator (5.41) can be reduced to an equivalent problem

$$\begin{pmatrix}A\\T\end{pmatrix} : H^s(\mathrm{int}\,X,E) \longrightarrow \begin{matrix}H^{s-\mu}(\mathrm{int}\,X,F)\\\oplus\\P_+H^{s-\mu}(Y,\bigoplus_{0\le k\le\mu-1}E')\end{matrix} \tag{5.42}$$

for the trace operator $T := \mathcal{R}_0^{-1}\mathcal{P}_+\mathbb{T}^\mu$. In fact, since $\mathcal{P}_+ = \mathcal{R}_0 P_+\mathcal{R}_0^{-1}$, the operator

$$\mathcal{P}_+\mathbb{T}^\mu : H^s(\mathrm{int}\,X,E) \to \mathcal{P}_+C^{s,\mu}(Y,E')$$

is equivalent to

$$T = \mathcal{R}_0^{-1}\mathcal{P}_+\mathbb{T}^\mu : H^s(\mathrm{int}\,X,E) \to (\mathcal{R}_0^{-1}\mathcal{P}_+\mathcal{R}_0)\mathcal{R}_0^{-1}C^{s,\mu}(Y,E')$$

$$= P_+\bigoplus_{k=0}^{\mu-1} H^{s-\mu}(Y,E').$$

Now in the modified notation

$$P_2 := P_+, \quad \mathbb{L}_2 := (P_2, J_2, L_2) \in \mathbb{P}(Y)$$

for $J_2 := \bigoplus_{k=0}^{\mu-1} E'$, $L_2 := \operatorname{im} \sigma_\psi(P_2)$ our operator (5.42) takes the form

$$\mathcal{A} := \begin{pmatrix} A \\ T \end{pmatrix} : H^s(\operatorname{int} X, E) \longrightarrow \begin{array}{c} H^{s-\mu}(\operatorname{int} X, F) \\ \oplus \\ H^{s-\mu}(Y, \mathbb{L}_2) \end{array}. \tag{5.43}$$

$\square$

The operator $\sigma_\partial(\mathcal{P}_+ \mathbb{T}^\mu)(y, \eta) : \ker \sigma_\partial(A)(y, \eta) \to L_+(y, \eta)$ is an isomorphism for every $(y, \eta) \in T^*Y \setminus 0$, cf. also Remark 5.2.6. Since the order reductions over Y are isomorphisms on the level of (y, η)-dependent principal symbols, we also have an isomorphism $\sigma_\partial(T)(y, \eta) : \ker \sigma_\partial(A)(y, \eta) \to L_+(y, \eta)$. By notation the bundles L_2 and L_+ coincide. Thus the operator (5.43) is elliptic in the sense of Definition 4.2.1. Hence, by virtue of Theorem 4.2.8 (i), it is Fredholm. Then the same is true of (5.41).

5.3   Projections of Calderón–Seeley type

Let us now discuss an alternative way of constructing pseudo-differential projections to Cauchy data spaces of solutions to elliptic differential equations contributed in communication with Tarkhanov. Consider again a differential operator

$$A : C^\infty(X, E) \to C^\infty(X, F)$$

of order μ on a smooth compact manifold X with boundary Y, and $E, F \in \operatorname{Vect}(X)$. Similarly as in Section 5.1, we assume that A is the restriction to $\Omega := \operatorname{int} X$ of a differential operator

$$D : C^\infty(M, V) \to C^\infty(M, W),$$

for a smooth Riemannian manifold M containing X in its interior, and $V, W \in \operatorname{Vect}(M)$. This time it is not essential that M is the double of X; for convenience we set $M = X \cup U$, where U is a tubular neighbourhood of Y in $2X$, and $A = D|_\Omega$, $E = V|_\Omega$, $F = W|_\Omega$. The characteristic function χ_Ω will be interpreted as an operator of multiplication

$$\chi_\Omega : H^s_{\operatorname{loc}}(M, V) \to \mathcal{D}'(M, V)$$

for $s \geq \mu$ and any $V \in \operatorname{Vect}(M)$. Observe that the distributional kernel of the commutator

$$[D, \chi_\Omega] = D\chi_\Omega - \chi_\Omega D : C_0^\infty(M, V) \to \mathcal{D}'(M, W)$$

is supported by $Y \times Y$. Moreover, $f \in H^s_{\mathrm{loc}}(M, V)$, $s \geq \mu$, and $\mathbb{T}^\mu f = 0$ entails $[D, \chi_\Omega] f = 0$. In the rest of this section we assume that there is a $Q \in L^{-\mu}_{\mathrm{cl}}(M; W, V)$ such that

$$DQf = f, \quad QDu = u, \tag{5.44}$$

for all distributional sections f and u in the respective bundles that are supported in an ε-neighbourhood of X in M for some $\varepsilon > 0$.

Now with e^+ and r^+ of analogous meaning as the corresponding operators in Section 5.1, we have $A = \mathrm{r}^+ D \mathrm{e}^+$, and we set $P := \mathrm{r}^+ Q \mathrm{e}^+$.

Proposition 5.3.1. *Under the above-mentioned assumptions on the elliptic differential operator A, we have*

$$G := 1 - PA \in B^{0,\mu}_G(X; E, F)$$

(as an operator on $H^s(\mathrm{int}\, X, E)$, $s \geq \mu$), and

$$G^2 = G.$$

Proof. First we have $G \in B^{0,\mu}(X; E, F)$, more precisely, $G \in B^{0,\mu}_G(X; E, F)$. Now the first of the identities (5.44) and the fact that $\tilde{A}$ is a differential operator gives $AP = 1$. Thus $P : H^{s-\mu}(\mathrm{int}\, X, F) \to H^s(\mathrm{int}\, X, E)$ is a right inverse of A, and hence PA is a projection. Then G is just the complementary projection. □

Remark 5.3.2. Let $s \geq \mu$, and set

$$\ker_s A = \{u \in H^s(\mathrm{int}\, X, E) : Au = 0\}.$$

Then

$$G : H^s(\mathrm{int}\, X, E) \to \ker_s A$$

is a projection to $\ker_s A$.

In fact, $u \in \ker_s A$ yields $Gu = u$, Moreover, for arbitrary $f \in H^s(\mathrm{int}\, X, E)$ we have $AGf = A(1 - PA)f = Af - (AP)Af = 0$, i.e., $Gf \in \ker_s A$.

Proposition 5.3.3. *The operator $G = 1 - PA$ of Proposition 5.3.1 can be written as*

$$G = \sum_{j=0}^{\mu-1} K_j \circ T_j \tag{5.45}$$

for unique potential operators $K_j \in B^{0,-j-1/2}(X; \boldsymbol{v})$ for $\boldsymbol{v} := (0, F; E', 0)$ and $T_j u = \partial_t^j u|_{t=0}$. Thus $f \in H^s(\mathrm{int}\, X, E)$ and $\mathbb{T}^\mu f = 0$ entails $Gf = 0$.

Proof. Because of $G = 1 - PA \in B^{0,\mu}_G(X; E, F)$ and Proposition 2.4.17, we have a representation

$$G = G_0 + \sum_{j=0}^{\mu-1} K_j \circ T_j,$$

with a $G_0 \in B_G^{0,0}(X; E, F)$ and K_j, T_j as indicated. However, $PA\varphi = \varphi$ for every $\varphi \in C_0^\infty(\Omega, E)$, which is a consequence of the second relation of (5.44) yields $G\varphi = (1 - PA)\varphi = 0$ for all $\varphi \in C_0^\infty(\Omega, E)$. Since $G\varphi = G_0\varphi$ for all those φ it follows that $G_0\varphi = 0$ for all $\varphi \in C_0^\infty(\Omega)$. This yields the representation (5.45), and we finally obtain $Gf = 0$ when $\mathbb{T}^\mu f = 0$. $\qquad\square$

Theorem 5.3.4. *Let A be an elliptic differential operator satisfying the above-mentioned assumptions. Let $\mathbb{T}^{-\mu}$ denote a right inverse of the map*

$$\mathbb{T}^\mu : H^s(\mathrm{int}\, X, E) \to \mathcal{C}^{s,\mu}(Y, E'),$$

cf. (5.38). Then

$$\Pi := \mathbb{T}^\mu G \mathbb{T}^{-\mu} : \mathcal{C}^{s,\mu}(Y, E') \to \mathcal{C}^{s,\mu}(Y, E')$$

is a projection to the Cauchy data space of solutions to $Au = 0$, $u \in H^s(\mathrm{int}\, X, E)$, called the Calderón–Seeley projection.

Proof. Let $g = \mathbb{T}^\mu u$ for some $u \in \ker_s A$. Then it follows that $\mathbb{T}^\mu(u - \mathbb{T}^{-\mu}g) = 0$. Thus Proposition 5.3.3 gives $G(u - \mathbb{T}^{-\mu}g) = 0$. Using also Remark 5.3.2, we obtain $G\mathbb{T}^{-\mu}g = Gu = u$, and hence $\Pi g = \mathbb{T}^\mu u = g$. Moreover, for arbitrary $h \in \mathcal{C}^{s,\mu}(Y, E')$ we have $\mathbb{T}^{-\mu}h \in H^s(Y, E)$ and $G\mathbb{T}^{-\mu}h \in \ker_s A$ by Remark 5.3.2. This yields $\Pi h \in \mathbb{T}^\mu \ker_s A$. $\qquad\square$

Remark 5.3.5. (i) The operator Π is independent of the choice of $\mathbb{T}^{-\mu}$. In fact, if $\mathbb{T}_1^{-\mu}$ is another right inverse of $\mathbb{T}^\mu$, then

$$\mathbb{T}^\mu G \mathbb{T}^{-\mu} - \mathbb{T}^\mu G \mathbb{T}_1^{-\mu} = \mathbb{T}^\mu G(\mathbb{T}^{-\mu} - \mathbb{T}_1^{-\mu}) = 0$$

since $G(\mathbb{T}^{-\mu} - \mathbb{T}_1^{-\mu})$ vanishes because of $\mathbb{T}^\mu(\mathbb{T}^{-\mu} - \mathbb{T}_1^{-\mu}) = 0$ and the second assertion of Proposition 5.3.3.

(ii) The operator Π is independent of the specific choice of Q, because for another Q_1 and the corresponding projection $G_1 = 1 - P_1 A$, the function $G_1 \mathbb{T}^{-\mu}g \in \ker_s A$ has the Cauchy data g. Since the same is true of $G\mathbb{T}^{-\mu}g$, it follows that $\mathbb{T}^\mu(G - G_1)\mathbb{T}^{-\mu} = 0$.

Remark 5.3.6. Assume that the operator A has the unique continuation property of solutions. Then

$$\mathbb{T}^\mu : \ker_s A \to \mathcal{C}^{s,\mu}(Y, E')$$

is surjective, i.e., there is a unique solution $u \in \ker_s A$ for every $g \in \Pi\mathcal{C}^{s,\mu}(Y, E')$ such that $\mathbb{T}^\mu u = g$.

Part II

Edge Operators with Global Projection Conditions

Chapter 6

The cone algebra

6.1 Mellin transform and weighted spaces on the half-line

A manifold with smooth boundary can be regarded as a manifold with edge, and boundary value problems can be interpreted as specific edge problems. In the edge case the inner normal turns into the axial variable of a cone transverse to the edge, and the boundary symbol calculus is now replaced by a calculus on the respective cone. In order to treat edge problems with global projection conditions, we first outline the calculus of operators on a manifold with conical singularities.

Let N be a smooth closed manifold, and form the cone

$$N^\Delta := (\overline{\mathbb{R}}_+ \times N)/(\{0\} \times N)$$

where $\{0\} \times N$ in the quotient space represents the vertex v_0. Moreover, set $N^\wedge := \mathbb{R}_+ \times N$. A manifold M with conical singularity v is a topological space such that $M \setminus \{v\}$ is a smooth manifold, and v has a neighbourhood V in M such that there is a homeomorphism

$$\chi^\Delta : V \to N^\Delta \qquad (6.1)$$

with $\chi^\Delta(v) = v_0$, and (6.1) restricts to a diffeomorphism

$$\chi^\wedge : V \setminus \{v\} \to N^\wedge. \qquad (6.2)$$

The choice of (6.2) determines a splitting of (local) variables (r, x) on $V \setminus \{v\}$ near v. For practical reasons we keep it fixed. Another choice of pair

$$\widetilde{\chi}^\Delta : V \to N^\Delta, \quad \widetilde{\chi}^\wedge : V \setminus \{v\} \to N^\wedge$$

is said to define an equivalent conical structure of M near v if

$$\widetilde{\chi}^\wedge \circ (\chi^\wedge)^{-1} : \mathbb{R}_+ \times N \to \mathbb{R}_+ \times N$$

© Springer Nature Switzerland AG 2018
X. Liu, B.-W. Schulze, *Boundary Value Problems with Global Projection Conditions*,
Operator Theory: Advances and Applications 265, https://doi.org/10.1007/978-3-319-70114-1_6

extends to a diffeomorphism $\overline{\mathbb{R}}_+ \times N \to \overline{\mathbb{R}}_+ \times N$ in the sense of the respective manifolds with smooth boundary. Such a notion of equivalence allows us to distinguish the chosen conical structure from various cuspidal configurations relative to the given conical one.

For simplicity we consider here the case of a single conical singularity. This automatically covers the case of finitely many conical singularities, since N may have finitely many connected components.

From M we can pass to the so-called stretched manifold $\mathbb{M}$, which is a smooth manifold with boundary $\partial\mathbb{M} \cong N$, obtained by invariantly attaching a copy of N at $M \setminus \{v\}$. Then $\partial\mathbb{M}$ has a collar neighbourhood $\mathbb{V}$ in $\mathbb{M}$ such that there is a diffeomorphism

$$\chi^{\|} : \mathbb{V} \to \overline{\mathbb{R}}_+ \times N \qquad (6.3)$$

between the corresponding manifolds with smooth boundary.

A simple example is $M := N^{\Delta}$. Then $\mathbb{M} = \overline{\mathbb{R}}_+ \times N$.

Note that the quotient map $\overline{\mathbb{R}}_+ \times N \to N^{\Delta}$ allows us to define a continuous map

$$\pi : \mathbb{M} \to M$$

such that $\pi|_{\mathbb{M}\setminus\partial\mathbb{M}} : \mathbb{M} \setminus \partial\mathbb{M} \to M \setminus \{v\}$ is a diffeomorphism and $\pi|_{\partial\mathbb{M}} : \partial\mathbb{M} \to v$ collapses $\partial\mathbb{M}$ to the point v.

It is often convenient to interpret M as a stratified space, here as a disjoint union

$$M = s_0(M) \cup s_1(M)$$

with the strata $s_0(M) := M \setminus \{v\}$, $s_1(M) := \{v\}$.

Let $\mathrm{Diff}^{\mu}(\Omega)$ for an open manifold Ω denote the space of all differential operators of order μ with smooth coefficients in local coordinates.

If M is a manifold with conical singularity v, an operator $A \in \mathrm{Diff}^{\mu}(M \setminus \{v\})$ is said to be of Fuchs type if locally close to v in the above-mentioned splitting of variables $(r, x) \in N^{\wedge}$ it has the form

$$A = r^{-\mu} \sum_{j=0}^{\mu} a_j(r)(-r\partial_r)^j, \qquad (6.4)$$

with coefficients $a_j \in C^{\infty}\big(\overline{\mathbb{R}}_+, \mathrm{Diff}^{\mu-j}(N)\big)$. The general program of the cone calculus is to construct a subalgebra of $\bigcup_{\mu \in \mathbb{R}} L^{\mu}_{\mathrm{cl}}(M \setminus \{v\})$ that contains all Fuchs type differential operators together with the parametrices of elliptic elements. Similarly as in BVPs, the ellipticity in this case is not only determined by the principal interior symbol σ_{ψ} over $s_0(M)$, but by a pair $(\sigma_{\psi}, \sigma_M)$, where σ_M, the so-called principal conormal symbol, is contributed by the singular stratum $s_1(M)$. In the case of a Fuchs type differential operator we have $\sigma_M(A)$ as an operator function

$$\sigma_M(A) := \sum_{j=0}^{\mu} a_j(0)z^j : H^s(N) \to H^{s-\mu}(N) \qquad (6.5)$$

depending on the complex variable z, dual to $r \in \mathbb{R}_+$ coming from the Mellin transform. For future references we now recall some material on the Mellin transform and associated weighted Sobolev spaces. We consider the Mellin transform

$$Mu(z) := \int_0^\infty r^z u(r) \frac{dr}{r},$$

first for $u \in C_0^\infty(\mathbb{R}_+)$. Set

$$\Gamma_\beta := \{z \in \mathbb{C} : \operatorname{Re} z = \beta\},$$

$\beta \in \mathbb{R}$, for any $\beta \in \mathbb{R}$. It can be easily verified that M induces a continuous operator $M : C_0^\infty(\mathbb{R}_+) \to \mathcal{A}(\mathbb{C})$, and we have $Mu|_{\Gamma_\beta} \in \mathcal{S}(\Gamma_\beta)$ for every β, uniformly in compact β-intervals. That means the Schwartz space referring to $\operatorname{Im} z$ for $z \in \Gamma_\beta$ is interpreted as

$$\mathcal{S}(\Gamma_\beta) := \psi_\beta^* \mathcal{S}(\mathbb{R})$$

for

$$\psi_\beta : \Gamma_\beta \to \mathbb{R}, \quad z \mapsto \operatorname{Im} z. \tag{6.6}$$

We distinguish the spaces $\mathcal{S}(\Gamma_\beta)$ for different β, and here "uniform" means that $(\psi_\beta^*)^{-1} Mu|_{\Gamma_\beta}$ is a bounded set in $\mathcal{S}(\mathbb{R})$ when β varies over a compact interval. More generally, if $E(\mathbb{R})$ is a function or distribution space on $\mathbb{R}$ with a given, say, locally convex topology, then we set

$$E(\Gamma_\beta) := \psi_\beta^*(E(\mathbb{R})),$$

and the uniformity terminology is used with a corresponding meaning.

Observe that for

$$(\iota_\lambda u)(r) := u(\lambda r), \quad \lambda \in \mathbb{R}_+, \tag{6.7}$$

we have

$$(M\iota_\lambda u)(z) = \lambda^{-z} Mu(z). \tag{6.8}$$

Other simple properties are the relations

$$M((-r\partial_r)u(r))(z) = z Mu(z), \tag{6.9}$$

and

$$M(r^\beta u(r))(z) = Mu(z + \beta), \quad \beta \in \mathbb{C}, \tag{6.10}$$

first on $C_0^\infty(\mathbb{R}_+)$, and then extended to weighted distribution spaces.

Remark 6.1.1. For every $n \in \mathbb{N}$ we have the relations

$$r^n \partial_r^n = \sum_{k=0}^n (-1)^k s_{n,k} (-r\partial_r)^k, \quad (-r\partial_r)^n = (-1)^n \sum_{k=0}^n S_{n,k} r^k \partial_r^k \tag{6.11}$$

where $s_{n,k}$ and $S_{n,k}$ are the Stirling numbers of first and second kind, respectively. They are determined by the recursive formulas

$$s_{n+1,k} = s_{n,k-1} - n s_{n,k}, \quad S_{n+1,k} = S_{n,k-1} + k S_{n,k}, \quad n \in \mathbb{N}, \; k \in \mathbb{N} \setminus \{0\}.$$

A proof may be found in [13].

The operator

$$M_\gamma : C_0^\infty(\mathbb{R}_+) \mapsto Mu|_{\Gamma_{1/2-\gamma}}, \quad M_\gamma : C_0^\infty(\mathbb{R}_+) \to \mathcal{S}(\Gamma_{1/2-\gamma}),$$

is called the weighted Mellin transform and $\gamma \in \mathbb{R}$ the corresponding weight. The transformation

$$S_\gamma u(\boldsymbol{r}) := e^{-(1/2-\gamma)\boldsymbol{r}} u(e^{-\boldsymbol{r}}), \quad \boldsymbol{r} \in \mathbb{R}, \tag{6.12}$$

induces an isomorphism

$$S_\gamma : C_0^\infty(\mathbb{R}_+) \to C_0^\infty(\mathbb{R}) \tag{6.13}$$

with the inverse

$$S_\gamma^{-1} v(r) = r^{\gamma-1/2} v(-\log r).$$

Lemma 6.1.2. *We have*

$$(M_\gamma u)\left(\frac{1}{2} - \gamma + i\rho\right) = (F S_\gamma u)(\rho), \tag{6.14}$$

for the Fourier transform $(Fv)(\rho) = \int_{-\infty}^\infty e^{-i\rho \boldsymbol{r}} v(\boldsymbol{r}) d\boldsymbol{r}$ *on the real axis.*

Proof. The assertion follows from

$$\begin{aligned}
(Mu)\left(\frac{1}{2} - \gamma + i\rho\right) &= \int_0^\infty r^{\frac{1}{2}-\gamma+i\rho} u(r)\frac{dr}{r} \\
&= \int_0^\infty e^{(\frac{1}{2}-\gamma+i\rho)\log r} u(r)\frac{dr}{r} \\
&= \int_{-\infty}^\infty e^{-i\rho \boldsymbol{r}} e^{-(\frac{1}{2}-\gamma)\boldsymbol{r}} u(e^{-\boldsymbol{r}}) d\boldsymbol{r}. \quad\square
\end{aligned}$$

Thus

$$M_{\gamma,r\to z} = \psi_{1/2-\gamma}^* \circ F_{\boldsymbol{r}\to\rho} \circ S_\gamma. \tag{6.15}$$

From the Fourier inversion formula

$$(F_{\rho\to r}^{-1} g)(\boldsymbol{r}) = \int e^{i\boldsymbol{r}\rho} g(\rho) \bar{d}\rho, \quad \bar{d}\rho = (2\pi)^{-1} d\rho,$$

together with (6.15), i.e., $M_{\gamma,z\to r}^{-1} = S_\gamma^{-1} \circ F_{\rho\to r}^{-1} \circ (\psi_{1/2-\gamma}^{-1})^*$, it follows that

$$\left(M_{\gamma,z\to r}^{-1} f\right)(r) = \int_{\Gamma_{1/2-\gamma}} r^{-z} f(z) \bar{d}z. \tag{6.16}$$

for $\bar{d}z := (2\pi i)^{-1} dz$. Because of the isomorphisms

$$(\psi_{1/2-\gamma}^{-1})^* : \mathcal{S}(\Gamma_{1/2-\gamma}) \to \mathcal{S}(\mathbb{R}_\rho), \quad F^{-1} : \mathcal{S}(\mathbb{R}_\rho) \to \mathcal{S}(\mathbb{R}_{\boldsymbol{r}})$$

it makes sense to interpret

$$T^{\gamma}(\mathbb{R}_+) := S_{\gamma}^{-1} \mathcal{S}(\mathbb{R}_r) \tag{6.17}$$

as the weighted Mellin analogue of the Schwartz space on $\mathbb{R}_+$. Then

$$M_{\gamma, r \to z} : T^{\gamma}(\mathbb{R}_+) \to \mathcal{S}(\Gamma_{1/2-\gamma}) \tag{6.18}$$

is an isomorphism. Observe that the operator of multiplication by $r^{\beta}, \beta \in \mathbb{R}$, induces an isomorphism

$$r^{\beta} : T^{\gamma}(\mathbb{R}_+) \to T^{\gamma+\beta}(\mathbb{R}_+). \tag{6.19}$$

Moreover, we have isomorphisms

$$(\psi_{1/2}^{-1})^* : L^2(\Gamma_{1/2}) \to L^2(\mathbb{R}_\rho), \quad F^{-1} : L^2(\mathbb{R}_\rho) \to L^2(\mathbb{R}_r), \quad S_0^{-1} : L^2(\mathbb{R}_r) \to L^2(\mathbb{R}_+).$$

Thus $M = M_0$ gives rise to an isomorphism

$$M : L^2(\mathbb{R}_+) \to L^2(\Gamma_{1/2}). \tag{6.20}$$

More generally, M_γ induces an isomorphism

$$M_\gamma : L^{2,\gamma}(\mathbb{R}_+) := r^\gamma L^2(\mathbb{R}_+) \to L^2(\Gamma_{1/2-\gamma})$$

for every $\gamma \in \mathbb{R}$. The inverse has again the form (6.16).

Relation (6.10) allows us to shift the involved weights. In particular, it is often convenient to consider $\gamma = 1/2$. In this case

$$(S_{1/2}u)(r) := u(e^{-r}) = ((\chi^{-1})^* u)(r)$$

is the function pull-back under χ^{-1}, for the diffeomorphism $\chi : \mathbb{R}_+ \to \mathbb{R}, \chi : r \mapsto r := -\log r$. Let us set

$$C_b^\infty(\mathbb{R}) := \left\{ v(r) \in C^\infty(\mathbb{R}) : \sup_{r \in \mathbb{R}} |\partial_r^k v(r)| < \infty \text{ for all } k \in \mathbb{N} \right\},$$

$$C_B^\infty(\mathbb{R}_+) := \left\{ u(r) \in C^\infty(\mathbb{R}_+) : \sup_{r \in \mathbb{R}_+} |(r\partial_r)^j u(r)| < \infty \text{ for all } j \in \mathbb{N} \right\}.$$

Remark 6.1.3. The transformation

$$S_{1/2} : C_0^\infty(\mathbb{R}_+) \xrightarrow{\cong} C_0^\infty(\mathbb{R}) \tag{6.21}$$

extends to an isomorphism

$$S_{1/2} : C_B^\infty(\mathbb{R}_+) \to C_b^\infty(\mathbb{R}).$$

In fact, for every $k \in \mathbb{N}$ we have

$$S_{1/2}(-r\partial_r)^k S_{1/2}^{-1} = \partial_r^k,$$

or, equivalently, $S_{1/2}(-r\partial_r)^k u(r) = \partial_r^k v(r)$ for $v(r) = u(e^{-r})$. This yields

$$\sup_{r \in \mathbb{R}_+} \left| (r\partial_r)^k u(r) \right| = \sup_{r \in \mathbb{R}} \left| \partial_r^k v(r) \right|$$

for $v(r) = (S_{1/2} u)(r)$.

Let us endow the weighted space $L^{2,1/2}(\mathbb{R}_+) = r^{1/2} L^2(\mathbb{R}_+) =: L^2(\mathbb{R}_+, dr/r)$ with the scalar product $(f,g)_{L^{2,1/2}(\mathbb{R}_+)} = \int_0^\infty f(r)\overline{g}(r) dr/r$. Then we have

$$\left(S_{1/2} f, S_{1/2} g \right)_{L^2(\mathbb{R}_r)} = (f,g)_{L^{2,1/2}(\mathbb{R}_+)},$$

for all $f, g \in L^{2,1/2}(\mathbb{R}_+)$. Analogously to the case of the Fourier transform, it follows that

$$(f,g)_{L^{2,1/2}(\mathbb{R}_+)} = \left(S_{1/2} f, S_{1/2} g \right)_{L^2(\mathbb{R})}$$
$$= \int_{\mathbb{R}} (FS_{1/2}f)(\rho)(\overline{FS_{1/2}g})(\rho)đ\rho = \int_{\Gamma_0} (M_{1/2}f)(i\rho)(\overline{M_{1/2}g})(i\rho)đz$$

i.e.,

$$\int_0^\infty f(r)\overline{g}(r)\frac{dr}{r} = \int_{\Gamma_0} (M_{1/2}f)(z)(\overline{M_{1/2}g})(z)đz, \qquad (6.22)$$

the Mellin version of Parseval's formula, and

$$\|f\|_{L^2(\mathbb{R}_+, dr/r)} = (2\pi)^{-1/2}\|M_{1/2}f\|_{L^2(\Gamma_0)},$$

the Mellin version of Plancherel's formula. In terms of the Mellin transform, the convolution of functions $f, g \in T^{1/2}(\mathbb{R}_+)$ becomes

$$\int_0^\infty f(r)g\left(\frac{t}{r}\right)\frac{dr}{r} =: (f *_M g)(t).$$

Then we have

$$M_{1/2}(f *_M g)(z) = (M_{1/2}f)(z)(M_{1/2}g)(z). \qquad (6.23)$$

In fact, after some simple substitutions and changing order of integration we obtain

$$\int_0^\infty t^z \left\{ \int f(r)g\left(\frac{t}{r}\right)\frac{dr}{r} \right\}\frac{dt}{t} = \int_0^\infty f(r)\left\{ \int_0^\infty (rs)^z g(s)\frac{ds}{s} \right\}\frac{dr}{r}$$
$$= \int_0^\infty r^z f(r)\frac{dr}{r}\int_0^\infty s^z g(s)\frac{ds}{s}.$$

Definition 6.1.4. We denote by $\mathcal{H}^{s,\gamma}(\mathbb{R}_+)$ the completion of $C_0^\infty(\mathbb{R}_+)$ with respect to the norm

$$\|u\|_{\mathcal{H}^{s,\gamma}(\mathbb{R}_+)} := \left\{ \int_{\Gamma_{1/2-\gamma}} \left(1+|z|^2\right)^s |Mu(z)|^2 \, dz \right\}^{1/2}, \quad dz = (2\pi i)^{-1} dz. \quad (6.24)$$

Proposition 6.1.5.   (i) *The transformation* (6.13) *extends to an isomorphism*

$$S_\gamma : \mathcal{H}^{s,\gamma}(\mathbb{R}_+) \to H^s(\mathbb{R})$$

for every $s, \gamma \in \mathbb{R}$. In particular, for $s = 0$ we have an isomorphism $S_\gamma :$ $r^\gamma L^2(\mathbb{R}_+) \to L^2(\mathbb{R})$.

(ii) *The weighted Mellin transform* (6.18) *extends to an isomorphism*

$$M_{\gamma, r\to z} : \mathcal{H}^{s,\gamma}(\mathbb{R}_+) \to \widehat{H}^s(\Gamma_{1/2-\gamma}) \quad (6.25)$$

for every $s, \gamma \in \mathbb{R}$. Here

$$\widehat{H}^s(\Gamma_{1/2-\gamma}) := (\psi_{1/2-\gamma}^{-1})^* \left(F_{t\to\rho} H^s(\mathbb{R}_t) \right).$$

(iii) *The operator of multiplication by r^β induces an isomorphism*

$$r^\beta : \mathcal{H}^{s,\gamma}(\mathbb{R}_+) \to \mathcal{H}^{s,\gamma+\beta}(\mathbb{R}_+) \quad (6.26)$$

for every $s, \gamma, \beta \in \mathbb{R}$.

Proof. For (i) we use (6.14) and write $\left(M_{\gamma, r\to z} u\right)(1/2 - \gamma + i\rho) = \left(F_{r\to\rho} S_\gamma u\right)(\rho)$. Thus the norm

$$\|u\|_{\mathcal{H}^{s,\gamma}(\mathbb{R}_+)} = \left\{ \frac{1}{2\pi} \int \langle 1/2 - \gamma + i\rho \rangle^{2s} \left| \left(F_{r\to\rho} S_\gamma u\right)(\rho)\right|^2 d\rho \right\}^{1/2} \quad (6.27)$$

is equivalent to $\|S_\gamma u\|_{H^s(\mathbb{R}_r)}$.

(ii) follows immediately follows from expression (6.27) together with (6.15).

For (iii) we assume $u \in \mathcal{H}^{s,\gamma}(\mathbb{R}_+)$ and show that $r^\beta u \in \mathcal{H}^{s,\gamma'}(\mathbb{R}_+)$ for $\gamma' = \gamma + \beta$. In fact, the assumption means that the inverse Fourier transform $F_{\rho\to r}^{-1}$ of $F_{r\to\rho}(S_\gamma u(r)) = F_{r\to\rho}(e^{\gamma r} S_0 u(r))$ as a function of $r \in \mathbb{R}$ belongs to $H^s(\mathbb{R}_r)$. Then the inverse Fourier transform of $F_{r\to\rho}(S_{\gamma'}(r^\beta u)(r)) = F_{r\to\rho}(e^{\gamma' r} e^{-\beta r} S_0 u(r))$ belongs to $H^s(\mathbb{R}_t)$ if and only if $\gamma' - \beta = \gamma$, i.e., $\gamma' = \gamma + \beta$. $\qquad\square$

Since $H_{\text{loc}}^s(\mathbb{R}_+)$-spaces are invariant under diffeomorphisms $\mathbb{R}_+ \to \mathbb{R}_+$ we have $\mathcal{H}^{s,\gamma}(\mathbb{R}_+) \subset H_{\text{loc}}^s(\mathbb{R}_+)$. More precisely,

$$\varphi \mathcal{H}^{s,\gamma}(\mathbb{R}_+) = \varphi H^s(\mathbb{R}_+)$$

for every $\varphi \in C_0^\infty(\mathbb{R}_+)$.

By a cut-off function ω on the half-line we mean any $\omega \in C_0^\infty(\overline{\mathbb{R}}_+)$ such that $\omega \equiv 1$ close to zero. We introduce the Kegel spaces with weight $\gamma \in \mathbb{R}$ at zero by

$$\mathcal{K}^{s,\gamma}(\mathbb{R}_+) := \left\{ u = \omega u_0 + (1 - \omega)u_\infty : u_0 \in \mathcal{H}^{s,\gamma}(\mathbb{R}_+), u_\infty \in H^s(\mathbb{R}) \right\}. \qquad (6.28)$$

Those spaces are independent of the choice of ω. Moreover, we define the Kegel spaces with weight $e \in \mathbb{R}$ at infinity by

$$\mathcal{K}^{s,\gamma;e}(\mathbb{R}_+) := [r]^{-e} \mathcal{K}^{s,\gamma}(\mathbb{R}_+),$$

and set

$$\mathcal{S}^\gamma(\mathbb{R}_+) := \varprojlim_{l \in \mathbb{N}} \mathcal{K}^{l,\gamma;l}(\mathbb{R}_+). \qquad (6.29)$$

Remark 6.1.6. (i) The operators ∂_r^j and $(r\partial_r)^j$ induce continuous operators

$$\partial_r^j : \mathcal{K}^{s,\gamma;e}(\mathbb{R}_+) \to \mathcal{K}^{s-j,\gamma-j;e}(\mathbb{R}_+)$$

and

$$(r\partial_r)^j : \mathcal{K}^{s,\gamma;e}(\mathbb{R}_+) \to \mathcal{K}^{s-j,\gamma;e-j}(\mathbb{R}_+),$$

respectively, $s, \gamma, e \in \mathbb{R}$.

(ii) The operators $\iota_\lambda : u(r) \mapsto u(\lambda r)$, $\lambda \in \mathbb{R}_+$, induce continuous operators

$$\iota_\lambda : \mathcal{K}^{s,\gamma;e}(\mathbb{R}_+) \to \mathcal{K}^{s,\gamma;e}(\mathbb{R}_+),$$

and we have

$$\partial_r^j = \lambda^j \iota_\lambda \partial_r^j \iota_\lambda^{-1}, \quad (r\partial_r)^j = \iota_\lambda (r\partial_r)^j \iota_\lambda^{-1},$$

$\lambda \in \mathbb{R}_+$, $s, \gamma, e \in \mathbb{R}$.

6.2 Weighted spaces on a manifold with conical singularities

Let $\mathcal{H}^{s,\gamma}(\mathbb{R}_+ \times \mathbb{R}^n)$ be the completion of $C_0^\infty(\mathbb{R}_+ \times \mathbb{R}^n)$ with respect to the norm

$$\left\{ \int_{\Gamma_{(n+1)/2-\gamma}} \int_{\mathbb{R}^n} \langle z, \xi \rangle^{2s} \left| (M_{r \to z} F_{x \to \xi} u)(z, \xi) \right|^2 dz d\xi \right\}^{1/2}. \qquad (6.30)$$

Then $u(r, x) \in \mathcal{H}^{s,\gamma}(N^\wedge)$ is defined to be the space of all $u \in H_{\mathrm{loc}}^s(\mathbb{R}_+ \times N)$ such that for any chart $\chi : U \to \mathbb{R}^n$ on N and

$$1 \times \chi : \mathbb{R}_+ \times U \to \mathbb{R}_+ \times \mathbb{R}^n, \quad (1 \times \chi)(r, x) := (r, \chi(x)),$$

we have

$$(\varphi u) \circ (1 \times \chi)^{-1} \in \mathcal{H}^{s,\gamma}(\mathbb{R}_+ \times \mathbb{R}^n)$$

for every $\varphi \in C_0^\infty(U)$.

Let $R^s(\lambda) \in L_{cl}^s(N; \mathbb{R}_\lambda)$ be a parameter-dependent elliptic family of operators of order $s \in \mathbb{R}$ that induces isomorphisms

$$R^s(\lambda) : H^m(N) \to H^{m-s}(N)$$

for all $m \in \mathbb{R}$, $\lambda \in \mathbb{R}$. Then $\mathcal{H}^{s,\gamma}(N^\wedge)$ can be equivalently defined as the completion of $C_0^\infty(N^\wedge)$ with respect to the norm

$$\left\{ \int_{\Gamma_{(n+1)/2-\gamma}} \left\| R^s(\operatorname{Im} z)(M_{r \to z}u)(z) \right\|_{L^2(N)}^2 dz \right\}^{1/2} .$$

Note that another equivalent definition is as follows. The space $\mathcal{H}^{s,\gamma}(N^\wedge)$, first for $s \in \mathbb{N}$, $\gamma \in \mathbb{R}$, is the set of all $u(r,x) \in r^{\gamma - n/2} L^2(N^\wedge)$, $n = \dim N$, (with $L^2(N^\wedge)$ referring to the measure $drdx$), such that

$$(r\partial_r)^j D_x^\alpha u(r,x) \in r^{\gamma - n/2} L^2(N^\wedge)$$

for all $D_x^\alpha \in \operatorname{Diff}^{|\alpha|}(N)$, $j \in \mathbb{N}$, $j + |\alpha| \le s$. The space $L^2(N^\wedge)$ is equipped with the scalar product

$$(u,v)_{L^2(N^\wedge)} = \int_N \int_0^\infty u(r,x) \overline{v}(r,x) dr dx,$$

where dx refers to the fixed Riemannian metric on N. In particular,

$$\mathcal{H}^{0,0}(N^\wedge) = r^{-n/2} L^2(N^\wedge). \tag{6.31}$$

By duality with respect to the scalar product of $\mathcal{H}^{0,0}(N^\wedge) = r^{-n/2} L^2(N^\wedge)$ we define $\mathcal{H}^{s,\gamma}(N^\wedge)$ for $-s \in \mathbb{N}$, $\gamma \in \mathbb{R}$. Then complex interpolation gives us the spaces for all $s, \gamma \in \mathbb{R}$.

We also employ cylindrical Sobolev spaces $H^s(\mathbb{R}^q \times N)$ for any $q \in \mathbb{N}$, defined as the completion of $C_0^\infty(\mathbb{R}^q \times N)$ with respect to the norm

$$\|u\|_{H^s(\mathbb{R}^q \times N)} := \left\{ \sum_{j=1}^N \left\| (\varphi_j u) \circ (\operatorname{id}_{\mathbb{R}^q} \times \chi_j^{-1}) \right\|_{H^s(\mathbb{R}^q \times \mathbb{R}^n)}^2 \right\}^{1/2} . \tag{6.32}$$

Here $\chi_j : U_j \to \mathbb{R}^n$ are charts for an open covering $(U_j)_{j=1,\dots,N}$ of N by coordinate neighbourhoods, and $(\varphi_j)_{j=1,\dots,N}$ is a subordinate partition of unity.

Remark 6.2.1. (i) From (6.30) it follows that $M_{\gamma-n/2}u \in \widehat{H}^s(\Gamma_{(n+1)/2-\gamma} \times N)$ for $u \in \mathcal{H}^{s,\gamma}(N^\wedge)$; the hat indicates the image under the Fourier transform of the cylindrical Sobolev space $H^s(\Gamma_{(n+1)/2-\gamma} \times N)$ with respect to the one-dimensional variable $\operatorname{Im} z$.

(ii) Under the assumption of (i), for any cut-off function ω the Mellin transform $M_{\gamma-n/2}\,\omega\,u$ extends to $\operatorname{Re} z > (n+1)/2 - \gamma$ as a holomorphic function, and we have

$$M_{\gamma-n/2}\,\omega\,u\big|_{\Gamma_\delta \times N} \in \widehat{H}^s(\Gamma_\delta \times N)$$

for every $\delta > (n+1)/2 - \gamma$, uniformly in compact subintervals of

$$\left[\frac{n+1}{2} - \gamma, \frac{n+1}{2} - \gamma + \beta\right).$$

An analogous statement is true for $(1-\omega)u$ with respect to the half-plane $\operatorname{Re} z < (n+1)/2 - \gamma$.

Definition 6.2.2. Let M be a compact manifold with conical singularity v, locally near v modelled on N^Δ as described at the beginning. Then

$$H^{s,\gamma}(M) \quad \text{for } s, \gamma \in \mathbb{R} \tag{6.33}$$

denotes the subspace of all $u \in H^s_{\mathrm{loc}}(M \setminus \{v\})$ such that for (6.2) we have

$$\omega\left(u\big|_{V\setminus\{v\}} \circ (\chi^\wedge)^{-1}\right) \in \mathcal{H}^{s,\gamma}(N^\wedge) \tag{6.34}$$

for any cut-off function ω.

By a cut-off function ω_1 on M we understand any $\omega_1 \in C(M)$ with compact support in V such that $\omega_1 \equiv 1$ close to v and $\omega_1 \in C^\infty(V \setminus \{v\})$.

Note that the spaces $\mathcal{H}^{s,\gamma}(N^\wedge)$ encode some control with respect to the dependence of distributions on r not only for $r \to 0$, but also for $r \to \infty$. Accordingly, we distinguish between the notation $\mathcal{H}^{s,\gamma}$ and $H^{s,\gamma}$.

On $M \setminus \{v\}$ we fix a strictly positive function r^1 that is equal to r close to v in a fixed splitting of variables $(r, x) \in \mathbb{R}_+ \times X$. For any $\beta \in \mathbb{R}$ we call r^β a weight function with weight β. The operator of multiplication by r^β, $\beta \in \mathbb{R}$, induces an isomorphism

$$\mathrm{r}^\beta : H^{s,\gamma}(M) \to H^{s,\gamma+\beta}(M). \tag{6.35}$$

Let us now turn to other spaces defined globally on the infinite stretched cone $N^\wedge$. Let $\chi : U \to \mathbb{R}^n$ be a chart on N and form

$$\chi_{\mathrm{cone}} : \mathbb{R}_+ \times U \to \mathbb{R}^{n+1}_+, \quad \chi_{\mathrm{cone}}(r, x) := (r, r\chi(x)).$$

Then $H^s_{\mathrm{cone}}(N^\wedge)$ is defined to be the set of all

$$u \in H^s_{\mathrm{loc}}(\mathbb{R} \times N)\big|_{N^\wedge} \text{ such that } ((1-\omega)\varphi u) \circ \chi^{-1}_{\mathrm{cone}} \in H^s(\mathbb{R}^{n+1}) \tag{6.36}$$

for every such chart χ_{cone}, every $\varphi \in C^\infty_0(U)$, and some cut-off function ω. Observe that for $N = S^n$ (the unit sphere in $\mathbb{R}^{n+1}$) the condition (6.36) is equivalent to

$$u \in H^s_{\mathrm{loc}}(\mathbb{R} \times N)\big|_{N^\wedge} \text{ such that } (1-\omega)u \in H^s(\mathbb{R}^{n+1}).$$

In the latter formula the cut-off function is used in the meaning $\omega = \omega(|\tilde{x}|)$, $\tilde{x} \in \mathbb{R}^{n+1}$. Incidentally, we also employ spaces with weight $e \in \mathbb{R}$ at infinity:

$$H_{\text{cone}}^{s;e}(N^{\wedge}) := [r]^{-e} H_{\text{cone}}^{s}(N^{\wedge}).$$

Analogously to (6.28) and (6.29), we also define the weighted Kegel spaces

$$\mathcal{K}^{s,\gamma}(N^{\wedge}) := \{\omega u_0 + (1-\omega)u_\infty : u_0 \in \mathcal{H}^{s,\gamma}(N^{\wedge}),\ u_\infty \in H_{\text{cone}}^{s}(N^{\wedge})\}, \quad (6.37)$$

and

$$\mathcal{K}^{s,\gamma;e}(N^{\wedge}) := [r]^{-e} \mathcal{K}^{s,\gamma}(N^{\wedge}), \qquad (6.38)$$

for $s, \gamma, e \in \mathbb{R}$.

Remark 6.2.3. There are continuous embedding

$$\mathcal{K}^{s',\gamma';e'}(N^{\wedge}) \hookrightarrow \mathcal{K}^{s,\gamma;e}(N^{\wedge}) \qquad (6.39)$$

whenever $s' \geq s$, $\gamma' \geq \gamma$, $e' \geq e$, and (6.39) is compact for $s' > s$, $\gamma' > \gamma$, $e' > e$.

We have natural identifications

$$\mathcal{K}^{0,0}(N^{\wedge}) = \mathcal{H}^{0,0}(N^{\wedge}) = r^{-n/2} L^2(N^{\wedge}), \qquad (6.40)$$

and there are non-degenerate sesquilinear pairings

$$(\cdot,\cdot)_{\mathcal{K}^{0,0}(N^{\wedge})} : \mathcal{K}^{s,\gamma;e}(N^{\wedge}) \times \mathcal{K}^{-s,-\gamma;-e}(X^{\wedge}) \to \mathbb{C}, \qquad (6.41)$$

for all $s, \gamma, e \in \mathbb{R}$.

Let $\mathrm{k}(r) \in C_0^\infty(\mathbb{R}_+)$ be a strictly positive function and $\mathrm{k}(r) = r$ for $0 < r < \varepsilon_0$, $\mathrm{k}(r) = 1$ for $r > \varepsilon_1$ for some $0 < \varepsilon_0 < \varepsilon_1$. Then the operator of multiplication by k^β induces an isomorphism

$$\mathrm{k}^\beta : \mathcal{K}^{s,\gamma;e}(N^{\wedge}) \to \mathcal{K}^{s,\gamma+\beta;e}(N^{\wedge}). \qquad (6.42)$$

for every $s, \gamma, e \in \mathbb{R}$.

Proposition 6.2.4. *Let ω be a cut-off function.*

(i) *Given $s, \gamma \in \mathbb{R}$, for every $L \in \mathbb{R}$ there is an $e = e(s, \gamma, L) \in \mathbb{R}$ such that the multiplication by $(1 - \omega(r))r^{-L}$ defines a continuous operator*

$$(1 - \omega(r))r^{-L} : \mathcal{H}^{s,\gamma}(N^{\wedge}) \to \mathcal{K}^{s,\gamma';e}(N^{\wedge})$$

for every $\gamma' \in \mathbb{R}$.

(ii) *Given $s, \gamma \in \mathbb{R}$, for every $\gamma', e \in \mathbb{R}$ there is exists an $L = L(s, \gamma, \gamma', e) \geq 0$ such that the multiplication by $(1 - \omega(r))r^{-L}$ defines a continuous operator*

$$(1 - \omega(r))r^{-L} : \mathcal{K}^{s,\gamma;e}(N^{\wedge}) \to \mathcal{H}^{s,\gamma'}(N^{\wedge}).$$

(iii) *For any $s, \gamma', e' \in \mathbb{R}$, the multiplication by $\omega(r)r^L$ defines a continuous operator*

$$\omega(r)r^L : \mathcal{H}^{s,\gamma'}\left(N^{\wedge}\right) \to \mathcal{K}^{s,\gamma'+L;e'}\left(N^{\wedge}\right)$$

for every $L \in \mathbb{R}$.

(iv) *For any $s, \gamma, e \in \mathbb{R}$, the multiplication by $\omega(r)r^L$ defines a continuous operator*

$$\omega(r)r^L : \mathcal{K}^{s,\gamma;e}\left(N^{\wedge}\right) \to \mathcal{H}^{s,\gamma}\left(N^{\wedge}\right)$$

for every $L \geq 0$.

A proof of this result is given in [16]; see also [55] in a more general context.

6.3   Mellin pseudo-differential operators

The cone calculus is based to a large extent on Mellin pseudo-differential operators on $\mathbb{R}_+$. In the case of scalar amplitude functions

$$f(r, r', z) \in S^{\mu}_{(cl)}(\mathbb{R}_+ \times \mathbb{R}_+ \times \Gamma_{1/2-\gamma}), \quad z = 1/2 - \gamma + i\rho,$$

we set

$$\mathrm{op}^{\gamma}_M(f)u(r) := \int_{\mathbb{R}} \int_{\mathbb{R}_+} \left(\frac{r}{r'}\right)^{1/2-\gamma+i\rho} f\left(r, r', 1/2 - \gamma + i\rho\right)u(r')\frac{dr'}{r'}\,\dj\rho,$$

$\dj\rho = (2\pi)^{-1}d\rho$, first for $u \in C_0^{\infty}(\mathbb{R}_+)$ and then extended to larger distribution spaces. In the case of an r'-independent Mellin symbol f we have

$$\mathrm{op}^{\gamma}_M(f) = M_{\gamma}^{-1}fM_{\gamma}$$

for the weighted Mellin transform M_{γ}.

Observe that

$$\mathrm{op}^{\gamma}_M(f) = r^{\gamma}\mathrm{op}_M(T^{-\gamma}f)r^{-\gamma} \tag{6.43}$$

for

$$\mathrm{op}_M := \mathrm{op}^0_M$$

and

$$(T^{\beta}f)(r, r', z) := f(r, r', z + \beta),$$

for any real β. More generally we have

$$\mathrm{op}^{\gamma+\delta}_M(f) = r^{\delta}\mathrm{op}^{\gamma}_M(T^{-\delta}f)r^{-\delta} \tag{6.44}$$

for every $\delta \in \mathbb{R}$.

Remark 6.3.1. For $(\iota_\lambda u)(r, x) = u(\lambda r, x)$ for $\lambda \in \mathbb{R}_+$, we have

$$\iota_\lambda \mathrm{op}_M^\gamma(f)\iota_\lambda^{-1} = \mathrm{op}_M^\gamma(f_\lambda)$$

for $f_\lambda(r, r', z) := f(\lambda r, \lambda r', z)$. In particular, for $f = f(z)$, the case with constant coefficients, we have

$$\iota_\lambda \mathrm{op}_M^\gamma(f)\iota_\lambda^{-1} = \mathrm{op}_M^\gamma(f) \tag{6.45}$$

for all $\lambda \in \mathbb{R}_+$.

Remark 6.3.2. We have

$$\mathrm{op}_M^\gamma(f) = S_\gamma^{-1}\mathrm{Op}_{\boldsymbol{r}}(f_\gamma)S_\gamma$$

for $f_\gamma(\boldsymbol{r}, \boldsymbol{r}', \rho) = f(e^{-\boldsymbol{r}}, e^{-\boldsymbol{r}'}, 1/2 - \gamma + i\rho)$ and S_γ defined by (6.12), where

$$\mathrm{Op}_{\boldsymbol{r}}(f_\gamma)u(\boldsymbol{r}) = \iint e^{i(\boldsymbol{r}-\boldsymbol{r}')\rho} f_\gamma(\boldsymbol{r}, \boldsymbol{r}', \rho)u(\boldsymbol{r}')d\boldsymbol{r}'\bar{d}\rho.$$

We can interpret $\mathrm{op}_M^\gamma(f)$ in terms of a kernel

$$k_M(f)(t, t', \theta) := \int \theta^{-(1/2-\gamma+i\tau)} f(t, t', 1/2 - \gamma + i\tau)\bar{d}\tau, \tag{6.46}$$

namely,

$$\mathrm{op}_M^\gamma(f)u(t) = \int_0^\infty k_M(f)\left(t, t', \frac{t}{t'}\right) u(t')\frac{dt'}{t'} \tag{6.47}$$

which is a Mellin convolution with respect to the $\mathbb{R}_+$-variable in the kernel

$$k_M(f)(t, t', \theta) = \left(M_{\gamma, z \to \theta}^{-1}f\right)(t, t', \theta).$$

Mellin operators will also be necessary for operator-valued symbols

$$f(r, r', z) \in C^\infty\big(\mathbb{R}_+ \times \mathbb{R}_+, L_{(\mathrm{cl})}^\mu\big(N; \Gamma_{(n+1)/2-\gamma}\big)\big).$$

The choice of the weight line in connection with n only has a normalising effect in our calculus. Most of the generalities are valid both for the classical and non-classical case. However, in the cone calculus there will be a specific point where the property of being classical is essential. Therefore, we will often concentrate on the classical case. Let us first consider the case with constant coefficients.

Proposition 6.3.3. *The operator* $\mathrm{op}_M^{\gamma-n/2}(f)$ *for* $f \in L^\mu\big(N; \Gamma_{(n+1)/2-\gamma}\big)$ *induces a continuous operator*

$$\mathrm{op}_M^{\gamma-n/2}(f) : \mathcal{H}^{s,\gamma}(N^\wedge) \to \mathcal{H}^{s-\mu,\gamma}(N^\wedge)$$

for every $s \in \mathbb{R}$.

Proof. Straightforward.   □

Given a Fréchet space E with the semi-norm system $\{\pi_j\}_{j\in\mathbb{N}}$, by $C_B^\infty(\mathbb{R}_+ \times \mathbb{R}_+, E)$ we denote the set of all $f \in C^\infty(\mathbb{R}_+ \times \mathbb{R}_+, E)$ such that

$$\sup_{r,r'\in\mathbb{R}_+} \pi_j\left(\left(r\partial_r\right)^k\left(r'\partial_{r'}\right)^{k'} f(r,r')\right) < \infty$$

for every k, k', for all $j \in \mathbb{N}$. In a similar manner we define $C_B^\infty(\mathbb{R}_+, E)$.

Example 6.3.4. Let $\omega(r) \prec \omega'(r)$ be cut-off functions and $N \in \mathbb{N}, N \geq 1$. Then we have

$$C_B^\infty(\mathbb{R}_+ \times \mathbb{R}_+) \ni \begin{cases} f_0(r,r') := \omega(r)\left(\log(r/r')\right)^{-N}\left(1 - \omega'(r')\right), \\ f_1(r,r') := \left(1 - \omega'(r)\right)\left(\log(r/r')\right)^{-N}\omega(r'). \end{cases}$$

Theorem 6.3.5. *For any* $f(r,r',z) \in C_B^\infty\big(\mathbb{R}_+ \times \mathbb{R}_+, L^\mu(N;\Gamma_{(n+1)/2-\gamma})\big)$, *the map* $\mathrm{op}_M^{\gamma-n/2}(f) : C_0^\infty(N^\wedge) \to C^\infty(N^\wedge)$ *induces a continuous operator*

$$\mathrm{op}_M^{\gamma-n/2}(f) : \mathcal{H}^{s,\gamma}(N^\wedge) \to \mathcal{H}^{s-\mu,\gamma}(N^\wedge)$$

for every $s \in \mathbb{R}$.

Remark 6.3.6. The transformation $\iota_\lambda : u(r,x) \mapsto u(\lambda r, x)$, $\lambda \in \mathbb{R}_+$, defines a continuous operator

$$\iota_\lambda : \mathcal{H}^{s,\gamma}(N^\wedge) \to \mathcal{H}^{s,\gamma}(N^\wedge)$$

for every $s, \gamma \in \mathbb{R}$, and relation (6.45) holds for every $f \in L^\mu(N;\Gamma_{(n+1)/2-\gamma})$.

Let us now compute the formal adjoint of $\mathrm{op}_M^{\gamma-n/2}(f)$ with respect to the non-degenerate sesquilinear pairing

$$(\cdot,\cdot)_{\mathcal{H}^{0,0}(N^\wedge)} : \mathcal{H}^{s,\gamma}(N^\wedge) \times \mathcal{H}^{-s,-\gamma}(N^\wedge) \to \mathbb{C}.$$

Here we take into account that $\mathcal{H}^{0,0}(N^\wedge) = r^{-n/2}L^2(\mathbb{R}_+ \times N)$ carries the scalar product

$$(\cdot,\cdot)_{\mathcal{H}^{0,0}(N^\wedge)} = \int_{\mathbb{R}_+\times N} u(r,x)\overline{v(r,x)}r^n dr dx. \tag{6.48}$$

Proposition 6.3.7. *For* $f(r,r',z) \in C^\infty\big(\mathbb{R}_+ \times \mathbb{R}_+, L^\mu(N;\Gamma_{(n+1)/2-\gamma})\big)$ *the formal adjoint of* $\mathrm{op}_M^{\gamma-n/2}(f)$ *in the sense that*

$$\left(\mathrm{op}_M^{\gamma-n/2}(f)u, v\right)_{\mathcal{H}^{0,0}(N^\wedge)} = \left(u, \left(\mathrm{op}_M^{\gamma-n/2}(f)\right)^* v\right)_{\mathcal{H}^{0,0}(N^\wedge)}$$

for $u, v \in C_0^\infty(N^\wedge)$, *can be written as*

$$\left(\mathrm{op}_M^{\gamma-n/2}(f)\right)^* = \mathrm{op}_M^{-\gamma-n/2}(f^{[*]}) \quad \text{for } f^{[*]}(r,r',z) := f^{(*)}(r',r,n+1-\bar{z}) \tag{6.49}$$

where $(*)$ *indicates the pointwise formal adjoint of operators over* N.

Proof. Let $u, v \in C_0^\infty(X^\wedge)$. By virtue of (6.48), the left-hand side of (6.49) is equal to

$$\int \left(\mathrm{op}_M^{\gamma-n/2}(f)u, v\right)_{L^2(N)} r^n \, dr$$

$$= \int \left\{ \iint \left((r/r')^{-(n+1)/2+\gamma-i\rho} f\left(r, r', \frac{n+1}{2} - \gamma + i\rho\right) u(r') \frac{dr'}{r'} đ\rho, v(r)\right)_{L^2(X)} \right\} r^n \, dr$$

$$= \int \left\{ \iint (r/r')^{-(n+1)/2+\gamma-i\rho} (r/r')r^{-1}(r/r')^n (r')^n \right.$$
$$\cdot \left. \left(f\left(r, r', \frac{n+1}{2} - \gamma + i\rho\right) u(r'), v(r)\right)_{L^2(X)} dr' đ\rho \right\} dr$$

$$= \int \left(u(r'), \iint (r/r')^{(n+1)/2+\gamma+i\rho} f^{(*)}\left(r, r', \frac{n+1}{2} - \gamma + i\rho\right) v(r) \frac{dr}{r} đ\rho\right)_{L^2(N)} (r')^n \, dr'$$

$$= \int \left(u(r'), \iint (r'/r)^{-(n+1)/2-\gamma-i\rho} f^{(*)}\left(r, r', \frac{n+1}{2} - \gamma + i\rho\right) v(r) \frac{dr}{r} đ\rho\right)_{L^2(N)} (r')^n \, dr'$$

$$= \int \left(u(r), \iint (r/r')^{-(n+1)/2-\gamma-i\rho} f^{(*)}\left(r', r, \frac{n+1}{2} - \gamma + i\rho\right) v(r') \frac{dr'}{r'} đ\rho\right)_{L^2(N)} r^n \, dr$$

$$= \left(u, \mathrm{op}_M^{-\gamma-n/2}(f^{[*]})v\right)_{\mathcal{H}^{0,0}(N^\wedge)}.$$

In order to check that the claimed $f^{[*]}$ is of the right shape we take into account that

$$\mathrm{op}_M^{-\gamma-n/2}(f^{[*]})v = (r')^{-\gamma-n/2}\mathrm{op}_M\left(T^{\gamma+n/2}f^{[*]}\right)r^{\gamma+n/2}v$$

$$= \iint (r'/r)^{-(n+1)/2-\gamma-i\rho} f^{[*]}\left(r, r', \frac{n+1}{2} + \gamma + i\rho\right) v(r) \, dr$$

$$= \iint (r'/r)^{-(n+1)/2-\gamma-i\rho} f^{(*)}(r', r, n+1-\bar{z}) v(r) \, dr$$

for $z = (n+1)/2 + \gamma + i\rho \in \Gamma_{(n+1)/2+\gamma}$.   □

Definition 6.3.8. The space $M_\mathcal{O}^\mu(N)$, $\mu \in \mathbb{R}$, is defined to be the set of all $h \in \mathcal{A}(\mathbb{C}, L_{\mathrm{cl}}^\mu(N))$ such that

$$h|_{\Gamma_\beta} \in L_{\mathrm{cl}}^\mu(N; \Gamma_\beta)$$

for every real β, uniformly in compact intervals.

The space $M_\mathcal{O}^\mu(N)$ is (nuclear) Fréchet with a semi-norm system that results directly from the definition.

Remark 6.3.9. Observe that for $h \in M_\mathcal{O}^\mu(N)$ we have

$$\omega \, \mathrm{op}_M^{\gamma-n/2}(h) \, \omega' = \omega \, \mathrm{op}_M^{\beta-n/2}(h) \, \omega'$$

as operators $C_0^\infty(N^\wedge) \to C_0^\infty(N^\wedge)$, for arbitrary $\gamma, \beta \in \mathbb{R}$.

Definition 6.3.10. (i) Let N be a compact C^∞ manifold of dimension $n > 0$. The space of holomorphic parameter-dependent Mellin symbols of the cone and edge calculus,

$$M_O^\mu(N; \mathbb{R}^l),$$

is defined to be the set of all $h(z, \lambda) \in \mathcal{A}(\mathbb{C}_z, L_{\mathrm{cl}}^\mu(N; \mathbb{R}_\lambda^l))$ such that

$$h(z, \lambda)|_{\Gamma_\beta \times \mathbb{R}^l} \in L_{\mathrm{cl}}^\mu(N; \Gamma_\beta \times \mathbb{R}_\lambda^l)$$

for every $\beta \in \mathbb{R}$, uniformly for $c \le \beta \le c'$ for every $c \le c'$. The case $l = 0$ corresponds to $M_O^\mu(N)$.

(ii) If N consists of a single point, then $M_O^\mu(\mathbb{R}^l)$ means the set of all $h(z, \lambda) \in \mathcal{A}(\mathbb{C}_z, S_{\mathrm{cl}}^\mu(\mathbb{R}_\lambda^l))$ such that

$$h(z, \lambda)|_{\Gamma_\beta \times \mathbb{R}^l} \in S_{\mathrm{cl}}^\mu(\Gamma_\beta \times \mathbb{R}_\lambda^l)$$

for every $\beta \in \mathbb{R}$, uniformly for $c \le \beta \le c'$ for every $c \le c'$. In particular, for $l = 0$ we obtain the space M_O^μ as the subspace of all $h \in \mathcal{A}(\mathbb{C}_z)$ such that

$$h(z)|_{\Gamma_\beta} \in S_{\mathrm{cl}}^\mu(\Gamma_\beta)$$

for every $\beta \in \mathbb{R}$, uniformly for $c \le \beta \le c'$ for every $c \le c'$.

The space $M_O^\mu(N; \mathbb{R}_\lambda^l)$ is a Fréchet space with respect to the system of semi-norms

$$\sup_{z \in K} \pi_k^{(l)}(h(z)), \quad K \Subset \mathbb{C}, \quad \sup_{c \le \beta \le c'} \pi_k^{(l+1)}\left(h|_{\Gamma_\beta}\right)$$

where $\{\pi_k^{(l)}\}_{k \in \mathbb{N}}$ and $\{\pi_k^{(l+1)}\}_{k \in \mathbb{N}}$ are semi-norm systems for the Fréchet topology of $L_{\mathrm{cl}}^\mu(X; \mathbb{R}^l)$ and $L_{\mathrm{cl}}^\mu(X; \mathbb{R}^{1+l})$, respectively, with $\mathbb{R}$ being identified with Γ_β.

We set

$$M_O^{-\infty}(N; \mathbb{R}^l) := \bigcap_{\mu \in \mathbb{R}} M_O^\mu(N; \mathbb{R}^l).$$

Remark 6.3.11. For every $m \in \mathbb{N}$, $\alpha \in \mathbb{N}^l$, we have

$$\frac{\partial^m}{\partial z^m} D_\lambda^\alpha M_O^\mu(N; \mathbb{R}^l) \subseteq M_O^{\mu-(m+|\alpha|)}(N; \mathbb{R}^l).$$

Let us now recall the definition of the kernel cut-off operator based on the Mellin transform, first applied to symbols $S_{\mathrm{cl}}^\mu(\Gamma_0)$ where $\rho \in \mathbb{R}$ for $i\rho \in \Gamma_0$ is regarded as the covariable. First, let

$$k_M(f)(s) := \int_{-\infty}^\infty s^{-i\rho} f(i\rho) \, đ\rho,$$

cf. the formula (6.46). For any $\psi(s) \in C_0^\infty(\mathbb{R}_+), \psi(s) \equiv 1$ in a neighbourhood of $s = 1$, we set

$$\mathcal{V}(\psi)f(z) := M_{1/2, s \to z}(\psi(s) k_M(f)(s)) \qquad (6.50)$$

where $M_{1/2}$ is the weighted Mellin transform for the weight $1/2$. Then $\mathcal{V}(\psi)f \in M_{\mathcal{O}}^{\mu}$, and $\mathcal{V}(\psi)f|_{\Gamma_0} = f \bmod S^{-\infty}(\Gamma_0)$. Alternatively, we may form

$$\mathcal{V}(\psi)f(i\rho) := \iint e^{-i\theta\tilde{\rho}}\psi(e^{-\theta})f(i(\rho - \tilde{\rho}))\, d\theta\, d\tilde{\rho}. \tag{6.51}$$

Then (6.50) is the analytic extension of (6.51) from Γ_0 to $\mathbb{C}$. Moreover, $\mathcal{V}(\psi)$ induces a continuous operator

$$\mathcal{V}(\psi) : S_{\mathrm{cl}}^{\mu}(\Gamma_0) \to M_{\mathcal{O}}^{\mu}. \tag{6.52}$$

More generally, applying $\mathcal{V}(\psi)$ to symbols

$$f(r, i\rho) \in S_{\mathrm{cl}}^{\mu}(\overline{\mathbb{R}}_+ \times \Gamma_0) = C^{\infty}(\overline{\mathbb{R}}_+, S_{\mathrm{cl}}^{\mu}(\Gamma_0))$$

it follows a continuous operator

$$\mathcal{V}(\psi) : S_{\mathrm{cl}}^{\mu}(\overline{\mathbb{R}}_+ \times \Gamma_0) \to C^{\infty}(\overline{\mathbb{R}}_+, M_{\mathcal{O}}^{\mu}).$$

Let us also observe that $h \in C^{\infty}(\overline{\mathbb{R}}_+, M_{\mathcal{O}}^{\mu})$ and $f := h|_{\Gamma_0}$ gives rise to

$$h - \mathcal{V}(\psi)f \in C^{\infty}(\overline{\mathbb{R}}_+, M_{\mathcal{O}}^{-\infty}).$$

Note that, $M_{\mathcal{O}}^{-\infty}(X; \mathbb{R}^l) = M_{\mathcal{O}}^{-\infty} \widehat{\otimes}_{\pi} C^{\infty}(X \times X) \widehat{\otimes}_{\pi} \mathcal{S}(\mathbb{R}^l)$, where $\widehat{\otimes}_{\pi}$ is the (completed) projective tensor product of the respective Fréchet spaces. The latter relation follows from the nuclearity and the completeness of the involved spaces. The following theorem, also referred to as the kernel cut-off theorem, gives an impression on the nature the spaces $M_{\mathcal{O}}^{\mu}(X; \mathbb{R}_{\lambda}^l)$. Thus it makes sense to form spaces of smooth functions in $\overline{\mathbb{R}}_+ \times \Omega$ or Ω with values in $M_{\mathcal{O}}^{\mu}(X; \mathbb{R}^l)$ for any open set $\Omega \subseteq \mathbb{R}^q$.

Theorem 6.3.12. *For every $f(r, z) \in C^{\infty}(\overline{\mathbb{R}}_+, L^{\mu}(N; \Gamma_{\beta}))$ there exists an*

$$h(r, z) \in C^{\infty}(\overline{\mathbb{R}}_+, M_{\mathcal{O}}^{\mu}(N)),$$

namely, $h(r, z) = \mathcal{V}(\psi)(f)(r, z)$, such that

$$h|_{\Gamma_{\beta}} = f \bmod L^{-\infty}(\mathbb{R}_+ \times N),$$

and h is unique modulo $C^{\infty}(\mathbb{R}_+, M_{\mathcal{O}}^{-\infty}(N))$.

Theorem 6.3.12 is a consequence of the construction on the kernel cut-off outlined earlier.

Theorem 6.3.13. *Let $h(z, \lambda) \in M_{\mathcal{O}}^{\mu}(N; \mathbb{R}^l)$. Then $h|_{\Gamma_{\beta} \times \mathbb{R}^l} \in L_{\mathrm{cl}}^{\mu-1}(N; \Gamma_{\beta} \times \mathbb{R}^l)$ for some fixed $\beta \in \mathbb{R}$ entails $h(z, \lambda) \in M_{\mathcal{O}}^{\mu-1}(N; \mathbb{R}^l)$.*

Proof. The values of h at $\Gamma_\delta \times \mathbb{R}^l$ can be obtained from Taylor's formula of length $N \in \mathbb{N}$. The terms are differentiated at $\Gamma_\beta \times \mathbb{R}^l$ and have orders less than $\mu - j$ for $j = 0, \ldots, N+1$, with a remainder which contains differentiation in z up to order $N+1$. Then it suffices to apply Remark 6.3.11. $\qquad\square$

Theorem 6.3.14. *Given any* $f(z, \lambda) \in L^\mu_{cl}(N; \Gamma_\beta \times \mathbb{R}^l)$ *for some fixed* $\beta \in \mathbb{R}$ *there exists an* $h(z, \lambda) \in M^\mu_\mathcal{O}(N; \mathbb{R}^l)$ *such that* $h|_{\Gamma_\beta \times \mathbb{R}^l} = f \bmod L^{-\infty}(N; \Gamma_\beta \times \mathbb{R}^l)$.

Proof. It suffices to apply the kernel cut-off operator to f with respect to the weight line Γ_β. $\qquad\square$

Theorem 6.3.15. *For any sequence* $h_j(z, \lambda) \in M^{\mu-j}_\mathcal{O}(N; \mathbb{R}^l)$, $j \in \mathbb{N}$, *there exists an asymptotic sum* $h \sim \sum_{j=0}^\infty h_j$, *i.e., an* $h(z, \lambda) \in M^\mu_\mathcal{O}(N; \mathbb{R}^l)$ *such that* $h - \sum_{j=0}^N h_j \in M^{\mu-(N+1)}_\mathcal{O}(X; \mathbb{R}^l)$ *for every* $N \in \mathbb{N}$, *and* h *is unique modulo* $M^{-\infty}_\mathcal{O}(N; \mathbb{R}^l)$.

Proof. We first restrict h_j to $\Gamma_\beta \times \mathbb{R}^l$ for any fixed $\beta \in \mathbb{R}$ and form the asymptotic sum of the resulting elements of $L^{\mu-j}_{cl}(N; \Gamma_\beta \times \mathbb{R}^l)$. We then obtain an $f \in L^\mu_{cl}(N; \Gamma_\beta \times \mathbb{R}^l)$ in the standard sense. Then, applying Theorem 6.3.14 we obtain an $h(z, \lambda) \in M^\mu_\mathcal{O}(N; \mathbb{R}^l)$ with the claimed property. $\qquad\square$

Remark 6.3.16. Let $\Omega \subseteq \mathbb{R}^q$ be open. Theorem 6.3.14 admits a variant for functions $f(r, y, z, \lambda) \in C^\infty(\overline{\mathbb{R}}_+ \times \Omega, L^\mu_{cl}(N; \Gamma_\beta \times \mathbb{R}^l))$ which gives a resulting $h(r, y, z, \lambda) \in C^\infty(\overline{\mathbb{R}}_+ \times \Omega, M^\mu_\mathcal{O}(N; \mathbb{R}^l))$. Analogously, we can generalise Theorem 6.3.15 to a result on asymptotic expansions of sequences

$$h_j(r, y, z, \lambda) \in C^\infty(\overline{\mathbb{R}}_+ \times \Omega, M^{\mu-j}_\mathcal{O}(N; \mathbb{R}^l)),$$

which yields an $h(r, y, z, \lambda) \in C^\infty(\overline{\mathbb{R}}_+ \times \Omega, M^\mu_\mathcal{O}(N; \mathbb{R}^l))$.

6.4   Oscillatory integrals based on the Mellin transform

For our calculus we formulate an analogue of oscillatory integrals in terms of the Mellin transform. In this section we first develop the basics analogously as in Section 1.4. After that we pass to more specific material, especially, extensions of oscillatory integral constructions for holomorphic amplitude functions. Moreover, we study several interesting variants such as amplitude functions taking values in pseudo-differential operators on a manifold, and we establish Mellin operator conventions. More details are given in Seiler's thesis [58].

Definition 6.4.1. Let V be a Fréchet space with the system of semi-norms $(\pi_\iota)_{\iota \in \mathbb{N}}$.

(i) By $S^{\mu;\nu}(\mathbb{R}_+ \times \Gamma_\beta, V)$ for fixed $\beta \in \mathbb{R}$ and sequences of reals $\boldsymbol{\mu} := (\mu_\iota)_{\iota \in \mathbb{N}}$, $\boldsymbol{\nu} := (\nu_\iota)_{\iota \in \mathbb{N}}$ we defined the set of all $a(r, \beta + i\rho) \in C^\infty(\mathbb{R}_+ \times \Gamma_\beta, V)$ such that

$$\sup_{(r,\rho) \in \mathbb{R}_+ \times \Gamma_\beta} \langle \rho \rangle^{-\mu_\iota} \langle \log r \rangle^{-\nu_\iota} \pi_\iota\big(\big((r\partial_r)^k \partial_\rho^l a\big)(r, \beta + i\rho)\big) < \infty$$

for all $k, l \in \mathbb{N}$, $\iota \in \mathbb{N}$.
Moreover, we set

$$S^{\infty;\infty}(\mathbb{R}_+ \times \Gamma_\beta, V) = \bigcup_{\mu;\nu} S^{\mu,\nu}(\mathbb{R}_+ \times \Gamma_\beta, V).$$

(ii) By $S^{\mu;\nu}(\mathbb{R}_+ \times \mathbb{C}, V)$ for sequences μ, ν of reals we denote the space of all $a(r, z) \in C^\infty(\mathbb{R}_+, \mathcal{A}(\mathbb{C}, V))$ such that

$$\sup_{r \in \mathbb{R}_+, c \leq \beta \leq c'} \langle \rho \rangle^{-\mu_\iota} e^{-\nu_\iota \langle \log r \rangle} \pi_\iota \big(((r\partial_r)^k \partial_z^l a)(r, \beta + i\rho) \big) < \infty$$

for all $k, l \in \mathbb{N}$, $c \leq c'$, $\iota \in \mathbb{N}$.
Moreover, we set

$$S^{\infty;\infty}(\mathbb{R}_+ \times \mathbb{C}, V) = \bigcup_{\mu;\nu} S^{\mu;\nu}(\mathbb{R}_+ \times \mathbb{C}, V).$$

Remark 6.4.2. The transformation $\chi : \mathbb{R} \to \mathbb{R}_+$, $\chi(y) := e^{-y} = r$, induces by function pull back an isomorphism

$$\chi^* : S^{\mu;\nu}(\mathbb{R}_+ \times \Gamma_0, V) \to S^{\mu;\nu}(\mathbb{R} \times \Gamma_0, V), \tag{6.53}$$

where $S^{\mu;\nu}(\mathbb{R} \times \Gamma_0, V)$ corresponds to the space $S^{\mu;\nu}(\mathbb{R} \times \mathbb{R}, V)$ in (1.130) for $q = 1$ under the identification of Γ_0 with $\mathbb{R}$.

There is an analogue of Proposition 1.4.4 for the spaces of Definition 6.4.1 that we do not formulate here in detail. Similarly to Definition 1.4.7, we also need the notion of regularising functions.

For a Fréchet space E we denote by

$$\mathcal{T}(\mathbb{R}_+ \times \mathbb{R}, E)$$

the space of all $u(r, \rho) \in C^\infty(\mathbb{R}_+ \times \mathbb{R}, E)$ such that $u(e^{-y}, \rho) \in \mathcal{S}(\mathbb{R} \times \mathbb{R}, E)$.

Definition 6.4.3. (a) A function $\chi_\varepsilon(r, z) : (0, 1] \times \mathbb{R}_+ \times \Gamma_0 \to \mathbb{C}$ is called regularising if

(i) $\chi_\varepsilon \in \mathcal{T}(\mathbb{R}_+ \times \Gamma_0)$ for each ε;

(ii) $\sup \big\{ |(r\partial_r)^k \partial_z^l \chi_\varepsilon(r, z)| : 0 < \varepsilon \leq 1, (r, z) \in \mathbb{R}_+ \times \Gamma_0 \big\} < \infty$ for all $k, l \in \mathbb{N}$;

(iii) $(r\partial_r)^k \partial_z^l \chi_\varepsilon(r, z) \to \begin{cases} 1 & \text{for } k + l = 0 \\ 0 & \text{for } k + l \neq 0 \end{cases}$ pointwise on $\mathbb{R}_+ \times \Gamma_0$ as $\varepsilon \to 0$.

(b) A function $\chi_\varepsilon(r, z) : (0, 1] \times \mathbb{R}_+ \times \mathbb{C} \to \mathbb{C}$ is called holomorphically regularising if

(i) $(\varepsilon, r, i\rho) \to \chi_\varepsilon(r, \beta + i\rho)$ is regularising in the sense of (a) for every $\beta \in \mathbb{R}$;

(ii) $z \mapsto \chi_\varepsilon(r, z)$ is entire and $\rho \mapsto \chi_\varepsilon(r, \beta + i\rho)$ belongs to $\mathcal{S}(\mathbb{R})$ uniformly for β in compact intervals;

(iii) for every ε there is a compact set $K_\varepsilon \subset \mathbb{R}_+$ such that $\chi_\varepsilon(r, z) = 0$ when $r \notin K_\varepsilon$.

Example 6.4.4. (i) Let $\chi \in \mathcal{T}(\mathbb{R}_+ \times \Gamma_0)$, $\chi(1) = 1$, and $\chi_\varepsilon(r, i\rho) := \chi(r^\varepsilon, i\varepsilon\rho)$. Then χ_ε is regularising in the sense of Definition 6.4.3 (a).

(ii) Assume $\varphi \in C_0^\infty(\mathbb{R}_+)$, $\varphi(1) = (M\varphi)(0) = 1$, with M being the Mellin transform. Then $\chi_\varepsilon(r, z) := \varphi(r^\varepsilon)M\varphi(\varepsilon z)$ is regularising in the sense of Definition 6.4.3 (b).

Similarly as in Section 1.4, we intend to regularise the Mellin oscillatory integral

$$\mathrm{Os}[a] := \iint_0^\infty r^{i\rho} a(r, i\rho) \frac{dr}{r} \dbar\rho \tag{6.54}$$

first for $a(r, z) \in S^{\mu;\nu}(\mathbb{R}_+ \times \Gamma_0, V)$, and then for $a(r, z) \in S^{\mu;\nu}(\mathbb{R}_+ \times \mathbb{C}, V)$. Observe that the diffeomorphism $\mathbb{R} \to \mathbb{R}_+$, $y \mapsto r = e^{-y}$, mentioned in Remark 6.4.2, gives us a relationship between (6.54) and (1.132) for the one-dimensional case. In fact, under the substitution

$$r^{i\rho} = e^{-i\rho \log r} = e^{i\rho y}, \qquad \frac{dr}{r} = dy,$$

the expression (6.54) becomes

$$\iint_0^\infty e^{i\rho y} a(e^{-y}, i\rho) \, dy \, \dbar\rho$$

Thus upon setting $a_F(y, \rho) := a(e^{-y}, i\rho)$ expression (6.54) takes the form $\mathrm{Os}[a_F]$ in the sense of (1.132).

Theorem 6.4.5. (i) *Let $a(r, z) \in S^{\mu;\nu}(\mathbb{R}_+ \times \Gamma_0, V)$ and $\chi_\varepsilon(r, z)$ be a regularising function. Then the oscillatory integral*

$$\mathrm{Os}[a] = \iint_0^\infty r^{i\rho} a(r, i\rho) \frac{dr}{r} \dbar\rho = \lim_{\varepsilon \to 0} \iint_0^\infty r^{i\rho} \chi_\varepsilon(r, i\rho) a(r, i\rho) \frac{dr}{r} \dbar\rho \tag{6.55}$$

defines an element in V which is independent of the regularising function. Moreover, $a \mapsto \mathrm{Os}[a]$ induces a continuous map $\mathrm{Os}[\cdot] : S^{\mu;\nu}(\mathbb{R}_+ \times \Gamma_0, V) \to V$ for every $\mu; \nu$.

(ii) *Let $a(r, z) \in S^{\mu;\nu}(\mathbb{R}_+ \times \mathbb{C}, V)$ and $\chi_\varepsilon(r, z)$ a holomorphically regularising function. Then the oscillatory integral (6.55) defines an element in V which* *is independent of the regularising function, and* Os[·] *induces a continuous map*

$$\mathrm{Os}[\,\cdot\,] : S^{\mu;\nu}(\mathbb{R}_+ \times \mathbb{C}, V) \to V \tag{6.56}$$

for all $\boldsymbol{\mu}$ *and* $\boldsymbol{\nu}$. *For* $a(r,z) \in S^{\mu;\nu}(\mathbb{R}_+ \times \Gamma_0, V) \cap S^{\mu,\nu}(\mathbb{R}_+ \times \mathbb{C}, V)$ *the maps* (6.55) *and* (6.56) *coincide.*

Proof. (i) It is convenient to reduce the proof to a corresponding statement on oscillatory integrals for the Fourier transform, here denoted for the moment by $\mathrm{Os}_F[\,\cdot\,]$. The substitution $r = e^{-y}$ gives us the isomorphism (6.53) and then (6.55) turns to

$$\mathrm{Os}_F[a] = \int_{\mathbb{R}} \int_{\mathbb{R}} e^{-iy\eta} a(y,\eta) dy d\!\!\!/\eta = \lim_{\varepsilon \to 0} \int_{\mathbb{R}} \int_{\mathbb{R}} e^{-iy\eta} \chi_\varepsilon(y, i\eta) h(y,\eta) dy d\!\!\!/\eta \tag{6.57}$$

for $\chi_\varepsilon(y, i\eta) := \chi_\varepsilon(e^{-y}, i\eta)$, $h(y,\eta) := a(e^{-y}, i\eta)$. Thus Theorem 6.4.5 (i) turns into a special case of Theorem 1.4.5.

(ii) We also reduce the statement to an oscillatory integral for the Fourier transform. To this end we slightly modify Definition 6.4.1 (ii) and denote by $S^{\mu;\nu}(\mathbb{R}_+ \times \mathbb{C}, V)$ the space of all $a(y,z) \in C^\infty(\mathbb{R}, \mathcal{A}(\mathbb{C}, V))$ such that

$$\pi_\iota\big(\big(D_y^k D_z^l a\big)(y, \beta + i\eta)\big) \le c \langle \eta \rangle^{\mu_\iota} e^{\nu_\iota \langle y \rangle} \tag{6.58}$$

for all $(y, \beta + i\eta) \in \mathbb{R} \times \{\beta \in \mathbb{R} : c \le \beta \le c'\}$, for all $k, l \in \mathbb{N}$, $c \le c'$, $\iota \in \mathbb{N}$, for a constant that depends on k, l, c, c', ι. Now we write $h(y, z) := a(e^{-y}, z)$ and set

$$A_\varepsilon := \iint e^{-iy\eta} \chi_\varepsilon(y, i\eta) h(y, i\eta) dy d\!\!\!/\eta$$

for a regularising function $\chi_\varepsilon(y, z) := \chi_\varepsilon(e^{-y}, z)$ where $\chi_\varepsilon(r, z)$ is holomorphically regularising in the sense of Definition 6.4.3 (ii). We first consider the η-integration and pass to an integration over weight lines in the complex plane. In other words we write

$$\int_{\mathbb{R}} e^{-iy\eta} \chi_\varepsilon(y, i\eta) h(y, i\eta) d\!\!\!/\eta = \int_{\Gamma_0} e^{-yz} \chi_\varepsilon(y, z) h(y, z) d\!\!\!/z \tag{6.59}$$

for $d\!\!\!/z = (2\pi i)^{-1} dz$. Choose a $\varphi \in C^\infty(\mathbb{R})$ such that $\varphi = 0$ for $y < c_0$ and $\varphi = 1$ for $y > c_1$ for a $c_0 < c_1$. Then write the right-hand side of (6.59) as

$$\int_{\Gamma_0} e^{-yz} \chi_\varepsilon(y, z) \varphi(y) h(y, z) d\!\!\!/z + \int_{\Gamma_0} e^{-yz} \chi_\varepsilon(y, z)(1 - \varphi(y)) h(y, z) d\!\!\!/z$$

$$= \int_{\Gamma_\alpha} e^{-yz} \chi_\varepsilon(y, z) \varphi(y) h(y, z) d\!\!\!/z + \int_{\Gamma_\beta} e^{-yz} \chi_\varepsilon(y, z)(1 - \varphi(y)) h(y, z) d\!\!\!/z$$

for any reals α, β. Here we used the fact that the integrals are holomorphic in $z = \beta + i\eta$ and Schwartz functions on Γ_δ for every δ, uniformly in compact δ-

intervals. Thus

$$
\begin{aligned}
A_\varepsilon &= \iint e^{-iy\eta} \chi_\varepsilon(y, \alpha + i\eta) e^{-\alpha y} \varphi(y) h(y, \alpha + i\eta) \, dy \, \text{\dj}\eta \\
&\quad + \iint e^{-iy\eta} \chi_\varepsilon(y, \beta + i\eta) e^{-\beta y} (1 - \varphi(y)) h(y, \beta + i\eta) \, dy \, \text{\dj}\eta \\
&= \iint e^{-iy\eta} \big(\varphi(y) h_{\varepsilon,\alpha}(y, \eta) + (1 - \varphi(y)) h_{\varepsilon,\beta}(y, \eta) \big) \, dy \, \text{\dj}\eta
\end{aligned}
$$

for

$$
\begin{aligned}
h_{\varepsilon,\alpha}(y, \eta) &:= \chi_\varepsilon(y, \alpha + i\eta) e^{-\alpha y} h(y, \alpha + i\eta), \\
h_{\varepsilon,\beta}(y, \eta) &:= \chi_\varepsilon(y, \beta + i\eta) e^{-\beta y} h(y, \beta + i\eta).
\end{aligned}
$$

The estimate (6.58) gives

$$
\begin{aligned}
\pi_\iota \big((D_y^k D_\eta^l \varphi(y) h_{\varepsilon,\alpha}(y, \eta)) \big) &\leq c \, \langle \eta \rangle^{\mu_\iota} e^{\nu_\iota \langle y \rangle} e^{-\alpha y}, \\
\pi_\iota \big((D_y^k D_\eta^l (1 - \varphi(y)) h_{\varepsilon,\beta}(y, \eta)) \big) &\leq c \, \langle \eta \rangle^{\mu_\iota} e^{\nu_\iota \langle y \rangle} e^{-\beta y}.
\end{aligned}
$$

The choice of α, β is arbitrary, and we set $\alpha := \nu_\iota$, $\beta := -\nu_\iota$. Then, recalling what the supports of φ and $1 - \varphi$ are, we obtain the estimates

$$
\pi_\iota \big((D_y^k D_\eta^l \varphi(y) h_{\varepsilon,\nu_\iota}(y, \eta)) \big) \leq c \, \langle \eta \rangle^{\mu_\iota}, \quad \pi_\iota \big((D_y^k D_\eta^l (1 - \varphi(y)) h_{\varepsilon,-\nu_\iota}(y, \eta)) \big) \leq c \, \langle \eta \rangle^{\mu_\iota}.
$$

In fact, we need the estimates

$$
e^{\nu_i(\langle y \rangle - y)} \leq c \quad \text{for } y > 0, \tag{6.60}
$$

$$
e^{\nu_i(\langle y \rangle + y)} \leq c \quad \text{for } y < 0. \tag{6.61}
$$

In the case $\nu_\iota \geq 0$ we employ the relation $\langle y \rangle - y = (\langle y \rangle + y)^{-1}$, which shows that $e^{\nu_i(\langle y \rangle - y)} = e^{\nu_i(\langle y \rangle + y)^{-1}} \leq c$ for $y > 0$. The estimate (6.61) follows in an analogous manner when we replace y by $-x$. In the case $\nu \leq 0$ the estimate (6.60) is obvious, since $\langle y \rangle - y \geq 0$ for $x > 0$, while (6.61) is a consequence of $\langle y \rangle + y \geq 0$ for $y < 0$. This allows us to pass to the limit in A_ε for $\varepsilon \to 0$ for $\alpha = \nu_\iota$, $\beta = -\nu_\iota$ by the regularisations of oscillatory integrals as in (i) separately for the summands with the factors φ and $1 - \varphi$. It is evident that the limit is altogether independent of the choice of φ. The final statement of (ii) is also clear. □

We employ the following notation. If E is a Fréchet space with the semi-norm system $(e_j)_{j \in \mathbb{N}}$, we set

$$
C_{\mathrm{B}}^\infty(\mathbb{R}_+, E) := \Big\{ u \in C^\infty(\mathbb{R}_+, E) : \sup_{r \in \mathbb{R}_+} e_j \big(((r\partial_r)^k u)(r) \big) < \infty \ \forall \, k, k', j \in \mathbb{N} \Big\}
$$

and

$$
C_{\mathrm{B}}^\infty(\mathbb{R}_+ \times \mathbb{R}_+, E) := \Big\{ u \in C^\infty(\mathbb{R}_+ \times \mathbb{R}_+, E) :
$$

$$
\sup_{r, r' \in \mathbb{R}_+} e_j \big(((r\partial_r)^k (r'\partial_{r'})^k u)(r) \big) < \infty \ \forall \, k, k', j \in \mathbb{N} \Big\}.
$$

Definition 6.4.6. Let $H, \widetilde{H}$ be Hilbert spaces with group actions κ and $\widetilde{\kappa}$, respectively. We define

$$S^{\mu}_{(\mathrm{cl})}\big(\mathbb{R}_+ \times \Gamma_0; H, \widetilde{H}\,\big)_{\mathrm{B}} := C^{\infty}_{\mathrm{B}}\big(\mathbb{R}_+, S^{\mu}_{(\mathrm{cl})}\big(\Gamma_0; H, \widetilde{H}\,\big)\big)$$

and, analogously,

$$S^{\mu}_{(\mathrm{cl})}\big(\mathbb{R}_+ \times \mathbb{R}_+ \times \Gamma_0; H, \widetilde{H}\,\big)_{\mathrm{B}} := C^{\infty}_{\mathrm{B}}\big(\mathbb{R}_+ \times \mathbb{R}_+, S^{\mu}_{(\mathrm{cl})}\big(\Gamma_0; H, \widetilde{H}\,\big)\big)$$

If necessary, we write $S^{\mu}_{(\mathrm{cl})}\big(\mathbb{R}_+ \times \mathbb{R}_+ \times \Gamma_0; H, \widetilde{H}\,\big)_{\kappa, \widetilde{\kappa}, \mathrm{B}}$, etc.

Next for any $a(r, r', z) \in S^{\mu}_{(\mathrm{cl})}\big(\mathbb{R}_+ \times \mathbb{R}_+ \times \Gamma_0; H, \widetilde{H}\,\big)_{\mathrm{B}}$ and $u(r) \in C^{\infty}_{\mathrm{B}}(\mathbb{R}_+, H)$ we define

$$\mathrm{op}^{1/2}_M(a)u(r) := \iint_0^\infty \left(\frac{r}{r'}\right)^{-i\rho} a(r, r', i\rho)u(r')\frac{dr'}{r'}\,đ\rho. \tag{6.62}$$

The expression (6.62) is interpreted as an oscillatory integral with the variables r', ρ in the sense of (6.4.5), for every fixed $r \in \mathbb{R}_+$. Clearly,

$$\mathrm{op}^{1/2}_M(a) \in L^{\mu}_{(\mathrm{cl})}\big(\mathbb{R}_+, H, \widetilde{H}\,\big)$$

in the sense of Definition 1.3.30. Hence, we obtain a continuous map

$$\mathrm{op}^{1/2}_M(a) : C^{\infty}_0(\mathbb{R}_+, H) \to C^{\infty}_0(\mathbb{R}_+, \widetilde{H}\,). \tag{6.63}$$

Concerning the operator $B = \mathrm{Op}_y(b)$ in the Fourier set-up,

$$b(y, y', \eta) \in S^{\mu}_{(\mathrm{cl})}\big(\mathbb{R}^{2q}_{y,y'} \times \mathbb{R}^q_\eta; H, \widetilde{H}\,\big),$$

we have

$$\mathrm{op}^{1/2}_M(a) = S^{-1}_{1/2}\,\mathrm{Op}_y(b)S_{1/2} \tag{6.64}$$

for $a(r, r', i\rho) \in S^{\mu}_{(\mathrm{cl})}\big(\mathbb{R}_+ \times \mathbb{R}_+ \times \Gamma_0; H, \widetilde{H}\,\big)$ and

$$b(y, y', \rho) = a\big(e^{-y}, e^{-y'}, i\rho\big). \tag{6.65}$$

Observe that (6.65) gives rise to an isomorphism

$$S^{\mu}_{(\mathrm{cl})}\big(\mathbb{R}_+ \times \mathbb{R}_+ \times \Gamma_0; H, \widetilde{H}\,\big)_{\mathrm{B}} \to S^{\mu}_{(\mathrm{cl})}\big(\mathbb{R} \times \mathbb{R} \times \mathbb{R}; H, \widetilde{H}\,\big)_{\mathrm{B}}, \tag{6.66}$$

with a and b being related via (6.65). Analogously to (6.17), we define the weighted Mellin–Schwartz spaces with values in a Hilbert space as

$$T^\gamma(\mathbb{R}_+, H) := S^{-1}_\gamma S(\mathbb{R}_+, H) \tag{6.67}$$

with the transformation S_γ from (6.13). Note that $T^{\gamma+\delta}(\mathbb{R}_+, H) = r^\delta T^\gamma(\mathbb{R}_+, H)$ for every $\gamma, \delta \in \mathbb{R}$.

Theorem 6.4.7. *The operator* (6.63) *extends to a continuous map*

$$\mathrm{op}_M^{1/2}(a) : \mathcal{T}^{1/2}(\mathbb{R}_+, H) \to \mathcal{T}^{1/2}(\mathbb{R}_+, \widetilde{H}).$$

Proof. The proof is a consequence of Theorem 1.4.11 and of relations (6.67), (6.66), (6.64).   □

We set

$$M_{1/2} L_{(\mathrm{cl})}^{\mu}(\mathbb{R}_+; H, \widetilde{H})_{\mathrm{B}} := \left\{ \mathrm{op}_M^{1/2}(a) : a(r, r', i\rho) \in S_{(\mathrm{cl})}^{\mu}(\mathbb{R}_+ \times \mathbb{R}_+ \times \Gamma_0; H, \widetilde{H})_{\mathrm{B}} \right\}.$$

Remark 6.4.8. The mapping

$$\mathrm{op}_M^{1/2} : S_{(\mathrm{cl})}^{\mu}(\mathbb{R}_+ \times \Gamma_0; H, \widetilde{H})_{\mathrm{B}} \to M_{1/2} L_{(\mathrm{cl})}^{\mu}(\mathbb{R}_+; H, \widetilde{H})_{\mathrm{B}}$$

is an isomorphism for every μ, and $a(r, i\rho)$ can be recovered from $\mathrm{op}_M^{1/2}(a)$ via

$$B := S_{1/2} \,\mathrm{op}_M^{1/2}(a) S_{1/2}^{-1} \in L_{(\mathrm{cl})}^{\mu}(\mathbb{R}; H, \widetilde{H})_{\mathrm{B}}$$

by Proposition 1.4.12.

Theorem 6.4.9. *Let H and $\widetilde{H}$ be Hilbert spaces with group action, and let $A = \mathrm{op}_M^{1/2}(a)$ for $a(r, r', i\rho) \in S_{(\mathrm{cl})}^{\mu}(\mathbb{R}_+ \times \mathbb{R}_+ \times \Gamma_0; H, \widetilde{H})_{\mathrm{B}}$. Then*

(i) *there exist unique left or right symbols*

$$a_{\mathrm{L}}(r, i\rho), a_{\mathrm{R}}(r', i\rho) \in S_{(\mathrm{cl})}^{\mu}(\mathbb{R}_+ \times \Gamma_0; H, \widetilde{H})_{\mathrm{B}}$$

   such that

$$\mathrm{op}_M^{1/2}(a) = \mathrm{op}_M^{1/2}(a_{\mathrm{L}}) = \mathrm{op}_M^{1/2}(a_{\mathrm{R}}); \tag{6.68}$$

(ii) *the symbols $a_{\mathrm{L}}(r, i\rho), a_{\mathrm{R}}(r, i\rho)$ can be expressed by oscillatory integrals*

$$a_{\mathrm{L}}(r, i\rho) = \iint t^{i\xi} a(r, rt, i(\rho + \xi)) \frac{dt}{t} d\xi,$$

$$a_{\mathrm{R}}(r', i\rho) = \iint t^{i\xi} a(tr', r', i(\rho - \xi)) \frac{dt}{t} d\xi,$$

which have asymptotic expansions

$$a_{\mathrm{L}}(r, i\rho) \sim \sum_{k \in \mathbb{N}} \frac{1}{k!} \left((-r' \partial_{r'})^k \partial_z^k a \right)(r, r', i\rho) \big|_{r'=r},$$

$$a_{\mathrm{R}}(r', i\rho) \sim \sum_{k \in \mathbb{N}} \frac{1}{k!} (-1)^k \left((-r \partial_r)^k \partial_z^k a \right)(r, r', i\rho) \big|_{r=r'};$$

(iii) *the mappings $a \mapsto a_{\mathrm{L}}$ and $a \mapsto a_{\mathrm{R}}$ are continuous;*

(iv) *writing*

$$a_{\mathrm{L}}(r, i\rho) = \sum_{k=0}^{N} \frac{1}{k!} \left((-r'\partial_{r'})^k \partial_z^k a \right)(r, r', i\rho)\big|_{r'=r} + r_{\mathrm{L},N+1}(r, i\rho),$$

$$a_{\mathrm{R}}(r', i\rho) = \sum_{k=0}^{N} \frac{1}{k!} (-1)^k \left((-r'\partial_{r'})^k \partial_z^k a \right)(r, r', i\rho)\big|_{r=r'} + r_{\mathrm{R},N+1}(r', i\rho),$$

for any $N \in \mathbb{N}$ we have $r_{\mathrm{L},N+1}, r_{\mathrm{R},N+1} \in S^{\mu-(N+1)}\big(\mathbb{R}_+ \times \Gamma_0; H, \widetilde{H}\big)_{\mathrm{B}}$, and

$$r_{\mathrm{L},N+1}(r, i\rho) = \int_0^1 \frac{(1-\theta)^N}{N!}$$
$$\cdot \iint t^{i\xi} \left((-r'\partial_{r'})^{N+1} \partial_z^{N+1} a \right)(r, rt, i(\rho + \theta\xi)) \frac{dt}{t} d\xi d\theta,$$

$$r_{\mathrm{R},N+1}(r', i\rho) = \int_0^1 \frac{(1-\theta)^N}{N!} (-1)^{N+1}$$
$$\cdot \iint t^{i\xi} \left((-r\partial_r)^{N+1} \partial_z^{N+1} a \right)(tr', r', i(\rho - \theta\xi)) \frac{dt}{t} d\xi d\theta;$$

(v) *the mappings $a \mapsto r_{\mathrm{L},N+1}$ and $a \mapsto r_{\mathrm{R},N+1}$ are continuous.*

Clearly, on functions in $z = i\rho$ we also write $\partial_z = -i\partial_\rho$.

Theorem 6.4.10. *Let $H, \widetilde{H}$ and H_0 be Hilbert spaces with group actions $\kappa, \widetilde{\kappa}$ and κ_0, respectively.*

(i) *If $A = \mathrm{op}_M^{1/2}(a) \in M_{1/2}L_{(\mathrm{cl})}^\mu\big(\mathbb{R}_+; H_0, \widetilde{H}\big)_{\mathrm{B}}$ for $a(r, i\rho) \in S_{(\mathrm{cl})}^\mu\big(\mathbb{R}_+ \times \Gamma_0; H_0, \widetilde{H}\big)_{\mathrm{B}}$ and $B = \mathrm{op}_M^{1/2}(b) \in M_{1/2}L_{(\mathrm{cl})}^\mu\big(\mathbb{R}_+; H, H_0\big)_{\mathrm{B}}$ for $b(r, i\rho) \in S_{(\mathrm{cl})}^\mu(\mathbb{R}_+ \times \Gamma_0; H, H_0)_{\mathrm{B}}$ then $AB \in M_{1/2}L_{(\mathrm{cl})}^{\mu+\nu}\big(\mathbb{R}_+; H, \widetilde{H}\big)_{\mathrm{B}}$, and we have*

$$AB = \mathrm{Op}(a\#b),$$

with

$$(a\#b)(r, i\rho) = \iint t^{i\xi} a(r, i(\rho + \xi)) b(rt, i\rho) \frac{dt}{t} d\xi.$$

There is an asymptotic expansion

$$(a\#b)(r, i\rho) \sim \sum_{k=0}^{\infty} \frac{1}{k!} \left(\partial_z^k a(r, i\rho) \right) (-r\partial_r)^k b(r, i\rho).$$

(ii) *The mapping $(a, b) \mapsto a\#b$ is continuous bilinear between the respective spaces of symbols.*

(iii) *Writing*

$$(a\#b)(r,i\rho) = \sum_{k=0}^{N} \frac{1}{k!} (\partial_z^k a(r,i\rho))(-r\partial_r)^k b(r,i\rho) + r_{N+1}(r,i\rho), \quad N \in \mathbb{N},$$

we have $r_{N+1}(r,i\rho) \in S_{(\mathrm{cl})}^{\mu+\nu-(N+1)}\big(\mathbb{R}_+ \times \Gamma_0; H, \widetilde{H}\big)_{\mathrm{B}}$ *and*

$$r_{N+1}(r,i\rho) = \int_0^1 \frac{(1-\theta)^N}{N!}$$
$$\cdot \iint t^{i\xi} \big(\partial_z^{N+1} a\big)(r,i(\rho+\theta\xi))\big((-r\partial_r)^{N+1} b\big)(rt,i\rho) \frac{dt}{t} \bar{d}\xi d\theta.$$

(iv) *The mapping* $(a,b) \mapsto r_{N+1}$ *is continuous bilinear between the respective symbol spaces.*

Definition 6.4.11. The space $\mathcal{H}^{s,\gamma}(\mathbb{R}_+, H)$ for a Hilbert space H with group action $\kappa = \{\kappa_\lambda\}_{\lambda \in \mathbb{R}_+}$ and $s, \gamma \in \mathbb{R}$ is defined to be the completion of $\mathcal{T}^\gamma(\mathbb{R}_+, H)$ with respect to the norm

$$\|u\|_{\mathcal{H}^{s,\gamma}(\mathbb{R}_+, H)} := \left\{ \int_{\Gamma_{1/2-\gamma}} \langle z \rangle^{2s} \big\|\kappa_{\langle z \rangle}^{-1} M_\gamma u(z)\big\|_H^2 \bar{d}z \right\}^{1/2}.$$

Here $\langle z \rangle = \big(1 + |z|^2\big)^{1/2}$.

Theorem 6.4.12. *Let* $H, \widetilde{H}$ *be Hilbert spaces with group action. Then* $\mathrm{op}_M^{1/2}(a)$ *for* $a(r,z) \in S_{(\mathrm{cl})}^\mu\big(\mathbb{R}_+ \times \Gamma_0; H, \widetilde{H}\big)_{\mathrm{B}}$ *extends the operator from Theorem 6.4.7 to a continuous operator*

$$\mathrm{op}_M^{1/2} : \mathcal{H}^{s,1/2}(\mathbb{R}_+, H) \to \mathcal{H}^{s-\mu,1/2}\big(\mathbb{R}_+, \widetilde{H}\big)$$

for every $s \in \mathbb{R}$. *The map* $a \mapsto \mathrm{op}_M^{1/2}$ *induces a continuous operator*

$$S_{(\mathrm{cl})}^\mu\big(\mathbb{R}_+ \times \Gamma_0; H, \widetilde{H}\big)_{\mathrm{B}} \to \mathcal{L}\big(\mathcal{H}^{s,1/2}(\mathbb{R}_+, H), \mathcal{H}^{s-\mu,1/2}(\mathbb{R}_+, \widetilde{H})\big) \qquad (6.69)$$

for every $s \in \mathbb{R}$.

Proof. We have
$$\mathcal{H}^{s,\gamma}(\mathbb{R}_+, H) = S_\gamma^{-1} \mathcal{H}^{s,\gamma}(\mathbb{R}, H) \qquad (6.70)$$

for every $s, \gamma \in \mathbb{R}$. In particular, by virtue of the relation (6.64) and Theorem 1.4.15, one obtains the asserted continuity (6.69). Moreover, (6.70) is a consequence of (6.66). $\square$

The spaces $\mathcal{H}^{s,\gamma}(\mathbb{R}_+, H)$ are Hilbert spaces in a natural way. In particular, if κ is unitary in H, then $\mathcal{H}^{0,1/2}(\mathbb{R}_+, H)$ may be identified with $L^{2,1/2}(\mathbb{R}_+, H) = r^{1/2}L^2(\mathbb{R}_+, H)$ with the scalar product

$$(u, v)_{L^{2,1/2}(\mathbb{R}_+, H)} = \int_0^\infty (u(r), v(r))_H \frac{dr}{r}.$$

Let us now pass to formal adjoints of Mellin operators. Analogously to how we proceeded in Theorem 1.4.16, we consider Hilbert space triples $\{H, H_0, H'; \kappa\}$ and $\{\widetilde{H}, \widetilde{H}_0, \widetilde{H}'; \widetilde{\kappa}\}$ with group action. The pointwise application of adjoints $\mathcal{L}(H, \widetilde{H}) \to \mathcal{L}(\widetilde{H}', H')$ yields an antilinear isomorphism

$$(*) : S^\mu_{(\text{cl})}(\mathbb{R}_+ \times \Gamma_0; H, \widetilde{H})_{\text{B}} \to S^\mu_{(\text{cl})}(\mathbb{R}_+ \times \Gamma_0; \widetilde{H}', H')_{\text{B}}.$$

Theorem 6.4.13. *Let* $A \in M_{1/2}L^\mu_{(\text{cl})}(\mathbb{R}_+; H, \widetilde{H})_{\text{B}}$, *and consider the formal adjoint* A^*, *defined by*

$$(Au, v)_{\mathcal{H}^{0,1/2}(\mathbb{R}_+, \widetilde{H}_0)} = (u, A^*v)_{\mathcal{H}^{0,1/2}(\mathbb{R}_+, H_0)}$$

for all $u \in T^{1/2}(\mathbb{R}_+, H)$, $v \in T^{1/2}(\mathbb{R}_+, \widetilde{H}')$.

(i) *We have* $A^* \in M_{1/2}L^\mu_{(\text{cl})}(\mathbb{R}_+; \widetilde{H}', \widetilde{H})_{\text{B}}$. *Moreover, for* $A = \text{Op}(a)$, $a(r, z) \in S^\mu_{(\text{cl})}(\mathbb{R}_+ \times \Gamma_0; H, \widetilde{H})_{\text{B}}$ *we have*

$$A^* = \text{Op}(a^*) \quad \text{for } a^*(r, z) \in S^\mu_{(\text{cl})}(\mathbb{R}_+ \times \Gamma_0; \widetilde{H}', \widetilde{H})_{\text{B}},$$

$$a^*(r, i\rho) = \iint t^{i\xi} a^{(*)}(rt, i(\rho + \xi)) \frac{dt}{t} \dbar\xi,$$

$$a^*(r, z) \sim \sum_{k=0}^\infty \frac{1}{k!}(-r\partial_r)^k \partial_z^k a^{(*)}(r, z).$$

(ii) *The mapping* $a \mapsto a^*$ *is continuous antilinear between the respective spaces of symbols.*

(iii) *Writing*

$$a^*(r, z) = \sum_{k=0}^N \frac{1}{k!}(-r\partial_r)^k \partial_z^k a^{(*)}(r, z) + r^*_{N+1}(r, z), \quad N \in \mathbb{N},$$

we have $r^*_{N+1} \in S^{\mu-(N+1)}(\mathbb{R}_+ \times \Gamma_0; \widetilde{H}', \widetilde{H})_{\text{B}}$ *and*

$$r^*_{N+1}(r, z) = \int_0^1 \frac{(1-\theta)^N}{N!} \iint t^{i\xi}\left((-r\partial_r)^{N+1}\partial_z^{N+1}a^{(*)}\right)(rt, i(\rho+\theta\xi)) \frac{dt}{t} \dbar\xi d\theta.$$

(iv) *The mapping* $a \mapsto r^*_{N+1}$ *is continuous antilinear between the respective spaces of symbols.*

Let us also establish the Mellin variant of the kernel cut-off constructions of Section 1.4. We start again with symbols with constant coefficients. Let $a(z, \eta) \in S^\mu_{(\mathrm{cl})}(\Gamma_0 \times \mathbb{R}^q_\eta; H, \widetilde{H})$ for Hilbert spaces H and $\widetilde{H}$ with group actions κ and $\widetilde{\kappa}$, respectively. Let us set

$$k_M(a)(\theta, \eta) := \int_{\Gamma_0} \theta^{-z} a(z, \eta) \, d\!\!\!^{-}z.$$

For any cut-off function $\psi \in C^\infty_0(\mathbb{R}_+)$ such that $\psi(\theta) = 1$ close to $\theta = 1$ and for $\chi(\theta) := 1 - \psi(\theta)$, we obtain a decomposition

$$a(z, \eta) = M_{1/2, \theta \to z} k_M(a)(\theta, \eta) = a_0(z, \eta) + c(z, \eta)$$

for

$$a_0(z, \eta) = M_{1/2, \theta \to z}(\psi(\theta) k_M(a)(\theta, \eta)) \in S^\mu_{(\mathrm{cl})}(\Gamma_0 \times \mathbb{R}^q_\eta; H, \widetilde{H})$$

and

$$c(z, \eta) = M_{1/2, \theta \to z}(\chi(\theta) k_M(a)(\theta, \eta)) \in \mathcal{S}(\Gamma_0 \times \mathbb{R}^q_\eta, \mathcal{L}(H, \widetilde{H})).$$

This gives us a Mellin kernel cut-off operator

$$S^\mu_{(\mathrm{cl})}(\Gamma_0 \times \mathbb{R}^q_\eta; H, \widetilde{H}) \to S^\mu_{(\mathrm{cl})}(\Gamma_0 \times \mathbb{R}^q_\eta; H, \widetilde{H}), \quad a(z, \eta) \mapsto a_0(z, \eta).$$

For similar reasons as in the Fourier set-up of kernel cut-off we admit operators for arbitrary $\varphi \in C^\infty_0(\mathbb{R}_+)$ or $C^\infty_B(\mathbb{R}_+)$,

$$\mathcal{V}_M(\varphi) a(z, \eta) = M_{1/2, \theta \to z}(\psi(\theta) k_M(a)(\theta, \eta)) = \iint \theta^{i\widetilde{\rho}} \varphi(\theta) a(i(\rho - \widetilde{\rho}), \eta) \frac{d\theta}{\theta} d\!\!\!^{-}\widetilde{\rho}$$

which is a Mellin oscillatory integral in the sense of (6.54) for every fixed ρ, η. Let for the moment $\varphi_M := \varphi$; then we may also write

$$\mathcal{V}_M(\varphi_M) a = M_{1/2} \varphi_M M^{-1}_{1/2} a.$$

Moreover, the Fourier kernel cut-off operator has the form

$$\mathcal{V}_F(\varphi_F) b := F \varphi_F F^{-1} b.$$

We obtain the relationship between $\mathcal{V}_M(\varphi_M)$ and $\mathcal{V}_F(\varphi_F)$, i.e.,

$$\mathcal{V}_M(\varphi_M) a = M_{1/2} \varphi_M M^{-1}_{1/2} a$$
$$= \Psi^{-1}_{1/2} F S_{1/2} \varphi_M S^{-1}_{1/2} F^{-1} \Psi_{1/2} a = \Psi^{-1}_{1/2} \mathcal{V}_F(\varphi_F) \Psi_{1/2} b$$

for $\varphi_F = S_{1/2} \varphi_M S^{-1}_{1/2}$.

Theorem 6.4.14. *The kernel cut-off operator $(\varphi, a) \mapsto \mathcal{V}_M(\varphi)a$ defines a continuous bilinear mapping*

$$\mathcal{V}_M : C_{\mathrm{B}}^{\infty}(\mathbb{R}_+) \times S_{(\mathrm{cl})}^{\mu}(\Gamma_0 \times \mathbb{R}_{\eta}^q; H, \widetilde{H}) \to S_{(\mathrm{cl})}^{\mu}(\Gamma_0 \times \mathbb{R}_{\eta}^q; H, \widetilde{H}), \qquad (6.71)$$

and $\mathcal{V}_M(\varphi)a(z, \eta)$ admits an asymptotic expansion

$$\mathcal{V}_M(\varphi)a(z, \eta) \sim \sum_{k=0}^{\infty} \frac{(-1)^k}{k!} ((-\theta\partial_\theta)^k \varphi)(1)\partial_z^k a(z, \eta). \qquad (6.72)$$

In particular, if $\varphi = \psi$ is a cut-off function, then

$$\mathcal{V}_M(\psi)a(z, \eta) = a(z, \eta) \mod S^{-\infty}(\Gamma_0 \times \mathbb{R}_{\eta}^q; H, \widetilde{H}).$$

Proof. The proof can be reduced to the proof of Theorem 1.4.17. $\qquad \square$

Theorem 6.4.15. *Let $\psi(\theta)$ be a cut-off function in the above-mentioned sense, and set $\psi_\varepsilon(\theta) := \psi(\theta^\varepsilon)$ for $0 < \varepsilon \le 1$. Then for every $a(z, \eta) \in S_{(\mathrm{cl})}^{\mu}(\Gamma_0 \times \mathbb{R}^q; H, \widetilde{H})$ we have*

$$\lim_{\varepsilon \to 0} \mathcal{V}_M(\psi_\varepsilon)a(z, \eta) = a(z, \eta),$$

with convergence in $S_{(\mathrm{cl})}^{\mu}(\Gamma_0 \times \mathbb{R}^q; H, \widetilde{H})$.

Proof. The proof follows from the continuity of (6.71) and $\mathcal{V}_M(1)a = a$. $\qquad \square$

Remark 6.4.16. For $\varphi \in C_{\mathrm{B}}^{\infty}(\mathbb{R}_+)$ and $\partial_\theta^k \varphi(1) = 0$ for all $0 \le k \le N$, we obtain a continuous operator

$$\mathcal{V}_M(\varphi) : S_{(\mathrm{cl})}^{\mu}(\Gamma_0 \times \mathbb{R}_{\eta}^q; H, \widetilde{H}) \to S_{(\mathrm{cl})}^{\mu-(N+1)}(\Gamma_0 \times \mathbb{R}_{\eta}^q; H, \widetilde{H}).$$

Moreover, for $\chi = 1 - \psi$ for a cut-off function ψ we have a continuous operator

$$\mathcal{V}_M(\chi) : S_{(\mathrm{cl})}^{\mu}(\Gamma_0 \times \mathbb{R}_{\eta}^q; H, \widetilde{H}) \to \mathcal{S}(\Gamma_0 \times \mathbb{R}_{\eta}^q, \mathcal{L}(H, \widetilde{H})).$$

Next we pass to amplitude functions with holomorphic dependence on the covariable.

Definition 6.4.17. Define

$$S_{(\mathrm{cl}),\mathcal{O}}^{\mu}(\mathbb{R}_{\eta}^q; H, \widetilde{H})$$

to be the set of all $h(z, \eta) \in \mathcal{A}(\mathbb{C}, S_{(\mathrm{cl})}^{\mu}(\mathbb{R}_{\eta}^q; H, \widetilde{H}))$ such that

$$h(z, \eta)|_{\Gamma_\beta} \in S_{(\mathrm{cl})}^{\mu}(\Gamma_0 \times \mathbb{R}_{\eta}^q; H, \widetilde{H}) \quad \text{for all } \beta \in \mathbb{R},$$

uniformly in compact β-intervals.

Theorem 6.4.18. *The kernel cut-off operator* $\mathcal{V}_M : (\varphi, a) \mapsto \mathcal{V}_M(\varphi)a$ *induces a separately continuous mapping*

$$\mathcal{V}_M : C_0^\infty(\mathbb{R}_+) \times S_{(cl)}^\mu(\Gamma_0 \times \mathbb{R}^q; H, \widetilde{H}) \to S_{(cl), \mathcal{O}}^\mu(\mathbb{R}_\eta^q; H, \widetilde{H}).$$

In particular, $a(z, \eta) \in S^{-\infty}(\Gamma_0 \times \mathbb{R}^q; H, \widetilde{H})$ *entails* $\mathcal{V}_M(\varphi)a \in S_{\mathcal{O}}^{-\infty}(\mathbb{R}^q; H, \widetilde{H})$ *for every* $\varphi \in C_0^\infty(\mathbb{R}_+)$.

Proof. The proof is analogous to the proof of Theorem 1.4.21. The expression $\mathcal{V}_M(\varphi)a$ has the form

$$\mathcal{V}_M(\varphi)a(i\rho, \eta) = \int \theta^{i\rho} \left\{ \varphi(\theta) \int \theta^{-i\rho'} a(i\rho', \eta) \d\rho' \right\} \frac{d\theta}{\theta}.$$

The inverse Mellin transform of a may be interpreted as an element

$$k_M(a)(\theta, \eta) \in (\mathcal{T}^{1/2})'(\mathbb{R}_+ \times \mathbb{R}^q, \mathcal{L}(H, \widetilde{H})) \left(= \mathcal{L}(\mathcal{T}^{1/2}(\mathbb{R}_+ \times \mathbb{R}^q, \mathcal{L}(H, \widetilde{H}))) \right).$$

Then $\varphi(\theta)k_M(a)(\theta, \eta)$ has compact support in θ on $\mathbb{R}_+$. Thus

$$\int \theta^{i\rho} \varphi(\theta) k_M(a)(\theta, \eta) \frac{d\theta}{\theta}$$

extends from $\Gamma_0 \ni z$ to a function

$$(\mathcal{V}_M(\varphi)a)(\beta + i\rho) = \int \theta^{\beta + i\rho} \varphi(\theta) k_M(a)(\theta, \eta) \frac{d\theta}{\theta} = \int \theta^{i\rho} \theta^\beta \varphi(\theta) k_M(a)(\theta, \eta) \frac{d\theta}{\theta},$$

which is holomorphic in $z = \beta + i\rho$ and $S_{(cl)}^\mu(\mathbb{R}^q; H, \widetilde{H})$-valued, and we have $\mathcal{V}_M(\varphi)a(\beta + i\rho) \in S_{(cl)}^\mu(\Gamma_\beta \times \mathbb{R}^q; H, \widetilde{H})$ for every $\beta \in \mathbb{R}$, since the map $\beta \mapsto \varphi_\beta(\theta) := \theta^\beta \varphi(\theta)$, $\mathbb{R}_+ \to C_0^\infty(\mathbb{R}_+)$, is uniformly continuous on compact β-intervals, which completes the proof. $\square$

Proposition 6.4.19. *For every* $a(z, \eta) \in S_{(cl)}^\mu(\Gamma_0 \times \mathbb{R}^q; H, \widetilde{H})$ *there exists an* $h(z, \eta) \in S_{(cl), \mathcal{O}}^\mu(\mathbb{R}^q; H, \widetilde{H})$ *such that*

$$a(z, \eta) = h(z, \eta)\big|_{\Gamma_0} \mod S^{-\infty}(\Gamma_0 \times \mathbb{R}^q; H, \widetilde{H}).$$

Proof. It suffices to set $h = \mathcal{V}_M(\psi)$ for a cut-off function ψ and to employ the asymptotic expansion (6.72). $\square$

The following results are also analogous to what we did before in the context of kernel cut-off for the Fourier transform.

Remark 6.4.20. The kernel cut-off operator $\mathcal{V}_M$ can also be introduced with respect to any other line Γ_β rather than Γ_0.

Proposition 6.4.21. *Let* $a(z, \eta) \in S^{\mu}_{(\mathrm{cl})}\big(\Gamma_0 \times \mathbb{R}^q; H, \widetilde{H}\big)$ *and* $h = \mathcal{V}_M(\varphi)$ *for a* $\varphi \in C^{\infty}_0(\mathbb{R}_+)$. *Then for any* $\beta \in \mathbb{R}$ *we have the asymptotic expansion*

$$h(\beta + i\rho, \eta) \sim \sum_{k=0}^{\infty} \frac{(-1)^k}{k!} \big((-\theta\partial_\theta)^k \varphi_\beta\big)(1)(\partial_z^k a)(i\rho, \eta),$$

where $\varphi_\beta(\theta) = \theta^\beta \varphi(\theta)$.

Proposition 6.4.22. *Let* $h(z, \eta) \in S^{\mu}_{(\mathrm{cl}),\mathcal{O}}\big(\mathbb{R}^q; H, \widetilde{H}\big)$. *Then*

$$h(\beta + i\rho, \eta) \sim \sum_{k=0}^{\infty} \frac{(-1)^k}{k!} \big((-\theta\partial_\theta)^k \varphi_\beta\big)(1)(\partial_z^k h)(i\rho, \eta)$$

for every $\beta \in \mathbb{R}$.

The Mellin kernel cut-off operator $\mathcal{V}_M$ has been defined so far with respect to the weighted Mellin transform $M_{1/2}$. If necessary, we will write $\mathcal{V}_{M_{1/2}}$ instead of $\mathcal{V}_M$. It makes sense to consider also the case of kernel cut-off with respect to arbitrary weight $\gamma \in \mathbb{R}$, namely,

$$\mathcal{V}_{M_\gamma} : C^{\infty}_{\mathrm{B}}(\mathbb{R}_+) \times S^{\mu}_{(\mathrm{cl})}\big(\Gamma_{1/2-\gamma} \times \mathbb{R}^q; H, \widetilde{H}\big) \to S^{\mu}_{(\mathrm{cl})}\big(\Gamma_{1/2-\gamma} \times \mathbb{R}^q; H, \widetilde{H}\big)$$

or

$$\mathcal{V}_{M_\gamma} : C^{\infty}_{\mathrm{B}}(\mathbb{R}_+) \times S^{\mu}_{(\mathrm{cl})}\big(\Gamma_{1/2-\gamma} \times \mathbb{R}^q; H, \widetilde{H}\big) \to S^{\mu}_{(\mathrm{cl}),\mathcal{O}}\big(\mathbb{R}^q; H, \widetilde{H}\big). \qquad (6.73)$$

The map (6.73) is defined by

$$\mathcal{V}_{M_\gamma}(\varphi) := T^{1/2-\gamma} \mathcal{V}_{M_{1/2}}(\varphi) T^{-1/2+\gamma},$$

where $(T^\beta f)(z, \eta) := f(z+\beta, \eta)$ for any real β. Clearly, all results and observations on $\mathcal{V}_{M_{1/2}}$ are valid in analogous form for $\mathcal{V}_{M_\gamma}$. If the weight is clear from the context, e.g., in the sense of (6.73), for brevity we also write $\mathcal{V}_M$ rather than $\mathcal{V}_{M_\gamma}$.

Proposition 6.4.23. *Let* $a(z, \eta) \in S^{\mu}_{(\mathrm{cl}),\mathcal{O}}\big(\mathbb{R}^q; H, \widetilde{H}\big)$; *then*

$$h(\beta_0 + i\rho, \eta) \in S^{\mu-\varepsilon}_{(\mathrm{cl})}\big(\Gamma_{\beta_0} \times \mathbb{R}^q; H, \widetilde{H}\big)$$

for any fixed $\beta_0 \in \mathbb{R}$ *and* $0 < \varepsilon \leq 1$ ($\varepsilon = 1$ *in the classical case*) *implies* $h(z, \eta) \in S^{\mu-\varepsilon}_{(\mathrm{cl}),\mathcal{O}}\big(\mathbb{R}^q; H, \widetilde{H}\big)$.

Corollary 6.4.24. $a, b \in S^{\mu}_{(\mathrm{cl}),\mathcal{O}}\big(\mathbb{R}^q; H, \widetilde{H}\big)$ *and* $(a-b)|_{\Gamma_\beta} \in S^{-\infty}\big(\Gamma_\beta \times \mathbb{R}^q; H, \widetilde{H}\big)$ *for some fixed* $\beta \in \mathbb{R}$ *implies* $a = b \bmod S^{-\infty}_{\mathcal{O}}\big(\mathbb{R}^q; H, \widetilde{H}\big)$.

Theorem 6.4.25. *For every sequence* $a_j \in S^{\mu_j}_{(\mathrm{cl}),\mathcal{O}}\big(\mathbb{R}^q; H, \widetilde{H}\big)$, $\mu_j \to -\infty$ *as* $j \to \infty$ ($\mu_j = \mu - j$ *in the classical case*) *there is an asymptotic sum* $a \in S^{\mu}_{(\mathrm{cl}),\mathcal{O}}\big(\mathbb{R}^q; H, \widetilde{H}\big)$ *for* $\mu = \{\mu_j\}_{j \in \mathbb{N}}$, *i.e., for every* $N \in \mathbb{N}$ *there is a sequence* $\nu_N \in \mathbb{R}$, $\nu_N \to -\infty$ *as* $N \to \infty$, *such that* $a - \sum_{j=0}^{N} a_j \in S^{\nu_N}_{\mathcal{O}}\big(\mathbb{R}^q; H, \widetilde{H}\big)$. *If* $b \in S^{\mu}_{(\mathrm{cl}),\mathcal{O}}\big(\mathbb{R}^q; H, \widetilde{H}\big)$ *is another such symbol then* $a = b \bmod S^{-\infty}_{\mathcal{O}}\big(\mathbb{R}^q; H, \widetilde{H}\big)$.

6.5   Spaces and operators with asymptotics

In this section we discuss asymptotics of weighted distributions on a manifold with conical singularity.

A sequence

$$\mathcal{P} := \{(p_j, m_j)\}_{j=0,\ldots,J} \subset \mathbb{C} \times \mathbb{N}, \quad J = J(\mathcal{P}) \in \mathbb{N} \cup \{\infty\},$$

is called a (discrete) asymptotic type associated with the weight data (γ, Θ), $\Theta = (\vartheta, 0]$, $-\infty \leq \vartheta < 0$, $\gamma \in \mathbb{R}$, if

$$\pi_{\mathbb{C}}\mathcal{P} := \{p_j\}_{j=0,\ldots,J} \subset \left\{\frac{n+1}{2} - \gamma + \vartheta < \operatorname{Re} z < \frac{n+1}{2} - \gamma\right\};$$

moreover, $\pi_{\mathbb{C}}\mathcal{P}$ is finite if Θ is finite, otherwise we assume $\operatorname{Re} p_j \to -\infty$ as $j \to \infty$.

Singular functions with asymptotics of type $\mathcal{P}$ on $N^\wedge = \mathbb{R}_+ \times N \ni (r, x)$ for a smooth closed manifold N are of the form

$$\omega(r)c(x)r^{-p_j} \log^k r, \quad 0 \leq k \leq m_j, \quad c \in C^\infty(N),$$

for $(p_j, m_j) \in \mathcal{P}$ and a cut-off function ω. For finite Θ and a fixed cut-off function ω we set

$$\mathcal{E}_{\mathcal{P}}(N^\wedge) := \left\{\sum_{j=0}^{J}\sum_{k=0}^{m_j} \omega(r)c_{jk}(x)r^{-p_j} \log^k r : 0 \leq k \leq m_j, c_{jk} \in C^\infty(N)\right\}.$$

Observe that $\pi_{\mathbb{C}}\mathcal{P} \subset \left\{\operatorname{Re} z < \frac{n+1}{2} - \gamma\right\}$ entails $\mathcal{E}_{\mathcal{P}}(N^\wedge) \subset \mathcal{K}^{\infty,\gamma}(N^\wedge)$. Moreover, we introduce spaces of functions of flatness Θ with respect to the weight γ, namely,

$$\mathcal{K}_{\Theta}^{s,\gamma}(N^\wedge) := \varprojlim_{i \in \mathbb{N}} \mathcal{K}^{s,\gamma-\vartheta-(1+i)^{-1}}(N^\wedge).$$

For $\Theta = (-\infty, 0]$ we simply set $\mathcal{K}_{\Theta}^{s,\gamma}(N^\wedge) := \mathcal{K}^{s,\infty}(N^\wedge) = \bigcap_{\delta \in \mathbb{R}} \mathcal{K}^{s,\delta}(N^\wedge)$. If Θ is finite, the space $\mathcal{E}_{\mathcal{P}}(N^\wedge)$ is isomorphic to a finite direct sum of copies of $C^\infty(N)$, in particular, a Fréchet space. Then we set

$$\mathcal{K}_{\mathcal{P}}^{s,\gamma}(N^\wedge) := \mathcal{K}_{\Theta}^{s,\gamma}(N^\wedge) + \mathcal{E}_{\mathcal{P}}(N^\wedge)$$

as a direct sum of Fréchet spaces. In the case $\Theta = (-\infty, 0]$ we first form $\Theta_l := (-(l+1), 0]$, $l \in \mathbb{N}$, $\mathcal{P}_l := \{(p, m) \in \mathcal{P} : \operatorname{Re} p > \frac{n+1}{2} - \gamma - (l+1)\}$. We then have continuous embeddings $\mathcal{K}_{\mathcal{P}_{l+1}}^{s,\gamma}(N^\wedge) \hookrightarrow \mathcal{K}_{\mathcal{P}_l}^{s,\gamma}(N^\wedge)$ for all l, and we set

$$\mathcal{K}_{\mathcal{P}}^{s,\gamma}(N^\wedge) := \varprojlim_{l \in \mathbb{N}} \mathcal{K}_{\mathcal{P}_l}^{s,\gamma}(N^\wedge).$$

Furthermore, let

$$\mathcal{K}_{\mathcal{P}}^{s,\gamma;e}(N^\wedge) := \{\omega u_0 + (1-\omega)u_\infty : u_0 \in \mathcal{K}_{\mathcal{P}}^{s,\gamma}(N^\wedge), u_\infty \in \mathcal{K}^{s,\gamma;e}(N^\wedge)\}, \quad (6.74)$$

and

$$S_{\mathcal{P}}^{\gamma}(N^{\wedge}) := \mathcal{K}_{\mathcal{P}}^{\infty,\gamma;\infty}(N^{\wedge}) \qquad (6.75)$$

for any asymptotic type $\mathcal{P}$ associated (γ, Θ), $\Theta = (\vartheta, 0]$, $-\infty \leq \vartheta < 0$. In particular, we set

$$S_{\Theta}^{\gamma}(N^{\wedge}) := \mathcal{K}_{\Theta}^{\infty,\gamma;\infty}(N^{\wedge}). \qquad (6.76)$$

For a manifold M with conical singularities we set

$$H_{\mathcal{P}}^{s,\gamma}(M) := \left\{ u \in H^{s,\gamma}(M) : \omega \big(u|_{V\setminus\{v\}} \circ (\chi^{\wedge})^{-1}\big) \in \mathcal{K}_{\mathcal{P}}^{s,\gamma}(N^{\wedge}) \right\},$$

for any cut-off function ω, cf. also (6.34).

In the future we will often identify $V \setminus \{v\}$ with $N^{\wedge}$ and for brevity drop the pull-backs under $\chi^{\wedge}$ when we talk about functions or operators in a neighbourhood of the conical singularity v. In that sense we have the simpler notation

$$H_{\mathcal{P}}^{s,\gamma}(M) = \left\{ u \in H^{s,\gamma}(M) : \omega u \in \mathcal{K}_{\mathcal{P}}^{s,\gamma}(N^{\wedge}) \right\}.$$

Remark 6.5.1. Let us fix a cut-off function $\omega(r)$. Then we have continuous embeddings

(i)

$$[\omega]r^{j}C^{m}(\overline{\mathbb{R}}_{+}) \hookrightarrow [\omega]\mathcal{H}^{\widetilde{m},j}(\mathbb{R}_{+})$$

for every $j, m \in \mathbb{N}$ and $\widetilde{m} = \widetilde{m}(m) \to \infty$ as $m \to \infty$.

(ii)

$$[\omega]\mathcal{H}_{\mathcal{P}_{k}}^{n,0}(\mathbb{R}_{+}) \hookrightarrow [\omega]C^{\widetilde{n}}(\overline{\mathbb{R}}_{+})$$

for every $n \in \mathbb{N}$ and $\widetilde{n} = \widetilde{n}(n) \to \infty$ as $n \to \infty$, and $k \to \infty$; here $\mathcal{P}_{k} := \{(-j, 0) : j = 0, \dots, k\}$ is associated with $\Theta = (-(k+1), 0]$.

(iii) Also,

$$\mathcal{S}(\overline{\mathbb{R}}_{+}) \subset \mathcal{H}_{\mathcal{T}}^{\infty,0}(\mathbb{R}_{+}) \quad \text{for} \quad \mathcal{T} = \{(-j, 0) : j \in \mathbb{N}\}.$$

Definition 6.5.2. (i) An operator

$$G \in \bigcap_{s\in\mathbb{R}} \mathcal{L}\big(\mathcal{K}^{s,\gamma}(N^{\wedge}), \mathcal{K}^{\infty,\gamma-\mu}(N^{\wedge})\big)$$

is called a Green operator on $N^{\wedge}$ for a closed smooth manifold N, if there are (G-dependent) asymptotic types $\mathcal{P}$ and $\mathcal{Q}$ such that

$$G \in \bigcap_{s,e\in\mathbb{R}} \mathcal{L}\big(\mathcal{K}^{s,\gamma;e}(N^{\wedge}), \mathcal{K}_{\mathcal{P}}^{\infty,\gamma-\mu;\infty}(N^{\wedge})\big),$$

$$G^{*} \in \bigcap_{s,e\in\mathbb{R}} \mathcal{L}\big(\mathcal{K}^{s,-\gamma+\mu;e}(N^{\wedge}), \mathcal{K}_{\mathcal{Q}}^{\infty,-\gamma;\infty}(N^{\wedge})\big).$$

(ii) An operator $G \in \bigcap_{s\in\mathbb{R}} \mathcal{L}(H^{s,\gamma}(M), H^{\infty,\gamma-\mu}(M))$ is called a Green operator on the (compact) manifold M with conical singularity if there are asymptotic types $\mathcal{P}$ and $\mathcal{Q}$ such that

$$G \in \bigcap_{s\in\mathbb{R}} \mathcal{L}(H^{s,\gamma}(M), H^{\infty,\gamma-\mu}_{\mathcal{P}}(M)), \quad G^* \in \bigcap_{s\in\mathbb{R}} \mathcal{L}(H^{s,-\gamma+\mu}(M), H^{\infty,-\gamma}_{\mathcal{Q}}(M)).$$

By $L_G(\cdot, \boldsymbol{g})$ for $\boldsymbol{g} := (\gamma, \gamma - \mu, \Theta)$ we denote the space of Green operators on $N^\wedge$ and M, respectively. The subspaces with fixed $\mathcal{P}$, $\mathcal{Q}$ will be denoted by $L_G(\cdot, \boldsymbol{g})_{\mathcal{P},\mathcal{Q}}$. In the case $\pi_{\mathbb{C}}\mathcal{P} = \emptyset$ or $\pi_{\mathbb{C}}\mathcal{Q} = \emptyset$ we write

$$L_G(\cdot, \boldsymbol{g})_{O,\mathcal{Q}}, \quad \text{respectively } L_G(\cdot, \boldsymbol{g})_{\mathcal{P},O}. \tag{6.77}$$

Green operators will play the role of smoothing operators in the cone calculus below. Those are compact as operators

$$G : \mathcal{K}^{s,\gamma}(N^\wedge) \to \mathcal{K}^{s-\mu,\gamma-\mu}(N^\wedge)$$

and

$$G : H^{s,\gamma}(M) \to H^{s-\mu,\gamma-\mu}(M),$$

respectively, for every $s \in \mathbb{R}$.

Remark 6.5.3. $L_G(M, \boldsymbol{g})_{\mathcal{P},\mathcal{Q}}$ and $L_G(N^\wedge, \boldsymbol{g})_{\mathcal{P},\mathcal{Q}}$ have natural structures of Fréchet spaces.

In fact, the conditions of Definition 6.5.2 (i) are equivalent to

$$G : \mathcal{K}^{s,\gamma;e}(N^\wedge) \to \mathcal{K}^{s',\gamma-\mu;e'}_{\mathcal{P}}(N^\wedge), \quad G^* : \mathcal{K}^{s,-\gamma+\mu;e}(N^\wedge) \to \mathcal{K}^{s',-\gamma;e'}_{\mathcal{Q}}(N^\wedge) \tag{6.78}$$

for arbitrary $s, s', e, e' \in \mathbb{Z}$. Then by virtue of interpolation properties, cf. [25], the continuities hold for all $s, s', e, e' \in \mathbb{R}$. This allows us to characterise Green operators by a countable number of conditions. Similar observations hold for Green operators over M.

Remark 6.5.4. It can be proved, cf. [60], that the operators $G \in L_G(N^\wedge, \boldsymbol{g})_{\mathcal{P},\mathcal{Q}}$, $\boldsymbol{g} = (\gamma, \gamma - \mu, \Theta)$, admit a characterisation by means of kernels in

$$\mathcal{S}^{\gamma-\mu}_{\mathcal{P}}(N^\wedge) \widehat{\otimes}_\Gamma \mathcal{S}^{-\gamma}_{\mathcal{Q}}(N^\wedge) := \mathcal{S}^{\gamma-\mu}_{\mathcal{P}}(N^\wedge) \widehat{\otimes}_\pi \mathcal{S}^{-\gamma}_0(N^\wedge) \bigcap \mathcal{S}^{\gamma-\mu}_0(\mathbb{R}_+) \widehat{\otimes}_\pi \mathcal{S}^{-\gamma}_{\mathcal{Q}}(N^\wedge)$$

for

$$\mathcal{S}^\beta_0(N^\wedge) := \{u \in \mathcal{K}^{\infty,\beta;\infty}(N^\wedge) : \log^k r\, u \in \mathcal{K}^{\infty,\beta;\infty}(N^\wedge) \text{ for all } k \in \mathbb{N}\}, \quad \beta \in \mathbb{R}.$$

Remark 6.5.5. Definition 6.5.2 immediately extends to the case of arbitrary weights $\delta \in \mathbb{R}$ instead of $\gamma - \mu$, and we tacitly employ the notation $L_G(M, \boldsymbol{g})$ and $L_G(N^\wedge, \boldsymbol{g})$ for the corresponding spaces with weight data $\boldsymbol{g} = (\gamma, \delta, \Theta)$ as well as for the respective subspaces $L_G(M, \boldsymbol{g})_{\mathcal{P},\mathcal{Q}}$ and $L_G(N^\wedge, \boldsymbol{g})_{\mathcal{P},\mathcal{Q}}$ with asymptotic types $\mathcal{P}$, $\mathcal{Q}$.

Observe that

$$L_G\big(N^\wedge,(\gamma,\delta,\Theta)\big) = r^{-\gamma+\mu+\delta}L_G\big(N^\wedge,(\gamma,\gamma-\mu,\Theta)\big) \qquad (6.79)$$

for any $\gamma,\mu,\delta \in \mathbb{R}$, or

$$r^\beta L_G\big(N^\wedge,(\gamma,\delta,\Theta)\big)r^{-\alpha} = L_G\big(N^\wedge,(\gamma+\beta,\delta+\alpha,\Theta)\big).$$

More precisely, $G \in L_G\big(N^\wedge,(\gamma,\delta,\Theta)\big)_{\mathcal{P},\mathcal{Q}}$ is characterised by the of continuity of

$$G : \mathcal{K}^{s,\gamma;e}\big(N^\wedge\big) \to \mathcal{K}_{\mathcal{P}}^{\infty,\delta;\infty}\big(N^\wedge\big) \quad \text{and} \quad G^* : \mathcal{K}^{s,-\delta;e}\big(N^\wedge\big) \to \mathcal{K}_{\mathcal{Q}}^{\infty,-\gamma;\infty}\big(N^\wedge\big)$$
$$(6.80)$$

for all $s,e \in \mathbb{R}$. Then we have

$$r^\beta G r^{-\alpha} \in L_G\big(N^\wedge,(\gamma+\alpha,\delta+\beta,\Theta)\big)_{T^{-\beta}\mathcal{P},T^\alpha\mathcal{Q}}. \qquad (6.81)$$

Note that

$$L_G\big(N^\wedge,(\gamma,\delta,\Theta)\big) \subseteq L_G\big(N^\wedge,(\gamma',\delta',\Theta')\big) \qquad (6.82)$$

for all

$$\gamma \le \gamma', \ \delta \ge \delta', \ \Theta \supseteq \Theta'$$

where $\Theta \supseteq \Theta'$ for $\Theta = (\vartheta,0]$, $\Theta' = (\vartheta',0]$ means $\vartheta \le \vartheta'$.

In the case $\Theta = (-\infty,0]$ the latter class is independent of γ or δ and we denote it simply by

$$L_G\big(N^\wedge\big)_{O,O}. \qquad (6.83)$$

The following operator spaces will be employed below. Set

$$\mathbb{G}\big(N^\wedge\big) := \bigcup_{\gamma,\delta\in\mathbb{R}} L_G\big(N^\wedge,(\gamma,\delta,(-\infty,0])\big), \qquad (6.84)$$

and let

$$\mathbb{G}\big(N^\wedge\big)_{\mathcal{P},\mathcal{Q}}$$

be the space of all $G \in \mathbb{G}(N^\wedge)$ that induce continuous operators

$$G : \mathcal{K}^{s,\gamma;e}\big(N^\wedge\big) \to \mathcal{K}_{\mathcal{P}}^{\infty,\delta;\infty}\big(N^\wedge\big) \quad \text{and} \quad G^* : \mathcal{K}^{s,-\delta;e}\big(N^\wedge\big) \to \mathcal{K}_{\mathcal{Q}}^{\infty,-\gamma;\infty}\big(N^\wedge\big)$$

for all $s,e \in \mathbb{R}$, for asymptotic types $\mathcal{P}$ and $\mathcal{Q}$ associated with the weight data $(\delta,(-\infty,0])$ and $(-\gamma,(-\infty,0])$, respectively.

Example 6.5.6. Any finite linear combination $G_{\gamma,\delta}$ of products $f(r,x)g(r',x')$ for $f \in \mathcal{K}_{\mathcal{P}}^{\infty,\delta;\infty}\big(N^\wedge\big)$, $g \in \mathcal{K}_{\mathcal{Q}}^{\infty,-\gamma;\infty}\big(N^\wedge\big)$, and $\mathcal{P},\mathcal{Q}$ associated with $(\delta,(-\infty,0])$, $(-\gamma,(-\infty,0])$, respectively, defines kernels of operators in $L_G\big(N^\wedge,(\gamma,\delta,(-\infty,0])\big)$ of finite dimension.

Remark 6.5.7. Operators in $L_G(N^\wedge)_{O,O}$ have kernels in $\mathcal{S}_O(N^\wedge) \,\widehat{\otimes}_\pi\, \mathcal{S}_O(N^\wedge)$ where

$$\mathcal{S}_O(N^\wedge) := \{u(r,x) \in \mathcal{S}(\mathbb{R}, C^\infty(N)) : \operatorname{supp} u \subset \overline{\mathbb{R}}_+ \times N\}.$$

The latter space coincides with $\mathcal{K}^{\infty,\infty;\infty}(N^\wedge)$.

Another essential ingredient of the cone calculus are smoothing Mellin operators. Those are, in general, not compact in such a sense.

Definition 6.5.8. A discrete asymptotic type for Mellin symbols is a sequence

$$\mathcal{R} := \{(r_j, n_j)\}_{j \in \mathbb{J}} \subset \mathbb{C} \times \mathbb{N}$$

for some subset $\mathbb{J} \subseteq \mathbb{Z}$ such that for $\pi_\mathbb{C} \mathcal{R} := \{r_j\}_{j \in \mathbb{J}}$ the set

$$\pi_\mathbb{C} \mathcal{R} \cap \{c \le \operatorname{Re} z \le c'\}$$

is finite for every $c < c'$.

In order to have a convenient expression for operations between Mellin symbols with different asymptotic types $\mathcal{R} = \{(r_j, n_j)\}_{j \in \mathbb{J}}$, $\mathcal{Q} = \{(q_l, m_l)\}_{l \in \mathbb{L}}$, where $\mathbb{J}, \mathbb{L} \subseteq \mathbb{Z}$, we introduce $\mathcal{R} \cup \mathcal{Q}$ which is simply the set-theoretic union of the respective sequences. Moreover, we define the addition

$$\mathcal{R} + \mathcal{Q}, \tag{6.85}$$

between asymptotic types $\mathcal{R}, \mathcal{Q}$ as a sequence of pairs $(p, s) \subset \mathbb{C} \times \mathbb{N}$ where

$$\pi_\mathbb{C}(\mathcal{R} + \mathcal{Q}) = \pi_\mathbb{C} \mathcal{R} \bigcup \pi_\mathbb{C} \mathcal{Q}$$

and

$$s = \begin{cases} n_j + m_l + 1 & \text{if } p = r_j = q_l \text{ for a pair } j, l, \\ n_j & \text{if } p = r_j \text{ for a } j \text{ and } p \ne q_l \text{ for all } l, \\ m_l & \text{if } p = q_l \text{ for an } l \text{ and } p \ne r_j \text{ for all } j. \end{cases}$$

Moreover, we set

$$\begin{aligned} \mathcal{R}^* &:= \{(n + 1 - \bar{r}_j, n_j) : (r_j, n_j) \in \mathcal{R}\}_{j \in \mathbb{J}}, \\ T^{-\beta} \mathcal{R} &:= \{(r_j - \beta, n_j) : (r_j, n_j) \in \mathcal{R}\}_{j \in \mathbb{J}}, \end{aligned} \tag{6.86}$$

where $n = \dim N$, $\beta \in \mathbb{R}$. Observe that $T^\gamma(\mathcal{R} + \mathcal{Q}) = T^\gamma \mathcal{R} + T^\gamma \mathcal{Q}$.

Definition 6.5.9. Let $M_\mathcal{R}^{-\infty}(N)$ be the subspace of all $f \in \mathcal{A}(\mathbb{C} \setminus \pi_\mathbb{C} \mathcal{R}, L^{-\infty}(N))$ that are meromorphic with poles at the points r_j of multiplicity $n_j + 1$, where the Laurent coefficients at $(z - r_j)^{-(k+1)}$, $0 \le k \le n_j$, are operators in $L^{-\infty}(N)$ of finite rank, and for any $\pi_\mathbb{C} \mathcal{R}$-cut-off function χ we have $\chi f|_{\Gamma_\beta} \in \mathcal{S}(\Gamma_\beta, L^{-\infty}(N))$ for every $\beta \in \mathbb{R}$, uniformly in finite β-intervals. Moreover, set

$$M_\mathcal{R}^\mu(N) := M_\mathcal{O}^\mu(N) + M_\mathcal{R}^{-\infty}(N), \tag{6.87}$$

endowed with the Fréchet topology of the non-direct sum.

Observe that we also have a non-direct sum decomposition

$$C^\infty\big(\overline{\mathbb{R}}_+, M_{\mathcal{R}}^\mu(N)\big) = C^\infty\big(\overline{\mathbb{R}}_+, M_{\mathcal{O}}^\mu(N)\big) + C^\infty\big(\overline{\mathbb{R}}_+, M_{\mathcal{R}}^{-\infty}(N)\big).$$

Let us now extend notation (6.87) to the case of operators between distributional sections of vector bundles over N. Using the spaces $L^{-\infty}(N; E, F)$ for $E, F \in \mathrm{Vect}(N)$ we denote by

$$M_{\mathcal{R}}^{-\infty}(N; E, F) \tag{6.88}$$

the set of all $f \in \mathcal{A}\big(\mathbb{C} \setminus \pi_{\mathbb{C}}\mathcal{R}, L^{-\infty}(N; E, F)\big)$ that are meromorphic with poles at the points r_j of multiplicity $n_j + 1$ and finite-rank Laurent coefficients at $(z - r_j)^{-(k+1)}$, $0 \le k \le n_j$, and $\chi f|_{\Gamma_\beta} \in \mathcal{S}\big(\Gamma_\beta, L^{-\infty}(N; E, F)\big)$ for every $\beta \in \mathbb{R}$, uniformly in finite β-intervals. In addition we have a straightforward generalisation of the notation $M_{\mathcal{O}}^\mu(N)$ to $M_{\mathcal{O}}^\mu(N; E, F)$, cf. also formula (1.19) and its extension to the parameter-dependent case. Then we set again

$$M_{\mathcal{R}}^\mu(N; E, F) := M_{\mathcal{O}}^\mu(N; E, F) + M_{\mathcal{R}}^{-\infty}(N; E, F) \tag{6.89}$$

as a non-direct sum of Fréchet spaces.

For the remaining considerations on spaces of Mellin symbols we return to the case of trivial bundles of fibre dimension 1. Simple generalisations to the bundle case will be tacitly used below.

Remark 6.5.10. (i) If $f \in M_{\mathcal{R}}^{-\infty}(N)$ and $g \in M_{\mathcal{Q}}^{-\infty}(N)$ then $f + g \in M_{\mathcal{R}\cup\mathcal{Q}}^{-\infty}(N)$, $fg \in M_{\mathcal{R}+\mathcal{Q}}^{-\infty}(N)$;

(ii) $f \in M_{\mathcal{R}}^{-\infty}(N)$ implies for the z-wise formal adjoint $f^{(*)}(n+1-\bar{z}) \in M_{\mathcal{R}^*}^{-\infty}(N)$;

(iii) $f \in M_{\mathcal{R}}^{-\infty}(N)$ implies $T^\beta f \in M_{T^{-\beta}\mathcal{R}}^{-\infty}(N)$ for $T^\beta f(z) := f(z + \beta)$;

(iv) More generally, $f \in M_{\mathcal{R}}^\mu(N)$, $g \in M_{\mathcal{Q}}^\nu(N)$ implies $f + g \in M_{\mathcal{R}\cup\mathcal{Q}}^{\max\{\mu,\nu\}}(N)$ when $\mu - \nu \in \mathbb{Z}$, $fg \in M_{\mathcal{R}+\mathcal{Q}}^{\mu+\nu}(N)$; moreover, $f^{(*)}(n+1-\bar{z}) \in M_{\mathcal{R}^*}^{-\infty}(N)$ and $T^\beta f \in M_{T^{-\beta}\mathcal{R}}^{-\infty}(N)$.

Theorem 6.5.11. *Let* $f \in M_{\mathcal{R}}^{-\infty}(N)$ *for some Mellin asymptotic type* $\mathcal{R}$*, and fix cut-off functions* ω, ω'*. Then for any* $\beta, \gamma \in \mathbb{R}$ *with*

$$\pi_{\mathbb{C}}\mathcal{R} \cap \Gamma_{(n+1)/2-\gamma} = \pi_{\mathbb{C}}\mathcal{R} \cap \Gamma_{(n+1)/2-(\gamma+\beta)} = \emptyset$$

we have

$$\omega \, \mathrm{op}_M^{\gamma-n/2}(f) r^\beta \omega' - \omega r^\beta \, \mathrm{op}_M^{\gamma-n/2}(T^{-\beta}f)\omega'$$
$$=: G \in L_G\big(N^\wedge, (\max\{\gamma, \gamma - \beta\}, \min\{\gamma, \gamma + \beta\}, (-\infty, 0])\big)$$

or, alternatively,

$$\omega \, \mathrm{op}_M^{\gamma-n/2}(f)\omega' = \omega \, \mathrm{op}_M^{\gamma+\beta-n/2}(f)\omega'$$
$$\mathrm{mod}\ L_G\big(N^\wedge, (\max\{\gamma, \gamma - \beta\}, \min\{\gamma, \gamma + \beta\}, (-\infty, 0])\big). \tag{6.90}$$

Proof. For $u \in C_0^\infty(\mathbb{R}_+, C^\infty(N))$ we write $v := \omega'u$. Then

$$\omega r^\beta \, \mathrm{op}_M^{\gamma-n/2}(T^{-\beta}f)\omega'u = \omega \int_{\Gamma_{(n+1)/2-\gamma}} r^{-z+\beta} f(z-\beta) \left\{ \int_0^\infty (r')^{z-1} v(r')dr' \right\} đz$$

$$= \omega \int_{\Gamma_{(n+1)/2-(\gamma+\beta)}} r^{-z} f(z) \left\{ \int_0^\infty (r')^{z+\beta-1} v(r')dr' \right\} đz$$

$$= \omega \int_{\Gamma_{(n+1)/2-\gamma}} r^{-z} f(z) \left\{ \int_0^\infty (r')^{z-1}(r')^\beta v(r')dr' \right\} đz + Gu$$

$$= \omega \, \mathrm{op}_M^{\gamma-n/2}(f) r^\beta \omega'u + Gu$$

for

$$Gu(r) = \omega \int_{\Delta_{\beta\gamma}} r^{-z} f(z) \left\{ \int_0^\infty (r')^{z-1}(r')^\beta v(r')dr' \right\} đz,$$

where the orientation in

$$\Delta_{\beta\gamma} := \Gamma_{(n+1)/2-\gamma} \cup \Gamma_{(n+1)/2-(\gamma+\beta)}$$

is from $\mathrm{Im}\, z = \infty$ to $\mathrm{Im}\, z = -\infty$ on $\Gamma_{(n+1)/2-\gamma}$ and in the opposite direction on $\Gamma_{(n+1)/2-(\gamma+\beta)}$. Since $\int_0^\infty (r')^{z-1}(r')^\beta v(r')dr' =: g(z)$ is an entire $C^\infty(N)$-valued function, strongly decreasing for $|\mathrm{Im}\, z| \to \infty$, uniformly in strips of finite width, by Cauchy's theorem the integral over $\Delta_{\beta\gamma}$ may be replaced by an integral

$$\omega \int_{C_{\beta\gamma}} r^{-z}(fg)(z)đz \tag{6.91}$$

for any smooth compact curve $C_{\beta\gamma}$ surrounding

$$\pi_\mathbb{C} \mathcal{R} \cap S_{\beta,\gamma} =: \left\{ (p_j, m_j) \right\}_{j=0,\dots,N} = \mathcal{P},$$

where $S_{\beta,\gamma}$ is the open strip between $\Gamma_{(n+1)/2-\gamma}$ and $\Gamma_{(n+1)/2-(\gamma+\beta)}$, no matter whether $\gamma \leq \beta$ or $\gamma \geq \beta$. The integrand in (6.91) is meromorphic and hence (6.91) has the form

$$\omega(r) \sum_{j=0}^N \sum_{k=0}^{m_j} c_{jk} r^{-p_j} \log^k r \in \mathcal{K}_\mathcal{P}^{\infty,\delta;\infty}(N^\wedge),$$

for some $c_{jk} \in C^\infty(N)$. Because of the position of $S_{\beta,\gamma}$ in the complex plane, the operator G extends to $\mathcal{K}^{s,\gamma;-\infty}(N^\wedge)$ when $\beta \geq 0$, and to $\mathcal{K}^{s,\gamma-\beta;-\infty}(N^\wedge)$ when $\beta \leq 0$ (where $G = 0$ for $\beta = 0$). In fact, for $\beta > 0$ the Mellin transform of $\omega r^\beta u$ for $u \in \mathcal{K}^{s,\gamma;-\infty}(N^\wedge)$ is holomorphic in $\mathrm{Re}\, z > (n+1)/2 - (\alpha+\beta)$, cf. Remark 6.2.1 (i), and has the asserted growth properties as $|\mathrm{Im}\, z| \to \infty$. Thus fg is meromorphic in $S_{\beta,\gamma}$ and strongly decreasing as $|\mathrm{Im}\, z| \to \infty$, so Cauchy's theorem is still applicable. In other words, in this case we obtain a continuous extension

$$G : \mathcal{K}^{s,\gamma;-\infty}(N^\wedge) \to \mathcal{E}_\mathcal{P}(N^\wedge),$$

for every $s, e \in \mathbb{R}$, cf. the notation (6.74). For $\beta < 0$ and $u \in \mathcal{K}^{s,\gamma-\beta;-\infty}(N^\wedge)$ we have $\omega r^\beta u \in \mathcal{K}^{s,\gamma;-\infty}(N^\wedge)$. Again because of Remark 6.2.1 (i) the product fg is meromorphic in $S_{\beta,\gamma}$ which now lies in $\{\operatorname{Re} z > (n+1)/2 - \gamma\}$ and we may argue as before. Thus, in this case we obtain a continuous operator

$$G : \mathcal{K}^{s,\gamma-\beta;-\infty}(N^\wedge) \to \mathcal{E}_{\mathcal{P}}(N^\wedge),$$

where the poles including multiplicities of f in $S_{\beta,\gamma}$ just furnish the asymptotic type $\mathcal{P}$. In order to complete the characterisation of G we look at the adjoint

$$G^* = \omega' r^\beta \big(\operatorname{op}_M^{\gamma-n/2}(f)\big)^* \omega - \omega' \big(\operatorname{op}_M^{\gamma-n/2}(T^{-\beta}f)\big)^* r^\beta \omega.$$

By Proposition 6.3.7, we have $\big(\operatorname{op}_M^{\gamma-n/2}(f)\big)^* = \operatorname{op}_M^{-\gamma-n/2}(f^{[*]})$, $f^{[*]} \in M_{\mathcal{R}^*}^{-\infty}(N)$, cf. (6.86), and $\big(\operatorname{op}_M^{\gamma-n/2}(T^{-\beta}f)\big)^* = \operatorname{op}_M^{-\gamma-n/2}\big((T^{-\beta}f)^{[*]}\big)$, where $f^{[*]}(z) = f^{(*)}(n+1-\bar{z})$, $(T^{-\beta}f)^{[*]}(z) = f^{(*)}(n+1-\beta-\bar{z}) = T^\beta f^{[*]}(z)$. Thus

$$-G^* = \omega' \operatorname{op}_M^{-\gamma-n/2}(T^\beta f^{[*]}) r^\beta \omega - \omega' r^\beta \operatorname{op}_M^{-\gamma-n/2}(f^{[*]}) \omega$$
$$= \omega' \operatorname{op}_M^{-\gamma-n/2}(g^{[*]}) r^\beta \omega - \omega' r^\beta \operatorname{op}_M^{-\gamma-n/2}(T^{-\beta}g^{[*]}) \omega$$

for $g^{[*]} := T^\beta f^{[*]}$. This operator is of exactly the same form as G, with the only change from γ to $-\gamma$. It follows that G defines continuous operators

$$G^* : \mathcal{K}^{s,-\gamma;e}(N^\wedge) \to \mathcal{E}_{\mathcal{Q}}(N^\wedge) \quad \text{for } \beta > 0$$

and

$$G^* : \mathcal{K}^{s,-\gamma-\beta;e}(N^\wedge) \to \mathcal{E}_{\mathcal{Q}}(N^\wedge) \quad \text{for } \beta < 0$$

for every $s, e \in \mathbb{R}$. Similarly as before, $\mathcal{Q}$ is determined by the poles including multiplicities of $g^{[*]}$ in the strip $S_{\beta,-\gamma}$. $\square$

Theorem 6.5.11 determines two maps, operating on Mellin asymptotic type $\mathcal{R}$ and producing the asymptotic types $\mathcal{P}$ and $\mathcal{Q}$ of the resulting Green operator G obtained by commuting r^β through the weighted Mellin operator $\omega \operatorname{op}_M^{\gamma-n/2}(f)\omega'$ with a symbol $f \in M_{\mathcal{R}}^{-\infty}(M)$. Let us set

$$\mathcal{P} := \mathrm{m}_{\beta,\gamma}(\mathcal{R}), \quad \mathcal{Q} := \mathrm{m}_{\beta,\gamma}^*(\mathcal{R}).$$

Corollary 6.5.12. *Let* $f \in M_{\mathcal{R}}^{-\infty}(N)$, $\mathcal{R} = \{(r_j, m_j)\}_{j \in \mathbb{J}}$, $\gamma, \delta \in \mathbb{R}$, *and assume*

$$\pi_{\mathbb{C}}\mathcal{R} \cap \Gamma_{(n+1)/2-\gamma} = \pi_{\mathbb{C}}\mathcal{R} \cap \Gamma_{(n+1)/2-\delta} = \emptyset.$$

Then

$$\omega \operatorname{op}_M^{\gamma-n/2}(f)\omega' - \omega \operatorname{op}_M^{\delta-n/2}(f)\omega' =: G$$

belongs to $L_G(N^\wedge, \boldsymbol{g})_{\mathcal{P},\mathcal{Q}}$ *for* $\boldsymbol{g} = (\delta, \gamma, (-\infty, 0])$,

$$\mathcal{P} = \left\{ (r, m) \in \mathcal{R} : \frac{n+1}{2} - \delta < \operatorname{Re} r < \frac{n+1}{2} - \gamma \right\}, \tag{6.92}$$

$$\mathcal{Q} = \left\{ (n+1-\bar{r}, m) : (r, m) \in \mathcal{R},\ \frac{n+1}{2} - \delta < \operatorname{Re} r < \frac{n+1}{2} - \gamma \right\} \qquad (6.93)$$

in the case $\gamma \leq \delta$, *while for* $\gamma \geq \delta$ *we have* $L_G(N^\wedge, \boldsymbol{g})_{\mathcal{P}, \mathcal{Q}}$ *for weight data* $\boldsymbol{g} = (\gamma, \delta, (-\infty, 0])$, *and we interchange the roles of* γ *and* δ *in* (6.92)–(6.93).

In fact, we have

$$Gu(r) = \int_{\Delta_{\gamma\delta}} r^{-z} f(z) \left\{ \int_0^\infty (r')^z u(r') \frac{dr'}{r'} \right\} \bar{d}z$$

for $\Delta_{\gamma\delta} = \Gamma_{(n+1)/2-\gamma} \cup \Gamma_{(n+1)/2-\delta}$ and integration from $\operatorname{Re} z = -\infty$ to $\operatorname{Re} z = +\infty$ on $\Gamma_{(n+1)/2-\gamma}$ and in opposite direction on $\Gamma_{(n+1)/2-\delta}$. This gives us the relation (6.92). In order to compute $\mathcal{Q}$ we employ that

$$\left(\omega \operatorname{op}_M^{\gamma-n/2}(f) \omega' - \omega \operatorname{op}_M^{\delta-n/2}(f) \omega' \right)^* = \omega' \operatorname{op}_M^{-\gamma-n/2}\left(f^{[*]} \right) \omega - \omega' \operatorname{op}_M^{-\delta-n/2}\left(f^{[*]} \right) \omega,$$

$f^{[*]} \in M_{\mathcal{R}^*}^{-\infty}(N)$, and

$$\mathcal{R}^* = \left\{ (n+1-\bar{r}_j, m_j) \right\}_{j \in \mathbb{J}}. \qquad (6.94)$$

Thus, similarly as we did for $\mathcal{P}$, we obtain

$$\mathcal{Q} = \left\{ (r^*, m^*) \in \mathcal{R}^* : \frac{n+1}{2} + \gamma < \operatorname{Re} r^* < \frac{n+1}{2} + \delta \right\}.$$

This relation is equivalent to

$$\frac{n+1}{2} + \gamma < \operatorname{Re}(n+1-\bar{r}) < \frac{n+1}{2} + \delta$$

for $(r, m) \in \mathcal{R}$, and implies immediately the characterisation (6.93).

The rules $\mathcal{R} \to \mathcal{P}$, $\mathcal{R} \to \mathcal{Q}$ in (6.92)–(6.93) will be denoted by

$$\mathcal{P} := \mathrm{a}_{\gamma, \delta}(\mathcal{R}), \quad \mathcal{Q} := \mathrm{a}_{\gamma, \delta}^*(\mathcal{R}).$$

in the case $\gamma \leq \delta$.

We now formulate smoothing Mellin operators of the cone algebra, first for a finite weight interval $\Theta := (-(k+1), 0]$, $k \in \mathbb{N}$ and weight data $\boldsymbol{g} = (\gamma, \gamma - \mu, \Theta)$. Starting with a sequence

$$f_j \in M_{\mathcal{R}_j}^{-\infty}(N), \quad j = 0, \ldots, k,$$

of Mellin symbols for Mellin asymptotic types $\mathcal{R}_j$ we form operators

$$M := r^{-\mu} \sum_{j=0}^k \omega r^j \operatorname{op}_M^{\gamma_j-n/2}(f_j) \omega' \qquad (6.95)$$

for weights $\gamma_j \in \mathbb{R}$ such that

$$\gamma - j \leq \gamma_j \leq \gamma, \quad \Gamma_{(n+1)/2-\gamma_j} \cap \pi_{\mathbb{C}} \mathcal{R}_j = \emptyset, \quad j = 0, \ldots, k, \qquad (6.96)$$

for cut-off functions ω, ω'.

Remark 6.5.13. The Mellin symbols f_j, $j = 0, \ldots, k$, in (6.95) are uniquely determined by the operator M as a map $C_0^\infty(N^\wedge) \to C^\infty(N^\wedge)$. More precisely, if $\widetilde{M}$ is another operator with the same Mellin symbol but other weights $\widetilde{\gamma}_j$ satisfying analogous conditions as (6.96), and other cut-off functions $\widetilde{\omega}$, $\widetilde{\omega}'$ rather than ω, ω', then we have

$$M - \widetilde{M} \in L_{\mathrm{G}}(N^\wedge, \boldsymbol{g}) \tag{6.97}$$

for $\boldsymbol{g} = (\gamma, \gamma - \mu, \Theta)$.

In fact, the uniqueness of the sequence of f_j is proved in [44, Subsection 1.3.1, Theorem 4]. The fact that changing cut-off functions causes Green remainders is straightforward, while appearance of Green remainders under changing weights is a consequence of Theorem 6.5.11.

Definition 6.5.14. (i) We define $L_{\mathrm{M+G}}(\cdot, \boldsymbol{g})$, $\boldsymbol{g} = (\gamma, \gamma - \mu, \Theta)$, $\Theta := (-(k+1), 0]$, as the set of all operators $M + G$ for arbitrary $G \in L_{\mathrm{G}}(\cdot, \boldsymbol{g})$ and M of the form (6.95); here $\cdot$ stands for $N^\wedge$ or M.

(ii) For $\Theta = (-\infty, 0]$ we put $\Theta_m := (-(m + 1), 0]$, $m \in \mathbb{N}$, and then define

$$L_{\mathrm{M+G}}(\cdot, \boldsymbol{g}) := \bigcap_{m \in \mathbb{N}} L_{\mathrm{M+G}}(\cdot, \boldsymbol{g}_m)$$

for $\boldsymbol{g}_m := (\gamma, \gamma - \mu, \Theta_m)$.

The operators in $L_{\mathrm{M+G}}(\cdot, \boldsymbol{g})$ are called smoothing Mellin plus Green operators of the cone calculus. Set

$$\sigma_1^{\mu-j}(M + G)(z) := f_j(z), \quad j = 0, \ldots, k. \tag{6.98}$$

The operator functions (6.98) will be called conormal symbols of the operator $M + G$, of order $\mu - j$. If we only need the highest component, we will often write

$$\sigma_1(M + G) := \sigma_1^\mu(M + G). \tag{6.99}$$

Remark 6.5.15. For every sequence of f_j in (6.95) belonging to $L_{\mathrm{M+G}}(N^\wedge, \boldsymbol{g})$, no matter where the singularities might lie; the weights $\gamma_j \neq \gamma$ for $j > 0$ may be necessary for the conditions (6.96).

Proposition 6.5.16. *Let* $f(r, z) \in C^\infty(\overline{\mathbb{R}}_+, M_{\mathcal{R}}^{-\infty}(N))$, $\mathcal{R} := \{(r_j, m_j)\}_{j \in \mathbb{J}}$, *and* $\gamma \in \mathbb{R}$, $\pi_{\mathbb{C}}\mathcal{R} \cap \Gamma_{(n+1)/2-\gamma} = \emptyset$; *then*

$$A := \omega r^{-\mu} \mathrm{op}_M^{\gamma-n/2}(f)\omega' \in L_{\mathrm{M+G}}(N^\wedge, \boldsymbol{g})$$

for $\boldsymbol{g} = (\gamma, \gamma - \mu, (-\infty, 0])$. *In this case* $\sigma_1^{\mu-j}(A)(z) = (\partial_r^j f)(0, z)/j!$, $j \in \mathbb{N}$.

Proof. It suffices to show that $A \in L_{\mathrm{M+G}}(N^\wedge, \boldsymbol{g})$ for $\boldsymbol{g} = (\gamma, \gamma - \mu, (-(k+1), 0])$, for any $k \in \mathbb{N}$. By Taylor's formula,

$$f(r, z) = \sum_{j=0}^{k} r^j \frac{1}{j!} (\partial_r^j f)(0, z) + r^{k+1} f_{(k+1)}(r, z),$$

where $(\partial_r^j f)(0, z) \in M_{\mathcal{R}}^{-\infty}(N)$ and $f_{(k+1)}(r, z) \in C^\infty(\overline{\mathbb{R}}_+, M_{\mathcal{R}}^{-\infty}(N))$, it suffices to observe that

$$r^{-\mu} \omega \operatorname{op}_M^{\gamma - n/2}(f_{(k+1)}(r, z)) \omega' \in L_{\mathrm{G}}(N^\wedge, \boldsymbol{g}). \qquad \square$$

Theorem 6.5.17. (i) *An operator $A \in L_{\mathrm{M+G}}(M, \boldsymbol{g})$ induces a continuous operator*

$$A : H^{s,\gamma}(M) \to H^{\infty, \gamma - \mu}(M) \qquad (6.100)$$

for every $s \in \mathbb{R}$. Moreover, for every asymptotic type associated with the weight data (γ, Θ) there is a $\mathcal{Q}$ associated with $(\gamma - \mu, \Theta)$ such that (6.100) restricts to a continuous operator

$$A : H_{\mathcal{P}}^{s,\gamma}(M) \to H_{\mathcal{Q}}^{\infty, \gamma - \mu}(M)$$

for any $s \in \mathbb{R}$.

(ii) *An operator $A \in L_{\mathrm{M+G}}(N^\wedge, \boldsymbol{g})$ induces a continuous operator*

$$A : \mathcal{K}^{s,\gamma;e}(N^\wedge) \to \mathcal{K}^{\infty, \gamma - \mu; e'}(N^\wedge) \qquad (6.101)$$

for any $s, e, e' \in \mathbb{R}$. Moreover, analogously as in (i), for every $\mathcal{P}$ there is a $\mathcal{Q}$ such that (6.101) restricts to a continuous operator

$$A : \mathcal{K}_{\mathcal{P}}^{s,\gamma;e}(N^\wedge) \to \mathcal{K}_{\mathcal{Q}}^{\infty, \gamma - \mu; e'}(N^\wedge)$$

for any $s, e, e' \in \mathbb{R}$.

The result is known from the cone pseudo-differential calculus, cf. [44, Subsection 2.1.2, Theorem 8], or [46, Theorem 2.3.55]. Details can also be found in the article [27].

In the following proposition we express formal adjoints, referring to the non-degenerate sesquilinear pairing (6.41) for arbitrary $s, \gamma \in \mathbb{R}$.

Proposition 6.5.18. *Having $A \in L_{\mathrm{M+G}}(M, \boldsymbol{g})$ for $\boldsymbol{g} = (\gamma, \gamma - \mu, \Theta)$ entails $A^* \in L_{\mathrm{M+G}}(M, \boldsymbol{g}^*)$ for $\boldsymbol{g}^* = (-\gamma + \mu, -\gamma, \Theta)$. An analogous result holds for $A \in L_{\mathrm{M+G}}(N^\wedge, \boldsymbol{g})$.*

Proof. In view of Definition 6.5.2, it suffices to consider operators A of the form (6.95) for $f := f_j \in M_{\mathcal{R}}^{-\infty}(N)$. In this case we have

$$A^* = \omega' \left(\operatorname{op}_M^{\gamma_j - n/2}(f) \right)^* r^{-\mu + j} \omega$$

and it suffices to observe that $\left(\mathrm{op}_M^{\gamma_j-n/2}(f)\right)^* = \mathrm{op}_M^{-\gamma_j-n/2}(f^{[*]})$, where $f^{[*]}$ is known from Theorem 6.3.7 to be of the form $f^{[*]}(z) = f^{(*)}(n+1-\bar{z})$, where upper $(*)$ means the pointwise adjoint of the respective smoothing operator over N with respect to the scalar product $(\cdot,\cdot)_{L^2(N)}$. Next, we write

$$A^* = \omega' \,\mathrm{op}_M^{-\gamma_j-n/2}(f^{[*]}) r^{-\mu+j}\omega = \omega' r^{-\mu}\,\mathrm{op}_M^{-\gamma_j+\mu-n/2}(T^\mu f^{[*]}) r^j\omega. \qquad (6.102)$$

This shows that the Mellin action on the right-hand side of (6.102) makes sense. Moreover, if we wish we may commute the r^j-terms to the left when $j > 0$, however, at the expense of Green operators and a translation of Mellin symbols as they are generated in Theorem 6.5.11, and under a modification of the weights as soon as the poles of $T^{\mu-j}f^{[*]}$ do not intersect the weight line. Clearly, in forming adjoints there is no extra weight condition for $f^{[*]}$ with respect to the weight line $\Gamma_{\frac{n+1}{2}+\gamma_j}$. In fact, in A we assume that

$$\pi_\mathbb{C}\mathcal{R} \cap \Gamma_{(n+1)/2-\gamma_j} = \emptyset. \qquad (6.103)$$

Then, using $f^{[*]} \in M_{\mathcal{R}^*}^{-\infty}(N)$ for $\mathcal{R}^*$ as in (6.94), it suffices to observe that

$$\pi_\mathbb{C}\mathcal{R} \cap \Gamma_{(n+1)/2+\gamma_j} = \emptyset.$$

The condition (6.103) means that for every $(r,m) \in \mathcal{R}$ we have

$$\mathrm{Re}\, r \neq \frac{n+1}{2} - \gamma_j \iff \mathrm{Re}\,(n+1-\bar{r}) \neq \frac{n+1}{2} + \gamma_j. \qquad \square$$

Theorem 6.5.19. (i) *Let M be a compact manifold with conical singularity. Then*

$$A \in L^\mu_{\mathrm{M+G}}(M,\boldsymbol{g}), \quad B \in L^\nu_{\mathrm{M+G}}(M,\boldsymbol{h})$$

for $\boldsymbol{g} = (\gamma-\nu, \gamma-(\mu+\nu), \Theta)$, $\boldsymbol{h} = (\gamma, \gamma-\nu, \Theta)$ implies

$$AB \in L^{\mu+\nu}_{\mathrm{M+G}}(M,\boldsymbol{g}\circ\boldsymbol{h})$$

for $\boldsymbol{g}\circ\boldsymbol{h} = (\gamma, \gamma-(\mu+\nu), \Theta)$, and we have

$$\sigma_1^{\mu+\nu-j}(AB) = \sum_{j=l+m} \left(T^{\nu-l}\sigma_1^{\mu-m}(A)(z)\right)\sigma_1^{\nu-l}(B)(z); \qquad (6.104)$$

(ii) *If $A \in L^\mu_{\mathrm{M+G}}(N^\wedge,\boldsymbol{g})$, $B \in L^\mu_{\mathrm{M+G}}(N^\wedge,\boldsymbol{h})$ for $\boldsymbol{g},\boldsymbol{h}$ as in (i), then $AB \in L^{\mu+\nu}_{\mathrm{M+G}}(N^\wedge,\boldsymbol{g}\circ\boldsymbol{h})$ and (6.104) holds.*

For the proof we refer to [46, Theorem 2.3.84]. The details essentially rely on commutations of powers of r through Mellin actions, as in Theorem 6.5.11.

6.6   Explicit computation of asymptotic types

For the following considerations we fix cut-off functions $\omega, \omega', \widetilde{\omega}, \widetilde{\omega}'$.

Proposition 6.6.1.   (i) *Let* $f \in M_{\mathcal{R}}^{-\infty}(N)$, $\gamma - j \leq \gamma_j \leq \gamma$, $\gamma - j \leq \delta_j \leq \gamma$, $j \in \mathbb{N}$, *and, say,* $\gamma_j \leq \delta_j$. *Then*

$$\omega r^{-\mu+j} \operatorname{op}_M^{\gamma_j - n/2}(f_j)\omega' - \omega r^{-\mu+j} \operatorname{op}_M^{\delta_j - n/2}(f_j)\omega' = G$$

for $G \in L_{\mathrm{G}}\big(N^\wedge, (\gamma, \gamma - \mu, (-\infty, 0])\big)_{\mathcal{P}_j, \mathcal{Q}_j}$, *where*

$$\mathcal{P}_j = T^{\mu-j} \mathrm{a}_{\gamma_j, \delta_j}(\mathcal{R}), \quad \mathcal{Q}_j = \mathrm{a}_{\gamma_j, \delta_j}^*(\mathcal{R}).$$

(ii) *Let* $f \in M_{\mathcal{R}}^{-\infty}(N)$ *and* $\gamma - j \leq \gamma_j \leq \gamma$. *Then*

$$\omega r^{-\mu+j} \operatorname{op}_M^{\gamma_j - n/2}(f)\omega' - \widetilde{\omega} r^{-\mu+j} \operatorname{op}_M^{\gamma_j - n/2}(f)\widetilde{\omega}' = \widetilde{G}$$

for $\widetilde{G} \in L_{\mathrm{G}}\big(N^\wedge, (\gamma, \gamma - \mu, (-(k+1), 0])\big)_{\mathcal{P}_j, \mathcal{Q}_j}$ *and some other* $\mathcal{P}_j, \mathcal{Q}_j$.

Proof. (i) immediately follows from Corollary 6.5.12 and formula (6.81).
(ii) Let us write $F := \operatorname{op}_M^{\gamma_j - n/2}(f)$. Then

$$\omega r^{-\mu+j} F\omega' - \widetilde{\omega} r^{-\mu+j} F\widetilde{\omega}' = (\omega - \widetilde{\omega}) r^{-\mu+j} F\omega' + \widetilde{\omega} r^{-\mu+j} F(\omega' - \widetilde{\omega}')$$

gives rise to

$$(\omega - \widetilde{\omega}) r^{-\mu+j} F\omega' : \mathcal{K}^{s,\gamma;e}(N^\wedge) \to \mathcal{K}_{\Theta}^{\infty, \gamma-\mu;\infty}(N^\wedge),$$
$$\widetilde{\omega} r^{-\mu+j} F(\omega' - \widetilde{\omega}') : \mathcal{K}^{s,\gamma;e}(N^\wedge) \to \mathcal{K}_{\mathcal{P}_j}^{\infty, \gamma-\mu;\infty}(N^\wedge)$$

for $\mathcal{P}_j = T^{\mu-j-\gamma_j+n/2}\widetilde{\mathcal{R}}_j$, $\widetilde{\mathcal{R}}_j := \{(r, m) \in \mathcal{R} : \operatorname{Re} r + \gamma_j < (n+1)/2\}$. In order to compute $\mathcal{Q}_j$, we form

$$\widetilde{G}^* = \omega' F^* r^{-\mu+j}(\omega - \widetilde{\omega}) + (\omega' - \widetilde{\omega}')F^* r^{-\mu+j}\widetilde{\omega}$$

where $F^* = \operatorname{op}_M^{-\gamma_j - n/2}(f^{[*]})$, $f^{[*]} \in M_{\mathcal{R}^*}^{-\infty}(N)$, $\mathcal{R}^* = \{(n+1-\overline{r}_i, m_i)\}_{i \in \mathbb{J}}$.

Here it suffices to consider the first summand in $\widetilde{G}^*$, since the second one maps to functions with trivial asymptotics Θ. Similarly as for $\widetilde{G}$, we have

$$\widetilde{G}^* : \mathcal{K}^{s, -\gamma+\mu; e}(N^\wedge) \to \mathcal{K}_{\mathcal{Q}_j}^{\infty, -\gamma;\infty}(N^\wedge)$$

for $\mathcal{Q}_j = T^{\gamma_j + n/2}\widetilde{\mathcal{R}}_j^*$ for

$$\widetilde{\mathcal{R}}_j^* := \left\{ \left(\frac{n}{2} + 1 - \gamma_j - \overline{r}, m\right) : (r, m) \in \mathcal{R}, \ \operatorname{Re} r + \gamma_j > \frac{n+1}{2} \right\}.$$

In fact, we have

$$F^* = r^{-\gamma_j - n/2} \operatorname{op}_M\left(T^{\gamma_j + n/2} f^{[*]}\right) r^{\gamma_j + n/2}$$

and $T^{\gamma_j + n/2} f^{[*]} \in M_{T^{-\gamma_j - n/2}\mathcal{R}^*}^{-\infty}(N)$,

$$T^{-\gamma_j - n/2}\mathcal{R}^* = \left\{ \left(\frac{n}{2} + 1 - \gamma_j - \bar{r}, m \right) : (r, m) \in \mathcal{R} \right\}.$$

The asymptotic type in the image under the first summand follows from the fact that the image belongs to $\mathcal{K}_{T^{\gamma_j + n/2}\mathcal{R}^*}^{\infty, -\gamma; \infty}(N^\wedge)$. $\qquad\square$

Proposition 6.6.2. (i) *There is a* $0 < \tilde{\varepsilon} < 1$ *such that for arbitrary* $0 < \varepsilon < \tilde{\varepsilon}$, *the operator* (6.95) *has the form*

$$M = r^{-\mu} \left\{ \omega \operatorname{op}_M^{\gamma - n/2}(f_0)\omega' + \sum_{j=1}^k r^j \omega \operatorname{op}_M^{\gamma - \varepsilon - n/2}(f_j)\omega' \right\} + G$$

for some $G \in L_G(N^\wedge, \boldsymbol{g})$ *which is independent of* ε.

(ii) *The space* $L_{M+G}(N^\wedge, \boldsymbol{g})$ *for* $\boldsymbol{g} = (\gamma, \gamma - \mu, \Theta)$, $\Theta = (-(k+1), 0]$, *can be equivalently defined as the space of all operators* $M + G$ *for* $G \in L_G(N^\wedge, \boldsymbol{g})$ *and*

$$M := r^{-\mu} \omega \sum_{j=0}^k \operatorname{op}_M^{\gamma_j - n/2}(f_j) r^j \omega'$$

with Mellin symbols $f_j \in M_{\mathcal{R}_j}^{-\infty}(N)$, *such that* $\pi_{\mathbb{C}} \mathcal{R}_j \cap \Gamma_{(n+1)/2 - \gamma_j} = \emptyset$ *for all* j, *and weights* γ_j *satisfying the conditions*

$$\gamma + j \geq \gamma_j \geq \gamma, \quad \text{for all } j = 0, \ldots, k. \tag{6.105}$$

Proof. (i) follows from Proposition 6.6.1, applied to the summands of (6.95) for every $j \neq 0$. In fact, it suffices to observe that there is an $0 < \tilde{\varepsilon} < 1$ such that $\Gamma_{n+1/2 - (\gamma - \varepsilon)} \cap \pi_{\mathbb{C}} \mathcal{R}_j = \emptyset$ for all $0 < \varepsilon < \tilde{\varepsilon}$ and all $j = 1, \ldots, k$. In order to make the resulting asymptotic type of G more explicit, we consider the summands separately, e.g., those

$$N_j := r^{-\mu} \omega r^j \operatorname{op}_M^{\gamma_j - n/2}(f_j)\omega', \quad f_j \in M_{\mathcal{R}_j}^{-\infty}(N),$$

where $\gamma_j < \gamma$. As noted before, we can pass to the operator

$$\tilde{N}_j := r^{-\mu} \omega r^j \operatorname{op}_M^{\gamma - \varepsilon - n/2}(f_j)\omega'.$$

Then it follows that

$$N_j - \tilde{N}_j \in L_G\big(N^\wedge, (\gamma, \gamma - \mu, (-(k+1), 0])\big)_{\mathcal{P}_j, \mathcal{Q}_j}$$

for $\mathcal{P}_j = T^{\mu - j} a_{\gamma_j, \gamma - \varepsilon}(\mathcal{R}_j)$, $\mathcal{Q}_j = a^*_{\gamma_j, \gamma - \varepsilon}(\mathcal{R}_j)$.

Assertion (ii) is also a consequence of Proposition 6.6.1 (i). The relations (6.105) are necessary for the continuity between weighted spaces in (6.101). $\qquad\square$

Proposition 6.6.3.   (i) *Fix weight data* $\boldsymbol{g} = (\gamma, \gamma - \mu, (-(k+1), 0])$ *for* $k \in \mathbb{N}$, *and let* $f \in C^\infty(\overline{\mathbb{R}}_+, M_\mathcal{R}^{-\infty}(N))$, $\mathcal{R} = \{(r_l, m_l)\}_{l \in \mathbb{J}}$. *Let* $\gamma_j \in \mathbb{R}$,

$$\pi_\mathbb{C} \mathcal{R} \cap \Gamma_{(n+1)/2-\gamma_j} = \emptyset, \quad \gamma - j \le \gamma_j \le \gamma,$$

and assume $j > k$. *Then we have*

$$r^{-\mu+j} \omega \operatorname{op}_M^{\gamma_j - n/2}(f) \omega' \in L_G(N^\wedge, \boldsymbol{g})_{\mathcal{P}, \mathcal{Q}}$$

for $\mathcal{P} = T^{\mu-j} \mathsf{a}_{\gamma_j, \gamma-\varepsilon}(\mathcal{R})$, *where* ε *is as in Proposition 6.6.2, and*

$$\mathcal{Q} = \left\{ (n+1-\bar{r}_l, m_l) : \gamma_j + \mu - j < \frac{n+1}{2} - \operatorname{Re} r_l < \gamma_j, l \in \mathbb{J} \right\} \quad \text{for } \mu \le j,$$

$$\mathcal{Q} = \left\{ (n+1-\bar{r}_l, m_l) : \gamma_j < \frac{n+1}{2} - \operatorname{Re} r_l < \gamma_j + \mu - j, l \in \mathbb{J} \right\} \quad \text{for } \mu \ge j.$$

(ii) *For any* $\beta \in \mathbb{R}$, $\pi_\mathbb{C} \mathcal{R} \cap \Gamma_{(n+1)/2-\beta} = \emptyset$, *there exists an* $l > 0$ *such that*

$$G := r^{-\mu+j} \omega \operatorname{op}_M^{\beta - n/2}(f) r^{j'} \omega' \in L_G(N^\wedge, \boldsymbol{g})$$

for every $j, j' \ge l$.

Proof. (i) From the continuity of the operator

$$G := r^{-\mu+j} \omega \operatorname{op}_M^{\gamma_j - n/2}(f) \omega' : \mathcal{K}^{s,\gamma;e}(N^\wedge) \to \mathcal{K}_\Theta^{\infty, \gamma-\mu;\infty}(N^\wedge)$$

for all $s, e \in \mathbb{R}$ we see that G satisfies the first condition of Definition 6.5.2, where $\pi_\mathbb{C} \mathcal{P} = \emptyset$, cf. the notation (6.77). In order to determine the asymptotic type $\mathcal{Q}$, we first consider the case of r-independent $f \in M_\mathcal{R}^{-\infty}(N)$, $\mathcal{R} = \{(r_l, m_l)\}_{l \in \mathbb{J}}$. Then we have $G^* = \omega' \operatorname{op}_M^{-\gamma_j - n/2}(f^{[*]}) r^{-\mu+j} \omega$ for $f^{[*]}(z) = f^{(*)}(n+1-\bar{z}) \in M_{\mathcal{R}^*}^{-\infty}(N)$, cf. Remark 6.5.10 (ii), and

$$\mathcal{R}^* = \{(n+1-\bar{r}_l, m_l)\}_{l \in \mathbb{J}}. \tag{6.106}$$

By Theorem 6.5.11, $G^* = \omega' r^{-\mu+j} \operatorname{op}_M^{-\gamma_j - n/2}(T^{\mu-j} f^{[*]}) \omega + L$ for an $L \in \mathbb{G}(N^\wedge)$, or, alternatively, $\omega' \operatorname{op}_M^{-\gamma_j - n/2}(f^{[*]}) \omega - \omega' \operatorname{op}_M^{-\gamma_j - n/2-\mu+j}(f^{[*]}) \omega = L r^{\mu-j} =: N$. Corollary 6.5.12 gives

$$N \in L_G(N^\wedge, \boldsymbol{g}_1)_{\widetilde{\mathcal{P}}_1, \widetilde{\mathcal{Q}}_1} \text{ for } \mu \le j, \quad N \in L_G(N^\wedge, \boldsymbol{g}_2)_{\widetilde{\mathcal{P}}_2, \widetilde{\mathcal{Q}}_2} \text{ for } \mu \ge j$$

where $\boldsymbol{g}_1 = (-\gamma_j - \mu + j, -\gamma_j, (-\infty, 0])$, $\boldsymbol{g}_2 = (-\gamma_j, -\gamma_j - \mu + j, (-\infty, 0])$ and

$$\widetilde{\mathcal{P}}_1 = \left\{ (r, m) \in \mathcal{R}^* : \frac{n+1}{2} - (-\gamma_j - \mu + j) < \operatorname{Re} r < \frac{n+1}{2} + \gamma_j \right\} \quad \text{for } \mu \le j,$$

$$\widetilde{\mathcal{P}}_2 = \left\{ (r, m) \in \mathcal{R}^* : \frac{n+1}{2} + \gamma_j < \operatorname{Re} r < \frac{n+1}{2} - (-\gamma_j - \mu + j) \right\} \quad \text{for } \mu \ge j.$$

Using (6.106) we obtain

$$\widetilde{\mathcal{P}}_1 = \left\{ (n+1-\bar{r}_l,,m_l) : \gamma_j + \mu - j < \frac{n+1}{2} - \operatorname{Re} r_l < \gamma_j, l \in \mathbb{J} \right\} \quad \text{for } \mu \le j,$$

$$\widetilde{\mathcal{P}}_2 = \left\{ (n+1-\bar{r}_l,,m_l) : \gamma_j < \frac{n+1}{2} - \operatorname{Re} r_l < \gamma_j + \mu - j, l \in \mathbb{J} \right\} \quad \text{for } \mu \ge j.$$

(ii) According to (6.43), we can write

$$\omega \operatorname{op}_M^{\beta-n/2}(f) r^{j'} \omega' = \omega r^{-j'} \operatorname{op}_M^{\beta-n/2}(T^{j'} T^{-j'} f) r^{j'} \omega'$$
$$= \omega \operatorname{op}_M^{\beta-j'-n/2}(T^{-j'} f) \omega', \tag{6.107}$$

and so $G = r^{-\mu+j} \omega \operatorname{op}_M^{\beta-j'-n/2}(T^{-j'} f) \omega'$. This operator defines a continuous map

$$G : \mathcal{K}^{s,\beta-j';e}(N^\wedge) \to \mathcal{K}^{s',\beta-j'+j-\mu;e'}(N^\wedge).$$

Since we have continuous embeddings $\mathcal{K}^{s,\gamma;e}(N^\wedge) \hookrightarrow \mathcal{K}^{s,\beta-j';e}(N^\wedge)$ for $j' \ge \beta - \gamma$ and $\mathcal{K}^{s',\beta-j'+j-\mu;e'}(N^\wedge) \hookrightarrow \mathcal{K}^{s',\gamma-\mu;e'}(N^\wedge)$ for $j \ge \gamma+j'-\beta+(k+1)$ for arbitrary $s, e, s', e' \in \mathbb{R}$, it follows that G induces a continuous operator

$$G : \mathcal{K}^{s,\gamma;e}(N^\wedge) \to \mathcal{K}_\Theta^{s',\gamma-\mu;e'}(N^\wedge),$$

see (6.78) for flatness in the image. For the formal adjoint we may argue in an analogous manner. $\qquad\square$

Given $A \in L_{M+G}(N^\wedge, \boldsymbol{g})$ or $A \in L_{M+G}(M, \boldsymbol{g})$, we set

$$\sigma_1^{\mu-j}(A)(z) = f_j(z) + l_j(z), \quad j = 0, \dots, k.$$

By definition, we have

$$\sigma_1^{\mu-j}(A) \in M_{\mathcal{R}_j}^{-\infty}(N) \tag{6.108}$$

for certain asymptotic types $\mathcal{R}_j$. Let $\mathcal{R} := (\mathcal{R}_j)_{j=0,\dots,k}$ and denote for the moment by

$$L_{M+G}(N^\wedge, \boldsymbol{g})_{\mathcal{R}}$$

the subspace of those $A \in L_{M+G}(N^\wedge, \boldsymbol{g})$ such that (6.108) holds. In a similar manner we define $L_{M+G}(M, \boldsymbol{g})_{\mathcal{P}}$.

Remark 6.6.4. (i) $A \in L_{M+G}(N^\wedge, \boldsymbol{g})$ and $\sigma_1^{\mu-j}(A) = 0$, $j = 0, \dots, k$, implies $A \in L_G(N^\wedge, \boldsymbol{g})$.

(ii) The map

$$L_{M+G}(N^\wedge, \boldsymbol{g})_{\mathcal{R}} \to \mathop{\mathrm{X}}_{j=0}^{k} M_{\mathcal{R}_j}^{-\infty}(N)$$

is surjective.
An analogous result holds for $L_{M+G}(M, \boldsymbol{g})_{\mathcal{P}}$.

Theorem 6.6.5. *Let $f \in M_{\mathcal{R}}^{\mu}(N)$, $\mu \in \mathbb{R}$, for some Mellin asymptotic type $\mathcal{R}$, and fix cut-off functions ω, ω'. Then for any $\beta, \gamma \in \mathbb{R}$ with*

$$\pi_{\mathbb{C}} \mathcal{R} \cap \Gamma_{(n+1)/2-\gamma} = \pi_{\mathbb{C}} \mathcal{R} \cap \Gamma_{(n+1)/2-(\gamma+\beta)} = \emptyset,$$

we have

$$\omega \operatorname{op}_M^{\gamma-n/2}(f) r^{\beta} \omega' = \omega r^{\beta} \operatorname{op}_M^{\gamma-n/2}(T^{-\beta}f) \omega' \tag{6.109}$$

as operators $C_0^{\infty}(\mathbb{R}_+, C^{\infty}(N)) \to C_0^{\infty}(\mathbb{R}_+, C^{\infty}(N))$, up to a remainder $G \in \mathbb{G}(N^{\wedge})$. For $\pi_{\mathbb{C}} \mathcal{R} \cap S_{\gamma, \beta} = \emptyset$ (which is always the case for $f \in M_{\mathcal{O}}^{\mu}(N)$) we have $G \equiv 0$. Instead of (6.6.5), we can also write

$$\omega \operatorname{op}_M^{\gamma-n/2}(f) \omega' = \omega \operatorname{op}_M^{\gamma+\beta-n/2}(f) \omega' \tag{6.110}$$

up to such a G.

Proof. Applying Remark 6.6.8, we write $f = h + l$ for $h \in M_{\mathcal{O}}^{\mu}(N), l \in M_{\mathcal{R}}^{-\infty}(N)$. This gives a decomposition $G = G_h + G_l$ where G_h (resp. G_l) stands for the operator (6.109) for h (resp. l) rather than f. By Theorem 6.5.11, $G_l \in \mathbb{G}$. As for G_h, we have $G_h = 0$. In fact, we use a modification of the proof Theorem 6.5.11. The only difference is that $h(z)$ is not strongly decreasing as $|\operatorname{Im} z| \to \infty$. However, since in the Mellin transform of $r^{\beta} v(r)$ for $v = \omega' u$, $u \in C_0^{\infty}(\mathbb{R}_+, C^{\infty}(N))$ is strongly decreasing as $|\operatorname{Im} z| \to \infty$, we obtain the same for the composition with h and hence we can apply Cauchy's theorem, this time with a holomorphic function, and hence the result is zero. □

Remark 6.6.6. Observe that for $h \in M_{\mathcal{O}}^{\mu}(N)$ we have

$$\omega \operatorname{op}_M^{\gamma-n/2}(h) \omega' = \omega \operatorname{op}_M^{\beta-n/2}(h) \omega'$$

as operators on $C_0^{\infty}(\mathbb{R}_+, C^{\infty}(N))$, for arbitrary $\gamma, \beta \in \mathbb{R}$.

Proposition 6.6.7. (i) *Let $h \in C^{\infty}(\overline{\mathbb{R}}_+, M_{\mathcal{O}}^{\mu}(N))$, $\mu \in \mathbb{R}$. Then (6.109) vanishes.*

(ii) *For $f \in C^{\infty}(\overline{\mathbb{R}}_+, M_{\mathcal{R}}^{-\infty}(N))$, equation (6.109) holds up to a remainder in $\mathbb{G}$.*

Remark 6.6.8. (i) Let $l \in M_{\mathcal{R}}^{\mu}(N)$ and fix a real β such that $\Gamma_{\beta} \cap \pi_{\mathbb{C}} \mathcal{R} = \emptyset$. Then $f := l|_{\Gamma_{\beta}} \in L_{\mathrm{cl}}^{\mu}(N; \Gamma_{\beta})$. Applying Theorem 6.3.12 to f, we find an $h \in M_{\mathcal{O}}^{\mu}(N)$ such that $f - h|_{\Gamma_{\beta}} \in L^{-\infty}(N; \Gamma_{\beta})$. It follows that $l - h \in M_{\mathcal{R}}^{-\infty}(N)$, i.e., we recover in this way a decomposition

$$l = h + (l - h) \quad \text{for } h \in M_{\mathcal{O}}^{\mu}(N), \; l - h \in M_{\mathcal{R}}^{-\infty}(N).$$

(ii) Let $l \in C^{\infty}(\overline{\mathbb{R}}_+ \times \Omega, M_{\mathcal{R}}^{\mu}(N))$, $\Omega \subseteq \mathbb{R}^q$ open, and let $\Gamma_{\beta} \cap \pi_{\mathbb{C}} \mathcal{R} = \emptyset$ for some β. Then the kernel cut-off argument that proves an analogue of Theorem 6.3.12 in the r-dependent case yields a decomposition

$$l = l_0 + l_1 \quad \text{for } l_0 \in C^{\infty}(\overline{\mathbb{R}}_+ \times \Omega, M_{\mathcal{O}}^{\mu}(N)), \quad l_1 \in C^{\infty}(\overline{\mathbb{R}}_+ \times \Omega, M_{\mathcal{R}}^{-\infty}(N)).$$

6.7   Kernel characterisations of Green operators

In this section we refer to material in Seiler's article [60]. Green operators in the cone calculus can be characterised in terms of kernels. The requirements on their mapping properties can be considerably simplified; the conditions so far have been formulated for convenience. We first consider the case of operators on $\mathbb{R}_+$.

Definition 6.7.1. Let $L_G(\mathbb{R}_+, \boldsymbol{g})_{\mathcal{P}, \mathcal{Q}}$ for $\boldsymbol{g} := (\gamma, \delta, \Theta, \Phi)$, $\Theta = (\vartheta, 0]$, $\Phi = (\varphi, 0]$, and asymptotic types $\mathcal{P}$ and $\mathcal{Q}$ associated with (δ, Φ) and $(-\gamma, \Theta)$, respectively, denote the space of all $G \in \mathcal{L}\big(\mathcal{K}^{0,\gamma}(\mathbb{R}_+), \mathcal{K}^{0,\delta}(\mathbb{R}_+)\big)$ that induce continuous operators

$$G : \mathcal{K}^{s,\gamma}(\mathbb{R}_+) \to \mathcal{S}_{\mathcal{P}}^{\delta}(\mathbb{R}_+), \quad G^* : \mathcal{K}^{s',-\delta}(\mathbb{R}_+) \to \mathcal{S}_{\mathcal{Q}}^{-\gamma}(\mathbb{R}_+) \tag{6.111}$$

for all $s, s' \in \mathbb{R}$, where G^* is the formal adjoint with respect to the $\mathcal{K}^{0,0}(\mathbb{R}_+) = L^2(\mathbb{R}_+)$-scalar product. For $\Phi = \Theta$ we simply write $\boldsymbol{g} := (\gamma, \delta, \Theta)$.

From (6.111) it follows that there is a kernel function

$$k_G(r, r') \in \mathcal{S}_{\mathcal{P}}^{\delta}(\mathbb{R}_+) \, \widehat{\otimes}_{\pi} \, \mathcal{K}^{-s,-\gamma}(\mathbb{R}_+) \bigcap \mathcal{K}^{s',\delta}(\mathbb{R}_+) \, \widehat{\otimes}_{\pi} \, \mathcal{S}_{\mathcal{Q}}^{-\gamma}(\mathbb{R}_+) \tag{6.112}$$

such that

$$Gu(r) = \int_0^{\infty} k_G(r, r') u(r') dr' \tag{6.113}$$

for any $u \in \mathcal{K}^{s,\gamma}(\mathbb{R}_+)$. Similarly as in Remark 6.5.4, we set

$$\mathcal{S}_0^{\gamma}(\mathbb{R}_+) := \big\{ u \in \mathcal{K}^{\infty,\gamma}(\mathbb{R}_+) : (1 - \omega) \, u \in \mathcal{S}(\mathbb{R}),$$

$$\log^k r \, \omega(r) u(r) \in \mathcal{K}^{\infty,\gamma}(\mathbb{R}_+) \text{ for all } k \in \mathbb{N} \big\}, \tag{6.114}$$

where ω is an arbitrary cut-off function.

Lemma 6.7.2. *For arbitrary* $\Theta = (\vartheta, 0]$, $-\infty \le \vartheta \le 0$, *we have*

$$L^2(\mathbb{R}_+) \, \widehat{\otimes}_{\pi} \, \mathcal{S}_{\Theta}^0(\mathbb{R}_+) \bigcap \mathcal{S}_0^0(\mathbb{R}_+) \, \widehat{\otimes}_{\pi} \, L^2(\mathbb{R}_+) = \mathcal{S}_0^0(\mathbb{R}_+) \, \widehat{\otimes}_{\pi} \, \mathcal{S}_{\Theta}^0(\mathbb{R}_+), \tag{6.115}$$

where $\mathcal{S}_{\Theta}^{\gamma}(\mathbb{R}_+) = \mathcal{K}_{\Theta}^{\infty,\gamma;\infty}(\mathbb{R}_+)$, *cf.* (6.76).

Proof. For convenience, in the following computations we work with $\theta := -\vartheta$; then $\Theta = (-\theta, 0]$. It is obvious that

$$L^2(\mathbb{R}_+) \, \widehat{\otimes}_{\pi} \, \mathcal{S}_{\Theta}^0(\mathbb{R}_+) \bigcap \mathcal{S}_0^0(\mathbb{R}_+) \, \widehat{\otimes}_{\pi} \, L^2(\mathbb{R}_+) \supseteq \mathcal{S}_0^0(\mathbb{R}_+) \, \widehat{\otimes}_{\pi} \, \mathcal{S}_{\Theta}^0(\mathbb{R}_+).$$

Thus it suffices to prove the converse inclusion. Writing a $g = g(r, t)$ in the left-hand side as

$$g = \omega(r)\omega(t)g + \omega(r)(1 - \omega(t))g + (1 - \omega(r))\omega(t)g + (1 - \omega(r))(1 - \omega(t))g$$

and applying the transformation $(S_{1/2}f)(x) = e^{-x/2}f(e^{-x})$, the assertion reduces to

$$L^2(\mathbb{R}) \, \widehat{\otimes}_\pi \, \mathcal{S}_\Theta(\mathbb{R}) \bigcap \mathcal{S}(\mathbb{R}) \, \widehat{\otimes}_\pi \, L^2(\mathbb{R}) = \mathcal{S}(\mathbb{R}) \, \widehat{\otimes}_\pi \, \mathcal{S}_\Theta(\mathbb{R}), \qquad (6.116)$$

where $\mathcal{S}_\Theta(\mathbb{R}) := \mathcal{S}(\mathbb{R})$ for $\theta = 0$ and

$$\mathcal{S}_\Theta(\mathbb{R}) := \left\{ u \in \mathcal{S}(\mathbb{R}) : e^{\tilde{\theta}x}u(x) \in \mathcal{S}(\mathbb{R}) \text{ for all } 0 \le \tilde{\theta} < \theta \right\}.$$

The result is known for $\theta = 0$. The proof in this case is simpler than that for $\theta < 0$, which will be given below. It suffices to show for some $g = g(x,y)$ belonging to the left-hand side of (6.116) that

$$g(x,y)_{\tilde{\theta}} := e^{\tilde{\theta}y}g(x,y) \in \mathcal{S}(\mathbb{R}) \, \widehat{\otimes}_\pi \, \mathcal{S}(\mathbb{R}) = \mathcal{S}(\mathbb{R} \times \mathbb{R})$$

for any fixed $0 < \tilde{\theta} < \theta$. Thus we have to show that

$$\langle x \rangle^{k'} \langle y \rangle^{l'} \langle D_x \rangle^k \langle D_y \rangle^l e^{\tilde{\theta}y} g(x,y) \in L^2(\mathbb{R} \times \mathbb{R})$$

for all $k, k', l, l' \in \mathbb{N}$, where $\langle D_x \rangle^\mu := \text{Op}(\langle \xi \rangle^\mu)$ for $\mu \in \mathbb{R}$. By repeatedly applying the inequality $\alpha\beta \le \alpha^2 + \beta^2$ and Plancherel's formula, for $\| \cdot \| := \| \cdot \|_{L^2(\mathbb{R} \times \mathbb{R})}$ we easily obtain

$$\begin{aligned}\left\| \langle x \rangle^{k'} \langle y \rangle^{l'} \langle D_x \rangle^k \langle D_y \rangle^l g_{\tilde{\theta}} \right\| &\le \left\| \langle x \rangle^{4k} \langle D_x \rangle^k g_{\tilde{\theta}} \right\| + \left\| \langle D_x \rangle^k \langle D_y \rangle^{2l} g_{\tilde{\theta}} \right\| \\ &\quad + \left\| \langle D_x \rangle^{2k} \langle D_y \rangle^l g_{\tilde{\theta}} \right\| + \left\| \langle y \rangle^{4l'} \langle D_y \rangle^l g_{\tilde{\theta}} \right\|. \end{aligned} \qquad (6.117)$$

The fourth term on the right-hand side is finite, since $g_{\tilde{\theta}} \in L^2(\mathbb{R}) \, \widehat{\otimes}_\pi \, \mathcal{S}(\mathbb{R})$. In order to treat the other summands, we choose a $p > 1$ such that $\tilde{\theta} < p\tilde{\theta} < p^2\tilde{\theta} < \theta$, and define p' by $1/p + 1/p' = 1$. Then we have $\alpha\beta \le \alpha^p + \beta^{p'}$ for all $\alpha, \beta \ge 0$. By passing to the image under the Fourier transform, we first obtain

$$\left\| \langle D_x \rangle^k \langle D_y \rangle^l g_{\tilde{\theta}} \right\| \le \left\| \langle D_x \rangle^{2k} g_{\tilde{\theta}} \right\| + \left\| \langle D_y \rangle^{2l} g_{\tilde{\theta}} \right\|.$$

The second summand is finite. For the first we set $\hat{g}(\xi,y) := (F_{x\to\xi}g)(\xi,y)$. It follows that

$$\begin{aligned}\left\| \langle D_x \rangle^{2k} g_{\tilde{\theta}} \right\| &= \left\| \langle \xi \rangle^{2k} e^{\tilde{\theta}y} \hat{g}(\xi,y) \right\| \\ &\le \left\| \langle \xi \rangle^{2p'k} \hat{g}(\xi,y) \right\| + \left\| e^{p\tilde{\theta}y} \hat{g}(\xi,y) \right\| \\ &= \left\| \langle D_x \rangle^{2p'k} g \right\| + \left\| e^{p\tilde{\theta}y} g \right\| < \infty. \end{aligned} \qquad (6.118)$$

Thus the second and third terms on the right-hand side of (6.117) are finite. It remains to note that

$$\begin{aligned}\left\| \langle x \rangle^{k'} \langle D_x \rangle^k g_{\tilde{\theta}} \right\| &\le \left\| \langle x \rangle^{p'k'} \langle D_x \rangle^k g \right\| + \left\| \langle D_x \rangle^k e^{p\tilde{\theta}y} g \right\| \\ &\le \left\| \langle x \rangle^{p'k'} \langle D_x \rangle^k g \right\| + \left\| e^{p^2\tilde{\theta}y} g \right\| < \infty \end{aligned} \qquad (6.119)$$

which shows that the right-hand side of (6.117) is finite. This completes the proof of (6.116). $\qquad\square$

Similarly as before, we set $\phi := -\varphi$, i.e., $\Phi = (-\phi, 0]$.

Theorem 6.7.3. *Assume* $G \in \mathcal{L}\big(\mathcal{K}^{s,\gamma}(\mathbb{R}_+), \mathcal{K}^{s',\delta}(\mathbb{R}_+)\big)$ *for some* $s, s', \gamma, \delta \in \mathbb{R}$ *induces continuous operators*

$$G : \mathcal{K}^{s,\gamma}(\mathbb{R}_+) \to \mathcal{S}_{\mathcal{P}}^{\delta}(\mathbb{R}_+) \quad and \quad G^* : \mathcal{K}^{s',-\delta}(\mathbb{R}_+) \to \mathcal{S}_{\mathcal{Q}}^{-\gamma}(\mathbb{R}_+) \qquad (6.120)$$

for asymptotic types $\mathcal{P}$ *and* $\mathcal{Q}$ *associated with* (δ, Φ) *and* $(-\gamma, \Theta)$, *respectively. Then* G *has an integral kernel* $k_G(r, r')$ *representing the operator by relation* (6.113), *where*

(i)

$$k_G \in \mathcal{S}_{\mathcal{P}}^{\delta}(\mathbb{R}_+) \,\widehat{\otimes}_{\Gamma}\, \mathcal{S}_{\overline{\mathcal{Q}}}^{-\gamma}(\mathbb{R}_+) := \mathcal{S}_{\mathcal{P}}^{\delta}(\mathbb{R}_+) \,\widehat{\otimes}_{\pi}\, \mathcal{S}_{0}^{-\gamma}(\mathbb{R}_+) \bigcap \mathcal{S}_{0}^{\delta}(\mathbb{R}_+) \,\widehat{\otimes}_{\pi}\, \mathcal{S}_{\overline{\mathcal{Q}}}^{-\gamma}(\mathbb{R}_+);$$
$$(6.121)$$

(ii) *let* $p, q > 1$ *be related by* $p^{-1} + q^{-1} = 1$ *and set*

$$\mathcal{P}_p := \big\{ (z, m) \in \mathcal{P} : \operatorname{Re} z > 1/2 - \delta - \phi/p \big\},$$
$$\mathcal{Q}_q := \big\{ (w, l) \in \mathcal{Q} : \operatorname{Re} w > 1/2 + \gamma + \theta/q \big\};$$

then we have

$$k_G \in \mathcal{S}_{\mathcal{P}_p}^{\delta}(\mathbb{R}_+) \,\widehat{\otimes}_{\pi}\, \mathcal{S}_{\overline{\mathcal{Q}_q}}^{-\gamma}(\mathbb{R}_+).$$

In particular, for $\Theta = \Phi = (-\infty, 0]$ *we have*

$$k_G \in \mathcal{S}_{\mathcal{P}}^{\delta}(\mathbb{R}_+) \,\widehat{\otimes}_{\pi}\, \mathcal{S}_{\overline{\mathcal{Q}}}^{-\gamma}(\mathbb{R}_+).$$

Proof. (i) First let Θ and Φ be finite. After applying suitable reductions of order we may assume $s = s' = \gamma = \delta = 0$. Choose an $A \in \mathcal{L}(L^2(\mathbb{R}_+))$ that restricts to isomorphisms $\mathcal{S}_{\mathcal{P}}^{0}(\mathbb{R}_+) \to \mathcal{S}_{\Theta}^{0}(\mathbb{R}_+)$ and $\mathcal{S}_{0}^{0}(\mathbb{R}_+) \to \mathcal{S}_{0}^{0}(\mathbb{R}_+)$. In a similar manner we define an operator B corresponding to $\mathcal{Q}$. Then $H := BGA^*$ has the mapping properties $H : L^2(\mathbb{R}_+) \to \mathcal{S}_{\Theta}^{0}(\mathbb{R}_+)$ and $H^* : L^2(\mathbb{R}_+) \to \mathcal{S}_{\Phi}^{0}(\mathbb{R}_+)$. By (6.112) and Lemma 6.7.2, the operator H has a kernel $h \in \mathcal{S}_{\Phi}^{0}(\mathbb{R}_+) \,\widehat{\otimes}_{\Gamma}\, \mathcal{S}_{\Theta}^{0}(\mathbb{R}_+)$. Then the assertion follows by representing $G = (A^{-1}(B^{-1}H)^*)^*$ in terms of kernels. The corresponding result for infinite Θ, Φ follows from the finite case for $\Theta_k := (-(k+1), 0]$, $\Phi_k := (-(k+1), 0]$, and then passing to the limit $k \to \infty$.

(ii) The proof is similar to that of (i), taking into account that

$$\mathcal{S}_{\Phi}^{0}(\mathbb{R}_+) \,\widehat{\otimes}_{\Gamma}\, \mathcal{S}_{\Theta}^{0}(\mathbb{R}_+) = \bigcap_{1 \leq p \leq \infty} \mathcal{S}_{\Phi_q}^{0}(\mathbb{R}_+) \,\widehat{\otimes}_{\pi}\, \mathcal{S}_{\Theta_p}^{0}(\mathbb{R}_+),$$

cf. also [60]. $\qquad\qquad\qquad\qquad\qquad\qquad\qquad\qquad\qquad\qquad\qquad\qquad\qquad\qquad\quad$ $\square$

6.8   The cone calculus

The cone calculus on a manifold M with conical singularity v will be formulated as a substructure of $L_{cl}^\mu(M \setminus \{v\})$, $\mu \in \mathbb{R}$, consisting of pseudo-differential operators with the typical Fuchs type degenerate behaviour near v, locally in the splitting of variables $(r, x) \in N^\wedge$. We consider here two cases. First we discuss the infinite cone N^Δ, where we talk about the calculus over the open stretched cone $N^\wedge$ which involves (by notation) also a specific control of operators for $r \to \infty$. Moreover, we consider the case of a compact M with conical singularities.

Definition 6.8.1.   (i) The space $L^\mu(N^\wedge, g)$ for $g = (\gamma, \gamma - \mu, \Theta)$, $\Theta = (-(k+1), 0]$, $k \in \mathbb{N} \cup \{\infty\}$, is defined as the set of all operators

$$A = \omega r^{-\mu} \mathrm{op}_M^{\gamma - n/2}(h)\omega' + (1 - \omega)A_\infty(1 - \omega'') + M + G$$

where $\omega'' \prec \omega \prec \omega'$ are cut-off functions and $h(r, z) \in C^\infty(\overline{\mathbb{R}}_+, M_O^\mu(N))$, cf. Definition 6.3.8, $M + G \in L_{M+G}(N^\wedge, g)$, cf. Definition 6.5.14, $A_\infty \in L_{cl}^{\mu;0}(N^{\overset{\times}{}})|_{N^\wedge}$, cf. Section 1.4.

(ii) The space $L^\mu(M, g)$ for g as in (i) is defined to be the subspace of all $A \in L_{cl}^\mu(M \setminus \{v\})$ of the form

$$A = \omega r^{-\mu} \mathrm{op}_M^{\gamma - n/2}(h)\omega' + (1 - \omega)A_{int}(1 - \omega'') + M + G,$$

where h and the cut-off functions are as before, $M + G \in L_{M+G}(M, g)$, cf. Definition 6.5.14, while $A_{int} \in L_{cl}^\mu(M \setminus \{v\})$.

Note that (6.5.16) implies embeddings

$$L^\mu(N^\wedge, g_{k+1}) \hookrightarrow L^\mu(N^\wedge, g_k)$$

for $g_k = (\gamma, \gamma - \mu, (-(k + 1), 0])$, $k \in \mathbb{N}$, and, similarly, for the operators over M.

Remark 6.8.2. There is a straightforward generalisation of Definition 6.8.1 to the case of arbitrary weight intervals

$$h = (\gamma, \delta, \Theta), \quad \gamma, \delta \in \mathbb{R}.$$

It suffices to set $L^\mu(M, h) := r^{-\gamma + \mu + \delta} L^\mu(M, g)$, cf. the notation in (6.35). In the case $N^\wedge$ instead of M we set $L^\mu(N^\wedge, h) := k^{-\gamma + \mu + \delta} L^\mu(N^\wedge, g)$, cf. the notation in (6.42).

Remark 6.8.3. For $(\iota_\lambda u)(r, x) := u(\lambda r, x)$, $\lambda \in \mathbb{R}_+$ we have

$$A \in L^\mu(N^\wedge, h) \implies \iota_\lambda A \iota_\lambda^{-1} \in L^\mu(N^\wedge, h), \quad \lambda \in \mathbb{R}_+.$$

Conjugation with ι_λ also leaves the subspaces $L_{M+G}(N^\wedge, h)$ and $L_G(N^\wedge, h)$ unchanged.

Theorem 6.8.4. *Let $A \in L^{\mu}(N^{\wedge}, (\gamma, \gamma - \mu, (-\infty, 0]))$. Then for every $\tilde{\gamma} \in \mathbb{R}$ there exists an $\tilde{A} \in L^{\mu}(N^{\wedge}, (\tilde{\gamma}, \tilde{\gamma} - \mu, (-\infty, 0]))$ such hat*

$$A = \tilde{A} \mod \mathbb{G}(N^{\wedge}).$$

Proof. If we have an operator $A \in L^{\mu}(N^{\wedge}, (\gamma, \gamma - \mu, \Theta))$ then we can first omit the Green term which belongs to $\mathbb{G}(N^{\wedge})$ and formulate the smoothing Mellin term with respect to the modified weight data $(\tilde{\gamma}, \tilde{\gamma} - \mu, (-\infty, 0])$, modulo another element in $\mathbb{G}(N^{\wedge})$, cf. Theorem 6.5.11. This shift of weight data does not affect the other ingredients. $\qquad \square$

Remark 6.8.5. There is an analogue of relation (6.90) for symbols

$$f(r, z) \in C^{\infty}(\overline{\mathbb{R}}_+, M_{\mathcal{R}}^{-\infty}(N)).$$

Moreover, (6.110) holds for $f(r, z) \in C^{\infty}(\overline{\mathbb{R}}_+, M_{\mathcal{O}}^{\mu}(N))$, $\mu \in \mathbb{R}$.

This is a consequence of (6.87). Let us now define the principal symbol structure

$$\sigma(A) = (\sigma_0(A), \sigma_1(A)) \tag{6.122}$$

of operators $A \in L^{\mu}(M, \boldsymbol{g})$ coming from the stratification $M = s_0(M) \cup s_1(M)$. By virtue of the inclusion $L^{\mu}(M, \boldsymbol{g}) \subset L_{\text{deg}}^{\mu}(M)$, we already have the first component $\sigma_0(A)$ including the reduced symbol $\tilde{\sigma}_0(A)$ close to $s_1(M)$ in the splitting of variables (r, x). Further, we set

$$\sigma_1(A)(z) := h(0, z) + \sigma_1(M + G)(z),$$

and call it the (principal) conormal symbol of the operator A. If we consider also the lower-order conormal symbols, we write $\sigma_1^{\mu}(A)(z) := \sigma_1(A)(z)$, and, more generally,

$$\sigma_1^{\mu - j}(A)(z) := \frac{1}{j!}\left(\frac{\partial^j}{\partial r^j} h\right)(0, z) + \sigma_1^{\mu - j}(M + G)(z), \quad j = 0, 1, \dots, k.$$

In the case $A \in L^{\mu}(N^{\wedge}, \boldsymbol{g})$ we define $\sigma_0(A)$ and $\sigma_1(A)$ as well as $\sigma_1^{\mu - j}(A)$ as before. In addition at $r = \infty$ we have exit symbols $\sigma_E(A)$ as in Section 1.5.

Let $A \in L^{\mu}(M, \boldsymbol{g})$ and set for the moment $\sigma_0^{\mu}(A) := \sigma_0(A)$, and $\sigma^{\mu}(A) := (\sigma_0^{\mu}(A), \sigma_1^{\mu}(A))$. The space $\{\sigma^{\mu}(A) : A \in L^{\mu}(M, \boldsymbol{g})\}$ can be characterised as a vector space of pairs with a suitable compatibility property between the components, left to the reader. Setting

$$L^{\mu - 1}(M, \boldsymbol{g}) := \{A \in L^{\mu}(M, \boldsymbol{g}) : \sigma^{\mu}(A) = 0\}$$

we obtain the space

$$L^{\mu - 1}(M, \boldsymbol{g}) = L^{\mu - 1}(M, \boldsymbol{g}^{\mu - 1}) + L_G(M, \boldsymbol{g}) \tag{6.123}$$

with $g^{\mu-1} := (\gamma, \gamma - (\mu - 1), (-k, 0])$, since $\sigma^\mu(G) = 0$ for $G \in L_G(M, g)$.

For A in the first space on the right of (6.123) we have the pair of the principal symbols $(\sigma_0^{\mu-1}(A), \sigma_1^{\mu-1}(A)) = \sigma^{\mu-1}(A)$. Successively we can define

$$L^{\mu-j}(M, g) := \{A \in L^{\mu-(j-1)}(M, g) : \sigma^{\mu-(j-1)}(A) = 0\}$$

for all $j \in \mathbb{N}$.

The operators in $L^{\mu-k}(M, g)$ for $\Theta = (-(k+1), 0]$ still contain one smoothing Mellin term while $L^{\mu-(k+1)}(M, g)$ only contains Green operators in $L_G(M, g)$.

Proposition 6.8.6. *The principal symbol map σ belongs to the exact sequence*

$$0 \to L^{\mu-1}(M, g) \overset{i}{\hookrightarrow} L^\mu(M, g) \overset{\sigma}{\to} \sigma(L^\mu(M, g)) \to 0,$$

where i is the canonical embedding, and there is a right inverse

$$\mathrm{op} : \sigma(L^\mu(M, g)) \to L^\mu(M, g)$$

of σ.

The statement immediately follows from the definition of symbol and operator spaces.

Theorem 6.8.7. *Let $A_j \in L^{\mu-j}(M, g)$, $j \in \mathbb{N}$, be an arbitrary sequence. Then there is an asymptotic sum $A \sim \sum_{j=0}^\infty A_j$ in $L^\mu(M, g)$ in the sense that $A - \sum_{j=0}^N A_j \in L^{\mu-(N+1)}(M, g)$ for every $N \in \mathbb{N}$, and A is unique modulo $L^{-\infty}(M, g) = L_G(M, g)$.*

The result follows from the fact that the involved holomorphic Mellin symbols of decreasing orders can be asymptotically summed up, see Theorem 6.3.15.

Theorem 6.8.8. (i) *Let M be a compact manifold with conical singularity and $A \in L^m(M, g)$, $g = (\gamma, \gamma - \mu, \Theta)$, $\mu - m \in \mathbb{N}$; then $A : C_0^\infty(s_0(M)) \to C^\infty(s_0(M))$ extends to a continuous operator*

$$A : H^{s,\gamma}(M) \to H^{s-m,\gamma-\mu}(M) \tag{6.124}$$

for every $s \in \mathbb{R}$. Moreover, for every asymptotic type $\mathcal{P}$ associated with the weight data (γ, Θ) there is a type $\mathcal{Q}$ associated with data $(\gamma - \mu, \Theta)$ such that (6.124) restricts to a continuous operator

$$A : H_\mathcal{P}^{s,\gamma}(M) \to H_\mathcal{Q}^{s-m,\gamma-\mu}(M)$$

for every $s \in \mathbb{R}$.

(ii) *In the case $A \in L^m(N^\wedge, g)$, $g = (\gamma, \gamma - \mu, \Theta)$, $\mu - m \in \mathbb{N}$ the map $A : C_0^\infty(N^\wedge) \to C^\infty(N^\wedge)$ extends to a continuous operator*

$$A : \mathcal{K}^{s,\gamma}(N^\wedge) \to \mathcal{K}^{s-m,\gamma-\mu}(N^\wedge)$$

and this restricts to a continuous operator

$$A : \mathcal{K}_{\mathcal{P}}^{s,\gamma}(N^\wedge) \to \mathcal{K}_{\mathcal{Q}}^{s-m,\gamma-\mu}(N^\wedge)$$

for every $\mathcal{P}$ with a resulting $\mathcal{Q}$, for every $s \in \mathbb{R}$. In addition we have corresponding continuous maps between the respective spaces with the same weights $e \in \mathbb{R}$ at infinity, i.e.,

$$A : \mathcal{K}^{s,\gamma;e}(N^\wedge) \to \mathcal{K}^{s-m,\gamma-\mu;e}(N^\wedge), \quad A : \mathcal{K}_{\mathcal{P}}^{s,\gamma;e}(N^\wedge) \to \mathcal{K}_{\mathcal{Q}}^{s-m,\gamma-\mu;e}(N^\wedge).$$

The proof is an immediate consequence of Definition 6.8.1 and the specific properties of the involved distribution spaces.

Proposition 6.8.9. *Let $h \in M_{\mathcal{O}}^\mu(N)$, $\mu \in \mathbb{R}$, and fix cut-off functions $\omega \prec \omega'$. Then the operators*

$$\omega \, \mathrm{op}_M^{\gamma-n/2}(h)(1-\omega'), \quad (1-\omega') \, \mathrm{op}_M^{\gamma-n/2}(h)\omega$$

belong to $L_{\mathrm{G}}(N^\wedge, \boldsymbol{g})_{O,O}$, cf. notation (6.83), for $\boldsymbol{g} = (\gamma, \gamma - \mu, (-\infty, 0])$ for all $\gamma \in \mathbb{R}$ (recall that these operators do not depend on γ at all).

Proof. Let us set

$$G_0 := \omega \, \mathrm{op}_M^{\gamma-n/2}(h)(1-\omega'), \quad G_1 := (1-\omega') \, \mathrm{op}_M^{\gamma-n/2}(h)\omega.$$

We need to show, in particular, that G_0, G_1 induce continuous operators

$$G_0, G_1 : \mathcal{K}^{s,\beta;e}(N^\wedge) \to \mathcal{K}^{s',\beta';e'}(N^\wedge) \tag{6.125}$$

for arbitrary $s, s', \beta, \beta', e, e' \in \mathbb{R}$. The corresponding mapping behaviour for G_0^*, G_1^* is then a direct consequence, because G_1 has the structure of the formal adjoint of an operator like G_0 and vice versa.

In order to show (6.125) for G_0, we pass to a symbol h_N via the relation

$$\omega \, \mathrm{op}_M^{\gamma-n/2}(h)(1-\omega') = \mathrm{op}_M^{\gamma-n/2}(h_N) \tag{6.126}$$

for

$$h_N(r, r', z) := f_0(r, r') \partial_z^N h(z), \quad f_0(r, r') := \widetilde{\omega}(r) \log(r/r')^{-N}(1 - \widetilde{\omega}'(r')),$$

cf. Example 6.3.4. The relation (6.126) follows by integration by parts in

$$\omega(r) \, \mathrm{op}_M^{\gamma-n/2}(h)(1-\omega'(r'))u(r) = \omega(r) \int_{\Gamma_{(n+1)/2-\gamma}} \int_0^\infty \left(\frac{r}{r'}\right)^{-z} h(z)u(r') \frac{dr'}{r'} \, d\!\!\!/ z$$

where $\widetilde{\omega}, \widetilde{\omega}'$ are cut-off functions such that $\widetilde{\omega} \succ \omega$, $1 - \widetilde{\omega}' \succ 1 - \omega'$. We now employ the relation

$$\mathrm{op}_M^{\gamma-n/2}(h_N) = r^L \, \mathrm{op}_M^{\beta_0-n/2}\big(f_0(r, r')T^{-L}\partial_z^N h(z)\big)(r')^{-L} \tag{6.127}$$

that is obtained first for $\beta_0 = \gamma$ from (6.126) by commuting r-powers through the Mellin action, cf. Theorem 6.6.5. Recall that the operators in consideration are taken on argument functions in $C_0^\infty(N^\wedge)$. Then we may replace γ by β_0, cf. Remark 6.6.6. We have

$$f_0(r,r')T^{-L}\partial_z^N h(z)\big|_{\Gamma_{(n+1)/2-\beta_0}} \in C_B^\infty\big(\mathbb{R}_+ \times \mathbb{R}_+, L^{\mu-N}\big(N; \Gamma_{(n+1)/2-\beta_0}\big)\big).$$

Thus, Theorem 6.3.5 yields a continuous operator

$$\mathrm{op}_M^{\beta_0-n/2}\big(f_0 T^{-L}\partial_z^N h\big) : \mathcal{H}^{s,\beta_0}\big(N^\wedge\big) \to \mathcal{H}^{s-\mu+N,\beta_0}\big(N^\wedge\big)$$

for every $s \in \mathbb{R}$. Let us write

$$G_0 = \mathrm{op}_M^{\beta_0-n/2}(h_N) = \widetilde{\widetilde{\omega}}(r)r^L \mathrm{op}_M^{\beta_0-n/2}\big(f_0 T^{-L}\partial_z^N h\big)\big(1 - \widetilde{\omega}'(r')\big)(r')^{-L}$$

for cut-off functions $\widetilde{\widetilde{\omega}}, \widetilde{\omega}'$ such that $\widetilde{\widetilde{\omega}} \succ \widetilde{\omega}$, $\big(1 - \widetilde{\omega}''\big) \succ (1 - \widetilde{\omega}')$. Applying Proposition 6.2.4 (ii), (iii) we have continuous operators

$$\big(1 - \widetilde{\omega}'(r')\big)(r')^{-L} : \mathcal{K}^{s,\beta;e}\big(N^\wedge\big) \to \mathcal{H}^{s,\beta_0}\big(N^\wedge\big) \tag{6.128}$$

for any given s, β, β_0, e for a suitable $L(s, \beta, \beta_0, e) > 0$ and

$$\widetilde{\widetilde{\omega}}(r)r^L : \mathcal{H}^{s-\mu+N,\beta_0}\big(N^\wedge\big) \to \mathcal{K}^{s-\mu+N,\beta_0+L;e'}\big(N^\wedge\big) \tag{6.129}$$

for every s, β_0, e', for arbitrary $L \in \mathbb{R}$, especially, $L = L(s, \beta, \beta_0, e)$ from (6.128). Taking into account the relations (6.126) and (6.39) we see altogether that G_0 induces a continuous map (6.125) for arbitrary $s, s', \beta, \beta', e, e' \in \mathbb{R}$.

Next we show the same for G_1. To this end introduce the function

$$f_1(r,r') := \big(1 - \widetilde{\omega}'(r)\big)\log(r/r')^{-N}\widetilde{\omega}(r')$$

for cut-off functions $\widetilde{\omega}', \widetilde{\omega}$ such that $\big(1 - \widetilde{\omega}'\big) \succ (1 - \omega')$, $\widetilde{\omega} \succ \omega$. Observe that $f_0(r,r') = (-1)^N f_1(r',r)$. Then we have

$$G_1 = \big(1 - \widetilde{\omega}'(r)\big)r^{-L}\mathrm{op}_M^{\beta_0-n/2}\big(f_1 T^{-L}\partial_z^N h\big)\widetilde{\omega}(r')(r')^L$$

and the remaining part of the proof is as before. The only change is that now we employ f_1 in Example 6.3.4 and Proposition 6.2.4 (i), (iv).   □

Theorem 6.8.10.   (i) $A \in L^\mu(M, \boldsymbol{g})$ for $\boldsymbol{g} = (\gamma, \gamma - \mu, \Theta)$, *implies for the formal adjoint* $A^* \in L^\mu(M, \boldsymbol{g}^*)$ for $\boldsymbol{g}^* = (-\gamma + \mu, -\gamma, \Theta)$, *and we have*

$$\sigma_0(A^*) = \overline{\sigma_0(A)}, \quad \sigma_1(A^*)(z) = \big(T^\mu \sigma_1(A)\big)(n + 1 - \bar{z}). \tag{6.130}$$

(ii) $A \in L^{\mu}(N^{\wedge}, \boldsymbol{g})$ for $\boldsymbol{g} = (\gamma, \gamma - \mu, \Theta)$, implies $A^* \in L^{\mu}(N^{\wedge}, \boldsymbol{g}^*)$ for $\boldsymbol{g}^* = (-\gamma + \mu, -\gamma, \Theta)$, again with (6.130) and the symbolic rules for the conical exit of $N^{\wedge}$ to infinity.

Proof. According to Definition 6.8.1, we write the operator $A \in L^{\mu}(M, \boldsymbol{g})$ in the form

$$A = H + J + M + G$$

for $H = \omega r^{-\mu} \operatorname{op}_M^{\gamma - n/2}(h)\omega'$, $J = (1 - \omega)A_{\mathrm{int}}(1 - \omega'')$ and $M + G \in L_{\mathrm{M+G}}(M, \boldsymbol{g})$, cf. Definition 6.5.14. The formal adjoint refers to the scalar product of $\mathcal{H}^{0,0}(M) = \omega \mathcal{K}^{0,0}(N^{\wedge}) + (1 - \omega)H^0_{\mathrm{loc}}(\mathrm{int}\, M)$, where M is locally close to the conical singularity modelled on $N^{\triangle}$ for a closed manifold N. We assume that N is compact. The interior part structure of the formal adjoint is analogous to that known from the standard pseudo-differential calculus. For the Green part G, the formal adjoint is of analogous kind, as a part of the definition. Concerning H and M we can apply Proposition 6.3.7 and deduce, in particular, the formula for $\sigma_1(A^*)(z)$. The first expression in (6.130) is again known from the standard pseudo-differential calculus, applied to A as an element of $L^{\mu}_{\mathrm{cl}}(M \setminus \{v\})$. The arguments for $N^{\wedge}$ are simple as well and left to the reader. $\qquad\square$

Theorem 6.8.11. (i) *Let M be a compact manifold with conical singularity, $A \in L^{\mu}(M, \boldsymbol{g})$ for $\boldsymbol{g} = (\gamma - \nu, \gamma - (\mu + \nu), \Theta)$, $B \in L^{\nu}(M, \boldsymbol{h})$ for $\boldsymbol{h} = (\gamma, \gamma - \nu, \Theta)$. Then $AB \in L^{\mu+\nu}(M, \boldsymbol{g} \circ \boldsymbol{h})$ for $\boldsymbol{g} \circ \boldsymbol{h} = (\gamma, \gamma - (\mu + \nu), \Theta)$ and we have*

$$\sigma_0(AB) = \sigma_0(A)\sigma_0(B), \quad \sigma_1(AB)(z) = \big(T^{\nu}\sigma_1(A)\big)(z)\sigma_1(B)(z). \quad (6.131)$$

(ii) *If $A \in L^{\mu}(N^{\wedge}, \boldsymbol{g})$, $B \in L^{\mu}(N^{\wedge}, \boldsymbol{h})$ for $\boldsymbol{g}, \boldsymbol{h}$ as in (i) then we have $AB \in L^{\mu+\nu}(N^{\wedge}, \boldsymbol{g} \circ \boldsymbol{h})$ with (6.131), and the composition rule from the conical exit of $N^{\wedge}$ to infinity.*

Proof. (i) Let $A = H + J + M + G$ as in the preceding proof and analogously, $B = L + I + N + K$. Then, what concerns the symbolic multiplicative rule for σ_0 in (6.131) holds for similar reasons as before. The second identity is a well-known rule for composing Mellin pseudo-differential operators multiplied together with a weight shift $r^{-\nu}$ in front of the second factor. Then the argument shift T^{ν} is generated from the rule on how such r-powers commute through the first Mellin action, see Theorem 6.5.11. For the items in the composition

$$AB = (H + J + M + G)(L + I + N + K)$$

we consider the various terms separately. Since $A \in L^{\mu}_{\mathrm{cl}}(M \setminus \{v\})$, $B \in L^{\nu}_{\mathrm{cl}}(M \setminus \{v\})$, and since all summands of A and B are of this kind, the structure of J and I shows that all products in $J(L + I + N + K)$ and $(H + J + M + G)I$ are of interior type. Moreover, all products of any such operator with G or K is again Green, because all items have corresponding mapping properties, see Theorem 6.8.8 (i). It remains to consider operators which are not multiplied by J or I and G or K. Thus we have to identify HL, ML, HN, MN as elements of $L^{\mu+\nu}(M, \boldsymbol{g} \circ \boldsymbol{h})$. Since all those operators are of Mellin type and localized close to the conical singularity, we may apply the fact that those Mellin compositions are again as desired. Note that the second factors are multiplied from the left by weight factors $r^{-\nu}$. Those can be through the first Mellin actions on the expense of a corresponding translation of the complex argument in the first factors. This leads, except for the resulting weight factor $r^{-(\mu+\nu)}$, to Mellin compositions with integrations over the same weight line, namely, $\Gamma_{(n+1)/2-(\mu+\nu)}$, and such composition are of known structure, again as Mellin operators, either with non-smoothing holomorphic symbols for HL, and smoothing meromorphic ones for ML, HN and MN. Assertion (ii) can be proved in an analogous manner. This completes the proof. $\qquad\square$

Proposition 6.8.12. *Let ω, ω_0, ω' be cut-off functions, $\omega_0 \succ \omega'$ and consider $f \in C^\infty\big(\overline{\mathbb{R}}_+, M_{\mathcal{R}}^\mu(N)\big)$, $g \in C^\infty\big(\overline{\mathbb{R}}_+, M_{\mathcal{Q}}^\nu(N)\big)$ for Mellin asymptotic types $\mathcal{R}$, $\mathcal{Q}$. Then*

$$\omega \operatorname{op}_M(f)(1 - \omega_0) \operatorname{op}_M(g)\, \omega' \in L_{\mathrm{G}}\big(N^\wedge, (0, 0, (-\infty, 0])\big).$$

Proof. We may apply the mapping properties of Definition 6.5.2 (i), characterizing Green operators. We can decompose $g(r, z)$ as a sum $g(r, z) = g_\mathcal{O}(r, z) + g_\mathcal{Q}(r, z)$ for some $g_\mathcal{O}(r, z) \in C^\infty\big(\overline{\mathbb{R}}_+; M_\mathcal{O}^\nu(N)\big)$, $g_\mathcal{Q}(r, z) \in C^\infty\big(\overline{\mathbb{R}}_+; M_\mathcal{Q}^{-\infty}(N)\big)$. To this end it suffices to apply Mellin kernel cut-off with respect to a weight line which does not intersect $\pi_\mathbb{C}\mathcal{R}$. Then from Proposition 6.8.9 we obtain

$$(1 - \omega_0) \operatorname{op}_M(g_\mathcal{O})\omega' \in L_{\mathrm{G}}\big(N^\wedge, \boldsymbol{g}\big)_{\mathcal{O},\mathcal{O}}.$$

Thus, $\operatorname{op}_M(f)(1 - \omega_0) \operatorname{op}_M(g_\mathcal{O})\omega' \in L_{\mathrm{G}}(N^\wedge, (0, 0, (-\infty, 0]))$ since Green operators form a two-sided ideal in the cone algebra. Moreover, we have

$$(1 - \omega_0) \operatorname{op}_M(g_\mathcal{Q})\omega' \in L_{\mathrm{G}}\big(N^\wedge, (0, 0, (-\infty, 0])\big)$$

because of the mapping properties. This shows altogether that $\omega \operatorname{op}_M(f)(1 - \omega_0) \operatorname{op}_M(g)\omega'$ is of Green type. $\qquad\square$

Theorem 6.8.13. *Let $A \in L^\mu(M, \boldsymbol{g})$, $B \in L^\nu(M, \boldsymbol{c})$ for $\boldsymbol{g} = (\gamma - \nu, \gamma - (\mu + \nu), \Theta)$, $\boldsymbol{c} = (\gamma, \gamma - \nu, \Theta)$, $\Theta = (-(k+1), 0]$, $k \in \mathbb{N}$. Then for*

$$\sigma_1^{\mu-p}(A)(z) = \frac{1}{p!} \frac{\partial^p h}{\partial r^p}(0, z) + \sigma_1^{\mu-p}(M)(z),$$

$$\sigma_1^{\nu-q}(B)(z) = \frac{1}{q!} \frac{\partial^q l}{\partial r^q}(0, z) + \sigma_1^{\nu-q}(L)(z)$$

where h, M are as in Definition 6.8.1 (i) (for $\gamma - \nu$ rather than γ) and l, L of analogous meaning with respect to the operator B, we have the Mellin translation product

$$\sigma_1^{\mu+\nu-l}(AB) = \sum_{p+q=l} \big(T^{\nu-q}\sigma_1^{\mu-p}(A)\big)\sigma_1^{\nu-q}(B), \quad l = 0, \dots, k.$$

Proof. The proof is analogous to that of Theorem 6.5.19. $\qquad\square$

6.9 Ellipticity in the cone calculus

As noted at the beginning, the singular algebra and especially the cone algebra contain the typical degenerate differential operators together with the parametrices of elliptic elements. The ellipticity refers to the principal symbol hierarchy of an operator A, in this case $\sigma(A) = (\sigma_0(A), \sigma_1(A))$, the components of which are contributed by the stratification $s(M) = (s_0(M), s_1(M))$. In the following Θ will be employed in the meaning $(-(k+1), 0]$ for some $k \in \mathbb{N} \cup \{\infty\}$.

Definition 6.9.1. Let M be a manifold with conical singularity v. An element $A \in L^\mu(M, \boldsymbol{g})$ for $\boldsymbol{g} = (\gamma, \gamma - \mu, \Theta)$ is called elliptic (of order μ) if

(i) $\sigma_0(A) \neq 0$ as a function on $T^*((s_0(M)) \setminus 0$ and if close to $s_1(M)$ in the splitting of variables and covariables into (r, x, ρ, ξ), the reduced symbol

$$\widetilde{\sigma}_0(A)(r, x, \rho, \xi) = r^\mu \sigma_0(A)(r, x, r^{-1}\rho, \xi)$$

does not vanish for $(\rho, \xi) \neq 0$, including $r = 0$.

(ii) The conormal symbol $\sigma_1(A)$ induces a family of bijective operators

$$\sigma_1(A) : H^s(N) \to H^{s-\mu}(N) \tag{6.132}$$

for all $z \in \Gamma_{(n+1)/2-\gamma}$, $n = \dim N$, with N being the base of the local cone near v.

An $A \in L^\mu(N^\wedge, \boldsymbol{g})$ for $\boldsymbol{g} = (\gamma, \gamma - \mu, \Theta)$ is called elliptic (of order μ) if

(iii) $\sigma_0(A) \neq 0$ as a function on $T^*(X^\wedge \setminus 0)$ and if close to zero the reduced symbol $\widetilde{\sigma}_0(A)$ is as in (i).

(iv) The conormal symbol $\sigma_1(A)$ satisfies condition (ii).

(v) The operator $A \in L^{\mu;0}_{\mathrm{cl}}(X^\succeq)|_{N^\wedge}$ in the notation at the end of Section 1.5 is exit elliptic for $r \to \infty$.

Recall that $\sigma_1(A)(z) \in M^\mu_{\mathcal{R}}(N)$ for some Mellin asymptotic type $\mathcal{R}$. Since

$$M^\mu_{\mathcal{R}}(N) = M^\mu_{\mathcal{O}}(N) + M^{-\infty}_{\mathcal{R}}(N),$$

it follows that $f := \sigma_1(A)$ can be written as $f = h + l$ for an $h \in M^\mu_{\mathcal{O}}(N)$, $l \in M^{-\infty}_{\mathcal{R}}(N)$. Under condition (6.132), the restriction

$$h|_{\Gamma_{(n+1)/2-\gamma}} \in L^\mu_{\mathrm{cl}}(N; \Gamma_{(n+1)/2-\gamma})$$

is parameter-dependent elliptic of order μ. Let us give a brief explanation on why in the cone calculus we meet Mellin asymptotic types $\mathcal{R}$ and finite rank Laurent coefficients in $L^{-\infty}(N)$ of elements in $M^{-\infty}_{\mathcal{R}}(N)$. The reason comes from the construction of parametrices within the cone calculus, which requires inverting the principal symbolic components. Condition (6.132) implies the property $\sigma_1(A)^{-1} \in M^{-\mu}_{\mathcal{Q}}(N)$ for some resulting asymptotic type.

Theorem 6.9.2. (i) *Let* $A \in L^\mu(M, \boldsymbol{g})$, $\boldsymbol{g} = (\gamma, \gamma - \mu, \Theta)$, *for a manifold* M *with conical singularity be elliptic in the sense of* Definition 6.9.1 (i), (ii). *Then there is a (properly supported) parametrix* $P \in L^{-\mu}(M, \boldsymbol{g}^{-1})$, $\boldsymbol{g}^{-1} = (\gamma - \mu, \gamma, \Theta)$, *such that*

$$1 - PA \in L_G(M, \boldsymbol{g}_L), \quad 1 - AP \in L_G(M, \boldsymbol{g}_R),$$

$\boldsymbol{g}_L = (\gamma, \gamma, \Theta)$, $\boldsymbol{g}_R = (\gamma - \mu, \gamma - \mu, \Theta)$, *and we have*

$$\sigma_0(P) = \sigma_0(A)^{-1}, \quad \sigma_1(P) = T^{-\mu}\sigma_1(A)^{-1}. \tag{6.133}$$

(ii) *Let* $A \in L^\mu(N^\wedge, \boldsymbol{g})$, $\boldsymbol{g} = (\gamma, \gamma - \mu, \Theta)$, N *closed, compact , be elliptic in the sense of* Definition 6.9.1 (iii), (iv), (v). *Then there is a parametrix* $P \in L^{-\mu}(N^\wedge, \boldsymbol{g}^{-1})$ *such that*

$$1 - PA \in L_G(N^\wedge, \boldsymbol{g}_L) \quad and \quad 1 - AP \in L_G(X^\wedge, \boldsymbol{g}_R),$$

where $\boldsymbol{g}^{-1}$, $\boldsymbol{g}_L$, $\boldsymbol{g}_R$ *are as in* (i), *and we have an analogue of* (6.133) *together with the symbolic rule under parametrix construction from the exit calculus in* Section 1.5.

Proof. We consider operators on M; the case $N^\wedge$ is left to the reader. According to Definition 6.8.1 (ii), the operator A is of the form

$$A = \omega r^{-\mu} \operatorname{op}_M^{\gamma-n/2}(h)\omega' + (1 - \omega)A_{\text{int}}(1 - \omega'') + M + G. \tag{6.134}$$

The Green operator G can be completely neglected, so we drop it. The choice of cut-off functions $\omega'' \prec \omega \prec \omega'$ is inessential. For the parametrix we choose cut-off functions $\widetilde{\omega}'' \prec \widetilde{\omega} \prec \widetilde{\omega}'$ such that $\widetilde{\omega}'\omega \succ \widetilde{\omega}$. We consider a neighbourhood of $r = 0$ and write

$$P = \widetilde{\omega} r^\mu \operatorname{op}_M^{\gamma-\mu-n/2}(l)\widetilde{\omega}' + (1 - \widetilde{\omega})P_{\text{int}}(1 - \widetilde{\omega}'') + L$$

for a smoothing Mellin operator L. Close to $r = 0$ we compute PA and write

$$\widetilde{\omega} r^\mu \operatorname{op}_M^{\gamma-\mu-n/2}(l)\widetilde{\omega}'\omega r^{-\mu} \operatorname{op}_M^{\gamma-n/2}(h)\omega' = \widetilde{\omega} \operatorname{op}_M^{\gamma-n/2}(T^{-\mu}l) \operatorname{op}_M^{\gamma-n/2}(h)\omega' + G, \tag{6.135}$$

where $G = \widetilde{\omega} \operatorname{op}_M^{\gamma-n/2}(T^{-\mu}l)(\widetilde{\omega}'\omega - 1) \operatorname{op}_M^{\gamma-n/2}(h)\omega'$. By Proposition 6.8.9, the operator G is of Green type. So we may focus on the first summand on the right-hand side of (6.135). We determine the Mellin symbol $m(r, z) := (T^{-\mu}l)(r, z) = l(r, z - \mu)$ by applying the Mellin Leibniz product for the composition of Mellin operators

$$\operatorname{op}_M^{\gamma-n/2}(m) \operatorname{op}_M^{\gamma-n/2}(h) = \operatorname{op}_M^{\gamma-n/2}(m \sharp h)$$

for

$$m \sharp h \sim \sum_{j=0}^\infty \frac{1}{j!} \partial_z^j m(r, z) \cdot (r\partial_r)^j h(r, z) = 1. \tag{6.136}$$

From the invertibility of $\sigma_1(A)(z) = h(0, z) + \sigma_1(M)(z)$ in $L_{\mathrm{cl}}^\mu(N; \Gamma_{(n+1)/2-\gamma})$ it follows that

$$h(r, z) + \sigma_1(M)(z) \in C^\infty\big(\overline{\mathbb{R}}_+, L_{\mathrm{cl}}^\mu(N; \Gamma_{(n+1)/2-\gamma})\big) \tag{6.137}$$

has an inverse in

$$C^\infty\big([0, \varepsilon), L_{\mathrm{cl}}^\mu(N; \Gamma_{(n+1)/2-\gamma})\big) \tag{6.138}$$

for some $\varepsilon > 0$. Applying Mellin kernel cut-off with respect to the weight line $\Gamma_{(n+1)/2-\gamma}$ for a sufficiently small support of ψ to this element gives us an $m_1(r, z) \in C^\infty([0, \varepsilon), M_{\mathcal{O}}^{-\mu}(N))$ which is invertible for all $r \in [0, \varepsilon)$. We may assume that all cut-off functions we are using in (6.134) or (6.135) are supported in $[0, \varepsilon)$. We have

$$m_1(r, z)h(r, z) = 1 \bmod C^\infty\big([0, \varepsilon), M_{\mathcal{O}}^{-\infty}(N)\big) \tag{6.139}$$

since $m_1(r, z)(h(r, z) + \sigma_1(M)(r, z)) = 1 \bmod C^\infty([0, \varepsilon), L^{-\infty}(N))$. We solve the asymptotic identity for $m_1(r, z)$ rather than $m(r, z)$. Then starting with

$$m_1 \sharp h = 1 + \sum_{j=1}^\infty \frac{1}{j!} \partial_z^j m_1(r, z)(r\partial_r)^j h(r, z) \tag{6.140}$$

we carry out the asymptotic summation in $C^\infty([0, \varepsilon), M_{\mathcal{O}}^{-1}(N))$ and obtain an $m_2(r, z)$ in this space. By a formal Neumann series argument we find an $m_3(r, z) \in C^\infty([0, \varepsilon), M_{\mathcal{O}}^{-1}(N))$ such that

$$(1 + m_2)^{\sharp - 1} = 1 + m_3(r, z) \bmod C^\infty\big([0, \varepsilon), M_{\mathcal{O}}^{-\infty}(N)\big).$$

Thus (6.140) gives

$$\{(1 + m_3(r, z)\sharp m_1(r, z)\}\sharp h(r, z) = 1 \bmod C^\infty\big([0, \varepsilon), M_{\mathcal{O}}^{-\infty}(N)\big),$$

and hence, setting $m(r, z) = 1 + m_3(r, z) + m_1(r, z)$ we get a solution of (6.136) mod $C^\infty([0, \varepsilon), M^{-\infty}(N))$. We then have

$$m(r, z)\sharp(h(r, z) + \sigma_1(M)(z)) = 1 + n(r, z)$$

for some $n(r, z) \in C^\infty([0, \varepsilon), M_{\mathcal{R}}^{-\infty}(N))$. The remainder $\mathcal{R}$ in the relation

$$\widetilde{\omega} \operatorname{op}_M^{\gamma - n/2}(m) \operatorname{op}_M^{\gamma - n/2}(h + \sigma_1(A))\omega' = 1 + R$$

belongs to $L_{\mathrm{M+G}}(M, \boldsymbol{g}_{\mathrm{L}})$ and $1 + R$ is invertible modulo Green operators, i.e., we can add 1 plus a smoothing Mellin operator to $\operatorname{op}_M^{\gamma - n/2}(m)$ which gives a $f(r, z) := m(r, z) + \sigma_1(L)(z)$ such that for $T^\mu f(r, z) = f(r, z + \mu)$ it follows that

$$\widetilde{\omega} r^\mu \operatorname{op}_M^{\gamma - \mu - n/2}(T^\mu f)\widetilde{\omega}' \omega r^{-\mu} \operatorname{op}_M^{-\mu}(h + \sigma_1(M))\omega' = 1 + G_1$$

for some Green operator G_1. Thus we constructed a left parametrix of A. In a similar manner we get a right parametrix of A close to $r = 0$. Since our operator A is elliptic far from $r = 0$ as a classical pseudo-differential operator in the standard sense, it has a standard parametrix there, belonging to $L_{cl}^{-\mu}(M \setminus \{v\})$, where v is the vertex of M. By applying a suitable partition of unity on M, we can glue the latter parametrix together with the one close to $r = 0$ to a parametrix in the sense of definition.   □

Theorem 6.9.3.   (i) *Let M be a compact manifold with conical singularity, and let $A \in L^\mu(M, \boldsymbol{g})$, $\boldsymbol{g} = (\gamma, \gamma - \mu, \Theta)$. Then the following conditions are equivalent:*

(a) *The operator*

$$A : H^{s,\gamma}(M) \to H^{s-\mu,\gamma-\mu}(M) \tag{6.141}$$

is Fredholm for some $s = s_0 \in \mathbb{R}$.

(b) *The operator A is elliptic in the sense of* Definition 6.9.1 (i), (ii).

(c) *The Fredholm property of* (6.141) *for some $s = s_0 \in \mathbb{R}$ entails the Fredholm property for all $s \in \mathbb{R}$.*

(ii) *Let $A \in L^\mu(N^\wedge, \boldsymbol{g})$, $\boldsymbol{g} = (\gamma, \gamma - \mu, \Theta)$. Then the following conditions are equivalent:*

(d) *The operator*

$$\Lambda : \mathcal{K}^{s,\gamma}(N^\wedge) \to \mathcal{K}^{s-\mu,\gamma-\mu}(N^\wedge) \tag{6.142}$$

is Fredholm for some $s = s_0 \in \mathbb{R}$.

(e) *The operator A is elliptic in the sense of* Definition 6.9.1 (iii), (iv), (v).

(f) *The Fredholm property of* (6.142) *for some $s = s_0 \subset \mathbb{R}$ entails the Fredholm property for all $s \in \mathbb{R}$.*

Proof. We content ourselves with (i) (a) and (b). Assertion (i) (c) will not be employed in the exposition. A proof of necessity of ellipticity for the Fredholm property in the cone algebra is given in Schrohe and Seiler [41], see also corresponding material in [44]. The Fredholm property of an elliptic operator $A \in L^\mu(M, \boldsymbol{g})$ for compact M follows from the existence of a parametrix P which leaves remainders of Green operators that are compact in space $H^{s,\gamma}(M)$, and $H^{s-\mu,\gamma-\mu}(M)$, respectively. The fact that the Fredholm property of (6.141) for $s = s_0$ gives rise to the Fredholm property for all s is a consequence of elliptic regularity of solutions to $Au = f$ and $A^*v = g$ for the formal adjoint A^*. Solutions to the respective homogeneous equations, i.e., $\ker A$ and $\operatorname{coker} A$, are independent of s. The scheme of the proof for (ii) is similar to (i).   □

Theorem 6.9.4.   (i) *Let M be a compact manifold with conical singularity, and $A \in L^\mu(M, \boldsymbol{g})$, $\boldsymbol{g} = (\gamma, \gamma - \mu, \Theta)$, and assume that*

$$A : H^{s,\gamma}(M) \to H^{s-\mu,\gamma-\mu}(M) \tag{6.143}$$

is an isomorphism for some $s = s_0 \in \mathbb{R}$. Then (6.143) is an isomorphism for all $s \in \mathbb{R}$, and we have $A^{-1} \in L^{-\mu}(M, \boldsymbol{g}^{-1})$.

(ii) Let $A \in L^\mu(N^\wedge, \boldsymbol{g})$, $\boldsymbol{g} = (\gamma, \gamma - \mu, \Theta)$, and assume that

$$A : \mathcal{K}^{s,\gamma}(N^\wedge) \to \mathcal{K}^{s-\mu, \gamma-\mu}(N^\wedge) \tag{6.144}$$

is an isomorphism for some $s = s_0 \in \mathbb{R}$. Then (6.144) is an isomorphism for all $s \in \mathbb{R}$, and we have $A^{-1} \in L^{-\mu}(N^\wedge, \boldsymbol{g}^{-1})$.

Proof. We focus again on (i). The proof of (ii) is similar. Assuming that (6.143) is an isomorphism then, by Theorem 6.9.3, A is necessarily elliptic in the cone algebra. Therefore, by Theorem 6.9.2, it has a parametrix P in the cone algebra which has index zero, since A has index zero and the Green remainders are compact. For standard functional analytic reasons, in the relation $PA = 1 + G_L$ for a Green operator G_L we always find a Green operator H of finite rank such that $P + H$ is again an isomorphism. Then also $1 + G_L$ is an isomorphism and its inverse $(1 + G_L)^{-1}$ has the form $1 + C_L$ for another Green operator C_L. To conclude this formally one proceeds in much the same way as in Proposition 2.2.39 for Green operators in BVPs. Then $P_1 := (1 + C_L)P$ is obviously a left inverse of A, and it belongs to the cone calculus, because of its algebra property. In a similar manner we find a right inverse, i.e., P_1 is a two-sided inverse of A, and this holds for all $s \in \mathbb{R}$, see Theorem 6.9.3 (i). □

Theorem 6.9.5. (i) *The space* $L^\mu(M, \boldsymbol{g})$, $\boldsymbol{g} = (\gamma, \gamma - \mu, \Theta)$, *contains an element* A *such that* (6.143) *is an isomorphism for every* $s \in \mathbb{R}$.

(ii) *The space* $L^\mu(N^\wedge, \boldsymbol{g})$, $\boldsymbol{g} = (\gamma, \gamma - \mu, \Theta)$, *contains an element* A *such that* (6.144) *is an isomorphism for all* $s \in \mathbb{R}$.

Proof. (i) First note that for any $\mu \in \mathbb{R}$ there exists an elliptic operator $A \in L^\mu(M, \boldsymbol{g})$. In fact, it suffices to choose a parameter-dependent elliptic element $f \in L^\mu_{\mathrm{cl}}(X; \Gamma_\lambda)$, pass via kernel cut-off to an $h(z) \in M^\mu_{\mathcal{O}}(X)$, and form the Mellin operator

$$r^{-\mu} \mathrm{op}_M^{\gamma - n/2}(h) \tag{6.145}$$

for a prescribed weight. As we know, the set of non-bijectivity points of the conormal symbol

$$h(z) : H^s(X) \to H^{s-\mu}(X) \tag{6.146}$$

is discrete in the complex plane. Therefore, after a translation in z in the real direction we find h in such a way that $\Gamma_{(n+1)/2-\gamma}$ does not touch the non-bijectivity points of (6.146). We can base the construction of h on an $f(\lambda)$ for local symbols of the form $\left(1 + |\xi|^2 + |\lambda|^2\right)^{\mu/2}$. Then, if we assume that the compact manifold M with conical singularity $\{v\}$ is locally, close to $\{v\}$, modelled on X^Δ for closed X, then we find an extension of (6.145) to $M \setminus \{v\}$ to an elliptic operator A_{int} of order μ belonging to $L^\mu_{\mathrm{cl}}(M \setminus \{v\})$. Then

$$A_1 = \omega r^{-\mu} \mathrm{op}_M^{\gamma - n/2}(h)\omega' + (1 - \omega)A_{\mathrm{int}}(1 - \omega'')$$

for cut-off functions $\omega'' \prec \omega \prec \omega'$ is elliptic in $L^\mu(M, \boldsymbol{g})$. Because of Theorem 6.141 (i), A_1 is Fredholm in weighted Sobolev spaces of some index $k \in \mathbb{N}$. Now we may use a result on the space $L_{\mathrm{M+G}}(M, \boldsymbol{g}_0)$ for $\boldsymbol{g}_0 = (\gamma - \mu, \gamma - \mu, \Theta)$ asserting that there is an element $L_{-k} \in L_{\mathrm{M+G}}(M, \boldsymbol{g}_0)$ such that

$$1 + L_{-k} : H^{s-\mu,\gamma-\mu}(M) \to H^{s-\mu,\gamma-\mu}(M)$$

is of index $-k$. Then $L_{-k}A_1 : H^{s,\gamma}(M) \to H^{s-\mu,\gamma-\mu}(M)$ has index zero as an operator in $L^\mu(M, \boldsymbol{g})$. Proceeding in much the same way as in the proof of Theorem 6.9.4, we can choose a finite rank Green operator H such that

$$A := L_{-k}A_1 + H : H^{s,\gamma}(M) \to H^{s-\mu,\gamma-\mu}(M)$$

is an isomorphism. This is then the case for all $s \in \mathbb{R}$, and hence the desired order-reducing isomorphism is constructed.   □

Remark 6.9.6. Definition 6.9.1 and Theorems 6.9.2–6.9.5 have a straightforward generalisation to $L^\mu(M, \boldsymbol{g})$ and $L^\mu(X^\wedge, \boldsymbol{h})$, respectively, for $\boldsymbol{h} = (\gamma, \delta, \Theta)$, $\gamma, \delta \in \mathbb{R}$ arbitrary, cf. Remark 6.8.2.

6.10 Interpretation of standard Sobolev spaces as weighted spaces on a cone

We first compare the spaces $H^s(\mathbb{R}^{n+1})$ and $\mathcal{H}^{s,s}((S^n)^\wedge)$ for $s \in \mathbb{N}$.

Variables in $\mathbb{R}^{n+1}$ are denoted by $\widetilde{x}$ and polar coordinates in $\mathbb{R}^{n+1} \setminus \{0\}$ by (r, φ), and we identify $L^2(\mathbb{R}^{n+1}) = H^0(\mathbb{R}^{n+1})$ with $r^{-n/2}L^2(\mathbb{R}_+ \times S^n)$, where $L^2(\mathbb{R}_+ \times S^n)$ is based on the measure $dr\,d\varphi$. For convenience we denote general operators in $\mathrm{Diff}^m(S^n)$ by D_φ^m.

Lemma 6.10.1. *For every $s \in \mathbb{N}$ we have*

$$H^s(\mathbb{R}^{n+1}) = \Big\{ u(r, \varphi) \in r^{-n/2}L^2(\mathbb{R}_+ \times S^n) :$$

$$r^{-s}\Big(r\frac{\partial}{\partial r}\Big)^k D_\varphi^{|\alpha|}u(r, \varphi) \in r^{-n/2}L^2(\mathbb{R}_+ \times S^n) \qquad (6.147)$$

$$\text{for all } D_\varphi^{|\alpha|} \in \mathrm{Diff}^{|\alpha|}(S^n),\ k + |\alpha| = s \Big\}.$$

Proof. We have

$$H^s(\mathbb{R}^{n+1}) = \{ u \in L^2(\mathbb{R}^{n+1}) : D_{\widetilde{x}}^\alpha \in L^2(\mathbb{R}^{n+1}) \text{ for all } \alpha \in \mathbb{N}^{n+1}, |\alpha| = s \}, \tag{6.148}$$

or, equivalently,

$$H^s(\mathbb{R}^{n+1}) = \{ u \in L^2(\mathbb{R}^{n+1}) : D_{\widetilde{x}}^\alpha u \in L^2(\mathbb{R}^{n+1}) \text{ for all } \alpha \in \mathbb{N}^{n+1}, |\alpha| \le s \}.$$

In polar coordinates,

$$D_{\tilde{x}}^{\alpha} = r^{-|\alpha|} \sum_{j=0}^{|\alpha|} a_{j,\alpha} \left(r \frac{\partial}{\partial r} \right)^j$$

for suitable $a_{j,\alpha} \in C^{\infty}\left(\overline{\mathbb{R}}_+, \operatorname{Diff}^{|\alpha|-j}(S^n) \right)$. Because of the homogeneity

$$D_{\tilde{x}}^{\alpha} = \iota_{\delta} D_{\tilde{x}}^{\alpha} \iota_{\delta^{-1}} \quad \text{and} \quad \left(r \frac{\partial}{\partial r} \right)^j = \iota_{\delta} \left(r \frac{\partial}{\partial r} \right)^j \iota_{\delta^{-1}} \quad \text{for all } \delta \in \mathbb{R}_+,$$

where $(\iota_{\delta} u)(\tilde{x}) := u(\delta \tilde{x})$, $(\iota_{\delta} v)(r, \varphi) := v(\delta r, \varphi)$, and so the coefficients $a_{j,\alpha}$ are independent of r, i.e., $a_{j,\alpha} \in \operatorname{Diff}^{|\alpha|-j}(S^n)$. Thus the characterisation (6.148) takes the form

$$H^s\left(\mathbb{R}^{n+1}\right) = \left\{ u \in r^{-n/2} L^2(\mathbb{R}_+ \times S^n) : r^{-s} \sum_{j=0}^{s} a_{j,\alpha} \left(r \frac{\partial}{\partial r} \right)^j u \in L^2(\mathbb{R}_+ \times S^n) \right.$$

$$\left. \text{for all } \alpha \in \mathbb{N}^{n+1}, |\alpha| = s \right\}.$$

$$(6.149)$$

$\square$

For $m \in \mathbb{N}$ and $s > m/2$ we set

$$H_0^s(\mathbb{R}^m) := \left\{ u \in H^s(\mathbb{R}^m) : (D_x^{\alpha} u)(0) = 0 \text{ for all } |\alpha| < s - m/2 \right\}.$$

Remark 6.10.2. The operator $r_{x_{m+1}=0} : u(x_1, \ldots, x_m, x_{m+1}) \mapsto u(x_1, \ldots, x_m, 0)$, which induces a continuous and surjective operator

$$r_{x_{m+1}=0} : H^s(\mathbb{R}^{m+1}) \to H^{s-1/2}(\mathbb{R}^m)$$

for every $s > 1/2$, restricts to a surjective operator

$$r_{x_{m+1}=0} : H_0^s(\mathbb{R}^{m+1}) \to H_0^{s-1/2}(\mathbb{R}^m)$$

for every $s > (m+1)/2$.

Lemma 6.10.3. *For $s > m/2$ we have*

$$H_0^s(\mathbb{R}^m) = \left\{ u \in H^s(\mathbb{R}^m) : (\partial_r^k u)(0) = 0 \text{ for all } 0 \le k < s - m/2 \right\}$$

for $r = |x|$, $x \in \mathbb{R}^m$.

Proof. Every $u \in H^s(\mathbb{R}^m)$ can be written in the form $u(x) = u_0(x) + p(x)$ for some $u_0 \in H_0^s(\mathbb{R}^m)$ and $p(x) = \sum_{0 \le |\beta| < s - m/2} \omega(x) c_{\beta} x^{\beta}$, where $\omega \in C_0^{\infty}(\mathbb{R}^m)$ is a cut-off function in $\mathbb{R}^m$ (i.e., $\omega \equiv 1$ close to 0), and the coefficients $c_{\beta} \in \mathbb{C}$ are uniquely determined by u. In polar coordinates $x = (r, \varphi) \in \mathbb{R}_+ \times S^{m-1}$ (and for $\omega = \omega(r)$) we have

$$p(x) = \left\{ \omega(r) \sum_{l=0}^{[s-m/2]} r^l \sum_{|\beta|=l} c_\beta g_\beta(\varphi) \right\},$$

with coefficients $g_\beta(\varphi) \in C^\infty(S^{m-1})$. Since $u \in H_0^s(\mathbb{R}^m)$ is equivalent to

$$(D_x^\alpha p)(0) = 0 \quad \text{for all } 0 \le |\alpha| < s - m/2, \tag{6.150}$$

it suffices to show that (6.150) is equivalent to

$$\left. (\partial_r^k p) \right|_{r=0} = 0 \quad \text{for all } 0 \le k < s - m/2. \tag{6.151}$$

Clearly, (6.150) entails (6.151). Conversely, (6.151) for $k = 0$ gives us $c_0 g_0(\varphi) = 0$. Inductively it follows that $\sum_{|\beta|=l} c_\beta g_\beta(\varphi) = 0$ for every $0 \le l < m/2$. This yields $D_\varphi^\gamma \partial_r^k p|_{r=0} = 0$ for all $0 \le |\gamma| + k < s - m/2$, and hence $u \in H_0^s(\mathbb{R}^m)$. $\square$

Remark 6.10.4. Let $x = (x_1, \ldots, x_{m-1}, x_m)$, $x' := (x_1, \ldots, x_{m-1})$, and define the restriction $\mathrm{r} : u(x) \mapsto u(x', 0)$. Then

$$H^s(\mathbb{R}^m) \to H^{s-1/2}(\mathbb{R}^{m-1}) \quad \text{for } s > 1/2 \tag{6.152}$$

restricts to a surjective operator

$$H_0^s(\mathbb{R}^m) \to H_0^{s-1/2}(\mathbb{R}^{m-1}) \quad \text{for } s > m/2. \tag{6.153}$$

Theorem 6.10.5. *We have canonical isomorphisms*

$$H^s(\mathbb{R}^{1+n}) \cong \mathcal{K}^{s,s}\big((S^n)^\wedge\big) \quad \text{for } -(1+n)/2 < s < (1+n)/2,$$

and

$$H_0^s(\mathbb{R}^{1+n}) \cong \mathcal{K}^{s,s}\big((S^n)^\wedge\big) \quad \text{for } s - (1+n)/2 \notin \mathbb{N}, \ s > (1+n)/2.$$

A proof may be found in [25, Theorem 2.1.32]. The role of Theorem 6.10.5 is to establish a relationship between standard Sobolev spaces and weighted spaces of the edge calculus.

Remark 6.10.6. Note that weighted Sobolev spaces have natural interpolation properties both with respect to smoothness and to weights, see [25] and the article [24] of Hirschmann.

6.11   Examples and remarks

We now consider some special cases of the above cone algebra representations of truncated operators $\mathrm{op}^+(a)$. First note that the operators $\mathrm{op}^+(l_+^\alpha)$, $\alpha \in \mathbb{R}$, induce isomorphisms

$$\mathrm{op}^+(l_+^\alpha) : H_0^s(\overline{\mathbb{R}}_+) \to H_0^{s-\alpha}(\overline{\mathbb{R}}_+)$$

for every $s \in \mathbb{R}$. Let us consider the operator

$$\delta + \partial_r = r^{-1}\{\mathrm{op}_M^\gamma(-z) + r\delta\} : C_0^\infty(\mathbb{R}_+) \to C_0^\infty(\mathbb{R}_+),$$

which has the Fourier amplitude functions $\delta + i\rho$; it extends to an isomorphism

$$r^{-1}\{\mathrm{op}_M^\gamma(-z) + r\delta\} : \mathcal{H}^{s-1,\gamma-1}(\mathbb{R}_+) \to \mathcal{H}^{s-1,\gamma-1}(\mathbb{R}_+)$$

for $s \in \mathbb{R}, \gamma \neq 1/2$, with the inverse $\{\mathrm{op}_M^\gamma(-z^{-1}) + r^{-1}\delta\}r$. In particular, we have isomorphisms

$$\delta + \partial_r = r^{-1}\{\mathrm{op}_M^1(-z) + r\delta\} : \mathcal{H}^{1,1}(\mathbb{R}_+) \to \mathcal{H}^{0,0}(\mathbb{R}_+) = L^2(\mathbb{R}_+)$$

or

$$\delta + \partial_r^j = r^{-1}\{\mathrm{op}_M^j(-z) + r\delta\} : \mathcal{H}^{j,j}(\mathbb{R}_+) \to \mathcal{H}^{j-1,j-1}(\mathbb{R}_+),$$

for any $j \in \mathbb{N}, j \geq 1$, which by iteration gives

$$\begin{aligned}\delta + \partial_r^j \\ = \left(r^{-1}\{\mathrm{op}_M^1(-z) + r\delta\}\right) \cdots \left(r^{-1}\{\mathrm{op}_M^j(-z) + r\delta\}\right) : \mathcal{H}^{j,j}(\mathbb{R}_+) \to L^2(\mathbb{R}_+).\end{aligned}$$
$$(6.154)$$

Since $\delta + \partial_r^j$ is equal to $\mathrm{op}^+(l_+^j)$ on the half-line, we may expect that also $\mathrm{op}^+(l_+^{-j})$ has a Mellin representation.

Remark 6.11.1. On the Mellin side we can, of course, compute the inverse of (6.154), namely, as

$$\begin{aligned}(\delta + \partial_r^j)^{-1} \\ = \{\mathrm{op}_M^j(-z^{-1}) + r^{-1}\delta\}) \cdots (\{\mathrm{op}_M^1(-z^{-1}) + r^{-1}\delta\})) : L^2(\mathbb{R}_+) \to \mathcal{H}^{j,j}(\mathbb{R}_+)\end{aligned}$$

which is an expression without any remainder, but it is not so automatic to connect it with something like $\mathrm{op}^+((\delta + i\rho)^{-j})$.

Choose cut-off functions $\omega'' \prec \omega \prec \omega'$, and write the operator $\mathrm{op}^+(\delta + i\rho)$ in the form

$$A := \omega r^{-1}\{\mathrm{op}_M^1(-z) + r\delta\}\omega' + (1 - \omega)\,\mathrm{op}(l_+)(1 - \omega'') : \mathcal{K}^{1,1}(\mathbb{R}_+) \to \mathcal{K}^{0,0}(\mathbb{R}_+).$$
$$(6.155)$$

We have $H_0^1(\overline{\mathbb{R}}_+) = \mathcal{K}^{1,1}(\mathbb{R}_+), \mathcal{K}^{0,0}(\mathbb{R}_+) = L^2(\mathbb{R}_+)$, and

$$\mathrm{op}^+(l_+) : H_0^1(\overline{\mathbb{R}}_+) \to L^2(\mathbb{R}_+)$$

is known to be an isomorphism. In other words, (6.155) is an isomorphism. This gives us altogether the following result.

Proposition 6.11.2. *For every* $j \in \mathbb{N}$ *the operator*

$$A_{-j} := \mathrm{op}^+\left(l_+^{-j}\right) : L^2(\mathbb{R}_+) \to H_0^j(\overline{\mathbb{R}}_+)$$

belongs to $L^{-j}(\mathbb{R}_+, \boldsymbol{g_j})$ *for* $\boldsymbol{g_j} = (0, j, (-\infty, 0])$ *and induces an isomorphism*

$$A_{-j} : L^2(\mathbb{R}_+) \to \mathcal{K}^{j,j}(\mathbb{R}_+). \tag{6.156}$$

In particular, for any cut-off function ω *we have*

$$\omega A_{-j} = \omega r^j A_{-j,0}$$

for an operator $A_{-j,0} \in L^{-j}(\mathbb{R}_+, \boldsymbol{g_0})$, *cf. the notation in Remark 6.8.2.*

6.12   Theorems of Paley–Wiener type

We first recall some generalities on the Fourier transform in $\mathbb{R}^n$

$$(Fu)(\xi) = \int e^{-ix\xi} u(x) dx,$$

also denoted by $\widehat{u}(\xi)$. The inverse has the form

$$(F^{-1}g)(x) = (2\pi)^{-n} \int e^{ix\xi} g(\xi) d\xi.$$

We often set $d\!\!\!\bar{}\,\xi = (2\pi)^{-n} d\xi$. The Fourier transform induces an isomorphism

$$F : \mathcal{S}(\mathbb{R}_x^n) \to \mathcal{S}(\mathbb{R}_\xi^n), \tag{6.157}$$

which comes from the relations

$$F\left((-x)^\beta D_x^\alpha u\right)(\xi) = D_\xi^\beta \xi^\alpha Fu(\xi) \tag{6.158}$$

for all $\alpha, \beta \in \mathbb{N}^n$, and extends to a isomorphism $F : \mathcal{S}'(\mathbb{R}_x^n) \to \mathcal{S}'(\mathbb{R}_\xi^n)$.

Observe that $(\kappa_\lambda u)(x) := \lambda^{n/2} u(\lambda x)$, $\lambda \in \mathbb{R}_+$, induces a group of isomorphisms $\kappa_\lambda : \mathcal{S}(\mathbb{R}^n) \to \mathcal{S}(\mathbb{R}^n)$, and we have

$$\kappa_\lambda F = F \kappa_\lambda^{-1}, \quad \lambda \in \mathbb{R}_+. \tag{6.159}$$

There are different equivalent choices of countable systems of semi-norms in $\mathcal{S}(\mathbb{R}^n)$ that turn $\mathcal{S}(\mathbb{R}^n)$ into a Schwartz space. Instead of the semi-norms

$$u \mapsto \sup_{x \in \mathbb{R}^n} \left| x^\alpha D_x^\beta u(x) \right|, \quad \alpha, \beta \in \mathbb{N}^n,$$

we may also take $\pi_m : u \mapsto \sup_{x \in \mathbb{R}^n} (1 + |x|)^m \sum_{|\alpha| \leq m} |D_x^\alpha u(x)|$, $m \in \mathbb{N}$. From (6.158) we have $\xi^\alpha D_\xi^\beta \widehat{u}(\xi) = \int_{\mathbb{R}^n} e^{-ix\xi} D_x^\alpha ((-x)^\beta u(x)) dx$, which yields

$$\left| \xi^\alpha D_\xi^\beta \widehat{u}(\xi) \right| = \int_{\mathbb{R}^n} \left| D_x^\alpha ((-x)^\beta u(x)) \right| dx. \tag{6.160}$$

Thus for $b(\xi) := (1 + |\xi|)^m / \sum_{|\alpha| \leq m} |\xi^\alpha|$,

$$\begin{aligned}
\pi_m(\widehat{u}\,) &= \sup_{\xi \in \mathbb{R}^n} (1 + |\xi|)^m \sum_{|\beta| \leq m} \left| D_\xi^\beta \widehat{u}(\xi) \right| \\
&= \sup_{\xi \in \mathbb{R}^n} b(\xi) \sum_{|\alpha| \leq m} |\xi^\alpha| \sum_{|\beta| \leq m} \left| D_\xi^\beta \widehat{u}(\xi) \right| \\
&\leq c \sum_{|\alpha| \leq m} \sum_{|\beta| \leq m} \int \left| D_x^\alpha x^\beta u(x) \right| dx \\
&\leq c \sum_{|\alpha| \leq m+n+1} \sum_{|\beta| \leq m} \int |x^\beta| \left| D_x^\alpha u(x) \right| dx.
\end{aligned} \tag{6.161}$$

Using $\sum_{|\beta| \leq m} |x^\beta| \leq c(1 + |x|)^{m+n+1} / (1 + |x|)^{n+1}$ and $\int (1 + |x|)^{-(n+1)} dx < \infty$ we have

$$\begin{aligned}
\pi_m(\widehat{u}\,) &\leq c \sum_{|\alpha| \leq m+n+1} \int (1 + |x|)^{m+n+1} \left| D_x^\alpha u(x) \right| (1 + |x|)^{-(n+1)} dx \\
&\leq \sup_{x \in \mathbb{R}^n} (1 + |x|)^{m+n+1} \sum_{|\alpha| \leq m+n+1} \left| D_x^\alpha u(x) \right| \int (1 + |x|)^{-(n+1)} dx \\
&= \text{const } \pi_{m+n+1}(u).
\end{aligned}$$

This establishes the continuity of (6.157). Similar estimates are valid for F^{-1} instead of F, which shows that (6.157) is an isomorphism.

Let us also mention other standard properties of the Fourier transform. The operator (6.157) extends by continuity to an isomorphism

$$F : L^2(\mathbb{R}_x^n) \to L^2(\mathbb{R}_\xi^n),$$

and we have

$$\int \widehat{u}(x) v(x) dx = \int u(x) \widehat{v}(x) dx$$

$$\int u(x) \overline{v}(x) dx = \int \widehat{u}(\xi) \overline{\widehat{v}(\xi)} \, d\xi \quad \text{(Parseval's formula)},$$

and in particular

$$\int |u(x)|^2 dx = \int |\widehat{u}(\xi)|^2 \bar{d}\xi \qquad \text{(Plancherel's formula)}.$$

Concerning the convolution

$$u * v(x) = \int u(y)v(x-y)dy = \int u(x-y)v(y)dx,$$

first for $u, v \in \mathcal{S}(\mathbb{R}^n)$ we have

$$\widehat{u * v}(\xi) = \widehat{u}(\xi)\widehat{v}(\xi), \quad \widehat{uv}(\xi) = (2\pi)^{-n}\widehat{u}(\xi) * \widehat{v}(\xi).$$

The convolution extends to functions $u \in L^1(\mathbb{R}^n)$, $v \in L^p(\mathbb{R}^n)$, $p \geq 1$, we have $u * v \in L^p(\mathbb{R}^n)$, and

$$\|u * v\|_{L^p(\mathbb{R}^n)} \leq \|u\|_{L^1(\mathbb{R}^n)}\|v\|_{L^p(\mathbb{R}^n)}.$$

We now recall some material on Sobolev spaces and the action of pseudo-differential operators. Concerning well-known properties of Sobolev distributions we refer to standard textbooks. More specific results will be given with proofs.

We define the Sobolev space $H^s(\mathbb{R}^n)$ of smoothness $s \in \mathbb{R}$ to be the completion of $C_0^\infty(\mathbb{R}^n)$ with respect to the norm

$$\|u\|_{H^s(\mathbb{R}^n)} = \left\{ \int \langle\xi\rangle^{2s} |\widehat{u}(\xi)|^2 \bar{d}\xi \right\}^{1/2}, \qquad (6.162)$$

with $Fu = \widehat{u}$ being the Fourier transform in $\mathbb{R}^n$. Equivalently, we may define $H^s(\mathbb{R}^n)$ as the subspace of all $u \in \mathcal{S}'(\mathbb{R}^n)$ such that $\widehat{u}$ belongs to $L^1_{\text{loc}}(\mathbb{R}^n)$ and (6.162) is finite; $H^s(\mathbb{R}^n)$ is a Hilbert space with the scalar product

$$(u, v)_{H^s(\mathbb{R}^n)} = \int \langle\xi\rangle^{2s} \widehat{u}(\xi) \overline{\widehat{v}}(\xi) \, \bar{d}\xi.$$

In particular, we identify $H^0(\mathbb{R}^n)$ with $L^2(\mathbb{R}^n)$ with the standard scalar product

$$(u, v)_{L^2(\mathbb{R}^n)} = \int u(x) \overline{v(x)} \, dx.$$

We also write $H^\infty(\mathbb{R}^n) := \bigcap_{s \in \mathbb{R}} H^s(\mathbb{R}^n)$; the superscript "$\infty$" will be used with analogous meaning also for other variants of Sobolev spaces.

Remark 6.12.1. The space $C_0^\infty(\mathbb{R}^n)$ is dense in $H^s(\mathbb{R}^n)$.

Example 6.12.2. (i) For $s \in \mathbb{N}$ the space $H^s(\mathbb{R}^n)$ can be equivalently be defined as

$$H^s(\mathbb{R}^n) = \left\{ u \in L^2(\mathbb{R}^n) : D_x^\alpha u \in L^2(\mathbb{R}^n) \text{ for all } |\alpha| = s \right\}$$

(which is also equivalent to an analogous condition for all $|\alpha| \leq s$).

(ii) Every $u \in \mathcal{D}'(\mathbb{R}^n)$ with compact support belongs to $H^s(\mathbb{R}^n)$ for a suitable $s = s(u) \in \mathbb{R}$. In particular, for the Dirac distribution δ_0 we have $\delta_0 \in H^s(\mathbb{R}^n)$ for every $s < -n/2$.

For reference, we recall below a few steps of the proof. For $u \in H^s(\mathbb{R}^n)$ we pass to an approximation of the form $u_\varepsilon = u * \varphi_\varepsilon \in C^\infty(\mathbb{R}^n)$, where $*$ is the convolution and $\varphi_\varepsilon(x) = \varepsilon^{-n}\varphi(x/\varepsilon)$, $\varepsilon > 0$, for a $\varphi \in C_0^\infty(\mathbb{R}^n)$, $\varphi \geq 0$, $\varphi = 0$ for $|x| \geq 1$, $\int \varphi(x)dx = 1$. It follows that $\widehat{u}_\varepsilon = \widehat{\varphi}_\varepsilon \widehat{u}$. Substituting $x = \varepsilon y$ we obtain

$$\widehat{\varphi}_\varepsilon(\xi) = \int \varepsilon^n \varphi\left(\frac{x}{\varepsilon}\right) e^{-ix\xi} dx = \int \varphi(y) e^{-iy\varepsilon\xi} dy = \widehat{\varphi}(\varepsilon\xi).$$

This yields $|\widehat{\varphi}_\varepsilon(\xi)| \leq \int \varphi(y)dy = \widehat{\varphi}(0) = 1$. Since $\widehat{\varphi}_\varepsilon(\xi) \in \mathcal{S}(\mathbb{R}_\xi^n)$, we have $\widehat{u}_\varepsilon \in \widehat{H}^\infty(\mathbb{R}^n)$. It follows that

$$\|u - u_\varepsilon\|_{H^s(\mathbb{R}^n)}^2 = \int \langle\xi\rangle^{2s} \left|\widehat{u}(\xi) - \widehat{u}_\varepsilon(\xi)\right|^2 d\xi$$

$$\leq c \int |\widehat{u}(\xi)|^2 \left|1 - \widehat{\varphi}(\varepsilon\xi)\right|^2 (1 + |\xi|)^{2s} d\xi \to 0 \tag{6.163}$$

by applying Lebesgue's theorem. In a second step, one shows that u_ε can be approximated by functions in $C_0^\infty(\mathbb{R}^n)$ with respect to $H^N(\mathbb{R}^n)$, for every $N \in \mathbb{N}$.

Let $\Omega \subseteq \mathbb{R}^n$ be open; then $H_{\text{loc}}^s(\Omega)$ is the subspace of all $u \in \mathcal{D}'(\Omega)$ such that $\varphi u \in H^s(\mathbb{R}^n)$ for every $\varphi \in C_0^\infty(\Omega)$. Moreover, $H_{\text{comp}}^s(\Omega)$ denotes the subspace of all $u \in H^s(\mathbb{R}^n)$ such that $\text{supp}\, u$ is compact and contained in Ω. If Ω has a smooth boundary we set

$$H_0^s(\overline{\Omega}) := \{u \in H^s(\mathbb{R}^n) : \text{supp}\, u \subseteq \overline{\Omega}\}$$

and

$$H^s(\Omega) := \{u|_\Omega : u \in H^s(\mathbb{R}^n)\}.$$

Clearly $H_0^s(\overline{\Omega})$ is a closed subspace in $H^s(\Omega)$, since convergence of a sequence with respect to $H^s(\mathbb{R}^n)$ entails the convergence of the values as distributions on $C_0^\infty(\mathbb{R}^n)$, and vanishing on $C_0^\infty(\mathbb{R}^n \setminus \overline{\Omega})$ entails vanishing of the limit.

We consider the case $\Omega = \overline{\mathbb{R}}_+^n = \{(x_1, \ldots, x_n) \in \mathbb{R}^n : x_n \gtrless 0\}$. We also write $x = (y, t)$ for $y = (x_1, \ldots, x_{n-1})$, and $\xi = (\eta, \tau)$. Observe that

$$H^s(\mathbb{R}_\pm^n) = H^s(\mathbb{R}^n)/H_0^s(\overline{\mathbb{R}}_\mp^n)$$

and $H^s(\mathbb{R}_\pm^n)$ may be identified with the orthogonal complement of $H_0^s(\overline{\mathbb{R}}_\mp^n)$ in $H^s(\mathbb{R}^n)$.

Remark 6.12.3. The space $C_0^\infty(\mathbb{R}_\pm^n)$ is dense in $H_0^s(\overline{\mathbb{R}}_\pm^n)$.

The proof employs the fact that $u \in H_0^s(\overline{\mathbb{R}}_+^n)$ entails

$$u_\varepsilon(x) := u(y, t - \varepsilon) \in H_0^s(\overline{\mathbb{R}}_+^n) \quad \text{for any } \varepsilon > 0$$

and $u_\varepsilon \to u$ as $\varepsilon \to 0$ in $H_0^s(\overline{\mathbb{R}}_+^n)$. As noted before, u_ε can be approximated in $H^s(\mathbb{R}^n)$ by functions in $C_0^\infty(\mathbb{R}_+^n)$. This yields altogether a sequence of approximating functions for u itself. The arguments for $H_0^s(\overline{\mathbb{R}}_-^n)$ are analogous.

Theorem 6.12.4. *Let* $u_+(t) \in H_0^s(\overline{\mathbb{R}}_+)$, $s \in \mathbb{R}$. *Then the Fourier transform*

$$\widehat{u}_+(\tau) = \int e^{-it\tau} u_+(t) dt$$

extends to a function

$$h_+(\zeta) = \int e^{-it\tau}(e^{t\vartheta} u_+(t)) dt \in C(\operatorname{Im}\zeta \leq 0) \cap \mathcal{A}(\operatorname{Im}\zeta < 0), \quad \zeta = \tau + i\vartheta, \quad (6.164)$$

such that

$$\int (1 + |\tau| + |\vartheta|)^{2s} |h_+(\tau + i\vartheta)|^2 d\tau \leq C \tag{6.165}$$

for $\vartheta \leq 0$ *and some constant* $C > 0$ *independent of* ϑ.

Conversely, let $h_+(\tau + i\vartheta)$ be a locally integrable function in $-\infty < \vartheta < 0$ satisfying the estimate (2.22) for some $C > 0$ independent of ϑ and belonging to $\mathcal{A}(\operatorname{Im}\zeta < 0)$. Then there is an $u_+(t) \in H_0^s(\overline{\mathbb{R}}_+), s \in \mathbb{R}$, such that

$$h_+(\zeta) = \int e^{-it\tau}(e^{t\vartheta} u_+(t)) dt.$$

A proof of this version of Paley–Wiener Theorem can be found in Eskin's book [14].

The following considerations will prepare for the Paley–Wiener theorem for Sobolev spaces.

Lemma 6.12.5. *Let* $u(x)$ *be a locally integrable function in* $x \in \mathbb{R}^n$; *set* $x = (y, t)$, $y \in \mathbb{R}^{n-1}$, $t \in \mathbb{R}$, *and assume that* $\frac{\partial u}{\partial t} = 0$ *in the distributional sense, i.e.,*

$$\left(u, \frac{\partial \varphi}{\partial t}\right) = \int_{\mathbb{R}^n} u(y, t) \frac{\partial \varphi}{\partial t}(y, t) dy dt = 0 \tag{6.166}$$

for all $\varphi \in C_0^\infty(\mathbb{R}^n)$. *Then* $u(x) = u_0(y)$ *in the distributional sense, for a locally integrable* $u_0(y)$ *in* $\mathbb{R}^{n-1}$.

Proof. We set

$$u_0(y) := \int u(y, t) \varphi_0(t) dt \tag{6.167}$$

for a function $\varphi_0 \in C_0^\infty(\mathbb{R}_t)$ such that $\int \varphi_0(t) dt = 1$, and verify the relation

$$\iint u(y, t) \varphi(y, t) dy dt = \iint u_0(y) \varphi(y, t) dy dt \tag{6.168}$$

for every $\varphi \in C_0^\infty(\mathbb{R}^n)$. To this end we set $\psi(y,t) := \varphi(y,t) - \varphi_1(y)\varphi_0(t)$ for $\varphi_1(y) := \int \varphi(y,t)dt$. Then $\int \psi(y,t)dt = 0$ and hence $\varphi_2(y,t) := \int_{-\infty}^t \psi(y,t)dt \in C_0^\infty(\mathbb{R}^n)$, where $\frac{\partial \varphi_2}{\partial t} = \psi(y,t)$. Because of (6.166) it follows that

$$(u,\varphi) = (u, \psi + \varphi_1\varphi_0) = \left(u, \frac{\partial \varphi_2}{\partial t} + \varphi_1(y)\varphi_0(t) \right)$$

$$= (u, \varphi_1(y)\varphi_0(t)) = \iint u(y,t)\varphi_1(y)\varphi_0(t)dydt.$$

Using (6.167) this finally yields

$$\iint u(y,t)\varphi(y,t)dydt = \int u_0(y)\varphi_1(y)dy = \iint u_0(y)\varphi(y,t)dydt$$

which is just (6.168).   □

Lemma 6.12.6. Let $\mathcal{S}_0(\overline{\mathbb{R}}_+^n)$ denote the subspace of all $u \in \mathcal{S}(\mathbb{R}_+^n)$ that vanish for $t < 0$. Then $F\mathcal{S}_0(\overline{\mathbb{R}}_+^n)$ consists of the space of functions $f(\eta, \tau)$ that admit an analytic extension in $\zeta = \tau + i\sigma$ in $\sigma < 0$, C^∞ in $(\eta, \tau) \in \mathbb{R}^n$, $\sigma \leq 0$, and satisfy the estimates

$$(1 + |\eta| + |\tau| + |\sigma|)^m |(D_\eta^{\kappa'} D_\zeta^{\kappa''} f)(\eta, \zeta)| \leq c_{m,\kappa} \qquad (6.169)$$

for all $m \geq 0$, $\kappa = (\kappa', \kappa'') \in \mathbb{N}^n$.

Proof. Let $u(y,t) \in \mathcal{S}_0(\overline{\mathbb{R}}_+^n)$; then

$$\widehat{u}(\eta, \tau) = \int_0^\infty \int_{\mathbb{R}^{n-1}} u(y,t) e^{-iy\eta - it\tau} dydt.$$

Since the integration in t is over the positive half-line, we obtain an analytic extension to the negative ζ half-plane $\sigma = \operatorname{Im} \zeta < 0$ as

$$\widehat{u}(\eta, \tau) = \int_0^\infty \int_{\mathbb{R}^{n-1}} u(y,t) e^{-iy\eta - it\zeta} dydt.$$

The proof of (6.169) follows in an analogous manner as that of (6.161) for every $m \in \mathbb{N}$ and every derivative in the covariables, now carried out in (η, ζ) for $\operatorname{Im} \zeta \leq 0$. The constant on the right-hand side is independent of σ, while the relation (6.160) is valid in (η, ζ) on the left-hand side which leads to the presence of $|\sigma|$ on the left of (6.169).   □

Let A_+^μ denote the space of functions $a_+(\eta, \tau + i\sigma)$ that are continuous in (η, τ, σ) for $\eta \neq 0$, $\sigma \leq 0$, and holomorphic in $\zeta = \tau + i\sigma$ in $\sigma < 0$, where

$$|a_+(\eta, \tau + i\sigma)| \leq c(1 + |\eta| + |\tau| + |\sigma|)^\mu \qquad (6.170)$$

for a constant $c > 0$. In an analogous manner we define the space A_-^μ of continuous functions in (η, τ, σ) for $\eta \neq 0$, $\sigma \geq 0$, holomorphic in ζ for $\sigma > 0$ and satisfying the estimate (6.170).

Proposition 6.12.7. *Let $a_\pm \in A_\pm^\mu$ and $p_\pm(\eta,\tau) := \left(a_\pm\big|_{\sigma=0}\right)(\eta,\tau)$. Then the operator of multiplication by $p_\pm$ induces a continuous operator*

$$\mathcal{M}_{p_\pm} : FH_0^s\left(\overline{\mathbb{R}}_\pm^n\right) \to FH_0^{s-\mu}\left(\overline{\mathbb{R}}_\pm^n\right)$$

for every $s \in \mathbb{R}$.

Proof. Let us consider the case of $FH_0^s\left(\overline{\mathbb{R}}_+^n\right)$; for $FH_0^s\left(\overline{\mathbb{R}}_-^n\right)$ the arguments are analogous. First, it is clear that multiplication by a function $p(\xi)$ satisfying the estimate $|p(\xi)| \le c\,\langle\xi\rangle^\mu$ induces a continuous operator

$$\mathcal{M}_p : FH^s(\mathbb{R}^n) \to FH^{s-\mu}(\mathbb{R}^n).$$

In particular, we have a continuous operator

$$\mathcal{M}_{p_+} : FH_0^s\left(\overline{\mathbb{R}}_+^n\right) \to FH^{s-\mu}(\mathbb{R}^n). \tag{6.171}$$

Thus it remains to show that $\mathcal{M}_{p_+}$ is continuous with values in $FH_0^{s-\mu}\left(\overline{\mathbb{R}}_+^n\right)$. Due to Remark 6.12.3, the space $C_0^\infty(\mathbb{R}_+^n)$ is dense in $H_0^s\left(\overline{\mathbb{R}}_+^n\right)$, so we first assume $u \in C_0^\infty(\mathbb{R}_+^n)$. Applying Lemma 6.12.6 for every $N \ge 0$ we have the estimate

$$(1 + |\eta| + |\tau| + |\sigma|)^N \left|\hat{u}(\eta,\tau + i\sigma)\right| \le c_N \tag{6.172}$$

for every $\sigma < 0$. Let $\hat{v}(\eta,\tau) := p_+(\eta,\tau)\hat{u}(\eta,\tau)$ and form

$$v(y,t) = \int e^{iy\eta+it\tau} p_+(\eta,\tau)\,\hat{u}(\eta,\tau)\, d\eta\, d\tau.$$

Because of (6.170) and (6.172), we have $v(y,t) \in C^\infty(\mathbb{R}^n)$. Applying the holomorphy of $a_+(\eta,\tau+i\sigma)\hat{u}(\eta,\tau+i\sigma)$ for $\sigma < 0$ and continuity up to $\sigma = 0$ for $\eta \ne 0$, Cauchy's theorem yields

$$v(y,t) = \int e^{iy\eta+it(\tau+i\sigma)} a_+(\eta,\tau+i\sigma)\,\hat{u}(\eta,\tau+i\sigma)\, d\eta\, d\tau,$$

and $|v(y,t)| \le ce^{-t\sigma}$ for a constant $c > 0$ independent of σ. For $t < 0$ we can let $\sigma \to -\infty$ and obtain $v(y,t) = 0$ for $t < 0$. Therefore, $\mathrm{supp}\, v \subset \overline{\mathbb{R}}_+^n$ and hence $\hat{v}(\eta,\tau) \in FH_0^{s-\mu}\left(\overline{\mathbb{R}}_+^n\right)$ (at this point we even have $\hat{v}(\eta,\tau) \in FH_0^\infty\left(\overline{\mathbb{R}}_+^n\right)$).

Now for arbitrary $u \in H_0^s\left(\overline{\mathbb{R}}_+^n\right)$ there exists a sequence $\varphi_j \in C_0^\infty(\mathbb{R}_+^n)$ tending to u in $H_0^s\left(\overline{\mathbb{R}}_+^n\right)$ as $j \to \infty$, cf. Remark 6.12.3. From the first part we have $p_+(\eta,\tau)\hat{\varphi}_j(\eta,\tau) \in H_0^{s-\mu}\left(\overline{\mathbb{R}}_+^n\right)$. By virtue of (6.171) the sequence $p_+(\eta,\tau)\hat{\varphi}_j(\eta,\tau)$ converges in $FH^{s-\mu}(\mathbb{R}^n)$. Since $FH_0^{s-\mu}\left(\overline{\mathbb{R}}_+^n\right)$ is closed in $FH^{s-\mu}(\mathbb{R}^n)$ (cf. the above observation on the corresponding spaces in the Fourier preimage), we obtain $p_+(\eta,\tau)\hat{u}_j(\eta,\tau) \in FH_0^{s-\mu}\left(\overline{\mathbb{R}}_+^n\right)$. $\qquad\square$

In other words, we proved the continuity of the pseudo-differential operators

$$F^{-1}p_\pm F : H_0^s\left(\overline{\mathbb{R}}_\pm^n\right) \to H_0^{s-\mu}\left(\overline{\mathbb{R}}_\pm^n\right). \tag{6.173}$$

Proposition 6.12.8. *For every $\mu \in \mathbb{R}$ there exist functions $a_\pm \in A_\pm^\mu$ such that for $p_\pm$ in the notation of* **Proposition** *6.12.7 the operators (6.173) are isomorphisms for all $s \in \mathbb{R}$.*

Proof. Let us consider the plus case, and choose a^+ in the form

$$a_+(\eta, \tau + i\sigma) := b_+(\eta, \tau + i(\sigma - 1)),$$

for

$$b_+(\eta, \tau + i\sigma) := (-i|\eta| + \tau + i\sigma)^\mu = e^{\mu[\log |(-i|\eta|+\tau+i\sigma)|+i \arg(-i|\eta|+\tau+i\sigma)]}, \quad \sigma \leq 0,$$

where we take the branch of the logarithm with $\arg(-i|\eta|+\tau+i\sigma) \to 0$ for $\tau \to \infty$. This function is holomorphic in $\zeta = \tau + i\sigma$ for $\sigma < 0$. In addition, a_+ itself satisfies the estimate (6.170), a_+ is obviously invertible, and the inverse has analogous properties with $-\mu$ instead of μ. Then it suffices to apply Proposition 6.12.7 to observe that $F^{-1}p_+^{-1}F$ is the inverse of $F^{-1}p_+F$. In the minus case we take $\sigma \geq 0$, for $a_-(\eta, \tau + i\sigma) := b_-(\eta, \tau + i(\sigma + 1))$ where $b_-(\eta, \tau + i\sigma) := (i|\eta| + \tau + i\sigma)^\mu = e^{\mu[\log |(i|\eta|+\tau+i\sigma)|+i \arg(i|\eta|+\tau+i\sigma)]}$; the consideration is then analogous. $\qquad \square$

Let $u \in H_0^s(\overline{\mathbb{R}}_+^n)$ which belongs to $\mathcal{S}'(\mathbb{R}^n)$ where $\operatorname{supp} u \subseteq \overline{\mathbb{R}}_+^n$. Thus the pairing $\langle u, \varphi \rangle$ for test functions $\varphi \in \mathcal{S}'(\mathbb{R}^n)$ is well-defined, and the support condition tells us that $\langle u, \varphi \rangle = 0$ whenever $\varphi \in C_0^\infty(\mathbb{R}_-^n)$. Now let us give the product $e^{t\sigma}u$ a meaning for $t < 0$, $\sigma \leq 0$, as an element of $\mathcal{S}'(\mathbb{R}^n)$ supported by $\overline{\mathbb{R}}_+^n$. To this end we choose a function $e \in C^\infty(\mathbb{R}_r)$ defined by

$$e(r) = \begin{cases} e^r & \text{for } r \leq 0, \\ 0 & \text{for } r > 1. \end{cases} \tag{6.174}$$

Then for $\sigma \leq 0$ we have

$$e(t\sigma) = e^{t\sigma} \quad \text{for } t \geq 0, \tag{6.175}$$

and the fact that $e(t\sigma)\varphi \in \mathcal{S}(\mathbb{R}^n)$ for $\varphi \in \mathcal{S}(\mathbb{R}^n)$ shows that $e(t\sigma)u \in \mathcal{S}'(\mathbb{R}^n)$. Moreover, $\langle e(t\sigma)u, \varphi \rangle = \langle u, e(t\sigma)\varphi \rangle$ vanishes for arbitrary $\varphi \in C_0^\infty(\mathbb{R}_-^n)$ since $e(t\sigma)\varphi \in C_0^\infty(\mathbb{R}_-^n)$ and $\operatorname{supp} u \subseteq \overline{\mathbb{R}}_+^n$. It can be easily verified that $e(t\sigma)u \in H^s(\mathbb{R}^n)$ and hence $e(t\sigma)u \in H_0^s(\overline{\mathbb{R}}_+^n)$ because of $\operatorname{supp}(e(t\sigma)u) \subseteq \overline{\mathbb{R}}_+^n$. Moreover, we have $e^{t\sigma}f \in \mathcal{D}'(\mathbb{R}^n)$ for any $\sigma \in \mathbb{R}$. Finally, let us show that $(e(t\sigma)u)|_{\mathbb{R}_+^n} = (e^{t\sigma}f)|_{\mathbb{R}_+^n}$ for $\sigma \leq 0$ for any choice of the function (6.174). Here it suffices to observe that $\langle e(t\sigma)u, \varphi \rangle = \langle e^{t\sigma}u, \varphi \rangle$ for any $\varphi \in C_0^\infty(\mathbb{R}_+^n)$, which is an immediate consequence of the relation (6.175). Summing up, it is justified to write $e^{t\sigma}u$ rather than $e(t\sigma)u$ for $\sigma \leq 0$, $u \in H_0^s(\overline{\mathbb{R}}_+^n)$.

Lemma 6.12.9. *In the case $s = \mu$, i.e., $p_+(\eta, \tau) := (-i|\eta| + \tau - i)^s$, $s \in \mathbb{R}$, for $u \in H_0^s(\overline{\mathbb{R}}_+^n)$, $v := F^{-1}p_+Fu \in H_0^0(\overline{\mathbb{R}}_+^n) = L^2(\mathbb{R}_+^n)$ we have*

$$F(ve^{t\sigma}) = a_+(\eta, \tau + i\sigma)F(ue^{t\sigma}), \quad \sigma \leq 0, \tag{6.176}$$

where $F = F_{(y,t) \to (\eta,\tau)}$.

Proof. We first assume $u \in C_0^\infty(\mathbb{R}_+^n)$. Then by Lemma 6.12.6, the function

$$(Fu)(\eta, \tau + i\sigma) = F(ue^{t\sigma})$$

is holomorphic in $\zeta = \tau + i\sigma$ for $\sigma < 0$ and satisfies the estimate (6.169). From Cauchy's theorem it follows that

$$\int e^{iy\eta + it(\tau + i\sigma)} a_+(\eta, \tau + i\sigma)(Fu)(\eta, \tau + i\sigma) \bar{d}\eta \bar{d}\tau = v,$$

which yields $F_{(\eta,\tau) \to (y,t)}^{-1}(a_+(\eta, \tau + i\sigma)(Fu)(\eta, \tau + i\sigma) = e^{t\sigma}v(y,t)$ which is equivalent to (6.176).

If $u \in H_0^s(\overline{\mathbb{R}}_+^n)$ is arbitrary there is a sequence $u_k \in C_0^\infty(\mathbb{R}_+^n)$, such that $u_k \to u$ in $H_0^s(\overline{\mathbb{R}}_+^n)$ for $k \to \infty$, cf. Remark 6.12.3. Thus $v_k := F^{-1}p_+Fu_k \to v$ as well as $v_k e^{t\sigma} \to ve^{t\sigma}$ in $H_0^s(\overline{\mathbb{R}}_+^n)$ for $k \to \infty$. Consequently, $a_+^{-1}(\eta, \tau + i\sigma)F(v_k e^{t\sigma}) \to a_+^{-1}(\eta, \tau + i\sigma)F(ve^{t\sigma})$ in $FH_0^s(\overline{\mathbb{R}}_+^n)$ for every $\sigma < 0$. On the other hand, $u_k \to u$ in $H_0^s(\overline{\mathbb{R}}_+^n)$ entails $u_k e^{t\sigma} \to ue^{t\sigma}$ in $\mathcal{S}'(\mathbb{R}^n)$ and so $F(u_k e^{t\sigma}) \to F(ue^{t\sigma})$ in $\mathcal{S}'(\mathbb{R}^n)$. Thus, passing in the equation

$$a_+^{-1}(\eta, \tau + i\sigma)F(v_k e^{t\sigma}) = F(u_k e^{t\sigma})$$

to the limit in $\mathcal{S}'(\mathbb{R}^n)$ for $k \to \infty$ we finally obtain (6.176). $\square$

Theorem 6.12.10 (Paley–Wiener theorem). (i) *For any* $u(y,t) \in H_0^s(\overline{\mathbb{R}}_+^n)$, $s \in \mathbb{R}$, *the Fourier transform* $f(\eta, \tau) = (F_{(y,t) \to (\eta,\tau)}u)(\eta, \tau)$ *extends to a function*

$$f(\eta, \tau + i\sigma) = F_{(y,t) \to (\eta,\tau)}(ue^{t\sigma})(\eta, \tau) \tag{6.177}$$

belonging to $C(\overline{\mathbb{R}}_{-,\sigma}, \hat{H}^s(\mathbb{R}^{n-1}))$ *that is holomorphic in* $\tau + i\sigma$ *for* $\sigma < 0$ *for almost every* $\eta \in \mathbb{R}^{n-1}$ *and satisfies the estimate*

$$\int_{\mathbb{R}^n} (1 + |\eta| + |\tau| + |\sigma|)^{2s}|f(\eta, \tau + i\sigma)|^2 d\eta d\tau \leq C \quad \text{for } \sigma \leq 0 \tag{6.178}$$

and for a constant $C > 0$ *which is independent of* σ.

(ii) *Conversely, suppose the function* $f(\eta, \tau + i\sigma)$ *is locally integrable in* $(\eta, \tau) \in \mathbb{R}^n$ *for every* $\sigma < 0$, *holomorphic in* $\tau + i\sigma$ *for almost all* $\eta \in \mathbb{R}^{n-1}$ *and satisfies the estimate* (6.178) *for a* $C > 0$ *independent of* σ. *Then there exists a function* $u \in H_0^s(\overline{\mathbb{R}}_+^n)$ *such that* (6.177) *holds.*

Proof. (i) $\implies$ (ii) Let $u(y,t) \in H_0^s(\overline{\mathbb{R}}_+^n)$. Applying Proposition 6.12.8 to the plus case and $\mu = s$ we obtain an isomorphism $F^{-1}p_+F : H_0^s(\overline{\mathbb{R}}_+^n) \to H_0^0(\overline{\mathbb{R}}_+^n)$. Set $v = F^{-1}p_+Fu$; then $\hat{v}(\eta, \tau + i\sigma) = F(ve^{t\sigma})$, cf. the relation (6.176). We show that the function $g(y,t)e^{t\sigma}$ is continuous in $\sigma \in (-\infty, 0]$ with values in $L^2(\mathbb{R}_+^n) = H_0^0(\overline{\mathbb{R}}_+^n)$. To this end we verify that

$$\int_0^\infty \int_{\mathbb{R}^{n-1}} \left|v(y,t)e^{t\sigma'} - v(y,t)e^{t\sigma''}\right|^2 dy dt \to 0 \tag{6.179}$$

for $\sigma' - \sigma'' \to 0$, $\sigma' \le 0$, $\sigma'' \le 0$ belonging to a relatively compact neighbourhood of any fixed $\sigma_0 \in (-\infty, 0]$. The convergence (6.179) follows from Lebesgue's theorem. In fact, we easily see the pointwise convergence of the integrand to zero, and we have an integrable function, namely, $4|v(y,t)|^2 \ge |v(y,t)(e^{it\sigma'} - e^{it\sigma''})|$. Therefore, thanks to the Parseval's identity the function $\widehat{v}(\eta, \tau + i\sigma)$ is continuous in $\sigma \in (-\infty, 0]$ with values in $\widehat{H}_0^0(\mathbb{R}_+^n)$ where

$$\int_{\mathbb{R}^n} |\widehat{v}(\eta, \tau + i\sigma)|^2 d\eta d\tau = (2\pi)^n \int_0^\infty \int_{\mathbb{R}^{n-1}} |v(y,t)e^{t\sigma}|^2 dy dt \le (2\pi)^n \|v\|_{L^2(\mathbb{R}_+^n)}^2.$$

Let $h(\eta, t) = F_{y \to \eta} v(y, t)$; then Plancherel's theorem yields $h(\eta, t) \in L^2(\mathbb{R}_+^n)$. For almost every $\eta \in \mathbb{R}^{n-1}$ we have $\int_{-\infty}^\infty |h(\eta, t)|^2 dt < \infty$. For arbitrary η it follows that

$$\widehat{v}(\eta, \tau + i\sigma) = F_{t \to \tau}(h(\eta, t)e^{t\sigma}) = \int h(\eta, t)e^{-it(\tau + i\sigma)} dt$$

is holomorphic in $\zeta = \tau + i\sigma$ for $\sigma < 0$. By (6.176), it follows that

$$f(\eta, \tau + i\sigma) = a_+^{-1}(\eta, \tau + i\sigma)\,\widehat{v}(\eta, \tau + i\sigma)$$

which gives the estimate (6.178). Thus $f(\eta, \tau + i\sigma)$ has the asserted properties.

(ii) $\Longrightarrow$ (i) Let us assume now that we have a locally integrable function $f(\eta, \tau + i\sigma)$ that is holomorphic in $\zeta = \tau + i\sigma$ for $\sigma < 0$, for almost every $\eta \in \mathbb{R}^{n-1}$. Furthermore, assume that (6.178) holds for $\sigma < 0$. Set

$$\widehat{v}(\eta, \tau + i\sigma) = a_+(\eta, \tau + i\sigma)f(\eta, \tau + i\sigma).$$

Since $\widehat{v}(\eta, \tau + i\sigma)$ is holomorphic in $\sigma < 0$ for almost every $\eta \in \mathbb{R}^{n-1}$, we have

$$\frac{\partial \widehat{v}(\eta, \tau + i\sigma)}{\partial \sigma} = i\frac{\partial \widehat{v}(\eta, \tau + i\sigma)}{\partial \tau}. \tag{6.180}$$

If we interpret $\widehat{v}(\eta, \tau + i\sigma)$ as a regular functional on the space

$$\mathcal{S}(\mathbb{R}^n \times (-\infty, 0)) := C_0^\infty(\mathbb{R}_{-,\sigma}, \mathcal{S}(\mathbb{R}^n)),$$

then (6.180) can be regarded as the equality of the derivatives of that functional. Throughout this proof C will denote different suitable constants. Let us verify that

$$\left\|\widehat{v}(\eta, \tau + i\sigma)\right\|_{L^2(\mathbb{R}_{\eta,\tau}^n)} \le C \tag{6.181}$$

for some $C > 0$ independent of $\sigma \in (-\infty, 0)$. In fact,

$$\int |\widehat{v}(\eta, \tau + i\sigma)|^2 d\eta d\tau \tag{6.182}$$

$$= \int |a_+(\eta, \tau + i\sigma) f(\eta, \tau + i\sigma)|^2 d\eta d\tau$$

$$= \int \frac{|a_+(\eta, \tau + i\sigma)|^2}{(1 + |\eta| + |\tau| + |\sigma|)^{2s}} (1 + |\eta| + |\tau| + |\sigma|)^{2s} |f(\eta, \tau + i\sigma)|^2 d\eta d\tau$$

$$\leq C \int (1 + |\eta| + |\tau| + |\sigma|)^{2s} |f(\eta, \tau + i\sigma)|^2 d\eta d\tau \leq C \tag{6.183}$$

for a $C > 0$ independent of $\sigma \in (-\infty, 0)$. In (6.182) we employed (6.170). Set

$$v(y, t, \sigma) := \int e^{iy\eta + it\tau} \widehat{v}(\eta, \tau + i\sigma) d\eta d\tau, \tag{6.184}$$

the inverse Fourier transform of $\widehat{v}(\eta, \tau + i\sigma)$ with respect to (η, τ). From (6.181) it follows that

$$\int_{\mathbb{R}^n} |v(y, t, \sigma)|^2 dy dt \leq C \tag{6.185}$$

for a C independent of σ. This is an immediate consequence of Plancherel's identity, namely,

$$\int_{\mathbb{R}^n} |v(y, t, \sigma)|^2 dy dt = \int_{\mathbb{R}^n} |\widehat{v}(\eta, \tau + i\sigma)|^2 d\eta d\tau \leq C. \tag{6.186}$$

Then for $v(y, t, \sigma)$ defined in (6.184) we consider

$$(v(y, t, \sigma), \varphi(y, t, \sigma)) = \int_{\mathbb{R}_-} \int_{-\infty}^{\infty} \int_{\mathbb{R}^{n-1}} v(y, t, \sigma) \overline{\varphi}(y, t, \sigma) dy dt d\sigma$$

$$= (2\pi)^{-n} \left(\widehat{v}(\eta, \tau + i\sigma), F_{(y,t) \to (\eta, \tau)} \varphi(y, t, \sigma) \right) \tag{6.187}$$

for all $\varphi \in C_0^\infty(\mathbb{R}^n \times (-\infty, 0))$. From (6.178), (6.187) and Parseval's identity it follows that

$$|(v(x, \sigma), \varphi(x, \sigma))| = (2\pi)^{-n} \iint \widehat{v}(\xi, \sigma) \widehat{\varphi}(\xi, \sigma) d\xi d\sigma$$

$$\leq C \int_{-\infty}^{0} \|\widehat{v}(\cdot, \sigma)\|_{L^2(\mathbb{R}^n_\xi)} \|\widehat{\varphi}(\cdot, \sigma)\|_{L^2(\mathbb{R}^n_\xi)} d\sigma \tag{6.188}$$

$$\leq C \int_{-\infty}^{0} \|\varphi(\cdot, \sigma)\|_{L^2(\mathbb{R}^n_x)} d\sigma.$$

The expression $\int_{\mathbb{R}_-} \|v(\cdot, \sigma)\|_{L^2(\mathbb{R}^n_x)} d\sigma$ is a norm on the space $L^1\big((-\infty, 0), L^2(\mathbb{R}^n)\big)$ of locally integrable functions in σ with values in $L^2(\mathbb{R}^n)$. Because of (6.188) and since $C_0^\infty(\mathbb{R}^n \times (-\infty, 0))$ is dense in $L^1((-\infty, 0), L^2(\mathbb{R}^n))$, the function g represents a linear continuous functional on $L^1((-\infty, 0), L^2(\mathbb{R}^n))$. Thus $g(y, t, \sigma)$ is a locally integrable function in $\mathbb{R}^n \times (-\infty, 0)$ satisfying the estimates (6.186) for a C independent of σ.

Applying (6.180) with derivatives $i\frac{\partial}{\partial \tau} = \frac{\partial}{\partial \sigma}$ in the distributional sense under the integral (6.184), we obtain

$$\frac{\partial v(y, t, \sigma)}{\partial \sigma} = \int e^{iy\eta + it\tau} \frac{\partial}{\partial \sigma} \widehat{v}(\eta, \tau + i\sigma) d\eta d\tau$$

$$= \int e^{iy\eta + it\tau} \frac{1}{i} \frac{\partial}{\partial \tau} \widehat{v}(\eta, \tau + i\sigma) d\eta d\tau$$

$$= \int \left(\left(-\frac{1}{i} \frac{\partial}{\partial \tau} \right) e^{iy\eta + it\tau} \right) \widehat{v}(\eta, \tau + i\sigma) d\eta d\tau$$

$$= \int -t e^{iy\eta + it\tau} \widehat{v}(\eta, \tau + i\sigma) d\eta d\tau$$

$$= -t v(y, t, \sigma).$$

It follows that $\frac{\partial v(y,t,\sigma)}{\partial \sigma} + t v(y, t, \sigma) = 0$ as a distribution on $\mathbb{R}^n \times (-\infty, 0)$. The evaluation on any test function $\varphi \in C_0^\infty(\mathbb{R}^n \times (-\infty, 0))$ gives

$$0 = \left(\frac{\partial v(y, t, \sigma)}{\partial \sigma} + t v(y, t, \sigma), \varphi(y, t, \sigma) \right)$$

$$= \left(v(y, t, \sigma), -\frac{\partial \varphi(y, t, \sigma)}{\partial \sigma} + t \varphi(y, t, \sigma) \right).$$

In particular, for $\varphi(y, t, \sigma) := e^{t\sigma} \psi(y, t, \sigma)$ for a $\psi \in C_0^\infty(\mathbb{R}^n \times (-\infty, 0))$ we obtain

$$0 = \left(v(y, t, \sigma), -\frac{\partial}{\partial \sigma} (e^{t\sigma} \psi(y, t, \sigma)) + t e^{t\sigma} \psi(y, t, \sigma) \right)$$

$$= \left(v(y, t, \sigma), e^{t\sigma} \frac{\partial}{\partial \sigma} \psi(y, t, \sigma) \right)$$

$$= \left(e^{-t\sigma} v(y, t, \sigma), \frac{\partial}{\partial \sigma} \psi(y, t, \sigma) \right)$$

$$= \left(-\frac{\partial}{\partial \sigma} (e^{-t\sigma} v(y, t, \sigma)), \psi(y, t, \sigma) \right),$$

for any test function ψ. Lemma 6.12.5 shows that vanishing of the σ-derivative of $e^{-t\sigma} v(y, t, \sigma)$ has the consequence that the respective distribution is independent of σ, and so it follows that $v(y, t, \sigma) = e^{t\sigma} v(y, t)$ for a locally integrable function $v(y, t)$. From (6.186) we obtain

$$\int_{\mathbb{R}^n} |v(y, t)|^2 e^{2t\sigma} dy dt \leq C$$

for all $\sigma < 0$, for a constant C independent of σ. Therefore, if we let $\sigma \to -\infty$ we see that $v(y, t) = 0$ for $t < 0$. Moreover, letting $\sigma \to 0$ we conclude that $v(y, t) \in L^2(\mathbb{R}^n_+)$. Therefore, $\hat{v}(\eta, \tau + i\sigma) = F(ve^{t\sigma})$. Finally, setting $u = F^{-1}p_+^{-1}Fv$, Proposition 6.12.8 shows that $u \in H_0^s(\overline{\mathbb{R}}^n_+)$ and $\hat{u}(\eta, \tau + i\sigma) = F(ue^{t\sigma})$ according to (6.176). $\qquad\square$

We now formulate a Mellin analogue of the Paley–Wiener Theorem 6.12.10. Let $a \in \mathbb{R}_+$ and set

$$\mathcal{H}^{s,\gamma}_{(0,a)}(\mathbb{R}_+ \times \mathbb{R}^n) := \{u \in \mathcal{H}^{s,\gamma}(\mathbb{R}_+ \times \mathbb{R}^n) : u = 0 \text{ for } r > a\}, \tag{6.189}$$

$$\mathcal{H}^{s,\gamma}_{(a,\infty)}(\mathbb{R}_+ \times \mathbb{R}^n) := \{u \in \mathcal{H}^{s,\gamma}(\mathbb{R}_+ \times \mathbb{R}^n) : u = 0 \text{ for } 0 < r < a\}. \tag{6.190}$$

Observe that the operator δ_a, cf. (6.7), induces isomorphisms

$$\begin{aligned}
\delta_a : \mathcal{H}^{s,\gamma}_{(0,a)}(\mathbb{R}_+ \times \mathbb{R}^n) &\to \mathcal{H}^{s,\gamma}_{(0,1)}(\mathbb{R}_+ \times \mathbb{R}^n), \\
\delta_a : \mathcal{H}^{s,\gamma}_{(a,\infty)}(\mathbb{R}_+ \times \mathbb{R}^n) &\to \mathcal{H}^{s,\gamma}_{(1,\infty)}(\mathbb{R}_+ \times \mathbb{R}^n)
\end{aligned} \tag{6.191}$$

for every $a \in \mathbb{R}_+$. From Definition 6.30 (i) it follows that $u(r, x) \in \mathcal{H}^{s,\gamma}(\mathbb{R}_+ \times \mathbb{R}^n)$ implies that

$$(M_{\gamma-n/2,r\to z}F_{x\to\xi}u)\left(\frac{n+1}{2} - \gamma + i\rho, \xi\right),$$

as a function of $z = (n+1)/2 - \gamma + i\rho \in \Gamma_{(n+1)/2-\gamma}$, belongs to the space

$$\{(F_{t\to\tau, x\to\xi}v)(\tau, \xi) : v(t, x) \in H^s(\mathbb{R}_t \times \mathbb{R}^n_x)\}.$$

For purposes below we denote the latter space by $\hat{H}^s(\Gamma_{(n+1)/2-\gamma} \times \mathbb{R}^n)$.

Theorem 6.12.11. (i) *For $u(r, x) \in \mathcal{H}^{s,\gamma}_{(0,a)}(\mathbb{R}_+ \times \mathbb{R}^n)$, $s, \gamma \in \mathbb{R}$, $a \in \mathbb{R}_+$, the function*

$$(M_{\gamma-n/2,r\to z}F_{x\to\xi}u)((n+1)/2 - \gamma + i\rho, \xi) =: \tilde{v}(z, \xi), \tag{6.192}$$

$z = (n+1)/2 - \gamma + i\rho$ *extends to a holomorphic function in*

$$\beta + i\rho \in \{z \in \mathbb{C} : \beta > (n+1)/2 - \gamma\}$$

with values in $\hat{H}^s(\mathbb{R}^n_\xi)$, belonging to $C(\beta \geq (n+1)/2 - \gamma, \hat{H}^s(\mathbb{R}^n_\xi))$, such that

$$\iint (1 + |\beta| + |\rho| + |\xi|)^{2s}|\tilde{v}(\beta + i\rho, \xi)|^2 d\rho d\xi \leq Ca^\beta \tag{6.193}$$

for a $C > 0$, independent of $\beta \geq (n+1)/2 - \gamma$.

(ii) *Conversely, let $\tilde{v}(\beta + i\rho, \xi)$ be a function that is locally integrable in (ρ, ξ), holomorphic in $\beta + i\rho$ for $\beta > (n+1)/2 - \gamma$ for almost all $\xi \in \mathbb{R}^n$, and satisfies the estimate (6.193) for a constant $C > 0$ independent of β. Then there exists a function $u(r, x) \in \mathcal{H}^{s,\gamma}_{(0,a)}(\mathbb{R}_+ \times \mathbb{R}^n)$ such that the relation (6.192) holds.*

Proof. Without loss of generality we may assume $a = 1$. In fact, by virtue of the first relation of (6.191), for $u \in \mathcal{H}^{s,\gamma}_{(0,a)}(\mathbb{R}_+ \times \mathbb{R}^n)$ we have $u_a := \delta_a u \in \mathcal{H}^{s,\gamma}_{(0,1)}(\mathbb{R}_+ \times \mathbb{R}^n)$. Assume our theorem is proved for $a = 1$. Then using (6.8) we obtain from (6.193), applied to u_a, the estimate

$$\iint (1 + |\beta| + |\rho| + |\xi|)^{2s} a^{-(\beta + i\rho)} |\widetilde{v}(\beta + i\rho, \xi)|^2 d\rho d\xi \leq C,$$

where $\widetilde{v} = M_{r \to z} F_{x \to \xi} u$, and this entails (6.193). Moreover, it suffices to assume $\gamma = (n+1)/2$, since a translation in the complex plane shifts $\Gamma_{(n+1)/2-\gamma}$ to Γ_0. We reduce the proof of (i) to Theorem 6.12.10. We first employ the isomorphism

$$S_{1/2} : \mathcal{H}^{s,(n+1)/2}(\mathbb{R}_+ \times \mathbb{R}^n) \to H^s(\mathbb{R} \times \mathbb{R}^n) \tag{6.194}$$

and observe that (6.194) induces an isomorphism

$$S_{1/2} : \mathcal{H}^{s,(n+1)/2}_{(0,1)}(\mathbb{R}_+ \times \mathbb{R}^n) \to H^s_0(\overline{\mathbb{R}}_+ \times \mathbb{R}^n).$$

This is true since $u = 0$ in $r > 1$ is equivalent to $(S_{1/2}u)(t, \cdot) = 0$ for $t < 0$. From (6.14) it follows that for $u(r, x) \in \mathcal{H}^{s,(n+1)/2}_{(0,1)}(\mathbb{R}_+ \times \mathbb{R}^n)$ we have

$$\big(M_{1/2, r \to z} F_{x \to \xi} u\big)(i\rho, \xi) = \big(F_{t \to \tau} S_{1/2} F_{x \to \xi} u\big)(\tau, \xi) = \big(F_{t \to \tau} F_{x \to \xi} f\big)(\tau, \xi),$$

where $f(t, x) := (S_{1/2}u)(t, x)$. Now Theorem 6.12.10 (here (x, ξ) plays the role of (y, η)) tells us that $\widehat{f}(\tau, \xi)$ extends to a holomorphic function in $\zeta := \tau + i\sigma$ for $\sigma < 0$ with values in $\widehat{H}^s(\mathbb{R}^n_\xi)$ which belongs to $C\big(\sigma \leq 0, \widehat{H}^s(\mathbb{R}^n_\xi)\big)$, such that

$$\iint_{\mathbb{R}^{n+1}} (1 + |\tau| + |\sigma| + |\xi|)^{2s} \big|\widehat{f}(\tau + i\sigma, \xi)\big|^2 d\tau d\xi \leq C$$

for $\sigma \leq 0$, for a $C > 0$ independent of σ. Returning to the original covariables it follows that

$$\big(M_{1/2} F_{x \to \xi} u\big)(z, \xi) = \widetilde{v}(z, \xi)$$

for $z = \beta + i\rho$, obtained from $\zeta = \tau + i\sigma$ by the rotation $-i : \mathbb{C}_\zeta \to \mathbb{C}_z$ by $-\pi/2$, is holomorphic in $\operatorname{Re} z > 0$, with values in $\widehat{H}^s(\mathbb{R}^n_\xi)$, continuous up to $\beta = 0$, and such that (6.193) holds for a $C > 0$, independent of β.

The proof of (ii) follows in an analogous manner from Theorem 6.12.10 (ii). $\qquad \square$

Let us define the spaces

$$\mathcal{H}^{s,\gamma}_{(0,a)}(X^\wedge) \quad \text{and} \quad \mathcal{H}^{s,\gamma}_{(a,\infty)}(X^\wedge)$$

in an analogous manner as (6.189) and (6.190), respectively. Then we have the following modification of Theorem 6.12.11.

Theorem 6.12.12. (i) *For* $u(r,x) \in \mathcal{H}_{(0,a)}^{s,\gamma}(X^\wedge)$ *the function*

$$\big(M_{r \to z}u\big)\big((n+1)/2 - \gamma + i\rho, x\big) := \widetilde{w}(z,x), \quad z = (n+1)/2 - \gamma + i\rho, \quad (6.195)$$

extends to a holomorphic function in the set $\{\beta > (n+1)/2 - \gamma\}$ *with values in* $H^s(X)$, *belonging to* $C\big(\beta \geq (n+1)/2 - \gamma, H^s(X)\big)$, *such that*

$$\int \big\|\widetilde{w}(\beta + i\rho, \cdot)\big\|_{H_{\beta,\rho}^s(X)}^2 d\rho \leq Ca^\beta \qquad (6.196)$$

for a $C > 0$, *independent of* $\beta \geq (n+1)/2 - \gamma$.

(ii) *Conversely, let* $\widetilde{w}(\beta + i\rho, \cdot)$ *be an* $H^s(X)$-*valued function for* $\beta > (n+1)/2 - \gamma$ *satisfying the estimate* (6.196) *for a* $C > 0$ *independent of* β. *Then there is a function* $u(r,x) \in \mathcal{H}_{(0,a)}^{s,\gamma}(X^\wedge)$ *such that the relation* (6.195) *holds.*

Proof. We reduce the proof to Theorem 6.12.11, by observing that $u(r,x) \in \mathcal{H}_{(0,a)}^{s,\gamma}(X^\wedge)$ is equivalent to $\varphi_j u \in \mathcal{H}_{(0,a)}^{s,\gamma}(X^\wedge)$ for every $\varphi_j \in C_0^\infty(U_j)$ for a coordinate neighbourhood U_j on X. Then using any chart $\kappa_j : U_j \to \mathbb{R}^n$ we can pass to functions in $\mathbb{R}_+ \times \mathbb{R}^n$. In this form we immediately see that it is allowed to pass to the x-variables in the preimage under the Fourier transform, i.e., to talk everywhere about $H^s(\mathbb{R}_x^n)$-valued functions instead of $\widehat{H}^s(\mathbb{R}_\xi^n)$-valued ones. In this case it is also clear that Sobolev norms may be replaced by the parameter-dependent version of $H^s(\mathbb{R}_x^n)$-norms, namely, $H_{\beta,\rho}^s(\mathbb{R}^n)$. The step to the case of a compact manifold X is straightforward. $\qquad \square$

Corollary 6.12.13. *There are analogues of* Theorems 6.12.11 *and* 6.12.12 *for the spaces* $\mathcal{H}_{(a,\infty)}^{s,\gamma}\big(\mathbb{R}_+ \times \mathbb{R}^n\big)$ *and* $\mathcal{H}_{(a,\infty)}^{s,\gamma}(X^\wedge)$, *respectively, where we only replace the former condition on* β *by* $\beta < (n+1)/2 - \gamma$.

Theorem 6.12.14. *Let* M_β, $\beta \in \mathbb{R}$, *be the weighted Mellin transform*

$$M_\beta : u(t) \to L^2\big(\Gamma_{1/2 - \beta}, H_1\big)$$

for $u(t) \in t^\beta L^2(\mathbb{R}_+, H_1)$. *Then the following conditions are equivalent.*

(i) $h(v) = M_\beta u(v)$ *for some* $u(t) \in t^\beta L^2(\mathbb{R}_+, H_1)$, $\operatorname{supp} u \subseteq [0,a], a > 0$;

(ii) h *is holomorphic in* $\operatorname{Re} v > 1/2 - \beta$; *moreover,* $h_\delta(\tau) := h(1/2 - \beta + \delta + i\tau)$ *belongs to* $L^2(\mathbb{R}_\tau, H_1)$ *for every* $\delta > 0$, *and there are constants* $a, c > 0$ *such that*

$$\|h_\delta\|_{L^2(\mathbb{R}_\tau, H_1)} \leq c\, a^\delta$$

for all $\delta \in \mathbb{R}_+$.

Under the conditions (i), (ii) *we have*

$$\lim_{\delta \to 0} h_\delta = h_0$$

in $L^2(\mathbb{R}_\tau, H_1)$. *Moreover,*

$$\|h_\delta\|_{L^2(\mathbb{R}_\tau, H_1)} = \sqrt{2\pi}\,\big\|t^\delta u(t)\big\|_{t^\beta L^2(\mathbb{R}_t, H_1)} = \sqrt{2\pi}\,\big\|t^{\delta - \beta} u(t)\big\|_{L^2(\mathbb{R}_t, H_1)}, \quad \delta \in \overline{\mathbb{R}}_+.$$

Chapter 7

The edge algebra

7.1 Weighted spaces on a manifold with edge

Let N be a smooth closed manifold. The wedge $W := N^\Delta \times \Omega$ for any open $\Omega \subseteq \mathbb{R}^q$ is an example of a manifold with edge Ω. The space W is at the same time a trivial N^Δ-bundle over Ω where the fibre is just the cone $N^\Delta = (\overline{\mathbb{R}}_+ \times N)/(\{0\} \times N)$. As before, we often set $N^\wedge = \mathbb{R}_+ \times N$. Let us also illustrate some other notation in this example, e.g., the stretched manifold $\mathbb{W} := \overline{\mathbb{R}}_+ \times N \times \Omega$ associated with W, which is a smooth manifold with boundary $\partial \mathbb{W} = N \times \Omega$ (identified with $\{0\} \times N \times \Omega$), which in turn is a trivial N-bundle over Ω. There is then a canonical projection

$$\pi : \mathbb{W} \to W$$

which restricts to the bundle projection $\pi|_{\partial \mathbb{W}} : \partial \mathbb{W} \to \Omega$ and to a diffeomorphism

$$\pi\big|_{\text{int } \mathbb{W}} : \text{int } \mathbb{W} \to W \setminus \Omega.$$

We interpret W as a stratified space with the strata

$$s_0(W) := W \setminus \Omega = N^\wedge \times \Omega, \quad s_1(W) := \Omega;$$

both are smooth manifolds, and we have

$$W = s_0(W) \cup s_1(W)$$

as a disjoint union.

More generally, a topological space M is a manifold with edge $Y := s_1(M)$ if both $s_1(M)$ and $s_0(M) := M \setminus s_1(M)$ are smooth manifolds, and Y has a neighbourhood V in M which has the structure of a (locally trivial) N^Δ-bundle for a smooth closed manifold N. Denoting an arbitrary trivialisation by

$$\chi_G^\Delta : V\big|_G \to N^\Delta \times \Omega \tag{7.1}$$

© Springer Nature Switzerland AG 2018
X. Liu, B.-W. Schulze, *Boundary Value Problems with Global Projection Conditions*,
Operator Theory: Advances and Applications 265, https://doi.org/10.1007/978-3-319-70114-1_7

for some coordinate neighbourhood G on Y which restricts to a chart $\chi_G : G \to \Omega$ on Y, we have a restriction

$$\chi_G^\wedge : V\big|_G \setminus G \to N^\wedge \times \Omega. \tag{7.2}$$

Let $(r, x, y) \in N^\wedge \times \Omega$ denote the corresponding local splitting of variables. Another choice of a trivialisation

$$\widetilde{\chi}_G^\wedge : V\big|_G \to N^\Delta \times \widetilde{\Omega}$$

is said to belong to an equivalent wedge structure if

$$\widetilde{\chi}_G^\wedge \circ \left(\chi_G^\wedge\right)^{-1} : \mathbb{R}_+ \times N \times \Omega \to \mathbb{R}_+ \times N \times \widetilde{\Omega}, \quad (r, x, y) \mapsto (\widetilde{r}, \widetilde{x}, \widetilde{y}),$$

is the restriction of a transition map $\overline{\mathbb{R}}_+ \times N \times \Omega \to \overline{\mathbb{R}}_+ \times N \times \widetilde{\Omega}$ to $\mathbb{R}_+ \times N \times \Omega$ which belongs to a corresponding $\overline{\mathbb{R}}_+ \times N$ bundle on the edge. This gives us a cocycle of transition maps $N \times \Omega \to N \times \widetilde{\Omega}$ of a corresponding N-bundle over Y. Similarly as in the case of conical singularities, we fix a system of trivialisations (not necessarily a maximal one) and assume for simplicity that the transition maps $\widetilde{r} = \widetilde{r}(r, x, y)$, $\widetilde{x} = \widetilde{x}(r, x, y)$, $\widetilde{y} = \widetilde{y}(r, x, y)$ are independent of r in a small neighbourhood of 0 (we have $\widetilde{r}(0, x, y) = 0$ and $\widetilde{y}(0, x, y)$ is independent of x).

By invariantly attaching the above-mentioned N-bundle over Y to $M \setminus Y$ we obtain a smooth manifold $\mathbb{M}$ with boundary, called the stretched manifold $\mathbb{M}$ associated with M. The boundary $\partial\mathbb{M}$ is just our N-bundle, and we have a canonical projection

$$\pi : \mathbb{M} \to M \tag{7.3}$$

which restricts to the bundle projection

$$\pi_{\partial\mathbb{M}} : \partial\mathbb{M} \to Y$$

and to a diffeomorphism $\pi_{\text{int}\,\mathbb{M}} : \text{int}\,\mathbb{M} \to W \setminus Y$. Throughout this exposition, for notational convenience, we assume that the bundle $\partial\mathbb{M}$ is trivial, i.e.,

$$\partial\mathbb{M} = N \times Y. \tag{7.4}$$

The generalisation of the considerations to the general case is straightforward. As in BVPs, it makes sense to study operators between distributional sections of vector bundles over the manifold M with edge Y. It will be convenient to admit bundles $E, F \in \text{Vect}(\mathbb{M})$, not only on M itself. The case of bundles over M is a special case; we can pass to pull-backs to $\mathbb{M}$ under the projection (7.3).

For any $E \in \text{Vect}(\mathbb{M})$ we have the restriction $E_\partial := E\big|_{\partial\mathbb{M}} \in \text{Vect}(\partial\mathbb{M})$ and restrictions to the fibres N_y of $\partial\mathbb{M}$ over $y \in Y$, denoted by $E'_y := E_\partial\big|_{N_y}$. If $\partial\mathbb{M}$ is locally described by $N \times \Omega$, then we also write E' rather than E'_y. For simplicity, by E', etc., we also denote elements of $\text{Vect}(N)$. This should not cause confusion, since the former E' can be seen as the pull-back of a bundle on N under the projection $N \times \Omega \to N$.

Finally, for any $E' \in \text{Vect}(N)$ by $E'^{\wedge}$ denote the pull-back of E' under the canonical projection $N^{\wedge} \to N$, $(r, x) \mapsto x$. For brevity we denote the restriction of an $E \in \text{Vect}(\mathbb{M})$ to $M \setminus \partial \mathbb{M} \cong M \setminus Y$ again by E.

Let $\text{Diff}^{\mu}(\,\cdot\,; J, G)$ denote the space of differential operators of order μ between the C^{∞} sections of bundles J, G over the respective smooth manifold, indicated by "$\cdot$".

If M is a manifold with edge Y, and E, $F \in \text{Vect}(\mathbb{M})$, an operator $A \in \text{Diff}^{\mu}(M \setminus Y; E, F)$ is said to be edge-degenerate if close to Y in the above-mentioned splitting of variables $(r, x, y) \in N^{\wedge} \times \Omega$ it is of the form

$$A = r^{-\mu} \sum_{j + |\alpha| \le \mu} a_{j\alpha}(r, y)(-r\partial_r)^j (rD_y)^{\alpha}, \tag{7.5}$$

with coefficients $a_{j\alpha} \in C^{\infty}\left(\overline{\mathbb{R}}_+ \times \Omega, \text{Diff}^{\mu - (j + |\alpha|)}(N; E', F')\right)$. The general idea of the edge calculus is to construct a subalgebra of $\bigcup_{\mu \in \mathbb{R}} L^{\mu}_{\text{cl}}(M \setminus Y; E, F)$ that contains all edge-degenerate differential operators together with the parametrices of elliptic elements.

The ellipticity in this situation is again determined by a principal symbol hierarchy, here denoted by

$$\sigma(A) = \big(\sigma_{\psi}(A), \sigma_{\wedge}(A)\big),$$

where the components are contributed by the strata $M \setminus Y$ and Y, respectively. The first component $\sigma_{\psi}(A) : \pi^*_{M \setminus Y} E \to \pi^*_{M \setminus Y} F$ is the homogeneous principal symbol of A over $M \setminus Y$. The so-called principal edge symbol $\sigma_{\wedge}$ in the pseudo-differential case will be defined below. The situation is similar to boundary value problems; the boundary symbol σ_{∂} is an analogue of $\sigma_{\wedge}$. In the case of differential operators (7.5) it is defined as

$$\sigma_{\wedge}(A)(y, \eta) := r^{-\mu} \sum_{j + |\alpha| \le \mu} a_{j\alpha}(0, y)(-r\partial_r)^j (r\eta)^{\alpha},$$

as an operator function

$$\sigma_{\wedge}(A)(y, \eta) : \mathcal{K}^{s, \gamma}\big(N^{\wedge}\big) \otimes E'^{\wedge}_y \to \mathcal{K}^{s - \mu, \gamma - \mu}\big(N^{\wedge}\big) \otimes F'^{\wedge}_y, \tag{7.6}$$

$(y, \eta) \in \Omega \times (\mathbb{R}^q \setminus \{0\})$, where in this case $E'^{\wedge}$, $F'^{\wedge}$ are interpreted as pull-backs from N to $N^{\wedge}$ of the former bundles E', F' under the canonical projection $N^{\wedge} \to N$, and the spaces $\mathcal{K}^{s, \gamma}\big(N^{\wedge}\big) \otimes E'^{\wedge}_y$ have the meaning of $\mathcal{K}^{s, \gamma}\big(N^{\wedge}_y\big)$-distributional sections in $E'^{\wedge}_y$.

The smoothness s is arbitrary, while the weight γ will be fixed in the discussion of ellipticity. From BVPs we already know that ellipticity requires additional conditions, here edge conditions in place of boundary conditions. This will be the topic in Chapter 8. In the present section we introduce weighted Sobolev spaces that play a major role in the edge algebra.

Recall that in boundary value problems the presence of a boundary introduces anisotropic aspects to the calculus. More precisely, we distinguish between the tangential variables and a variable transverse to the boundary. Nevertheless, in the case of operators with the transmission property we could argue in terms of the negative counterpart of the configuration with boundary, and the Sobolev spaces over a neighbouring manifold do not "feel" the presence of the boundary. However, in the case of edge-degenerate operators and, as we shall see below, in associated distribution spaces, a more significant difference arises between variables on the edge and the transversal cone.

In order to define weighted Sobolev spaces, first locally on an open (stretched) wedge $N^\wedge \times \mathbb{R}^q$, we go back to the material of Section 1.3.

Remark 7.1.1. The space $H := \mathcal{K}^{s,\gamma;e}(N^\wedge)$ for any fixed s, γ, $e \in \mathbb{R}$, cf. (6.37), is a Hilbert space with group action $\kappa = \{\kappa_\lambda\}_{\lambda \in \mathbb{R}_+}$,

$$(\kappa_\lambda u)(r,x) = \lambda^{(n+1)/2} u(\lambda r, x), \quad \lambda \in \mathbb{R}_+, \ n = \dim N.$$

It induces a group action on the Fréchet subspace $\mathcal{K}^{s,\gamma;e}_{\mathcal{P}}(N^\wedge)$, cf. (6.74), for any asymptotic type $\mathcal{P}$ associated with the weight data (γ, Θ).

This allows us to apply Definition 1.3.31. For $e = 0$ we obtain so-called weighted edge spaces

$$\mathcal{W}^s(\mathbb{R}^q, \mathcal{K}^{s,\gamma}(N^\wedge)) \tag{7.7}$$

of smoothness $s \in \mathbb{R}$ and weight $\gamma \in \mathbb{R}$, and also subspaces $\mathcal{W}^s(\mathbb{R}^q, \mathcal{K}^{s,\gamma}_{\mathcal{P}}(N^\wedge))$ with edge asymptotics of type $\mathcal{P}$.

Note that Theorem 1.3.33 (ii) has the consequence that

$$H^s_{\mathrm{comp}}(N^\wedge \times \mathbb{R}^q) \subset \mathcal{W}^s(\mathbb{R}^q, \mathcal{K}^{s,\gamma}(N^\wedge)) \subset H^s_{\mathrm{loc}}(N^\wedge \times \mathbb{R}^q) \tag{7.8}$$

for every s, $\gamma \in \mathbb{R}$. This allows us to define weighted spaces modelled on (7.7) on a manifold M with edge Y. For convenience, we first assume that M is compact.

Let $G_1, \dots, G_N$ be an open covering of the edge Y by coordinate neighbourhoods and

$$\chi_j : G_j \to \mathbb{R}^q, \quad j = 1, \dots, N,$$

be charts such that the transition maps $\chi_j \chi_l^{-1} : \mathbb{R}^q \to \mathbb{R}^q$ satisfying the same boundedness condition for large $|y|$ as in [46]. Let $\{\varphi_1, \dots, \varphi_N\}$ be partition of unity subordinate to the open covering, and set

$$\mathcal{W}^s(Y, \mathcal{K}^{s,\gamma}(N^\wedge)) = \left\{ \sum_{j=1}^N \varphi_j \chi_j^* u_j : u_j \in \mathcal{W}^s(\mathbb{R}^q, \mathcal{K}^{s,\gamma}(N^\wedge)) \right\}$$

with pull-backs $\chi_j^* u_j(u, \cdot) = u_j(\chi_j(y), \cdot)$, $j = 1, \dots, N$. This is a coordinate invariant definition, and we have

$$\mathcal{W}^s(Y, \mathcal{K}^{s,\gamma}(N^\wedge)) \subseteq H^s_{\mathrm{loc}}(N^\wedge \times Y).$$

Moreover, by a cut-off function θ on M we understand any $\theta \in C^\infty(M \setminus Y)$ which is equal to 1 in a small neighbourhood of Y and 0 outside some other small neighbourhood of Y.

Definition 7.1.2. Let M be a compact manifold with edge Y, locally near Y modelled on $N^\Delta \times \mathbb{R}^q$ as described in the beginning. Then we set

$$H^{s,\gamma}(M) = \theta \mathcal{W}^s\big(Y, \mathcal{K}^{s,\gamma}(N^\wedge)\big) + (1-\theta)H^s_{\mathrm{loc}}(M \setminus Y) \quad \text{for } s,\gamma \in \mathbb{R}. \tag{7.9}$$

There is a straightforward extension of the definition of our spaces to distributional sections of vector bundles $E \in \mathrm{Vect}(\mathbb{M})$. Remark 7.1.1 is valid in analogous form for the spaces $\mathcal{K}^{s,\gamma;e}(N^\wedge, E^\wedge)$, cf. also the notation in formula (7.6), which gives us the spaces

$$\mathcal{W}^s\big(\mathbb{R}^q, \mathcal{K}^{s,\gamma}(N^\wedge, E'^\wedge)\big). \tag{7.10}$$

For a compact manifold M with edge Y and $E \in \mathrm{Vect}(\mathbb{M})$ we define (by abuse of notation)

$$H^{s,\gamma}(M, E) \quad \text{for } s,\gamma \in \mathbb{R} \tag{7.11}$$

as the subspace of all $u \in H^s_{\mathrm{loc}}(M \setminus Y, E)$ such that an analogue of (7.9) holds.

Remark 7.1.3. Note that there are also straightforward generalisations of the spaces

$$\mathcal{K}^{s,\gamma;e}(N^\wedge, E'^\wedge) = [r]^{-e} \mathcal{K}^{s,\gamma}(N^\wedge, E'^\wedge)$$

to spaces $\mathcal{K}^{s,\gamma;e}_{\mathcal{P}}(N^\wedge, E'^\wedge)$ for asymptotic types $\mathcal{P}$, and we can also form spaces $\mathcal{W}^s\big(Y, \mathcal{K}^{s,\gamma}_{\mathcal{P}}(N^\wedge)\big)$ as well as

$$H^{s,\gamma}_{\mathcal{P}}(M) := \theta \mathcal{W}^s\big(Y, \mathcal{K}^{s,\gamma}_{\mathcal{P}}(N^\wedge)\big) + (1-\theta)H^s_{\mathrm{loc}}(M \setminus Y),$$

or

$$H^{s,\gamma}_{\mathcal{P}}(M, E) := \theta \mathcal{W}^s\big(Y, \mathcal{K}^{s,\gamma}_{\mathcal{P}}(N^\wedge, E'^\wedge)\big) + (1-\theta)H^s_{\mathrm{loc}}(M \setminus Y, E).$$

Clearly, in the latter spaces the bundle E means a restriction to $M \setminus Y$ and $E'^\wedge$ is interpreted as before.

7.2 Green and smoothing Mellin edge symbols

Edge amplitude functions will be specific operator-valued symbols in the sense of Definition 1.3.8. The values of the symbols will belong to the cone algebra $L^\mu(N^\wedge, \boldsymbol{g}; E'^\wedge, F'^\wedge)$ for $\boldsymbol{g} = (\gamma, \gamma-\mu, \Theta)$ or $\boldsymbol{g} = (\gamma, \gamma-\mu)$, cf. Definition 6.8.1, and we take into account also the Green and smoothing Mellin plus Green subclasses. Concerning the Green symbols, we admit from the very beginning 2×2 block matrix symbols of a similar meaning as the Green symbols in Definition 2.4.6, which also contain entries of trace and potential type. The following definition takes into account Remark 7.1.1.

Definition 7.2.1. By $\mathcal{R}_G^\nu(\Omega \times \mathbb{R}^q, \boldsymbol{g}; \boldsymbol{w})$ for an open set $\Omega \subseteq \mathbb{R}^q$, weight data $\boldsymbol{g} :=$ $(\gamma, \gamma - \mu, \Theta)$ of similar meaning as in Definition 6.5.2, $\mu, \nu, \gamma \in \mathbb{R}$ and $\boldsymbol{w} :=$ $(E', F'; j_1, j_2)$, we denote the space of all operator-valued symbols

$$g(y, \eta) \in \bigcap_{s,e \in \mathbb{R}} S_{\mathrm{cl}}^\nu \left(\Omega \times \mathbb{R}^q; \mathcal{K}^{s,\gamma;e}\left(N^\wedge, E'^\wedge\right) \oplus \mathbb{C}^{j_1}, \mathcal{K}^{\infty,\gamma-\mu;\infty}\left(N^\wedge, F'^\wedge\right) \oplus \mathbb{C}^{j_2}\right)$$

such that for (g-dependent) asymptotic types $\mathcal{P}$ and $\mathcal{Q}$

$$g(y, \eta) \in \bigcap_{s,e \in \mathbb{R}} S_{\mathrm{cl}}^\nu \left(\Omega \times \mathbb{R}^q; \mathcal{K}^{s,\gamma;e}\left(N^\wedge, E'^\wedge\right) \oplus \mathbb{C}^{j_1}, \mathcal{K}_{\mathcal{P}}^{\infty,\gamma-\mu;\infty}\left(N^\wedge, F'^\wedge\right) \oplus \mathbb{C}^{j_2}\right),$$

$$g^*(y, \eta) \in \bigcap_{s,e \in \mathbb{R}} S_{\mathrm{cl}}^\nu \left(\Omega \times \mathbb{R}^q; \mathcal{K}^{s,-\gamma+\mu;e}\left(N^\wedge, F'^\wedge\right) \oplus \mathbb{C}^{j_2}, \mathcal{K}_{\mathcal{Q}}^{\infty,-\gamma;\infty}\left(N^\wedge, E'^\wedge\right) \oplus \mathbb{C}^{j_1}\right).$$

The pointwise formal adjoint refers to sesquilinear pairings

$$\mathcal{K}^{s,\gamma;e}\left(N^\wedge, E'^\wedge\right) \times \mathcal{K}^{-s,-\gamma;-e}\left(N^\wedge, E'^\wedge\right) \to \mathbb{C},$$

$s, \gamma, e \in \mathbb{R}$, induced by the $\mathcal{K}^{0,0;0}\left(N^\wedge, E'^\wedge\right) (= r^{-n/2} L^2(\mathbb{R}_+ \times N, E'^\wedge))$-scalar product. The elements of

$$\mathcal{R}_G^\nu(\Omega \times \mathbb{R}^q, \boldsymbol{g}; \boldsymbol{w})$$

are called Green symbols of the edge calculus. Moreover, $\mathcal{R}_G^\nu(\Omega \times \mathbb{R}^q, \boldsymbol{g}; \boldsymbol{w})_{\mathcal{P},\mathcal{Q}}$ is the subspace of Green symbols with fixed asymptotic types $\mathcal{P}$, $\mathcal{Q}$.

In the case of $\boldsymbol{w} := (E', F'; 0, 0)$, i.e., Green symbols in top left corners, we also use the notation

$$\mathcal{R}_G^\nu(\Omega \times \mathbb{R}^q, \boldsymbol{g}; E', F') \quad \text{and} \quad \mathcal{R}_G^\nu(\Omega \times \mathbb{R}^q, \boldsymbol{g}; E', F')_{\mathcal{P},\mathcal{Q}},$$

respectively. If $\mathcal{P} = O$ and $\mathcal{Q} = O$ are the trivial asymptotic types (i.e., $\pi_{\mathbb{C}}\mathcal{P} = \emptyset$, $\pi_{\mathbb{C}}\mathcal{Q} = \emptyset$) and $\Theta = (-\infty, 0]$, then $\mathcal{R}_G^\nu(\Omega \times \mathbb{R}^q, \boldsymbol{g}; E', F')_{O,O}$ is independent of $\boldsymbol{g}$ and we write

$$\mathcal{R}_G^\nu(\Omega \times \mathbb{R}^q; E', F')_{O,O}. \tag{7.12}$$

In the general edge calculus it is not always necessary to control the asymptotic types as in Definition 7.2.1; it may be reasonable to pass to continuous asymptotic types anyway. Therefore, it makes sense to introduce the space

$$\mathcal{R}_G^\nu(\Omega \times \mathbb{R}^q, \boldsymbol{g}; \boldsymbol{w}) \quad \text{for } \boldsymbol{g} := (\gamma, \gamma - \mu) \tag{7.13}$$

(i.e., for omitted Θ) of all $g(y, \eta)$ such that

$$g(y, \eta) \in \bigcap_{s,e \in \mathbb{R}} S_{\mathrm{cl}}^\nu \left(\Omega \times \mathbb{R}^q; \mathcal{K}^{s,\gamma;e}\left(N^\wedge, E'^\wedge\right) \oplus \mathbb{C}^{j_1}, \mathcal{K}^{\infty,\gamma-\mu+\varepsilon;\infty}\left(N^\wedge, F'^\wedge\right) \oplus \mathbb{C}^{j_2}\right),$$

$$g^*(y, \eta) \in \bigcap_{s,e \in \mathbb{R}} S_{\mathrm{cl}}^\nu \left(\Omega \times \mathbb{R}^q; \mathcal{K}^{s,-\gamma+\mu;e}\left(N^\wedge, F'^\wedge\right) \oplus \mathbb{C}^{j_2}, \mathcal{K}^{\infty,-\gamma+\varepsilon;\infty}\left(N^\wedge, E'^\wedge\right) \oplus \mathbb{C}^{j_1}\right)$$

for some $\varepsilon = \varepsilon(g) > 0$.

Theorem 7.2.2. *Let $g_j(y, \eta) \in \mathcal{R}_G^{\nu-j}(\Omega \times \mathbb{R}^q, \boldsymbol{g})_{\mathcal{P},\mathcal{Q}}$, $j \in \mathbb{N}$, be a sequence of Green symbols. Then there is a $g(y, \eta) \in \mathcal{R}_G^\nu(\Omega \times \mathbb{R}^q, \boldsymbol{g})_{\mathcal{P},\mathcal{Q}}$ such that $g \sim \sum_{j=0}^\infty g_j$, and g is unique modulo $\mathcal{R}_G^{-\infty}(\Omega \times \mathbb{R}^q, \boldsymbol{g})_{\mathcal{P},\mathcal{Q}}$.*

There is the following generalisation of Definition 7.2.1 referring to different weight intervals Θ and Θ'. For simplicity, we often omit the variants with vector bundles.

Definition 7.2.3. Fix weight data $(\gamma, \Theta; \gamma', \Theta')$ for reals γ, γ' and $\Theta = (\theta, 0]$, $\Theta' = (\theta', 0]$. Then

$$\mathcal{R}_G^\nu\big(\Omega \times \mathbb{R}^q, (\gamma, \Theta; \gamma', \Theta')\big)$$

denotes the space of all $g(y, \eta)$ such that

$$g(y, \eta) \in S_{\mathrm{cl}}^\nu\big(\Omega \times \mathbb{R}^q; \mathcal{K}^{s,\gamma;e}(N^\wedge), \mathcal{K}_{\mathcal{P}'}^{\infty,\gamma';\infty}(N^\wedge)\big),$$

$$g^*(y, \eta) \in S_{\mathrm{cl}}^\nu\big(\Omega \times \mathbb{R}^q; \mathcal{K}^{s,-\gamma';e}(N^\wedge), \mathcal{K}_{\mathcal{P}}^{\infty,-\gamma;\infty}(N^\wedge)\big)$$

for all $s, e \in \mathbb{R}$, for g-dependent asymptotic types $\mathcal{P}'$ and $\mathcal{P}$ associated with the weight data (γ', Θ') and $(-\gamma, \Theta)$, respectively. If $\mathcal{P}', \mathcal{P}$ are fixed, we also write

$$\mathcal{R}_G^\nu\big(\Omega \times \mathbb{R}^q, (\gamma, \Theta; \gamma', \Theta')\big)_{\mathcal{P}',\mathcal{P}}.$$

For $\Theta = \Theta'$ we replace $(\gamma, \Theta; \gamma', \Theta')$ by $(\gamma, \gamma', \Theta)$, i.e., we have the spaces

$$\mathcal{R}_G^\nu\big(\Omega \times \mathbb{R}^q, (\gamma, \gamma', \Theta)\big) \quad \text{and} \quad \mathcal{R}_G^\nu\big(\Omega \times \mathbb{R}^q, (\gamma, \gamma', \Theta)\big)_{\mathcal{P}',\mathcal{P}},$$

respectively.

Remark 7.2.4. Analogously to Remark 1.3.20, the following properties hold. Let

$$g(y, \eta) \in C^\infty\big(\Omega \times \mathbb{R}^q, L_G\big(N^\wedge, (\gamma, \gamma', \Theta)\big)\big)_{\mathcal{P}',\mathcal{P}}$$

be a family of operators such that

$$g(y, \lambda\eta) = \lambda^\nu \kappa_\lambda g(y, \eta) \kappa_\lambda^{-1}$$

for all $\lambda \geq 1$, $|\eta| \geq C$ for some $C > 0$. Then $g(y, \eta) \in \mathcal{R}_G^\nu\big(\Omega \times \mathbb{R}^q, (\gamma, \gamma', \Theta)\big)_{\mathcal{P}',\mathcal{P}}$.

Remark 7.2.5. The operator $\mathcal{M}_\varphi$ of multiplication by

$$\varphi(r, x, y) \in C_0^\infty\big(\overline{\mathbb{R}}_+ \times N \times \Omega\big), \quad \varphi(r, x, y) = 0 \quad \text{for } r > c_0,$$

for some $c_0 > 0$, represents elements

$$\mathcal{M}_\varphi \in S^0\big(\Omega \times \mathbb{R}^q; \mathcal{K}^{s,\gamma}(N^\wedge), \mathcal{K}^{s,\gamma}(N^\wedge)\big), \quad \mathcal{M}_\varphi \in S^0\big(\Omega \times \mathbb{R}^q; \mathcal{K}_{\mathcal{P}}^{s,\gamma}(N^\wedge), \mathcal{K}_{\mathcal{Q}}^{s,\gamma}(N^\wedge)\big)$$

for every $s, \gamma \in \mathbb{R}$ and every asymptotic type $\mathcal{P}$ associated with the weight data (γ, Θ), and some resulting $\mathcal{Q}$.

Proposition 7.2.6. *Let* $g(y,\eta) \in \mathcal{R}_G^\nu(\Omega \times \mathbb{R}^q, \boldsymbol{g})$, $\boldsymbol{g} = (\gamma, \gamma - \mu, \Theta)$, $\mu, \nu \in \mathbb{R}$, *and* $\varphi(r,x,y) \in C^\infty(\overline{\mathbb{R}}_+ \times X \times \Omega)$, $\varphi(r,x,y) = 0$ *for* $r > c_0$. *Then, denoting* $\mathcal{M}_\varphi$ *also by* φ, *we have*

$$g\varphi, \varphi g \in \mathcal{R}_G^\nu(\Omega \times \mathbb{R}^q, \boldsymbol{g}).$$

Moreover, for every $j \in \mathbb{N}$ *we have*

$$r^j g(y,\eta), \ g(y,\eta) r^j \in \mathcal{R}_G^{\nu-j}(\Omega \times \mathbb{R}^q, \boldsymbol{g}). \tag{7.14}$$

Proof. Let us first show (7.14). The continuity of

$$g(y,\eta) : \mathcal{K}^{s,\gamma;e}(N^\wedge) \to \mathcal{K}_{\mathcal{P}}^{\infty,\gamma-\mu;\infty}(N^\wedge)$$

implies an analogous mapping property of $g_j := r^j g(y,\eta)$, namely,

$$g_j(y,\eta) : \mathcal{K}^{s,\gamma;e}(N^\wedge) \to \mathcal{K}_{\mathcal{P}_j}^{\infty,\gamma-\mu;\infty}(N^\wedge), \tag{7.15}$$

it is continuous, where $\mathcal{P}_j$ is obtained by a translation of $\pi_{\mathbb{C}}\mathcal{P}$ by j to the left in the complex plane. More precisely, we have

$$g_j(y,\eta) \in S_{\mathrm{cl}}^{\nu-j}(\Omega \times \mathbb{R}^q; \mathcal{K}^{s,\gamma;e}(N^\wedge) \to \mathcal{K}_{\mathcal{P}_j}^{\infty,\gamma-\mu-j;\infty}(N^\wedge))$$

for every $s, e \in \mathbb{R}$. The resulting order $\nu - j$ and the property of being classical follows, by evaluating the homogeneous components, which are just

$$g_{j,(\nu-j-k)}(y,\eta) = r^j g_{(\nu-k)}(y,\eta).$$

In fact, we have $g_{(\nu-k)}(y,\lambda\eta) = \lambda^{\nu-k}\kappa_\lambda g_{(\nu-k)}(y,\eta)\kappa_\lambda^{-1}$, whence

$$r^j g_{(\nu-k)}(y,\lambda\eta) = r^j \lambda^{\nu-k}\kappa_\lambda g_{(\nu-k)}(y,\eta)\kappa_\lambda^{-1} = \lambda^{\nu-j-k}\kappa_\lambda r^j g_{(\nu-k)}(y,\eta)\kappa_\lambda^{-1},$$

and so

$$g_{j,(\nu-j-k)}(y,\lambda\eta) = \lambda^{\nu-j-k}\kappa_\lambda(r^j g_{(\nu-k)}(y,\eta))\kappa_\lambda^{-1}$$

for all $\lambda \in \mathbb{R}_+$, $j, k \subset \mathbb{N}$. For the adjoints we may argue in an analogous manner. Here the factor r^j is applied from the right, i.e., we have treated also the case $g(y,\eta) r^j$.

Next we consider the multiplication by φ. For convenience, we assume $n = \dim N = 0$, $\varphi(r) \in C^\infty(\overline{\mathbb{R}}_+)$; the general case is completely analogous. We apply Taylor's formula:

$$\varphi(r) = \sum_{j=0}^{N} \frac{1}{j!} r^j \frac{\partial^j \varphi}{\partial r^j}(0) + r_N(r), \quad r_N(r) = \frac{r^{N+1}}{N!} \int_0^1 (1-N)^N \frac{\partial^{N+1}\varphi}{\partial r^{N+1}}(tr)dt.$$

This yields

$$\varphi g(y,\eta) = \sum_{j=0}^{N} c_j g_j(y,\eta) + r^{N+1} g_{(N)}(y,\eta) \tag{7.16}$$

for $c_j = \frac{1}{j!}(\partial_r^j)(0)$ and $g_{(N)}(y,\eta) = \psi(r)g(y,\eta)$ with

$$\psi(r) = \frac{1}{N!}\int_0^1 (1-N)^N (\partial_r^{N+1}\varphi)(tr)dt \in C_b^\infty(\overline{\mathbb{R}}_+).$$

The operator of multiplication by ψ represents a symbol

$$\mathcal{M}_\psi \in S^0(\mathbb{R}^q; \mathcal{K}^{s,\gamma;e}(\mathbb{R}_+), \mathcal{K}^{s,\gamma;e}(\mathbb{R}_+)) \tag{7.17}$$

for every $s, \gamma, e \in \mathbb{R}$. It suffices to discuss the case $s \in \mathbb{N}$; the assertion in general follows by interpolation. For $s \in \mathbb{N}$ the space $\mathcal{K}^{s,\gamma;e}(\mathbb{R}_+)$ can be characterised by the system of conditions

$$r^l \partial_r^l (r^{-\gamma}\omega(r)u(r)), \ \partial_r^l(\langle r\rangle^{-e}(1-\omega(r))u(r)) \in L^2(\mathbb{R}_+)$$

for $l = 0, \ldots, s$, where $\omega(r)$ is a cut-off function. In order to show the symbol estimates it suffices to verify that

$$\int \left| r^l \partial_r^l (r^{-\gamma}\omega(r)(r\langle\eta\rangle^{-1})u(r))\right|^2 dr \le c\|u\|_{L^2(\mathbb{R}_+)}^2,$$

$$\int \left| \partial_r^l(\langle r\rangle^{-e}(1-\omega(r))(r\langle\eta\rangle^{-1})u(r))\right|^2 dr \le c\|u\|_{L^2(\mathbb{R}_+)}^2$$

for all $\eta \in \mathbb{R}^q$, for some constant $c > 0$, $l = 0, \ldots, s$. However, this is evident. From

$$\left\|\kappa_{\langle\eta\rangle}^{-1}\{D_y^\alpha D_\eta^\beta g(y,\eta)\}\kappa_{\langle\eta\rangle}u\right\|_{\mathcal{K}^{s',\gamma-\mu;e'}(\mathbb{R}_+)} \le c\,\langle\eta\rangle^{\nu-|\beta|}\|u\|_{\mathcal{K}^{s,\gamma;e}(\mathbb{R}_+)}$$

for every $s, s', \gamma \in \mathbb{R}$ together with (7.17) we obtain

$$\left\|\kappa_{\langle\eta\rangle}^{-1}\{D_y^\alpha D_\eta^\beta \psi(r)g_N(y,\eta)\}\kappa_{\langle\eta\rangle}u\right\|_{\mathcal{K}^{s',\gamma-\mu;e'}(\mathbb{R}_+)} \le c\,\langle\eta\rangle^{\nu-|\beta|}\|u\|_{\mathcal{K}^{s,\gamma;e}(\mathbb{R}_+)}.$$

This yields

$$\left\|\kappa_{\langle\eta\rangle}^{-1}\{D_y^\alpha D_\eta^\beta r^{N+1}\psi(r)g_N(y,\eta)\}\kappa_{\langle\eta\rangle}u\right\|_{\mathcal{K}^{s',\gamma-\mu+(N+1);e'}(\mathbb{R}_+)}$$
$$\le c\,\langle\eta\rangle^{\nu-|\beta|-(N+1)}\|u\|_{\mathcal{K}^{s,\gamma;e}(\mathbb{R}_+)},$$

and hence

$$r^{N+1}\psi(r)g_N(y,\eta) \in S^{\nu-|\beta|-(N+1)}(\Omega \times \mathbb{R}^q; \mathcal{K}^{s,\gamma;e}(\mathbb{R}_+), \mathcal{K}^{s',\gamma-\mu;e'}(\mathbb{R}_+)).$$

Thus (7.16) shows that $\varphi g(y,\eta)$ is a classical symbol in the sense of the first part of Definition 7.2.1 for the weight interval $\Theta = (-(N+1), 0]$. Since N is arbitrary, we obtain the desired property for all Θ, in particular $(-\infty, 0]$ if the original asymptotics refer to the infinite weight interval. For the adjoints we can argue in an analogous manner. $\qquad\square$

The term Green symbol is motivated by Green functions in boundary value problems. For similar reasons Green symbols play a crucial role also in the edge algebra below. In the framework of the edge calculus we will encounter many specific examples of Green symbols, also with a holomorphic dependence on covariables.

As classical operator-valued symbols with twisted symbol estimates the Green symbols have a sequence of homogeneous components

$$g_{(\nu-j)}(y,\eta) \in S^{(\nu-j)}\big(\Omega \times (\mathbb{R}^q \setminus \{0\}); \mathcal{K}^{s,\gamma;e}(N^\wedge), \mathcal{K}^{\infty,\gamma-\mu;\infty}(N^\wedge)\big), \quad j \in \mathbb{N},$$

cf. the notation in Definition 6.5.2.

Example 7.2.7. Examples of Green symbols are provided by families of operators

$$g(y,\eta)u(r,x) = \iint_{\mathbb{R}_+} c(r[\eta], x, r'[\eta], x', y, \eta) u(r', x')(r')^n \, dr' dx'$$

for $c(\widetilde{r}, x, \widetilde{r}', x', y, \eta) = b(y,\eta)\omega(\widetilde{r})\widetilde{r}^{-p} c(x,x') \log \widetilde{r}^k (\widetilde{r}')^{-p'} \log \widetilde{r}'^{k'} \omega'(\widetilde{r}')$, with $p, p' \in \mathbb{C}$, $\operatorname{Re} p < (n+1)/2 - (\gamma - \mu)$, $\operatorname{Re} p' < (n+1)/2 + \gamma$, $k, k' \in \mathbb{N}$, $c(x,x') \in C^\infty(N \times N)$, $b(y,\eta) \in S^{\nu+n+1}_{\mathrm{cl}}(\Omega \times \mathbb{R}^q)$.

Denote for the moment by $\mathcal{R}^\nu_{\mathrm{G}}(\Omega \times \mathbb{R}^q, \boldsymbol{g})$ the space of top left corners of Green symbols for trivial bundles of fibre dimension 1. Then the operator-valued symbols as in Definition 7.2.1 are considered with respect to the group action

$$(\kappa_\lambda u)(r,x) := \lambda^{(n+1)/2} u(\lambda r, x), \quad \lambda \in \mathbb{R}_+, \quad \text{for } n = \dim N. \tag{7.18}$$

In an analogous manner we can also define Green symbols based on id, the trivial group action in the involved spaces.

Let $\mathcal{R}^\nu_{\mathrm{G}}(\Omega \times \mathbb{R}^q, \boldsymbol{g})_{\mathrm{id,id}}$ denote the corresponding space.

Remark 7.2.8. The map

$$g_1(y,\eta) \mapsto \kappa_{[\eta]} g_1(y,\eta) \kappa_{[\eta]}^{-1}$$

defines an isomorphism

$$\mathcal{R}^\nu_{\mathrm{G}}(\mathbb{R}^q \times \mathbb{R}^q, \boldsymbol{g})_{\mathrm{id,id}} \to \mathcal{R}^\nu_{\mathrm{G}}(\mathbb{R}^q \times \mathbb{R}^q, \boldsymbol{g}).$$

For a Green symbol g in the notation of Definition 7.2.1 we form the (twisted) homogeneous principal symbol of order μ

$$\sigma_\wedge(g)(y,\eta): \quad \begin{matrix} \mathcal{K}^{s,\gamma}(N^\wedge) \otimes E_y'^\wedge \\ \oplus \\ \mathbb{C}^{j_1} \end{matrix} \quad \longrightarrow \quad \begin{matrix} \mathcal{K}^{\infty,\gamma-\mu}(N^\wedge) \otimes F_y'^\wedge \\ \oplus \\ \mathbb{C}^{j_2} \end{matrix}, \tag{7.19}$$

according to the homogeneous component of order μ of the corresponding classical symbol.

Let us now turn to smoothing Mellin symbols of the edge calculus. These are obtained in terms of smoothing Mellin operators of the cone algebra. First consider a finite weight interval $\Theta = (-(k+1), 0]$, $k \in \mathbb{N}$, and start with a sequence

$$f_{j\alpha} \in C^\infty\big(\Omega, M_{\mathcal{R}_{j\alpha}}^{-\infty}(N; E', F')\big), \quad \alpha \in \mathbb{N}^q, \ |\alpha| \leq j, \ j = 0, \ldots, k, \qquad (7.20)$$

cf. (6.88). For any cut-off function $\omega(r)$ we write $\omega_\eta(r) := \omega(r[\eta])$; here $\eta \to [\eta]$ is any strictly positive function in $C^\infty(\mathbb{R}^q)$ such that $[\eta] = |\eta|$ for $|\eta| \geq c$ for some $c > 0$.

Definition 7.2.9. Let $\mathcal{R}_{\mathrm{M+G}}^\mu(\Omega \times \mathbb{R}^q, \boldsymbol{g}; E', F')$ for $\boldsymbol{g} = (\gamma, \gamma - \mu, \Theta)$ with $\Theta = (-(k+1), 0]$, $k \in \mathbb{N}$, be the space of all operator functions $m(y, \eta) + g(y, \eta)$ with $g(y, \eta) \in \mathcal{R}_{\mathrm{G}}^\mu(\Omega \times \mathbb{R}^q, \boldsymbol{g}; E', F')$, and

$$m(y, \eta) := r^{-\mu} \sum_{j=0}^k r^j \sum_{|\alpha| \leq j} \omega_\eta \mathrm{op}_M^{\gamma_{j\alpha} - n/2}(f_{j\alpha}) \eta^\alpha \omega_\eta', \qquad (7.21)$$

for cut-off functions ω, ω' and weights $\gamma_{j\alpha}$ such that

$$\gamma - j \leq \gamma_{j\alpha} \leq \gamma, \quad \Gamma_{(n+1)/2 - \gamma_{j\alpha}} \cap \pi_\mathbb{C} \mathcal{R}_{j\alpha} = \emptyset \text{ for all } j, \alpha, \qquad (7.22)$$

where $\omega_\eta(r) = \omega(r[\eta])$, $\omega_\eta'(r) = \omega'(r[\eta])$.

Remark 7.2.10. For any $m(y, \eta)$ of the form (7.21) and $l \in \mathbb{N} \setminus \{0\}$, we have $r^l m(y, \eta) \in \mathcal{R}_{\mathrm{G}}^{\mu - l}(\Omega \times \mathbb{R}^q, \boldsymbol{g}; E', F')$.

Remark 7.2.11. We have

$$m(y, \eta) \in \bigcap_{s, e \in \mathbb{R}} S_{\mathrm{cl}}^\mu\big(\Omega \times \mathbb{R}^q; \mathcal{K}^{s, \gamma; e}(N^\wedge, E'^\wedge), \mathcal{K}^{\infty, \gamma - \mu; \infty}(N^\wedge, F'^\wedge)\big)$$

and

$$m(y, \eta) \in \bigcap_{s, e \in \mathbb{R}} S_{\mathrm{cl}}^\mu\big(\Omega \times \mathbb{R}^q; \mathcal{K}_\mathcal{P}^{s, \gamma; e}(N^\wedge, E'^\wedge), \mathcal{K}_\mathcal{Q}^{\infty, \gamma - \mu; \infty}(N^\wedge, F'^\wedge)\big),$$

for every asymptotic type $\mathcal{P}$ and some resulting $\mathcal{Q}$.

Remark 7.2.12. Let $\widetilde{m}(y, \eta)$ be of analogous form as (7.21) for the same Mellin symbol (7.20), but with other cut-off functions $\widetilde{\omega}, \widetilde{\omega}'$, other $\eta \to [\widetilde{\eta}]$, and weights $\widetilde{\gamma}_{j\alpha}$ satisfying conditions like (7.22). Then

$$m(y, \eta) - \widetilde{m}(y, \eta) \in \mathcal{R}_{\mathrm{G}}^\mu(\Omega \times \mathbb{R}^q, \boldsymbol{g}; E', F')$$

(the notation on the right means the space of Green symbols of the form of top left corners).

Remark 7.2.13. Let $f \in C^\infty(\Omega, M_{\mathcal{R}}^{-\infty}(N; E', F'))$, $\mathcal{R} = \{(r_j, m_j)\}_{j \in \mathbb{J}}$, $\gamma, \delta \in \mathbb{R}$, and assume $\pi_{\mathbb{C}}\mathcal{R} \cap \Gamma_{(n+1)/2-\gamma} = \pi_{\mathbb{C}}\mathcal{R} \cap \Gamma_{(n+1)/2-\delta} = \emptyset$, and let $j \in \mathbb{N}$, $\alpha \in \mathbb{R}^q$, $|\alpha| \leq j$. Then

$$g(y, \eta) := \omega_\eta r^j \operatorname{op}_M^{\gamma - n/2}(f)(y)\eta^\alpha \omega'_\eta - \omega_\eta r^j \operatorname{op}_M^{\delta - n/2}(f)(y)\eta^\alpha \omega'_\eta$$
$$\in \mathcal{R}_G^{-j+|\alpha|}(\Omega \times \mathbb{R}^q, (\max\{\gamma, \delta\}, j + \min\{\gamma, \delta\}, (-\infty, 0])).$$

Definition 7.2.14. By $\mathcal{R}_{M+G}^\mu(\Omega \times \mathbb{R}^q, \boldsymbol{g}; \boldsymbol{w})$ for $\boldsymbol{w} = (E', F'; j_1, j_2)$ we denote the space of all operator-valued functions of the form

$$\begin{pmatrix} m(y, \eta) & 0 \\ 0 & 0 \end{pmatrix} + g(y, \eta)$$

for arbitrary $m(y, \eta)$ of the form (7.21) and $g(y, \eta) \in \mathcal{R}_G^\mu(\Omega \times \mathbb{R}^q, \boldsymbol{g}; \boldsymbol{w})$.

We set

$$\sigma_\wedge(m)(y, \eta) := r^{-\mu} \sum_{j=0}^k r^j \sum_{|\alpha|=j} \omega_{|\eta|} \operatorname{op}_M^{\gamma_{j\alpha} - n/2}(f_{j\alpha})\eta^\alpha \omega'_{|\eta|}, \qquad (7.23)$$

$\omega_{|\eta|}(r) := \omega(r|\eta|)$, etc.

7.3 Edge amplitude functions

In order to describe the structure of full edge amplitude functions, we need some auxiliary material on the Mellin quantisation of edge-degenerate operator functions.

Let

$$\widetilde{p}(r, y, \widetilde{\rho}, \widetilde{\eta}) \in C^\infty(\overline{\mathbb{R}}_+ \times \Omega, L_{\mathrm{cl}}^\mu(N; E', F'; \mathbb{R}_{\widetilde{\rho}, \widetilde{\eta}}^{1+q})), \qquad (7.24)$$

$\Omega \subseteq \mathbb{R}^q$ open, $E', F' \in \mathrm{Vect}(N)$, and form

$$p(r, y, \rho, \eta) := \widetilde{p}(r, y, r\rho, r\eta). \qquad (7.25)$$

Definition 7.3.1. The space $M_{\mathcal{O}}^\mu(N; E', F'; \mathbb{R}_\eta^q)$, $\mu \in \mathbb{R}$, is defined to be the set of all $h(z, \eta) \in \mathcal{A}(\mathbb{C}, L_{\mathrm{cl}}^\mu(N; E', F'; \mathbb{R}_\eta^q))$ such that

$$h(z, \eta)|_{\Gamma_\beta} \in L_{\mathrm{cl}}^\mu(N; E', F'; \Gamma_\beta \times \mathbb{R}^q)$$

for every $\beta \in \mathbb{R}$, uniformly in compact β-intervals.

Similarly as $M_{\mathcal{O}}^\mu(X)$, cf. Definition 6.3.8, the space $M_{\mathcal{O}}^\mu(N; E', F'; \mathbb{R}^q)$ is a (nuclear) Fréchet space.

We frequently employ the following Mellin quantisation result.

Theorem 7.3.2. *For every $p(r, y, \rho, \eta)$ of the form (7.25) with (7.24) there is an*

$$\widetilde{h}(r, y, z, \widetilde{\eta}) \in C^\infty\big(\overline{\mathbb{R}}_+ \times \Omega, M^\mu_{\mathcal{O}}(N; E', F'; \mathbb{R}^q_{\widetilde{\eta}})\big) \tag{7.26}$$

such that for

$$h(r, y, z, \eta) := \widetilde{h}(r, y, z, r\eta) \tag{7.27}$$

we have

$$\mathrm{Op}_r(p)(y, \eta) = \mathrm{op}^\beta_M(h)(y, \eta) \ \mathrm{mod} \ C^\infty\big(\Omega, L^{-\infty}(N^\wedge; E'^\wedge, F'^\wedge; \mathbb{R}^q_\eta)\big), \tag{7.28}$$

for every $\beta \in \mathbb{R}$, cf. notation (1.28).

If p and h are related to each other as in Theorem 7.3.2, we say that h is a Mellin quantisation of p. Mellin quantisations have been known for a long time, cf. [42], [46, Theorem 2.1.3]. For more details and references, see also [22], or [12, Section 3.3]. Note that Theorem 7.3.2 can also be obtained from (the vector bundle-analogue of) Theorem 7.3.3 below.

In order to further interpret the Mellin operator convention for edge-degenerate families of pseudo-differential operators, we consider for convenience the case of trivial bundles of fibre dimension 1. Choose an

$$a(r, \tau, \lambda) \in C^\infty\big(\overline{\mathbb{R}}_+, L^\mu_{\mathrm{cl}}(N; \mathbb{R}^{1+l}_{\tau, \lambda})\big).$$

For any $\varphi \in C^\infty_0(\mathbb{R}_+)$ the oscillatory integral

$$(Q(\varphi)a)(r, z, \lambda) := \int_{\mathbb{R}} \int_{\mathbb{R}} e^{i(1-t)\tau} t^{-z} \varphi(t) a(r, \tau, \lambda) dt \, d\tau, \quad z \in \Gamma_0,$$

defines an element of $C^\infty\big(\overline{\mathbb{R}}_+, L^\mu_{\mathrm{cl}}(N; \Gamma_0 \times \mathbb{R}^l_\lambda)\big)$. The following theorem tells us, in particular, that $Q(\varphi)a$ extends to a holomorphic operator-valued function in $z \in \mathbb{C}$, for convenience again denoted by $Q(\varphi)a(r, z, \lambda)$.

Theorem 7.3.3. (i) *Under the above-mentioned assumptions we have*

$$Q(\varphi)(a)(r, z, \lambda) \in C^\infty\big(\overline{\mathbb{R}}_+, M^\mu_{\mathcal{O}}(N; \mathbb{R}^l_\lambda)\big).$$

(ii) *If $\varphi \in C^\infty_0(\mathbb{R}_+)$ is equal to 1 in a neighbourhood of 1, then*

$$Q(\varphi)(a)(r, z, \lambda) := h(r, z, \lambda)$$

in $C^\infty\big(\overline{\mathbb{R}}_+, M^\mu_{\mathcal{O}}(N; \mathbb{R}^l_\lambda)\big)$ has the property that

$$\mathrm{Op}_r(a)(\lambda) - \mathrm{op}^\gamma_M(h)(\lambda) = \mathrm{Op}_r((1 - \varphi(r'/r))a)(\lambda) \in L^{-\infty}(N^\wedge; \mathbb{R}^l_\lambda)$$

for every $\gamma \in \mathbb{R}$.

References for Theorem 7.3.3 are [15, Theorem 2.3], or [46, Theorem 3.2.7].

Corollary 7.3.4. *Let φ be as in* Theorem 7.3.3 (ii). *For every*

$$a(r,\rho,\lambda) = \tilde{a}(r,r\rho,r\lambda), \quad \tilde{a}(r,\tilde{\rho},\tilde{\lambda}) \in C^{\infty}\left(\overline{\mathbb{R}}_+, L_{\mathrm{cl}}^{\mu}\left(N;\mathbb{R}_{\tilde{\rho},\tilde{\lambda}}^{1+l}\right)\right)$$

there exists an $\tilde{a}\left(r,z,\tilde{\lambda}\right)$ such that for $h(r,z,\lambda) := \tilde{h}(r,z,r\lambda)$ and for every $\gamma \in \mathbb{R}$ we have

$$\mathrm{Op}_r(a)(\lambda) - \mathrm{op}_M^{\gamma}(h)(\lambda) = \mathrm{Op}((1-\varphi(r'/r))a)(\lambda) \in L^{-\infty}\left(N^{\wedge};\mathbb{R}_{\lambda}^{l}\right).$$

Observe that when we set

$$p_0(r,y,\rho,\eta) := \tilde{p}(0,y,r\rho,r\eta), \quad h_0(r,y,z,\eta) := \tilde{h}(0,y,z,r\eta), \tag{7.29}$$

also h_0 is a Mellin quantisation of p_0.

Let us fix cut-off functions $\omega'' \prec \omega \prec \omega'$, set $\chi := 1 - \omega$, $\chi' := 1 - \omega''$, write $\omega_\eta(r) := \omega(r[\eta])$, etc., for some strictly positive function $\eta \mapsto [\eta]$ in $C^{\infty}(\mathbb{R}^q)$ such that $[\eta] = |\eta|$ for $|\eta| > C$ for some $C > 0$. The following observation is a consequence of pseudo-locality:

Remark 7.3.5. We have

$$r^{-\mu}\mathrm{Op}_r(p)(y,\eta) = r^{-\mu}\omega_\eta\mathrm{Op}_r(p)(y,\eta)\omega'_\eta + r^{-\mu}\chi_\eta\mathrm{Op}_r(p)(y,\eta)\chi'_\eta + c(y,\eta), \tag{7.30}$$

where $c(y,\eta) \in C^{\infty}\left(\Omega, L^{-\infty}\left(N^{\wedge};E'^{\wedge},F'^{\wedge};\mathbb{R}_\eta^q\right)\right)$.

Thus Theorem 7.3.2 allows us to write

$$r^{-\mu}\mathrm{Op}_r(p)(y,\eta) = r^{-\mu}\omega_\eta\,\mathrm{op}_M^{\gamma-n/2}(h)(y,\eta)\omega'_\eta + r^{-\mu}\chi_\eta\mathrm{Op}_r(p)(y,\eta)\chi'_\eta \tag{7.31}$$

mod $C^{\infty}\left(\Omega, L^{-\infty}\left(N^{\wedge};E'^{\wedge},F'^{\wedge};\mathbb{R}^q\right)\right)$. Now let us choose cut-off functions

$$\epsilon'' \prec \epsilon \prec \epsilon'.$$

Similarly as (7.31) we can decompose once again the left-hand side of (7.31) as

$$r^{-\mu}\mathrm{Op}_r(p)(y,\eta) = r^{-\mu}\epsilon\,\mathrm{Op}_r(p)(y,\eta)\epsilon' + r^{-\mu}(1-\epsilon)\mathrm{Op}_r(p)(y,\eta)(1-\epsilon'')$$

mod $C^{\infty}\left(\Omega, L^{-\infty}\left(N^{\wedge};E'^{\wedge},F'^{\wedge};\mathbb{R}^q\right)\right)$. Combined with (7.30) we obtain

$$r^{-\mu}\mathrm{Op}_r(p)(y,\eta) = w(y,\eta) + b(y,\eta) + c(y,\eta) \tag{7.32}$$

for

$$w(y,\eta) := r^{-\mu}\epsilon\{\omega_\eta\,\mathrm{op}_M^{\gamma-n/2}(h)(y,\eta)\omega'_\eta + \chi_\eta\mathrm{Op}_r(p)(y,\eta)\chi'_\eta\}\epsilon', \tag{7.33}$$

$$b(y,\eta) := r^{-\mu}(1-\epsilon)\,\mathrm{Op}_r(p)(y,\eta)(1-\epsilon''), \tag{7.34}$$

and $c(y,\eta) \in C^{\infty}\left(\Omega, L^{-\infty}\left(N^{\wedge};E'^{\wedge},F'^{\wedge};\mathbb{R}^q\right)\right)$.

Remark 7.3.6. The operator function $b(y, \eta)$ can be written in the form

$$b(y, \eta) = \mathrm{Op}_r(p_{\mathrm{int}})(y, \eta) \bmod C^\infty(\Omega, L^{-\infty}(N^\wedge; E'^\wedge, F'^\wedge; \mathbb{R}^q)) \qquad (7.35)$$

for some $p_{\mathrm{int}}(r, y, \rho, \eta) \in C^\infty(\mathbb{R}_+ \times \Omega, L^\mu_{\mathrm{cl}}(N; E', F'; \mathbb{R}^{1+q}_{\rho, \eta}))$.

Remark 7.3.7. For (7.33) we have

$$w(y, \eta) \in S^\mu(\Omega \times \mathbb{R}^q; \mathcal{K}^{s,\gamma}(N^\wedge, E'^\wedge), \mathcal{K}^{s-\mu,\gamma-\mu}(N^\wedge, F'^\wedge)), \qquad (7.36)$$

$$w(y, \eta) \in S^\mu(\Omega \times \mathbb{R}^q; \mathcal{K}^{s,\gamma}_\mathcal{P}(N^\wedge, E'^\wedge), \mathcal{K}^{s-\mu,\gamma-\mu}_\mathcal{Q}(N^\wedge, F'^\wedge)), \qquad (7.37)$$

for every asymptotic type $\mathcal{P}$ and some resulting $\mathcal{Q}$, associated with the respective weight data. Moreover, if $b(y, \eta)$ is defined by (7.35) it follows that

$$\varphi b(y, \eta) \varphi' \in S^\mu(\Omega \times \mathbb{R}^q; \mathcal{K}^{s,\gamma}(N^\wedge, E'^\wedge), \mathcal{K}^{s-\mu,\gamma-\mu}(N^\wedge, F'^\wedge)) \qquad (7.38)$$

for every $\varphi, \varphi' \in C_0^\infty(\mathbb{R}_+)$.

The way to construct from (7.25), (7.24) close to $r = 0$ an operator function $a(y, \eta)$ of the form (7.33) which has the property (7.36) is interpreted as a quantisation (also referred to as edge quantisation) of the edge-degenerate operator family (7.25). The summand $b(y, \eta)$ in (7.32) is localised off $r = 0$ and only contributes to the interior part of the operators.

Definition 7.3.8. The space $\mathcal{R}^\mu(\Omega \times \mathbb{R}^q, \boldsymbol{g}; E', F')$ for $\boldsymbol{g} = (\gamma, \gamma - \mu, \Theta)$, $\Theta = (-(k+1), 0]$, $k \in \mathbb{N} \cup \{\infty\}$, and $E', F' \in \mathrm{Vect}(N)$ is defined to be the set of all operator functions of the form

$$
\begin{aligned}
a(y, \eta) = {}& \epsilon r^{-\mu}\{\omega_\eta \, \mathrm{op}_M^{\gamma-n/2}(h)(y, \eta)\omega'_\eta + \chi_\eta \mathrm{Op}_r(p)(y, \eta)\chi'_\eta\}\epsilon' \\
& + \varphi \, \mathrm{Op}_r(p_{\mathrm{int}})(y, \eta)\varphi' + (m+g)(y, \eta)
\end{aligned}
\qquad (7.39)
$$

for arbitrary p and h, where h is a Mellin quantisation of p, and cut-off functions as in (7.33) more precisely, arbitrary $\epsilon \prec \epsilon'$, $\omega'' \prec \omega \prec \omega'$, $\chi = 1 - \omega$, $\chi' = 1 - \omega''$, moreover, $(m+g)(y, \eta) \in R^\mu_{\mathrm{M}+\mathrm{G}}(\Omega \times \mathbb{R}^q, \boldsymbol{g}; E', F')$, and

$$p_{\mathrm{int}}(r, y, \rho, \eta) \in C^\infty(\mathbb{R}_+ \times \Omega, L^\mu_{\mathrm{cl}}(N; \mathbb{R}^{1+q}_{\rho, \eta}; E', F')), \quad \varphi \prec \varphi' \in C_0^\infty(\mathbb{R}_+).$$

Remark 7.3.9. The space $\mathcal{R}^\mu(\Omega \times \mathbb{R}^q, \boldsymbol{g}; E', F')$ can be equivalently defined as the set of all operator functions (7.39) for arbitrary data as in Definition 7.3.8, but for $p_{\mathrm{int}} \equiv 0$. In fact, the term $\varphi \, \mathrm{Op}_r(p_{\mathrm{int}})(y, \eta)\varphi'$ can always be integrated under $\epsilon r^{-\mu}\{\cdots\}\epsilon'$ by a suitable choice of ϵ, ϵ' and modified h, p with the indicated properties.

Remark 7.3.10. An alternative to the edge quantisation $r^{-\mu}\mathrm{Op}_r(p)(y, \eta) \to w(y, \eta)$ is to set

$$v(y, \eta) := r^{-\mu}\epsilon \, \mathrm{op}_M^{\gamma-n/2}(h)(y, \eta)\epsilon'. \qquad (7.40)$$

It can be proved, cf. [16], that

$$g_{\mathrm{rem}}(y, \eta) := w(y, \eta) - v(y, \eta) \in \mathcal{R}^\mu_\mathrm{G}(\Omega \times \mathbb{R}^q; E', F')_{O,O},$$

cf. formula (7.12).

Definition 7.3.11. Let $\mathcal{R}^\mu(\Omega \times \mathbb{R}^q; \boldsymbol{g}; \boldsymbol{w})$, $\boldsymbol{g} = (\gamma, \gamma - \mu, \Theta)$, $\boldsymbol{w} = (E', F'; j_1, j_2)$, denote the set of all operator functions

$$a(y, \eta) = \begin{pmatrix} w(y, \eta) + m(y, \eta) & 0 \\ 0 & 0 \end{pmatrix} + g(y, \eta) \tag{7.41}$$

with arbitrary $w(y, \eta)$ like (7.33), $m(y, \eta)$ of the form (7.21), and $g(y, \eta) \in \mathcal{R}^\mu_G(\Omega \times \mathbb{R}^q; \boldsymbol{g}; \boldsymbol{w})$.

Remark 7.3.12. Below we will employ the notation $\mathcal{R}^\mu(\Omega \times \mathbb{R}^q, \boldsymbol{g}; \boldsymbol{w})$ also for more general spaces of operator functions. When we write $\boldsymbol{g} = (\gamma, \gamma - \mu)$ rather than $\boldsymbol{g} = (\gamma, \gamma - \mu, \Theta)$ we mean the set of all operator functions (7.41) where $w(y, \eta)$ is as before, but

$$m(y, \eta) := r^{-\mu} \omega_\eta \, \mathrm{op}_M^{\gamma - n/2}(f) \omega'_\eta$$

for some $f \in C^\infty(\Omega, M_{\mathcal{R}}^{-\infty}(N; E', F'))$, $\Gamma_{(n+1)/2-\gamma} \cap \pi_\mathbb{C} \mathcal{R} = \emptyset$, certain ω, ω', and $g \in \mathcal{R}^\mu_G(\Omega \times \mathbb{R}^q; \boldsymbol{g}, \boldsymbol{w})$ for $\boldsymbol{g} = (\gamma, \gamma - \mu)$, as explained after Definition 7.2.1.

For a function $a(y, \eta) \in \mathcal{R}^\mu(\Omega \times \mathbb{R}^q, \boldsymbol{g}; \boldsymbol{w})$ written in the form (7.41) we set

$$\sigma_\wedge(a)(y, \eta) = \begin{pmatrix} \sigma_\wedge(w + m)(y, \eta) & 0 \\ 0 & 0 \end{pmatrix} + \sigma_\wedge(g)(y, \eta), \tag{7.42}$$

where $\sigma_\wedge(g)(y, \eta)$ is given by (7.19), $\sigma_\wedge(m)(y, \eta)$ by (7.23) for $\boldsymbol{g} = (\gamma, \gamma - \mu, \Theta)$, and

$$\sigma_\wedge(w)(y, \eta) := r^{-\mu} \omega_{|\eta|} \, \mathrm{op}_M^{\gamma - n/2}(h_0)(y, \eta) \omega'_{|\eta|} + r^{-\mu} \chi_{|\eta|} \, \mathrm{Op}_r(p_0)(y, \eta) \chi'_{|\eta|}. \tag{7.43}$$

If we represent our element $a(y, \eta) \in \mathcal{R}^\mu(\Omega \times \mathbb{R}^q, \boldsymbol{g}; \boldsymbol{w})$ in the top left corner by (7.40) rather than $w(y, \eta)$, we replace $\sigma_\wedge(w)(y, \eta)$ in the top left corner of (7.42) equivalently by

$$\sigma_\wedge(w)(y, \eta) = \sigma_\wedge(v)(y, \eta) + \sigma_\wedge(g_{\mathrm{rem}})(y, \eta), \tag{7.44}$$

where

$$\sigma_\wedge(v)(y, \eta) = r^{-\mu} \, \mathrm{op}_M^{\gamma - n/2}(h_0)(y, \eta), \tag{7.45}$$

cf. Remark 7.3.10.

With elements $a(y, \eta) \in \mathcal{R}^\mu(\Omega \times \mathbb{R}^q, \boldsymbol{g}; \boldsymbol{w})$ we also associate the homogeneous principal interior symbol $\sigma_\psi(a)$ of order μ. This only concerns the family $w(y, \eta)$ in the upper left corner of (7.41). The operator function $\widetilde{p}(r, y, \widetilde{\rho}, \widetilde{\eta})$ in (7.24) has a parameter-dependent homogeneous principal symbol

$$\widetilde{p}_{(\mu)}(r, x, y, \widetilde{\rho}, \xi, \widetilde{\eta})$$

in $(\widetilde{\rho}, \xi, \widetilde{\eta}) \neq 0$. Then we define

$$\sigma_\psi(a)(r, x, y, \rho, \xi, \eta) \sigma_\psi(w)(r, x, y, \rho, \xi, \eta) := \epsilon(r) \widetilde{p}_{(\mu)}(r, x, y, r\rho, \xi, r\eta). \tag{7.46}$$

In addition, we introduce the reduced symbol

$$\widetilde{\sigma}_\psi(a)(r, x, y, \rho, \xi, \eta) \, \widetilde{\sigma}_\psi(w)(r, x, y, \rho, \xi, \eta) := r^\mu \sigma_\psi(w)(r, x, y, r^{-1}\rho, \xi, r^{-1}\eta). \tag{7.47}$$

7.4 The edge calculus

Let M be a compact manifold with edge Y, let E, $F \in \mathrm{Vect}(\mathbb{M})$, J_1, $J_2 \in \mathrm{Vect}(Y)$, and fix weight data $\boldsymbol{g}$, either as $\boldsymbol{g} = (\gamma, \gamma - \mu, \Theta)$ or $\boldsymbol{g} = (\gamma, \gamma - \mu)$, which yields different variants of edge operators.

Before we give a definition of the edge algebra, we have a look at general pseudo-differential operators globally on $M \setminus Y$, acting between distributional sections of E, F. Let $(G_j)_{j=1,\ldots,N}$ be an open covering of Y by coordinate neighbourhoods, $(\varphi_j)_{j=1,\ldots,N}$ a subordinate partition of unity, and $(\varphi_j')_{j=1,\ldots,N}$ a system of functions $\varphi_j' \in C_0^\infty(G_j)$, $\varphi_j' \prec \varphi_j$ for all j. Further, let $\theta \prec \theta'$ be cut-off functions on the half-line, and choose $\psi \prec \psi'$ in $C_0^\infty(M \setminus Y)$ such that $\mathrm{dist}(\mathrm{supp}\,\psi, Y)$ is small enough. Then every $A \in L_{\mathrm{cl}}^\mu(M \setminus Y; E, F)$ can be written in the form

$$A = \sum_{j=1}^N \theta \varphi_j B_j \theta' \varphi_j' + \psi A_{\mathrm{int}} \psi' + C$$

with the following ingredients: $C \in L^{-\infty}(M \setminus Y; E, F)$, $A_{\mathrm{int}} \in L_{\mathrm{cl}}^\mu(M \setminus Y; E, F)$, cf. Remark 1.1.11, and $B_j = \left((\chi_{G_j}^\wedge)^{-1}\right)_* \mathrm{Op}_{r,y}(b_j)$, cf. the notation (7.2),

$$b_j(r, y, \rho, \eta) \in C^\infty\left(\mathbb{R}_+ \times \mathbb{R}^q, L_{\mathrm{cl}}^\mu\left(N; E', F'; \mathbb{R}_{\rho,\eta}^{1+q}\right)\right),$$

where the operator push-forwards also take into account the transition maps of the bundles E, F close to Y.

Let $L^{-\infty}(M, \boldsymbol{g}; \boldsymbol{v})$ for $\boldsymbol{v} := (E, F; J_1, J_2)$ and $\boldsymbol{g} := (\gamma, \gamma - \mu, \Theta)$ be defined as the set of all continuous operators

$$\mathcal{C}: \quad \begin{matrix} H^{s,\gamma}(M, E) \\ \oplus \\ H^s(Y, J_1) \end{matrix} \quad \longrightarrow \quad \begin{matrix} H_{\mathcal{P}}^{\infty, \gamma - \mu}(M, F) \\ \oplus \\ H^\infty(Y, J_2) \end{matrix} \tag{7.48}$$

such that

$$\mathcal{C}^*: \quad \begin{matrix} H^{s,-\gamma+\mu}(M, F) \\ \oplus \\ H^s(Y, J_2) \end{matrix} \quad \longrightarrow \quad \begin{matrix} H_{\mathcal{Q}}^{\infty, -\gamma}(M, E) \\ \oplus \\ H^\infty(Y, J_1) \end{matrix} \tag{7.49}$$

for all $s \in \mathbb{R}$, for certain asymptotic types $\mathcal{P}$ and $\mathcal{Q}$, associated with the weight data $(\gamma - \mu, \Theta)$ and $(-\gamma, \Theta)$, respectively. In the case $\boldsymbol{g} := (\gamma, \gamma - \mu)$ we denote by $L^{-\infty}(M, \boldsymbol{g}; \boldsymbol{v})$ the set of all continuous operators

$$\mathcal{C}: \quad \begin{matrix} H^{s,\gamma}(M, E) \\ \oplus \\ H^s(Y, J_1) \end{matrix} \quad \longrightarrow \quad \begin{matrix} H^{\infty, \gamma - \mu + \varepsilon}(M, F) \\ \oplus \\ H^\infty(Y, J_2) \end{matrix} \tag{7.50}$$

such that

$$\mathcal{C}^*: \quad \begin{matrix} H^{s,-\gamma+\mu}(M, F) \\ \oplus \\ H^s(Y, J_2) \end{matrix} \quad \longrightarrow \quad \begin{matrix} H^{\infty, -\gamma + \varepsilon}(M, E) \\ \oplus \\ H^\infty(Y, J_1) \end{matrix} \tag{7.51}$$

for all $s \in \mathbb{R}$, for some $\varepsilon(\mathcal{C}) > 0$.

Moreover, let $g(y, \eta) \in \mathcal{R}_G^\mu(\mathbb{R}^q \times \mathbb{R}^q, \boldsymbol{g}; \boldsymbol{w})$ for $\boldsymbol{w} = (E', F'; j_1, j_2)$ be a Green symbol. For the global operators below it suffices to assume that $g(y, \eta)$ is independent of y for large $|y|$. Then we obtain a continuous operator

$$
\text{Op}_y(g): \quad
\begin{matrix}
\mathcal{W}^s\big(\mathbb{R}^q, \mathcal{K}^{s,\gamma}(N^\wedge, E'^\wedge)\big) \\
\oplus \\
H^s(\mathbb{R}^q, \mathbb{C}^{j_1})
\end{matrix}
\quad \longrightarrow \quad
\begin{matrix}
\mathcal{W}^s\big(\mathbb{R}^q, \mathcal{K}_{\mathcal{P}}^{\infty,\gamma-\mu}(N^\wedge, F'^\wedge)\big) \\
\oplus \\
H^{s-\mu}(\mathbb{R}^q, \mathbb{C}^{j_2})
\end{matrix}
$$

for every $s \in \mathbb{R}$ and an asymptotic type $\mathcal{P}$ which is determined by the involved Green symbol $g(y, \eta)$. For charts $\chi_{G_i}: G_i \to \mathbb{R}^q$ on the edge Y and

$$
\chi_{G_i}^\wedge: (V \setminus Y)\big|_{G_i} \to N^\wedge \times \mathbb{R}^q,
$$

cf. (7.2), using bundle pull-backs $E^\wedge := p^* E'^\wedge$, $F^\wedge := p^* F'^\wedge$ under $p: N^\wedge \times \mathbb{R}^q \to N^\wedge$, and natural identifications

$$
(\chi_{G_i}^\wedge)^*(E^\wedge) = E\big|_{(V\setminus Y)|_{G_i}}, \quad (\chi_{G_i}^\wedge)^*(F^\wedge) = F\big|_{(V\setminus Y)|_{G_i}}, \quad \chi_{G_i}^*(\mathbb{C}^{j_l}) = J_l\big|_{G_i}, \quad l = 1, 2,
$$

we introduce the operators

$$
\mathcal{G}_i := \begin{pmatrix} \theta\varphi_i & 0 \\ 0 & \varphi_i \end{pmatrix} \begin{pmatrix} (\chi_{G_i})_*^{-1} & 0 \\ 0 & (\chi_{G_i})_*^{-1} \end{pmatrix} \text{Op}_y(g_i) \begin{pmatrix} \theta'\varphi_i' & 0 \\ 0 & \varphi_i' \end{pmatrix}, \tag{7.52}
$$

$$
\mathcal{G}_i := \quad
\begin{matrix}
H^{s,\gamma}(M, E) \\
\oplus \\
H^s(Y, J_1)
\end{matrix}
\quad \longrightarrow \quad
\begin{matrix}
H_{\mathcal{P}_i}^{s-\mu,\gamma-\mu}(M, F) \\
\oplus \\
H^{s-\mu}(Y, J_2)
\end{matrix}
\quad , \quad i = 1, \dots, N. \tag{7.53}
$$

We define $L_G^\mu(M, \boldsymbol{g}; \boldsymbol{v})$, $\boldsymbol{v} := (E, F; J_1, J_2)$, as the set of all operators of the form

$$
\mathcal{G} := \sum_{j=1}^N \mathcal{G}_i + \mathcal{C} \tag{7.54}
$$

for arbitrary $\mathcal{G}_i$ of the form (7.52) and $\mathcal{C} \in L^{-\infty}(M, \boldsymbol{g})$.

The elements of $L_G^\mu(M, \boldsymbol{g}; \boldsymbol{v})$ are called Green operators of order μ of the edge calculus on M. From the local (twisted) homogeneous principal symbols of the Green amplitude functions we get an invariantly defined homogeneous principal symbol

$$
\sigma_\wedge(\mathcal{G})(y, \eta): \quad
\begin{matrix}
\mathcal{K}^{s,\gamma}(N^\wedge) \otimes E_y'^\wedge \\
\oplus \\
J_{1,y}
\end{matrix}
\quad \longrightarrow \quad
\begin{matrix}
\mathcal{K}^{\infty,\gamma-\mu}(N^\wedge) \otimes F_y'^\wedge \\
\oplus \\
J_{2,y}
\end{matrix}
\quad , \quad (y, \eta) \in T^*Y \setminus 0,
$$

for a Green operator $\mathcal{G} \in L_G^\mu(M, \boldsymbol{g}; \boldsymbol{v})$. Homogeneity means that

$$
\sigma_\wedge(\mathcal{G})(y, \lambda\eta) = \lambda^\mu \begin{pmatrix} \kappa_\lambda & 0 \\ 0 & 1 \end{pmatrix} \sigma_\wedge(\mathcal{G})(y, \eta) \begin{pmatrix} \kappa_\lambda^{-1} & 0 \\ 0 & 1 \end{pmatrix}
$$

for every $(y, \eta) \in T^*Y \setminus 0$, $\lambda \in \mathbb{R}_+$.

Definition 7.4.1. Let M be a compact manifold with edge Y. The space $L^\mu(M, \boldsymbol{g}; \boldsymbol{v})$ for $\boldsymbol{v} := (E, F; J_1, J_2)$ is defined to be the set of all operators

$$\mathcal{A} = \begin{pmatrix} A & 0 \\ 0 & 0 \end{pmatrix} + \mathcal{G} \tag{7.55}$$

with $\mathcal{G} \in L^\mu_{\mathrm{G}}(M, \boldsymbol{g}; \boldsymbol{v})$ and

$$A = \sum_{j=1}^{N} \theta \varphi_j \left(\chi_{G_j}\right)_*^{-1} \mathrm{Op}_y(a_j) \theta' \varphi_j' + \psi A_{\mathrm{int}} \psi' \tag{7.56}$$

for $A_{\mathrm{int}} \in L^\mu_{\mathrm{cl}}(M \setminus Y; E, F)$, $\psi, \psi' \in C_0^\infty(M \setminus Y)$, $a_j \in \mathcal{R}^\mu(\mathbb{R}^q \times \mathbb{R}^q, \boldsymbol{g}; E', F')$.

Remark 7.4.2. There is a straightforward extension of Definition 7.4.1 to the case of a non-compact manifold M with edge Y, under some standard assumptions, paracompact, etc. Instead of a finite sum (7.56) we then have a countable sum, referring to a locally finite system of interior and singular charts, indicated at the beginning of this section. Instead of (weighted) Sobolev spaces and subspaces with asymptotics we then have comp- and loc-versions, similarly as standard comp- and loc-Sobolev spaces. In particular, the smoothing operators of the class

$$L^{-\infty}(M, \boldsymbol{g}; \boldsymbol{v}) = L^{-\infty}_{\mathrm{G}}(M, \boldsymbol{g}; \boldsymbol{v})$$

are defined in terms of analogues of the above-mentioned mapping properties, here between corresponding comp- and loc-spaces. In this context, the analogues of (7.56), as well as those involved in $L^\mu_{\mathrm{G}}(M, \boldsymbol{g}; \boldsymbol{v})$, are locally finite sums, and they may be taken as properly supported representatives in a similar meaning as properly supported elements of standard pseudo-differential operators on a smooth non-compact manifold. We say that the operator (7.55) is properly supported if $\mathcal{C}$ in (7.54) vanishes and if A_{int} is properly supported in the standard sense.

Theorem 7.4.3. *An $\mathcal{A} \in L^\mu(M, \boldsymbol{g}; \boldsymbol{v})$ with compact M and $\boldsymbol{g} := (\gamma, \gamma - \mu, \Theta)$, $\boldsymbol{v} := (E, F; J_1, J_2)$ induces continuous operators*

$$\mathcal{A} : \begin{array}{c} H^{s,\gamma}(M, E) \\ \oplus \\ H^s(Y, J_1) \end{array} \longrightarrow \begin{array}{c} H^{s-\mu,\gamma-\mu}(M, F) \\ \oplus \\ H^{s-\mu}(Y, J_2) \end{array} \tag{7.57}$$

and

$$\mathcal{A} : \begin{array}{c} H^{s,\gamma}_{\mathcal{P}}(M, E) \\ \oplus \\ H^s(Y, J_1) \end{array} \longrightarrow \begin{array}{c} H^{s-\mu,\gamma-\mu}_{\mathcal{Q}}(M, F) \\ \oplus \\ H^{s-\mu}(Y, J_2) \end{array} \tag{7.58}$$

for $s \in \mathbb{R}$ and every asymptotic type $\mathcal{P}$ for some resulting $\mathcal{Q}$. In the case $\boldsymbol{g} := (\gamma, \gamma - \mu)$ we have (7.57), $s \in \mathbb{R}$.

Proof. The claimed continuity of $\mathcal{A} \in L^{-\infty}(M, \boldsymbol{g}; \boldsymbol{v})$ holds by definition, the one of Green operators by (7.53). Moreover, we have $\psi A_{\mathrm{int}} \psi' \in L^{\mu}_{\mathrm{cl}}(M \setminus Y)$, and these summands are continuous as operators $H^{s}_{\mathrm{loc}}(M \setminus Y) \to H^{s-\mu}_{\mathrm{comp}}(M \setminus Y)$, cf. (7.8). It remains to consider the summands $\theta\varphi_j\left(\chi_{G_j}^{-1}\right)_* \mathrm{Op}_y(a_j)\theta'\varphi'_j$ contained in (7.56). It suffices to note that corresponding continuity properties of $\mathrm{Op}_y(a_j)$ hold between local wedge spaces over $\mathbb{R}^q$. Those are a consequence of Theorem 1.3.34 and Remarks 7.2.11, 7.3.7. $\qquad\square$

Every $\mathcal{A} \in L^{\mu}(M, \boldsymbol{g}; \boldsymbol{v})$ has a pair of principal symbols, namely,

$$\sigma(\mathcal{A}) = \big(\sigma_\psi(\mathcal{A}), \sigma_\wedge(\mathcal{A})\big)$$

with $\sigma_\psi(\mathcal{A}) := \sigma_\psi(A)$ being the standard homogeneous principal symbol of $A \in L^{\mu}_{\mathrm{cl}}(M \setminus Y; E, F)$ of order μ,

$$\sigma_\psi(\mathcal{A}) : \pi^*_{M\setminus Y} E \to \pi^*_{M\setminus Y} F, \tag{7.59}$$

$\pi_{M\setminus Y} : T^*(M\setminus Y)\setminus 0 \to M\setminus Y$. The homogeneous principal edge symbol represents a family of continuous operators

$$\sigma_\wedge(\mathcal{A})(y,\eta) : \quad \begin{array}{c} \mathcal{K}^{s,\gamma}(N^\wedge) \otimes E'^\wedge_y \\ \oplus \\ J_{1,y} \end{array} \quad \longrightarrow \quad \begin{array}{c} \mathcal{K}^{s-\mu,\gamma-\mu}(N^\wedge) \otimes F'^\wedge_y \\ \oplus \\ J_{2,y} \end{array}, \quad (y,\eta) \in T^*Y \setminus 0.$$

This follows from the definition of the various amplitude functions involved in $\sigma_\wedge(\cdot)$.

Let us set for the moment $\sigma(\mathcal{A}) =: \sigma^{\mu}(\mathcal{A}) = \big(\sigma^{\mu}_0(\mathcal{A}), \sigma^{\mu}_1(\mathcal{A})\big)$, and form

$$L^{\mu-1}(M, \boldsymbol{g}; \boldsymbol{v}) := \big\{\mathcal{A} \in L^{\mu}(M, \boldsymbol{g}; \boldsymbol{v}) : \sigma^{\mu}(\mathcal{A}) = 0\big\}. \tag{7.60}$$

Every $\mathcal{A} \in L^{\mu-1}(M, \boldsymbol{g}; \boldsymbol{v})$ has again a pair of principal symbols of order $\mu - 1$, now denoted by $\sigma^{\mu-1}(\mathcal{A}) = (\sigma^{\mu-1}_0(\mathcal{A}), \sigma^{\mu-1}_1(\mathcal{A}))$. This gives a subspace

$$L^{\mu-2}(M, \boldsymbol{g}; \boldsymbol{v}) := \big\{\mathcal{A} \in L^{\mu-1}(M, \boldsymbol{g}; \boldsymbol{v}) : \sigma^{\mu-1}(\mathcal{A}) = 0\big\}, \tag{7.61}$$

and then, successively,

$$L^{\mu-j}(M, \boldsymbol{g}; \boldsymbol{v}) := \big\{\mathcal{A} \in L^{\mu-(j-1)}(M, \boldsymbol{g}; \boldsymbol{v}) : \sigma^{\mu-(j-1)}(\mathcal{A}) = 0\big\}, \tag{7.62}$$

for every $j \in \mathbb{N}$. We then have

$$L^{-\infty}(M, \boldsymbol{g}; \boldsymbol{v}) := \bigcap_{j\in\mathbb{N}} L^{\mu-j}(M, \boldsymbol{g}; \boldsymbol{v}). \tag{7.63}$$

Theorem 7.4.4. *Every* $\mathcal{A} \in L^{\mu-1}(M, \boldsymbol{g}; \boldsymbol{v})$ *for compact M is compact as an operator* (7.57) *for every $s \in \mathbb{R}$.*

Proof. Vanishing of $\sigma^\mu(\mathcal{A})$ means that the interior order of the top left corner is equal to $\mu - 1$. In addition, close to the edge the holomorphic Mellin symbol of order $\mu-1$ has a weight factor $r^{-\mu+1}$, i.e., it improves weights by 1. The smoothing Mellin symbol together with the Green symbol has twisted order $\leq \mu-1$ and maps to spaces $\mathcal{K}^{s-(\mu-1),\gamma-\mu+\varepsilon}(N^\wedge) \otimes F_y^{\prime\wedge} \oplus J_{2,y}$, for some $\varepsilon > 0$. We have altogether that $\mathcal{A}$ locally close to the edge has a symbol of improved order and maps to spaces with improved weights. Such operators are necessarily compact. $\qquad\square$

Theorem 7.4.5. *Let $\mathcal{A}_j \in L^{\mu-j}(M, \boldsymbol{g}; \boldsymbol{v})$, $j \in \mathbb{N}$, be an arbitrary sequence, $\boldsymbol{g} := (\gamma, \gamma - \mu, (-(k+1), 0])$ for a finite k, or $\boldsymbol{g} := (\gamma, \gamma - \mu)$, and assume that the asymptotic types involved in the Green operators are independent of j. Then there exists an $\mathcal{A} \in L^\mu(M, \boldsymbol{g}; \boldsymbol{v})$ such that*

$$\mathcal{A} - \sum_{j=0}^{N} \mathcal{A}_j \in L^{\mu-(N+1)}(M, \boldsymbol{g}; \boldsymbol{v})$$

for every $N \in \mathbb{N}$, and $\mathcal{A}$ is unique mod $L^{-\infty}(M, \boldsymbol{g}; \boldsymbol{v})$.

Proof. The operators $\mathcal{A}_j$ can be decomposed in interior parts, i.e., far from the edge, of order $\mu - j$, and locally close to the edge into operators in $y \in \mathbb{R}^q$ with amplitude functions in $\mathcal{R}^{\mu-j}(\mathbb{R}_y^q \times \mathbb{R}_\eta^q, \boldsymbol{g}; \boldsymbol{w})$, cf. Definition 7.3.11. The interior parts can be asymptotically summed up, according to the standard pseudo-differential calculus, and the amplitude functions as symbols in $S^{\mu-j}(\mathbb{R}_y^q \times \mathbb{R}_\eta^q; H, \widetilde{H})$ for the respective Hilbert spaces H, $\widetilde{H}$ with group action can be asymptotically summed up as well. This gives rise to the claimed asymptotic summation, and uniqueness obviously holds mod $L^{-\infty}(M, \boldsymbol{g}; \boldsymbol{v})$, since remainders as well as their formal adjoints approximate the right mapping properties for $N \to \infty$. $\qquad\square$

Theorem 7.4.6. (i) $\mathcal{A} \in L^\mu(M, \boldsymbol{g}_0; \boldsymbol{v}_0)$, $\mathcal{B} \in L^\rho(M, \boldsymbol{g}_1; \boldsymbol{v}_1)$ for $\boldsymbol{g}_0 = (\gamma - \rho, \gamma - (\mu + \rho), \Theta)$, $\boldsymbol{g}_1 = (\gamma, \gamma - \rho, \Theta)$, or $\boldsymbol{g}_0 = (\gamma - \rho, \gamma - (\mu + \rho))$, $\boldsymbol{g}_1 = (\gamma, \gamma - \rho)$, *implies $\mathcal{A}\mathcal{B} \in L^{\mu+\nu}(M, \boldsymbol{g}_0 \circ \boldsymbol{g}_1; \boldsymbol{v}_0 \circ \boldsymbol{v}_1)$ (when the bundle data in the middle fit together so that $\boldsymbol{v}_0 \circ \boldsymbol{v}_1$ makes sense and when $\mathcal{A}$ or $\mathcal{B}$ is properly supported), and we have $\sigma(\mathcal{A}\mathcal{B}) = \sigma(\mathcal{A})\sigma(\mathcal{B})$ with component wise multiplication.*

(ii) *$\mathcal{A} \in L^\mu(M, \boldsymbol{g}; \boldsymbol{v})$ for $\boldsymbol{g} = (\gamma, \gamma-\mu, \Theta)$ or $\boldsymbol{g} = (\gamma, \gamma-\mu)$ and $\boldsymbol{v} = (V_1, V_2; J_1, J_2)$ implies $\mathcal{A}^* \in L^\mu(M, \boldsymbol{g}^*; \boldsymbol{v}^*)$ for $\boldsymbol{g}^{\wedge *} = (\gamma - \mu, \gamma, \theta)$ or $\boldsymbol{g}^{\wedge *} = (\gamma - \mu, \gamma)$ and $\boldsymbol{v}^* = (V_2, V_1; J_2, J_1)$, where $\mathcal{A}^*$ is the formal adjoint in the sense*

$$(u, \mathcal{A}^* v)_{H^{0,0}(M, V_1) \oplus H^0(Y, J_1)} = (\mathcal{A}u, v)_{H^{0,0}(M, V_2) \oplus H^0(Y, J_2)}$$

for all $u \in H^{\infty,\infty}(M, V_1) \oplus H^\infty(Y, J_1)$, $v \in H^{\infty,\infty}(M, V_2) \oplus H^\infty(Y, J_2)$, and we have $\sigma(\mathcal{A}^*) = \sigma(\mathcal{A})^*$ with componentwise formal adjoint.

Proof. (i) The composition of properly supported operators in L_G^μ-classes, i.e., when the smoothing summands as on the right of (7.54) are vanishing, is elementary and left to the reader. So we may focus on operators in the respective top left corners. In this case, in local representations of amplitude functions a_j as in (7.56) , $a_j \in \mathcal{R}^\mu(\mathbb{R}^q \times \mathbb{R}^q, \boldsymbol{g}; E', F')$, cf. Definition 7.3.8, we employ in the first summands on the right of (7.39) the alternative edge quantisation (7.40), which is of a relatively simple composition behaviour, cf. the article [16]. Then apart from elementary Leibnitz product effects and (y, η)-dependent analogues of arguments of the proof of Theorem 6.8.11 we can easily deduce the desired composition behaviour.

(ii) The behaviour under formal adjoints can also be analysed for the matrix-valued Green part and for the top left corner entry, separately. While the Green part is again elementary we may concentrate on the top left corner. In this case we refer to the alternative edge quantisation (7.40) and results from [16], see also Seiler's thesis [58].   □

Chapter 8

Edge-ellipticity

8.1 An edge analogue of the Atiyah–Bott obstruction

In the preceding section for operators $\mathcal{A}$ in the edge calculus we defined a pair

$$\sigma(\mathcal{A}) = (\sigma_\psi(\mathcal{A}), \sigma_\wedge(\mathcal{A})) \tag{8.1}$$

of principal symbols. Both components will take part in the notion of ellipticity. The ellipticity with respect to σ_ψ, also called interior ellipticity, is a condition on the top left corner A of the 2×2 block matrix operator $\mathcal{A}$.

Definition 8.1.1. Let M be a manifold with edge Y. Then an operator $A \in L^\mu(M, \boldsymbol{g}; E, F)$, $E, F \in \mathrm{Vect}(\mathbb{M})$, is called σ_ψ-elliptic if

(i) the standard homogeneous principal symbol

$$\sigma_\psi(A) : \pi^*_{M\setminus Y} E \to \pi^*_{M\setminus Y} F,$$

$\pi_{M\setminus Y} : T^*(M \setminus Y) \setminus 0 \to M \setminus Y$, is an isomorphism (here E and F mean the restrictions $E|_{M\setminus Y}$ and $F|_{M\setminus Y}$, respectively, $M \setminus Y \cong \mathbb{M} \setminus \partial \mathbb{M}$);

(ii) locally near $\partial \mathbb{M}$ in the splitting of variables $(r, x, y) \in \overline{\mathbb{R}}_+ \times \Sigma \times \Omega$ with covariables (ρ, ξ, η) the reduced symbol

$$\tilde{\sigma}_\psi(A)(r, x, y, \rho, \xi, \eta) := r^\mu \sigma_\psi(A)\big(r, x, y, r^{-1}\rho, \xi, r^{-1}\eta\big)$$

induces isomorphisms

$$\tilde{\sigma}_\psi(A)(r, x, y, \rho, \xi, \eta) : E_{(r,x,y)} \to F_{(r,x,y)}$$

for all $(r, x, y, \rho, \xi, \eta) \in \overline{\mathbb{R}}_+ \times \Sigma \times (\mathbb{R}^{1+n+q}_{\rho,\xi,\eta} \setminus \{0\})$ (here $\Sigma \subseteq \mathbb{R}^n$ and $\Omega \subseteq \mathbb{R}^q$ correspond to charts on N and Y, respectively, and $E_{(r,x,y)}, \ldots$, mean the fibres of $E, \ldots$, over the corresponding points in local coordinates).

© Springer Nature Switzerland AG 2018
X. Liu, B.-W. Schulze, *Boundary Value Problems with Global Projection Conditions*,
Operator Theory: Advances and Applications 265, https://doi.org/10.1007/978-3-319-70114-1_8

Recall that $\sigma_\wedge(A)(y,\eta)$ for $A = P + M + G$, with P being the non-smoothing pseudo-differential part and $M + G$ the smoothing Mellin plus Green part, defines an operator family

$$\sigma_\wedge(A)(y,\eta) : \mathcal{K}^{s,\gamma}(N^\wedge) \otimes E_y'^\wedge \to \mathcal{K}^{s-\mu,\gamma-\mu}(N^\wedge) \otimes F_y'^\wedge \qquad (8.2)$$

of the form $\sigma_\wedge(A) = \sigma_\wedge(P) + \sigma_\wedge(M) + \sigma_\wedge(G)$, where

$$\sigma_\wedge(P)(y,\eta) = r^{-\mu}\{\omega_{|\eta|}\operatorname{op}_M^{\gamma-n/2}(h_0)(y,\eta)\omega'_{|\eta|} + \chi_{|\eta|}\operatorname{Op}_r(p_0)(y,\eta)\chi'_{|\eta|}\},$$

$$\sigma_\wedge(M)(y,\eta) = r^{-\mu}\omega_{|\eta|}\operatorname{op}_M^{\gamma-n/2}(f)(y,\eta)\omega'_{|\eta|}$$

(in the case $\boldsymbol{g} = (\gamma,\gamma-\mu)$, otherwise with extra smoothing Mellin summands, cf. (7.23), and the principal homogeneous component $\sigma_\wedge(G)(y,\eta) = g_{(\mu)}(y,\eta)$ of the Green symbol $g(y,\eta)$ of the operator G, $(y,\eta) \in T^*Y \setminus 0$. We have

$$\sigma_\wedge(A)(y,\eta) \in L^\mu(N^\wedge, \boldsymbol{g}; E'^\wedge, F'^\wedge),$$

cf. Definition 6.8.1. The cone calculus also has a symbolic structure, in the case of an infinite stretched cone a tuple

$$\sigma = (\sigma_\psi, \sigma_M, \sigma_E),$$

cf. Section 1.5.

Remark 8.1.2. Let $A \in L^\mu(M, \boldsymbol{g}; E, F)$ be σ_ψ-elliptic in the sense of Definition 8.1.1. Then $\sigma_\wedge(A)(y,\eta)$ is σ_ψ-elliptic in the sense of the cone calculus and σ_E-elliptic in the sense of Definition 1.5.18 for every fixed $(y,\eta) \in T^*Y \setminus 0$. Moreover, for every $y \in Y$ there is a discrete set $D_A(y) \subset \mathbb{C}$ such that

$$\sigma_M\sigma_\wedge(A)(y,z) : H^s(N, E') \to H^{s-\mu}(N, F')$$

is an isomorphism, $s \in \mathbb{R}$, if and only if $z \notin D_A(y)$.

This observation yields the following result.

Proposition 8.1.3. Let $A \in L^\mu(M, \boldsymbol{g}; E, F)$ be σ_ψ-elliptic. Then (8.2) is a Fredholm operator for $y \in Y$, $\eta \neq 0$, $s \in \mathbb{R}$, if and only if $\gamma \in \mathbb{R}$ satisfies the condition

$$\Gamma_{(n+1)/2-\gamma} \cap D_A(y) = \emptyset. \qquad (8.3)$$

In the following we assume that our operator $A \in L^\mu(M, \boldsymbol{g}; E, F)$ is σ_ψ-elliptic and satisfies the condition (8.3) for every $y \in Y$. Because of the homogeneity

$$\sigma_\wedge(A)(y,\lambda\eta) = \lambda^\mu \kappa_\lambda \sigma_\wedge(A)(y,\eta)\kappa_\lambda^{-1}$$

we have

$$\operatorname{ind}\sigma_\wedge(A)(y,\eta) = \operatorname{ind}\sigma_\wedge(A)(y,\eta/|\eta|).$$

More precisely, the dimensions of $\ker \sigma_\wedge(A)(y,\eta)$ and $\operatorname{coker} \sigma_\wedge(A)(y,\eta)$ only depend on $\eta/|\eta|$. Therefore, (8.2) may be regarded as a family of Fredholm operators depending on the parameters $(y,\eta) \in S^*Y$, the unit cosphere bundle of Y, which is a compact topological space. This gives rise to an index element

$$\operatorname{ind}_{S^*Y} \sigma_\wedge(A) \in K(S^*Y).$$

The property

$$\operatorname{ind}_{S^*Y} \sigma_\wedge(A) \in \pi^* K(Y), \tag{8.4}$$

$\pi : S^*Y \to Y$, is of analogous meaning for the edge calculus as the corresponding condition in Theorem 3.2.2. If A satisfies the relation (8.4) we say that the Atiyah–Bott obstruction vanishes.

Theorem 8.1.4. *Let* $A \in L^\mu(M, \boldsymbol{g}; E, F)$ *be* σ_ψ-*elliptic, and* (8.3) *be satisfied for some* $\gamma \in \mathbb{R}$ *and all* $y \in Y$, *and denote the family of Fredholm operators* (8.2) *for the moment by*

$$\sigma_\wedge(A)^\gamma(y,\eta) : \mathcal{K}^{s,\gamma}(N^\wedge) \otimes E_y'^\wedge \to \mathcal{K}^{s-\mu,\gamma-\mu}(N^\wedge) \otimes F_y'^\wedge,$$

$(y,\eta) \in S^*Y$. *Then if* $\tilde{\gamma} \in \mathbb{R}$ *is another weight satisfying*

$$\Gamma_{(n+1)/2 - \tilde{\gamma}} \cap D_A(y) = \emptyset$$

for all $y \in Y$, *we have*

$$\operatorname{ind}_{S^*Y} \sigma_\wedge(A)^\gamma \in \pi^* K(Y) \iff \operatorname{ind}_{S^*Y} \sigma_\wedge(A)^{\tilde{\gamma}} \in \pi^* K(Y).$$

In addition, if $\tilde{A} \in L^\mu(M, \boldsymbol{g}; E, F)$ *satisfies* $\tilde{A} = A \bmod L^\mu_{M+G}(M, \boldsymbol{g}; E, F)$, *then*

$$\Gamma_{(n+1)/2 - \gamma} \cap D_{\tilde{A}}(y) = \emptyset, \quad \Gamma_{(n+1)/2 - \gamma} \cap D_A(y) = \emptyset$$

for some $\gamma \in \mathbb{R}$ *and all* $y \in Y$ *implies that*

$$\operatorname{ind}_{S^*Y} \sigma_\wedge(\tilde{A})^\gamma \in \pi^* K(Y) \iff \operatorname{ind}_{S^*Y} \sigma_\wedge(A)^\gamma \in \pi^* K(Y).$$

Proof. The assertions are a consequence of the fact that both $\sigma_\wedge(A)^\gamma$ and $\sigma_\wedge(A)^{\tilde{\gamma}}$, as well as $\sigma_\wedge(A)^\gamma$ and $\sigma_\wedge(\tilde{A})^\gamma$, differ only by Green operator-valued families. Their pointwise compactness shows that the respective index elements coincide, in particular, are at the same time pull-back under $\pi : S^*Y \to Y$. $\qquad \square$

8.2 Construction of elliptic edge conditions

Let us now study ellipticity with respect to both components of (8.1) from the point of view of an analogue of the Shapiro–Lopatinskii ellipticity in the edge calculus, again referred to as SL-ellipticity.

Definition 8.2.1. An operator $\mathcal{A} \in L^\mu(M, \boldsymbol{g}; \boldsymbol{v})$ for $\boldsymbol{v} := (E, F; J_1, J_2)$ is called elliptic if

(i) its top left corner $A \in L^\mu(M, \boldsymbol{g}; E, F)$ is σ_ψ-elliptic in the sense of Definition 8.1.1;

(ii) $\mathcal{A}$ is $\sigma_\wedge$-elliptic, i.e.,

$$\sigma_\wedge(\mathcal{A})(y, \eta) : \quad \begin{matrix} \mathcal{K}^{s,\gamma}(N^\wedge) \otimes E_y'^\wedge \\ \oplus \\ J_{1,y} \end{matrix} \quad \longrightarrow \quad \begin{matrix} \mathcal{K}^{s-\mu,\gamma-\mu}(N^\wedge) \otimes F_y'^\wedge \\ \oplus \\ J_{2,y} \end{matrix} \quad (8.5)$$

is a family of isomorphisms for all $(y, \eta) \in T^*Y \setminus 0$ and some $s \in \mathbb{R}$.

Remark 8.2.2. (i) If (8.5) is a family of isomorphisms for some $s \in \mathbb{R}$, then the same holds for all $s \in \mathbb{R}$, and $\ker \sigma_\wedge(\mathcal{A})$ as well as $\dim \operatorname{coker} \sigma_\wedge(\mathcal{A})$ are independent of s.

(ii) The condition of isomorphisms (8.5) for all $(y, \eta) \in T^*Y \setminus 0$ is equivalent to the one for all $(y, \eta) \in S^*Y$.

(iii) The condition of isomorphisms (8.5) for all $(y, \eta) \in S^*Y$ has the consequence that

$$\sigma_\wedge(A)(y, \eta) : \mathcal{K}^{s,\gamma}(N^\wedge) \otimes E_y'^\wedge \to \mathcal{K}^{s-\mu,\gamma-\mu}(N^\wedge) \otimes F_y'^\wedge \quad (8.6)$$

is a family of Fredholm operators, and we have

$$\operatorname{ind}_{S^*Y} \sigma_\wedge(A) = [J_2] - [J_1] \in \pi^* K(Y).$$

Theorem 8.2.3. *Let $A \in L^\mu(M, \boldsymbol{g}; E, F)$ be σ_ψ-elliptic, and let (8.6) be a family of Fredholm operators. Then the property*

$$\operatorname{ind}_{S^*Y} \sigma_\wedge(A) \in \pi^* K(Y)$$

is equivalent to the existence of an SL-elliptic operator $\mathcal{A} \in L^\mu(M, \boldsymbol{g}; \boldsymbol{v})$, $\boldsymbol{v} := (E, F; J_1, J_2)$ for suitable $J_1, J_2 \in \operatorname{Vect}(Y)$ which contains A as the top left corner.

Proof. Theorem 8.2.3 is an analogue of Theorem 3.2.2 and can be proved in an analogous manner. $\qquad \square$

Theorem 8.2.4. *Let $A \in L^\mu(M, \boldsymbol{g}; E, F)$ be σ_ψ-elliptic, cf. Definition 8.1.1, and let*

$$\sigma_\wedge(A)(y, \eta) : \mathcal{K}^{s,\gamma}(N^\wedge) \otimes E_y'^\wedge \to \mathcal{K}^{s-\mu,\gamma-\mu}(N^\wedge) \otimes F_y'^\wedge, \quad (y, \eta) \in S^*Y, \quad (8.7)$$

be a family of Fredholm operators. Choose $L_1, L_2 \in \operatorname{Vect}(T^*Y \setminus 0)$ such that

$$\operatorname{ind}_{S^*Y} \sigma_\wedge(A) = [L_2|_{S^*Y}] - [L_1|_{S^*Y}]. \quad (8.8)$$

Then there exists an element $G \in L_G^\mu(M, \boldsymbol{g}; E, F)$ such that

$$\ker_{S^*Y} \sigma_\wedge(A + G) \cong L_2|_{S^*Y}, \quad \operatorname{coker}_{S^*Y} \sigma_\wedge(A + G) \cong L_1|_{S^*Y}. \quad (8.9)$$

If $J_1, J_2 \in \mathrm{Vect}(Y)$ are bundles such that L_i are subbundles of $\pi_Y^* J_i$ for π_Y : $T^*Y \setminus 0 \to Y$, $i = 1, 2$, there exists an $\mathcal{A} \in L^\mu(M, \boldsymbol{g}; \boldsymbol{v})$ for $\boldsymbol{v} = (E, F; J_1, J_2)$ with $A + G$ as the top left corner of $\mathcal{A}$, such that $\sigma_\wedge(\mathcal{A})$ induces an isomorphism

$$
\sigma_\wedge(\mathcal{A}): \quad
\begin{matrix}
\mathcal{K}^{s,\gamma}(N^\wedge) \otimes \pi_Y^* E'^\wedge \\
\oplus \\
L_1
\end{matrix}
\quad \longrightarrow \quad
\begin{matrix}
\mathcal{K}^{s-\mu,\gamma-\mu}(N^\wedge) \otimes \pi_Y^* F'^\wedge \\
\oplus \\
L_2
\end{matrix} . \tag{8.10}
$$

Proof. Since (8.7) is Fredholm, we find a potential symbol

$$
k_{(\mu)}(y, \eta) : \mathbb{C}^{N_-} \to \mathcal{K}^{s-\mu,\gamma-\mu}(N^\wedge) \otimes F_y'^\wedge
$$

in the sense of a top right corner of a Green symbol in Definition 7.2.1 of order μ with weight data $\boldsymbol{g}$, and for a suitable N_-, such that

$$
\big(\sigma_\wedge(A)(y, \eta) \quad k_{(\mu)}(y, \eta) \big) :
\begin{matrix}
\mathcal{K}^{s,\gamma}(N^\wedge) \otimes E_y'^\wedge \\
\oplus \\
\mathbb{C}^{N_-}
\end{matrix}
\to \mathcal{K}^{s-\mu,\gamma-\mu}(N^\wedge) \otimes F_y'^\wedge
$$

is surjective for all $(y, \eta) \in S^*Y$. This holds for all s. Let

$$
p_2 : \mathcal{K}^{0,\gamma-\mu}(N^\wedge) \otimes \pi^* F'^\wedge \to \mathrm{im}_{S^*Y} k_{(\mu)}
$$

be the orthogonal projection with respect to the scalar product in the fibres $\mathcal{K}^{0,\gamma-\mu}(N^\wedge) \otimes \pi^* F_y'^\wedge$. Since its kernel is smooth, p_2 extends to $\mathcal{K}^{s,\gamma-\mu}(N^\wedge) \otimes \pi^* F'^\wedge$ for every s. There are subbundles

$$
\widetilde{L}_1, \widetilde{L}_1^\perp \subset \mathcal{K}^{\infty,\gamma}(N^\wedge) \otimes \pi^* E'^\wedge, \quad \widetilde{L}_2, \widetilde{L}_2^\perp \subset \mathcal{K}^{\infty,\gamma-\mu}(N^\wedge) \otimes \pi^* F'^\wedge
$$

such that $\widetilde{L}_{1,2} \cong L_{1,2}$, $\widetilde{L}_2^\perp \cong \widetilde{L}_1^\perp$, and

$$
\mathrm{im}_{S^*Y}\big((1 - p_2)\sigma_\wedge(A)\big) \cong \widetilde{L}_2 \oplus \widetilde{L}_2^\perp, \quad \mathrm{im}_{S^*Y} k_{(\mu)} \cong \widetilde{L}_1 \oplus \widetilde{L}_1^\perp.
$$

Let $p_1 : \mathcal{K}^{s,\gamma}(N^\wedge) \otimes \pi^* E'^\wedge \to \widetilde{L}_2^\perp$ be induced by the corresponding orthogonal projection for $s = 0$. Further, let $\lambda : \widetilde{L}_2^\perp \to \widetilde{L}_1^\perp$ be any smooth isomorphism, and $\iota : \widetilde{L}_1^\perp \to \mathcal{K}^{s-\mu,\gamma-\mu}(N^\wedge) \otimes \pi^* F'^\wedge$ be the canonical embedding. Set $q := \iota \circ \lambda \circ p_1$, and form

$$
g_1 := -p_2 \sigma_\wedge(A) + q : \mathcal{K}^{s,\gamma}(N^\wedge) \otimes \pi^* E'^\wedge \to \mathcal{K}^{s,\gamma-\mu}(N^\wedge) \otimes \pi^* F'^\wedge.
$$

Then g_1 can be regarded as the restriction to S^*Y of the homogeneous principal symbol $g_{(\mu)}$ of a Green operator $G \in L_G^\mu(M, \boldsymbol{g}; E, F)$, and by construction (8.9) holds. In order to construct the operator $\mathcal{A}$, it suffices to define its principal edge symbol $\sigma_\wedge(\mathcal{A}) := (\sigma_\wedge(\mathcal{A})_{ij})_{i,j=1,2}$ for $\sigma_\wedge(\mathcal{A})_{11} := \sigma_\wedge(A + G)$. For the remaining entries we choose arbitrary $J_{1,2} \in \mathrm{Vect}(Y)$ such that $L_{1,2}$ are subbundles of $\pi_Y^* J_{1,2}$.

Similarly as in the standard calculus of pseudo-differential BVPs outlined before, see also in Chapter 2 and Section 3.1, there are a potential symbol

$$\sigma_\wedge(\mathcal{A})_{12} : \pi_Y^* J_1 \to \mathcal{K}^{s-\mu,\gamma-\mu}(N^\wedge) \otimes \pi_Y^* F'^\wedge$$

and a trace symbol

$$\sigma_\wedge(\mathcal{A})_{21} : \mathcal{K}^{s,\gamma}(N^\wedge) \otimes \pi_Y^* E'^\wedge \to \pi_Y^* J_2$$

such that, if we set $\sigma_\wedge(\mathcal{A})_{22} := 0$, the matrix $\sigma_\wedge(\mathcal{A})$ induces an isomorphism (8.10). □

8.3 Parametrices and the Fredholm property

Let M be a manifold with edge.

Definition 8.3.1. Let $\mathcal{A} \in L^\mu(M, \boldsymbol{g}; \boldsymbol{v})$, $\boldsymbol{v} := (E, F; J_1, J_2)$, $\mathcal{P} \in L^{-\mu}(M, \boldsymbol{g}^{-1}; \boldsymbol{v}^{-1})$, $\boldsymbol{v}^{-1} := (F, E; J_2, J_1)$, and let $\mathcal{A}$ or $\mathcal{P}$ be properly supported, cf. Remark 7.4.2. Then $\mathcal{P}$ is called a parametrix of $\mathcal{A}$ if

$$\mathcal{C}_L := \mathcal{I} - \mathcal{P}\mathcal{A} \in L^{-\infty}(M, \boldsymbol{g}_L; \boldsymbol{v}_L), \quad \mathcal{C}_R := \mathcal{I} - \mathcal{A}\mathcal{P} \in L^{-\infty}(M, \boldsymbol{g}_R; \boldsymbol{v}_R) \quad (8.11)$$

for $\boldsymbol{v}_L := (E, E; J_1, J_1)$, $\boldsymbol{v}_R := (F, F; J_2, J_2)$, with $\mathcal{I}$ denoting the respective identity operators.

Theorem 8.3.2. *Any elliptic operator* $\mathcal{A} \in L^\mu(M, \boldsymbol{g}; \boldsymbol{v})$, $\boldsymbol{v} := (E, F; J_1, J_2)$, *has a properly supported parametrix* $\mathcal{P} \in L^{-\mu}(M, \boldsymbol{g}^{-1}; \boldsymbol{v}^{-1})$. *If M is compact, the following conditions are equivalent:*

(i) $\mathcal{A}$ *is elliptic in the sense of* Definition 8.2.1.

(ii) *The operator*

$$\mathcal{A} : H^{s,\gamma}(M, E) \oplus H^s(Y, J_1) \to H^{s-\mu,\gamma-\mu}(M, F) \oplus H^{s-\mu}(Y, J_2) \quad (8.12)$$

is Fredholm for some $s = s_0 \in \mathbb{R}$.

Proof. Without loss of generality we may assume that the operators in consideration are properly supported. Then the existence of a parametrix to $\mathcal{A}$ means that the relations (8.11) hold and hence

$$1 = \sigma(\mathcal{P})\sigma(\mathcal{A}), \; 1 = \sigma(\mathcal{A})\sigma(\mathcal{P}), \quad (8.13)$$

with (8.13) being valid both for σ_ψ and for $\sigma_\wedge$. Hence, both components are invertible and $\mathcal{A}$ is elliptic. Conversely, the ellipticity of $\mathcal{A}$ has the consequence (8.13), and using $\sigma_\psi(\mathcal{P})$ and $\sigma_\wedge(\mathcal{P})$ we can construct a parametrix $\mathcal{P}$ of $\mathcal{A}$. □

Remark 8.3.3. Let $\mathcal{A} \in L^\mu(M, \boldsymbol{g}; \boldsymbol{v})$, $\boldsymbol{v} := (E, F; J_1, J_2)$, be elliptic.

(i) The operator (8.12) is Fredholm for every $s \in \mathbb{R}$.

(ii) $\mathcal{V} := \ker_s \mathcal{A} = \{u \in H^{s,\gamma}(M,E) \oplus H^s(Y,J_1) : \mathcal{A}u = 0\}$ is a finite-dimensional subspace of $H^{\infty,\gamma}(M,E) \oplus H^\infty(Y,J_1)$ independent of s, and there is a finite-dimensional $\mathcal{W} \subset H^{\infty,\gamma-\mu}(M,F) \oplus H^\infty(Y,J_2)$ independent of s such that

$$\mathrm{im}_s\, \mathcal{A} + \mathcal{W} = H^{s-\mu,\gamma-\mu}(M,F) \oplus H^{s-\mu}(Y,J_2)$$

for every s; here $\mathrm{im}_s\, \mathcal{A} = \{\mathcal{A}u : u \in H^{s,\gamma}(M,E) \oplus H^s(Y,J_1)\}$.

Theorem 8.3.4. *Let M be compact, and $\mathcal{A} \in L^\mu(M,\boldsymbol{g};\boldsymbol{v})$, $\boldsymbol{v} := (E,F;J_1,J_2)$, be elliptic. Then there is a parametrix $\mathcal{B} \in L^{-\mu}(M,\boldsymbol{g}^{-1};\boldsymbol{v}^{-1})$, $\boldsymbol{v}^{-1} := (F,E;J_2,J_1)$, such that the operators $\mathcal{C}_\mathrm{L}$ and $\mathcal{C}_\mathrm{R}$ in the relation (8.11) are projections*

$$\mathcal{C}_\mathrm{L} : H^{s,\gamma}(M,E) \oplus H^s(Y,J_1) \to \mathcal{V}, \quad \mathcal{C}_\mathrm{R} : H^{s-\mu,\gamma-\mu}(M,F) \oplus H^{s-\mu}(Y,J_2) \to \mathcal{W}$$

for all $s \in \mathbb{R}$.

Proof. The formal aspects of the proof are standard, i.e., Fredholmness follows from the existence of a parametrix with remainder terms as projections of finite rank. $\qquad\square$

8.4 Edge calculus with parameters

We now establish an analogue of the edge calculus for operators depending on a parameter $\lambda \in \mathbb{R}^l$. More details in this direction can be found in Dorschfeldt [13]. The idea is similar to the case of operators on a smooth manifold with boundary, cf. Section 3.3. Since the generalisation of the edge calculus to the parameter-dependent case is to a large extent straightforward, we simply list a number corresponding modifications.

Parameter-dependent Green symbols

$$g(y,\eta,\lambda) \in \mathcal{R}^\nu_\mathrm{G}\big(\Omega \times \mathbb{R}^{q+l}_{\eta,\lambda}\big)$$

are obtained from Definition 7.2.1 by replacing $\eta \in \mathbb{R}^q$ by $(\eta,\lambda) \in \mathbb{R}^{q+l}$. Then Theorem 7.2.2 has an obvious analogue in the parameter-dependent case. Moreover, replacing η by (η,λ) we obtain parameter-dependent Mellin plus Green symbols, analogously as Definition 7.2.9. This gives us the spaces

$$\mathcal{R}^\mu_\mathrm{M+G}\big(\Omega \times \mathbb{R}^q, \boldsymbol{g}; E', F'; \mathbb{R}^l\big).$$

As for the non-smoothing contributions, we first consider parameter-dependent versions of (7.24) and (7.25), namely,

$$\widetilde{p}\big(r,y,\widetilde{\rho},\widetilde{\eta},\widetilde{\lambda}\big) \in C^\infty\Big(\overline{\mathbb{R}}_+ \times \Omega, L^\mu_\mathrm{cl}\big(N; E', F'; \mathbb{R}^{1+q+l}_{\widetilde{\rho},\widetilde{\eta},\widetilde{\lambda}}\big)\Big), \qquad (8.14)$$

and $\Omega \subseteq \mathbb{R}^q$ open, and form

$$p(r, y, \rho, \eta, \lambda) := \widetilde{p}(r, y, r\rho, r\eta, r\lambda). \tag{8.15}$$

There is then an analogue of Theorem 7.3.2 which yields

$$\widetilde{h}(r, y, z, \widetilde{\eta}, \widetilde{\lambda}) \in C^\infty\left(\overline{\mathbb{R}}_+ \times \Omega, M_{\mathcal{O}}^\mu\left(N; E', F'; \mathbb{R}_{\widetilde{\eta}, \widetilde{\lambda}}^{q+l}\right)\right) \tag{8.16}$$

and the associated

$$h(r, y, z, \eta, \lambda) := \widetilde{h}(r, y, z, r\eta, r\lambda) \tag{8.17}$$

such that

$$\begin{aligned}
\mathrm{Op}_r(p)(y, \eta, \lambda) &= \mathrm{op}_M^\beta(h)(y, \eta, \lambda) \\
&\quad \mathrm{mod}\, C^\infty\left(\Omega, L^{-\infty}\left(N^\wedge; E'^\wedge, F'^\wedge; \mathbb{R}_{\eta, \lambda}^{q+l}\right)\right),
\end{aligned} \tag{8.18}$$

for every $\beta \in \mathbb{R}$.

The space of edge amplitude functions

$$\mathcal{R}^\mu(\Omega \times \mathbb{R}^q, \boldsymbol{g}; E', F'; \mathbb{R}^l)$$

is the set of all $a(y, \eta, \lambda)$, of the form

$$a(y, \eta, \lambda) - \epsilon r^{-\mu}\left\{\omega_{\eta, \lambda}\mathrm{Op}_M^{\gamma - n/2}(h)(y, \eta, \lambda)\omega'_{\eta, \lambda} + \chi_{\eta, \lambda}\mathrm{Op}_r(p)(y, \eta, \lambda)\chi'_{\eta, \lambda}\right\}\epsilon'$$
$$+ (m + g)(y, \eta, \lambda)$$

for some $\epsilon'' \prec \epsilon \prec \epsilon'$ and arbitrary (8.15), (8.17) with (8.16), such that h is a Mellin quantisation of p, cf. notation after Theorem 7.3.2, and

$$(m + g)(y, \eta, \lambda) \in \mathcal{R}_{\mathrm{M+G}}^\mu(\Omega \times \mathbb{R}^q, \boldsymbol{g}; E', F'; \mathbb{R}^l).$$

Analogously to Definition 7.3.11, we then also obtain the space

$$\mathcal{R}^\mu(\Omega \times \mathbb{R}^q, \boldsymbol{g}; \boldsymbol{w}; \mathbb{R}^l) \quad \text{for } \boldsymbol{w} = (E', F'; j_1, j_2).$$

For a compact manifold M with edge Y and $\boldsymbol{v} = (E, F; J_1, J_2)$ for $E, F \in \mathrm{Vect}(\mathbb{M})$, $J_1, J_2 \in \mathrm{Vect}(Y)$ we have the space

$$L^{-\infty}(M, \boldsymbol{g}; \boldsymbol{v}; \mathbb{R}^l) := \mathcal{S}(\mathbb{R}^l, L^{-\infty}(M, \boldsymbol{g}; \boldsymbol{v})) \tag{8.19}$$

of parameter-dependent smoothing operators. In (8.19) we use the fact that $L^{-\infty}(M, \boldsymbol{g}; \boldsymbol{v})$ is a union of Fréchet spaces $L^{-\infty}(M, \boldsymbol{g}; \boldsymbol{v})_{\mathcal{P}, \mathcal{Q}}$, determined by the mapping properties (7.48)–(7.49) for $\boldsymbol{g} = (\gamma, \gamma - \mu, \Theta)$, otherwise by $L^{-\infty}(M, \boldsymbol{g}; \boldsymbol{v})_\varepsilon$ when we require (7.50)–(7.51) for an $\varepsilon > 0$.

Using that

$$(m + g)(y, \eta, \lambda) \in \mathcal{R}_{\mathrm{M+G}}^\mu(\Omega \times \mathbb{R}^q, \boldsymbol{g}; \boldsymbol{w}; \mathbb{R}^l)$$

for $\boldsymbol{w} = (E', F'; j_1, j_2)$ implies $(m + g)(y, \eta, \lambda_0) \in \mathcal{R}^\mu_{M+G}(\Omega \times \mathbb{R}^q, \boldsymbol{g}; \boldsymbol{w})$ and that

$$a(y, \eta, \lambda) \in \mathcal{R}^\mu(\Omega \times \mathbb{R}^q, \boldsymbol{g}; \boldsymbol{w}; \mathbb{R}^l)$$

implies $a(y, \eta, \lambda) \in \mathcal{R}^\mu(\Omega \times \mathbb{R}^q, \boldsymbol{g}; \boldsymbol{w})$ for every $\lambda_0 \in \mathbb{R}^l$, it follows that Definition 7.4.1 yields operators

$$\mathcal{A}(\lambda) = \begin{pmatrix} A(\lambda) & 0 \\ 0 & 0 \end{pmatrix} + \mathcal{G}(\lambda) \tag{8.20}$$

for every fixed $\lambda = \lambda_0$ when we apply the procedure to formulate operators in terms of amplitude functions. The set of all operator families (8.20) then defines the space of all parameter-dependent edge operators

$$L^\mu(M, \boldsymbol{g}; \boldsymbol{v}; \mathbb{R}^l). \tag{8.21}$$

By virtue of

$$\mathcal{A}(\lambda) \in L^\mu(M, \boldsymbol{g}; \boldsymbol{v}; \mathbb{R}^l) \implies \mathcal{A}(\lambda_0) \in L^\mu(M, \boldsymbol{g}; \boldsymbol{v})$$

for every $\lambda_0 \in \mathbb{R}^l$ it follows that every $\mathcal{A}(\lambda) \in L^\mu(M, \boldsymbol{g}; \boldsymbol{v}; \mathbb{R}^l)$ induces continuous operators (7.57)–(7.58), for every $\lambda \in \mathbb{R}^l$.

An $\mathcal{A}(\lambda) \in L^\mu(M, \boldsymbol{g}; \boldsymbol{v}; \mathbb{R}^l)$ has a pair

$$\sigma(\mathcal{A}(\lambda)) = (\sigma_\psi(\mathcal{A}(\lambda)), \sigma_\wedge(\mathcal{A}(\lambda))) \tag{8.22}$$

of parameter-dependent homogeneous principal symbols, where

$$\sigma_\psi(\mathcal{A}(\lambda)) = \sigma_\psi(A(\lambda))$$

is the parameter-dependent homogeneous principal symbol of the top left corner $A(\lambda)$, which is similar in meaning to (7.59), now for the projection

$$\pi_{M \setminus Y} : T^* M \setminus Y \times \mathbb{R}^l \setminus 0 \to M \setminus Y,$$

where 0 means $(\xi, \lambda) = 0$, with ξ being the covariables of $T^* M \setminus Y$. Moreover, the parameter-dependent homogeneous principal edge symbol represents a family of continuous operators

$$\sigma_\wedge(\mathcal{A}(\lambda))(y, \eta, \lambda) : \begin{array}{c} \mathcal{K}^{s,\gamma}(N^\wedge) \otimes E_y'^\wedge \\ \oplus \\ J_{1,y} \end{array} \longrightarrow \begin{array}{c} \mathcal{K}^{s-\mu,\gamma-\mu}(N^\wedge) \otimes F_y'^\wedge \\ \oplus \\ J_{2,y} \end{array}, \tag{8.23}$$

for $(y, \eta, \lambda) \in T^* Y \times \mathbb{R}^l \setminus 0$; in this case 0 indicates $(\eta, \lambda) = 0$.

The filtration of (8.21) is similar in structure to that defined by (7.62). In addition, similarly as (7.63) we have

$$L^{-\infty}(M, \boldsymbol{g}; \boldsymbol{v}; \mathbb{R}^l) := \bigcap_{j \in \mathbb{N}} L^{\mu-j}(M, \boldsymbol{g}; \boldsymbol{v}; \mathbb{R}^l). \tag{8.24}$$

Remark 8.4.1. We have parameter-dependent analogues of Theorems 7.4.5 and 7.4.6 (i), (ii).

Definition 8.4.2. Let M be a (not necessarily compact) manifold with edge Y. An operator
$$A(\lambda) \in L^\mu(M, \boldsymbol{g}; E, F; \mathbb{R}^l)$$
is called parameter-dependent σ_ψ-elliptic, if

(i) the parameter-dependent homogeneous principal symbol
$$\sigma_\psi(A(\lambda)) : \pi^*_{M \backslash Y} E \to \pi^*_{M \backslash Y} F, \quad \pi_{M \backslash Y} : T^*(M \backslash Y) \times \mathbb{R}^l \backslash 0 \to M \backslash Y$$
is an isomorphism;

(ii) locally near ∂M in the splitting of variables $(r, x, y) \in \overline{\mathbb{R}}_+ \times \Sigma \times \Omega$ with covariables (ρ, ξ, η), the reduced parameter-dependent symbol
$$\tilde{\sigma}_\psi(A(\lambda))(r, x, y, \rho, \xi, \eta, \lambda) = r^\mu \sigma_\psi(A(\lambda))(r, x, y, r^{-1}\rho, \xi, r^{-1}\eta, r^{-1}\lambda)$$
induces isomorphisms
$$\tilde{\sigma}_\psi(A(\lambda))(r, x, y, \rho, \xi, \eta, \lambda) : E_{(r,x,y)} \to F_{(r,x,y)}$$
for all $(r, x, y, \rho, \xi, \eta, \lambda) \in \overline{\mathbb{R}}_+ \times \Sigma \times \Omega \times \left(\mathbb{R}^{1+n+q+l}_{\rho,\xi,\eta,\lambda} \backslash \{0\} \right)$.

An operator
$$\mathcal{A}(\lambda) \in L^\mu(M, \boldsymbol{g}; \boldsymbol{v}; \mathbb{R}^l), \quad \boldsymbol{v} = (E, F; J_1, J_2),$$
is called parameter-dependent elliptic, if

(iii) The top left corner $A(\lambda)$ of $\mathcal{A}(\lambda)$ is parameter-dependent σ_ψ-elliptic in the sense of (i), (ii);

(iv) the parameter-dependent homogeneous principal edge symbol (8.23) defines isomorphisms for all $(y, \eta, \lambda) \in T^*Y \times \mathbb{R}^l \backslash 0$.

Theorem 8.4.3. *Let $\mathcal{A}(\lambda) \in L^\mu(M, \boldsymbol{g}; \boldsymbol{v}; \mathbb{R}^l)$, $\boldsymbol{v} = (E, F; J_1, J_2)$, be parameter-dependent elliptic. Then $\mathcal{A}(\lambda)$ has a properly supported parametrix*
$$\mathcal{P}(\lambda) \in L^{-\mu}(M, \boldsymbol{g}^{-1}; \boldsymbol{v}^{-1}; \mathbb{R}^l),$$
such that analogues of relations (8.11) hold.

Proof. Straightforward.   □

Theorem 8.4.4. *Let M be a compact manifold with edge Y, and let $\mathcal{A}(\lambda) \in L^\mu(M, \boldsymbol{g}; \boldsymbol{v}; \mathbb{R}^l)$ be parameter-dependent elliptic. Then $\mathcal{A}(\lambda)$ defines a family of Fredholm operators*

$$\mathcal{A}(\lambda) \quad \begin{matrix} H^{s,\gamma}(M, E) \\ \oplus \\ H^s(Y, J_1) \end{matrix} \longrightarrow \begin{matrix} H^{s-\mu,\gamma-\mu}(M, F) \\ \oplus \\ H^{s-\mu}(Y, J_2) \end{matrix} \qquad (8.25)$$

of index zero, for every $s \in \mathbb{R}$, and there is a constant $C > 0$ such that the operators (8.25) are isomorphisms for all $|\lambda| > C$.

Proof. Straightforward. □

Remark 8.4.5. Let $A(\lambda) \in L^{\mu}(M, \boldsymbol{g}; E, F; \mathbb{R}^{l})$ be parameter-dependent σ_{ψ}-elliptic. Then, similarly as in the case without parameters, the edge symbol

$$\sigma_{\wedge}(A)(y, \eta) : \mathcal{K}^{s,\gamma}(N^{\wedge}) \otimes E_y'^{\wedge} \to \mathcal{K}^{s-\mu,\gamma-\mu}(N^{\wedge}) \otimes F_y'^{\wedge}$$

is a family of Fredholm operators, for all $(y, \eta, \lambda) \in T^*Y \times \mathbb{R}^{l} \setminus 0$. For every fixed $\lambda_0 \in \mathbb{R}^{l} \setminus \{0\}$ we have $A(\lambda_0) \in L^{\mu}(M, \boldsymbol{g}; E, F)$, and then

$$\sigma_{\wedge}(A(\lambda_0))(y, \eta) : \mathcal{K}^{s,\gamma}(N^{\wedge}) \otimes E_y'^{\wedge} \to \mathcal{K}^{s-\mu,\gamma-\mu}(N^{\wedge}) \otimes F_y'^{\wedge}$$

is a family of Fredholm operators for all $(y, \eta) \in T^*Y$, including the zero section. Thus we have

$$\mathrm{ind}_{S^*Y} \sigma_{\wedge}(A(\lambda_0)) \in \pi^* K(Y)$$

for every $\lambda_0 \neq 0$, cf. Theorem 8.2.3.

8.5 Order-reducing operators on a manifold with edge

Theorem 8.5.1. *Let M be a compact manifold with edge and $\mathbb{M}$ its stretched manifold. For every $\mu, \gamma \in \mathbb{R}, V \in \mathrm{Vect}(\mathbb{M})$ there exists a parameter-dependent elliptic element $R_V^{\mu}(\lambda) \in L^{\mu}(M, \boldsymbol{g}; V, V; \mathbb{R}^{l})$ for $\boldsymbol{g} = (\gamma, \gamma - \mu, \Theta)$, which induces isomorphisms*

$$R_V^{\mu}(\lambda) : H^{s,\gamma}(M, V) \to H^{s-\mu,\gamma-\mu}(M, V) \tag{8.26}$$

for all $\lambda \in \mathbb{R}^{l}, s \in \mathbb{R}$.

Proof. Let us prove the assertion for the case $V = \mathbb{C}$, i.e., the trivial bundle over $\mathbb{M}$ of fibre dimension 1. The simple modification of arguments in the general case is left to the reader. We start the construction with local edge-degenerate $L_{\mathrm{cl}}^{\mu}(N)$-valued symbols close to $\partial \mathbb{M}$ in the splitting of variables

$$(r, y) \in \mathbb{R}_+ \times \mathbb{R}^q$$

and covariables

$$(\rho, \eta) \in \mathbb{R}^{1+q},$$

depending on extra parameters $\theta \in \mathbb{R}$, $\lambda \in \mathbb{R}^{l}$. More precisely, we consider operator functions

$$\widetilde{p}\left(r, \widetilde{\rho}, \theta, \widetilde{\eta}, \widetilde{\lambda}\right) \in C^{\infty}\left(\overline{\mathbb{R}}_+, L_{\mathrm{cl}}^{\mu}\left(N; \mathbb{R}_{\widetilde{\rho}, \theta, \widetilde{\eta}, \widetilde{\lambda}}^{2+q+l}\right)\right),$$

for the moment without the variable y. Then we form

$$p(r, \rho, \theta, \eta, \lambda) := \widetilde{p}\left(r, r\rho, \theta, r\eta, r\lambda\right). \tag{8.27}$$

The operator function $\tilde{p}$ is constructed in terms of parameter-dependent pseudo-differential operators on N, with symbols in local coordinates $x \in \mathbb{R}^n$ and covariables $\xi \in \mathbb{R}^n$ of the form

$$\left(1 + |\tilde{\rho}|^2 + |\xi|^2 + \theta^2 + |\tilde{\eta}|^2 + |\tilde{\lambda}|^2\right)^{\mu/2}. \tag{8.28}$$

Then by applying an operator convention with respect to (x, ξ), using an open cover of N by coordinate neighbourhoods $(U_1, \ldots, U_N)$, a subordinate partition of unity $(\varphi_1, \ldots, \varphi_N)$, functions

$$(\psi_1, \ldots, \psi_N), \quad \psi_j \in C_0^\infty(U_j), \; \psi_j \succ \varphi_j,$$

and charts $\chi_j : U_j \to \mathbb{R}^n$, we obtain the element (8.27). Another step is the construction of local edge amplitude functions

$$a(y, \eta, \lambda) \in \mathcal{R}^\mu\left(\mathbb{R}^q \times \mathbb{R}^q, \boldsymbol{g}; \mathbb{R}^l\right)$$

for the future operator $R_{\mathbb{C}}^\mu(\lambda)$ ($\mathbb{C}$ stands for V, and $a(y, \eta, \lambda)$ also depends on ϑ). To this end we apply a Mellin quantisation, which gives an

$$\tilde{h}\left(r, z, \vartheta, \tilde{\eta}, \tilde{\lambda}\right) \in C^\infty\left(\overline{\mathbb{R}}_+, M_{\mathcal{O}}^\mu\left(N; \mathbb{R}_{\vartheta, \tilde{\eta}, \tilde{\lambda}}^{1+q+l}\right)\right),$$

such that for

$$h(r, z, \vartheta, \eta, \lambda) := \tilde{h}(r, z, \vartheta, r\eta, r\lambda)$$

we have

$$\mathrm{Op}_r(p)(\vartheta, \eta, \lambda) = \mathrm{Op}_M^\beta(h)(\vartheta, \eta, \lambda),$$

for every $\beta \in \mathbb{R}$, modulo a remainder in $L^{-\infty}\left(N^\wedge; \mathbb{R}_{\vartheta, \eta, \lambda}^{1+q+l}\right)$. Now we form

$$b(\vartheta, \eta, \lambda) := \omega(r) r^{-\mu} \mathrm{Op}_M^\beta(h)(\vartheta, \eta, \lambda) \tilde{\omega}(r)$$

for fixed cut-off functions $\omega, \tilde{\omega}$. In this way we obtain a family of operators in the cone algebra over $N^\wedge$. Its principal conormal symbol is a parameter-dependent elliptic operator family

$$\sigma_M(b)(\vartheta, z) \in L_{\mathrm{cl}}^\mu\left(N; \mathbb{R}_\vartheta \times \Gamma_{(n+1)/2-\gamma}\right)$$

belonging to $M_{\mathcal{O}}^\mu(N; \mathbb{R}_\vartheta)$. We find a $C > 0$ such that for $|\vartheta| > C$ the operators

$$\sigma_M(b)(\vartheta, z)(\vartheta, z) : H^s(N) \to H^{s-\mu}(N)$$

are isomorphisms for all $z \in \Gamma_{(n+1)/2-\gamma}$ and $s \in \mathbb{R}$. We obtain a family of Fredholm operators

$$b(\vartheta, \eta, \lambda) : \mathcal{K}^{s, \gamma}\left(N^\wedge\right) \to \mathcal{K}^{s-\mu, \gamma-\mu}\left(N^\wedge\right)$$

for $|\vartheta| > C$ and $(\eta, \lambda) \neq 0$, cf. Theorem 6.9.3, (ii). Since ϑ is fixed, $|\vartheta|$ sufficiently large, from now on we omit indicating ϑ, i.e., write $b(\eta, \lambda) := b(\vartheta, \eta, \lambda)$. Now we use the fact, cf. [45], [50], that there is an element $f(z) \in M_O^{-\infty}(N)$, such that for any fixed cut-off functions ω, ω' the operator

$$1 + \omega_{|\eta,\lambda|}\mathrm{Op}_M^{\gamma-n/2}(f)\omega'_{|\eta,\lambda|} : \mathcal{K}^{s,\gamma}(N^\wedge) \to \mathcal{K}^{s,\gamma}(N^\wedge)$$

is of index $-\mathrm{ind}\, b(\eta, \lambda)$. Recall that $\omega_{|\eta,\lambda|}(r) = \omega(r|\eta, \lambda|)$, etc. It follows that

$$b_0(\eta, \lambda) := b(\eta, \lambda)\Big(1 + \omega_{|\eta,\lambda|}\mathrm{Op}_M^{\gamma-n/2}(f)\omega'_{|\eta,\lambda|}\Big) : \mathcal{K}^{s,\gamma}(N^\wedge) \to \mathcal{K}^{s-\mu,\gamma-\mu}(N^\wedge)$$

is of index 0. There is then a Green symbol $g(\eta, \lambda) \in \mathcal{R}_G^\mu(\mathbb{R}_\eta^q, \boldsymbol{g}; \mathbb{R}_\lambda^l)$ of finite rank such that

$$b_0(\eta, \lambda) + \sigma_\wedge(g)(\eta, \lambda) : \mathcal{K}^{s,\gamma}(N^\wedge) \to \mathcal{K}^{s-\mu,\gamma-\mu}(N^\wedge)$$

is a family of isomorphisms for all $s \in \mathbb{R}$. We can refer this construction to local coordinates $y \in \mathbb{R}^q$ for any coordinate neighbourhood of the edge Y. For an open covering $G_1, \ldots, G_N$ of Y by coordinate neighbourhoods and charts $\chi_j : G_j \to \mathbb{R}^q$, $j = 1, \ldots, N$, together with a subordinate partition of unity φ_j and functions $\varphi'_j \succ \varphi_j$ in $C_0^\infty(G_j)$, we can form edge amplitude functions

$$a_j(y, \eta, \lambda) \in \mathcal{R}^\mu(\mathbb{R}_y^q \times \mathbb{R}_\eta^q, \boldsymbol{g}; \mathbb{R}_\lambda^l)$$

such that

$$\sigma_\wedge(a_j)(y, \eta, \lambda) = \varphi_j(y)\big(b_0(\eta, \lambda) + \sigma_\wedge(g)(\eta, \lambda)\big).$$

Similarly as in (7.56), we pass to the family of operators

$$R_\mathbb{C}^\mu(\lambda) := \theta\left\{\sum_{j=1}^N \varphi_j(\chi_j^{-1})_*\mathrm{Op}_y(a_j)(\lambda)\varphi'_j\right\}\theta' + (1 - \theta)A_{\mathrm{int}}(\lambda)(1 - \theta''), \quad (8.29)$$

where $\theta'' \prec \theta \prec \theta'$ are cut-off functions on M, $\equiv 1$ in a neighbourhood of Y. The element $A_{\mathrm{int}}(\lambda) \in L_{\mathrm{cl}}^\mu(M \setminus Y; \mathbb{R}^l)$ is chosen in such a way that (8.29) in $L^\mu(M, \boldsymbol{g}; \mathbb{R}^l)$ is parameter-dependent elliptic in the sense of Definition 8.4.2. This is possible because of the specific form of the local interior symbols (8.28). By virtue of Theorem 8.4.4, the operators $R_\mathbb{C}^\mu(\lambda) : H^{s,\gamma}(M) \to H^{s-\mu,\gamma-\mu}(M)$ are isomorphisms whenever $|\lambda| \geq C$ for some sufficiently large $C > 0$. If in this construction we replace for the moment the parameter $\lambda \in \mathbb{R}^l$ by $(\lambda, \tilde{\lambda}) \in \mathbb{R}^l \times \mathbb{R}^{\tilde{l}}$ for some $\tilde{l} > 0$, then $|\lambda, \tilde{\lambda}| \geq C$ is satisfied for $|\tilde{\lambda}| \geq C$ and arbitrary $\lambda \in \mathbb{R}^l$. Therefore, $R_\mathbb{C}^\mu(\lambda) := R_\mathbb{C}^\mu(\lambda, \tilde{\lambda})$ for any fixed $|\tilde{\lambda}| \geq C$ is as desired. $\qquad\square$

The operator functions (8.26) are referred to as parameter-dependent order-reducing operators of the edge calculus.

Remark 8.5.2. (i) Assume that operators $R_V^\mu(\lambda) \in L^\mu(M, \boldsymbol{g}; V, V; \mathbb{R}^l)$ are parameter-dependent order-reducing as in Theorem 8.5.1. Then for the family of inverses we have

$$\big(R_V^\mu(\lambda)\big)^{-1} \in L^{-\mu}(M, \boldsymbol{g}^{-1}; V, V; \mathbb{R}^l)$$

for $\boldsymbol{g}^{-1} = (\gamma - \mu, \gamma, \Theta)$.

(ii) If $R_V^\mu(\lambda) \in L^\mu\big(M, \boldsymbol{g}; V, V; \mathbb{R}^l\big)$ is parameter-dependent order-reducing, then for any $\lambda_0 \in \mathbb{R}^l$ the operator $R_V^\mu := R_V^\mu(\lambda_0) \in L^\mu(M, \boldsymbol{g}; V, V)$ is order reducing, in the sense that

$$R_V^\mu : H^{s,\gamma}(M, V) \to H^{s-\mu, \gamma-\mu}(M, V) \tag{8.30}$$

is an isomorphism for every $s \in \mathbb{R}$, and $\big(R_V^\mu\big)^{-1} \in L^{-\mu}\big(M, \boldsymbol{g}^{-1}; V, V\big)$.

Chapter 9

Toeplitz edge problems

9.1 Edge operators with global projection conditions

Let M be a compact manifold with edge Y.

Definition 9.1.1. Let $\mathbb{L}_i := (P_i, J_i, L_i) \in \mathbb{P}(Y)$ be projection data, cf. Definition 1.2.7, $V_1, V_2 \in \mathrm{Vect}(\mathbb{M})$, $i = 1, 2$, and set

$$\boldsymbol{v} := (V_1, V_2; J_1, J_2), \quad \boldsymbol{l} := (V_1, V_2; \mathbb{L}_1, \mathbb{L}_2).$$

Then $\mathcal{T}^\mu(M, \boldsymbol{g}; \boldsymbol{l})$, $\mu \in \mathbb{R}$, for $\boldsymbol{g} = (\gamma, \gamma - \mu, \Theta)$ or $\boldsymbol{g} = (\gamma, \gamma - \mu)$ is defined to be the set of all operators

$$\mathcal{A} := \mathcal{P}_2 \widetilde{\mathcal{A}} \mathcal{E}_1$$

for $\mathcal{P}_2 := \mathrm{diag}(1, P_2)$, $\mathcal{E}_1 := \mathrm{diag}(1, E_1)$, cf. formula (4.1), for arbitrary $\widetilde{\mathcal{A}} \in L^\mu(M, \boldsymbol{g}; \boldsymbol{v})$. The elements of $\mathcal{T}^\mu(M, \boldsymbol{g}; \boldsymbol{l})$ will be called edge problems of order μ with global projection conditions. Moreover, set

$$\mathcal{T}^{-\infty}(M, \boldsymbol{g}; \boldsymbol{l}) := \{\mathcal{P}_2 \widetilde{\mathcal{C}} \mathcal{E}_1 : \widetilde{\mathcal{C}} \in L^{-\infty}(M, \boldsymbol{g}; \boldsymbol{l})\}. \tag{9.1}$$

Observe that the space (9.1) can be equivalently characterised as the set of all $\mathcal{A} \in \mathcal{T}^\mu(M, \boldsymbol{g}; \boldsymbol{l})$, $\mathcal{A} := \mathcal{P}_2 \widetilde{\mathcal{A}} \mathcal{E}_1$ for some $\widetilde{\mathcal{A}} \in L^\mu(M, \boldsymbol{g}; \boldsymbol{v})$, such that $\mathcal{P}_2 \widetilde{\mathcal{A}} \mathcal{P}_1 \in L^{-\infty}(M, \boldsymbol{g}; \boldsymbol{v})$; then $\mathcal{A} = \mathcal{P}_2 (\mathcal{P}_2 \widetilde{\mathcal{A}} \mathcal{P}_1) \mathcal{E}_1$. Moreover,

$$\mathcal{P}_2 (\mathcal{P}_2 \widetilde{\mathcal{A}} \mathcal{P}_1) \mathcal{E}_1 \in \mathcal{T}^{-\infty}(M, \boldsymbol{g}; \boldsymbol{l}) \implies \mathcal{P}_2 \widetilde{\mathcal{A}} \mathcal{P}_1 \in L^{-\infty}(M, \boldsymbol{g}; \boldsymbol{v}).$$

Theorem 9.1.2. *Every $\mathcal{A} \in \mathcal{T}^\mu(M, \boldsymbol{g}; \boldsymbol{l})$ induces continuous operators*

$$\mathcal{A} : \begin{array}{c} H^{s,\gamma}(M, V_1) \\ \oplus \\ H^s(Y, \mathbb{L}_1) \end{array} \longrightarrow \begin{array}{c} H^{s-\mu,\gamma-\mu}(M, V_2) \\ \oplus \\ H^{s-\mu}(Y, \mathbb{L}_2) \end{array} \tag{9.2}$$

for every $s \in \mathbb{R}$.

© Springer Nature Switzerland AG 2018
X. Liu, B.-W. Schulze, *Boundary Value Problems with Global Projection Conditions*,
Operator Theory: Advances and Applications 265, https://doi.org/10.1007/978-3-319-70114-1_9

Proof. The proof is evident after Theorem 7.4.3. □

Remark 9.1.3. (i) Let $\mathbb{L}_{1,2} := (P_{1,2}, J_{1,2}, L_{1,2})$, $\widetilde{\mathbb{L}}_{1,2} := (\widetilde{P}_{1,2}, \widetilde{J}_{1,2}, \widetilde{L}_{1,2}) \in \mathbb{P}(Y)$, such that $J_{1,2}$ are subbundles of $\widetilde{J}_{1,2}$, and

$$\widetilde{P}_{1,2}\big|_{H^s(Y,J_{1,2})} = P_{1,2}. \tag{9.3}$$

Then we have a canonical isomorphism

$$\mathcal{T}^\mu(M, g; l) \cong \mathcal{T}^\mu(M, g; \widetilde{l})$$

for $\widetilde{l} := (V_1, V_2; \widetilde{\mathbb{L}}_1, \widetilde{\mathbb{L}}_2)$ $l := (V_1, V_2; \mathbb{L}_1, \mathbb{L}_2)$.

(ii) If $\mathbb{L}_{1,2} \in \mathbb{P}(Y)$ and $\widetilde{J}_{1,2} \in \mathrm{Vect}(Y)$ contain $J_{1,2}$ as subbundles, we find projections $\widetilde{P}_{1,2} \in L^0_{cl}(Y; \widetilde{J}_{1,2}, \widetilde{J}_{1,2})$ with the property (9.3).

Proposition 9.1.4. *Given $V_i \in \mathrm{Vect}(\mathbb{M})$, $i = 1, 2$, we have a canonical isomorphism*

$$\mathcal{T}^\mu(M, g; l) \to \left\{ \mathcal{P}_2 \widetilde{\mathcal{A}} \mathcal{P}_1 : \widetilde{\mathcal{A}} \in L^\mu(M, g; v) \right\}.$$

Proof. The proof is analogous to that of Proposition 1.2.16, cf. also Proposition 4.1.3. □

In other words, we have an identification

$$\mathcal{T}^\mu(M, g; l) = L^\mu(M, g; v)/\sim \tag{9.4}$$

with the equivalence relation

$$\widetilde{\mathcal{A}} \sim \widetilde{\mathcal{B}} \iff \mathcal{P}_2 \widetilde{\mathcal{A}} \mathcal{P}_1 = \mathcal{P}_2 \widetilde{\mathcal{B}} \mathcal{P}_1. \tag{9.5}$$

The space $\mathcal{T}^\mu(M, g; l)$ is equipped with the principal symbol structure

$$\sigma(\mathcal{A}) = (\sigma_\psi(\mathcal{A}), \sigma_\wedge(\mathcal{A}))$$

with the interior and the edge symbol component. Writing $\mathcal{A} = (\mathcal{A}_{ij})_{i,j=1,2}$, we first set $\sigma_\psi(\mathcal{A}) := \sigma_\psi(\mathcal{A}_{11})$, i.e.,

$$\sigma_\psi(\mathcal{A}) : \pi^*_{M\setminus Y} V_1 \to \pi^*_{M\setminus Y} V_2$$

for $\pi_{M\setminus Y} : T^*(M \setminus Y) \setminus 0 \to M \setminus Y$.

The edge symbol of $\mathcal{A}$, represented as $\mathcal{A} = \mathcal{P}_2 \widetilde{\mathcal{A}} \mathcal{E}_1$, is defined as

$$\sigma_\wedge(\mathcal{A}) = \mathrm{diag}\,(1, p_2)\, \sigma_\wedge(\widetilde{\mathcal{A}})\, \mathrm{diag}\,(1, e_1),$$

$$\sigma_\wedge(\mathcal{A})(y, \eta) : \begin{matrix} \mathcal{K}^{s,\gamma}(N^\wedge) \otimes V'^\wedge_{1,y} \\ \oplus \\ L_{1,(y,\eta)} \end{matrix} \longrightarrow \begin{matrix} \mathcal{K}^{s-\mu,\gamma-\mu}(N^\wedge) \otimes V'^\wedge_{2,y} \\ \oplus \\ L_{2,(y,\eta)} \end{matrix} \tag{9.6}$$

where $p_2(y, \eta)$ is the homogeneous principal symbol of order zero of the projection $P_2 \in L^0_{cl}(Y; J_2, J_2)$, and $e_1 : L_{1,(y,\eta)} \to (\pi^*_Y J_1)_{(y,\eta)}$ is the canonical embedding.

Remark 9.1.5. Identifying an operator $\mathcal{A} = \mathcal{P}_2\widetilde{\mathcal{A}}\mathcal{E}_1 \in \mathcal{T}^\mu(M, \boldsymbol{g}; \boldsymbol{l})$ with $\widetilde{\widetilde{\mathcal{A}}} :=$ $\mathcal{P}_2\widetilde{\mathcal{A}}\mathcal{P}_1 \in L^\mu(M, \boldsymbol{g}; \boldsymbol{v})$, cf. Proposition 9.1.4, then $\sigma\!\left(\widetilde{\widetilde{\mathcal{A}}}\right) = 0$ in the sense of $L^\mu(M, \boldsymbol{g}; \boldsymbol{v})$ is equivalent to $\sigma(\mathcal{A}) = 0$ in the sense of $\mathcal{T}^\mu(M, \boldsymbol{g}; \boldsymbol{l})$.

Analogously as (7.62) we define $\mathcal{T}^\nu(M, \boldsymbol{g}; \boldsymbol{l})$ for $\boldsymbol{g} = (\gamma, \gamma - \mu, \Theta)$ or $\boldsymbol{g} = (\gamma, \gamma - \mu)$, $\mu - \nu \in \mathbb{N}$,

$$\mathcal{T}^\nu(M, \boldsymbol{g}; \boldsymbol{l}) := \left\{ \mathcal{P}_2\widetilde{\mathcal{A}}\mathcal{E}_1 : \widetilde{\mathcal{A}} \in L^\nu(M, \boldsymbol{g}; \boldsymbol{v}) \right\}.$$

Remark 9.1.6. $\mathcal{A} \in \mathcal{T}^\mu(M, \boldsymbol{g}; \boldsymbol{l})$ and $\sigma(\mathcal{A}) = 0$ imply $\mathcal{A} \in \mathcal{T}^{\mu-1}(M, \boldsymbol{g}; \boldsymbol{l})$, and the operator (9.2) is compact for every $s \in \mathbb{R}$.

Theorem 9.1.7. *Let* $\mathcal{A}_j \in \mathcal{T}^{\mu-j}(M, \boldsymbol{g}; \boldsymbol{l})$, $j \in \mathbb{N}$, *be an arbitrary sequence,* $\boldsymbol{g} :=$ $(\gamma, \gamma - \mu, (-(k+1), 0])$ *for a finite* k, *or* $\boldsymbol{g} := (\gamma, \gamma - \mu)$, *and assume that the asymptotic types involved in the Green operators are independent of* j. *Then there exists an* $\mathcal{A} \in \mathcal{T}^\mu(M, \boldsymbol{g}; \boldsymbol{l})$ *such that*

$$\mathcal{A} - \sum_{j=0}^{N} \mathcal{A}_j \in \mathcal{T}^{\mu-(N+1)}(M, \boldsymbol{g}; \boldsymbol{l})$$

for every $N \in \mathbb{N}$, *and* $\mathcal{A}$ *is unique mod* $\mathcal{T}^{-\infty}(M, \boldsymbol{g}; \boldsymbol{l})$.

Proof. The proof is analogous to the proof of Theorem 4.1.8. $\square$

Theorem 9.1.8. (i) $\mathcal{A} \in \mathcal{T}^\mu(M, \boldsymbol{g}_0; \boldsymbol{l}_0)$, $\mathcal{B} \in \mathcal{T}^\rho(M, \boldsymbol{g}_1; \boldsymbol{l}_1)$ *for* $\boldsymbol{g}_0 = (\gamma - \rho, \gamma - (\mu + \rho), \Theta)$, $\boldsymbol{g}_1 = (\gamma, \gamma - \rho, \Theta)$, *or* $\boldsymbol{g}_0 = (\gamma - \rho, \gamma - (\mu + \rho))$, $\boldsymbol{g}_1 = (\gamma, \gamma - \rho)$, *implies* $\mathcal{A}\mathcal{B} \in \mathcal{T}^{\mu+\nu}(M, \boldsymbol{g}_0 \circ \boldsymbol{g}_1; \boldsymbol{l}_0 \circ \boldsymbol{l}_1)$ *(when the projection data in the middle fit together such that* $\boldsymbol{l}_0 \circ \boldsymbol{l}_1$ *makes sense), and we have* $\sigma(\mathcal{A}\mathcal{B}) = \sigma(\mathcal{A})\sigma(\mathcal{B})$ *with componentwise multiplication.*

(ii) $\mathcal{A} \in \mathcal{T}^\mu(M, \boldsymbol{g}; \boldsymbol{l})$ *for* $\boldsymbol{g} = (\gamma, \gamma - \mu, \Theta)$ *or* $\boldsymbol{g} = (\gamma, \gamma - \mu)$, *and* $\boldsymbol{l} = (V_1, V_2; \mathbb{L}_1, \mathbb{L}_2)$ *implies* $\mathcal{A}^* \in \mathcal{T}^\mu(M, \boldsymbol{g}^*; \boldsymbol{l}^*)$ *for* $\boldsymbol{l}^* = (V_2, V_1; \mathbb{L}_2^*, \mathbb{L}_1^*)$, *where* $\mathcal{A}^*$ *is the formal adjoint in the sense*

$$(u, \mathcal{A}^* v)_{H^{0,0}(M, V_1) \oplus H^0(Y, \mathbb{L}_1)} = (\mathcal{A}u, v)_{H^{0,0}(M, V_2) \oplus H^0(Y, \mathbb{L}_2)}$$

for all $u \in H^{\infty,\infty}(M, V_1) \oplus H^\infty(Y, \mathbb{L}_1)$, $v \in H^{\infty,\infty}(M, V_2) \oplus H^\infty(Y, \mathbb{L}_2)$, *and we have* $\sigma(\mathcal{A}^*) = \sigma(\mathcal{A})^*$ *with componentwise formal adjoint.*

Proof. The proof is formally analogous to the proof of Theorem 4.1.7 when we take into account the corresponding results from the edge calculus, see Section 7.4. $\square$

Remark 9.1.9. Similarly to the case of the Toeplitz algebras of Section 1.2 or 4.1 we have a natural notion of direct sum

$$\mathcal{T}^\mu(M, \boldsymbol{g}; \boldsymbol{l}) \oplus \mathcal{T}^\mu(M, \boldsymbol{g}; \boldsymbol{m}) = \mathcal{T}^\mu(M, \boldsymbol{g}; \boldsymbol{l} \oplus \boldsymbol{m})$$

where $\sigma(\mathcal{A} \oplus \mathcal{B}) = \sigma(\mathcal{A}) \oplus \sigma(\mathcal{B})$ with the componentwise direct sum of symbols.

9.2 Ellipticity, parametrices, and the Fredholm property

We now turn to ellipticity in the Toeplitz calculus of edge problems.

Definition 9.2.1. Let $\mathcal{A} \in T^\mu(M, \boldsymbol{g}; \boldsymbol{l})$ for $\boldsymbol{g} := (\gamma, \gamma - \mu, \Theta)$ or $\boldsymbol{g} := (\gamma, \gamma - \mu)$, $\mu \in \mathbb{R}$, and $\boldsymbol{l} := (V_1, V_2; \mathbb{L}_1, \mathbb{L}_2)$, $V_i \in \mathrm{Vect}(\mathbb{M})$, $\mathbb{L}_i = (P_i, J_i, L_i) \in \mathbb{P}(Y)$, $i = 1, 2$. The operator $\mathcal{A}$ is called elliptic if the top left corner $A \in L^\mu(M, \boldsymbol{g}; V_1, V_2)$ is σ_ψ-elliptic in the sense of Definition 8.1.1 and if the edge symbol (9.6) is an isomorphism for every $(y, \eta) \in T^*Y \setminus 0$ and some $s = s_0 \in \mathbb{R}$.

Remark 9.2.2. The bijectivity of (9.6) for some $s = s_0 \in \mathbb{R}$ is equivalent to the bijectivity of (9.6) for every $s \in \mathbb{R}$. The latter property is equivalent to the bijectivity

$$
\sigma_\wedge(\mathcal{A})(y, \eta) : \quad
\begin{array}{c}
\mathcal{K}^{\infty, \gamma; \infty}(N^\wedge) \otimes V'^\wedge_{1,y} \\
\oplus \\
L_{1,(y,\eta)}
\end{array}
\longrightarrow
\begin{array}{c}
\mathcal{K}^{\infty, \gamma - \mu; \infty}(N^\wedge) \otimes V'^\wedge_{2,y} \\
\oplus \\
L_{2,(y,\eta)}
\end{array}
\tag{9.7}
$$

Theorem 9.2.3. (i) *For every σ_ψ-elliptic $A \in L^\mu(M, \boldsymbol{g}; V_1, V_2)$ such that*

$$
\sigma_\wedge(A)(y, \eta) : \mathcal{K}^{s, \gamma}(N^\wedge) \otimes V'^\wedge_{1,y} \to \mathcal{K}^{s - \mu, \gamma - \mu}(N^\wedge) \otimes V'^\wedge_{2,y}, \quad (y, \eta) \in T^*Y \setminus 0,
\tag{9.8}
$$

is a family of Fredholm operators with

$$
\mathrm{ind}_{S^*Y} \sigma_\wedge(A) = \left[L_2 \big|_{S^*Y} \right] - \left[L_1 \big|_{S^*Y} \right] \text{ for some } L_1, L_2 \in \mathrm{Vect}(T^*Y \setminus 0),
$$

there exist a Green operator $G \in L_G^\mu(M, \boldsymbol{g}; V_1, V_2)$, projection data $\mathbb{L}_1, \mathbb{L}_2 \in \mathbb{P}(Y)$, and an elliptic $\mathcal{A} \in T^\mu(M, \boldsymbol{g}; \boldsymbol{l})$, $\boldsymbol{l} := (V_1, V_2; \mathbb{L}_1, \mathbb{L}_2)$ with $A + G$ as the top left corner, such that $\mathcal{A}$ is elliptic in the sense of Definition 9.2.1.

(ii) *For A as in (i) and suitable projection data $\mathbb{L}_i = (P_i, J_i, L_i) \in \mathbb{P}(Y)$, $i = 1, 2$, there exists an elliptic operator $\mathcal{A} \in T^\mu(M, \boldsymbol{g}; \boldsymbol{l})$ containing A as the upper left corner.*

Proof. (i) is a consequence of Theorem 8.2.4. For (ii) we choose $J_{1,2} \in \mathrm{Vect}(Y)$ of sufficiently large fibre dimension and a potential edge symbol

$$
\sigma_\wedge(K) : \pi_Y^* J_1 \to \pi_Y^* \mathcal{K}^{s - \mu, \gamma - \mu}(N^\wedge) \otimes V'^\wedge_2
$$

of κ_δ-homogeneity μ such that

$$
\left(\sigma_\wedge(A) \quad \sigma_\wedge(K) \right) : \pi_Y^*
\begin{pmatrix}
\mathcal{K}^{s, \gamma}(N^\wedge) \otimes V'^\wedge_1 \\
\oplus \\
J_1
\end{pmatrix}
\to \mathcal{K}^{s - \mu, \gamma - \mu}(N^\wedge) \otimes V'^\wedge_2
$$

is surjective; this is always possible, also when J_1 is trivial and of sufficiently large fibre dimension. Then

$$
\ker_{S^*Y} \left(\sigma_\wedge(A) \quad \sigma_\wedge(K) \right) =: L_2
$$

has the property $\ker_{S^*Y} = [L_2|_{S^*Y}] - [\pi^*J_1]$. This allows us to apply (i) for $\mathbb{L}_1 := (\mathrm{id}, J_1, J_1)$. □

Proposition 9.2.4. *For every $\mu, \gamma \in \mathbb{R}$, $\boldsymbol{g} := (\gamma, \gamma - \mu, \Theta)$, $V \in \mathrm{Vect}(\mathbb{M})$ and $\mathbb{L} \in \mathbb{P}(Y)$, there exists an elliptic element $\mathcal{R}_{V,\mathbb{L}}^{\mu} \in T^{\mu}(M, \boldsymbol{g}; l)$ for $l := (V, V; \mathbb{L}, \mathbb{L})$ which induces a Fredholm operator*

$$\mathcal{R}_{V,\mathbb{L}}^{\mu} : \begin{matrix} H^{s,\gamma}(M,V) \\ \oplus \\ H^s(Y, \mathbb{L}) \end{matrix} \longrightarrow \begin{matrix} H^{s-\mu,\gamma-\mu}(M,V) \\ \oplus \\ H^{s-\mu}(Y, \mathbb{L}) \end{matrix}$$

for every $s \in \mathbb{R}$.

Proof. It suffices to set

$$\mathcal{R}_{V,\mathbb{L}}^{\mu} := \mathrm{diag}\big(R_V^{\mu}, R_{\mathbb{L}}^{\mu}\big)$$

for R_V^{μ} from Theorem 8.5.1 and $R_{\mathbb{L}}^{\mu}$ from Remark 1.2.29. □

Theorem 9.2.5. *For every elliptic operator $\mathcal{A} \in T^{\mu}(M, \boldsymbol{g}; l)$, $l = (V_1, V_2; \mathbb{L}_1, \mathbb{L}_2)$, there exists an elliptic operator $\mathcal{B} \in T^{\mu}(M; \boldsymbol{g}; \boldsymbol{m})$, $\boldsymbol{m} := (V_2, V_1; \mathbb{M}_1, \mathbb{M}_2)$, for certain projection data $\mathbb{M}_1, \mathbb{M}_2 \in \mathbb{P}(Y)$ of the form $\mathbb{M}_i := (Q_i, \mathbb{C}^N, M_i)$, $i = 1, 2$, for some $N \in \mathbb{N}$, such that $\mathcal{A} \oplus \mathcal{B} \in L^{\mu}(M, \boldsymbol{g}; \boldsymbol{v})$ for $\boldsymbol{v} = \big(V_1 \oplus V_2, V_2 \oplus V_1; \mathbb{C}^N, \mathbb{C}^N\big)$, is SL-elliptic, cf. Definition 8.2.1.*

Proof. The operator $A \in L^{\mu}(M, \boldsymbol{g}; V_1, V_2)$ in the top left corner of $\mathcal{A}$ induces continuous maps

$$A : H^{s,\gamma}(M, V_1) \to H^{s-\mu,\gamma-\mu}(M, V_2) \tag{9.9}$$

for all $s \in \mathbb{R}$. This will be applied for $s = \gamma$. From Theorem 8.5.1 and Remark 8.5.2 (ii) we have order reducing isomorphisms

$$R_{V_1}^{\gamma} : H^{\gamma,\gamma}(M, V_1) \to H^{0,0}(M, V_1), \quad R_{V_2}^{\gamma-\mu} : H^{\gamma-\mu,\gamma-\mu}(M, V_2) \to H^{0,0}(M, V_2),$$

belonging to $L^{\gamma}(M, (\gamma, 0, \Theta); V_1, V_1)$ and $L^{\gamma-\mu}(M, (\gamma - \mu, 0, \Theta); V_2, V_2)$, respectively. These operators are elliptic in the sense of Definition 8.2.1. According to Theorem 7.4.6 (i), we can form

$$A_0 := R_{V_2}^{\gamma-\mu} A \big(R_{V_1}^{\gamma}\big)^{-1} \in L^0(M, (0, 0, \Theta); V_1, V_2),$$

$A_0 : H^{0,0}(M) \to H^{0,0}(M)$. It follows that $\sigma(A_0) = \sigma\big(R_{V_2}^{\gamma-\mu}\big)\sigma(A)\sigma\big((R_{V_1}^{\gamma})^{-1}\big)$. Applying this for the $\sigma_\wedge$-components we see that

$$\mathrm{ind}_{S^*Y}\sigma_\wedge(A_0) = \mathrm{ind}_{S^*Y}\sigma_\wedge(A) = [L_2|_{S^*Y}] - [L_1|_{S^*Y}]. \tag{9.10}$$

There is an N such that the bundles J_i contained in $\mathbb{L}_i$, $i = 1, 2$, are subbundles of $\mathbb{C}^N$. Because of Remark 9.1.3, without loss of generality we assume $\mathbb{L}_{1,2} = \big(P_{1,2}, \mathbb{C}^N, L_{1,2}\big)$. For complementary bundles $L_{1,2}^{\perp}$ of $L_{1,2}$ in $\mathbb{C}^N$ we have

$$[L_2|_{S^*Y}] - [L_1|_{S^*Y}] = [L_1^{\perp}|_{S^*Y}] - [L_2^{\perp}|_{S^*Y}]. \tag{9.11}$$

For the adjoint of A_0 we have $A_0^* \in L^0(M, (0,0,\Theta); V_2, V_1)$, and relations (9.10), (9.11) imply

$$\text{ind}_{S^*Y}\sigma_\wedge\left(A_0^*\right) = \left[L_2^\perp\big|_{S^*Y}\right] - \left[L_1^\perp\big|_{S^*Y}\right].$$

By Theorem 8.2.4, there is a Green operator $G_0 \in L^0(M, (0,0,\Theta); V_2, V_1)$ such that

$$\ker_{S^*Y}\sigma_\wedge(A_0^* + G_0) \cong L_2^\perp\big|_{S^*Y}, \quad \text{coker}_{S^*Y}\sigma_\wedge(A_0^* + G_0) \cong L_1^\perp\big|_{S^*Y}.$$

For $B := \left(R_{V_1}^{\gamma-\mu}\right)^{-1}(A_0^* + G_0)R_{V_2}^{\gamma} \in L^\mu(M, \boldsymbol{g}; V_2, V_1)$ we also have

$$\ker_{S^*Y}\sigma_\wedge(B) \cong L_2^\perp\big|_{S^*Y}, \quad \text{coker}_{S^*Y}\sigma_\wedge(B) \cong L_1^\perp\big|_{S^*Y}.$$

Let us set $\mathbb{M}_{1,2} := \left(P_{1,2}^\perp, \mathbb{C}^N, L_{1,2}^\perp\right)$, where $P_{1,2}^\perp$ are complementary projections to $P_{1,2}$. Because of Theorem 9.2.3 there is now an elliptic operator $\mathcal{B} \in \mathcal{T}^\mu(M, \boldsymbol{g}; \boldsymbol{m})$ for $\boldsymbol{m} := (V_2, V_1; \mathbb{M}_1, \mathbb{M}_2)$ containing B as upper left corner. From the construction it is then evident that $\mathcal{A} \oplus \mathcal{B}$ has the desired properties. $\qquad\square$

Definition 9.2.6. Let an operator $\mathcal{A} \in \mathcal{T}^\mu(M, \boldsymbol{g}; \boldsymbol{l})$ be as in Definition 9.2.1. A $\mathcal{P} \in \mathcal{T}^{-\mu}(M, \boldsymbol{g}^{-1}; \boldsymbol{l}^{-1})$ for $\boldsymbol{g}^{-1} = (\gamma - \mu, \gamma, \Theta)$ or $(\gamma - \mu, \gamma)$ and $\boldsymbol{l}^{-1} = (V_2, V_1; \mathbb{L}_2, \mathbb{L}_1)$ is called a parametrix of $\mathcal{A}$, if

$$\mathcal{C}_L := \mathcal{I} - \mathcal{P}\mathcal{A} \in \mathcal{T}^{-\infty}(M, \boldsymbol{g}_L; \boldsymbol{l}_L), \quad \mathcal{C}_R := \mathcal{I} - \mathcal{A}\mathcal{P} \in \mathcal{T}^{-\infty}(M, \boldsymbol{g}_R; \boldsymbol{l}_R) \quad (9.12)$$

for $\boldsymbol{g}_L = (\gamma, \gamma, \Theta)$ or $\boldsymbol{g}_R = (\gamma - \mu, \gamma - \mu, \Theta)$ and similarly without Θ, with $\mathcal{I}$ being the respective identity operators, and $\boldsymbol{l}_L := (V_1, V_1; \mathbb{L}_1, \mathbb{L}_1)$, $\boldsymbol{l}_R := (V_2, V_2; \mathbb{L}_2, \mathbb{L}_2)$.

Theorem 9.2.7. Let $\mathcal{A} \in \mathcal{T}^\mu(M, \boldsymbol{g}; \boldsymbol{l})$, $\mu \in \mathbb{R}$, $\boldsymbol{l} := (V_1, V_2; \mathbb{L}_1, \mathbb{L}_2)$ for V_1, $V_2 \in \text{Vect}(\mathbb{M})$, $\mathbb{L}_1$, $\mathbb{L}_2 \in \mathbb{P}(Y)$.

(i) *Let $\mathcal{A}$ be elliptic; then*

$$\mathcal{A}: \begin{matrix} H^{s,\gamma}(M, V_1) \\ \oplus \\ H^s(Y, \mathbb{L}_1) \end{matrix} \longrightarrow \begin{matrix} H^{s-\mu,\gamma-\mu}(X, V_2) \\ \oplus \\ H^{s-\mu}(Y, \mathbb{L}_2) \end{matrix} \qquad (9.13)$$

is a Fredholm operator for every $s \in \mathbb{R}$. Moreover, if (9.13) is Fredholm for $s = \gamma$, then the operator $\mathcal{A}$ is elliptic.

(ii) *If $\mathcal{A}$ is elliptic, (4.11) is Fredholm for all $s \in \mathbb{R}$, and $\dim \ker \mathcal{A}$ and $\dim \text{coker} \mathcal{A}$ are independent of s.*

(iii) *An elliptic operator $\mathcal{A} \in \mathcal{T}^\mu(M, \boldsymbol{g}; \boldsymbol{l})$ has a parametrix $\mathcal{P} \in \mathcal{T}^{-\mu}(M, \boldsymbol{g}^{-1}; \boldsymbol{l}^{-1})$ in the sense of Definition 4.2.7, and $\mathcal{P}$ can be chosen in such a way that the remainders in (4.10) are projections*

$$\mathcal{C}_L : H^{s,\gamma}(M, V_1) \oplus H^s(Y, \mathbb{L}_1) \to \mathcal{V}_1,$$
$$\mathcal{C}_R : H^{s-\mu,s-\mu}(M, V_2) \oplus H^{s-\mu}(Y, \mathbb{L}_2) \to \mathcal{V}_2$$

for all $s \in \mathbb{R}$, *for* $\mathcal{V}_1 = \ker \mathcal{A} \subset H^{\infty,\gamma}(M, V_1) \oplus H^{\infty}(Y, \mathbb{L}_1)$ *and a finite-dimensional subspace* $\mathcal{V}_2 \subset H^{\infty,\gamma-\mu}(M, V_2) \oplus H^{\infty}(Y, \mathbb{L}_2)$ *with the property*

$$\mathcal{V}_2 + \operatorname{im} \mathcal{A} = H^{s-\mu,\gamma-\mu}(M, V_2) \oplus H^{s-\mu}(Y, \mathbb{L}_2), \quad \mathcal{V}_2 \cap \operatorname{im} \mathcal{A} = \{0\},$$

for every $s \in \mathbb{R}$.

Proof. We first show that an elliptic operator $\mathcal{A} \in \mathcal{T}^{\mu}(M, \boldsymbol{g}; \boldsymbol{l})$ has a parametrix $\mathcal{P} \in \mathcal{T}^{-\mu}(M; \boldsymbol{g}^{-1}; \boldsymbol{l}^{-1})$. We apply Theorem 9.2.5 and choose a complementary operator

$$\mathcal{B} \in \mathcal{T}^{\mu,d}(M; \boldsymbol{g}; \boldsymbol{m}), \quad \boldsymbol{m} = (V_2, V_1; \mathbb{M}_1, \mathbb{M}_2)$$

such that $\widetilde{\mathcal{A}} := \mathcal{A} \oplus \mathcal{B} \in L^{\mu}(M, \boldsymbol{g}; \boldsymbol{v})$ for $\boldsymbol{v} = (V_1 \oplus V_2, V_2 \oplus V_1; \mathbb{C}^N, \mathbb{C}^N)$ is elliptic in the sense of Definition 8.2.1. Then

$$\mathcal{A} = \operatorname{diag}(1, P_2) \widetilde{\mathcal{A}} \operatorname{diag}(1, E_1). \tag{9.14}$$

From Theorem 8.4.3 we obtain a parametrix $\widetilde{\mathcal{P}} \in L^{-\mu}(M, \boldsymbol{g}^{-1}; \boldsymbol{v}^{-1})$ for $\boldsymbol{v}^{-1} := (V_2 \oplus V_1, V_1 \oplus V_2; \mathbb{C}^N, \mathbb{C}^N)$, where $\sigma(\widetilde{\mathcal{P}}) = \sigma(\widetilde{\mathcal{A}})^{-1}$. Let us set

$$\mathcal{P}_0 := \operatorname{diag}(1, P_1) \widetilde{\mathcal{P}} \operatorname{diag}(1, E_2) \in \mathcal{T}^{-\mu}(M, \boldsymbol{g}^{-1}; \boldsymbol{l}^{-1}),$$

where $E_2 : H^{s-\mu}(Y, \mathbb{L}_2) \to H^{s-\mu}(Y, J_2)$ is the canonical embedding and $P_1 : H^s(Y, J_1) \to H^s(Y, \mathbb{L}_1)$ the projection involved in $\mathbb{L}_1$. This yields

$$\mathcal{P}_0 \mathcal{A} = \operatorname{diag}(1, P_1) \widetilde{\mathcal{P}} \operatorname{diag}(1, P_2) \widetilde{\mathcal{A}} \operatorname{diag}(1, E_1).$$

Thus for $\mathcal{C}_{\mathrm{L}} := \mathcal{I} - \mathcal{P}_0 \mathcal{A} \in \mathcal{T}^0(M, \boldsymbol{g}_{\mathrm{L}}; \boldsymbol{v}_{\mathrm{L}})$ with $\boldsymbol{v}_{\mathrm{L}} = (V_1, V_1; \mathbb{L}_1, \mathbb{L}_1)$, we have $\sigma(\mathcal{C}_{\mathrm{L}}) = 0$, i.e., $\mathcal{C}_{\mathrm{L}} \in \mathcal{T}^{-1}(M, \boldsymbol{g}_{\mathrm{L}}; \boldsymbol{v}_{\mathrm{L}})$, cf. Remark 9.1.6. Applying Theorem 9.1.7 we find an operator $\mathcal{D}_{\mathrm{L}} \in \mathcal{T}^{-1}(M, \boldsymbol{g}_{\mathrm{L}}; \boldsymbol{v}_{\mathrm{L}})$ such that $(\mathcal{I} + \mathcal{D}_{\mathrm{L}})(\mathcal{I} - \mathcal{C}_{\mathrm{L}}) = \mathcal{I}$ mod $\mathcal{T}^{-\infty}(M, \boldsymbol{g}_{\mathrm{L}}; \boldsymbol{v}_{\mathrm{L}})$. We can define $\mathcal{D}_{\mathrm{L}}$ as an asymptotic sum $\sum_{j=1}^{\infty} \mathcal{C}_{\mathrm{L}}^j$. Thus $(\mathcal{I} + \mathcal{D}_{\mathrm{L}})\mathcal{P}_0 \mathcal{A} = \mathcal{I}$ mod $\mathcal{T}^{-\infty}(M, \boldsymbol{g}_{\mathrm{L}}; \boldsymbol{v}_{\mathrm{L}})$, and hence $\mathcal{P}_{\mathrm{L}} := \mathcal{I} + \mathcal{D}_{\mathrm{L}} \mathcal{P}_0 \in \mathcal{T}^{-\mu}(M, \boldsymbol{g}_{\mathrm{L}}; \boldsymbol{l}^{-1})$ is a left parametrix of $\mathcal{A}$. In a similar manner we find a right parametrix. Thus we may take $\mathcal{P} := \mathcal{P}_{\mathrm{L}}$.

The Fredholm property of (9.13) is a direct consequence of the compactness of the remainders $\mathcal{C}_{\mathrm{L}}, \mathcal{C}_{\mathrm{R}}$ in relation (9.12), cf. also Remark 9.1.6. The second part of (iii) is a consequence of general facts on elliptic operators that are always satisfied when elliptic regularity holds in the respective scales of spaces, see, for instance, [25, Subsection 1.2.7]. This confirms, in particular, assertion (ii).

It remains to show that the Fredholm property of (9.13) for $s = \gamma$ entails ellipticity. We reduce orders and weights to 0 by means of elliptic operators from Proposition 9.2.4, namely,

$$\mathcal{R}_{V_1,\mathbb{L}_1}^{-\gamma} : \begin{array}{c} L^2(M, V_1) \\ \oplus \\ H^0(Y, \mathbb{L}_1) \end{array} \longrightarrow \begin{array}{c} H^{\gamma,\gamma}(M, V_1) \\ \oplus \\ H^{\gamma}(Y, \mathbb{L}_1) \end{array},$$

$$\mathcal{R}_{V_2,\mathbb{L}_2}^{\gamma-\mu,\gamma-\mu} : \begin{array}{c} H^{\gamma-\mu,\gamma-\mu}(M, V_2) \\ \oplus \\ H^{\gamma-\mu}(Y, \mathbb{L}_2) \end{array} \longrightarrow \begin{array}{c} L^2(M, V_2) \\ \oplus \\ H^0(Y, \mathbb{L}_2) \end{array} \tag{9.15}$$

which are both Fredholm, according to the first part of the proof. The composition

$$\mathcal{A}_0 := \mathcal{R}^{\gamma-\mu}_{V_2,\mathbb{L}_2} \, \mathcal{A} \, \mathcal{R}^{-\gamma}_{V_1,\mathbb{L}_1} : \begin{array}{c} L^2(M,V_1) \\ \oplus \\ H^0(Y,\mathbb{L}_1) \end{array} \longrightarrow \begin{array}{c} L^2(M,V_2) \\ \oplus \\ H^0(Y,\mathbb{L}_2) \end{array} \tag{9.16}$$

is again a Fredholm operator. In addition it belongs to $T^0\big(M, \boldsymbol{g}_0; (V_1, V_2; \mathbb{L}_1, \mathbb{L}_2)\big)$ for $\boldsymbol{g}_0 = (0,0,\Theta)$. It suffices to show the ellipticity of $\mathcal{A}_0$. We now employ the fact that every $\mathbb{L} \in \mathbb{P}(Y)$ admits complementary projection data $\mathbb{L}^\perp \in \mathbb{P}(Y)$, cf. Proposition 1.2.8 (iii). In particular, for $\mathbb{L}_1 = (P_1, J_1, \sigma_\psi(P_1)J_1)$ we form $\mathbb{L}_1^\perp = (1 - P_1, J_1, \sigma_\psi(1 - P_1)J_1)$. Then $L^2(Y,J_1) = H^0(Y,\mathbb{L}_1) \oplus H^0(Y,\mathbb{L}_1^\perp)$. We define an operator

$$\mathcal{B} := \mathcal{I}_2 \mathcal{E} \mathcal{C} \mathcal{I}_1 : \begin{array}{c} L^2(M,V_1) \\ \oplus \\ L^2(Y,J_1) \end{array} \longrightarrow \begin{array}{c} L^2(M,V_2) \\ \oplus \\ L^2(Y,J_2 \oplus J_1) \end{array}$$

where

$$\mathcal{I}_1 : \begin{array}{c} L^2(M,V_1) \\ \oplus \\ L^2(Y,J_1) \end{array} \longrightarrow \begin{array}{c} L^2(M,V_1) \\ \oplus \\ H^0(Y,\mathbb{L}_1), \\ \oplus \\ H^0(Y,\mathbb{L}_1^\perp) \end{array} \quad \mathcal{I}_2 : \begin{array}{c} L^2(M,V_2) \\ \oplus \\ L^2(Y,J_2) \\ \oplus \\ L^2(Y,J_1) \end{array} \longrightarrow \begin{array}{c} L^2(M,V_2) \\ \oplus \\ L^2(Y,J_2 \oplus J_1) \end{array}$$

are canonical identifications, and

$$\mathcal{C} : \begin{array}{c} L^2(M,V_1) \\ \oplus \\ H^0(Y,\mathbb{L}_1) \\ \oplus \\ H^0(Y,\mathbb{L}_1^\perp) \end{array} \longrightarrow \begin{array}{c} L^2(M,V_2) \\ \oplus \\ H^0(Y,\mathbb{L}_2), \\ \oplus \\ H^0(Y,\mathbb{L}_1^\perp) \end{array} \quad \mathcal{E} : \begin{array}{c} L^2(M,V_2) \\ \oplus \\ H^0(Y,\mathbb{L}_2) \\ \oplus \\ H^0(Y,\mathbb{L}_1^\perp) \end{array} \hookrightarrow \begin{array}{c} L^2(M,V_2) \\ \oplus \\ L^2(Y,J_2) \\ \oplus \\ L^2(Y,J_1) \end{array}$$

with $\mathcal{E}$ being a canonical embedding, and $\mathcal{C} := \operatorname{diag}\big(\mathcal{A}_0, \operatorname{id}_{H^0(Y,\mathbb{L}_1^\perp)}\big)$. We obviously have $\dim \ker \mathcal{B} = \dim \ker \mathcal{A}_0 < \infty$. Moreover, $\ker \mathcal{B}^*\mathcal{B} = \ker \mathcal{B} = \operatorname{im}(\mathcal{B}^*\mathcal{B})^\perp$, and $\mathcal{B}^*\mathcal{B}$ has closed range, since $\mathcal{C}^*\mathcal{C}$ does. Therefore, $\mathcal{B}^*\mathcal{B} \in T^0\big(M, \boldsymbol{g}_0; (V_1, V_1; J_1, J_1)\big)$ for $\boldsymbol{g}_0 = (0,0,\Theta)$ is a Fredholm operator and hence elliptic by Theorem 8.3.2. Therefore, both $\sigma_\psi(\mathcal{A}_0)$ and $\sigma_\wedge(\mathcal{A}_0)$ are injective. Analogous arguments for adjoint operators show that $\sigma_\psi(\mathcal{A}_0)$ and $\sigma_\wedge(\mathcal{A}_0)$ are also surjective. $\qquad\square$

9.3   Reduction to the edge

The edge calculus, outlined in Chapters 7–8, furnished by operator spaces $L^\mu(M, \boldsymbol{g}; \boldsymbol{v})$ on a manifold M with edge Y is a substructure of the Toeplitz edge calculus, consisting of the spaces $T^\mu(M, \boldsymbol{g}; \boldsymbol{l})$ constructed in Chapter 9. Bundle data $\boldsymbol{v} = (V_1, V_2; J_1, J_2)$ can be regarded as a special case of tuples $\boldsymbol{l} = (V_1, V_2; \mathbb{L}_1, \mathbb{L}_2)$

for global projection data $\mathbb{L}_i = (P_i, J_i, L_i) \in \mathbb{P}(Y)$, $i = 1, 2$. Manifolds X with smooth boundary Y form a subcategory of manifolds with edge, and BVPs with the transmission property (of order and type 0) in $\mathcal{B}^{0,0}(X; \boldsymbol{v})$, cf. Chapter 2, as well as the Toeplitz BVPs, cf. Chapter 4, in $\mathcal{T}^{0,0}(X; \boldsymbol{l})$, are special cases of $L^\mu(M, \boldsymbol{g}; \boldsymbol{v})$ and $\mathcal{T}^\mu(M, \boldsymbol{g}; \boldsymbol{l})$, respectively (BVPs of order $\mu \in \mathbb{Z}$ and type $d \in \mathbb{N}$ are a slight modification of those of order and type 0). As we shall see in Part III below, BVPs without the transmission property at the boundary can also be subsumed under edge problems. In all those cases the idea of reducing boundary or edge problems to Y is an interesting aspect, cf. also Section 4.3.

Now let us consider elliptic operators

$$
\mathcal{A}_i = \begin{pmatrix} A \\ T_i \end{pmatrix} \in \mathcal{T}^\mu(M, \boldsymbol{g}; l_i),\ i = 0, 1,\quad \mathcal{A}_i : H^{s,\gamma}(M, V_1) \longrightarrow \begin{array}{c} H^{s-\mu,\gamma-\mu}(M, V_2) \\ \oplus \\ H^{s-\mu}(Y, L_i) \end{array}
$$
$$(9.17)$$

for $l_i = (V_1, V_2; \mathbb{O}, \mathbb{L}_i)$, $i = 0, 1$, $\mathbb{L}_i = (Q_i, J, L_i) \in \mathbb{P}(Y)$, where $\mathbb{O}$ indicates the case where the fibre dimension of the bundle in the middle is zero. For convenience we assume the trace operators to be of the same orders as A. However, a simple reduction of order allows us to pass to arbitrary orders, cf. Remark 1.2.29. By virtue of Theorem 9.2.7 (iii), the operators $\mathcal{A}_i$ have parametrices $\mathcal{P}_i \in \mathcal{T}^{-\mu}(M, \boldsymbol{g}^{-1}; l_i^{-1})$ for $l_i^{-1} = (V_2, V_1; \mathbb{L}_i, \mathbb{O})$, $i = 0, 1$, $\mathcal{P}_i =: (P_i \quad C_i)$, $i = 0, 1$. Since $\mathcal{A}_0 \mathcal{P}_0 = \operatorname{diag}\left(\operatorname{id}_{H^{s-\mu}(X, V_2)}, \operatorname{id}_{H^{s-\mu}(Y, \mathbb{L}_0)}\right) \bmod \mathcal{T}^{-\infty}(M, \boldsymbol{g}_L; (V_2, V_2; \mathbb{L}_0, \mathbb{L}_0))$ for $\boldsymbol{g}_L = (\gamma - \mu, \gamma - \mu, \Theta)$, it follows that

$$
\mathcal{A}_1 \mathcal{P}_0 = \begin{pmatrix} \operatorname{id}_{H^{s-\mu}(M; V_2)} & 0 \\ T_1 P_0 & T_1 C_0 \end{pmatrix} \bmod \mathcal{T}^{-\infty}(M, \boldsymbol{g}_L; (V_2, V_2; \mathbb{L}_0, \mathbb{L}_1)).
$$

Since the latter operator is elliptic, so is $R := T_1 C_0 \in \mathcal{T}^0(Y; \mathbb{L}_0, \mathbb{L}_1)$, now in the Toeplitz calculus on Y, developed in Section 1.2. In particular,

$$
R : H^{s-\mu}(Y, \mathbb{L}_0) \to H^{s-\mu}(Y, \mathbb{L}_1) \tag{9.18}
$$

is a Fredholm operator, and we have an analogue of the Agranovich–Dynin formula (4.16). Moreover, knowing a parametrix $\mathcal{P}_0$ of $\mathcal{A}_0$ we can easily express a parametrix $\mathcal{P}_1$ of $\mathcal{A}_1$ by applying the corresponding analogue of relation (4.17), here using a parametrix $R^{(-1)} \in \mathcal{T}^0(Y; \mathbb{L}_1, \mathbb{L}_0)$ of the operator R.

Let us extend the procedure of reducing operators to the edge to elliptic operators in block matrix form. For simplicity, we assume orders to be zero; the general case can be achieved by reduction of orders to 0. Consider

$$
\mathcal{A}_i = \begin{pmatrix} A & K_i \\ T_i & Q_i \end{pmatrix} \in \mathcal{T}^0(M, \boldsymbol{g}; l_i),\quad \mathcal{A}_i : \begin{array}{c} H^{0,\gamma}(M, V_1) \\ \oplus \\ H^0(Y, \mathbb{K}_i) \end{array} \longrightarrow \begin{array}{c} H^{0,\gamma}(M, V_2) \\ \oplus \\ H^0(Y, \mathbb{L}_i) \end{array},\quad i = 0, 1,
$$
$$(9.19)$$

for $l_i = (V_1, V_2; \mathbb{K}_i, \mathbb{L}_i)$, $i = 0, 1$, $\mathbb{K}_i = (P_i, J, K_i)$, $\mathbb{L}_i = (Q_i, J, L_i) \in \mathbb{P}(Y)$, where the top left corner is the same for $i = 1, 2$. In order to achieve an analogue of the Agranovich–Dynin formula for the Fredholm indices, we pass to the operators

$$
\widetilde{\mathcal{A}}_0 = \begin{pmatrix} A & K_1 & K_0 \\ T_0 & 0 & Q_0 \\ 0 & 1 & 0 \end{pmatrix} : \begin{matrix} H^{0,\gamma}(M, V_1) \\ \oplus \\ H^0(Y, \mathbb{K}_1) \\ \oplus \\ H^0(Y, \mathbb{K}_0) \end{matrix} \longrightarrow \begin{matrix} H^{0,\gamma}(M, V_2) \\ \oplus \\ H^0(Y, \mathbb{L}_0) \\ \oplus \\ H^0(Y, \mathbb{K}_1) \end{matrix} \tag{9.20}
$$

with $\widetilde{\mathcal{A}}_0 \in \mathcal{T}^\mu(M, \boldsymbol{g}; l_i)$, $\boldsymbol{g} = (\gamma, \gamma, \Theta)$, and

$$
\widetilde{\mathcal{A}}_1 = \begin{pmatrix} A & K_1 & K_0 \\ T_1 & Q_1 & 0 \\ 0 & 0 & 1 \end{pmatrix} : \begin{matrix} H^{0,\gamma}(M, V_1) \\ \oplus \\ H^0(Y, \mathbb{K}_1) \\ \oplus \\ H^0(Y, \mathbb{K}_0) \end{matrix} \longrightarrow \begin{matrix} H^{0,\gamma}(M, V_2) \\ \oplus \\ H^0(Y, \mathbb{L}_1) \\ \oplus \\ H^0(Y, \mathbb{K}_0) \end{matrix} \tag{9.21}
$$

with $\widetilde{\mathcal{A}}_1 \in \mathcal{T}^0(M, \boldsymbol{g}; l_i)$. If $\mathcal{P}_0 = \begin{pmatrix} P_0 & C_0 \\ B_0 & Q_0 \end{pmatrix} \in \mathcal{T}^0(M, \boldsymbol{g}; l_0^{-1})$ is a parametrix of $\mathcal{A}_0$ which exists by Theorem 9.2.7, we obtain a parametrix $\widetilde{\mathcal{P}}_0$ of $\widetilde{\mathcal{A}}_0$ in the form

$$
\widetilde{\mathcal{P}}_0 = \begin{pmatrix} P_0 & C_0 & -P_0 K_1 \\ 0 & 0 & 1 \\ B_0 & Q_0 & -B_0 K_1 \end{pmatrix}. \tag{9.22}
$$

It follows that

$$
\widetilde{\mathcal{A}}_1 \widetilde{\mathcal{P}}_0 = \begin{pmatrix} 1 & 0 & 0 \\ T_1 P_0 & T_1 C_0 & -T_1 P_0 K_1 + Q_1 \\ B_0 & Q_0 & -B_0 K_1 \end{pmatrix} \tag{9.23}
$$

mod $\mathcal{T}^{-\infty}(M, \boldsymbol{g}; \boldsymbol{n})$ for $\boldsymbol{n} = (V_2, V_2; \mathbb{L}_0 \oplus \mathbb{K}_1, \mathbb{L}_1 \oplus \mathbb{K}_0)$, where the bottom right corner

$$
\mathcal{R} = \begin{pmatrix} T_1 C_0 & -T_1 P_0 K_1 + Q_1 \\ Q_0 & -B_0 K_1 \end{pmatrix} : \begin{matrix} H^0(Y, \mathbb{L}_0) \\ \oplus \\ H^0(Y, \mathbb{K}_1) \end{matrix} \longrightarrow \begin{matrix} H^0(Y, \mathbb{L}_1) \\ \oplus \\ H^0(Y, \mathbb{K}_0) \end{matrix} \tag{9.24}
$$

is elliptic and belongs to $\mathcal{T}^0(Y; \boldsymbol{r})$ for $\boldsymbol{r} = (\mathbb{L}_0 \oplus \mathbb{K}_1, \mathbb{L}_1 \oplus \mathbb{K}_0)$. The analogue of the Agranovich–Dynin formula in this case is as follows.

Theorem 9.3.1. *For every two elliptic operators* (9.19) *the reduction to the edge* (9.23) *is elliptic, and we have*

$$
\operatorname{ind} \mathcal{A}_1 - \operatorname{ind} \mathcal{A}_0 = \operatorname{ind} \mathcal{R}.
$$

Proof. The result is a consequence of $\operatorname{ind} \mathcal{A}_1 - \operatorname{ind} \mathcal{A}_0 = \operatorname{ind} \mathcal{A}_1 \mathcal{P}_0 = \operatorname{ind} \widetilde{\mathcal{A}}_1 \widetilde{\mathcal{P}}_0 = \operatorname{ind} \mathcal{R}$. □

Part III

BVPs without the Transmission Property

Chapter 10

The edge approach to BVPs

10.1 Edge operators on a smooth manifold with boundary

BVPs without the transmission property on a manifold M with smooth boundary will be interpreted as specific edge problems. It is evident that such an M is a manifold with edge in the sense of Section 7.1, where now $\dim N = 0$. The above-mentioned neighbourhood V of $Y = \partial M$ in M will be chosen as a collar neighbourhood of Y, realised as $[0, 1) \times Y$, induced by the trivial normal bundle of Y. We adopt here notation and assumptions from the beginning of Section 2.3, including the order of variables. In Section 2.1 we wrote (y, t) with the half-line variable t normal to the boundary. However, here we prefer (r, y), since $r \in \mathbb{R}_+$ is interpreted as the cone axis variable of the edge calculus, cf. also the splitting of variables in (7.5). Clearly, r is now at the same time the variable normal to the boundary. Let $\widetilde{E}, \widetilde{F} \in \mathrm{Vect}(2M)$ be vector bundles on the double $2M$ of M, and $E := \widetilde{E}|_M, F := \widetilde{F}|_M$.

First we compare the degenerate operators of the edge calculus on M with the space

$$L_{\mathrm{cl}}^{\mu}(M; E, F)_{\mathrm{smooth}} := \big\{ A \in L_{\mathrm{cl}}^{\mu}(\mathrm{int}\, M; E, F) :$$
$$A = \widetilde{A}\big|_{\mathrm{int}\, M} + C, \ \widetilde{A} \in L_{\mathrm{cl}}^{\mu}\big(2M; \widetilde{E}, \widetilde{F}\big), \tag{10.1}$$
$$C \in L^{-\infty}(\mathrm{int}\, M; E, F)\big\}.$$

We will show that the operators in (10.1) are edge-degenerate modulo

$$L^{-\infty}(\mathrm{int}\, M; E, F),$$

though much more specific. They generate a proper substructure of the edge calculus which represents a calculus of pseudo-differential boundary value problems with or without the transmission property at the boundary.

© Springer Nature Switzerland AG 2018
X. Liu, B.-W. Schulze, *Boundary Value Problems with Global Projection Conditions*,
Operator Theory: Advances and Applications 265, https://doi.org/10.1007/978-3-319-70114-1_10

The double $2M$ in this definition is taken for convenience. We could equivalently define $L_{\mathrm{cl}}^{\mu}(M; E, F)_{\mathrm{smooth}}$ by replacing $2M$ by any other open smooth manifold of the same dimension containing M as an embedded manifold with boundary. We assume for a while that the involved bundles are trivial and of fibre dimension 1; then we simply write $L_{\mathrm{cl}}^{\mu}(M)_{\mathrm{smooth}}$ instead of (10.1). Analogous considerations will be valid (and later on tacitly employed) for arbitrary bundles.

Note that for $M := \overline{\mathbb{R}}_+ \times \Omega$, $\Omega \subseteq \mathbb{R}^q$ open, $q = \dim \partial M$, we have

$$L_{\mathrm{cl}}^{\mu}(\overline{\mathbb{R}}_+ \times \Omega)_{\mathrm{smooth}} = \Big\{ \mathrm{Op}_{r,y}(a) + C : a(r, y, \rho, \eta) \in S_{\mathrm{cl}}^{\mu}\big((\overline{\mathbb{R}}_+ \times \Omega) \times \mathbb{R}_{\rho,\eta}^{1+q}\big),$$

$$C \in L^{-\infty}(\mathbb{R}_+ \times \Omega)\Big\}.$$

Remark 10.1.1. Let $V = [0, 1) \times Y$ be the above-mentioned collar neighbourhood of $Y = \partial M$ and choose cut-off functions $\theta'' \prec \theta \prec \theta'$ on the r half-line, vanishing for $r > 1 - \varepsilon$ for some $0 < \varepsilon < 1/2$. Every $A \in L_{\mathrm{cl}}^{\mu}(M)_{\mathrm{smooth}}$ can be written in the form

$$A = \theta A_0 \theta' + (1 - \theta) A_{\mathrm{int}} (1 - \theta'') + C \qquad (10.2)$$

for some $A_{\mathrm{int}} \in L_{\mathrm{cl}}^{\mu}(\mathrm{int}\, M)$, $C \in L^{-\infty}(\mathrm{int}\, M)$ and an $A_0 \in L_{\mathrm{cl}}^{\mu}(M)_{\mathrm{smooth}}$ which is a locally finite sum of operators of the form $(\chi^{-1})_*(\varphi \mathrm{Op}_{r,y}(a)\varphi')$ for symbols $a(r, y, \rho, \eta) \in S_{\mathrm{cl}}^{\mu}\big((\overline{\mathbb{R}}_+ \times \mathbb{R}^q) \times \mathbb{R}_{\rho,\eta}^{1+q}\big)$, $\varphi, \varphi' \in C_0^{\infty}([0, 1) \times G)$, $G \subset Y$ a coordinate neighbourhood on Y and $\chi : [0, 1) \times G \to \overline{\mathbb{R}}_+ \times \mathbb{R}^q$ a chart near Y.

Theorem 10.1.2. *Let us fix $\mu, \gamma \in \mathbb{R}$, and set $\boldsymbol{g} := (\gamma, \gamma - \mu, (-\infty, 0])$. Then for every $A \in L_{\mathrm{cl}}^{\mu}(M; E, F)_{\mathrm{smooth}}$ there exists a $C_{\gamma} \in L^{-\infty}(\mathrm{int}\, M; E, F)$ such that*

$$A - C_{\gamma} \in L^{\mu}(M, \boldsymbol{g}; E, F). \qquad (10.3)$$

Proof. For simplicity we assume $E = F = \mathbb{C}$; the arguments in the general case are completely analogous. By virtue of Remark 10.1.1 it suffices to consider the case $A \in L_{\mathrm{cl}}^{\mu}(\overline{\mathbb{R}}_+ \times \Omega)_{\mathrm{smooth}}$, and we may assume $A := \mathrm{Op}_{r,y}(a)$ for some $a(r, y, \rho, \eta) \in S_{\mathrm{cl}}^{\mu}(\overline{\mathbb{R}}_+ \times \Omega \times \mathbb{R}_{\rho,\eta}^{1+q})$. The dependence of the symbol a on y does not affect the arguments; so we assume $a(r, \rho, \eta) \in S_{\mathrm{cl}}^{\mu}(\overline{\mathbb{R}}_+ \times \mathbb{R}_{\rho,\eta}^{1+q})$. Let $\chi(\rho, \eta)$ be an excision function, and write a as an asymptotic expansion

$$a(r, \rho, \eta) \sim \sum_{j=0}^{\infty} \chi(\rho, \eta) a_{(\mu-j)}(r, \rho, \eta)$$

in $S_{\mathrm{cl}}^{\mu}(\overline{\mathbb{R}}_+ \times \mathbb{R}_{\rho,\eta}^{1+q})$, where $a_{(\mu-j)}$ is the homogeneous component of a of order $\mu - j$. This allows us to write

$$a_{(\mu-j)}(r, \rho, \eta) = r^{-\mu}\big(r^j a_{(\mu-j)}(r, r\rho, r\eta)\big) \quad \text{for } r > 0. \qquad (10.4)$$

The functions $\widetilde{p}_{(\mu-j)}(r, \widetilde{\rho}, \widetilde{\eta}) := r^j a_{(\mu-j)}(r, \widetilde{\rho}, \widetilde{\eta})$ are homogeneous in $(\widetilde{\rho}, \widetilde{\eta}) \neq 0$ and smooth up to $r = 0$, and we can form an asymptotic expansion

$$\widetilde{p}\left(r,\widetilde{\rho},\widetilde{\eta}\right) \sim \sum_{j=0}^{\infty} \chi\left(\widetilde{\rho},\widetilde{\eta}\right) \widetilde{p}_{(\mu-j)}\left(r,\widetilde{\rho},\widetilde{\eta}\right)$$

in $S_{\mathrm{cl}}^{\mu}\left(\overline{\mathbb{R}}_{+} \times \mathbb{R}_{\widetilde{\rho},\widetilde{\eta}}^{1+q}\right)$. We have

$$p(r,\rho,\eta) := \widetilde{p}\left(r, r\rho, r\eta\right) \in S_{\mathrm{cl}}^{\mu}\left(\mathbb{R}_{+} \times \mathbb{R}_{\rho,\eta}^{1+q}\right)$$

and

$$a(r,\rho,\eta) = r^{-\mu} p(r,\rho,\eta) \ \mathrm{mod} \ S^{-\infty}\left(\mathbb{R}_{+} \times \mathbb{R}_{\rho,\eta}^{1+q}\right).$$

In fact, first we have

$$a(r,\rho,\eta) - \sum_{j=0}^{N} \chi(\rho,\eta) a_{(\mu-j)}(r,\rho,\eta) \in S_{\mathrm{cl}}^{-(N+1)}\left(\mathbb{R}_{+} \times \mathbb{R}_{\rho,\eta}^{1+q}\right), \qquad (10.5)$$

and, similarly,

$$\widetilde{p}\left(r,\widetilde{\rho},\widetilde{\eta}\right) - \sum_{j=0}^{N} \chi\left(\widetilde{\rho},\widetilde{\eta}\right) \widetilde{p}_{(\mu-j)}\left(r,\widetilde{\rho},\widetilde{\eta}\right) \in S^{-(N+1)}\left(\mathbb{R}_{+} \times \mathbb{R}_{\widetilde{\rho},\widetilde{\eta}}^{1+q}\right). \qquad (10.6)$$

This entails

$$r^{-\mu}\widetilde{p}\left(r, r\rho, r\eta\right) - r^{-\mu} \sum_{j=0}^{N} \chi(r\rho, r\eta) \widetilde{p}_{(\mu-j)}(r, r\rho, r\eta) \in S^{-(N+1)}\left(\mathbb{R}_{+} \times \mathbb{R}_{\rho,\eta}^{1+q}\right). \qquad (10.7)$$

From (10.5) and (10.7) it follows that

$$a(r,\rho,\eta) - r^{-\mu}\widetilde{p}\left(r, r\rho, r\eta\right) \in S^{-(N+1)}\left(\mathbb{R}_{+} \times \mathbb{R}_{\rho,\eta}^{1+q}\right)$$

since

$$\chi(\rho,\eta) a_{(\mu-j)}(r,\rho,\eta) - r^{-\mu}\chi(r\rho, r\eta) \widetilde{p}_{(\mu-j)}(r, r\rho, r\eta) \in S^{-\infty}\left(\mathbb{R}_{+} \times \mathbb{R}_{\rho,\eta}^{1+q}\right),$$

cf. relation (10.4). Thus we obtain

$$\mathrm{Op}_{r,y}(a) = r^{-\mu}\mathrm{Op}_{r,y}(p) \ \mathrm{mod} \ L^{-\infty}(\mathbb{R}_{+} \times \Omega). \qquad (10.8)$$

By Theorem 7.3.2, there is an $\widetilde{h}(r,z,\widetilde{\eta}) \in C^{\infty}\left(\overline{\mathbb{R}}_{+}, M_{\mathcal{O}}^{\mu}(\mathbb{R}_{\widetilde{\eta}}^{q})\right)$ such that for $h(r,z,\eta) := \widetilde{h}(r,z,r\eta)$ we have

$$\mathrm{Op}_{r,y}(p)(\eta) = \mathrm{Op}_{y}\mathrm{op}_{M}^{\beta}(h)(\eta) \ \mathrm{mod} \ L^{-\infty}(\mathbb{R}_{+} \times \Omega) \qquad (10.9)$$

for any real β. From Remark 7.3.10 we obtain that $f(\eta) := r^{-\mu}\omega\,\mathrm{op}_M^{\gamma-n/2}(h)(\eta)\omega'$ is an edge amplitude function for any choice of cut-off functions ω,ω' and that

$$r^{-\mu}\omega\,\mathrm{Op}_{r,y}(p)\omega' = \mathrm{Op}_y(f)\ \mathrm{mod}\ L^{-\infty}(\mathbb{R}_+ \times \Omega).$$

This gives us altogether a $C_\gamma \in L^{-\infty}(\mathbb{R}_+ \times \Omega)$ such that

$$\omega\,\mathrm{Op}_{r,y}(a)\omega' - C_\gamma \in L^\mu(\overline{\mathbb{R}}_+ \times \Omega, \boldsymbol{g}).$$

Since $\mathrm{Op}_{r,y}(a) = \omega\,\mathrm{Op}_{r,y}(a)\omega' + (1-\omega)\mathrm{Op}_{r,y}(a)(1-\omega'')\ \mathrm{mod}\ L^{-\infty}(\mathbb{R}_+ \times \Omega)$ for cut-off functions $\omega'' \prec \omega \prec \omega'$, we finally obtain $\mathrm{Op}_{r,y}(a) - C_\gamma \in L^\mu(\overline{\mathbb{R}}_+ \times \Omega, \boldsymbol{g})$ for a suitable $C_\gamma \in L^{-\infty}(\mathbb{R}_+ \times \Omega)$. $\qquad\square$

Remark 10.1.3. By definition, we have $L_{\mathrm{cl}}^\mu(M; E, F)_{\mathrm{smooth}} \subset L_{\mathrm{cl}}^\mu(\mathrm{int}\,M; E, F)$ and we first interpret the operators A in that space as continuous operators

$$A : C_0^\infty(\mathrm{int}\,M, E) \to C^\infty(\mathrm{int}\,M, F).$$

By Theorem 10.1.2, any choice of C_γ represents an edge quantisation of A depending on the weight γ. According to the results of the edge calculus we therefore have extensions as continuous operators

$$A - C_\gamma : H^{s,\gamma}(M, E) \to H^{s-\mu,\gamma-\mu}(M, F)$$

between weighted edge spaces for all $s \in \mathbb{R}$ (when M is compact, cf. the formula (7.11), otherwise between corresponding comp/loc-spaces).

Relation (10.4) suggests introducing the following space of edge-degenerate symbols:

Definition 10.1.4. Let $S_{\mathrm{cl}}^\mu\big(\mathbb{R}_+ \times \Omega \times \mathbb{R}_{\rho,\eta}^{1+q}\big)_{\mathrm{smooth}}$ denote the set of all $p(r, y, \rho, \eta) \in S_{\mathrm{cl}}^\mu\big(\mathbb{R}_+ \times \Omega \times \mathbb{R}_{\rho,\eta}^{1+q}\big)$ which are of the form

$$p(r, y, \rho, \eta) = \widetilde{p}(r, y, r\rho, r\eta)\ \text{ for some }\ \widetilde{p}(r, y, \widetilde{\rho}, \widetilde{\eta}) \in S_{\mathrm{cl}}^\mu\big(\overline{\mathbb{R}}_+ \times \Omega \times \mathbb{R}_{\widetilde{\rho},\widetilde{\eta}}^{1+q}\big).$$

such that the homogeneous components $\widetilde{p}_{(\mu-j)}(r, y, \widetilde{\rho}, \widetilde{\eta}), j \in \mathbb{N}$, have the property

$$\widetilde{p}_{(\mu-j)}(r, y, \widetilde{\rho}, \widetilde{\eta}) = r^j \widetilde{\widetilde{p}}_{(\mu-j)}(r, y, \widetilde{\rho}, \widetilde{\eta}),$$

$$\widetilde{\widetilde{p}}_{(\mu-j)}(r, y, \widetilde{\rho}, \widetilde{\eta}) \in S^{(\mu-j)}\Big(\overline{\mathbb{R}}_+ \times \Omega \times \big(\mathbb{R}_{\widetilde{\rho},\widetilde{\eta}}^{1+q} \setminus \{0\}\big)\Big).$$

Proposition 10.1.5. *For every* $a(r, y, \rho, \eta) \in S_{\mathrm{cl}}^\mu\big(\overline{\mathbb{R}}_+ \times \Omega \times \mathbb{R}_{\rho,\eta}^{1+q}\big)$ *there exists a* $p(r, y, \rho, \eta) \in S_{\mathrm{cl}}^\mu\big(\mathbb{R}_+ \times \Omega \times \mathbb{R}_{\rho,\eta}^{1+q}\big)_{\mathrm{smooth}}$ *satisfying the relation*

$$a(r, y, \rho, \eta) = r^{-\mu}p(r, y, \rho, \eta) \in S^{-\infty}\Big(\mathbb{R}_+ \times \Omega \times \mathbb{R}_{\rho,\eta}^{1+q}\Big). \qquad (10.10)$$

Conversely, for every $p(r, y, \rho, \eta) \in S_{\mathrm{cl}}^\mu\big(\mathbb{R}_+ \times \Omega \times \mathbb{R}_{\rho,\eta}^{1+q}\big)_{\mathrm{smooth}}$ *there exists an* $a(r, y, \rho, \eta) \in S_{\mathrm{cl}}^\mu\big(\overline{\mathbb{R}}_+ \times \Omega \times \mathbb{R}_{\rho,\eta}^{1+q}\big)$ *such that* (10.10) *holds.*

Proof. The first part of Proposition 10.1.5 is contained in the proof of Theorem 10.1.2. However, the relation between $r^{-\mu}p(r, y, \rho, \eta)$ and $a(r, y, \rho, \eta)$ can be established the other way around, using asymptotic summations in $S_{\mathrm{cl}}^{\mu}(\overline{\mathbb{R}}_+ \times \Omega \times \mathbb{R}_{\rho,\eta}^{1+q})$. $\qquad\square$

We have

$$L_{\mathrm{cl}}^{\mu}(\mathbb{R}_+ \times \Omega)_{\mathrm{smooth}} := \Big\{ \mathrm{Op}_{r,y}(b) + C : C \in L^{-\infty}(\mathbb{R}_+ \times \Omega),$$

$$b(r, y, \rho, \eta) \in r^{-\mu} S_{\mathrm{cl}}^{\mu}\Big(\mathbb{R}_+ \times \Omega \times \mathbb{R}_{\rho,\eta}^{1+q}\Big)_{\mathrm{smooth}} \Big\}.$$

By virtue of Proposition 10.1.5 every $A \in L_{\mathrm{cl}}^{\mu}(\mathbb{R}_+ \times \Omega)_{\mathrm{smooth}}$ has the form

$$A = \mathrm{Op}_{r,y}(a) + C \tag{10.11}$$

for an $a(r, y, \rho, \eta) \in S_{\mathrm{cl}}^{\mu}(\overline{\mathbb{R}}_+ \times \Omega \times \mathbb{R}_{\rho,\eta}^{1+q})$ and $C \in L^{-\infty}(\mathbb{R}_+ \times \Omega)$. Analogously as in Remark 10.1.3, operators of this kind are first interpreted as maps

$$A : C_0^{\infty}(\mathbb{R}_+ \times \Omega) \to C^{\infty}(\mathbb{R}_+ \times \Omega).$$

Every $A \in L_{\mathrm{cl}}^{\mu}(\mathbb{R}_+ \times \Omega)_{\mathrm{smooth}}$ can be written as $A = A_0 + C$ for a properly supported $A_0 \in L_{\mathrm{cl}}^{\mu}(\mathbb{R}_+ \times \Omega)_{\mathrm{smooth}}$ and a $C \in L^{-\infty}(\mathbb{R}_+ \times \Omega)$.

An $A \in L_{\mathrm{cl}}^{\mu}(\mathbb{R}_+ \times \Omega)_{\mathrm{smooth}}$ is called elliptic if the symbol $a(r, y, \rho, \eta) \in S_{\mathrm{cl}}^{\mu}(\overline{\mathbb{R}}_+ \times \Omega \times \mathbb{R}_{\rho,\eta}^{1+q})$ in the representation (10.11) is elliptic in the standard sense, more precisely, $a_{(\mu)}(r, y, \rho, \xi) \neq 0$ for all $(r, y, \rho, \xi) \in \overline{\mathbb{R}}_+ \times \Omega \times (\mathbb{R}_{\rho,\eta}^{1+q} \setminus \{0\})$.

Corollary 10.1.6. *Let $A \in L_{\mathrm{cl}}^{\mu}(\mathbb{R}_+ \times \Omega)_{\mathrm{smooth}}$, $B \in L_{\mathrm{cl}}^{\nu}(\mathbb{R}_+ \times \Omega)_{\mathrm{smooth}}$, and let A or B be properly supported. Then we have $AB \in L_{\mathrm{cl}}^{\mu+\nu}(\mathbb{R}_+ \times \Omega)_{\mathrm{smooth}}$. Moreover, an elliptic $A \in L_{\mathrm{cl}}^{\mu}(\mathbb{R}_+ \times \Omega)_{\mathrm{smooth}}$ has a properly supported parametrix $P \in L_{\mathrm{cl}}^{-\mu}(\mathbb{R}_+ \times \Omega)_{\mathrm{smooth}}$, where $AP - 1, PA - 1 \in L^{-\infty}(\mathbb{R}_+ \times \Omega)$.*

Let us recall notation (1.22) for a smooth manifold M with boundary, namely,

$$S^{(\mu)}(T^*M \setminus 0; E, F) \quad \text{for } E, F \in \mathrm{Vect}(M), \tag{10.12}$$

which is the set of all bundle morphisms

$$\sigma : \pi_M^* E \to \pi_M^* F, \tag{10.13}$$

$\pi_M : T^*M \setminus 0 \to M$, positively homogeneous of order μ in the sense (1.21). Note that

$$S^{(\mu)}(T^*M \setminus 0; E, F) = \Big\{ \widetilde{\sigma}\big|_{T^*(2M)\setminus 0} : \widetilde{\sigma} \in S^{(\mu)}(T^*(2M) \setminus 0; \widetilde{E}, \widetilde{F}) \Big\}$$

for bundles $\widetilde{E}, \widetilde{F} \in \mathrm{Vect}(2M)$ such that $E = \widetilde{E}\big|_M$, $F = \widetilde{F}\big|_M$.

Definition 10.1.7. Let the space $L^\mu(M, \boldsymbol{g}; \boldsymbol{v})_{\text{smooth}}$ for $\boldsymbol{g} = (\gamma, \gamma - \mu, (-\infty, 0])$, $\boldsymbol{v} = (E, F; J_1, J_2)$ be defined as the set of all

$$\mathcal{A} = \begin{pmatrix} A & K \\ T & Q \end{pmatrix} \in L^\mu(M, \boldsymbol{g}; \boldsymbol{v}) \qquad (10.14)$$

such that $A \in L_{\text{cl}}^\mu(M; E, F)_{\text{smooth}}$.

The 2×2-block matrix structure of operators (10.14) is analogous to that in the Boutet de Monvel's calculus outlined in Chapter 2, or, more generally, in the general edge calculus of Chapter 7. The subspace of Green operators

$$\mathcal{G} = \begin{pmatrix} G & K \\ T & Q \end{pmatrix} \in L_{\text{G}}^\mu(M, \boldsymbol{g}; \boldsymbol{v}) \qquad (10.15)$$

is the same as in the general edge calculus, here specified to a manifold with smooth boundary, with trace operators T, potential operators K and classical pseudo-differential operators Q on the boundary. For brevity we talk about Green operators also when we mean G in the upper left corner. More generally, we have the subspace of Mellin plus Green operators

$$\mathcal{G} = \begin{pmatrix} M + G & K \\ T & Q \end{pmatrix} \in L_{\text{M+G}}^\mu(M, \boldsymbol{g}; \boldsymbol{v}), \qquad (10.16)$$

which is also the same as in the general edge calculus on a manifold with boundary.

For purposes below we deepen the information on the space

$$\mathcal{R}_{\text{G}}^\mu(\Omega \times \mathbb{R}^q, \boldsymbol{g})_{\mathcal{P}, \mathcal{Q}} \qquad (10.17)$$

of Green symbols belonging to top left corners G in (10.15), cf. Definition 7.2.1. Recall that in the edge calculus we usually assume $\boldsymbol{g} := (\gamma, \gamma - \mu, \Theta)$. Nevertheless, similarly as in cone Green operators, we may replace $\gamma - \mu$ by another weight δ, cf. formula (6.80). Recall that for the model cone $\mathbb{R}_+$ an element $g(y, \eta)$ in (10.17) is characterised by the relations

$$g(y, \eta) \in S_{\text{cl}}^\mu\left(\Omega \times \mathbb{R}^q; \mathcal{K}^{s,\gamma}(\mathbb{R}_+), \mathcal{S}_{\mathcal{P}}^\delta(\mathbb{R}_+)\right), \qquad (10.18)$$

$$g^*(y, \eta) \in S_{\text{cl}}^\mu\left(\Omega \times \mathbb{R}^q; \mathcal{K}^{s,-\delta}(\mathbb{R}_+), \mathcal{S}_{\mathcal{Q}}^{-\gamma}(\mathbb{R}_+)\right), \qquad (10.19)$$

for all $s \in \mathbb{R}$ and asymptotic types $\mathcal{P}$ and $\mathcal{Q}$, associated with (δ, Θ) and $(-\gamma, \Theta)$, respectively.

Proposition 10.1.8. *Relation $g(y, \eta) \in \mathcal{R}_{\text{G}}^\mu(\Omega \times \mathbb{R}^q, \boldsymbol{g})_{\mathcal{P}, \mathcal{Q}}$ is equivalent to the existence of a kernel function*

$$k(y, \eta, r, r') \in S_{\text{cl}}^\mu\left(\Omega_y \times \mathbb{R}_\eta^q, \mathcal{S}_{\mathcal{P}}^\delta(\mathbb{R}_+) \,\widehat{\otimes}_\pi\, \mathcal{S}_{\overline{\mathcal{Q}}}^{-\gamma}(\mathbb{R}_+)\right)$$

such that

$$(g(y, \eta)u)(r) = \int_0^\infty k_g(y, \eta, r, r')u(r')dr', \qquad (10.20)$$

$u \in \mathcal{K}^{s,\gamma}(\mathbb{R}_+)$, *for*

$$k_g(y, \eta, r, r') = [\eta]k(y, \eta, r[\eta], r'[\eta]),$$

with $\hat{\otimes}_\Gamma$ *being the tensor product defined by*

$$\mathcal{K}^{0,-\gamma}(\mathbb{R}_+) \, \hat{\otimes}_\pi \, \mathcal{S}_{\mathcal{P}}^\delta(\mathbb{R}_+) \cup \mathcal{S}_{\mathcal{Q}}^{-\gamma} \, \hat{\otimes}_\pi \, \mathcal{K}^{0,\delta}(\mathbb{R}_+) = \mathcal{S}_{\mathcal{P}}^\delta(\mathbb{R}_+) \, \hat{\otimes}_\Gamma \, \mathcal{S}_{\mathcal{Q}}^{-\gamma}(\mathbb{R}_+),$$

cf. [36], [60].

Proof. The y-variables do not affect the ideas of the proof, so they will be dropped. First, it is clear that (10.20) defines a Green symbol as soon as the kernel is of the above-mentioned form. The homogeneous components $g_{(\mu-j)}$ are then given by the kernel functions

$$k_{g_{(\mu-j)}}(\eta, r, r') = |\eta|k_{(\mu-j)}(\eta, r|\eta|, r'|\eta|)$$

for $\eta \neq 0$. Conversely, if g is a Green symbol every homogeneous component $g_{(\mu-j)}$ has a kernel $k_{g_{(\mu-j)}} \in C^\infty(\mathbb{R}^q \setminus \{0\}, \mathcal{S}_{\mathcal{P}}^\delta(\mathbb{R}_+) \hat{\otimes}_\Gamma \mathcal{S}_{\mathcal{Q}}^{-\gamma}(\mathbb{R}_+))$. This is a consequence of Theorem 6.7.3. By virtue of twisted homogeneity of $g_{(\mu-j)}$ we have

$$k_{g_{(\mu-j)}}(\lambda\eta, r, r') = \lambda^{\mu-j+1}k_{g_{(\mu-j)}}(\eta, \lambda r, \lambda r')$$

for every $\lambda \in \mathbb{R}_+$. It follows that

$$k_{(\mu-j)}(\eta, r, r') = |\eta|^{-1}k_{g_{(\mu-j)}}(\eta, r|\eta|^{-1}, r'|\eta|^{-1})$$

is homogeneous of order $\mu - j$ in η and belongs to the space

$$C^\infty\left(\mathbb{R}^q \setminus \{0\}, \mathcal{S}_{\mathcal{P}}^\delta(\mathbb{R}_+) \hat{\otimes}_\Gamma \mathcal{S}_{\mathcal{Q}}^{-\gamma}(\mathbb{R}_+)\right).$$

We now choose a $\tilde{k} \in S_{\mathrm{cl}}^\mu\left(\mathbb{R}^q, \mathcal{S}_{\mathcal{P}}^\delta(\mathbb{R}_+) \hat{\otimes}_\Gamma \mathcal{S}_{\mathcal{Q}}^{-\gamma}(\mathbb{R}_+)\right)$ as an asymptotic sum

$$\sum_{j=0}^\infty \chi(\eta)k_{(\mu-j)}(\eta, r, r')$$

and pass to $\tilde{g}$ via the kernel $[\eta]\tilde{k}(\eta, r[\eta], r'[\eta])$. Then the first part of the proof shows $\tilde{g} \in \mathcal{R}_{\mathrm{G}}^\mu(\mathbb{R}^q, \boldsymbol{g})_{\mathcal{P},\mathcal{Q}}$. Moreover, we have $g - \tilde{g} \in \mathcal{R}_{\mathrm{G}}^{-\infty}(\mathbb{R}^q, \boldsymbol{g})_{\mathcal{P},\mathcal{Q}}$, by the construction of $\tilde{g}$ and $\tilde{k}$. It follows that

$$\tilde{\tilde{k}}(\eta, r, r') := k_g(\eta, r, r') - [\eta]\tilde{k}(\eta, r[\eta], r'[\eta]) \in \mathcal{S}\left(\mathbb{R}^q, \mathcal{S}_{\mathcal{P}}^\delta(\mathbb{R}_+) \hat{\otimes}_\Gamma \mathcal{S}_{\mathcal{Q}}^{-\gamma}(\mathbb{R}_+)\right).$$

Now we may set

$$k(\eta, r, r') := \widetilde{k}(\eta, r, r') + [\eta]^{-1}\widetilde{\widetilde{k}}(\eta, r[\eta]^{-1}, r'[\eta]^{-1})$$

since the second summand on the right-hand side also belongs to

$$\mathcal{S}\left(\mathbb{R}^q, \mathcal{S}_{\mathcal{P}}^{\delta}(\mathbb{R}_+) \widehat{\otimes}_{\Gamma} \mathcal{S}_{\mathcal{Q}}^{-\gamma}(\mathbb{R}_+)\right). \qquad \square$$

Corollary 10.1.9. *Let* $g \in \mathcal{R}_G^{\mu}(\Omega \times \mathbb{R}^q, \boldsymbol{g})_{\mathcal{P}, \mathcal{Q}}$*, and let* k *be a kernel function as in Proposition 10.1.8. Then we have*

$$k \in S_{\mathrm{cl}}^{\mu}\left(\Omega \times \mathbb{R}^q, \mathcal{S}_{\mathcal{P}_p}^{\delta}(\mathbb{R}_+) \widehat{\otimes}_{\pi} \mathcal{S}_{\mathcal{Q}_q}^{-\gamma}(\mathbb{R}_+)\right),$$

cf. notation in Theorem 6.7.3 (ii). *In particular, for* $\Theta = (-\infty, 0]$ *it follows that*

$$k \in S_{\mathrm{cl}}^{\mu}\left(\Omega \times \mathbb{R}^q, \mathcal{S}_{\mathcal{P}}^{\delta}(\mathbb{R}_+) \widehat{\otimes}_{\pi} \mathcal{S}_{\mathcal{Q}}^{-\gamma}(\mathbb{R}_+)\right).$$

Proposition 10.1.10. *Let* $a(r, \rho, \eta) \in S_{\mathrm{cl}}^{\mu}\left(\overline{\mathbb{R}}_+ \times \mathbb{R}_{\rho,\eta}^{1+q}\right)$*, and consider the operator family*

$$\mathrm{op}^+(a)(\eta) = \mathrm{r}^+\mathrm{op}(a)(\eta)\mathrm{e}^+ : C_0^{\infty}(\mathbb{R}_+) \to C^{\infty}(\mathbb{R}_+),$$

where e^+ *is the extension by* 0 *to* $\mathbb{R} \setminus \mathbb{R}_+$ *and* r^+ *the restriction to* $\mathbb{R}_+$*. Let* $T = \{(-j, 0) : j \in \mathbb{N}\}$ *be the Taylor asymptotic type.*

(i) *For cut-off functions* $\sigma \prec \sigma'$ *we have*

$$\sigma\mathrm{op}^+(a)(\eta)(1 - \sigma'), \ (1 - \sigma')\mathrm{op}^+(a)(\eta)\sigma \in \mathcal{R}_G^{-\infty}\left(\mathbb{R}^q, (0, 0, (-\infty, 0])\right)_{T,T}.$$

(ii) *Let* $\omega \prec \omega'$ *be cut-off functions; then*

$$\omega_{\eta}\,\mathrm{op}^+(a)(\eta)(1 - \omega'_{\eta}), \ (1 - \omega'_{\eta})\,\mathrm{op}^+(a)(\eta)\omega_{\eta} \in \mathcal{R}_G^{-\infty}\left(\mathbb{R}^q, (0, 0, (-\infty, 0])\right)_{T,T}.$$

Proof. (i) A straightforward computation shows that

$$g(\eta) := \sigma\,\mathrm{op}^+(a)(\eta)(1 - \sigma')$$

has a kernel $k_g(\eta, r, r') \in \mathcal{S}(\mathbb{R}^q, \mathcal{S}(\overline{\mathbb{R}}_+ \times \overline{\mathbb{R}}_+))$. Then it suffices to note that $\mathcal{S}(\overline{\mathbb{R}}_+ \times \overline{\mathbb{R}}_+) = \mathcal{S}(\overline{\mathbb{R}}_+) \widehat{\otimes}_{\pi} \mathcal{S}(\overline{\mathbb{R}}_+)$ and $\mathcal{S}(\overline{\mathbb{R}}_+) = \mathcal{S}_T^0(\mathbb{R}_+)$. The arguments for $(1 - \sigma')\mathrm{op}^+(a)(\eta)\sigma$ are analogous.

(ii) Define $\kappa(\eta) := \kappa_{[\eta]}$ for $(\kappa_{\delta}u)(r) = \delta^{1/2}u(\delta r), \delta \in \mathbb{R}_+$. Then, upon setting $\widetilde{g}(\eta) := \kappa^{-1}(\eta)g(\eta)\kappa(\eta)$ for $g(\eta) := \omega_{\eta}\,\mathrm{op}^+(a)(\eta)(1 - \omega'_{\eta})$ we obtain

$$\widetilde{g}(\eta) = \omega\,\mathrm{op}^+(\widetilde{a})(\eta)(1 - \omega')$$

for $\widetilde{a}(r, \rho, \eta) = a(r[\eta]^{-1}, \rho[\eta], \eta) \in S_{\mathrm{cl}}^{\mu}(\mathbb{R}_{\eta}^q, S_{\mathrm{cl}}^{\mu}(\overline{\mathbb{R}}_+ \times \mathbb{R}_{\rho}))$. By virtue of (i), the operators $\widetilde{g}(\eta)$ have a kernel function $\widetilde{k}(\eta, r, r') \in \mathcal{S}(\mathbb{R}_{\eta}^q, \mathcal{S}(\overline{\mathbb{R}}_+ \times \overline{\mathbb{R}}_+))$. Thus the assertion is a consequence of Proposition 10.1.8, since g has the kernel $k(\eta, r, r') = [\eta]\widetilde{k}(\eta, r[\eta], r'[\eta])$. The arguments for $(1-\omega'_{\eta})\,\mathrm{op}^+(a)(\eta)\omega_{\eta}$ are again analogous. $\square$

For $\mathcal{A} \in L^\mu(M, \boldsymbol{g}; \boldsymbol{v})_{\text{smooth}}$ we set

$$\sigma(\mathcal{A}) := (\sigma_\psi(\mathcal{A}), \sigma_\partial(\mathcal{A})), \tag{10.21}$$

where $\sigma_\psi(\mathcal{A}) := \sigma_\psi(\mathcal{A}_{11}) \in S^{(\mu)}(T^*M \setminus 0; E, F)$ is the (homogeneous principal) interior symbol of $\mathcal{A}$ of order μ,

$$\sigma_\psi(\mathcal{A}) : \pi_M^* E \to \pi_M^* F, \tag{10.22}$$

$\pi_M : T^*M \setminus 0 \to M$, while $\sigma_\partial(\mathcal{A}) := \sigma_\wedge(\mathcal{A})$ is the (homogeneous principal) boundary symbol of $\mathcal{A}$ of order μ, analogously as the edge symbol defined in Section 8.1,

$$\sigma_\partial(\mathcal{A}) : \pi_Y^* \begin{pmatrix} \mathcal{K}^{s,\gamma}(\mathbb{R}_+) \otimes E' \\ \oplus \\ J_1 \end{pmatrix} \to \pi_Y^* \begin{pmatrix} \mathcal{K}^{s-\mu,\gamma-\mu}(\mathbb{R}_+) \otimes F' \\ \oplus \\ J_2 \end{pmatrix}, \quad \pi_Y : T^*Y \setminus 0 \to Y. \tag{10.23}$$

Analogously to (7.60), we define

$$L^{\mu-j}(M, \boldsymbol{g}; \boldsymbol{v})_{\text{smooth}} := L^{\mu-j}(M, \boldsymbol{g}; \boldsymbol{v}) \cap L^{\mu-(j-1)}(M, \boldsymbol{g}; \boldsymbol{v})_{\text{smooth}} \tag{10.24}$$

for every $j \in \mathbb{N} \setminus \{0\}$. Set

$$L^{-\infty}(M, \boldsymbol{g}; \boldsymbol{v})_{\text{smooth}} := \bigcap_{j \in \mathbb{N}} L^{\mu-j}(M, \boldsymbol{g}; \boldsymbol{v})_{\text{smooth}}.$$

Many general properties of the general edge calculus outlined in Section 7.4 hold in analogous form also for the spaces in Definition 10.1.7 on a manifold with smooth boundary. For completeness, we recall here some important elements but add a number of essential observations. Let us first note that

$$L_G^{\mu-j}(M, \boldsymbol{g}; \boldsymbol{v})_{\text{smooth}} = L_G^{\mu-j}(M, \boldsymbol{g}; \boldsymbol{v}) \tag{10.25}$$

and

$$L_{M+G}^{\mu-j}(M, \boldsymbol{g}; \boldsymbol{v})_{\text{smooth}} = L_{M+G}^{\mu-j}(M, \boldsymbol{g}; \boldsymbol{v}). \tag{10.26}$$

Although the general aspects of the calculus of operators in $L^\mu(M, \boldsymbol{g}; \boldsymbol{v})_{\text{smooth}}$ are not the main topic of this exposition, this system of notions allows us to single out many substructures which are of independent interest.

Remark 10.1.11. We have

$$\mathcal{B}^{0,0}(M, \boldsymbol{v}) \subset L^0(M, \boldsymbol{g}; \boldsymbol{v}) \quad \text{for } \boldsymbol{g} = (0, 0, (-\infty, 0]), \tag{10.27}$$

cf. Definition 2.4.14 and

$$\mathcal{B}_G^{0,0}(M, \boldsymbol{v}) \subset L_G^0(M, \boldsymbol{g}; \boldsymbol{v}) \quad \text{for } \boldsymbol{g} = (0, 0, (-\infty, 0]), \tag{10.28}$$

cf. Definition 2.4.13.

In the following, for convenience, we assume M to be compact. The straight-forward generalisations to the non-compact case are left to the reader.

Theorem 10.1.12. *An element* $\mathcal{A} \in L^\mu(M, \boldsymbol{g}; \boldsymbol{v})_{\text{smooth}}$ *for* $\boldsymbol{g} := (\gamma, \gamma - \mu, \Theta)$ *and* $\boldsymbol{v} := (E, F; J_1, J_2)$ *induces continuous operators*

$$\mathcal{A}: \begin{array}{c} H^{s,\gamma}(M, E) \\ \oplus \\ H^s(Y, J_1) \end{array} \longrightarrow \begin{array}{c} H^{s-\mu,\gamma-\mu}(M, F) \\ \oplus \\ H^{s-\mu}(Y, J_2) \end{array} \tag{10.29}$$

and

$$\mathcal{A}: \begin{array}{c} H^{s,\gamma}_{\mathcal{P}}(M, E) \\ \oplus \\ H^s(Y, J_1) \end{array} \longrightarrow \begin{array}{c} H^{s-\mu,\gamma-\mu}_{\mathcal{Q}}(M, F) \\ \oplus \\ H^{s-\mu}(Y, J_2) \end{array} \tag{10.30}$$

for $s \in \mathbb{R}$ *and every asymptotic type* $\mathcal{P}$ *for some resulting* $\mathcal{Q}$*. In the case* $\boldsymbol{g} := (\gamma, \gamma - \mu)$ *we have (7.57),* $s \in \mathbb{R}$*.*

Remark 10.1.13. *Every* $\mathcal{A} \in L^{\mu-1}(M, \boldsymbol{g}; \boldsymbol{v})_{\text{smooth}}$ *is compact as an operator* (10.29) *for every* $s \in \mathbb{R}$*.*

10.2 Operator conventions

Theorem 10.2.1. *Let* M *be a smooth manifold with boundary* Y*. Then there exists a map*

$$\text{op}^\gamma : S^{(\mu)}(T^*M \setminus 0; E, F) \to L^\mu(M, \boldsymbol{g}; E, F)_{\text{smooth}} \tag{10.31}$$

for $E, F \in \text{Vect}(M)$*,* $\boldsymbol{g} = (\gamma, \gamma - \mu, \Theta)$ *or* $\boldsymbol{g} = (\gamma, \gamma - \mu)$*, such that*

$$\sigma_\psi \circ \text{op}^\gamma = \text{id}_{S^{(\mu)}(T^*M \setminus 0; E, F)}.$$

Theorem 10.2.1 is a direct consequence of Theorem 10.1.2. The construction shows that the operator $\text{op}^\gamma(\sigma)$ is determined by $\sigma \in S^\mu(T^*M \setminus 0; E, F)$, modulo $L^{\mu-1}(M, \boldsymbol{g}; E, F)_{\text{smooth}}$. The operator convention $\text{op}^\gamma(\cdot)$ is valid for all $\sigma \in S^{(\mu)}(T^*M \setminus 0; E, F)$, regardless of whether σ has the transmission property at the boundary or not, and for arbitrary weights γ. There are some reasons to take into account also other operator conventions, for instance, based on truncation, applied in Part I to symbols with the transmission property, cf. the formulas (2.16) or (2.139). In order to compare (10.31) with the truncation convention, we first formulate the operator convention suggested from the standard pseudo-differential calculus.

Proposition 10.2.2. *There is a map*

$$\text{op} : S^{(\mu)}(T^*M \setminus 0; E, F) \to L^\mu_{\text{cl}}(M; E, F)_{\text{smooth}} \tag{10.32}$$

such that

$$\sigma_\psi \circ \text{op} = \text{id}_{S^{(\mu)}(T^*M \setminus 0; E, F)}.$$

Proof. Given $\sigma \in S^{(\mu)}(T^*M \setminus 0; E, F)$, we first choose a

$$\tilde{\sigma} \in S^{(\mu)}(T^*(2M) \setminus 0; \tilde{E}, \tilde{F})$$

such that $\tilde{\sigma}|_{T^*M \setminus 0} = \sigma$. According to Proposition 1.1.13, we have an operator convention $\widetilde{\mathrm{op}} : S^{(\mu)}(T^*(2M) \setminus 0; \tilde{E}, \tilde{F}) \to L^\mu(2M; \tilde{E}, \tilde{F})$ on $2M$ such that $\tilde{\sigma}_\psi \circ \widetilde{\mathrm{op}} = \mathrm{id}_{S^{(\mu)}(T^*(2M) \setminus 0; \tilde{E}, \tilde{F})}$. Here $\tilde{\sigma}_\psi$ denotes the principal symbolic map on $2M$. It suffices then to set $\mathrm{op}(\sigma) := \widetilde{\mathrm{op}}(\tilde{\sigma})|_{\mathrm{int}\, M}$. $\square$

Setting $\tilde{A} := \widetilde{\mathrm{op}}(\tilde{\sigma})$ and $A := \mathrm{op}(\sigma)$, it is justified to write

$$A = \mathrm{r}^+ \tilde{A} \mathrm{e}^+ \in L^\mu_{\mathrm{cl}}(M; E, F)_{\mathrm{smooth}} \tag{10.33}$$

with e^+ being the operator of extension from int M to $2M$ by zero and r^+ the restriction of distributions on $2M$ to int M. Concerning the application of e^+, we have to specify the admitted distributions on int M, for instance, assume that they belong to $C_0^\infty(\mathrm{int}\, M, E)$. Assuming, for simplicity, that M is compact, then $\mathrm{e}^+, \mathrm{r}^+$ induce continuous operators

$$\mathrm{e}^+ : L^2(M, E) \to L^2(2M, \tilde{E}), \quad \mathrm{r}^+ : L^2(2M, \tilde{F}) \to L^2(M, F) \tag{10.34}$$

(the involved vector bundles are assumed to be equipped with Hermitean metrics). Then, for any $\tilde{A} \in L^0_{(\mathrm{cl})}(2M; \tilde{E}, \tilde{F})$ we obtain a continuous operator

$$\mathrm{r}^+ \tilde{A} \mathrm{e}^+ : L^2(M, E) \to L^2(M, F). \tag{10.35}$$

Theorem 10.2.3. *For every* $\tilde{A} \in L^0_{\mathrm{cl}}(2M; \tilde{E}, \tilde{F})$ *we have*

$$\mathrm{r}^+ \tilde{A} \mathrm{e}^+ \in L^0(M, \boldsymbol{g}; E, F)_{\mathrm{smooth}} \quad \text{for } \boldsymbol{g} = (0, 0, (-\infty, 0]). \tag{10.36}$$

Theorem 10.2.3 belongs to the results of [50]. Also the case of truncated operators for arbitrary orders and weights is treated there. Theorem 10.2.3 yields an operator convention

$$\mathrm{op}^+_{\boldsymbol{g}} : S^{(0)}(T^*M \setminus 0; E, F) \to L^0(M, \boldsymbol{g}; E, F)_{\mathrm{smooth}}, \tag{10.37}$$

through the following chain of maps:

$$\begin{aligned} S^{(0)}(T^*M \setminus 0; E, F) &\to S^{(0)}(T^*(2M) \setminus 0; \tilde{E}, \tilde{F}) \\ &\to L^0_{\mathrm{cl}}(2M; \tilde{E}, \tilde{F}) \to L^0(M, \boldsymbol{g}; E, F)_{\mathrm{smooth}}, \end{aligned} \tag{10.38}$$

namely, $\sigma \to \tilde{\sigma} \to \tilde{A} \to A = \mathrm{r}^+ \tilde{A} \mathrm{e}^+$. Clearly, the latter operator does not depend on the choice of $\tilde{\sigma}$ with $\sigma = \tilde{\sigma}|_{T^*M \setminus 0}$. In order to illustrate the difference between op^γ in (10.31) and $\mathrm{op}^+_{\boldsymbol{g}}$ for $\gamma = \mu = 0$, we consider the local situation in the half-space $M := \overline{\mathbb{R}}^n_+ = \overline{\mathbb{R}}_{+,r} \times \mathbb{R}^{n-1}_y$ for a symbol $\sigma := a_{(0)}(\rho, \eta) \in S^{(0)}(\mathbb{R}^n_{\rho, \eta} \setminus \{0\})$. Here in the case of constant coefficients we may choose $\tilde{\sigma} = \sigma$. Then for any excision function $\chi(\rho, \eta)$ in $\mathbb{R}^n$ we get an operator

$$\widetilde{A}u(r, y) := \mathrm{Op}(a)u(r, y) = \iint e^{i((r-r')\rho + (y-y')\eta)} a(\rho, \eta) u(r', y') dr' dy' đ\rho đ\eta$$

for $a(\rho, \eta) := (\chi a_{(0)})(\rho, \eta)$, and Theorem 10.2.3 shows that

$$\mathrm{r}^+ \mathrm{Op}(a) \mathrm{e}^+ \in L^0 \big(\overline{\mathbb{R}}^n_+, \boldsymbol{g} \big)_{\mathrm{smooth}}.$$

Note that because of $\mathrm{r}^+ \mathrm{Op}((1-\chi)a_{(0)}(\rho, \eta)) \mathrm{e}^+ \in L^{-\infty} \big(\overline{\mathbb{R}}^n_+, \boldsymbol{g} \big)$ we even have

$$\mathrm{r}^+ \mathrm{Op}(a_{(0)}) \mathrm{e}^+ \in L^0 \big(\overline{\mathbb{R}}^n_+, \boldsymbol{g} \big)_{\mathrm{smooth}}.$$

An inspection of the details of the proof of Theorem 10.2.3 shows that much more information on the original symbol is preserved under the truncation operator convention $\mathrm{op}_{\boldsymbol{g}}^+(\,\cdot\,)$ than under op^0. This concerns, in particular, the specific mero-morphic structure of the involved smoothing Mellin symbols. The effect is already visible in the case of the half-line, studied in [45], see also Eskin's book [14], cf. the remarks in Chapter 6. However, this leads to some restrictions on the weight when we want to pass from $\gamma = 0$ to an arbitrary γ. In addition, the truncation operator convention in general requires subtracting a Green operator G_γ, controlled in a more precise way than C_γ in Theorem 10.1.2. For arbitrary weights γ and orders μ, there are different generalisations of Theorem 10.2.3; we do not report all variants here. Let us content ourselves with the following cases; concerning the complete information, cf. [50].

Theorem 10.2.4. *For every $\widetilde{A} \in L^\mu_{\mathrm{cl}}\big(2M; \widetilde{E}, \widetilde{F}\big)$, $\mu \in \mathbb{R}$, and given $\boldsymbol{g} := (\gamma, \gamma - \mu, (-\infty, 0])$, for $\gamma \geq 0, \gamma - \mu \geq 0, \gamma - \mu \notin 1/2 + \mathbb{Z}$, there exists a $G_\gamma \in L^\mu_{\mathrm{G}}(M, \boldsymbol{b}; E, F)$ for $\boldsymbol{b} := (\gamma, 0, (-\infty, 0])$, such that*

$$\mathrm{r}^+ \widetilde{A} \mathrm{e}^+ - G_\gamma \in L^\mu(M, \boldsymbol{g}; E, F)_{\mathrm{smooth}}. \tag{10.39}$$

This result is similar in meaning with Theorem 10.2.3 and (10.38), yielding an operator convention

$$\mathrm{op}_{\boldsymbol{g}}^{\mathsf{l}} : S^{(\mu)}(T^*M \setminus 0; E, F) \to L^\mu(M, \boldsymbol{g}; E, F)_{\mathrm{smooth}}, \tag{10.40}$$

in this case by the chain of maps $\mathrm{op}_{\boldsymbol{g}}^+ : \sigma \to \tilde{\sigma} \to \widetilde{A} \to A := \mathrm{r}^+ \widetilde{A} \mathrm{e}^+ - G_\gamma$ where $\sigma_\psi \circ \mathrm{op}_{\boldsymbol{g}}^+ = \mathrm{id}_{S^{(\mu)}(T^*M \setminus 0; E, F)}$.

Remark 10.2.5. The image under (10.39) consists of edge operators that have con-stant discrete asymptotics (which is, of cause, also the case in (10.31)). Therefore, there is a discrete set $D_A \subset \mathbb{R}$ of weights such that for any $\delta \notin D_A$ there exists an $H_\delta \in L^\mu_{\mathrm{G}}(M, \boldsymbol{a}; E, F)$ with H_δ and $\boldsymbol{a} = (a_1, a_2, (-\infty, 0])$ depending on A, such that $A - H_\delta \in L^\mu(M, \boldsymbol{d}; E, F)_{\mathrm{smooth}}$ for $\boldsymbol{d} = (\delta, \delta - \mu, (-\infty, 0])$. In other words, for every such δ we have an operator convention

$$\mathrm{op}_{\boldsymbol{d}}^+ : S^{(\mu)}(T^*M \setminus 0; E, F) \to L^\mu(M, \boldsymbol{d}; E, F)_{\mathrm{smooth}}, \tag{10.41}$$

by $\mathrm{op}_{\boldsymbol{d}}^+ : \sigma \to \tilde{\sigma} \to \widetilde{A} \to A := \mathrm{r}^+ \widetilde{A} \mathrm{e}^+ - G_\gamma - H_\delta$ where $\sigma_\psi \circ \mathrm{op}_{\boldsymbol{d}}^+ = \mathrm{id}_{S^{(\mu)}(T^*M \setminus 0; E, F)}$.

10.3   The anti-transmission property

In this section we return to scalar symbols (for simplicity). Recall that the transmission property of a symbol $a(\rho) \in S^\mu_{\mathrm{cl}}(\mathbb{R})$ means the condition (2.6). In general, the curve

$$L(a) = \{a(\rho) \in \mathbb{C} : \rho \in \mathbb{R}\} \tag{10.42}$$

is not closed. Let $a(r, y, \rho, \eta) \in S^0_{\mathrm{cl}}(\Omega \times \overline{\mathbb{R}}_+ \times \mathbb{R}^n_{\rho,\eta})$ be an elliptic symbol, $a_{(0)}$ its homogeneous principal part, and $a(\rho) := a_{(0)}(0, y, \rho\eta)$ for fixed $(y, \eta) \in T^*\Omega \setminus 0$. Then, similarly as in elliptic BVPs with the transmission property, the task is to find a bijective 2×2 block matrix

$$\boldsymbol{a} = \begin{pmatrix} \mathrm{op}^+(a) & k \\ b & q \end{pmatrix} : \begin{matrix} L^2(\mathbb{R}_+) \\ \oplus \\ \mathbb{C}^{j-} \end{matrix} \quad \to \quad \begin{matrix} L^2(\mathbb{R}_+) \\ \oplus \\ \mathbb{C}^{j+} \end{matrix}$$

for suitable $j_\pm \in \mathbb{N}$. This is possible if and only if

$$\mathrm{op}^+(a) : L^2(\mathbb{R}_+) \to L^2(\mathbb{R}_+) \tag{10.43}$$

is a Fredholm operator. Set

$$M(a) := \{z \in \mathbb{C} : z = (1 - \lambda)a_0^+ + \lambda a_0^-, \ 0 \le \lambda \le 1\}. \tag{10.44}$$

The following result is well known.

Theorem 10.3.1. *The operator* (10.43) *is Fredholm if and only if*

$$L(a) \cup M(a) \subset \mathbb{C} \setminus \{0\}. \tag{10.45}$$

A proof of the Fredholm property of (10.43) under the condition (10.45) is given in Eskin's book [14]; it is also noted there that (10.42) is necessary. Details of that part of the proof may be found in [44, Theorem 2.1.180].

Corollary 10.3.2. *Let* $a(\rho) \in S^0_{\mathrm{cl}}(\mathbb{R})$ *be elliptic in the sense that* $L(a) \subset \mathbb{C} \setminus \{0\}$. *Then* (10.43) *is a Fredholm operator if and only if*

$$0 \notin M(a). \tag{10.46}$$

The union

$$C(a) := L(a) \cup M(a)$$

is a continuous and piecewise smooth curve which can be represented as the image of a continuous map $\gamma : [0, 1] \to \mathbb{C}$. If (10.45) holds, then we have a winding number wind $C(a)$, and there is the well-known relation

$$\mathrm{ind}\, \mathrm{op}^+(a) = \mathrm{wind}\, C(a).$$

Observe that

$$a_0^- = -a_0^+ \implies 0 \in M(a),$$

i.e., the operator (10.43) cannot be Fredholm in this case.

Definition 10.3.3. A symbol $a(\rho) \in S_{\text{cl}}^{\mu}(\mathbb{R})$ for $\mu \in \mathbb{Z}$ is said to have the anti-transmission property if the coefficients $a_j^{\pm}$ in the asymptotic expansion (2.4) satisfy the condition

$$a_j^+ = -a_j^- \quad \text{for all } j \in \mathbb{N}. \tag{10.47}$$

Let $S_{-\text{tr}}^{\mu}(\mathbb{R})$ denote the space of all symbols with the anti-transmission property.

Note that (10.47) is just the opposite of (2.6).

Proposition 10.3.4. *Every $a(\rho) \in S_{\text{cl}}^{\mu}(\mathbb{R})$ can be written in the form*

$$a(\rho) = \frac{1}{2}\big(a_{\text{tr}}(\rho) + a_{-\text{tr}}(\rho)\big) + c(\rho) \tag{10.48}$$

for suitable $a_{\text{tr}}(\rho) \in S_{\text{tr}}^{\mu}(\mathbb{R})$, $a_{-\text{tr}}(\rho) \in S_{-\text{tr}}^{\mu}(\mathbb{R})$, $c(\rho) \in \mathcal{S}(\mathbb{R})$.

Proof. We form a symbol

$$b(\rho) \sim \sum_{j=0}^{\infty} \chi(\rho)\big(a_j^- \theta^+(\rho) + a_j^+ \theta^-(\rho)\big)(i\rho)^{\mu-j}$$

belonging to $S_{\text{cl}}^{\mu}(\mathbb{R})$, where $\chi(\rho)$ is some excision function. Then we obviously have $a_{\text{tr}}(\rho) := a(\rho) + b(\rho) \in S_{\text{tr}}^{\mu}(\mathbb{R})$, $a_{-\text{tr}}(\rho) := a(\rho) - b(\rho) \in S_{-\text{tr}}^{\mu}(\mathbb{R})$, and we obtain the relation (10.48). $\qquad\square$

Remark 10.3.5. A symbol $a(\rho) \in S_{\text{cl}}^{\mu}(\mathbb{R})$ has the anti-transmission property precisely when

$$a_{(\mu-j)}(\rho) = (-1)^{\mu-j+1} a_{(\mu-j)}(-\rho) \tag{10.49}$$

for all $\rho \in \mathbb{R} \setminus \{0\}$ and all $j \in \mathbb{N}$.

In fact, the anti-transmission property means that

$$a_{(\mu-j)}(\rho) = \{c_j \theta^+(\rho) - c_j \theta^-(\rho)\}(i\rho)^{\mu-j}$$

for constants $c_j := a_j^+ \in \mathbb{C}$. This yields the relation

$$\begin{aligned}
a_{(\mu-j)}(-\rho) &= \{c_j \theta^+(-\rho) - c_j 0^-(-\rho)\}(-i\rho)^{\mu-j} \\
&= (-1)^{\mu-j}\{c_j \theta^-(\rho) - c_j \theta^+(\rho)\} = (-1)^{\mu-j+1} a_{(\mu-j)}(\rho),
\end{aligned}$$

where we used that $\theta^+(-\rho) = \theta^-(\rho), \theta^-(-\rho) = \theta^+(\rho)$.

Conversely, from (10.49) we obtain

$$\begin{aligned}
\{a_j^+ \theta^+(\rho) + a_j^- \theta^-(\rho)\}(i\rho)^{\mu-j} &= (-1)^{\mu-j+1}\{a_j^+ \theta^+(-\rho) + a_j^- \theta^-(-\rho)\}(-i\rho)^{\mu-j} \\
&= \{\theta^- a_j^+ \theta^-(\rho) + a_j^- \theta^+(\rho)\}(i\rho)^{\mu-j}.
\end{aligned}$$

This gives $a_j^+ = -a_j^-$, which are the conditions of Definition 10.3.3. Observe that there is also a higher-dimensional analogue of Definition 2.1.1 for symbols $p(r, y, \rho, \eta) \in S_{\text{cl}}^{\mu}(\overline{\mathbb{R}}_+ \times \Omega \times \mathbb{R}_{y,\rho})$, where instead of (2.1) we ask that

$$D_{r,y}^{\alpha} D_{\rho,\eta}^{\beta}\{p_{(\mu-j)}(r, y, \rho, \eta) - (-1)^{\mu-j+1} p_{(\mu-j)}(r, y, -\rho, -\eta)\} = 0$$

on $\{(r,y,\rho,\eta) : y \in \Omega, \ r = 0, \ \eta = 0, \ \rho \in \mathbb{R} \setminus \{0\}\}$ for all α, β, j. This gives us the symbol class $S^{\mu}_{-\mathrm{tr}}(\Omega \times \overline{\mathbb{R}}_+ \times \mathbb{R}^n)$. There is then a higher-dimensional analogue of Proposition 10.3.4. In fact, let $a(r,y,\rho,\eta) \in S^{\mu}_{\mathrm{cl}}(\overline{\mathbb{R}}_+ \times \Omega \times \mathbb{R}^n)$ be arbitrary, and define the homogeneous components

$$a_{\mathrm{tr},(\mu-j)}(r,y,\rho,\eta) := a_{(\mu-j)}(r,y,\rho,\eta) + (-1)^{\mu-j} a_{(\mu-j)}(r,y,-\rho,-\eta)$$

and

$$a_{-\mathrm{tr},(\mu-j)}(r,y,\rho,\eta) := a_{(\mu-j)}(r,y,\rho,\eta) - (-1)^{\mu-j} a_{(\mu-j)}(r,y,-\rho,-\eta)$$

for all j and $(r,y,\rho,\eta) \in \overline{\mathbb{R}}_+ \times \Omega \times (\mathbb{R}^n \setminus \{0\})$. Then we have

$$
\begin{aligned}
a_{\mathrm{tr},(\mu-j)}(r,y,\rho,\eta) &- (-1)^{\mu-j} a_{\mathrm{tr},(\mu-j)}(r,y,-\rho,-\eta) = \\
& a_{(\mu-j)}(r,y,\rho,\eta) + (-1)^{\mu-j} a_{(\mu-j)}(r,y,-\rho,-\eta) \\
& - (-1)^{\mu-j}\big\{ a_{(\mu-j)}(r,y,-\rho,-\eta) + (-1)^{\mu-j} a_{(\mu-j)}(y,t,\eta,\tau) \big\} = 0,
\end{aligned}
$$

and

$$
\begin{aligned}
a_{-\mathrm{tr},(\mu-j)}(r,y,\rho,\eta) &+ (-1)^{\mu-j} a_{-\mathrm{tr},(\mu-j)}(r,y,-\rho,-\eta) = \\
& a_{(\mu-j)}(r,y,\rho,\eta) - (-1)^{\mu-j} a_{(\mu-j)}(r,y,-\rho,-\eta) \\
& + (-1)^{\mu-j}\big\{ a_{(\mu-j)}(r,y,-\rho,-\eta) - (-1)^{\mu-j} a_{(\mu-j)}(r,y,\rho,\eta) \big\} = 0
\end{aligned}
$$

for all j and $(y,t,\eta,\tau) \in \overline{\mathbb{R}}_+ \times \Omega \times (\mathbb{R}^n \setminus \{0\})$. In other words, if we define

$$a_{\mathrm{tr}}(r,y,\rho,\eta) \sim \sum_{j=0}^{\infty} \chi(\rho,\eta) a_{\mathrm{tr},(\mu-j)}(r,y,\rho,\eta),$$

and

$$a_{-\mathrm{tr}}(r,y,\rho,\eta) \sim \sum_{j=0}^{\infty} \chi(\eta,\tau) a_{-\mathrm{tr},(\mu-j)}(r,y,\rho,\eta),$$

then a_{tr} has the transmission property, and $a_{-\mathrm{tr}}$ the anti-transmission property; here $\chi(\rho,\eta)$ is any excision function. Thus we have proved the following result.

Proposition 10.3.6. *Every symbol* $a(y,t,\eta,\tau) \in S^{\mu}_{\mathrm{cl}}(\overline{\mathbb{R}}_+ \times \Omega \times \mathbb{R}^n)$ *can be written in the form*

$$a(y,t,\eta,\tau) = \frac{1}{2}\big\{ a_{\mathrm{tr}}(y,t,\eta,\tau) + a_{-\mathrm{tr}}(y,t,\eta,\tau) \big\} + c(y,t,\eta,\tau)$$

with symbols $a_{\mathrm{tr}}(r,y,\rho,\eta) \in S^{\mu}_{\mathrm{tr}}(\overline{\mathbb{R}}_+ \times \Omega \times \mathbb{R}^n)$, $a_{-\mathrm{tr}}(r,y,\rho,\eta) \in S^{\mu}_{-\mathrm{tr}}(\overline{\mathbb{R}}_+ \times \Omega \times \mathbb{R}^n)$, *uniquely determined modulo* $S^{-\infty}(\overline{\mathbb{R}}_+ \times \Omega \times \mathbb{R}^n)$, *and*

$$c(r,y,\rho,\eta) \in S^{-\infty}(\overline{\mathbb{R}}_+ \times \Omega \times \mathbb{R}^n).$$

10.4 The calculus of BVPs

Recall that by definition we have

$$L^\mu(M, \boldsymbol{g}; \boldsymbol{v})_{\text{smooth}} \subset L^\mu(M, \boldsymbol{g}; \boldsymbol{v}) \tag{10.50}$$

for a smooth manifold M with boundary. The difference between these two operator spaces is in the nature of non-smoothing operators in the top left corners over $M \setminus Y$, while

$$L^\mu_{\text{G}}(M, \boldsymbol{g}; \boldsymbol{v})_{\text{smooth}} = L^\mu_{\text{G}}(M, \boldsymbol{g}; \boldsymbol{v}), \quad L^\mu_{\text{M+G}}(M, \boldsymbol{g}; \boldsymbol{v})_{\text{smooth}} = L^\mu_{\text{M+G}}(M, \boldsymbol{g}; \boldsymbol{v}).$$

Some elements of the calculus for $L^\mu(M, \boldsymbol{g}; \boldsymbol{v})_{\text{smooth}}$ are completely analogous to those for $L^\mu(M, \boldsymbol{g}; \boldsymbol{v})$. In particular, we have the spaces of operators of lower order (10.24), (10.25), and (10.26), respectively.

Theorem 10.4.1. *Let $\mathcal{A}_j \in L^{\mu-j}(M, \boldsymbol{g}; \boldsymbol{v})_{\text{smooth}}$, $j \in \mathbb{N}$, be an arbitrary sequence, $\boldsymbol{g} := (\gamma, \gamma - \mu, (-(k+1), 0])$ for a finite k, or $\boldsymbol{g} := (\gamma, \gamma - \mu)$, and assume that the asymptotic types involved in the Green operators are independent of j. Then there exists an asymptotic sum $\mathcal{A} \sim \sum_{j=0}^\infty \mathcal{A}_j$ in $L^\mu(M, \boldsymbol{g}; \boldsymbol{v})_{\text{smooth}}$, i.e., an $\mathcal{A} \in L^\mu(M, \boldsymbol{g}; \boldsymbol{v})_{\text{smooth}}$ such that*

$$\mathcal{A} - \sum_{j=0}^\infty \mathcal{A}_j \in L^{\mu-(N+1)}(M, \boldsymbol{g}; \boldsymbol{v})_{\text{smooth}}$$

for every $N \in \mathbb{N}$, and $\mathcal{A}$ is unique mod $L^{-\infty}(M, \boldsymbol{g}; \boldsymbol{v})$.

Theorem 10.4.2. (i) *Let $\mathcal{A} \in L^\mu(M, \boldsymbol{g}_0; \boldsymbol{v}_0)_{\text{smooth}}$, $\mathcal{B} \in L^\rho(M, \boldsymbol{g}_1; \boldsymbol{v}_1)_{\text{smooth}}$ for $\boldsymbol{g}_0 = (\gamma - \rho, \gamma - (\mu+\rho), \Theta)$, $\boldsymbol{g}_1 = (\gamma, \gamma - \rho, \Theta)$, or $\boldsymbol{g}_0 = (\gamma - \rho, \gamma - (\mu+\rho))$, $\boldsymbol{g}_1 = (\gamma, \gamma - \rho)$. Then $\mathcal{A}\mathcal{B} \in L^{\mu+\nu}(M, \boldsymbol{g}_0 \circ \boldsymbol{g}_1; \boldsymbol{v}_0 \circ \boldsymbol{v}_1)_{\text{smooth}}$ (when the bundle data in the middle fit together so that $\boldsymbol{v}_0 \circ \boldsymbol{v}_1$ makes sense), and we have $\sigma(\mathcal{A}\mathcal{B}) = \sigma(\mathcal{A})\sigma(\mathcal{B})$ with componentwise multiplication.*

(ii) *$\mathcal{A} \in L^\mu(M, \boldsymbol{g}; \boldsymbol{v})_{\text{smooth}}$ for $\boldsymbol{g} = (\gamma, \gamma - \mu, \Theta)$ or $\boldsymbol{g} = (\gamma, \gamma - \mu)$, and $\boldsymbol{v} = (V_1, V_2; J_1, J_2)$ implies $\mathcal{A}^* \in L^\mu(M, \boldsymbol{g}^*; \boldsymbol{v}^*)_{\text{smooth}}$ for $\boldsymbol{v}^* = (V_2, V_1; J_2, J_1)$, where $\mathcal{A}^*$ is the formal adjoint in the sense that*

$$(u, \mathcal{A}^* v)_{H^{0,0}(M, V_1) \oplus H^0(Y, J_1)} = (\mathcal{A}u, v)_{H^{0,0}(M, V_2) \oplus H^0(Y, J_2)}$$

for all $u \in H^{\infty,\infty}(M, V_1) \oplus H^\infty(Y, J_1)$, $v \in H^{\infty,\infty}(M, V_2) \oplus H^\infty(Y, J_2)$, and we have $\sigma(\mathcal{A}^*) = \sigma(\mathcal{A})^*$ with componentwise formal adjoint.

Theorems 10.4.1 and 10.4.2 can be proved in an analogous manner as Theorems 7.4.5 and 7.4.6, respectively.

Chapter 11

Boundary ellipticity

11.1 The Atiyah–Bott obstruction

Recall that the operators $\mathcal{A}$ in the calculus of boundary value problems without the transmission property have a pair

$$\sigma(\mathcal{A}) = \big(\sigma_\psi(\mathcal{A}), \sigma_\partial(\mathcal{A})\big) \tag{11.1}$$

of principal symbols. As in the general edge calculus, both components will take part in the notion of ellipticity. The ellipticity with respect to σ_ψ, also referred to as interior ellipticity, is a condition on the top left corner A of the 2×2 block matrix operator $\mathcal{A}$.

Definition 11.1.1. An operator $A \in L^\mu(M, \boldsymbol{g}; E, F)_{\mathrm{smooth}}$ on a manifold M with smooth boundary Y, for $E, F \in \mathrm{Vect}(M)$, is called σ_ψ-elliptic if the standard homogeneous principal symbol

$$\sigma_\psi(A) : \pi_M^* E \to \pi_M^* F,$$

where $\pi_M : T^*M \setminus 0 \to M$, defines an isomorphism.

Remark 11.1.2. Assume that $A \in L^\mu(M, \boldsymbol{g}; E, F)_{\mathrm{smooth}}$ is σ_ψ-elliptic. Then locally near Y, in the splitting of variables $(r, y) \in \overline{\mathbb{R}}_+ \times \Omega$ with the covariables (ρ, η), the reduced symbol

$$\widetilde{\sigma}_\psi(A)(r, y, \rho, \eta) := r^\mu \sigma_\psi(A)\big(r, y, r^{-1}\rho, r^{-1}\eta\big)$$

induces isomorphisms

$$\widetilde{\sigma}_\psi(A)(r, y, \rho, \eta) : E_{(r,y)} \to F_{(r,y)}$$

for all $(r, y, \rho, \eta) \in \overline{\mathbb{R}}_+ \times \Omega \times \big(\mathbb{R}^{1+q}_{\rho,\eta} \setminus \{0\}\big)$ (here $\Omega \subseteq \mathbb{R}^q$ correspond to charts on Y, and $E_{(r,y)}, \ldots$ stand for the fibres of $E, \ldots$ over the corresponding points in local coordinates).

© Springer Nature Switzerland AG 2018
X. Liu, B.-W. Schulze, *Boundary Value Problems with Global Projection Conditions*,
Operator Theory: Advances and Applications 265, https://doi.org/10.1007/978-3-319-70114-1_11

In fact, we have

$$\sigma_\psi(A)(r, y, r\rho, r\eta) = r^\mu \sigma_\psi(A)(r, y, \rho, \eta) \quad \text{for every } r > 0,$$

which shows that

$$\sigma_\psi(A)(r, y, \rho, \eta) = r^\mu \sigma_\psi(A)(r, y, r^{-1}\rho, r^{-1}\eta) = \widetilde{\sigma}_\psi(A)(r, y, \rho, \eta).$$

In other words, in this case the reduced symbol coincides with the standard homogeneous principal symbol.

Recall that $\sigma_\partial(A)(y, \eta)$ for $A = P + M + G$, with P being the non-smoothing pseudo-differential part and $M + G$ the smoothing Mellin plus Green part, defines an operator family

$$\sigma_\partial(A)(y, \eta) : \mathcal{K}^{s,\gamma}(\mathbb{R}_+) \otimes E'^\wedge_y \to \mathcal{K}^{s-\mu,\gamma-\mu}(\mathbb{R}_+) \otimes F'^\wedge_y \qquad (11.2)$$

of the form $\sigma_\partial(A) = \sigma_\partial(P) + \sigma_\partial(M) + \sigma_\partial(G)$ for $E'^\wedge := (\pi_+^* E')|_{\mathbb{R}_+ \times Y}$, etc., where $\pi_+ : \overline{\mathbb{R}}_+ \times Y \to Y$ is the canonical projection, and

$$\sigma_\partial(P)(y, \eta) = r^{-\mu} \{ \omega_{|\eta|} \operatorname{op}_M^\gamma(h_0)(y, \eta) \omega'_{|\eta|} + \chi_{|\eta|} \operatorname{Op}_r(p_0)(y, \eta) \chi'_{|\eta|} \},$$

$$\sigma_\partial(M)(y, \eta) = r^{-\mu} \omega_{|\eta|} \operatorname{op}_M^\gamma(f)(y, \eta) \omega'_{|\eta|}$$

(in the case $\boldsymbol{g} = (\gamma, \gamma - \mu)$, otherwise with extra smoothing Mellin summands, cf. (7.23), and the principal homogeneous component

$$\sigma_\partial(G)(y, \eta) = g_{(\mu)}(y, \eta)$$

of the Green symbol $g(y, \eta)$ of the operator G, $(y, \eta) \in T^*Y \setminus 0$. We have

$$\sigma_\partial(A)(y, \eta) \in L^\mu(\mathbb{R}_+, \boldsymbol{g}; E'^\wedge_y, F'^\wedge_y),$$

cf. Definition 6.8.1. The cone calculus also has a symbol structure, given in the case of an infinite stretched cone by a tuple

$$\sigma = (\sigma_\psi, \sigma_M, \sigma_E),$$

cf. Section 1.5.

Remark 11.1.3. Let $A \in L^\mu(M, \boldsymbol{g}; E, F)$ be σ_ψ-elliptic in the sense of Definition 11.1.1. Then $\sigma_\partial(A)(y, \eta)$ defines a family of Fredholm operators and its interior symbol on the half-line is σ_E-elliptic in the sense of Definition 1.5.18 for every fixed $(y, \eta) \in T^*Y \setminus 0$. Moreover, for every $y \in Y$ there is a discrete set $D_A(y) \subset \mathbb{C}$ such that

$$\sigma_M \sigma_\partial(A)(y, z) : E'_y \to F'_y$$

is an isomorphism if and only if $z \notin D_A(y)$.

Summing up we have the following result.

Proposition 11.1.4. *Let $A \in L^\mu(M, g; E, F)$ be σ_ψ-elliptic. Then (11.2) is a Fredholm operator for $\eta \neq 0$, $s \in \mathbb{R}$, and every $\gamma \in \mathbb{R}$, $y \in Y$, such that*

$$\Gamma_{1/2-\gamma} \cap D_A(y) = \emptyset. \tag{11.3}$$

In the following we assume that our operator $A \in L^\mu(M, g; E, F)$ satisfies the condition (11.3) for every $y \in Y$. Because of the homogeneity

$$\sigma_\partial(y, \lambda\eta) = \lambda^\mu \kappa_\lambda \sigma_\partial(A)(y, \eta) \kappa_\lambda^{-1},$$

we have

$$\operatorname{ind} \sigma_\partial(A)(y, \eta) = \sigma_\partial(A)(y, \eta/|\eta|).$$

More precisely, the dimensions of $\ker \sigma_\partial(A)(y, \eta)$ and $\operatorname{coker} \sigma_\partial(A)(y, \eta)$ only depend on $\eta/|\eta|$. Therefore, (11.2) may be regarded as a family of Fredholm operators depending on the parameters $(y, \eta) \in S^*Y$, the unit cosphere bundle of Y, which is a compact topological space. This gives rise to an index element

$$\operatorname{ind}_{S^*Y} \sigma_\partial(A) \in K(S^*Y).$$

The property

$$\operatorname{ind}_{S^*Y} \sigma_\partial(A) \in \pi^* K(Y), \tag{11.4}$$

where $\pi : S^*Y \to Y$, is an analogue of the corresponding condition in Theorem 3.2.2, and has a similar meaning in the edge calculus. If A satisfies relation (11.4), we say that the Atiyah–Bott obstruction vanishes.

The following theorem is a special case of Theorem 8.1.4.

Theorem 11.1.5. *Let $A \in L^\mu(M, g; E, F)$ be σ_ψ-elliptic, and (11.3) be satisfied for some $\gamma \in \mathbb{R}$ and all $y \in Y$. Denote the family of Fredholm operators (11.2) for the moment by*

$$\sigma_\partial(A)^\gamma(y, \eta) : \mathcal{K}^{s,\gamma}(\mathbb{R}_+) \otimes E'^\wedge \to \mathcal{K}^{s-\mu,\gamma-\mu}(\mathbb{R}_+) \otimes F'^\wedge, \quad (y, \eta) \in S^*Y.$$

Then if $\tilde{\gamma} \in \mathbb{R}$ is another weight satisfying

$$\Gamma_{1/2-\tilde{\gamma}} \cap D_A(y) = \emptyset \quad \text{for all } y \in Y,$$

we have

$$\operatorname{ind}_{S^*Y} \sigma_\partial(A)^\gamma \in \pi^* K(Y) \iff \operatorname{ind}_{S^*Y} \sigma_\partial(A)^{\tilde{\gamma}} \in \pi^* K(Y).$$

In addition, if $\tilde{A} \in L^\mu(M, g; E, F)_{\text{smooth}}$ satisfies

$$\tilde{A} = A \bmod L^\mu_{M+G}(M, g; E, F),$$

then

$$\Gamma_{1/2-\gamma} \cap D_{\tilde{A}}(y) = \emptyset, \quad \Gamma_{1/2-\gamma} \cap D_A(y) = \emptyset$$

for some $\gamma \in \mathbb{R}$ and all $y \in Y$ has the consequence that

$$\operatorname{ind}_{S^*Y} \sigma_\partial(\tilde{A})^\gamma \in \pi^* K(Y) \iff \operatorname{ind}_{S^*Y} \sigma_\partial(A)^\gamma \in \pi^* K(Y).$$

11.2   Elliptic boundary conditions

Let us now study ellipticity with respect to both components of (11.1) from the point of view of an analogue of Shapiro–Lopatinskii ellipticity, again referred to as SL-ellipticity.

Definition 11.2.1. An operator $\mathcal{A} \in L^\mu(M, \boldsymbol{g}; \boldsymbol{v})_{\mathrm{smooth}}$ for $\boldsymbol{v} := (E, F; J_1, J_2)$ is called elliptic if

(i) its top left corner $A \in L^\mu(M, \boldsymbol{g}; E, F)_{\mathrm{smooth}}$ is σ_ψ-elliptic in the sense of Definition 11.1.1;

(ii) $\mathcal{A}$ is σ_∂-elliptic, i.e.,

$$\sigma_\partial(\mathcal{A})(y, \eta) : \begin{array}{c} \mathcal{K}^{s,\gamma}(\mathbb{R}_+) \otimes E_y'^\wedge \\ \oplus \\ J_{1,y} \end{array} \longrightarrow \begin{array}{c} \mathcal{K}^{s-\mu,\gamma-\mu}(\mathbb{R}_+) \otimes F_y'^\wedge \\ \oplus \\ J_{2,y} \end{array} \qquad (11.5)$$

is a family of isomorphisms for all $(y, \eta) \in T^*Y \setminus 0$ and some $s \in \mathbb{R}$.

Remark 11.2.2. (i) If (11.5) is a family of isomorphisms for some $s \in \mathbb{R}$ then this is the case for all $s \in \mathbb{R}$, and $\ker \sigma_\partial(\mathcal{A})$ as well as $\dim \operatorname{coker} \sigma_\partial(\mathcal{A})$ are independent of s.

(ii) The isomorphism condition (11.5) for all $(y, \eta) \in T^*Y \setminus 0$ is equivalent to the one for all $(y, \eta) \in S^*Y$.

(iii) The isomorphism condition (11.5) for all $(y, \eta) \in S^*Y$ has the consequence that

$$\sigma_\partial(A)(y, \eta) : \mathcal{K}^{s,\gamma}(\mathbb{R}_+) \otimes E'^\wedge \to \mathcal{K}^{s-\mu,\gamma-\mu}(\mathbb{R}_+) \otimes F'^\wedge$$

is a family of Fredholm operators, and we have

$$\operatorname{ind}_{S^*Y} \sigma_\partial(A) = [J_2] - [J_1] \in \pi^* K(Y).$$

The following theorem is a special case of Theorem 8.2.3.

Theorem 11.2.3. *Let* $A \in L^\mu(M, \boldsymbol{g}; E, F)_{\mathrm{smooth}}$ *be* σ_ψ-*elliptic, and let*

$$\sigma_\wedge(A)(y, \eta) : \mathcal{K}^{s,\gamma}(\mathbb{R}_+) \otimes E_y'^\wedge \to \mathcal{K}^{s-\mu,\gamma-\mu}(\mathbb{R}_+) \otimes F_y'^\wedge \qquad (11.6)$$

be a family of Fredholm operators. Then the property

$$\operatorname{ind}_{S^*Y} \sigma_\partial(A) \in \pi^* K(Y)$$

is equivalent to the existence of an SL-*elliptic operator*

$$\mathcal{A} \in L^\mu(M, \boldsymbol{g}; \boldsymbol{v})_{\mathrm{smooth}}, \quad \boldsymbol{v} := (E, F; J_1, J_2),$$

for suitable $J_1, J_2 \in \operatorname{Vect}(Y)$ *which contains* A *as the top left corner.*

11.3 Parametrices and the Fredholm property

Let M be a smooth manifold with boundary.

Definition 11.3.1. Let $\mathcal{A} \in L^\mu(M, \boldsymbol{g}; \boldsymbol{v})_{\text{smooth}}$, $\boldsymbol{v} := (E, F; J_1, J_2)$, and

$$\mathcal{P} \in L^{-\mu}(M, \boldsymbol{g}^{-1}; \boldsymbol{v}^{-1})_{\text{smooth}}, \quad \boldsymbol{v}^{-1} := (F, E; J_2, J_1),$$

and let $\mathcal{A}$ or $\mathcal{P}$ be properly supported, cf. Remark 7.4.2. Then $\mathcal{P}$ is called a parametrix of $\mathcal{A}$ if

$$\mathcal{C}_L := \mathcal{I} - \mathcal{P}\mathcal{A} \in L^{-\infty}(M, \boldsymbol{g}_L; \boldsymbol{v}_L), \quad \mathcal{C}_R := \mathcal{I} - \mathcal{A}\mathcal{P} \in L^{-\infty}(M, \boldsymbol{g}_R; \boldsymbol{v}_R) \quad (11.7)$$

for $\boldsymbol{v}_L := (E, E; J_1, J_1)$, $\boldsymbol{v}_R := (F, F; J_2, J_2)$, with $\mathcal{I}$ denoting the respective identity operators.

The following results are a special case of Theorem 8.3.2, Remark 8.3.3 and Theorem 8.3.4, respectively.

Theorem 11.3.2. *An elliptic operator* $\mathcal{A} \in L^\mu(M, \boldsymbol{g}; \boldsymbol{v})_{\text{smooth}}$, $\boldsymbol{v} := (E, F; J_1, J_2)$, *has a properly supported parametrix* $\mathcal{P} \in L^{-\mu}(M, \boldsymbol{g}^{-1}; \boldsymbol{v}^{-1})_{\text{smooth}}$. *If M is compact, then for an operator* $\mathcal{A} \in L^\mu(M, \boldsymbol{g}; \boldsymbol{v})_{\text{smooth}}$, $\boldsymbol{v} := (E, F; J_1, J_2)$, *the following conditions are equivalent:*

(i) *$\mathcal{A}$ is elliptic in the sense of Definition 11.2.1.*

(ii) *The operator*

$$\mathcal{A} : H^{s,\gamma}(M, E) \oplus H^s(Y, J_1) \to H^{s-\mu,\gamma-\mu}(M, F) \oplus H^{s-\mu}(Y, J_2) \quad (11.8)$$

is Fredholm for some $s = s_0 \in \mathbb{R}$.

Remark 11.3.3. Let M be compact and $\mathcal{A} \in L^\mu(M, \boldsymbol{g}; \boldsymbol{v})_{\text{smooth}}$, $\boldsymbol{v} := (E, F; J_1, J_2)$, elliptic.

(i) The operator (11.8) is Fredholm for every $s \in \mathbb{R}$.

(ii) $\mathcal{V} := \ker_s \mathcal{A} := \{u \in H^{s,\gamma}(M, E) \oplus H^s(Y, J_1) : \mathcal{A}u = 0\}$ is a finite-dimensional subspace of $H^{\infty,\gamma}(M, E) \oplus H^\infty(Y, J_1)$ independent of s, and there is a finite-dimensional

$$\mathcal{W} \subset H^{\infty,\gamma-\mu}(M, F) \oplus H^\infty(Y, J_2)$$

independent of s such that

$$\operatorname{im}_s \mathcal{A} + \mathcal{W} = H^{s-\mu,\gamma-\mu}(M, F) \oplus H^{s-\mu}(Y, J_2)$$

for every s; here $\operatorname{im}_s \mathcal{A} := \{\mathcal{A}u : u \in H^{s,\gamma}(M, E) \oplus H^s(Y, J_1)\}$.

Theorem 11.3.4. *Let M be compact and $\mathcal{A} \in L^{\mu}(M, \boldsymbol{g}; \boldsymbol{v})_{\text{smooth}}$, $\boldsymbol{v} := (E, F; J_1, J_2)$ elliptic. Then there is a parametrix $\mathcal{B} \in L^{-\mu}(M, \boldsymbol{g}^{-1}; \boldsymbol{v}^{-1})_{\text{smooth}}$, with $\boldsymbol{v}^{-1} := (F, E; J_2, J_1)$, such that the operators $\mathcal{C}_{\text{L}}$ and $\mathcal{C}_{\text{R}}$ in the relation (11.7) are projections*

$$\mathcal{C}_{\text{L}} : H^{s,\gamma}(M, E) \oplus H^{s}(Y, J_1) \to \mathcal{V}, \quad \mathcal{C}_{\text{R}} : H^{s-\mu,\gamma-\mu}(M, F) \oplus H^{s-\mu}(Y, J_2) \to \mathcal{W}$$

for all $s \in \mathbb{R}$.

Chapter 12

Toeplitz boundary value problems without the transmission property

12.1 BVPs with global projection conditions

Let M be a compact manifold with boundary Y.

Definition 12.1.1. Let $\mathbb{L}_i := (P_i, J_i, L_i) \in \mathbb{P}(Y)$ be projection data (cf. Definition 1.2.7), $V_i \in \mathrm{Vect}(M)$, $i = 1, 2$, and set

$$\boldsymbol{v} := (V_1, V_2; J_1, J_2), \quad \boldsymbol{l} := (V_1, V_2; \mathbb{L}_1, \mathbb{L}_2).$$

Then $\mathcal{T}^\mu(M, \boldsymbol{g}; \boldsymbol{l})_{\mathrm{smooth}}$, $\mu \in \mathbb{R}$, for $\boldsymbol{g} = (\gamma, \gamma - \mu, \Theta)$ or $\boldsymbol{g} = (\gamma, \gamma - \mu)$ is defined to be the set of all operators

$$\mathcal{A} := \mathcal{P}_2 \widetilde{\mathcal{A}} \mathcal{E}_1 \quad \text{for } \mathcal{P}_2 := \mathrm{diag}\,(1, P_2), \ \mathcal{E}_1 := \mathrm{diag}\,(1, E_1),$$

cf. formula (4.1), for arbitrary $\widetilde{\mathcal{A}} \in L^\mu(M, \boldsymbol{g}; \boldsymbol{v})_{\mathrm{smooth}}$. The elements of the space $\mathcal{T}^\mu(M, \boldsymbol{g}; \boldsymbol{l})_{\mathrm{smooth}}$ will be called boundary value problems of order μ with global projection conditions. Moreover, set

$$\mathcal{T}^{-\infty}(M, \boldsymbol{g}; \boldsymbol{l}) := \{\mathcal{P}_2 \widetilde{\mathcal{C}} \mathcal{E}_1 : \widetilde{\mathcal{C}} \in L^{-\infty}(M, \boldsymbol{g}; \boldsymbol{l})\}. \tag{12.1}$$

Observe that the space (12.1) can be equivalently characterised as the set of all $\mathcal{A} \in \mathcal{T}^\mu(M, \boldsymbol{g}; \boldsymbol{l})_{\mathrm{smooth}}$, $\mathcal{A} := \mathcal{P}_2 \widetilde{\mathcal{A}} \mathcal{E}_1$ for a $\widetilde{\mathcal{A}} \in L^\mu(M, \boldsymbol{g}; \boldsymbol{v})_{\mathrm{smooth}}$, $\mathcal{P}_2 \widetilde{\mathcal{A}} \mathcal{P}_1 \in L^{-\infty}(M, \boldsymbol{g}; \boldsymbol{v})$; then $\mathcal{A} = \mathcal{P}_2 (\mathcal{P}_2 \widetilde{\mathcal{A}} \mathcal{P}_1) \mathcal{E}_1$. Moreover,

$$\mathcal{P}_2 (\mathcal{P}_2 \widetilde{\mathcal{A}} \mathcal{P}_1) \mathcal{E}_1 \in \mathcal{T}^{-\infty}(M, \boldsymbol{g}; \boldsymbol{l}) \implies \mathcal{P}_2 \widetilde{\mathcal{A}} \mathcal{P}_1 \in L^{-\infty}(M, \boldsymbol{g}; \boldsymbol{v}).$$

The remaining part of this section is a special case of Section 9.1.

© Springer Nature Switzerland AG 2018
X. Liu, B.-W. Schulze, *Boundary Value Problems with Global Projection Conditions*,
Operator Theory: Advances and Applications 265, https://doi.org/10.1007/978-3-319-70114-1_12

Theorem 12.1.2. *Every $\mathcal{A} \in \mathcal{T}^\mu(M, \boldsymbol{g}; \boldsymbol{l})_{\text{smooth}}$ induces continuous operators*

$$
\mathcal{A}: \begin{array}{c} H^{s,\gamma}(M, V_1) \\ \oplus \\ H^s(Y, \mathbb{L}_1) \end{array} \longrightarrow \begin{array}{c} H^{s-\mu,\gamma-\mu}(M, V_2) \\ \oplus \\ H^{s-\mu}(Y, \mathbb{L}_2) \end{array} \tag{12.2}
$$

for every $s \in \mathbb{R}$.

Proof. The proof is an obvious consequence of Theorem 7.4.3. □

Proposition 12.1.3. *Given $V_i \in \text{Vect}(M)$, $i = 1, 2$, we have a canonical isomorphism*

$$
\mathcal{T}^\mu(M, \boldsymbol{g}; \boldsymbol{l})_{\text{smooth}} \to \{ \mathcal{P}_2 \widetilde{\mathcal{A}} \mathcal{P}_1 : \widetilde{\mathcal{A}} \in L^\mu(M, \boldsymbol{g}; \boldsymbol{v})_{\text{smooth}} \}.
$$

Proof. The proof is analogous to that of Proposition 1.2.16, cf. also Proposition 4.1.3. □

In other words, we have an identification

$$
\mathcal{T}^\mu(M, \boldsymbol{g}; \boldsymbol{l})_{\text{smooth}} = L^\mu(M, \boldsymbol{g}; \boldsymbol{v})_{\text{smooth}}/\sim \tag{12.3}
$$

with the equivalence relation

$$
\widetilde{\mathcal{A}} \sim \widetilde{\mathcal{B}} \iff \mathcal{P}_2 \widetilde{\mathcal{A}} \mathcal{P}_1 = \mathcal{P}_2 \widetilde{\mathcal{B}} \mathcal{P}_1. \tag{12.4}
$$

The space $\mathcal{T}^\mu(M, \boldsymbol{g}; \boldsymbol{l})_{\text{smooth}}$ is equipped with the principal symbol structure

$$
\sigma(\mathcal{A}) = \big(\sigma_\psi(\mathcal{A}), \sigma_\partial(\mathcal{A}) \big)
$$

with the interior and the boundary symbol components. Writing $\mathcal{A} = (A_{ij})_{i,j=1,2}$ we first set $\sigma_\psi(\mathcal{A}) := \sigma_\psi(A_{11})$, i.e.,

$$
\sigma_\psi(\mathcal{A}) : \pi^*_{M \backslash Y} V_1 \to \pi^*_{M \backslash Y} V_2,
$$

where $\pi_{M \backslash Y} : T^*(M \backslash Y) \to M \backslash Y$.

The boundary symbol of $\mathcal{A}$, represented as $\mathcal{A} = \mathcal{P}_2 \widetilde{\mathcal{A}} \mathcal{E}_1$, is defined as

$$
\sigma_\partial(\mathcal{A}) = \text{diag}\,(1, p_2)\, \sigma_\partial\big(\widetilde{\mathcal{A}} \big)\, \text{diag}\,(1, e_1),
$$

$$
\sigma_\partial(\mathcal{A})(y, \eta) : \begin{array}{c} \mathcal{K}^{s,\gamma}\big(\mathbb{R}_+, V_{1,y}\big) \\ \oplus \\ L_{1,(y,\eta)} \end{array} \longrightarrow \begin{array}{c} \mathcal{K}^{s-\mu,\gamma-\mu}\big(\mathbb{R}_+, V_{2,y}\big) \\ \oplus \\ L_{2,(y,\eta)} \end{array} \tag{12.5}
$$

where $p_2(y, \eta)$ is the homogeneous principal symbol of order zero of the projection $P_2 \in L^0_{\text{cl}}(Y; J_2, J_2)$, and $e_1 : L_{1,(y,\eta)} \to (\pi^*_Y J_1)_{(y,\eta)}$ is the canonical embedding.

Remark 12.1.4. If we identify an operator $\mathcal{A} = \mathcal{P}_2 \widetilde{\mathcal{A}} \mathcal{E}_1 \in \mathcal{T}^\mu(M, \boldsymbol{g}; \boldsymbol{l})_{\text{smooth}}$ with $\widetilde{\widetilde{\mathcal{A}}} := \mathcal{P}_2 \widetilde{\mathcal{A}} \mathcal{P}_1 \in L^\mu(M, \boldsymbol{g}; \boldsymbol{v})_{\text{smooth}}$, cf. Proposition 12.1.3, then $\sigma\big(\widetilde{\widetilde{\mathcal{A}}} \big) = 0$ in the sense of $L^\mu(M, \boldsymbol{g}; \boldsymbol{v})_{\text{smooth}}$ is equivalent to $\sigma(\mathcal{A}) = 0$ in the sense of $\mathcal{T}^\mu(M, \boldsymbol{g}; \boldsymbol{l})_{\text{smooth}}$.

Analogously to Definition 9.1.1, we define $T^\nu(M, \boldsymbol{g}; \boldsymbol{l})_{\text{smooth}}$ for $\boldsymbol{g} = (\gamma, \gamma - \mu, \Theta)$ or $\boldsymbol{g} = (\gamma, \gamma - \mu)$, $\mu - \nu \in \mathbb{N}$, as

$$T^\nu(M, \boldsymbol{g}; \boldsymbol{l})_{\text{smooth}} := \{ \mathcal{P}_2 \widetilde{\mathcal{A}} \mathcal{E}_1 : \widetilde{\mathcal{A}} \in L^\nu(M, \boldsymbol{g}; \boldsymbol{v})_{\text{smooth}} \}.$$

Remark 12.1.5. If $\mathcal{A} \in T^\mu(M, \boldsymbol{g}; \boldsymbol{l})_{\text{smooth}}$ and $\sigma(\mathcal{A}) = 0$ then we have $\mathcal{A} \in T^{\mu-1}(M, \boldsymbol{g}; \boldsymbol{l})_{\text{smooth}}$, and the operator (12.2) is compact for every $s \in \mathbb{R}$.

Theorem 12.1.6. *Let $\mathcal{A}_j \in T^{\mu-j}(M, \boldsymbol{g}; \boldsymbol{l})_{\text{smooth}}$, $j \in \mathbb{N}$, be an arbitrary sequence, $\boldsymbol{g} := (\gamma, \gamma - \mu, (-(k+1), 0])$ for a finite k, or $\boldsymbol{g} := (\gamma, \gamma - \mu)$, and assume that the asymptotic types involved in the Green operators are independent of j. Then there exists an $\mathcal{A} \in T^\mu(M, \boldsymbol{g}; \boldsymbol{l})_{\text{smooth}}$ such that*

$$\mathcal{A} - \sum_{j=0}^{N} \mathcal{A}_j \in T^{\mu-(N+1)}(M, \boldsymbol{g}; \boldsymbol{l})_{\text{smooth}}$$

for every $N \in \mathbb{N}$, and $\mathcal{A}$ is unique mod $T^{-\infty}(M, \boldsymbol{g}; \boldsymbol{l})$.

Proof. The proof is analogous to the proof of Theorem 4.1.8. $\qquad\square$

Theorem 12.1.7.   (i) $\mathcal{A} \in T^\mu(M, \boldsymbol{g}_0; \boldsymbol{l}_0)_{\text{smooth}}$, $\mathcal{B} \in T^\rho(M, \boldsymbol{g}_1; \boldsymbol{l}_1)_{\text{smooth}}$ *for $\boldsymbol{g}_0 = (\gamma - \rho, \gamma - (\mu + \rho), \Theta)$, $\boldsymbol{g}_1 = (\gamma, \gamma - \rho, \Theta)$, or $\boldsymbol{g}_0 = (\gamma - \rho, \gamma - (\mu + \rho))$, $\boldsymbol{g}_1 = (\gamma, \gamma - \rho)$, implies $\mathcal{A}\mathcal{B} \in T^{\mu+\nu}(M, \boldsymbol{g}_0 \circ \boldsymbol{g}_1; \boldsymbol{l}_0 \circ \boldsymbol{l}_1)_{\text{smooth}}$ (when the projection data in the middle fit together so that $\boldsymbol{l}_0 \circ \boldsymbol{l}_1$ makes sense), and we have $\sigma(\mathcal{A}\mathcal{B}) = \sigma(\mathcal{A})\sigma(\mathcal{B})$ with componentwise multiplication.*

(ii) $\mathcal{A} \in T^\mu(M, \boldsymbol{g}; \boldsymbol{l})_{\text{smooth}}$ *for $\boldsymbol{g} = (\gamma, \gamma - \mu, \Theta)$ or $\boldsymbol{g} = (\gamma, \gamma - \mu)$, and $\boldsymbol{l} = (V_1, V_2; \mathbb{L}_1, \mathbb{L}_2)$ implies $\mathcal{A}^* \in T^\mu(M, \boldsymbol{g}^*; \boldsymbol{l}^*)_{\text{smooth}}$ for $\boldsymbol{l}^* = (V_2, V_1; \mathbb{L}_2^*, \mathbb{L}_1^*)$, where $\mathcal{A}^*$ is the formal adjoint in the sense that*

$$(u, \mathcal{A}^* v)_{H^{0,0}(M, V_1) \oplus H^0(Y, \mathbb{L}_1)} = (\mathcal{A}u, v)_{H^{0,0}(M, V_2) \oplus H^0(Y, \mathbb{L}_2)}$$

for all $u \in H^{\infty,\infty}(M, V_1) \oplus H^\infty(Y, \mathbb{L}_1)$, $v \in H^{\infty,\infty}(M, V_2) \oplus H^\infty(Y, \mathbb{L}_2)$, and we have $\sigma(\mathcal{A}^*) = \sigma(\mathcal{A})^*$ with componentwise formal adjoint.

Proof. The proof is formally analogous to the proof of Theorem 4.1.7 when we take into account the corresponding results from the edge calculus. $\qquad\square$

Remark 12.1.8. Similarly as in the Toeplitz algebras of Section 1.2 or 4.1, we have a natural notion of direct sum

$$T^\mu(M, \boldsymbol{g}; \boldsymbol{l})_{\text{smooth}} \oplus T^\mu(M, \boldsymbol{g}; \boldsymbol{m})_{\text{smooth}} = T^\mu(M, \boldsymbol{g}; \boldsymbol{l} \oplus \boldsymbol{m})_{\text{smooth}},$$

where $\sigma(\mathcal{A} \oplus \mathcal{B}) = \sigma(\mathcal{A}) \oplus \sigma(\mathcal{B})$ with the componentwise direct sum of symbols.

12.2   Ellipticity, parametrices, and the Fredholm property

We now turn to ellipticity in the Toeplitz calculus of edge problems. Definitions and results are analogous to those of Section 9.2.

Definition 12.2.1. Let $\mathcal{A} \in \mathcal{T}^\mu(M, \boldsymbol{g}; \boldsymbol{l})_{\text{smooth}}$ for $\boldsymbol{g} := (\gamma, \gamma{-}\mu, \Theta)$ or $\boldsymbol{g} := (\gamma, \gamma{-}\mu)$, $\mu \in \mathbb{R}$, and $\boldsymbol{l} := (V_1, V_2; \mathbb{L}_1, \mathbb{L}_2)$, $V_i \in \text{Vect}(\mathbb{M})$, $\mathbb{L}_i = (P_i, J_i, L_i) \in \mathbb{P}(Y)$, $i = 1, 2$. The operator $\mathcal{A}$ is called elliptic if the top left corner $A \in L^\mu(M, \boldsymbol{g}; V_1, V_2)$ is σ_ψ-elliptic in the sense of Definition 11.1.1 and if the boundary symbol (12.5) is an isomorphism for every $(y, \eta) \in T^*Y \setminus 0$ and some $s = s_0 \in \mathbb{R}$.

Remark 12.2.2. The bijectivity of (12.5) for some $s = s_0 \in \mathbb{R}$ is equivalent to the bijectivity of (12.5) for every $s \in \mathbb{R}$. The latter property is equivalent to the bijectivity of

$$\sigma_\partial(\mathcal{A})(y, \eta) : \begin{array}{c} \mathcal{K}^{\infty, \gamma; \infty}(\mathbb{R}_+) \otimes V_{1,y}^{\prime\wedge} \\ \oplus \\ L_{1,(y,\eta)} \end{array} \longrightarrow \begin{array}{c} \mathcal{K}^{\infty, \gamma-\mu; \infty}(\mathbb{R}_+) \otimes V_{2,y}^{\prime\wedge} \\ \oplus \\ L_{2,(y,\eta)} \end{array}. \tag{12.6}$$

Theorem 12.2.3.   (i) *For every σ_ψ-elliptic $A \in L^\mu(M, \boldsymbol{g}; V_1, V_2)$ such that*

$$\sigma_\partial(A)(y, \eta) : \mathcal{K}^{s, \gamma}(\mathbb{R}_+) \otimes V_{1,y}^{\prime\wedge} \to \mathcal{K}^{s-\mu, \gamma-\mu}(\mathbb{R}_+) \otimes V_{2,y}^{\prime\wedge}, \quad (y, \eta) \in T^*Y \setminus 0, \tag{12.7}$$

is a family of Fredholm operators with

$$\text{ind}_{S^*Y} \sigma_\partial(A) - \left[L_2 \big|_{S^*Y} \right] - \left[L_1 \big|_{S^*Y} \right] \quad \text{for some } L_1, L_2 \in \text{Vect}(T^*Y \setminus 0)$$

there exist a Green operator $G \in L_G^\mu(M, \boldsymbol{g}; V_1, V_2)$, projection data $\mathbb{L}_1, \mathbb{L}_2 \in \mathbb{P}(Y)$, and an elliptic $\mathcal{A} \in \mathcal{T}^\mu(M, \boldsymbol{g}; \boldsymbol{l})_{\text{smooth}}$, $\boldsymbol{l} := (V_1, V_2; \mathbb{L}_1, \mathbb{L}_2)$ with $A + G$ as the top left corner, such that $\mathcal{A}$ is elliptic in the sense of Definition 12.2.1.

(ii) *For A as in (i) and suitable projection data $\mathbb{L}_i = (P_i, J_i, L_i) \in \mathbb{P}(Y)$, $i = 1, 2$, there exists an elliptic operator $\mathcal{A} \in \mathcal{T}^\mu(M, \boldsymbol{g}; \boldsymbol{l})_{\text{smooth}}$ containing A as the top left corner.*

Proposition 12.2.4. *For every $\mu, \gamma \in \mathbb{R}$, $\boldsymbol{g} := (\gamma, \gamma - \mu, \Theta)$, $V \in \text{Vect}(\mathbb{M})$ and $\mathbb{L} \in \mathbb{P}(Y)$ there exists an elliptic element $\mathcal{R}_{V, \mathbb{L}}^\mu \in \mathcal{T}^\mu(M, \boldsymbol{g}; \boldsymbol{l})_{\text{smooth}}$ for $\boldsymbol{l} := (V, V; \mathbb{L}, \mathbb{L})$ which induces a Fredholm operator*

$$\mathcal{R}_{V, \mathbb{L}}^\mu : \begin{array}{c} H^{s, \gamma}(M, V) \\ \oplus \\ H^s(Y, \mathbb{L}) \end{array} \longrightarrow \begin{array}{c} H^{s-\mu, \gamma-\mu}(M, V) \\ \oplus \\ H^{s-\mu}(Y, \mathbb{L}) \end{array}$$

for every $s \in \mathbb{R}$.

Theorem 12.2.5. *For every elliptic operator* $\mathcal{A} \in \mathcal{T}^{\mu}(M, \boldsymbol{g}; \boldsymbol{l})_{\text{smooth}}$, *with* $\boldsymbol{l} = (V_1, V_2; \mathbb{L}_1, \mathbb{L}_2)$, *there exists an elliptic operator* $\mathcal{B} \in \mathcal{T}^{\mu}(M; \boldsymbol{g}; \boldsymbol{m})_{\text{smooth}}$, $\boldsymbol{m} := (V_2, V_1; \mathbb{M}_1, \mathbb{M}_2)$, *for certain projection data* $\mathbb{M}_1, \mathbb{M}_2 \in \mathbb{P}(Y)$ *of the form* $\mathbb{M}_i := (Q_i, \mathbb{C}^N, M_i)$, $i = 1, 2$, *for some* $N \in \mathbb{N}$, *such that* $\mathcal{A} \oplus \mathcal{B} \in L^{\mu}(M, \boldsymbol{g}; \boldsymbol{v})$ *for* $\boldsymbol{v} = (V_1 \oplus V_2, V_2 \oplus V_1; \mathbb{C}^N, \mathbb{C}^N)$, *is SL-elliptic, cf. Definition 11.2.1.*

Definition 12.2.6. Let $\mathcal{A} \in \mathcal{T}^{\mu}(M, \boldsymbol{g}; \boldsymbol{l})_{\text{smooth}}$ be as in Definition 12.2.1. A

$$\mathcal{P} \in \mathcal{T}^{-\mu}(M, \boldsymbol{g}^{-1}; \boldsymbol{l}^{-1})_{\text{smooth}}$$

for $\boldsymbol{g}^{-1} = (\gamma - \mu, \gamma, \Theta)$ or $(\gamma - \mu, \gamma)$ and $\boldsymbol{l}^{-1} = (V_2, V_1; \mathbb{L}_2, \mathbb{L}_1)$, is called a *parametrix* of $\mathcal{A}$, if

$$\mathcal{C}_{\text{L}} := \mathcal{I} - \mathcal{P}\mathcal{A} \in \mathcal{T}^{-\infty}(M, \boldsymbol{g}_{\text{L}}; \boldsymbol{l}_{\text{L}}), \quad \mathcal{C}_{\text{R}} := \mathcal{I} - \mathcal{A}\mathcal{P} \in \mathcal{T}^{-\infty}(M, \boldsymbol{g}_{\text{R}}; \boldsymbol{l}_{\text{R}}) \quad (12.8)$$

for $\boldsymbol{g}_{\text{L}} = (\gamma, \gamma, \Theta)$ or $\boldsymbol{g}_{\text{R}} = (\gamma - \mu, \gamma - \mu, \Theta)$ and similarly without Θ, with $\mathcal{I}$ being the respective identity operators, and $\boldsymbol{l}_{\text{L}} := (V_1, V_1; \mathbb{L}_1, \mathbb{L}_1)$, $\boldsymbol{l}_{\text{R}} := (V_2, V_2; \mathbb{L}_2, \mathbb{L}_2)$.

Theorem 12.2.7. *Let* $\mathcal{A} \in \mathcal{T}^{\mu}(M, \boldsymbol{g}; \boldsymbol{l})_{\text{smooth}}$, $\mu \in \mathbb{R}$, $\boldsymbol{l} := (V_1, V_2; \mathbb{L}_1, \mathbb{L}_2)$ *for* $V_1, V_2 \in \text{Vect}(\mathbb{M})$, $\mathbb{L}_1, \mathbb{L}_2 \in \mathbb{P}(Y)$.

(i) *Let* $\mathcal{A}$ *be elliptic; then*

$$\mathcal{A} : \begin{array}{c} H^{s,\gamma}(M, V_1) \\ \oplus \\ H^s(Y, \mathbb{L}_1) \end{array} \longrightarrow \begin{array}{c} H^{s-\mu,\gamma-\mu}(X, V_2) \\ \oplus \\ H^{s-\mu}(Y, \mathbb{L}_2) \end{array} \quad (12.9)$$

is a Fredholm operator for every $s \in \mathbb{R}$. *Moreover, if* (12.9) *is Fredholm for* $s = \gamma$, *then the operator* $\mathcal{A}$ *is elliptic.*

(ii) *If* $\mathcal{A}$ *is elliptic, then the operator* (12.9) *is Fredholm for all* $s \in \mathbb{R}$, *and* $\dim \ker \mathcal{A}$ *and* $\dim \operatorname{coker} \mathcal{A}$ *are independent of* s.

(iii) *An elliptic operator* $\mathcal{A} \in \mathcal{T}^{\mu}(M, \boldsymbol{g}; \boldsymbol{l})_{\text{smooth}}$ *has a parametrix*

$$\mathcal{P} \in \mathcal{T}^{-\mu}(M, \boldsymbol{g}^{-1}; \boldsymbol{l}^{-1})_{\text{smooth}}$$

in the sense of Definition 12.2.6, *and* $\mathcal{P}$ *can be chosen in such a way that the remainders in* (12.8) *are projections*

$$\mathcal{C}_{\text{L}} : H^{s,\gamma}(M, V_1) \oplus H^s(Y, \mathbb{L}_1) \to \mathcal{V}_1,$$
$$\mathcal{C}_{\text{R}} : H^{s-\mu, s-\mu}(M, V_2) \oplus H^{s-\mu}(Y, \mathbb{L}_2) \to \mathcal{V}_2$$

for all $s \in \mathbb{R}$, *for* $\mathcal{V}_1 = \ker \mathcal{A} \subset H^{\infty,\gamma}(M, V_1) \oplus H^{\infty}(Y, \mathbb{L}_1)$ *and a finite-dimensional subspace* $\mathcal{V}_2 \subset H^{\infty,\gamma-\mu}(M, V_2) \oplus H^{\infty}(Y, \mathbb{L}_2)$ *with the property* $\mathcal{V}_2 + \operatorname{im} \mathcal{A} = H^{s-\mu,\gamma-\mu}(M, V_2) \oplus H^{s-\mu}(Y, \mathbb{L}_2)$, $\mathcal{V}_2 \cap \operatorname{im} \mathcal{A} = \{0\}$ *for every* $s \in \mathbb{R}$.

Remark 12.2.8. The constructions concerned with reducing elliptic BVPs to the boundary are completely analogous to those in the case of a manifold with edge, cf. Section 9.3. In the case of a manifold M with boundary if suffices to note that $\mathcal{T}^\mu(M, \boldsymbol{g}; \boldsymbol{l})_{\text{smooth}} \subset \mathcal{T}^\mu(M, \boldsymbol{g}; \boldsymbol{l})$, and ellipticity in $\mathcal{T}^\mu(M, \boldsymbol{g}; \boldsymbol{l})_{\text{smooth}}$ entails ellipticity in $\mathcal{T}^\mu(M, \boldsymbol{g}; \boldsymbol{l})$. The constructions involved in reducing operators to the boundary can be carried out within the operator spaces with subscript "smooth", and hence conclusions such as the Agranovich–Dynin formula hold also in those spaces of BVPs.

Chapter 13

Examples, applications and remarks

The present section gives an abstract on additional results around the nature of cone operators and ellipticity. If proofs are dropped we refer to corresponding material in textbooks or articles.

13.1 Mellin expansions of truncated operators

By truncation of a pseudo-differential operator on the half-line

$$\mathrm{op}(a)\,u(r) = \iint e^{i(r-r')\rho} a(r,\rho) u(r') dr' đ\rho$$

for a symbol $a(r,\rho) \in S^{\mu}(\mathbb{R}_r \times \mathbb{R}_{\rho})$ we understand the operator

$$\mathrm{op}^+(a) := \mathrm{r}^+ \mathrm{op}(a)\, \mathrm{e}^+, \tag{13.1}$$

first in the sense $C_0^{\infty}(\mathbb{R}_+) \to C^{\infty}(\overline{\mathbb{R}}_+)$, later on extended to larger spaces, with e^+ being the extension operator by zero from $\mathbb{R}_+$ to $\mathbb{R}$ and r^+ the restriction of distributions to the open half-line, cf. Section 2.2. Clearly, as soon as we consider actions in $C_0^{\infty}(\mathbb{R}_+)$ then there is no problem of interpreting relation (13.1). However, if we intend to control argument functions up to $r = 0$, e.g., to have continuous operators

$$\mathrm{op}^+(a) : H^s(\mathbb{R}_+) \to H^{s-\mu}(\mathbb{R}_+) \tag{13.2}$$

for $s > -1/2$, or

$$\mathrm{op}^+(a) : \mathcal{S}(\overline{\mathbb{R}}_+) \to \mathcal{S}(\overline{\mathbb{R}}_+), \tag{13.3}$$

© Springer Nature Switzerland AG 2018
X. Liu, B.-W. Schulze, *Boundary Value Problems with Global Projection Conditions*,
Operator Theory: Advances and Applications 265, https://doi.org/10.1007/978-3-319-70114-1_13

then it is necessary to impose some conditions, and (13.3) will not be true in general. At the moment, for convenience, we look at symbols $a(r, \rho)$ which have the form

$$a(r, \rho) = p(r, \rho) + b(\rho)$$

for classical symbols p, b in ρ and $p(r, \rho) = 0$ for $|r| > C$ for some constant $C > 0$.

In Part I we saw that Boutet de Monvel's calculus employs symbols with the transmission property at the boundary, and then (13.2) and (13.3) for $\mu \in \mathbb{Z}$ are continuous extensions of (13.1). Recall that for $s = 0$ and $\mu = 0$ we also have continuous operators

$$\mathrm{op}^+(a) : L^2(\mathbb{R}_+) \to L^2(\mathbb{R}_+) \tag{13.4}$$

for arbitrary $a(r, \rho) \in S^0_{\mathrm{cl}}(\overline{\mathbb{R}}_+ \times \mathbb{R})$ under the above assumption on symbols for $r > \mathrm{const}$. This aspect has been studied in detail in Eskin's book [14]. From the point of view of "Mellin quantization", it is instructive to consider symbols $a(\rho) \in S^0_{\mathrm{cl}}(\mathbb{R}_\rho)$ with constant coefficients.

Composing

$$\mathrm{op}(a) : L^2(\mathbb{R}) \to L^2(\mathbb{R})$$

with the operators of extension by zero to the opposite side

$$e^+ : L^2(\mathbb{R}_+) \to L^2(\mathbb{R}), \quad e^- : L^2(\mathbb{R}_-) \to L^2(\mathbb{R})$$

and the operators $\mathrm{r}^\pm$ of restriction to $\mathbb{R}_\pm$, we obtain continuous operators

$$\mathrm{op}^+(a) := \mathrm{r}^+\mathrm{op}(a)\, e^+ : L^2(\mathbb{R}_+) \to L^2(\mathbb{R}_+),$$
$$\mathrm{op}^-(a) := \mathrm{r}^-\mathrm{op}(a)\, e^- : L^2(\mathbb{R}_-) \to L^2(\mathbb{R}_-),$$
$$\mathrm{r}^\pm\mathrm{op}(a)\, e^\mp : L^2(\mathbb{R}_\mp) \to L^2(\mathbb{R}_\pm).$$

More generally, we also have $\mathrm{op}^+(a) = \mathrm{r}^+\mathrm{op}(a)\, e^+$ for $a(r, \rho) \in S^\mu_{\mathrm{cl}}(\overline{\mathbb{R}}_+ \times \mathbb{R})$, $\mu \in \mathbb{R}$, first operating on $C_0^\infty(\mathbb{R}_+)$ and then extended to suitable distribution spaces. It makes sense also to form complementary projections $\mathrm{op}^+(\theta^\pm)$, where $\theta^\pm$ is the characteristic function of $\mathbb{R}_{\pm,\rho}$, which gives

$$\mathrm{op}^+(\theta^\pm) : L^2(\mathbb{R}_+) \to L^2(\mathbb{R}_+).$$

Note that although $\theta^\pm(\rho)$ is not a symbol in the standard sense, since it has a discontinuity at $\rho = 0$, we can write

$$\mathrm{op}(\theta^\pm) = \mathrm{op}(\chi\theta^\pm) + \mathrm{op}((1 - \chi)\theta^\pm) \tag{13.5}$$

for any excision function $\chi(\rho)$. Then $\chi(\rho)\theta^\pm(\rho) \in S^0_{\mathrm{cl}}(\mathbb{R})$ while $\mathrm{op}((1 - \chi)\theta^\pm)$ is a smoothing operator. Thus $\mathrm{op}(\theta^\pm)$ is an order-zero pseudo-differential operator on $\mathbb{R}$.

One of the major issues is to rephrase $\mathrm{op}^+(a)$ as a Mellin operator and to observe a relationship to cone algebras on $\mathbb{R}_+$. To this end we consider the following specific Mellin symbols, namely,

$$g^+(z) = \left(1 - e^{-2\pi i z}\right)^{-1}, \quad g^-(z) = \left(1 - e^{2\pi i z}\right)^{-1}, \tag{13.6}$$

cf. Eskin [14], where $g^{\pm}(z)$ appear in a similar context. The functions (13.6) are meromorphic with simple poles at the real integers. From (13.6) it follows that

$$g^{+}(z) + g^{-}(z) = 1, \quad g^{\pm}(z+k) = g^{\pm}(z) \text{ for every } k \in \mathbb{Z}.$$

Observe that $g^{\pm}(z) = 1 - g^{\mp}(z)$ gives rise to

$$g^{+}(z) = \frac{-e^{2\pi i}}{1 - e^{2\pi i}}, \quad g^{-}(z) = \frac{-e^{-2\pi i}}{1 - e^{-2\pi i}}.$$

Another useful identity is

$$\overline{g^{\pm}(1 - \bar{z})} = g^{\pm}(z). \tag{13.7}$$

In addition, if $\chi^{\pm}(\rho) \in C^{\infty}(\mathbb{R})$ are functions with

$$\chi^{+}(\rho) = \begin{cases} 1 & \text{for } \rho > 2\varepsilon, \\ 0 & \text{for } \rho < \varepsilon, \end{cases} \quad \chi^{-}(\rho) = \begin{cases} 1 & \text{for } \rho < -2\varepsilon, \\ 0 & \text{for } \rho > -\varepsilon, \end{cases}$$

for some $\varepsilon > 0$, then we have

$$\chi^{+}(\rho)g^{+}(\beta + i\rho), \ \chi^{-}(\tau)g^{-}(\beta + i\rho) \in \mathcal{S}(\mathbb{R}_{\rho})$$

for every $\beta \in \mathbb{R}$, uniformly in compact β-intervals. In particular, it follows that

$$g^{\pm}(z) \in M_{\mathcal{R}}^{0} \tag{13.8}$$

for the Mellin asymptotic type $\mathcal{R} = \{(l, 0)\}_{l \in \mathbb{Z}}$. Therefore, for every $\gamma \notin \mathbb{Z} + 1/2$ the operator

$$\mathrm{op}_{M}^{\gamma}(g^{\pm}(z)) : \mathcal{H}^{s,\gamma}(\mathbb{R}_{+}) \to \mathcal{H}^{s,\gamma}(\mathbb{R}_{+})$$

is continuous for every $s \in \mathbb{R}$. In particular, $\mathrm{op}_{M}(g^{\pm}(z)) : L^{2}(\mathbb{R}_{+}) \to L^{2}(\mathbb{R}_{+})$ is a continuous operator. According to the general properties of Mellin symbols, cf. Theorem 6.3.12, we have a non-direct sum decomposition

$$M_{\mathcal{R}}^{0} = M_{\mathcal{O}}^{0} + M_{\mathcal{R}}^{-\infty} \tag{13.9}$$

which allows us to write $g^{\pm}(z) = g_{0}^{\pm}(z) + g_{1}^{\pm}(z)$ for $g_{0}^{\pm}(z) \in M_{\mathcal{O}}^{0}$, $g_{1}^{\pm}(z) \in M_{\mathcal{R}}^{-\infty}$.

Remark 13.1.1. We have

$$\left(g^{\pm}(z)\right)^{2} - g^{\pm}(z), \ g^{+}(z)g^{-}(z) \in M_{\mathcal{Q}}^{-\infty} \text{ for } \mathcal{Q} = \{(k, 1)\}_{k \in \mathbb{Z}}, \tag{13.10}$$

$$g(z) := -e^{i\pi z}\left(1 - e^{2\pi i z}\right)^{-1} \in M_{\mathcal{R}}^{-\infty} \text{ for } \mathcal{R} = \{(k, 0)\}_{k \in \mathbb{Z}}, \tag{13.11}$$

and $g^{2}(z) = -g^{+}(z)g^{-}(z)$. Moreover,

$$g(z + k) = (-1)^{k} g(z) \quad \text{for every } k \in \mathbb{N}. \tag{13.12}$$

Proposition 13.1.2. *We have*

$$\mathrm{op}^+\left(\theta^\pm(\rho)\right) = \mathrm{op}_M\left(g^\pm(z)\right)$$

as continuous operators $L^2(\mathbb{R}_+) \to L^2(\mathbb{R}_+)$.

Proof. The multiplication by $\theta^\pm(\rho)$ defines complementary projections in $L^2(\mathbb{R}_\rho)$. Thus

$$\mathrm{op}(\theta^\pm) = F^{-1}\theta^\pm F$$

are complementary projections in $L^2(\mathbb{R}_r)$, where $F = F_{r\to\rho}$ is the Fourier transform on the real line. As we saw in Proposition 2.3.3, they have the form

$$\mathrm{op}(\theta^+)u(r) = \lim_{\varepsilon\to+0}\frac{1}{2\pi i}\int_{-\infty}^{\infty}\frac{u(s)}{r-i\varepsilon-s}ds, \tag{13.13}$$

$$\mathrm{op}(\theta^-)u(r) = -\lim_{\varepsilon\to+0}\frac{1}{2\pi i}\int_{-\infty}^{\infty}\frac{u(s)}{r+i\varepsilon-s}ds. \tag{13.14}$$

Let us consider the plus-case; the minus case is analogous. For $u \in L^2(\mathbb{R}_+)$ we have

$$M\left(\mathrm{r}^+\mathrm{op}(\theta^+)\mathrm{e}^+u\right)(z) = M\left(\lim_{\varepsilon\to+0}\frac{1}{2\pi i}\int_0^\infty\frac{u(s)}{r-i\varepsilon-s}ds\right)(z)$$

$$= \lim_{\varepsilon\to+0}\frac{1}{2\pi i}\int_0^\infty r^{z-1}\left(\int_0^\infty\frac{u(s)}{r-i\varepsilon-s}ds\right)dr$$

$$= \lim_{\varepsilon\to+0}\frac{1}{2\pi i}\int_0^\infty u(s)\left(\int_0^\infty\frac{r^{z-1}}{r-i\varepsilon-s}ds\right)dr.$$

Using Lemma 2.3.1 we obtain

$$M\left(\mathrm{r}^+\mathrm{op}(\theta^+)\mathrm{e}^+u\right)(z) = g^+(z)\lim_{\varepsilon\to+0}\frac{1}{2\pi i}\int_0^\infty u(s)(s+i\varepsilon)^{z-1}ds = g^-(z)Mu(z),$$

which is just the asserted relation. □

Proposition 13.1.2 is a simple example of a Mellin reformulation for a pseudo-differential operator on $\mathbb{R}_+$, where $\theta^\pm(\rho)$ are symbols with constant coefficients, i.e., independent of r, though discontinuous at $\rho = 0$, and $g^\pm(z)$ have constant coefficients. We will establish analogous reformulations for general symbols $a(\rho)$. It turns out that then the resulting Mellin symbols are r-dependent in general.

Proposition 13.1.3. *The operator*

$$Ku(r) := \frac{1}{2\pi i}\int_0^\infty\frac{u(s)}{r+s}ds,$$

$u \in C_0^\infty(\mathbb{R}_+)$, *can be written in the form*

$$Ku(r) = \mathrm{op}_M(g)u(r)$$

for $g(z)$ *as in* (13.11). *Thus* K *extends to a continuous operator* $K : L^2(\mathbb{R}_+) \to L^2(\mathbb{R}_+)$.

Proof. We have

$$M(Ku)(z) = \frac{1}{2\pi i} \int_0^\infty r^{z-1} \left(\int_0^\infty \frac{u(s)}{r+s} ds \right) dr = \int_0^\infty \left(\frac{1}{2\pi i} \int_0^\infty \frac{r^{z-1}}{r+s} dr \right) u(s) ds.$$

It remains to apply Lemma 2.3.1, i.e., the identity

$$\frac{1}{2\pi i} \int_0^\infty \frac{r^{z-1}}{r+s} dr = g^-(z) e^{(z-1)\log(-s)} = \frac{-e^{\pi iz} s^{z-1}}{1 - e^{2\pi iz}}. \qquad \square$$

Proposition 13.1.4. *Let* $(\varepsilon^* u)(r) := u(-r)$ *for* $u \in L^2(\mathbb{R}_+)$. *Then*

$$\varepsilon^* r^- \mathrm{op}(\theta^\pm) e^+ = \pm \mathrm{op}_M(g).$$

Proof. From (13.13) it follows that

$$\varepsilon^* r^- \mathrm{op}(\theta^+) e^+ u(r) = \varepsilon^* r^- \lim_{\delta \to +0} \frac{1}{2\pi i} \int_0^\infty \frac{u(s)}{r - i\delta - s} ds$$

$$= r^+ \lim_{\delta \to +0} \frac{-1}{2\pi i} \int_0^\infty \frac{u(s)}{r - i\delta + s} ds = -\mathrm{op}_M(g) u.$$

Moreover, from $\mathrm{op}(\theta^+) + \mathrm{op}(\theta^-) = 1$ we obtain $0 = \varepsilon^* r^- \big(\mathrm{op}(\theta^+) + \mathrm{op}(\theta^-)\big)$ which yields the assertion for θ^-. $\qquad \square$

Remark 13.1.1 and Propositions 13.1.2, 13.1.3, 13.1.4 can be found in Eskin's book [14].

Note that Euler's Γ-function

$$\Gamma(z) = \int_0^\infty r^{z-1} e^{-r} dr, \quad \mathrm{Re}\, z > 0, \tag{13.15}$$

is also a Mellin transform. For $z = n + 1$ we have $\Gamma(n+1) = \int_0^\infty r^n e^{-r} dr = n!$, for all $n \in \mathbb{N}$, $0! = \Gamma(1) = 1$. Integration by parts in (13.15) gives the functional equation of $\Gamma(z)$, namely

$$\Gamma(z+1) = z\Gamma(z), \quad z \in \mathbb{C}. \tag{13.16}$$

Consequently,

$$\Gamma(z+n) = (z+n-1)\Gamma(z+n-1) = (z+n-1)\cdots(z+1)z\Gamma(z) \tag{13.17}$$

for every $n \in \mathbb{N}$. Since the integral on the right-hand side of (13.15) converges uniformly in $\mathrm{Re}\, z > 0$, $\Gamma(z)$ is holomorphic; the proof is given in many textbooks on special functions (see [31]). Now write

$$\Gamma(z) = \int_0^1 r^{z-1} e^{-r} dr + \int_1^\infty r^{z-1} e^{-r} dr. \tag{13.18}$$

The second integral represents an entire function of z. In the first integral we substitute the series of the exponential function. Since this series convergence uniformly, we can interchange summation and integration when z is in the domain $0 < \beta \leq \operatorname{Re} z \leq \delta$, where β and δ are arbitrary reals. For these values of z we obtain the expansion due to Mittag-Leffler

$$\Gamma(z) = \sum_{n=0}^{\infty} \frac{(-1)^n}{n!(z+n)} + \int_1^{\infty} r^{z-1} e^{-r} dr,$$

which holds for all $z \neq -n$, $n = 0, 1, \ldots$. It follows that the Γ-function is meromorphic in $\mathbb{C}$ with simple poles at $0, -1, -2, \ldots$. Another well-known property of the Γ-function is the relation

$$\frac{\Gamma(z+a)}{\Gamma(z+b)} = z^{a-b} \left(1 + O\left(\frac{1}{z} \right) \right) \quad \text{for } |\arg(z+a)| < \pi.$$

Set

$$f_\rho(z) := \frac{\Gamma(1-z)}{\Gamma(1-z+\rho)}. \tag{13.19}$$

Then $f_\rho(z) \in M_{\mathcal{R}_\rho}^{-\rho}$, where

$$\mathcal{R}_\rho = \{(j,0)\}_{j=1,\ldots,\rho} \quad \text{for } \rho \in \mathbb{N}, \, \rho \geq 1; \quad \mathcal{R}_\rho = O \quad \text{for } \rho \in -\mathbb{N}.$$

From the functional equation for the Γ-function it follows that

$$\Gamma(1-z) = (-1)^k z(z+1) \cdots (z+k-1)\Gamma(1-z-k) \quad \text{for } k \in \mathbb{N} \setminus \{0\}.$$

This yields

$$f_k(z) = \prod_{q=1}^{k} (q-z)^{-1} \quad \text{for } k \in \mathbb{N} \setminus \{0\}, \tag{13.20}$$

and

$$f_{-k}(z) = \prod_{q=1}^{k} (1-q-z) \quad \text{for } k \in \mathbb{N} \setminus \{0\}.$$

In particular, $f_0 \equiv 1$. Note that

$$\overline{f_k(1-(\bar{z}-k))} = (-1)^k f_k(z). \tag{13.21}$$

Moreover, let us recall the binomial expansion

$$(1+z)^\alpha = \sum_{k=0}^{\infty} \binom{\alpha}{k} z^k, \quad -\pi < \arg z < \pi, \tag{13.22}$$

$$\binom{\alpha}{k} = \frac{(-1)^k}{k!} \frac{\Gamma(k-\alpha)}{\Gamma(-\alpha)}, \quad k \in \mathbb{N}, \, \alpha \in \mathbb{C}. \tag{13.23}$$

Lemma 13.1.5. $\tilde{a}(r,\rho) \in S^{\mu;0}(\mathbb{R}\times\mathbb{R})$, and $a(r,\rho) := \tilde{a}(r,\rho)\big|_{\overline{\mathbb{R}}_+\times\mathbb{R}}$. Then for every $\omega'' \prec \omega \prec \omega'$ we have for $\mathrm{op}^+(a)$ as an operator on $C_0^\infty(\mathbb{R}_+)$

$$\mathrm{op}^+(a) = \omega\,\mathrm{op}^+(a)\,\omega' + (1-\omega)\,\mathrm{op}^+(a)(1-\omega'') + G,$$

where

$$G := \omega\,\mathrm{op}^+(a)(1-\omega') + (1-\omega)\mathrm{op}^+(a)\,\omega'',$$

and G has a kernel in $\mathcal{S}(\overline{\mathbb{R}}_+\times\overline{\mathbb{R}}_+)$.

Proof. The result is a simple consequence of Corollary 1.1.3, applied first for $\mathrm{op}(\tilde{a})$ itself for functions $\tilde{\omega}' \prec \tilde{\omega} \prec \tilde{\omega}''$ in $C_0^\infty(\mathbb{R})$ and then truncated to $\mathbb{R}_+$. $\square$

In the following we set

$$l_\pm^\alpha(\delta,\rho) := (\delta \pm i\rho)^\alpha, \quad \alpha \in \mathbb{C}, \tag{13.24}$$

for some fixed $\delta \in \mathbb{R}_+$.

Proposition 13.1.6 ([45]). *Let $\alpha \in \mathbb{Z}$, and let ω, ω' be arbitrary cut-off functions. Then for every $m, n \in \mathbb{N}$ there exists an $N(m,n) \in \mathbb{N}$ such that the following relation holds:*

$$\omega\,\mathrm{op}^+\left(l_+^\alpha\right)(\delta)\omega' = \omega r^{-\alpha}\sum_{k=0}^{N}(r\delta)^k\binom{\alpha}{k}\mathrm{op}_M\left(f_{k-\alpha}\right)\omega' + G_N \tag{13.25}$$

on the space $C_0^\infty(\mathbb{R}_+)$ for every $N \geq N(m,n)$, where G_N is an integral operator with kernel in $C_0^m\left(\overline{\mathbb{R}}_+\times\overline{\mathbb{R}}_+\right)$ which extends to a continuous operator

$$G_N : L^2(\mathbb{R}_+) \to [\omega]r^n C^m\left(\overline{\mathbb{R}}_+\right), \tag{13.26}$$

cf. notation (1.84).

Corollary 13.1.7. *With notation of Proposition 13.1.6 for every $\gamma > -1/2$ on the space $C_0^\infty(\mathbb{R}_+)$ we have*

$$\omega\,\mathrm{op}^+(l_+^\alpha)\omega' = \omega r^{-\alpha}\sum_{k=0}^{N}(r\delta)^k\binom{\alpha}{k}\mathrm{op}_M^\gamma\left(f_{k-\alpha}\right)\omega' + G_N.$$

Let us now establish Mellin expansions of $\omega\,\mathrm{op}^+(a)\omega'$ for arbitrary $a(r,\rho) \in S_{\mathrm{cl}}^\mu(\mathbb{R}\times\mathbb{R})$. We shall see that the operators $\omega\,\mathrm{op}^+(a)\omega'$ belong to the cone algebra on $\mathbb{R}_+$ and then obtain explicit expansions for the sequence of conormal symbols.

For any $a(r,\rho) \in S_{\mathrm{cl}}^\mu\left(\overline{\mathbb{R}}_+\times\mathbb{R}\right) := S_{\mathrm{cl}}^\mu(\mathbb{R}\times\mathbb{R})\big|_{\overline{\mathbb{R}}_+\times\mathbb{R}}$ we have asymptotic expansions

$$a(r,\rho) \sim \sum_{j=0}^{\infty} a_j^\pm(r)(i\rho)^{\mu-j} \quad \text{for } \rho \to \pm\infty \tag{13.27}$$

for coefficients $a_j^\pm(r) \in C^\infty(\overline{\mathbb{R}}_+)$ that are uniquely determined by a. In fact, the function

$$a_{(-j)}(r, \rho) = \{a_j^+(r)\theta^+(\rho) + a_j^-(r)\theta^-(\rho)\}(i\rho)^{\mu-j}$$

is the (unique) homogeneous component of a of order $-j$. For purposes below we introduce the functions

$$\sigma_M^{\mu-j}(a)(r, z) := \{a_j^+(r)g^+(z + \mu) + a_j^-(r)g^-(z + \mu)\}f_{j-\mu}(z), \qquad (13.28)$$

cf. formula (13.19). Note that

$$\sigma_M^{\mu-j}(a)(r, z) \in C^\infty\left(\overline{\mathbb{R}}_+, M_{T^{-\mu}\mathcal{R}+\mathcal{R}_{j-\mu}}^{\mu-j}\right), \qquad (13.29)$$

for $\mathcal{R} = \{(l, 0)\}_{l \in \mathbb{Z}}$, $j \in \mathbb{N}$, taking into account that $T^\mu g^\pm \in M_{T^{-\mu}\mathcal{R}}^0$ and $f_{j-\mu} \in M_{\mathcal{R}_{j-\mu}}^{\mu-j}$, cf. also Remark 6.5.10. We have

$$T^{-\mu}\mathcal{R} = \mathcal{R} \quad \text{for } \mu \in \mathbb{Z}.$$

Moreover,

$$\mathcal{R}_{j-\mu} = \begin{cases} \{(n, 0)\}_{n=1,\dots,j-\mu} & \text{for } \mu \in \mathbb{Z}, \, j - \mu \geq 1, \\ O & \text{for } \mu \in \mathbb{Z}, \, j - \mu \leq 0. \end{cases} \qquad (13.30)$$

This gives

$$T^{-\mu}\mathcal{R} + \mathcal{R}_{j-\mu} = \begin{cases} \{(n, 1)\}_{n=1,\dots,j-\mu} \cup \mathcal{R} & \text{for } \mu \in \mathbb{Z}, \, j - \mu \geq 1, \\ \mathcal{R} & \text{for } \mu \in \mathbb{Z}, \, j - \mu \leq 0. \end{cases} \qquad (13.31)$$

for any $j \in \mathbb{N}$. Note that

$$\pi_\mathbb{C}\left(T^{-\mu}\mathcal{R} + \mathcal{R}_{j-\mu}\right) = \mathbb{Z} \quad \text{for } \mu \in \mathbb{Z}.$$

Remark 13.1.8. (i) For $a(\rho) \in S_{\mathrm{cl}}^\mu(\mathbb{R})$, $\mu \in \mathbb{R}$, the operator $\mathrm{op}_M^\gamma\left(\sigma_M^{\mu-j}(a)\right)$ induces a continuous map

$$\omega\, \mathrm{op}_M^\gamma\left(\sigma_M^{\mu-j}(a)\right)\omega' : \mathcal{H}^{s,\gamma}(\mathbb{R}_+) \to \mathcal{H}^{s-\mu+j,\gamma}(\mathbb{R}_+)$$

for $j \in \mathbb{N}$, $s \in \mathbb{R}$, $\gamma - \mu \notin 1/2 + \mathbb{Z}$, cf. Proposition 6.3.3.

(ii) The operator $\mathrm{op}_M(g^\pm)$ induces a continuous map

$$\mathrm{op}_M(g^\pm) : \mathcal{H}^{s,0}(\mathbb{R}_+) \to \mathcal{H}^{s,0}(\mathbb{R}_+), \quad s \in \mathbb{R}.$$

We have $r^{\gamma-\mu}\omega \in \mathcal{H}^{\infty,0}(\mathbb{R}_+)$ for $\gamma - \mu > -1/2$, whence

$$\mathrm{op}_M(g^\pm)r^{\gamma-\mu}\omega \in \mathcal{H}^{\infty,0}(\mathbb{R}_+).$$

Assume for the moment $a(\rho) \in S^0_{\mathrm{cl}}(\mathbb{R})$. It is instructive to consider the bounded set

$$\{a(\rho) \in \mathbb{C} : \rho \in \mathbb{R}\},$$

which is a curve (with finitely many self-intersections) in the complex plane and end points $a_0^\pm$. Recall that the symbol a has the transmission property if and only if this curve is closed and smooth including the end points, corresponding to $\rho \to \pm\infty$. In the present section we do not rule out this case, but we are mainly interested in symbols without the transmission property.

With a we associate the sequence of conormal symbols

$$\sigma_M^{-j}(a)(z) := \{a_j^+ g^+(z) + a_j^- g^-(z)\} f_j(z), \quad j \in \mathbb{N},$$

cf. formula (13.20). Let us set

$$\sigma_M(a) := \{\sigma_M^{-j}(a)\}_{j\in\mathbb{N}}. \tag{13.32}$$

Observe that in the special case $\sigma_M^{-j}(a) \in M_{\mathcal{R}_j}^{-j}$ for

$$\mathcal{R}_j = \{(k,0)\}_{k\in\mathbb{Z}\setminus\{1,\dots,j\}} \cup \{(k,1)\}_{k=1,\dots,j}.$$

Moreover, the points

$$\{a_j^+ g^+(z) + a_j^- g^-(z) : z \in \Gamma_{1/2}\}$$

form the straight connection in $\mathbb{C}$ between the points a_0^+ and a_0^-.

Example 13.1.9. Let A be a differential operator, i.e.,

$$A = \sum_{k=0}^{\mu} a_k \partial_r^k \tag{13.33}$$

$\mu \in \mathbb{N}$, $a_k \in \mathbb{C}$. We can easily express ∂_r^k by Fuchs type differentiations, i.e.,

$$\partial_r^k = r^{-k} \sum_{j=0}^{k} s_{k,j}(-1)^j(-r\partial_r)^j.$$

Thus (13.33) takes the form

$$A = \sum_{k=0}^{\mu} a_k r^{-k} \sum_{j=0}^{k} s_{k,j}(-1)^j(-r\partial_r)^j = r^{-\mu} \sum_{k=0}^{\mu} a_k r^{\mu-k} \sum_{j=0}^{k} s_{k,j}(-1)^j(-r\partial_r)^j.$$

This yields

$$\sigma_M^{\mu-k}(A)(z) = a_k \sum_{j=0}^{k} s_{k,j}(-1)^j z^j.$$

For $h(r,z) := \sum_{k=0}^{\mu} a_k r^{\mu-k} \sum_{j=0}^{k} s_{k,j}(-1)^j z^j$, $\gamma \in \mathbb{R}$, we have $A = r^{-\mu}\mathrm{op}_M^\gamma(h)$, cf. also Remark 6.3.9 (i).

Proposition 13.1.10. *Let* $a(\tau) \in S_{\mathrm{cl}}^0(\mathbb{R})$ *and* $a^* = \overline{a}(\tau)$. *Then*

$$\sigma_M^{-j}(a^*)(z) = \sigma_M^{-j}(a)^*(z-j) \tag{13.34}$$

for all $j \in \mathbb{N}$, *where the* $*$ *on the right is to be interpreted in the sense* $h^*(z) = \overline{h(1-\overline{z})}$.

Proof. The asymptotics of a^* has the coefficients $(-1)^j \overline{a_j^{\pm}}$ for $\tau \to \pm\infty$. Thus,

$$\sigma^{-j}(a^*)(z) = (-1)^j \{\overline{a_j^+} g^+(z) + \overline{a_j^-} g^-(z)\} f_j(z).$$

By (13.7) and (13.21), we have

$$\sigma^{-j}(a^*)(z) = (-1)^j \{\overline{a_j^+ g^+(1-(\overline{z}-j))} + \overline{a_j^- g^-(1-(\overline{z}-j))}\} \overline{f_j(1-(\overline{z}-j))}$$

which is just the right-hand side of (13.34). $\square$

There are other remarkable properties of the sequences (13.32). Those will be commented later on in detail after having developed Mellin expansions of operators $\mathrm{op}^+(a)$. The proofs of the following results are given in [45], see also [21]. For arbitrary $a(\rho) \in S_{\mathrm{cl}}^\mu(\mathbb{R})$ we have the following lemma.

Lemma 13.1.11. *The operator* $\mathrm{op}^+(a)$ *on the space* $C_0^\infty(\mathbb{R}_+)$ *for* $a(\rho) \in S_{\mathrm{cl}}^\mu(\mathbb{R})$, $\mu \in \mathbb{R}$, *admits the decomposition*

$$\mathrm{op}^+(a) = \sum_{l=0}^N \{A_l^+(\delta)\mathrm{op}_M(g^+) + A_l^-(\delta)\mathrm{op}_M(g^-)\}\mathrm{op}^+(l_+^{\mu-l})(\delta) + C_N, \tag{13.35}$$

$N \in \mathbb{N}$, *where* C_N *is an integral operator with kernel in* $C^m(\overline{\mathbb{R}}_+ \times \overline{\mathbb{R}}_+)$ *for any prescribed* $m \in \mathbb{N}$ *when* $N \geq N(m)$ *for* $N(m)$ *sufficiently large, and*

$$A_l^+(\delta) - \sum_{j+k=l} a_j^{\pm}\binom{\mu-j}{k}(-\delta)^k = a_l^{\pm}.$$

Proof. For any excision function $\chi(\rho)$ and $N \in \mathbb{N}$ the difference

$$r_N(\rho) := a(\rho) - \sum_{j=0}^N \chi(\rho)\{a_j^+\theta^+(\rho) + a_j^-\theta^-(\rho)\}(i\rho)^{\mu-j} \tag{13.36}$$

is a symbol in $S_{\mathrm{cl}}^{\mu-(N+1)}(\mathbb{R})$. Given $m \in \mathbb{N}$, there is an $N(m) \in \mathbb{N}$ such that the distributional kernel of $\mathrm{op}^+(r_N)$ belongs to $C^m(\overline{\mathbb{R}}_+ \times \overline{\mathbb{R}}_+)$. In order to analyse the finite sum on the right of (13.36), we consider the summand

$$\chi(\rho)a_j^+\theta^+(\rho)(i\rho)^{\mu-j};$$

the case with minus signs is completely analogous. Let us write

$$(i\rho)^{\mu-j} = \left(l_+(\delta,\rho) - \delta\right)^{\mu-j} = (1 - \delta l_+^{-1})^{\mu-j} l_+^{\mu-j}(\delta,\rho)$$

$$= \left\{ \sum_{k=0}^{\infty} \binom{\mu-j}{k} (-\delta)^k l_+^{-k}(\delta,\rho) \right\} l_+^{\mu-j}(\delta,\rho), \qquad (13.37)$$

cf. the formulas (13.22), (13.23). Then

$$\mathrm{op}^+\left(\chi(\rho)a_j^+\theta^+(\rho)(i\rho)^{\mu-j}\right)$$

$$= a_j^+ \mathrm{op}^+\left(\chi(\rho)\theta^+(\rho) \sum_{k=0}^{\infty} \binom{\mu-j}{k}(-\delta)^k l_+^{\mu-(j+k)}(\delta,\rho)\right) + R_N^+(\delta)$$

$$(13.38)$$

where the remainder $R_N(\delta)$ has a kernel in $C^m(\overline{\mathbb{R}}_+ \times \overline{\mathbb{R}}_+)$ for $N \geq N(m)$. Moreover, we have

$$\mathrm{op}^+\left(\chi(\rho)\theta^+(\rho)l_+^{\mu-l}(\delta,\rho)\right) = \mathrm{op}^+\left(\chi(\rho)\theta^+(\rho)l_+^{\mu-l}(\delta,\rho)\right) + G_l^+(\delta)$$

for $G_l^+(\delta)$ with kernel in $C^\infty(\overline{\mathbb{R}}_+ \times \overline{\mathbb{R}}_+)$. Since $l_+^{\mu-l}$ is a plus-symbol in ρ,

$$\mathrm{op}^+\left(\theta^+ l_+^{\mu-l}\right) = \mathrm{op}^+\left(\theta^+\right)\mathrm{op}^+\left(l_+^{\mu-l}\right). \qquad (13.39)$$

In fact, the plus-property has the consequence that

$$\mathrm{r}^- \mathrm{op}\left(l_+^{\mu-l}\right)\mathrm{e}^+ = 0$$

since $\mathrm{op}\left(l_+^{\mu-l}\right)$ preserves distributions supported by $\overline{\mathbb{R}}_+$. Then (13.39) follow from

$$\mathrm{op}^+\left(\theta^+\right)\mathrm{op}^+\left(l_+^{\mu-l}\right) = \mathrm{r}^+\mathrm{op}(\theta^+)\mathrm{e}^+\mathrm{r}^+\mathrm{op}(l_+^{\mu-l})\mathrm{e}^+$$

$$= \mathrm{r}^+\mathrm{op}(\theta^+)\mathrm{op}(l_+^{\mu-l})\mathrm{e}^+ = \mathrm{op}^+\left(\theta^+ l_+^{\mu-l}\right).$$

Replacing $\mathrm{op}^+(\theta^+)$ by $\mathrm{op}_M(g^+)$, cf. Proposition 13.1.2, for the plus-part of the sum in the expression (13.35) we obtain

$$\sum_{j=0}^{N}\sum_{k=0}^{N} a_j^+ \mathrm{op}_M(g^+)\binom{\mu-j}{k}(-\delta)^k \mathrm{op}^+\left(l_+^{\mu-(j+k)}\right)$$

$$= \sum_{l=0}^{N}\sum_{j+k=l} a_j^+ \binom{\mu-j}{k}(-\delta)^k \mathrm{op}_M(g^+)\mathrm{op}^+\left(l_+^{\mu-(j+k)}\right) + S_N^+$$

$$= \sum_{l=0}^{N} A_l^+(\delta)\,\mathrm{op}_M(g^+)\mathrm{op}^+\left(l_+^{\mu-l}\right) + S_N^+,$$

where

$$A_l^+(\delta) := \sum_{j+k=l} a_j^{\pm} \binom{\mu-j}{k}(-\delta)^k,$$

$$S_N^+ := \sum_{\substack{j+k>N \\ j,k\leq N}} a_j^+ \operatorname{op}_M(g^+)\binom{\mu-j}{k}(-\delta)^k \operatorname{op}^+\big(l_+^{\mu-(j+k)}\big),$$

and C_N in the assertion is the sum of all occurring remainders $R_N^\pm(\delta)$, $G_l^\pm$ and $S_N^\pm$. The analogous expressions with the minus sign can be treated in much the same way.   □

Lemma 13.1.12. *For $A_j(\delta) = \sum_{p+q=j} a_p \binom{\mu-p}{q}(-\delta)^q$ we have*

$$B_l = \sum_{l=j+k} A_j(\delta)\binom{\mu-j}{k}\delta^k = a_l.$$

Proof. Let us write

$$B_l(\delta) = \sum_{l=j+k}\left\{\sum_{p+q=j} a_p\binom{\mu-p}{q}(-\delta)^q\right\}\binom{\mu-j}{k}\delta^k. \tag{13.40}$$

We first observe that a_l occurs in the sum (13.40) exactly for $p = l$, $j = l$, $q = 0$, $k = 0$, which is the coefficient at δ^0 in the polynomial (13.40) of degree l in δ. Let us single out the coefficient of a_p in the expression (13.40) for $p < l$.
  Take first $p = l - 1$. Then the coefficient has the form

$$\sum_{l=j+k}\left\{\sum_{l-1+q=j} (-1)^q\binom{\mu-l+1}{q}\right\}\binom{\mu-j}{k}\delta^{k+q}. \tag{13.41}$$

From $l = j+k$, $l-1+q = j$ we get $m := k+q = 1$. There are two cases, namely,

$$k = 1, \quad q = 0, \quad \text{which implies } j+1 = l,$$

or

$$k = 0, \quad q = 1, \quad \text{which implies } j = l.$$

Thus (13.41) is equal to

$$(-1)^0\binom{\mu-l+1}{0}\binom{\mu-l+1}{1}\delta + (-1)^1\binom{\mu-l+1}{1}\binom{\mu-1}{0}\delta = 0.$$

Now let us treat the general case, i.e., determine the coefficient at a_{l-n} for $n = 2, \ldots, l$. This coefficient is equal to

$$\sum_{l=j+k} \left\{ \sum_{l-n+q=j} a_{l-n}(-1)^q \binom{\mu - l + n}{q} \right\} \binom{\mu - j}{k} \delta^{k+q}$$

$$= \sum_{l=j+k} \left\{ \sum_{l-n+q=j} a_{l-n}(-1)^q \binom{\mu - l + n}{q} \right\} \binom{\mu - j}{k} \delta^n.$$

Thus we have to show that

$$\sum_{l=j+k} \sum_{l-n+q=j} (-1)^q \binom{\mu - l + n}{q} \binom{\mu - j}{k} = 0.$$

Since $l = j + k = j + n - q$, i.e., $k + q = n$, this is equivalent to

$$\sum_{k+q=n} (-1)^q \binom{\mu - l + n}{q} \binom{\mu - l + n - q}{k} = 0,$$

which turns to $\sum_{k+q=n}(-1)^q \binom{\nu}{q}\binom{\nu-q}{k} = 0$, or, equivalently,

$$\sum_{k=0}^{n}(-1)^{n-k} \binom{\nu}{n-k}\binom{\nu - n + k}{k} = 0. \qquad (13.42)$$

In other words, it remains to show (13.42) for every $n \in \mathbb{N}$. For $n = 0, 1$ this is elementary, as we saw before. For general n we employ the definition (13.23) of the binomial coefficients, and write

$$\binom{\nu}{n-k} = \frac{(-1)^{n-k}}{(n-k)!} \frac{\Gamma(n-k-\nu)}{\Gamma(-\nu)}, \qquad \binom{\nu - n - k}{k} = \frac{(-1)^k}{k!} \frac{\Gamma(-\nu+n)}{\Gamma(-\nu+n-k)}.$$

Then (13.42) takes the form

$$\sum_{k=0}^{n} \frac{(-1)^k}{(n-k)!k!} \frac{\Gamma(-\nu+n)}{\Gamma(-\nu)} = \left\{ \sum_{k=0}^{n} \frac{(-1)^k}{(n-k)!k!} \right\} (-\nu)(-\nu+1) \cdots (-\nu+(n-1)).$$

The second factor on the right does not depend on k. For the first factor we write

$$\sum_{k=0}^{n} \frac{(-1)^k}{(n-k)!k!} = \frac{1}{n!} \sum_{k=0}^{n}(-1)^k \binom{n}{k} = (-1+1)^n = 0. \qquad \square$$

Remark 13.1.13. In the case $a(\rho) \in S_{\mathrm{cl}}^\mu(\mathbb{R})$ and $a_j^+ = a_j^-$ for all $j \in \mathbb{N}$, we have

$$\mathrm{op}^+(a)u(r) = \sum_{l=0}^{N} A_l^+(\delta)\mathrm{op}^+(l_+^{\mu-l})u(r) + C_N u(r),$$

for every $N \in \mathbb{N}$, $u \in C_0^\infty(\mathbb{R}_+)$, and C_N as in Lemma 13.1.11.

To formulate remainders in the following proposition, we denote by $\mathbb{G}^m(\mathbb{R}_+)$, $m \in \mathbb{N}$, the space of all operators $\mathcal{L}(L^2(\mathbb{R}_+))$ which define continuous operators

$$K : L^2(\mathbb{R}_+) \to \mathcal{S}^m(\overline{\mathbb{R}}_+), \quad K^* : L^2(\mathbb{R}_+) \to \mathcal{S}^m(\overline{\mathbb{R}}_+) + \mathcal{S}^0_{\mathcal{P}}(\mathbb{R}_+)$$

where

$$\mathcal{S}^m(\overline{\mathbb{R}}_+) := \left\{ u \in C^m(\overline{\mathbb{R}}_+) : \sup \left| \partial_r^k u(r) \right| \langle r \rangle^m < \infty \text{ for } k = 0, \ldots, m \right\}$$

and

$$\mathcal{S}^0_{\mathcal{P}}(\mathbb{R}_+) := \mathcal{K}^{\infty,0;\infty}_{\mathcal{P}}(\mathbb{R}_+)$$

for the asymptotic type $\mathcal{P} := \{(-j, 1) : j \in \mathbb{N}\}$, associated with the weight data $(0, (-\infty, 0])$. In the following theorems we fix cut-off functions ω, ω'.

Theorem 13.1.14. *Let $a(\rho) \in S^0_{\mathrm{cl}}(\mathbb{R})$. Then on $C_0^\infty(\mathbb{R}_+)$ we have the following representation:*

$$\omega \mathrm{op}^+(a)\omega' = \omega \sum_{k=0}^N r^k \mathrm{op}_M\left(\sigma_M^{-k}(a)\right)\omega' + R_N, \quad N \in \mathbb{N}, \tag{13.43}$$

for

$$\sigma_M^{-k}(a)(z) = \left\{ a_k^+ g^+(z) + a_k^- g^-(z) \right\} f_k(z),$$

and a remainder

$$R_N = C_N + G_N$$

for an operator C_N with kernel in $C_0^m(\overline{\mathbb{R}}_+ \times \overline{\mathbb{R}}_+)$, where $m = m(N) \to \infty$ as $N \to \infty$, and an operator G_N with kernel in $\mathcal{S}^0_{\mathcal{T}}(\mathbb{R}_+) \hat\otimes_\pi \mathcal{S}^0_{\mathcal{P}}(\mathbb{R}_+)$, where $\mathcal{T}$ indicates Taylor asymptotics and $\mathcal{P} := \{(-j, 1) : j \in \mathbb{N}\}$.

Theorem 13.1.15. *Let $a(\rho) \in S^\mu_{\mathrm{cl}}(\mathbb{R})$, $\mu \in \mathbb{Z}$, and assume*

$$\gamma \geq 0, \quad \gamma - \mu \geq 0, \quad \gamma - \mu \notin 1/2 + \mathbb{Z}. \tag{13.44}$$

Then on $C_0^\infty(\mathbb{R}_+)$ we have

$$\omega \, \mathrm{op}^+(a)\omega' = r^{-\mu}\omega \sum_{k=0}^N r^k \mathrm{op}_M^\gamma\left(\sigma_M^{\mu-k}(a)\right)\omega' + R_N, \quad N \in \mathbb{N}, \tag{13.45}$$

for

$$\sigma_M^{\mu-k}(a)(z) = \left\{ a_k^+ g^+(z+\mu) + a_k^- g^-(z+\mu) \right\} f_{k-\mu}(z)$$

and a remainder

$$R_N = C_N + G_N$$

where C_N has a kernel in $C_0^m(\overline{\mathbb{R}}_+ \times \overline{\mathbb{R}}_+)$, $m = m(N) \to \infty$ as $N \to \infty$, G_N is of finite rank, depending on N, and has the form

$$G_N = \sum_{i=1}^3 G_{N,i}$$

where $G_{N,i}$ has kernels in $\mathcal{S}_{\mathcal{P}_i}^{\delta_i}(\mathbb{R}_+) \widehat{\otimes}_\pi \mathcal{S}_{\mathcal{Q}_i}^{-\gamma_i}(\mathbb{R}_+)$ for asymptotic types $\mathcal{P}_i$ and $\mathcal{Q}_i$ associated with the weight data $(\delta_i, (-\infty, 0])$ and $(-\gamma_i, (-\infty, 0])$, respectively. For the pairs of weights we have

$$\gamma_1 = \gamma, \quad \delta_1 = 0, \quad \gamma_2 = \gamma_3 = \gamma, \quad \delta_2 = \delta_3 = \gamma - \mu,$$

and $\mathcal{P}_i$, $\mathcal{Q}_i$ are given by (13.66), (13.68), (13.70), (13.71), (13.77) in the proof below.

Proof of Theorem 13.1.14. We first assume $a(\rho) \in S_{\mathrm{cl}}^\mu(\mathbb{R})$. From Lemma 13.1.11 it follows that

$$\omega \operatorname{op}^+(a)\omega' = \omega \sum_{j=0}^{N} \left\{ A_j^+(\delta)\operatorname{op}_M(g^+) + A_j^-(\delta)\operatorname{op}_M(g^-) \right\} \operatorname{op}^+\left(l_+^{\mu-j}\right)\omega' + \omega C_N \omega'.$$

(13.46)

Since $\operatorname{op}^+\left(l_+^{\mu-j}\right)\omega' : H_0^s(\overline{\mathbb{R}}_+) \to H_0^{s-\mu+j}(\overline{\mathbb{R}}_+)$ for all s, we have, in particular,

$$\operatorname{op}^+\left(l_+^{\mu-j}\right)\omega' : C_0^\infty(\mathbb{R}_+) \to H_0^\infty(\overline{\mathbb{R}}_+) \subset L^2(\mathbb{R}_+),$$

and hence the first composition on the right of (13.46) makes sense. By Proposition 13.1.6 and Corollary 13.1.7,

$$\omega_0 \operatorname{op}^+\left(l_+^{\mu-j}\right)\omega' = \omega_0 r^{-\mu+j} \sum_{k=0}^{N} (r\delta)^k \binom{\mu-j}{k} \operatorname{op}_M^\gamma\left(f_{k+j-\mu}\right)\omega' + \omega_0 G_N \omega' \quad (13.47)$$

for cut-off functions ω_0, ω', where $\gamma \geq 0$. For convenience we assume $\omega_0 \succ \omega$, $\omega_0 \succ \omega'$. Let us write

$$\operatorname{op}^+\left(l_+^{\mu-j}\right)\omega' = \omega_0 \operatorname{op}^+\left(l_+^{\mu-j}\right)\omega' + (1-\omega_0)\operatorname{op}^+\left(l_+^{\mu-j}\right)\omega'.$$

Moreover, set

$$m_j := A_j^+ g^+ + A_j^- g^-, \quad m_j^\pm := A_j^\pm g^\pm, \quad n_j := T^{\mu-\gamma}m_j, \quad n_j^\pm := T^{\mu-\gamma}m_j^\pm,$$
$$M_j = \operatorname{op}_M(m_j), \quad N_j = \operatorname{op}_M(n_j), \quad (13.48)$$

for $A_j^\pm := A_j^\pm(\delta)$. Then (13.46) and (13.47) imply

$$\omega \operatorname{op}^+(a)\omega' = \omega \sum_{j=0}^{N} \sum_{k=0}^{N} M_j \omega_0 r^{-\mu+j} (r\delta)^k \binom{\mu-j}{k} \operatorname{op}_M^\gamma\left(f_{k+j-\mu}\right)\omega' + R_N^1 + R_N^2 + R_N^3,$$

(13.49)

where

$$R_N^1 := \omega C_N \omega', \quad R_N^2 := \omega \sum_{j=0}^{N} M_j \omega_0 G_N \omega', \quad R_N^3 := \omega \sum_{j=0}^{N} M_j (1-\omega_0)\operatorname{op}^+\left(l_+^{\mu-j}\right)\omega'.$$

(13.50)

We now recast the first summand on the right of (13.49) as

$$\omega \sum_{j=0}^{N} \sum_{k=0}^{N} M_j \omega_0 r^{-\mu+j+k} \delta^k \binom{\mu-j}{k} \mathrm{op}_M^\gamma \big(f_{k+j-\mu}\big) \omega'$$

$$= \omega \sum_{j=0}^{N} \sum_{k=0}^{N} M_j \omega_0 r^{\gamma-\mu+j+k} \delta^k \binom{\mu-j}{k} \mathrm{op}_M \big(T^{-\gamma} f_{k+j-\mu}\big) r^{-\gamma} \omega' \qquad (13.51)$$

$$= r^{-\mu} \omega \sum_{j=0}^{N} \sum_{k=0}^{N} r^{j+k+\gamma} N_j \omega_0 \delta^k \binom{\mu-j}{k} \mathrm{op}_M \big(T^{-\gamma} f_{k+j-\mu}\big) r^{-\gamma} \omega' + R_N^4$$

$$= r^{-\mu} \omega \sum_{j=0}^{N} \sum_{k=0}^{N} r^{j+k+\gamma} \mathrm{op}_M(n_j) \mathrm{op}_M \big(T^{-\gamma} f_{k+j-\mu}\big) r^{-\gamma} \delta^k \binom{\mu-j}{k} \omega' + R_N^4 + R_N^5$$

$$= r^{-\mu} \omega \sum_{l=0}^{N} r^l \big\{ B_l^+ \mathrm{op}_M^\gamma \big((T^\mu g^+) f_{l-\mu} \big) + B_l^- \mathrm{op}_M^\gamma \big((T^\mu g^-) f_{l-\mu} \big) \big\} \omega'$$

$$+ R_N^4 + R_N^5 + R_N^6, \qquad (13.52)$$

with $B_l^{\pm} := \sum_{l=j+k} A_j^{\pm}(\delta) \delta^k \binom{\mu-j}{k}$. Using Lemma 13.1.12, i.e., that $B_l^{\pm} = a_l^{\pm}$, we finally obtain the Mellin expansion (13.43) for

$$R_N = \sum_{j=1}^{6} R_N^j.$$

The function $M_j \omega_0 r^{\gamma-\mu+j+k}$ in (13.51) is well-defined as an element of $L^2(\mathbb{R}_+)$ for $\gamma - \mu > -1/2$, cf. Remark 13.1.8 (ii). Moreover, $\omega \, \mathrm{op}_M^\gamma \big(T^\mu g^{\pm}\big) f_{l-\mu}) \omega'$ in (13.52) is well-defined for $\gamma - \mu \notin 1/2 + \mathbb{Z}$, cf. Remark 13.1.8 (i). In this computation R_N^4 comes from the commutation of r^{j+k} through the Mellin actions $\mathrm{op}_M(g^+)$ and $\mathrm{op}_M(g^-)$, respectively, R_N^5 is generated by omitting ω_0, and R_N^6 is caused by the summands where $j + k > N$, i.e.,

$$R_N^4 := \omega \sum_{j=0}^{N} \sum_{k=0}^{N} \big\{ M_j \omega_0 r^{\gamma-\mu+j+k} - r^{\gamma-\mu+j+k} N_j \omega_0 \big\}$$

$$\times \delta^k \binom{\mu-j}{k} \mathrm{op}_M \big(T^{-\gamma} f_{k+j-\mu}\big) r^{-\gamma} \omega', \qquad (13.53)$$

$$R_N^5 := r^{-\mu} \omega \sum_{j=0}^{N} \sum_{k=0}^{N} r^{j+k+\gamma} N_j (1 - \omega_0) \delta^k \binom{\mu-j}{k} \mathrm{op}_M \big(T^{-\gamma} f_{k+j-\mu}\big) r^{-\gamma} \omega',$$

$$(13.54)$$

$$R_N^6 := \omega r^{-\mu} \sum_{\substack{j+k>N \\ j,k \leq N}} r^{j+k+\gamma} \mathrm{op}_M(n_j) \mathrm{op}_M \big(T^{-\gamma} f_{k+j-\mu}\big) r^{-\gamma} \delta^k \binom{\mu-j}{k} \omega'. \quad (13.55)$$

For the proof of Theorem 13.1.14 we now assume $\mu = \gamma = 0$. This means that for $a \in S^0_{\mathrm{cl}}(\mathbb{R})$ we have formula (13.43) for the remainder $R_N = \sum_1^6 R_N^j$, where R_N^j, $j = 1, 2, 3$, are given by (13.50).

In order to characterise $R_N^1 = \omega C_N \omega'$, we recall C_N from Lemma 13.1.11, namely, that C_N has the form

$$C_N u(r) = \int_0^\infty c_N(r, r') u(r') dr'$$

for a $c_N(r, r') \in C^m(\overline{\mathbb{R}}_+ \times \overline{\mathbb{R}}_+)$ and $m = m(N) \to \infty$ as $N \to \infty$. This shows immediately that R_N^1 and $(R_N^1)^*$ have the mapping properties

$$R_N : L^2(\mathbb{R}_+) \to [\omega] C^m(\overline{\mathbb{R}}_+), \tag{13.56}$$
$$R_N^* : L^2(\mathbb{R}_+) \to [\omega']\{C^m(\overline{\mathbb{R}}_+) + S^0_{\mathcal{P}}(\mathbb{R}_+)\}, \tag{13.57}$$

respectively. As for R_N^2, we employ from Proposition 13.1.6 that G_N has the form

$$G_N = \omega(r) r^n \widetilde{G}_N \quad \text{for} \quad \widetilde{G}_N u(r) := \int_0^\infty \widetilde{g}_N(r, r') u(r') dr'$$

for a $\widetilde{g}_N(r, r') \in C^m_0(\overline{\mathbb{R}}_+ \times \overline{\mathbb{R}}_+)$, where $n = n(N), m = m(N) \to \infty$ as $N \to \infty$. Now R_N^2 is a finite linear combination of operators of the form

$$\omega \operatorname{op}_M(g^\pm) \omega_0 G_N \omega' : L^2(\mathbb{R}_+) \to [\omega] C^m(\overline{\mathbb{R}}_+). \tag{13.58}$$

In fact, we may write the operator (13.58) as

$$\omega \operatorname{op}_M(g^\pm) \omega_0 r^n \widetilde{\omega}_0 \widetilde{G}_N \omega'$$

for any cut-off function $\widetilde{\omega}_0 \succ \omega_0$. Using Remark 6.5.1 (i) we have a continuous operator

$$r^n \widetilde{\omega}_0 \widetilde{G}_N \omega' : L^2(\mathbb{R}_+) \to [\widetilde{\omega}_0] \mathcal{H}^{\widetilde{m}, 0}_\Theta(\mathbb{R}_+)$$

for every fixed $\widetilde{m} \in \mathbb{N}$ and $k \in \mathbb{N}$, $\Theta = (-(k+1), 0]$. As a consequence of Theorem 6.8.8 (ii) we see that

$$\omega \operatorname{op}_M(g^\pm) \omega_0 : [\widetilde{\omega}_0] \mathcal{H}^{\widetilde{m}, 0}_\Theta(\mathbb{R}_+) \to [\omega] \mathcal{H}^{\widetilde{m}, 0}_{\mathcal{P}}(\mathbb{R}_+)$$

for $\mathcal{P} = \{(-j, 0) : j = 0, \ldots, k\}$. Applying Remark 6.5.1 (ii) we conclude that R_N^2 has the mapping property (13.56). For the adjoint of R_N^2 we have

$$\left(R_N^2\right)^* = \omega' G_N^* \omega_0 \operatorname{op}_M(g^\pm) \omega.$$

Here we employed that $\left(\operatorname{op}_M(g^\pm)\right)^* = \operatorname{op}_M(g^\pm)$, cf. the relations (13.7). Because of the continuity of the operators $\operatorname{op}_M(g^\pm) : L^2(\mathbb{R}_+) \to L^2(\mathbb{R}_+)$, cf. Proposition 13.1.2, and since G_N^* has a kernel in $C^m_0 = (\overline{\mathbb{R}}_+ \times \overline{\mathbb{R}}_+)$, we see that

$$\left(R_N^2\right)^* : L^2(\mathbb{R}_+) \to [\omega'] C^\infty_0(\overline{\mathbb{R}}_+)$$

is continuous.

For the characterisation of R_N^3 we employ the fact that $(1 - \omega_0)\mathrm{op}^+\big(l_+^{-j}\big)\omega'$ has a kernel $d(r, r') \in \mathcal{S}\big(\overline{\mathbb{R}}_+ \times \overline{\mathbb{R}}_+\big)$, cf. Lemma 13.1.5. Because of the factor $1 - \omega_0$ on the left-hand side, we have

$$(1 - \omega_0)\mathrm{op}^+\big(l_+^{-j}\big)\omega' : L^2(\mathbb{R}_+) \to (1 - \omega_0)\mathcal{H}^{\infty,0}(\mathbb{R}_+) \hookrightarrow \mathcal{H}_{(-\infty,0]}^{\infty,0}(\mathbb{R}_+),$$

see also Remark 6.5.1 (iii). Since $g^\pm \in M_{\mathcal{R}}^0$, cf. (13.8), the composition from the left by

$$\omega \, \mathrm{op}_M\big(g^\pm\big) : (1 - \omega_0)\mathcal{H}^{\infty,0}(\mathbb{R}_+) \to [\omega]\mathcal{H}_{\mathcal{T}}^{\infty,0}(\mathbb{R}_+)$$

for $\mathcal{T} := \{(-j, 0) : j \in \mathbb{N}\}$ yields

$$R_N^3 : L^2(\mathbb{R}_+) \to [\omega]\mathcal{H}_{\mathcal{T}}^{\infty,0}(\mathbb{R}_+) \hookrightarrow [\omega]C_0^\infty\big(\overline{\mathbb{R}}_+\big).$$

On the right-hand side we employed Remark 6.5.1 (ii). The adjoint is a linear combination of operators $\omega'\mathrm{op}^+\big(l_-^{-j}\big)(1 - \omega_0)\mathrm{op}_M\big(g^\pm\big)\omega$. By virtue of Proposition 13.1.2 and since $\omega'\mathrm{op}^+\big(l_-^{-j}\big)(1-\omega_0)$ has the kernel $\overline{d}(r, r') \in \mathcal{S}\big(\overline{\mathbb{R}}_+ \times \overline{\mathbb{R}}_+\big)$, we obtain the mapping property $\big(R_N^3\big)^* : L^2(\mathbb{R}_+) \to \mathcal{S}\big(\overline{\mathbb{R}}_+\big) \hookrightarrow [\omega']C^m\big(\overline{\mathbb{R}}_+\big)$.

The operator R_N^4 is a linear combination of expressions of the form

$$\big\{\omega \, \mathrm{op}_M\big(g^\pm\big)\omega_0 r^l - r^l\omega \, \mathrm{op}_M\big(g^\pm\big)\omega_0\big\}\mathrm{op}_M\big(f_l\big)\omega', \quad l \in \mathbb{N}.$$

Consider, for instance, the plus case, and set

$$D_l := \omega \, \mathrm{op}_M\big(g^+\big)\omega_0 r^l - r^l\omega \, \mathrm{op}_M\big(g^+\big)\omega_0. \tag{13.59}$$

By Theorem 6.6.5 and Definition 6.5.2 (ii), $D_l \in L_{\mathrm{G}}(\mathbb{R}_+, (0, 0, (-\infty, 0]))$, and D_l is of finite rank. The concrete asymptotic types follow as in the proof of Theorem 6.5.11, when we apply a decomposition $g^+ = g_0^+ + g_1^+$ for $g_0^+ \in M_{\mathcal{O}}^0$, $g_1^+ \in M_{\mathcal{R}}^{-\infty}$, cf. the notation in (13.8). From the continuity of

$$\mathrm{op}_M\big(f_l\big)\omega' : L^2(\mathbb{R}_+) = \mathcal{H}^{0,0}(\mathbb{R}_+) \to \mathcal{H}^{-l,0}(\mathbb{R}_+),$$

cf. Proposition 6.3.3, it follows that

$$F := D_l \, \mathrm{op}_M\big(f_l\big)\omega' : L^2(\mathbb{R}_+) \to \mathcal{H}_{\mathcal{T}}^{\infty,0}(\mathbb{R}_+) \hookrightarrow [\omega]C^m\big(\overline{\mathbb{R}}_+\big),$$

for every $m \in \mathbb{N}$, cf. Remark 6.5.1 (ii). It will be adequate to represent F as a sum $F_0 + F_1$ for

$$F_0 := D_l \, \mathrm{op}_M\big(f_{l,0}\big)\omega', \quad F_1 := D_l \, \mathrm{op}_M\big(f_{l,1}\big)\omega',$$

with $f_l = f_{l,0} + f_{l,1}$ being a decomposition into a holomorphic non-smoothing part $f_{l,0}$ and a meromorphic smoothing part $f_{l,1}$, according to Remark 6.6.8 (i). The adjoint of R_N^4 is a linear combination of operators

$$\omega'\mathrm{op}_M\big(f_l^{[*]}\big)D_l^* = F_0^* + F_1^*. \tag{13.60}$$

with D_l^* being of analogous structure as D_l, where $f_l^{[*]}(z) = \bar{f}_l(1 - \bar{z})$ belongs to $M_\mathcal{Q}^{-l}$ for $\mathcal{Q} := \{(i, 0) : i = 0, \ldots, l - 1\}$. Using Theorem 6.8.8 and the mapping property of D_l^*, we immediately see that the operator F_0^* is as required in the definition of C_N, while F_1^* maps to asymptotics of type $\mathcal{P}$ as indicated. Observe that F_1 is a Green operator with kernel $\mathcal{S}_T^0(\mathbb{R}_+) \,\widehat{\otimes}_\pi\, \mathcal{S}_O^0(\mathbb{R}_+)$.

The remainder R_N^5 is a linear combination of operators

$$\omega r^l \operatorname{op}_M(g^\pm)(1 - \omega_0) \operatorname{op}_M(f_l) \omega'$$

each of which is a Green operator in $L_G(\mathbb{R}_+, (0, 0, (-\infty, 0]))$ with Taylor asymptotic types. Here it suffices to apply Proposition 6.8.12 (ii). For $(R_N^5)^*$ we can argue in an analogous manner.

The operator R_N^6 is a linear combination of expressions of the form

$$\omega r^{j+k} \operatorname{op}_M(g^\pm f_{j+k}) \omega' \tag{13.61}$$

for $j + k > N$. Since $g^\pm f_{j+k} \in M_\mathcal{D}^{-N}$ for $\mathcal{D} = \{(n, 1) : n \in \mathbb{Z}\}$, it follows that

$$\omega r^{j+k} \operatorname{op}_M(g^\pm f_{j+k}) \omega' : L^2(\mathbb{R}_+) \to [\omega] \mathcal{H}^{N,N}(\mathbb{R}_+). \tag{13.62}$$

For every $m \in \mathbb{N}$ there is an N such that $[\omega] \mathcal{H}^{N,N}(\mathbb{R}_+) \hookrightarrow [\omega] C^m(\overline{\mathbb{R}}_+)$ is a continuous embedding. Thus (13.61) satisfies (13.56). The adjoint of (13.61) is equal to

$$\omega' \operatorname{op}_M\left(g^\pm \big(f_{j+k}\big)^{[*]} r^{j+k}\right) \omega, \tag{13.63}$$

and, of course, is Green again, but the asymptotic type in the image follows from Theorem 6.5.11. Combined with Remark 6.5.1 we finally obtain for (13.63) the required mapping property (13.57). Thus Theorem 13.1.14 is completely proved. $\qquad\square$

Proof of Theorem 13.1.15. Let us now finish the proof of Theorem 13.1.15. In other words, we employ the information up to formula (13.55). The remainders R_N^1, R_N^2 are as before, and it remains to characterise R_N^i, $i = 3, 4, 5, 6$.

For R_N^3 we may apply exactly the same arguments as in the case $\mu = 0$. The operator R_N^4 is a linear combination of expressions of the form

$$\left\{ \omega \operatorname{op}_M(g^\pm) \omega_0 r^{\gamma - \mu + l} - r^{\gamma - \mu + l} \operatorname{op}_M(T^{\mu - \gamma} g^\pm) \omega_0 \right\} \operatorname{op}_M(T^{-\gamma} f_{l-\mu}) r^{-\gamma} \omega', \quad l \in \mathbb{N}. \tag{13.64}$$

Again we content ourselves with the plus case and set

$$\begin{aligned}
D_l(\mu, \gamma) &:= \omega \operatorname{op}_M(g^+) \omega_0 r^{\gamma - \mu + l} - r^{\gamma - \mu + l} \operatorname{op}_M(T^{\mu - \gamma} g^+) \omega_0 \\
&= \omega \operatorname{op}_M(g^+) \omega_0 r^\beta - r^\beta \operatorname{op}_M(T^{-\beta} g^+) \omega_0
\end{aligned} \tag{13.65}$$

for $\beta := \gamma - \mu + l$, using $T^l g^+ = g^+$ for every $l \in \mathbb{N}$. Clearly, in the present case we have $D_l(\mu, \gamma) = D_l$ and this coincides with (13.59) for $\mu = \gamma = 0$. We apply Theorem 6.6.5 to $g^+ \in M_\mathcal{R}^0$, cf. relation (13.8), and look at the condition

$$\pi_\mathbb{C} \mathcal{R} \cap \Gamma_{1/2} = \pi_\mathbb{C} \mathcal{R} \cap \Gamma_{1/2 - \beta} = \emptyset,$$

which is satisfied because of the assumptions (13.44). Concerning the asymptotic types involved in D_l, we argue as in the proof of Theorem 6.5.11, since in a decomposition $g^+ = g_0^+ + g_1^+$ for $g_0^+ \in M_{\mathcal{O}}^0$, $g_1^+ \in M_{\mathcal{R}}^{-\infty}$, only the terms with g_1^+ contribute to the asymptotics. In the present case the difference in Theorem 6.5.11 defines a continuous map

$$D_l : \mathcal{K}^{s,0;-\infty}(\mathbb{R}_+) \to \mathcal{E}_{\mathcal{P}_{1,l}}$$

where $\mathcal{P}_{1,l} = \{(p,0) : p \in \mathbb{Z}, 1/2 - (\gamma - \mu + l) < p < 1/2\}$. Set

$$\mathcal{P}_1 := \bigcup_{l=0}^{2N} \mathcal{P}_{1,l} \tag{13.66}$$

This occurs for $l = 0, \ldots, 2N$, and the largest possible β is equal to $\gamma - \mu + 2N$. For analogous reasons, the adjoint defines a continuous map

$$D_l^* : \mathcal{K}^{s,0;-\infty}(\mathbb{R}_+) \to \mathcal{E}_{\mathcal{Q}_{1,l}} \subset \mathcal{K}^{\infty,0;\infty}(\mathbb{R}_+)$$

where $\mathcal{Q}_{1,l} = \{(k - \gamma + \mu, 0) : k \in \mathbb{Z}, 1/2 - (\gamma - \mu + l) < k - \gamma + \mu < 1/2\}$. Set

$$\mathcal{Q} := \bigcup_{l=0}^{2N} \mathcal{Q}_{1,l}. \tag{13.67}$$

The operator $F := \widetilde{\omega}_0 \mathrm{op}_M \left(T^{-\gamma} f_{l-\mu} \right) r^{-\gamma} \omega'$ for $\widetilde{\omega}_0 \succ \omega_0$ induces a continuous map

$$F : \mathcal{K}^{0,\gamma;-\infty}(\mathbb{R}_+) \to \mathcal{K}^{l-\mu,0;\infty}(\mathbb{R}_+).$$

This implies the continuity of $D_l F : \mathcal{K}^{0,\gamma;-\infty}(\mathbb{R}_+) \to [\omega]\mathcal{K}_{\mathcal{P}_1}^{\infty,0;\infty}(\mathbb{R}_+) \hookrightarrow \mathcal{S}_{\mathcal{P}_1}^0(\mathbb{R}_+)$. Moreover, $F^* = \omega' r^{-\gamma} \mathrm{op}_M \left(\left(T^{-\gamma} f_{l-\mu} \right)^{[*]} \right) \widetilde{\omega}_0$ induces a continuous map

$$F^* : \mathcal{K}^{0,0;-\infty}(\mathbb{R}_+) \to \mathcal{K}^{0,-\gamma;\infty}(\mathbb{R}_+)$$

and then $F^* D_l^* : \mathcal{K}^{s,0;-\infty}(\mathbb{R}_+) \to \mathcal{K}_{T^\gamma \mathcal{Q} + \mathcal{Q}_l'}^{\infty,-\gamma;\infty}(\mathbb{R}_+) \hookrightarrow \mathcal{S}_{T^\gamma \mathcal{Q} + \mathcal{Q}_l'}^{-\gamma}(\mathbb{R}_+)$ for

$$\mathcal{Q}_l' = \begin{cases} \{(-j,0)\}_{j=0,\ldots,l-\mu} & \text{for } l - \mu \in \mathbb{N}, \ l - \mu \geq 1, \\ O & \text{for } j - \mu \in -\mathbb{N}. \end{cases}$$

Set

$$\mathcal{Q}' := \bigcup_{l=0}^{2N} \mathcal{Q}_l', \quad \mathcal{Q}_1 := T^\gamma \mathcal{Q} + \mathcal{Q}'. \tag{13.68}$$

Thus R_N^4 is completely characterised as an operator with kernel in

$$\mathcal{S}_{\mathcal{P}_1}^0(\mathbb{R}_+) \widehat{\otimes}_\pi \mathcal{S}_{\mathcal{Q}_1}^{-\gamma}(\mathbb{R}_+). \tag{13.69}$$

Let us now pass to R_N^5, given as (13.55). In this case we have the mapping properties

$$R_N^5 : \mathcal{K}^{s,\gamma;-\infty}(\mathbb{R}_+) \to \mathcal{S}_{\mathcal{P}_2}^{\gamma-\mu}(\mathbb{R}_+)$$

for

$$\mathcal{P}_2 := \{(k,0) : k \in \mathbb{Z}, k < 1/2 - (\gamma - \mu)\}, \qquad (13.70)$$

and

$$\left(R_N^5\right)^* : \mathcal{K}^{s,-\gamma+\mu;-\infty}(\mathbb{R}_+) \to \mathcal{S}_{\mathcal{Q}_2}^{-\gamma}(\mathbb{R}_+)$$

for

$$\mathcal{Q}_2 := \mathcal{Q}' \text{ as in (13.68).} \qquad (13.71)$$

Thus R_N^5 has a kernel in $\mathcal{S}_{\mathcal{P}_2}^{\gamma-\mu}(\mathbb{R}_+) \widehat{\otimes}_\pi \mathcal{S}_{\mathcal{Q}_2}^{-\gamma}(\mathbb{R}_+)$.

Finally, consider the remainder R_N^6 where we content ourselves with the case involving symbol g^+. We then have a linear combination of operators

$$C := \omega r^{-\mu+l+\gamma} \mathrm{op}_M(d) r^{-\gamma} \omega'$$

for

$$d := \left(T^{\mu-\gamma} g^+\right)\left(T^{-\gamma} f_{l-\mu}\right), \quad l > N.$$

Let us write $d = d_0 + d_1$ for $d_0 \in M_{\mathcal{O}}^{\mu-l}$, $d_1 \in M_{\mathcal{D}}^{-\infty}$ for the Mellin asymptotic type $\mathcal{D} = \mathcal{G} + \mathcal{F}$ for $\mathcal{G} := \{(m - \mu + \gamma, 0) : m \in \mathbb{Z}\}$, $\mathcal{F} := T^\gamma \mathcal{R}_{l-\mu}$, cf. (13.30) and Remark 6.6.8 (i).

Writing $C_i := \omega r^{-\mu+l+\gamma} \mathrm{op}_M(d_i) r^{-\gamma} \omega'$, $i = 0, 1$, we have $C = C_0 + C_1$, and it suffices to consider C_0 and C_1 separately. We fix a finite weight interval $\Theta = (-(k+1), 0]$ and choose N larger than k, which is enough for our result. For C_i, $i = 0, 1$, we have continuity in the sense

$$C_0 : [\omega']\mathcal{H}^{s,\gamma}(\mathbb{R}_+) \to [\omega]\mathcal{H}_\Theta^{s+l-\mu,\gamma-\mu}(\mathbb{R}_+), \qquad (13.72)$$

$$C_1 : [\omega']\mathcal{H}^{s,\gamma}(\mathbb{R}_+) \to [\omega]\mathcal{H}_\Theta^{\infty,\gamma-\mu}(\mathbb{R}_+). \qquad (13.73)$$

The adjoint of C_i has the form

$$C_i^* := \omega' r^{-\gamma} \mathrm{op}_M\left(d_i^{[*]}\right) r^{-\mu+l+\gamma} \omega.$$

Applying Theorem 6.5.11 for C_0^* we may commute $r^{-\mu+l+\gamma}$ through the Mellin actions without remainder. Thus C_0^* induces a continuous operator

$$C_0^* : [\omega]\mathcal{H}^{s,-\gamma+\mu}(\mathbb{R}_+) \to [\omega']\mathcal{H}_\Theta^{s+l-\mu,-\gamma}(\mathbb{R}_+), \qquad (13.74)$$

$s \in \mathbb{R}$. Concerning G_1^*, we apply Theorem 6.5.11 to commute $r^{-\mu+l+\gamma}$ through the operator which leaves a Green remainder G_1^*, and

$$C_1^* - G_1^* : [\omega]\mathcal{H}^{s,-\gamma+\mu}(\mathbb{R}_+) \to [\omega']\mathcal{H}_\Theta^{\infty,-\gamma}(\mathbb{R}_+) \qquad (13.75)$$

is continuous, $s \in \mathbb{R}$. The Green operator G_1 has a kernel in

$$\mathcal{S}_{\Theta}^{\gamma-\mu}(\overline{\mathbb{R}}_+) \widehat{\otimes}_\pi \mathcal{S}_{\mathcal{N}_l}^{-\gamma}(\overline{\mathbb{R}}_+). \tag{13.76}$$

The trivial asymptotic type Θ relative to the chosen weight interval Θ comes from the corresponding mapping property of G_1 with a large weight in the image, caused by l in the exponent. The asymptotic type $\mathcal{N}_l$ may be obtained from the commutation result of Theorem 6.5.11. Since

$$C_1^* := \omega' \mathrm{op}_M^{-\gamma}\big(T^\gamma d_1^{[*]}\big) r^{-\mu+l}\omega,$$

where $T^\gamma d_1^{[*]} \in M_{T^{-\gamma}\mathcal{D}^*}^{-\infty}$, cf. the notation (6.86), we have

$$\mathcal{N}_l = \big\{(1-q-\gamma, m_q) : (q, m_q) \in \mathcal{D}, 1/2 + \mu - l < 1 - q < 1/2\big\},$$

since this corresponds to the pairs in $T^{-\gamma}\mathcal{D}^*$ where the first component lies in the strip $S_{\beta,\gamma}$ for $\beta = -\mu + l$. Set

$$\mathcal{P}_3 := \Theta, \quad \mathcal{Q}_3 := \bigcup_{l=N+1}^{2N} \mathcal{N}_l. \tag{13.77}$$

This shows altogether that R_N^6 has a kernel in $\mathcal{S}_{\Theta}^{\gamma-\mu}(\overline{\mathbb{R}}_+) \widehat{\otimes}_\pi \mathcal{S}_{\mathcal{Q}_3}^{-\gamma}(\overline{\mathbb{R}}_+)$.

A consequence of the mapping properties (13.72), (13.74) is that the mappings

$$C_0 : [\omega']\mathcal{H}^{0,\gamma}(\mathbb{R}_+) \to [\omega]\mathcal{H}^{M,M}(\mathbb{R}_+), \quad C_0^* : [\omega]\mathcal{H}^{0,-\gamma+\mu}(\mathbb{R}_+) \to [\omega']\mathcal{H}^{M,M}(\mathbb{R}_+) \tag{13.78}$$

are continuous for any given sufficiently large M. Similarly, setting $B_1 := C_1 - G_1$ from (13.73), (13.75), (13.76) one derives the continuity of the mappings

$$B_1 : [\omega']\mathcal{H}^{0,\gamma}(\mathbb{R}_+) \to [\omega]\mathcal{H}^{M,M}(\mathbb{R}_+), \quad B_1^* : [\omega]\mathcal{H}^{0,-\gamma+\mu}(\mathbb{R}_+) \to [\omega']\mathcal{H}^{M,M}(\mathbb{R}_+).$$

To simplify notation, set $A := C_0$ or $A := B_1$. Then A is a pseudo-differential operator of order $-M$, and we may chose M so large that its integral kernel $a(r, r')$ belongs to $C^n(\mathbb{R}_+ \times \mathbb{R}_+)$ for any prescribed $n \in \mathbb{N}$. Clearly, it has bounded support in r and r'.

The mapping properties (13.78) show that we may pass to a kernel

$$r^{\gamma-\mu} a(r, r')(r')^{-\gamma} =: a_0(r, r')$$

such that the associated operator A_0 induces continuous operators

$$A_0 : [\omega']\mathcal{H}^{0,0}(\mathbb{R}_+) \to [\omega]\mathcal{H}^{M,M}(\mathbb{R}_+), \quad A_0^* : [\omega]\mathcal{H}^{0,0}(\mathbb{R}_+) \to [\omega']\mathcal{H}^{M,M}(\mathbb{R}_+), \tag{13.79}$$

where M on the right-hand side can be taken as large as we want.

Now the relation

$$A_0 u = \int_0^\infty a_0(r, r') u(r') dr'$$

and (13.79) allow us to write

$$a_0(r, r') = r^{\widetilde{M}} a_1(r, r')(r')^{\widetilde{M}'}$$

for any fixed $\widetilde{M}, \widetilde{M}'$ such that when we define A_1 with the kernel a_1, the operators A_1 and A_1^* have properties analogous to (13.79), where for convenience we take again the same M that was allowed, as large as we want. It follows that

$$a(r, r') = r^{\widetilde{M}+\gamma-\mu} a_1(r, r')(r')^{\widetilde{M}'-\gamma}$$

and hence also $r^{-\widetilde{M}-\gamma+\mu} A(r')^{-\widetilde{M}'+\gamma}$ has the mapping properties (13.79). Now choose $\widetilde{M}$ and $\widetilde{M}'$ so that $-\widetilde{M}' + \gamma \le 0$, $-\widetilde{M} - \gamma + \mu \le 0$. Then A, A^* operate on $\mathcal{H}^{0,0}(\mathbb{R}_+)$ ($= L^2(\mathbb{R}_+)$) by restriction from the respective weighted spaces to $L^2(\mathbb{R}_+)$.

Concerning the operators C_0 and B_1, we are in the same position as in the proof of Theorem 13.1.14, formula (13.62), and we conclude that C_0, B_1 have kernels in $C_0^m(\overline{\mathbb{R}}_+ \times \overline{\mathbb{R}}_+)$ with $m = m(N) \to \infty$ as $N \to \infty$, i.e., C_0, B_1 contribute to the remainder C_N. □

Remark 13.1.16. Theorems 13.1.14 and 13.1.15 have natural generalisations to the case of symbols $a(r, \rho) \in S_{cl}^\mu(\overline{\mathbb{R}}_+ \times \mathbb{R})$, where in the formulas for $\sigma_M^{\mu-k}(a)$ we simply replace $a_k^\pm$ by corresponding functions in $C^\infty(\overline{\mathbb{R}}_+)$, cf. the asymptotic expansion (13.27). An inspection of the proofs shows that there are no essential modifications.

Remark 13.1.17. Theorem 13.1.14 is a refinement of a corresponding result in Eskin's book [14, §15] where the coefficients $a_0^\pm$ and the Mellin symbols $g^\pm$ are involved, while the other ingredients, namely Mellin convention operators plus Hilbert–Schmidt operators are the analogues of our smoothing Mellin plus Green operators. This source of information is also employed in [45].

Remark 13.1.18. Let $a(\rho) \in S_{cl}^\mu(\mathbb{R})$; then for every $m \in \mathbb{N}$ there exists a $\nu \le 0$ such that the distributional kernel of op(a) belongs to $C^m(\mathbb{R} \times \mathbb{R})$ for every $\mu \le \nu$. For any $\gamma \ge 0$ we then have $\gamma - \mu \ge 0$, and so $\omega \, \mathrm{op}^+(a)\omega'$ defines an operator

$$\omega \, \mathrm{op}^+(a)\omega' : \mathcal{K}^{s,\gamma}(\mathbb{R}_+) \to \mathcal{K}^{s-\mu,0}(\mathbb{R}_+)$$

and we cannot expect that the image will belong to $\mathcal{K}^{s-\mu,\gamma-\mu}(\mathbb{R}_+)$. In other words, when we intend to pass from $\omega \, \mathrm{op}^+(a)\omega'$ to an element in $L^\mu(\mathbb{R}_+, \boldsymbol{g})$ for $\boldsymbol{g} = (\gamma, \gamma - \mu, \Theta)$, then we necessarily have to subtract some remainder that belongs to $L_G^\mu(\mathbb{R}_+, (\gamma, 0, \Theta))$; hence, here we are dealing with a non-trivial Mellin operator convention.

Remark 13.1.19. Theorems 13.1.14 and 13.1.15 have analogues also for symbols

$$a(r, \rho) \in S_{\mathrm{cl}}^{\mu}(\overline{\mathbb{R}}_+ \times \mathbb{R})$$

where the indicated remainders have a similar form as before; however, the obtained asymptotic types $\mathcal{P}_i$, $\mathcal{Q}_i$ have to be respectively replaced by $\bigcup_{j \in \mathbb{N}} T^{-j} \mathcal{P}_i$ and $\bigcup_{j \in \mathbb{N}} T^{-j} \mathcal{Q}_i$.

Next we draw some important conclusions from Theorem 13.1.15 (or its special case Theorem 13.1.14). In the r-dependent case we have

$$\sigma_M^{\mu-l}(a)(r, z) \in C^{\infty}\left(\overline{\mathbb{R}}_+, M_{T^{-\mu}\mathcal{R} + \mathcal{R}_{l-\mu}}^{\mu-l}\right),$$

cf. (13.28), (13.29). Applying Remark 6.6.8 (ii) we can construct decompositions

$$\sigma_M^{\mu-l}(a)(r, z) = h_l(r, z) + b_l(r, z) \tag{13.80}$$

with $h_l(r, z) \in C^{\infty}\left(\overline{\mathbb{R}}_+, M_{\mathcal{O}}^{\mu-l}\right)$, $b_l(r, z) \in C^{\infty}\left(\overline{\mathbb{R}}_+, M_{T^{-\mu}\mathcal{R} + \mathcal{R}_{l-\mu}}^{-\infty}\right)$. By Theorem 6.3.15 and Remark 6.3.16, we can carry out the asymptotic summation

$$h \sim \sum_{l=0}^{\infty} h_l \quad \text{in } C^{\infty}\left(\overline{\mathbb{R}}_+, M_{\mathcal{O}}^{\mu}\right).$$

More precisely, as follows from the proof, there are decompositions

$$h_l = h_l' + h_l'' \quad \text{with } h_l' \in C^{\infty}\left(\overline{\mathbb{R}}_+, M_{\mathcal{O}}^{\mu-l}\right), \ h_l'' \in C^{\infty}\left(\overline{\mathbb{R}}_+, M_{\mathcal{O}}^{-\infty}\right),$$

such that

$$h = \sum_{l=0}^{\infty} r^l h_l' \quad \text{converges in } C^{\infty}\left(\overline{\mathbb{R}}_+, M_{\mathcal{O}}^{\mu}\right). \tag{13.81}$$

Since $r^l h_l \in C^{\infty}\left(\overline{\mathbb{R}}_+, M_{\mathcal{O}}^{\mu-l}\right)$, we in fact have even more, namely, that also

$$\sum_{l=M}^{\infty} r^{l-M} h_l' \quad \text{converges in } C^{\infty}\left(\overline{\mathbb{R}}_+, M_{\mathcal{O}}^{\mu-M}\right) \text{ for every } M \in \mathbb{N}. \tag{13.82}$$

Theorem 13.1.20. *Let* $a(r, \rho) \in S_{\mathrm{cl}}^{\mu}(\overline{\mathbb{R}}_+ \times \mathbb{R})$, $\mu, \gamma \in \mathbb{R}$, *and assume the relations* (13.44) *hold. Then there is a* $h(r, z) \in C^{\infty}\left(\overline{\mathbb{R}}_+, M_{\mathcal{O}}^{\mu}\right)$ *such that on the space* $C_0^{\infty}(\mathbb{R}_+)$ *we have*

$$\omega \operatorname{op}^+(a)\omega' = r^{-\mu} \omega \operatorname{op}_M^{\gamma}(h) \, \omega' + H$$

for a smoothing Mellin plus Green operator H *that will be characterised below.*

Proof. We start with the r-dependent analogue of Theorem 13.1.15, i.e., express $\omega \operatorname{op}^+(a)\omega'$ in the form

$$\omega \operatorname{op}^+(a)\omega' = r^{-\mu}\omega \sum_{l=0}^{N} r^l \operatorname{op}_M^\gamma\left(\sigma_M^{\mu-l}(a)\right)\omega' + R_N.$$

We employ the decomposition (13.80) including the subsequent information, which yields

$$\omega \operatorname{op}^+(a)\omega' = r^{-\mu}\omega \sum_{l=0}^{N} r^l \operatorname{op}_M^\gamma\left(h_l + b_l\right)\omega' + R_N$$

$$= r^{-\mu}\omega \sum_{l=0}^{N} r^l \operatorname{op}_M^\gamma\left(h_l'\right)\omega' + r^{-\mu}\omega \sum_{l=0}^{N} r^l \operatorname{op}_M^\gamma\left(h_l'' + b_l\right)\omega' + R_N$$

$$= r^{-\mu}\omega \operatorname{op}_M^\gamma(h)\omega' + H \tag{13.83}$$

with h as in (13.81), b_l as in (13.80), and

$$d_N(r,z) := -r^{N+1} \sum_{l=N+1}^{\infty} r^{l-(N+1)} h_l'(r,z) \in C^\infty\left(\overline{\mathbb{R}}_+, M_\mathcal{O}^{\mu-(N+1)}\right),$$

$$H := r^{-\mu}\omega \operatorname{op}_M^\gamma(d_N)\omega' + r^{-\mu}\omega \sum_{l=0}^{N} r^l \operatorname{op}_M^\gamma\left(h_l'' + b_l\right)\omega' + R_N.$$

Since both $\omega \operatorname{op}^+(a)\omega'$ and $r^{-\mu}\omega \operatorname{op}_M^\gamma(h)\omega'$ are independent of N, so is H.

By the characterisation of R_N in Theorem 13.1.15, the smoothness of the kernel of R_N tends to infinity as $N \to \infty$. Because of relation (13.82), also the smoothness of the kernel of $r^{-\mu}\omega \operatorname{op}_M^\gamma(d_N)\omega'$ tends to infinity as $N \to \infty$. This shows that H has a kernel in $C^\infty(\mathbb{R}_+ \times \mathbb{R}_+)$.

In order to represent H as a smoothing Mellin plus Green operator, we fix a weight interval $\Theta = (-(k+1), 0]$ for any $k \in \mathbb{N}$ and write $H = M + G$ for $M := r^{-\mu}\omega \sum_{l=0}^{k} r^l \operatorname{op}_M^\gamma\left(h_l'' + b_l\right)\omega'$ which is of smoothing Mellin type, while $G = S_N + R_N$ for $S_N := r^{-\mu}\omega \operatorname{op}_M^\gamma(d_N)\omega' + r^{-\mu}\omega \sum_{l=k+1}^{N} r^l \operatorname{op}_M^\gamma\left(h_l'' + b_l\right)\omega'$. For every choice of $s, e, s', e' \in \mathbb{R}$ we find an N such that

$$S_N : \mathcal{K}^{s,\gamma;e}(\mathbb{R}_+) \to \mathcal{K}_\mathcal{O}^{s',\gamma-\mu;e'}(\mathbb{R}_+).$$

Moreover, S_N^* induces a continuous operator

$$S_N^* : \mathcal{K}^{s,-\gamma+\mu;e}(\mathbb{R}_+) \to \mathcal{K}_\mathcal{B}^{s',-\gamma;e'}(\mathbb{R}_+),$$

where $\mathcal{B}$ is only induced by the mapping property of $\left(r^{-\mu}\sum_{l=k+1}^{N} r^l \operatorname{op}_M^\gamma(b_l)\omega'\right)^*$, while

$$\left(r^{-\mu}\operatorname{op}_M^\gamma(d_N)\omega'\right)^* \quad \text{and} \quad r^{-\mu}\sum_{l=k+1}^{N}\left(r^l \operatorname{op}_M^\gamma(h_l'')\omega'\right)^*$$

produce flatness in the image. The asymptotic type $\mathcal{B}$ is explicitly known from Theorem 6.5.11. For R_N we have

$$R_N = C_N + G_{N,1} + G_{N,2} + G_{N,3},$$

where from Theorem 13.1.15 we know that $G_{N,2} + G_{N,3} \in L_{\mathrm{G}}(\mathbb{R}_+, (\gamma, \gamma - \mu, \Theta))$ for every $k \in \mathbb{N}$ and $G_{N,1} \in L_{\mathrm{G}}(\mathbb{R}_+, (\gamma, 0, \Theta))$, cf. notation (6.79). Concerning C_N, we have the mapping properties

$$C_N, C_N^* : \mathcal{K}^{s,0;e}(\mathbb{R}_+) \to \mathcal{K}_{\mathcal{T}}^{s',0;e'}(\mathbb{R}_+)$$

for every $s, e, s', e' \in \mathbb{R}$ and suitable N sufficiently large, where $\mathcal{T}$ indicates the Taylor asymptotic type for the weight interval $(-(k+1), 0]$ (i.e., $\{(-j, 0)\}_{j=0,\dots,k}$). Thus, in the characterisation of H which is independent of N the term C_N contributes an operator with kernel in $\mathcal{S}(\overline{\mathbb{R}}_+) \widehat{\otimes}_\pi \mathcal{S}(\overline{\mathbb{R}}_+)$. $\qquad\square$

Corollary 13.1.21. *Let* $a(r, \rho) \in S_{\mathrm{cl}}^{0;0}(\overline{\mathbb{R}}_+ \times \mathbb{R})$, *which corresponds to Theorem 13.1.14. Then we have* $\mathrm{op}^+(a) \in L^0(\mathbb{R}_+, (0, 0, (-\infty, 0]))$.

In fact, because of Lemma 13.1.5 it suffices to state that

$$\omega\,\mathrm{op}^+(a)\omega' \in L^0(\mathbb{R}_+, (0, 0, (-\infty, 0])),$$

which is just a consequence of Theorem 13.1.20.

Remark 13.1.22. Condition (13.44) concerns arbitrary $\mu \in \mathbb{Z}$ and any sufficiently large γ is suitable, provided $\gamma - \mu \notin 1/2 + \mathbb{Z}$. The latter relation only requires that $\mathrm{op}_M^\gamma(\sigma_M^{\mu-k}(a))$ makes sense, i.e., $\Gamma_{1/2-\gamma} \cap \pi_{\mathbb{C}}(T^{-\mu}\mathcal{R} + \mathcal{R}_{k-\mu}) = \emptyset$ for all $k \in \mathbb{N}$, cf. (13.29).

Theorem 13.1.23. *Let* $a(r, \rho) \in S_{\mathrm{cl}}^{\mu;0}(\overline{\mathbb{R}}_+ \times \mathbb{R})$ *for fixed* $\gamma \in \mathbb{R}$, $\mu \in \mathbb{Z}$. *Then*

$$\mathrm{op}^+(a) \in L^\mu(\mathbb{R}_+, \boldsymbol{g}) \bmod \mathbb{G}(\mathbb{R}_+)$$

for $\boldsymbol{g} = (\gamma, \gamma - \mu, (-\infty, 0])$, *cf.* (6.84), *under the following conditions on* γ *and* μ:

$$\gamma \geq 0, \quad \gamma - \mu \geq 0, \quad \gamma - \mu \notin 1/2 + \mathbb{Z},$$

or

$$1/2 - \gamma \notin \mathbb{Z}.$$

Proof. By Lemma 13.1.5, we may focus on $\omega\,\mathrm{op}^+(a)\omega'$. Applying Theorem 13.1.20 for any fixed

$$\widetilde{\gamma} \geq 0, \quad \widetilde{\gamma} - \mu \geq 0, \quad \widetilde{\gamma} - \mu \notin 1/2 + \mathbb{Z}, \tag{13.84}$$

we have

$$\omega\,\mathrm{op}^+(a)\omega' = r^{-\mu}\omega\,\mathrm{op}_M^{\widetilde{\gamma}}(h)\omega' + \widetilde{H}$$

for an operator $\widetilde{H}$ of the form $\widetilde{M} + \widetilde{G}$, where $\widetilde{M} \in L_{\mathrm{M+G}}(\mathbb{R}_+, \widetilde{\boldsymbol{g}})$, $\widetilde{\boldsymbol{g}} = (\widetilde{\gamma}, \widetilde{\gamma} - \mu, -(k+1), 0])$ for arbitrary fixed $k \in \mathbb{N}$, and $\widetilde{G} \in \mathbb{G}(\mathbb{R}_+)$. The assertion is proved when we find operators $\widetilde{G}_1, \widetilde{G}_2 \in \mathbb{G}(\mathbb{R}_+)$ such that $(\widetilde{M} - \widetilde{G}_1) + (\widetilde{G} - \widetilde{G}_2) \in L_{M+G}(\mathbb{R}_+, \boldsymbol{g})$.

By the constructions in the proof of Theorem 13.1.20, we have $\widetilde{M} = \widetilde{M}_0 + \widetilde{M}_1$ for

$$\widetilde{M}_0 := r^{-\mu}\omega \sum_{l=0}^{k} r^l \mathrm{op}_M^{\widetilde{\gamma}}(h_l'')\omega', \quad \widetilde{M}_1 := r^{-\mu}\omega \sum_{l=0}^{k} r^l \mathrm{op}_M^{\widetilde{\gamma}}(b_l)\omega'.$$

Since $h_l'' \in C^\infty(\overline{\mathbb{R}}_+, M_{\mathcal{O}}^{-\infty})$ the operator $\widetilde{M}_0$ already belongs to $L_{M+G}(\mathbb{R}_+, \boldsymbol{g})$, since here we may write $\widetilde{M}_0 := r^{-\mu}\omega \sum_{l=0}^{k} r^l \mathrm{op}_M^{\gamma}(h_l'')\omega'$, applying Theorem 6.5.11 with vanishing remainder. As for $\widetilde{M}_1$, we have by commuting r-powers through the Mellin actions

$$\widetilde{M}_1 := r^{-\mu}\omega \sum_{l=0}^{k} r^l \mathrm{op}_M^{\gamma}(b_l)\omega' + \widetilde{G}_1$$

where $b_l(r, z) \in C^\infty\left(\overline{\mathbb{R}}_+, M_{T^{-\mu}\mathcal{R}+\mathcal{R}_{l-\mu}}^{-\infty}\right)$, and $\widetilde{G}_1 \in \mathbb{G}(\mathbb{R}_+)$ is the Green operator from Theorem 13.1.20. In fact,

$$\begin{aligned}
\omega\, \mathrm{op}_M^{\gamma}(b_l)\omega' &= r^\gamma \omega\, \mathrm{op}_M(T^{-\gamma}b_l)\omega' r^{-\gamma} \\
&= r^{-\widetilde{\gamma}\,|\,\gamma} r^{\widetilde{\gamma}} \omega\, \mathrm{op}_M(T^{-\widetilde{\gamma}}T^{\widetilde{\gamma}-\gamma}b_l)\omega' r^{-\widetilde{\gamma}} r^{\widetilde{\gamma}-\gamma} \qquad (13.85) \\
&= r^{-\widetilde{\gamma}+\gamma} \omega\, \mathrm{op}_M^{\widetilde{\gamma}}(T^{\widetilde{\gamma}-\gamma}b_l)\omega' r^{\widetilde{\gamma}-\gamma} = \omega\, \mathrm{op}_M^{\widetilde{\gamma}}(b_l)\omega' + G_0.
\end{aligned}$$

This works under the conditions

$$\Gamma_{1/2-\widetilde{\gamma}} \cap \pi_{\mathbb{C}}(T^{-\mu}\mathcal{R} + \mathcal{R}_{l-\mu}) = \emptyset, \quad \Gamma_{1/2-\gamma} \cap \pi_{\mathbb{C}}(T^{-\mu}\mathcal{R} + \mathcal{R}_{l-\mu}) = \emptyset. \quad (13.86)$$

The first one is satisfied through the assumptions of Theorem 13.1.20. So far the choice of $\widetilde{\gamma}$ is arbitrary. Now, in order to characterise the properties where only the weight γ and the order μ are involved, we specify $\widetilde{\gamma}$ as

$$\widetilde{\gamma} := \mu \quad \text{for } \mu \geq 0; \quad \widetilde{\gamma} := 0 \quad \text{for } \mu \leq 0. \quad (13.87)$$

According to the above description of the asymptotic type $T^{-\mu}\mathcal{R} + \mathcal{R}_{l-\mu}$, cf. formula (13.31), we first study the case (13.87) for $\mu \geq 0$, i.e., $\widetilde{\gamma} = \mu$. Relation (13.85) employs the formulas in (13.86), i.e.,

$$\Gamma_{1/2-\mu} \cap \pi_{\mathbb{C}}(T^{-\mu}\mathcal{R} + \mathcal{R}_{l-\mu}) = \emptyset, \quad \Gamma_{1/2-\gamma} \cap \pi_{\mathbb{C}}(T^{-\mu}\mathcal{R} + \mathcal{R}_{l-\mu}) = \emptyset,$$

which corresponds to the conditions

$$1/2 - \mu \notin \mathbb{Z}, \quad 1/2 - \gamma \notin \mathbb{Z}.$$

In the case $\mu \leq 0$, i.e., $\widetilde{\gamma} = 0$, the condition from (13.86) is

$$1/2 - \gamma \notin \mathbb{Z}. \qquad \square$$

Corollary 13.1.24. *For every* $a(r, \rho) \in S^{\mu}_{\mathrm{cl}}(\overline{\mathbb{R}}_+ \times \mathbb{R})$ *and* $\gamma \in \mathbb{R}$ *as in Theorem 13.1.23, the operator* $\omega \, \mathrm{op}^+(a) \, \omega'$ *induces continuous operators*

$$\omega \, \mathrm{op}^+(a)\omega' : \mathcal{K}^{s,\gamma}(\mathbb{R}_+) \to \mathcal{K}^{s-\mu,\gamma-\mu}(\mathbb{R}_+) + \mathcal{K}^{\infty,\delta}_{\mathcal{S}}(\mathbb{R}_+), \quad s \in \mathbb{R}, \qquad (13.88)$$

for some $\delta \in \mathbb{R}$ *and a discrete asymptotic type* $\mathcal{S}$ *associated with the weight data* $(\delta, (-\infty, 0])$ *(where both* δ *and* $\mathcal{S}$ *depend on* a*). Moreover, (13.88) restricts to continuous operators*

$$\omega \, \mathrm{op}^+(a)\omega' : \mathcal{K}^{s,\gamma}_{\mathcal{P}}(\mathbb{R}_+) \to \mathcal{K}^{s-\mu,\gamma-\mu}_{\mathcal{Q}}(\mathbb{R}_+) + \mathcal{K}^{\infty,\delta}_{\mathcal{S}}(\mathbb{R}_+),$$

$s \in \mathbb{R}$, *for every discrete asymptotic type* $\mathcal{P}$ *associated with* $(\gamma, (-\infty, 0])$ *and some resulting* $\mathcal{Q}$ *with* $(\gamma - \mu, (-\infty, 0])$.

13.2   The cone algebra of order zero on the half-line

We now discuss some observations on the cone algebra, here in most cases of order zero. The results are related to the boundary symbol calculus, outlined in Section 2.3 and the reformulation of

$$\mathrm{op}^+(a) : L^2(\mathbb{R}_+) \to L^2(\mathbb{R}_+)$$

as an operator in the cone algebra

$$L^0(\overline{\mathbb{R}}_+, \boldsymbol{g}) \qquad (13.89)$$

for weight data $\boldsymbol{g} = (0, 0, (-\infty, 0])$, which is the cone algebra on the infinite cone $\overline{\mathbb{R}}_+$ as a subalgebra of $\mathcal{L}(L^2(\mathbb{R}_+), L^2(\mathbb{R}_+))$, cf. [45]. In order to draw some interesting conclusions, we recall once again that on $\mathbb{R}_+$, interpreted as an open straight cone, there are the Kegel spaces

$$\mathcal{K}^{s,\gamma}(\mathbb{R}_+) = \omega \mathcal{H}^{s,\gamma}(\mathbb{R}_+) + (1 - \omega) H^s_{\mathrm{cone}}(\mathbb{R}_+)$$

for any $s, \gamma \in \mathbb{R}$ and some cut-off function ω on the half-line. Note that

$$(1 - \omega)H^s_{\mathrm{cone}}(\mathbb{R}_+) = (1 - \omega)H^s(\mathbb{R})\big|_{\mathbb{R}_+}.$$

In addition it makes sense to admit weights at $r = \infty$, i.e.,

$$\mathcal{K}^{s,\gamma;e}(\mathbb{R}_+) = [r]^{-e}\mathcal{K}^{s,\gamma}(\mathbb{R}_+)$$

for a strictly positive C^{∞} function $r \mapsto [r]$ on $\mathbb{R}_+$, which is equal to r for sufficiently large r. For instance, we could set

$$[r] = \omega(r) + r(1 - \omega).$$

Moreover, we have discrete asymptotic types $\mathcal{P}$, represented by sequences

$$\mathcal{P} := \big\{(p_j, m_j)\big\}_{j \in \mathbb{J}} \subset \mathbb{C} \times \mathbb{N} \tag{13.90}$$

for some index set $\mathbb{J} \subseteq \mathbb{N}$, where (13.90) is said to be associated with (γ, Θ) for $\Theta = (\nu, 0]$ for $\nu < 0$, when

$$\pi_{\mathbb{C}}\mathcal{P} = \{p_j\}_{j \in \mathbb{J}} \subset \{z \in \mathbb{C} : 1/2 - \gamma + \nu < \operatorname{Re} z < 1/2 - \gamma\},$$

and $\pi_{\mathbb{C}}\mathcal{P}$ is finite as soon as ν is finite, otherwise $\operatorname{Re} p_j \to -\infty$ as $j \to \infty$ if $\pi_{\mathbb{C}}\mathcal{P}$ is infinite. We then have spaces of singular functions for finite $\pi_{\mathbb{C}}\mathcal{P}$, namely,

$$\mathcal{E}_{\mathcal{P}}(\mathbb{R}_+) := \left\{ \omega(r) \sum_{j \in \mathbb{J}} \sum_{k=0}^{m_j} c_{jk} r^{-p_j} \log^k r : c_{jk} \in \mathbb{C} \right\}$$

for a fixed cut-off function. We set

$$\mathcal{K}_{\Theta}^{s,\gamma}(\mathbb{R}_+) := \varprojlim_{\varepsilon > 0} \mathcal{K}^{s, \gamma - \nu - \varepsilon}(\mathbb{R}_+)$$

and take the direct sum

$$\mathcal{K}_{\mathcal{P}}^{s,\gamma}(\mathbb{R}_+) := \mathcal{E}_{\mathcal{P}}(\mathbb{R}_+) \oplus \mathcal{K}_{\Theta}^{s,\gamma}(\mathbb{R}_+).$$

If $\Theta = (-\infty, 0]$ we set $\Theta_N := (-(N+1), 0]$, $N \in \mathbb{N}$, and

$$\mathcal{P}_N := \big\{(p, m) \in \mathcal{P} : \operatorname{Re} p > 1/2 - \gamma - (N+1)\big\},$$

and define

$$\mathcal{K}_{\mathcal{P}}^{s,\gamma}(\mathbb{R}_+) = \varprojlim_{N \in \mathbb{N}} \mathcal{K}_{\mathcal{P}_N}^{s,\gamma}(\mathbb{R}_+).$$

Setting

$$\mathcal{K}_{\mathcal{P}}^{s,\gamma;e}(\mathbb{R}_+) := [r]^{-e} \mathcal{K}_{\mathcal{P}}^{s,\gamma}(\mathbb{R}_+)$$

we have the space $L_{\mathrm{G}}\big(\overline{\mathbb{R}}_+, \boldsymbol{g}\big)$ of Green operators as the set of all continuous

$$G : \mathcal{K}^{s,\gamma;e}(\mathbb{R}_+) \to \mathcal{K}_{\mathcal{P}}^{\infty, \gamma - \mu; \infty}(\mathbb{R}_+)$$

such that

$$G^* : \mathcal{K}^{s, -\gamma + \mu; e}(\mathbb{R}_+) \to \mathcal{K}_{\mathcal{Q}}^{\infty, -\gamma; \infty}(\mathbb{R}_+)$$

for all $s, e \in \mathbb{R}$ and G-dependent asymptotic types $\mathcal{P}$ and $\mathcal{Q}$, with G^* being the formal adjoint with respect to the $\mathcal{K}^{0,0}(\mathbb{R}_+) = L^2(\mathbb{R}_+)$-scalar product.

Let us now consider $M_{\mathcal{O}}^{\mu}$ for $\mu \in \mathbb{R}$, defined as the space of all $h(z) \in \mathcal{A}(\mathbb{C})$ such that

$$h\big|_{\Gamma_\beta} \in S_{\mathrm{cl}}^{\mu}(\Gamma_\beta)$$

for any $\beta \in \mathbb{R}$, uniformly in compact β-intervals. A sequence

$$\mathcal{R} := \left\{ (p_j, m_j) \right\}_{j \in \mathbb{I}} \subset \mathbb{C} \times \mathbb{N}$$

for some index set $\mathbb{I}$ is called a discrete Mellin asymptotic type if $\pi_{\mathbb{C}} \mathcal{R} := \{p_j\}_{j \in \mathbb{I}}$ intersects any strip $\{z \in \mathbb{C} : c_1 \leq \operatorname{Re} z \leq c_2\}$ in a finite set for every $c_1 \leq c_2$. Then $M_{\mathcal{R}}^{-\infty}$ is defined as the set of all $f \in \mathcal{A}(\mathbb{C} \setminus \pi_{\mathbb{C}} \mathcal{R})$ such that f is meromorphic with poles at $p_j \in \pi_{\mathbb{C}} \mathcal{R}$ of multiplicity $m_j + 1$, and for any $\pi_{\mathbb{C}} \mathcal{R}$ -excision function χ (i.e., $\chi \equiv 0$ for $\operatorname{dist}(z, \pi_{\mathbb{C}} \mathcal{R}) < \varepsilon_0$, $\chi \equiv 1$ for $\operatorname{dist}(z, \pi_{\mathbb{C}} \mathcal{R}) > \varepsilon_1$, for some $0 < \varepsilon_0 < \varepsilon_1 < \infty$) we have

$$\chi f(z)\big|_{\Gamma_\beta} \in \mathcal{S}(\Gamma_\beta)$$

for every real β, uniformly for β in compact β-intervals. Both $M_{\mathcal{O}}^\mu$ and $M_{\mathcal{R}}^{-\infty}$ for any fixed $\mathcal{R}$ are Fréchet spaces, and we have non-direct sums

$$M_{\mathcal{R}}^\mu := M_{\mathcal{O}}^\mu + M_{\mathcal{R}}^{-\infty}$$

which are also Fréchet, where $M_{\mathcal{O}}^{-\infty}$ is the intersection of the two spaces.

For any symbol $p(r, \rho) \in S_{\mathrm{cl}}^\mu(\mathbb{R}_+ \times \mathbb{R})$ such that

$$p(r, \rho) = \widetilde{p}(r, r\rho) \text{ for some } \widetilde{p}(r, \widetilde{\rho}) \in S_{\mathrm{cl}}^\mu(\mathbb{R}_+ \times \mathbb{R}_{\widetilde{\rho}}) \tag{13.91}$$

where $\widetilde{p}(r, \widetilde{\rho})$ is independent of r for large r, there is via Mellin quantization an

$$h(r, z) \in C^\infty\left(\overline{\mathbb{R}}_+, M_{\mathcal{O}}^\mu\right), \tag{13.92}$$

which is also independent of r for large r, such that

$$\operatorname{Op}_r(p) = \operatorname{Op}_M^\beta(h) \bmod L^{-\infty}(\mathbb{R}_+) \tag{13.93}$$

for every $\beta \in \mathbb{R}$. Relation (13.93) is interpreted as an identity for operators $C_0^\infty\left(\overline{\mathbb{R}}_+\right) \to C^\infty(\mathbb{R}_+)$.

Remember that the symbol space $S^{\mu;\nu}(\mathbb{R}_r \times \mathbb{R}_\rho)$ for $\mu, \nu \in \mathbb{R}$ with exit property for $|r| \to \infty$ is determined by estimates

$$\left| D_r^l D_\rho^k p(r, \rho) \right| \leq C \langle r \rangle^{\nu-l} \langle \rho \rangle^{\mu-k}$$

for all $(r, \rho) \in \mathbb{R} \times \mathbb{R}$, $l, k \in \mathbb{N}$ and constants $C = C(l, k) > 0$. The corresponding subspace of classical symbols can be described by the projective tensor product

$$S_{\mathrm{cl}}^{\mu;\nu}(\mathbb{R} \times \mathbb{R}) := S_{\mathrm{cl}}^\nu(\mathbb{R}_r) \widehat{\otimes}_\pi S_{\mathrm{cl}}^\mu(\mathbb{R}_\rho).$$

It also makes sense to observe the specific exit property only on $r \in \mathbb{R}_+$ for $r \to \infty$. The corresponding spaces will be denoted by $S^{\mu;\nu}(\mathbb{R}_+ \times \mathbb{R})$ and $S_{\mathrm{cl}}^{\mu;\nu}(\mathbb{R}_+ \times \mathbb{R})$, respectively. Observe that the cone-degenerate symbols (13.91) just belong to $S_{\mathrm{cl}}^{\mu;\mu}(\mathbb{R}_+ \times \mathbb{R})$.

Let us also recall that we have a kernel cut-off operator which says that for any $f(r, z) \in C^\infty\left(\mathbb{R}_+, S^\mu_{\mathrm{cl}}(\Gamma_{1/2-\beta})\right)$, $\mu \in \mathbb{R}$, there is an $h(r, z) \in C^\infty\left(\mathbb{R}_+, M^\mu_\mathcal{O}\right)$ such that

$$\mathrm{Op}^\beta_M(f) = \mathrm{Op}^\gamma_M(h) \bmod L^{-\infty}(\mathbb{R}_+)$$

for any $\gamma \in \mathbb{R}$. Up to a translation in the complex plane, we may set $\beta = 1/2$. The correspondence

$$V : C^\infty\left(\mathbb{R}_+, S^\mu_{\mathrm{cl}}(\Gamma_0)\right) \to C^\infty\left(\mathbb{R}_+, M^\mu_\mathcal{O}\right)$$

can be chosen in terms of an excision function $\psi(\tau) \in C^\infty_0(\mathbb{R}_+)$ such that $\psi \equiv 1$ in a neighbourhood of $\tau = 1$. Then we have the corresponding weighted Mellin transform $M_{1/2, r \to \tau}$, and take first a symbol $f(z)$ with constant coefficients. This can be identified with $f(i\rho)$. The Mellin distributional kernel of $\mathrm{Op}^{1/2}_M(f)$ has the form $m(r/r')$ for

$$m(f)(\tau) = \left(M^{-1}_{1/2} f\right)(\tau\theta), \quad \tau \in \mathbb{R}_+,$$

and it follows that

$$\operatorname{sing\,supp} m(f)(\tau) \subseteq \{1\}.$$

Now $\psi(\tau) m(f)(\tau)$ has compact support in $\tau \in \mathbb{R}_+$, and it suffices to set

$$(Vf)(i\rho) := M_{1/2}\left(\psi(\tau) m(f)(\tau)\right)(i\rho).$$

This extends to the complex z-plane for $z = \beta + i\rho$ to an element of $M^\mu_\mathcal{O}$. Since

$$M_{1/2}\left((1 - \psi(\tau)) m(f)(\tau)\right)(z) \in S^{-\infty}(\Gamma_0),$$

the operator V is as desired. Since the kernel cut-off operator only acts on covariables, the procedure also works for r-dependent symbols $f(r, z)$.

Remark 13.2.1. The cut-off function ψ in this construction is arbitrary. We can also form

$$V := M_{1/2}\left(\phi(\tau) m(f)(\tau)\right)(i\rho)$$

for any

$$\phi(\tau) \in C^\infty(\mathbb{R}_+)_B := \left\{ u(\tau) \in C^\infty(\mathbb{R}_+) : \sup_{\tau \in \mathbb{R}_+} \left| \left(\tau \frac{\partial}{\partial \tau}\right)^j u(\tau) \right| < \infty \ \forall j \in \mathbb{N} \right\}.$$

Now using a family of cut-off functions ψ_ε, $0 < \varepsilon < 1$, such that $\psi_\varepsilon \to 1$ for $\varepsilon \to 0$ in the topology of $C^\infty(\mathbb{R}_+)_B$ we obtain that

$$(V(\psi_\varepsilon)f) \to f \quad \text{for } \varepsilon \to 0$$

in the topology of $S^\mu_{\mathrm{cl}}(\Gamma_0)$. Here we wrote for the moment $V = V(\psi_\varepsilon)$ for the cut-off operator associated with ψ_ε.

The space

$$L_{M+G}(\overline{\mathbb{R}}_+, \boldsymbol{g}) \tag{13.94}$$

for weight data $\boldsymbol{g} = (\gamma, \gamma - \mu, \Theta)$, $\Theta = (-(k+1), 0]$, $k \in \mathbb{N}$, is defined as the set of all operators $M + G$ with $G \in L_G(\overline{\mathbb{R}}_+, \boldsymbol{g})$ and

$$M = r^{-\mu} \omega \sum_{j=0}^{k} r^j \mathrm{Op}_M^{\gamma_j}(f_j) \omega'$$

for cut-off functions ω, ω' in the axial variable and Mellin symbols $f_j \in M_{\mathcal{R}_j}^{-\infty}$ for Mellin asymptotic types $\mathcal{R}_j$ and weights $\gamma - j \le \gamma_j \le \gamma$ such that

$$\Gamma_{1/2-\gamma_j} \bigcap \pi_{\mathbb{C}} \mathcal{R}_j = \emptyset \quad \text{for all } j.$$

The cone algebra on the half-line

$$L^\mu(\overline{\mathbb{R}}_+, \boldsymbol{g}) \tag{13.95}$$

is furnished by operator spaces

$$A = r^{-\mu}\{\omega \mathrm{Op}_M^\gamma(h)\omega' + (1-\omega)\mathrm{Op}_r(p)(1-\omega'')\} + M + G, \tag{13.96}$$

where $p(r, \rho)$ is a symbol as in (13.91) and $h(r, z)$ is as in (13.92) such that (13.93) holds, and $M + G$ lies in the space (13.94) and for cut-off functions $\omega'' \prec \omega \prec \omega'$.

Theorem 13.2.2. *Any $A \in L^\mu(\overline{\mathbb{R}}_+, \boldsymbol{g})$ for $\boldsymbol{g} = (\gamma, \gamma - \mu, \Theta)$ induces continuous operators*

$$A : \mathcal{K}^{s,\gamma}(\mathbb{R}_+) \to \mathcal{K}^{s-\mu,\gamma-\mu}(\mathbb{R}_+)$$

and

$$A : \mathcal{K}_{\mathcal{P}}^{s,\gamma}(\mathbb{R}_+) \to \mathcal{K}_{\mathcal{Q}}^{s-\mu,\gamma-\mu}(\mathbb{R}_+)$$

for every $s \in \mathbb{R}$ and asymptotic types $\mathcal{P}$ with A-dependent resulting asymptotic type $\mathcal{Q}$.

Operators in the cone calculus are controlled by several symbols. Thanks to the inclusion $L^\mu(\overline{\mathbb{R}}_+, \boldsymbol{g}) \subset L_{\mathrm{cl}}^\mu(\mathbb{R}_+)$ we have the homogeneous principal symbol

$$\sigma_\psi(A)(r, \rho) \in S^{(\mu)}(\mathbb{R}_+ \times (\mathbb{R} \setminus \{0\}))$$

with $S^{(\mu)}(\mathbb{R}_+ \times (\mathbb{R} \setminus \{0\}))$ being the set of all $a_{(\mu)}(r, \rho) \in C^\infty(\mathbb{R} \times (\mathbb{R} \setminus \{0\}))$ such that

$$a_{(\mu)}(r, \delta\rho) = \delta^\mu a_{(\mu)}(r, \rho)$$

for all $\delta \in \mathbb{R}_+$, $\rho \in \mathbb{R} \setminus \{0\}$. Moreover, there is a reduced symbol

$$\tilde{\sigma}_\psi(A)(r, \rho) := r^\mu \sigma_\psi(A)(r, r^{-1}\rho) \in S^{(\mu)}(\mathbb{R}_+ \times (\mathbb{R} \setminus \{0\})).$$

In addition, there is the principal homogeneous exit symbol $\sigma_e(A)$, which in the present case coincides with $\sigma_{\psi,e}(A)$ and satisfies

$$\sigma_\psi(A)(r,\rho)\big|_{r=\infty} \in S^{(\mu)}(\mathbb{R}_\rho \setminus \{0\}).$$

Finally, we have the conormal symbols

$$\sigma_M^\mu(A)(z) := h(0,z) + f_0(z), \tag{13.97}$$

$$\sigma_M^{\mu-j}(A)(z) := \frac{1}{j!}\frac{\partial^j}{\partial r^j}h(r,z)\big|_{r=0} + f_j(z), \quad j = 1,\ldots,k, \tag{13.98}$$

belonging to $M_{\mathcal{R}_j}^\mu$.

Definition 13.2.3. An $A \in L^\mu(\overline{\mathbb{R}}_+, \boldsymbol{g})$ for $\boldsymbol{g} = (\gamma, \gamma - \mu, \Theta)$ is called elliptic if

(i) $\sigma_\psi(A)(r,\rho) \neq 0$ for all $(r,\rho) \in \mathbb{R}_+ \times (\mathbb{R} \setminus \{0\})$, and $\tilde{\sigma}_\psi(A)(r,\rho) \neq 0$ for all $(r,\rho) \in \overline{\mathbb{R}}_+ \times (\mathbb{R} \setminus \{0\})$;

(ii) A is elliptic for $r \to \infty$ in the sense of the exit symbolic calculus, i.e., with respect to the symbols $\sigma_e(A)$ and $\sigma_{\psi,e}(A)$;

(iii) $\sigma_M(A)(z) \neq 0$ for all $z \in \Gamma_{1/2-\gamma}$.

Remark 13.2.4. Ellipticity as in Definition 13.2.3 depends on the weight γ. Since $\sigma_M(A) \in M_{\mathcal{R}_0}^\mu$, the conormal symbol $\sigma_M(A)$ is holomorphic in a strip $\{z \in \mathbb{C} : 1/2-\gamma-\varepsilon < \operatorname{Re} z < 1/2-\gamma+\varepsilon\}$ for some $\varepsilon > 0$. Moreover, we have $\sigma_M(A) \neq 0$ on Γ_β for every $\beta \in \mathbb{R}$ off some discrete set of reals. The pointwise inverse $\sigma_M(A)^{-1}(z)$ on any such Γ_β extends to an element in $M_{\mathcal{S}}^{-\mu}$ for some Mellin asymptotic type $\mathcal{S}$.

Theorem 13.2.5. *For an operator $A \in L^\mu(\overline{\mathbb{R}}_+, \boldsymbol{g})$ the following conditions are equivalent:*

(i) *A is elliptic in the sense of* Definition 13.2.3;

(ii) *A induces a Fredholm operator*

$$A : \mathcal{K}^{s,\gamma}(\mathbb{R}_+) \to \mathcal{K}^{s-\mu,\gamma-\mu}(\mathbb{R}_+) \tag{13.99}$$

for some $s = s_0 \in \mathbb{R}$. The Fredholm property of (13.99) for some $s = s_0$ entails the Fredholm property for all $s \in \mathbb{R}$ with index independent of s.

(iii) *If $A \in L^\mu(\overline{\mathbb{R}}_+, \boldsymbol{g})$ is elliptic, then it has a parametrix $A^{(-1)} \in L^{-\mu}(\overline{\mathbb{R}}_+, \boldsymbol{g}^{-1})$ for $\boldsymbol{g}^{-1} = (\gamma - \mu, \gamma, \Theta)$ and $A^{(-1)}A = 1 - G_L$, $AA^{(-1)} = 1 - G_R$ for some Green operators $G_L \in L_G(\mathbb{R}_+, \boldsymbol{g}_L)$, $G_R \in L_G(\mathbb{R}_+, \boldsymbol{g}_R)$ for $\boldsymbol{g}_L = (\gamma, \gamma, \Theta)$, $\boldsymbol{g}_R = (\gamma - \mu, \gamma - \mu, \Theta)$.*

Theorem 13.2.6. *Let (13.99) be an isomorphism and A^{-1} the inverse operator. Then we have $A^{-1} \in L^{-\mu}(\overline{\mathbb{R}}_+, \boldsymbol{g}^{-1})$ for $\boldsymbol{g}^{-1} = (\gamma - \mu, \gamma, \Theta)$.*

13.3 Homotopies between boundary symbols and Mellin operators

Let $a(\rho) \in S^0_{\mathrm{cl}}(\mathbb{R})$ be a symbol of order zero with constant coefficients. The case of symbols of arbitrary order belonging to $\mathbb{Z}$ can be obtained by reducing the order to 0 via composition with an elliptic element of opposite order. Then, as noted before, we have asymptotic expansions

$$a(\rho) \sim \sum_{j=0}^{\infty} a_j^{\pm}(i\rho)^{-j} \quad \text{for } \rho \to \pm\infty.$$

The imaginary unit i is only used for technical reasons. The coefficients $a_j^{\pm} \in \mathbb{C}$ are completely independent and arbitrary. However, symbols with the transmission property are characterized by the conditions

$$a_j^+ = a_j^- \quad \text{for all } j \in \mathbb{N}. \tag{13.100}$$

In that case $a(\rho)$ determines a curve

$$L_\psi(a) := \{z \in \mathbb{C} : z = a(\rho) \text{ for } \rho \in \mathbb{R}\}, \tag{13.101}$$

which is closed in the case (13.100), bounded, and smooth also at the point $z = a(-\infty) = a(+\infty) = a_0^{\pm}$. Otherwise $L_\psi(a)$ has two different end points a_0^- and a_0^+. In the general case (13.101) is a bounded set in $\mathbb{C}$, but it has different end points $a(-\infty) = a_0^-$, $a(+\infty) = a_0^+$. In any case we have

Theorem 13.3.1. *For every* $a(\rho) \in S^0_{\mathrm{cl}}(\mathbb{R})$,

$$\mathrm{op}^+(a) \in L^0(\overline{\mathbb{R}}_+, \boldsymbol{g}) \tag{13.102}$$

for $\boldsymbol{g} = (0, 0, (-\infty, 0])$.

In other words, truncated operators belong to the cone algebra on the half-line. A proof of Theorem 13.3.1 admits generalizations to symbols in $S^\mu_{\mathrm{cl}}(\overline{\mathbb{R}}_+ \times \mathbb{R})$ of different kind for the cone algebra described in the preceding section for arbitrary weight data, realized as operators in weighted Kegel spaces. It may be necessary in that case to allow remainders in Green operators which are more singular than those before. In addition, we have to impose some restrictions on the allowed weights.

Let us return to the situation of Theorem 13.3.1.

Remark 13.3.2. The conormal symbols of (13.102) are completely determined by $a(\rho) \in S^0_{\mathrm{cl}}(\mathbb{R})$. These have the form

$$\sigma_M^{-j}(a)(z) = \{a_j^+ g^+(z) + a_j^- g^-(z)\} f_j(z), \quad j \in \mathbb{N}, \tag{13.103}$$

for $g^+(z) = \left(1 - e^{-2\pi i z}\right)^{-1}$, $g^-(z) = 1 - g^+(z)$, and $f_j(z) = \displaystyle\prod_{p=1}^{j} (p - z)^{-1}$ for $j \geq 1$,

$f_0(z) = 1$, cf. [45, page 394].

The highest-order conormal symbol

$$\sigma_M^0(a)(z) = a_0^+ g^+(z) + a_0^- g^-(z)$$

defines a straight connection between the points $a_0^\pm \in \mathbb{C}$ when z varies over the weight line $\Gamma_{\frac{1}{2}}$, which is the reference weight line in the complex Mellin plane for $L^2(\Gamma_{\frac{1}{2}})$, the Mellin image of $L^2(\mathbb{R}_+)$.

The algebra in $\mathcal{L}(L^2(\mathbb{R}_+))$ generated by $\{\mathrm{op}^+(a) : a \in S_{\mathrm{cl}}^0(\mathbb{R})\}$ may be regarded as a subalgebra

$$\Psi^0(\overline{\mathbb{R}}_+, \boldsymbol{g}) \quad \text{of} \quad L^0(\overline{\mathbb{R}}_+, \boldsymbol{g}), \tag{13.104}$$

and the symbols behave multiplicatively under composition of operators. The composition

$$\mathrm{op}^+(a)\,\mathrm{op}^+(b) \quad \text{for } a, b \in S_{\mathrm{cl}}^0(\mathbb{R})$$

gives rise to the composition of leading conormal symbols

$$(a_0^+ g^+ + a_0^- g^-)(b_0^+ g^+ + b_0^- g^-) = a_0^+ b_0^+ g^+ + a_0^- b_0^- g^- + (a_0^+ b_0^- + a_0^- b_0^+) g^+ g^-.$$

While the first two right-hand side summands are meromorphic Mellin symbols as before, the third summand is a mixed term and is a smoothing meromorphic function. In fact, the functions $g^+(z) g^-(z)$ tend strongly to zero as $|\mathrm{Im}\, z| \to \infty$ along any weight line. This just corresponds to the smoothing Mellin contributions which are coded by $L_{M+G}^0(\overline{\mathbb{R}}_+, \boldsymbol{g})$. On the level of operators, we may compose the smoothing Mellin terms with cut-off functions from both sides, and the smoothing operators in this context are Green operators. In the following discussion on homotopies between boundary symbols and elements of the cone algebra we follow the work of Pirhayati [33, Chapter 5].

The homotopy to be outlined here concerns families of elliptic operators in (13.104) between $\mathrm{op}^+(a_0)$ for $a_0(\rho) \in S_{\mathrm{tr}}^0(\mathbb{R})$ and operators of the kind

$$1 + \omega \, \mathrm{Op}_M^0(f_1) \omega'$$

with $f_1(z) \in M_O^{-\infty}$ through elliptic elements of the algebra $\Psi^0(\overline{\mathbb{R}}_+, \boldsymbol{g})$, cf. formula (13.104).

Without loss of generality, we assume $a_0(\rho_1) = 1$ for some $\rho_1 \in \mathbb{R}$, otherwise we pass to the symbol $a_0(\rho_1)^{-1} a_0(\rho)$, and for the symbols $a_s(\rho) \in S_{\mathrm{cl}}^0(\mathbb{R})$, $0 < s < 1$, we do the same. In the proof of the following theorem we then look at the curve (13.101) and define intervals in this curve which include the point $z = 1$ where the end points are $a_s^- = a_s(-\infty)$ and $a_s^+ = a_s(+\infty)$ which tend from the respective sides on the curve to 1 as $s \to 1$.

Theorem 13.3.3. *For every operator*

$$A_0 := \mathrm{op}^+(a_0), \quad a_0(\rho) \in S^0_{\mathrm{tr}}(\mathbb{R}), \tag{13.105}$$

there exists a homotopy

$$A_s := \mathrm{op}^+(a_s) + \omega\,\mathrm{Op}_M(l_s)\omega', \quad 0 \le s \le 1, \tag{13.106}$$

for certain $a_s(\rho) \in S^0_{\mathrm{cl}}(\mathbb{R})$ (which need not have the transmission property) and smoothing Mellin symbols $l_s(z) \in M^{-\infty}_{\mathcal{O}}$ such that (13.106) is a homotopy though elliptic elements in $L^0(\overline{\mathbb{R}}_+, \boldsymbol{g})$ between (13.105) and an operator

$$A_1 := 1 + \omega\,\mathrm{Op}_M(l_1)\,\omega', \tag{13.107}$$

where $l_1 \in M^{-\infty}_{\mathcal{O}}$. Conversely, starting with an operator (13.107) that is elliptic in $L^0(\overline{\mathbb{R}}_+, \boldsymbol{g})$, we find such a homotopy to (13.105) for an elliptic $a_0(\rho) \in S^0_{\mathrm{tr}}(\mathbb{R})$.

Let us sketch the proof in terms of the pseudo-differential symbol curve (13.101) for $a(\rho) \in S^0_{\mathrm{cl}}(\mathbb{R})$ which is not closed and smooth, with end points a_0^+ and a_0^- in $\mathbb{C}$ where we can attach the Mellin symbol curve

$$L_M := \{ z \in \mathbb{C} : a_0^- g^-(z) + a_0^+ g^+(z) + l(z) \}; \tag{13.108}$$

here z varies along $\Gamma_{1/2} \ni 1/2 + i\rho$, from $\rho = -\infty$ to $\rho = +\infty$ and $z = z(\rho)$. The smoothing Mellin symbol $l(z)$ in (13.108) has the meaning of $f_0(z)$ in (13.97). The curves $L_\psi(s) := L_\psi(a_s)$ and $L_M(s)$ defined by (13.108) for $l(z) := l_s(z)$ depend on $0 \le s \le 1$ when we associate them with (13.106), and in the elliptic case

$$L(s) := L_\psi(s) \cup L_M(s) \tag{13.109}$$

is a closed (piecewise smooth) curve $L(s)$ which does not intersect the origin in $\mathbb{C}$. By small deformations of $l_s(z)$ near $\mathrm{Im}\,z = \pm\infty$ we may assume that the curves $L(s)$ are smooth including the points a_0^+ and a_0^-. The situation is then that $L_\psi(s)$ and $L_M(s)$ are complementary pieces of the closed curve $L(s)$ (possibly with self-intersection), and $L(0) = L_\psi(a_0)$, $L(1) = L_M(l_1)$.

Thus at the beginning of the homotopy we have only the closed smooth curve belonging to (13.105), at the end only the one of (13.107), and for $0 < s < 1$ non-trivial parts $L_\psi(s)$ and $L_M(s)$. The remarkable aspect is that we may assume that the curve $L(s)$ is the same for all s and for different s the subintervals $L_\psi(s)$ and $L_M(s)$ are varying. Then the interval belonging to $a_s(\rho)$ contracts to 1 when $s \to 1$. The curve itself determines the Fredholm index of

$$A_s : L^2(\mathbb{R}_+) \to L^2(\mathbb{R}_+)$$

for all s, which is its winding number, cf. relation (2.112) and (2.137).

Bibliography

[1] M. S. Agranovich and M. I. Vishik. Elliptic problems with parameter and parabolic problems of general type. *Uspekhi Mat. Nauk*, 19(3):53–161, 1964.

[2] M. F. Atiyah and D. W. Anderson. *K-theory*. Number 7 in Mathematics Lecture Notes. W. A. Benjamin, New York, Amsterdam, 1967. Lectures by Atiyah (Fall 1964), notes by Anderson. 2nd edition published in 1989.

[3] M. F. Atiyah and R. Bott. The index problem for manifolds with boundary. In *Differential Analysis: Papers presented at the international colloquium (Bombay, 7–14 January 1964)*, Tata Institute of Fundamental Research Studies in Mathematics 2, pages 175–186, London, 1964. Oxford University Press.

[4] M. F. Atiyah, V. Patodi, and I. M. Singer. Spectral asymmetry and Riemannian geometry I. *Math. Proc. Cambridge Philos. Soc.*, 77:43–69, 1975.

[5] M. F. Atiyah, V. Patodi, and I. M. Singer. Spectral asymmetry and Riemannian geometry II. *Math. Proc. Cambridge Philos. Soc.*, 78:405–432, 1976.

[6] M. F. Atiyah, V. Patodi, and I. M. Singer. Spectral asymmetry and Riemannian geometry III. *Math. Proc. Cambridge Philos. Soc.*, 79:315–330, 1976.

[7] M. S. Birman and M. Z. Solomjak. On the subspaces admitting a pseudodifferential projection. *Vestnik LGU*, 1:18–25, 1982.

[8] B. Booss-Bavnbek and K. Wojciechowski. *Elliptic boundary problems for Dirac operators*, volume 4 of *Mathematics: Theory & Applications*. Birkhäuser, Boston, Basel, Berlin, 1993.

[9] L. Boutet de Monvel. Comportement d'un opérateur pseudo-différentiel sur une variété à bord I–II. *J. d'Analyse Fonct.*, 17:241–304, 1966.

[10] L. Boutet de Monvel. Boundary problems for pseudo-differential operators. *Acta Math.*, 126:11–51, 1971.

[11] D.-C. Chang, N. Habal, and B.-W. Schulze. The edge algebra structure of the zaremba problem. *J. Pseudo-Differ. Oper. Appl.*, 5(1):69–155, 2014.

© Springer Nature Switzerland AG 2018
X. Liu, B.-W. Schulze, *Boundary Value Problems with Global Projection Conditions*,
Operator Theory: Advances and Applications 265, https://doi.org/10.1007/978-3-319-70114-1

[12] D.-C. Chang, X. Lyu, and B.-W. Schulze. Recent developments on pseudo-differential operators (ii). *Tamkang Journal of Mathematics*, 46(3):281–347, 2015.

[13] Ch. Dorschfeldt. *Algebras of pseudo-differential operators near edge and corner singularities*, volume 102 of *Mathematical Research*. Wiley-VCH, Berlin, Chichester, 1998.

[14] G.I. Eskin. *Boundary value problems for elliptic pseudodifferential equations*, volume 52 of *Translations of Math. Monographs*. AMS, Providence, RI, 1981. Translation of Kraevye zadachi dlia ellipticheskikh psevdodifferentsialnykh uravnenii.

[15] J. B. Gil, B.-W. Schulze, and J. Seiler. *Differential Equations, Asymptotic Analysis, and Mathematical Physics*, volume 100 of *Mathematical Research*, chapter Holomorphic operator-valued symbols for edge-degenerate pseudo-differential operators, pages 113–137. Akademie Verlag, Berlin, 1997.

[16] J. B. Gil, B.-W. Schulze, and J. Seiler. Cone pseudodifferential operators in the edge symbolic calculus. *Osaka J. Math.*, 37:221–260, 2000.

[17] B. Gramsch. Relative Inversion in der Störungstheorie von Operatoren und Ψ-Algebras. *Math. Ann.*, 269(1):27–71, 1984.

[18] G. Grubb. Pseudo-differential boundary value problems in L_p spaces. *Comm. Partial Differential Equations*, 15(3):289–340, 1990.

[19] G. Grubb. *Functional calculus of pseudo-differential boundary problems*, volume 65 of *Progress in Mathematics*. Birkhäuser, Boston, 2nd edition, 1996.

[20] G. Grubb and R. T. Seeley. Weakly parametric pseudodifferential operators and Atiyah–Patodi–Singer boundary problems. *Inventiones math.*, 121(1):481–529, 1995.

[21] N. Habal. *Operators on singular manifolds*. PhD thesis, University of Potsdam, 2013.

[22] N. Habal and B.-W. Schulze. *Operator Theory, Pseudo-Differential Equations, and Mathematical Physics*, volume 228 of *Oper. Theory: Adv. Appl.*, chapter Mellin quantisation in corner operators, pages 151–172. Birkhäuser, Basel, 2013.

[23] G. Harutjunjan and B.-W. Schulze. *Elliptic mixed, transmission and singular crack problems*, volume 4 of *EMS Tracks in Mathematics*. EMS, Zürich, 2008.

[24] T. Hirschmann. Functional analysis in cone and edge Sobolev spaces. *Ann. Global Anal. Geom.*, 8(2):167–192, 1990.

[25] D. Kapanadze and B.-W. Schulze. *Crack theory and edge singularities*. Kluwer Academic Publ., Dordrecht, 2003.

[26] H. Kumano-go. *Pseudo-differential operators.* The MIT Press, Cambridge, MA, 1981.

[27] X. Lyu. Asymptotics in weighted corner spaces. *Asian–European J. Math.*, 7(3), 2014.

[28] V. Nazaikinskij, A. Savin, B.-W. Schulze, and B. Ju. Sternin. Elliptic theory on manifolds with nonisolated singularities: Iv. obstructions to elliptic problems on manifolds with edges. preprint 24, University of Potsdam, Institute of Mathematics, Potsdam, 2002.

[29] V. Nazaikinskij, B.-W. Schulze, and B. Sternin. *The localization problem in index theory of elliptic operators,* volume 10 of *Pseudo-Differential Operators.* Birkhäuser, Basel, 2013.

[30] V. Nazaikinskij, B.-W. Schulze, B. Ju. Sternin, and V. Shatalov. Spectral boundary value problems and elliptic equations on singular manifolds. *Differentsial'nye Uravneniya,* 34(5):696–710, 1998. English transl.: Differential Equations.

[31] F. W. J. Olver. *Asymptotics and special functions.* Academic Press, New York, London, 1974.

[32] A. Pazy. *Semigroups of linear operators and applications to partial differential equations,* volume 44 of *Appl. Math. Sci.* Springer, Berlin, New York, Tokyo, 1983.

[33] M. Pirhayati. Boundary symbols in the cone algebra. *J. Pseudo-Differ. Oper. Appl.,* 6(3):307–339, 2015.

[34] S. Rempel and B.-W. Schulze. *Index theory of elliptic boundary problems.* Akademie-Verlag, Berlin, 1982.

[35] S. Rempel and B.-W. Schulze. Parametrices and boundary symbolic calculus for elliptic boundary problems without transmission property. *Math. Nachr.,* 105:45–149, 1982.

[36] S. Rempel and B.-W. Schulze. *Asymptotics for elliptic mixed boundary problems: pseudo-differential and M operators in spaces with conormal singularity,* volume 50 of *Math. Res.* Akademie-Verlag, Berlin, 1989.

[37] A. Savin and B. Sternin. Elliptic operators in even subspaces. *Math. Sbornik,* 190(8):125–160, 1999.

[38] H. Schröder. *Funktionalanalysis.* Akademie-Verlag, Berlin, 1997.

[39] E. Schrohe and B.-W. Schulze. *Pseudo-Differential Calculus and Mathematical Physics: Advances in Partial Differential Equations,* volume 5 of *Math. Topics,* chapter Boundary value problems in Boutet de Monvel's calculus for manifolds with conical singularities I, pages 97–209. Akademie-Verlag, Berlin, 1994.

[40] E. Schrohe and B.-W. Schulze. *Boundary Value Problems, Schrödinger Operators, Deformation Quantization*, volume 8 of *Math. Topics*, chapter Boundary value problems in Boutet de Monvel's calculus for manifolds with conical singularities II, pages 70–205. Akademie-Verlag, Berlin, 1995.

[41] E. Schrohe and J. Seiler. Ellipticity and invertibility in the cone algebra on L_p-Sobolev spaces. *Integral Equations Operator Theory*, 41(1):93–114, 2001.

[42] B.-W. Schulze. Corner Mellin operators and reduction of orders with parameters. *Ann. Sc. Norm. Sup. Pisa Cl. Sci.*, 16(1):1–81, 1989.

[43] B.-W. Schulze. *Symp. "Partial Differential Equations", Holzhau 1988*, volume 112 of *Teubner-Texte zur Mathematik*, chapter Pseudo-differential operators on manifolds with edges, pages 259–287. BSB Teubner, Leipzig, 1989.

[44] B.-W. Schulze. *Pseudo-differential operators on manifolds with singularities*, volume 24 of *Studies in Mathematics and its Applications*. North-Holland, Amsterdam, 1st edition, 1991.

[45] B.-W. Schulze. *Pseudo-differential boundary value problems, conical singularities, and asymptotics*, volume 4 of *Math. Topics*. Akademie-Verlag, Berlin, 1994.

[46] B.-W. Schulze. *Boundary value problems and singular pseudo-differential operators*, volume 4 of *Pure and Applied Mathematics: A Wiley Series of Texts, Monographs and Tracts*. Wiley, Chichester, 1st edition, 1998.

[47] B.-W. Schulze. An algebra of boundary value problems not requiring Shapiro–Lopatinskij conditions. *J. Funct. Anal.*, 179(2):374–408, 2001.

[48] B.-W. Schulze. *Aspects of Boundary Problems in Analysis and Geometry*, volume 151 of *Advances in Partial Differential Equations*, chapter Index defects in the theory of spectral boundary value problems, pages 324–429. Birkhäuser, Basel, 2004.

[49] B.-W. Schulze. *Pseudo-Differential Operators: Analysis, Applications and Computations*, volume 213 of *Oper. Theory: Adv. Appl.*, chapter The iterative structure of corner calculus, pages 79–103. Springer, Basel, 2011.

[50] B.-W. Schulze and J. Seiler. The edge algebra structure of boundary value problems. *Ann. Glob. Anal. Geom.*, 22(3):197–265, 2002.

[51] B.-W. Schulze and J. Seiler. Pseudodifferential boundary value problems with global projection conditions. *J. Funct. Anal.*, 206(2):449–498, 2004.

[52] B.-W. Schulze and J. Seiler. Edge operators with conditions of Toeplitz type. *J. of the Inst. Math. Jussieu*, 5(1):101–123, 2006.

[53] B.-W. Schulze and J. Seiler. Elliptic complexes with generalized Atiyah–Patodi–Singer boundary conditions. preprint arXiv:1510.02455[math. AP], 2015.

[54] B.-W. Schulze and J. Seiler. Elliptic complexes on manifolds with boundary. *J. Geometric Analysis*, 3(22):1–51, 2018.

[55] B.-W. Schulze and Y. Wei. The Mellin-edge quantisation for corner operators. *Complex Analysis and Operator Theory*, 8(4):803–841, 2014.

[56] B.-W. Schulze and M. W. Wong. Mellin and Green operators of the corner calculus. *J. Pseudo-Differ. Oper. Appl.*, 2(4):467–507, 2011.

[57] R. Seeley. Complex powers of an elliptic operator. In *Singular Integrals (Proc. Sympos. Pure Math., Chicago, 1966)*, volume 10, pages 288–307, Providence, RI, 1967. Amer. Math. Soc.

[58] J. Seiler. *Pseudodifferential calculus on manifolds with non-compact edges*. PhD thesis, University of Potsdam, 1997.

[59] J. Seiler. Continuity of edge and corner pseudo-differential operators. *Math. Nachr.*, 205(1):163–182, 1999.

[60] J. Seiler. *Approaches to Singular Analysis*, volume 125 of *Oper. Theory: Adv. Appl.*, chapter The cone algebra and a kernel characterization of Green operators, pages 1–29. Birkhäuser, Basel, 2001.

[61] M. Z. Solomjak. The Calderón projection. *Operator Theory and Function Theory*, 1:47–55, 1983. In Russian, University of Leningrad.

[62] M. I. Vishik and G. I. Eskin. Convolution equations in a bounded region. *Uspekhi Mat. Nauk*, 20(3):89–152, 1965.

[63] M. I. Vishik and G. I. Eskin. Convolution equations in bounded domains in spaces with weighted norms. *Math. Sbornik*, 69(111)(1):65–110, 1966.

Subject Index

Printed in the United States
By Bookmasters